PARASITOLOGY
A Conceptual Approach

PARASITOLOGY
A Conceptual Approach

Eric S. Loker

Bruce V. Hofkin

Garland Science
Taylor & Francis Group
NEW YORK AND LONDON

Vice President: Denise Schanck
Senior Editor: Elizabeth Owen
Editorial Assistant: Deepa Divakaran
Senior Production Editor: Georgina Lucas
Illustrator: Patrick Lane, ScEYEence Studios
Layout: Georgina Lucas
Cover Designer: Patrick Lane, ScEYEence Studio
Copyeditor: Marjorie Singer Anderson
Proofreader: Susan Wood
Indexer: Bill Johncocks

Eric S. Loker is Distinguished Professor of Biology, Curator of the Division of Parasitology of the Museum of Southwestern Biology, and Director of NIH COBRE Center for Evolutionary and Theoretical Immunology at the University of New Mexico where he has taught parasitology and related courses for many years. His research interests focus on the biology of schistosomes.

Bruce V. Hofkin received his PhD from the University of New Mexico where he is currently a principal lecturer in the Department of Biology. His primary research interest is the epidemiology and control of vector-borne and snail-borne diseases.

ISBN 978-0-8153-4473-5

Library of Congress Cataloging-in-Publication Data

Loker, Eric S.
Parasitology : a conceptual approach / Eric S. Loker, Bruce V. Hofkin.
pages cm
ISBN 978-0-8153-4473-5 (alk. paper)
1. Parasitology. I. Hofkin, Bruce V. II. Title.
QL757.L72 2015
616.9'6--dc23
201500060

Published by Garland Science, Taylor & Francis Group, LLC,
an informa business,
711 Third Avenue, New York, NY, 10017, USA,
and 3 Park Square, Milton Park, Abingdon, OX14 4RN, UK.

Printed in the United States of America

15 14 13 12 11 10 9 8 7 6 5 4 3 2 1

NEW YORK AND LONDON

Visit our web site at http://www.garlandscience.com

PREFACE

The world of parasites is becoming ever more fascinating. In *Parasitology: A Conceptual Approach*, we explore the absorbing and rapidly changing discipline of parasitology, and we do so by taking a distinctly different approach from other textbook treatments of the subject. We focus on the exciting ideas and concepts of parasitology, rather than follow a taxon-by-taxon approach. The organisms we emphasize are those traditionally covered in parasitology courses—namely, eukaryotic parasites—but the themes and concepts we explore apply to all infectious agents. Furthermore, we include current discussions of parasitic fungi and plants that expands the usual coverage of parasitic protozoans, helminths, and arthropods. There are three reasons for our approach.

First, as instructors who have long taught the subject, we have often been frustrated to find that after completing the obligatory survey of different parasite groups and life cycles, there was little class time left to discuss all the marvelous aspects of parasite biology that somehow don't fit the traditional course mold.

Second, as observed by Clark Read in 1972 in *Animal Parasitism*, his prescient textbook built around core concepts, the study of parasitism has greatly profited by integration with many other biological disciplines, so much so that it has become a major contributor of new insights in fields such as immunology, ecology, and evolutionary biology. This body of accomplishment deserves to be highlighted. It will only increase in stature and importance with time.

Third, many texts provide the traditional approach to parasitology, an approach that is also found with ease on the Internet. What is much harder to find online is a more synthetic discussion covering major thematic areas within parasitology. For example, an Internet search for "*Ascaris*" will immediately yield an avalanche of data on this nematode parasite, which include details of its life cycle, morphology, epidemiology, and treatment of infections. It is far more difficult to locate a succinct summary on how parasites in general affect the evolution of their hosts or an explanation of the impact of parasites on conservation of endangered species. Our book is structured to emphasize concepts ranging from overviews of parasite diversity to transmission biology, pathology, immunology, ecology, evolutionary biology, conservation biology, and control.

We have not abandoned the traditional organization of parasitology textbooks. We have provided in an easy to use appendix called the *Rogues' Gallery of Parasites*, a concise overview of the basic biology of major parasite groups. This appendix is cross-referenced throughout the book's ten chapters.

This book is intended for courses in parasitology for upper-level undergraduate students and graduate students with interests in parasitology. Students of human or veterinary medicine or of public health will also find the book of interest because it provides succinct coverage of parasites and their biology and discusses important topics such as pathology and control. Students interested more broadly in infectious diseases of all kinds will find that many of the examples, concepts, and ideas are valid even if the infectious disease is caused by a virus, bacterium, or tapeworm. Finally, for students interested primarily in immunology, ecology, evolutionary biology, or conservation biology, this book will provide more specific knowledge about a broad spectrum of parasites, giving them considerable perspective for integrating parasitology into their own specific disciplines.

We have deliberately kept the book of manageable length; with only ten chapters, instructors wishing to use the entire text should have no trouble doing so in a typical semester.

We begin with a chapter designed to delimit the field of parasitology and its relationships with other closely related and overlapping disciplines. The chapter also emphasizes some of the specialized vocabulary regarding parasites that is used throughout the book. Like all the chapters, it also features boxes that highlight topics requiring amplification or that are of particular interest or current relevance. Throughout this and the remaining chapters, the scientific names for those parasites we have identified as being of particular importance are designated in red type. This cue lets the reader know that the particular parasite in question is included in the *Rogues' Gallery* found at the end of the book. Here the reader can quickly find a concise summary of the parasite's taxonomic position, prevalence, distribution, mode of transmission, life cycle, pathological consequences, and means of diagnosis, treatment, and control. The *Rogues' Gallery* enables a parasitology instructor to provide students with traditional content on the basic features of parasite biology while focusing on major concepts. Additional features in the text include a summary that highlights the major points of each chapter and a set of questions to stimulate further student discussion and thought.

We hope that our text provides your students with an exciting and thought-provoking entry into the dynamic and endlessly intriguing study of parasites and their hosts. We are eager for specific and constructive comments regarding what works, what does not, and how we can improve it in the future.

ACKNOWLEDGMENTS

The reviewers who provided constructive feedback on early drafts of book chapters are especially deserving of our thanks: Abdel-aziz M Ahmed (University of Khartoum, Sudan), Deloris Alexander (Tuskegee University, USA), Gustavo Arrizabalaga (University of Idaho), Christopher Bayne (Oregon State University), Hebert de Matos Guedes (Universidade Federale do Rio de Janeiro, Brazil), Jillian Detwiler (Texas A&M University), Christian Doerig (Monash University, Australia), Érica dos Santos Martins Duarte (Universidade Federale do Rio de Janeiro, Brazil), Brian Lund Fredensborg (The University of Copenhagen, Denmark), Wendy Gibson (University of Bristol), Simon Harvey (Canterbury Christ Church University, UK), Petr Horák (Charles University in Prague, Czech Republic), Mariakuttikan Jayalakshmi (Madurai Kamaraj University, India), Armand Kuris (University of California, Santa Barbara), Kevin Lafferty (University of California, Santa Barbara), David Marcogliese (Environment Canada), Thomas Nolan (University of Pennsylvania), Kimberly Paul (Clemson University, USA), Tom Platt (St. Mary's College, USA), Michelle Power (Macquarie University, Australia), Greg Sandland (University of Wisconsin-La Crosse), Ravinder Sehgal (San Francisco State University), Jack Shurley (Idaho State University), Balbir Singh (Universiti Malaysia Sarawak), Photini Sinnis (NYU Langone Medical Center, USA), Mike Turner (University of Glasgow), Marcos André Vannier-Santos (Fundação Oswaldo Cruz, Brazil), Mark Wiser (Tulane University, USA).

Jennifer Kavka provided clerical and technical assistance on several occasions. We thank the COBRE Center for Evolutionary and Theoretical Immunology (CETI) for their infrastructure support. John T. Sullivan of the University of San Francisco provided useful advice and photographs. Gerald Mkoji of the Kenya Medical Research Institute, offered many helpful insights. We also wish to thank the community of parasitologists at the University of New Mexico for their expertise and feedback on key points and for providing needed photographs. Included here are Coen Adema, Sara Brant, Sarah Buddenborg, Charles Cunningham, Ben Hanelt, Martina Laidemitt, Melissa Sanchez, and Si-Ming Zhang. We also wish to acknowledge the financial assistance provided by the U.S. Agency for International Development, the National Institutes of Health, and the Bill and Melinda Gates Foundation for supporting our research careers in parasitology, thereby helping to make this book a reality in the process.

The authors would like to express their deep appreciation and thanks to several people who have helped make the publication of the book possible. We are especially appreciative of the hard work and professionalism of the staff at Garland Science. Senior Editor Elizabeth Owen was instrumental in guiding us through all stages of the project. Georgina Lucas, Senior Production Editor, worked with us on a daily basis for weeks to produce the final text. Deepa Divakaran worked diligently to identify and secure permissions for the photographs used throughout the book. Patrick Lane produced excellent new original artwork for the book and worked efficiently with us to modify the figures as needed. Adam Sendroff was in charge of both marketing and promotion. Denise Schanck, Vice President of Garland Science, supported the project from the start to the finish.

Finally, we cannot conclude without expressing our appreciation for the patience and understanding shown by Robin Loker and Leslie Kranz over the many months needed to complete this project.

ONLINE RESOURCES

Accessible from www.garlandscience.com, the Student and Instructor Resource Websites provide learning and teaching tools created for *Parasitology: A Conceptual Approach*. The Student Resource site is open to everyone, and users have the option to register in order to use book-marking and note-taking tools. The Instructor Resource site requires registration, and access is available only to qualified instructors. To access the Instructor Resource site, please contact your local sales representative or email science@garland.com. Below is an overview of the resources available for this book. On the Websites, the resources may be browsed by individual chapters and there is a search engine. You can also access the resources available for other Garland Science titles.

For students

Evolutionary Trees: A tutorial on constructing and using evolutionary trees, including molecular phylogenetics

Answers and Explanations: A guide to answering the end-of-chapter questions

Immunology: A review of basic immunology for parasitologists

Glossary: The complete glossary can be searched and browsed as a whole or sorted by chapter.

Flashcards: Each chapter contains a set of flashcards that allow students to review key terms from the text.

For instructors

Figures: The images from the book are available in two convenient formats: PowerPoint® and JPEG. They have been optimized for display on a computer.

Instructor's Guide: Guidance on how to structure a course using this text, how to integrate this text into a taxonomic parasitology course, and how to integrate the text with a laboratory course

CONTENTS

An Introduction to Parasitism

1

I don't mind a parasite. I object to a cut-rate one.
RICK BLAINE in the movie *Casablanca*

In This Chapter

Parasitism IS THE MODE OF EXISTENCE IN WHICH ONE ORGANISM, a **parasite**, infects another, the **host**, and the parasite does some measure of harm to the host while itself deriving a benefit. **Parasitism** is not rare—as we shall see in this book, parasitism is one of the most common lifestyles on earth. As such, the study of parasitism can teach us a great deal about life in general.

Parasitism can be studied at many different levels. For instance, parasites in the aggregate pose formidable problems for human health and well-being. Over 500,000 African children still die of malaria every year. Along with the parasites that cause malaria, there are many other parasites that further jeopardize the health of people, especially those living in disadvantaged conditions (**Figure 1.1**). A greater understanding of the biology of the responsible organisms could lead to development of new vaccines, drugs, or control strategies to abolish these long-standing scourges of humanity. Parallel considerations apply to veterinary medicine or production of food plants because parasites are also a constant menace to these endeavors. Some livestock parasites are resistant to virtually all known drugs and many of our essential food plants are shockingly uniform in their genetic makeup and thus vulnerable to diseases caused by newly emerging parasites of plants.

Other reasons to study parasitology are more subtle yet still important. For example, we are currently experiencing crashes of honeybee (*Apis mellifera*) populations in the Northern Hemisphere (**Box 1.1**). This situation is worrisome both because honeybees are an indicator of the overall health of the ecosystems in which we live and because they are indispensible pollinators of many of the plants we rely on for food. Many interacting factors, including parasite infections, lie behind these mysterious crashes in bee populations and are worthy of the attention of inquiring biologists. In general, the ongoing loss of biodiversity, in some cases aided and abetted by parasites, provides another strong motivation to know more about parasites and their biology. In the context of conservation biology, it is important to keep in mind that parasites represent as much as half of all species and, as we shall see, are valuable and imperiled forms of biodiversity in their own right.

The efforts of ecologists and evolutionary biologists have in recent years brought into much clearer focus the extent to which parasites have influenced our world. Parasites play key roles in food webs and energy flow within

Species names highlighted in red are included in the Rogues' Gallery, starting on page 429.

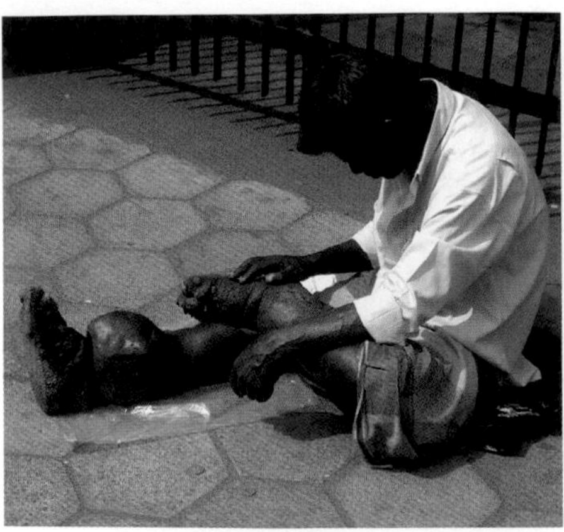

Figure 1.1 Parasites take a toll on human health. A man suffering from the disfiguring effects of filariasis, likely caused by *Wuchereria bancrofti*. A chronic and debilitating infection like this can impair physical and mental health.

ecosystems, and they have long acted as powerful selective agents influencing the population dynamics, competitive interactions, likelihood of predation, and mating behavior of the hosts they occupy, including humans. Revelation of the impact of parasitism in both ecological and evolutionary time frames is an exciting, and very much ongoing endeavor. Furthermore, it has a practical component, in helping us understand, for example, the underlying factors leading to emergence of new parasite problems, whether they afflict food plants, honeybees, lions, cattle, or humans.

Most students find parasites to be intrinsically fascinating and this book hopes to capture that interest. Whether the focus is on a parasite gene that encodes a new vaccine target or on images taken from satellites that help gauge the global responses of parasites to climate change, parasites have much to offer in improving our understanding of life on Earth.

In order to understand the concepts relating to parasites and the interfaces between the study of parasitism and several prominent biological disciplines, such as biodiversity studies, immunology, ecology, evolution, conservation biology, and disease control, that are the topics of the chapters to follow, the purpose of this chapter is to introduce the basic vocabulary needed to pursue the study of parasitism. This chapter also describes some of the nuances in definitions of parasitism and explores the boundaries between parasitism and other common intimate associations.

1.1 BUILDING AN UNDERSTANDING OF THE BASICS OF PARASITISM

Parasites live in or on their hosts and cause them harm

There is no single distinctive feature that cleanly separates all parasites from all nonparasites. Brooks and McLennan (1993) have said that "the only unambiguous definition is that parasites are those organisms studied by

BOX 1.1

Parasites and the Decline of Honeybees

The European, Western, or Common honeybee *Apis mellifera* (Figure 1.1A) is the most commonly domesticated honeybee species. It plays a vital role as a pollinator of both wild plants and important crop plants and in the production of hive products such as honey. Since late 2006, *A. mellifera* has suffered declines in its populations, especially in the Northern Hemisphere, a phenomenon called colony collapse disorder that is marked by the widespread and sudden disappearance of honeybee workers from colonies.

Honeybees and their health are instructive for parasitologists because they show that parasites can have far-reaching and unexpected impacts, that different kinds of parasites may have unforeseen synergistic effects, and that introductions of parasites into new environments often have disastrous consequences. Furthermore, they show that it is often not straightforward to ascribe a particular malady like colony collapse disorder to a single cause.

One of the players in this story is the infamous Varroa mite (*Varroa destructor*) (Figure 1.1B), a mite that attaches to the bee and sucks the bee's blood (hemolymph), weakening its host as it does so. In the process, the mite may also facilitate transmission of a range of RNA viruses such as the deformed wing virus (DWV) directly into the hemolymph of the honeybee. This bypasses more traditional routes of viral transmission, such as via oral or sexual contact. Although the mite alone has been implicated in collapse of colonies, the one–two punch of mite and virus have been considered to be major contributors to widespread advent of colony collapse disorder. In this case the mite serves as a host in which DWV can increase in abundance and persist for long periods of time. Studies in Hawaii where DWV and other viruses were present but mites were previously absent have shown that introduction of the mite has had the effect of increasing the prevalence of DWV from about 10% to 100% (Figure 1.1C). Furthermore, the titer, or amount, of the virus increased in individual bees by a millionfold and the genetic diversity of DWV decreased to a single strain. These trends suggest that the introduction of the mite has greatly favored transmission of a single highly contagious variant of DWV that is well-adapted to the Varroa mite, a partnership that poses grave threats for honeybees throughout the world.

Before concluding that colony collapse disorder can be attributed solely to Varroa mites, the honey bee viruses they habor, or both, note that many other causes also have been proposed including use of pesticides on crop plants that honeybees visit as sources of nectar and pollen and overall habitat degradation. Clearly a variety of stressors may be involved that interact with each other in complex and as yet imperfectly understood ways, unfortunately with considerable detriment to honeybees. The situation is further complicated by the use of chemicals called miticides to control Varroa mites, with attendant increases in mite populations resistant to the chemicals. Some feral honeybee populations appear to be developing increased resistance to mite attacks, offering hope for a brighter future for honeybees.

References

Henry M, Beguin M, Requier F et al (2012) A common pesticide decreases foraging success and survival in honey bees. *Science* 336:348–350 (doi:10.1126/science.1215039).

Martin SJ, Highfield AC, Brettell L et al (2012) Global Honey Bee Viral Landscape Altered by a Parasitic Mite. *Science* 336:1304–1306 (doi: 10.1126/science.1220941).

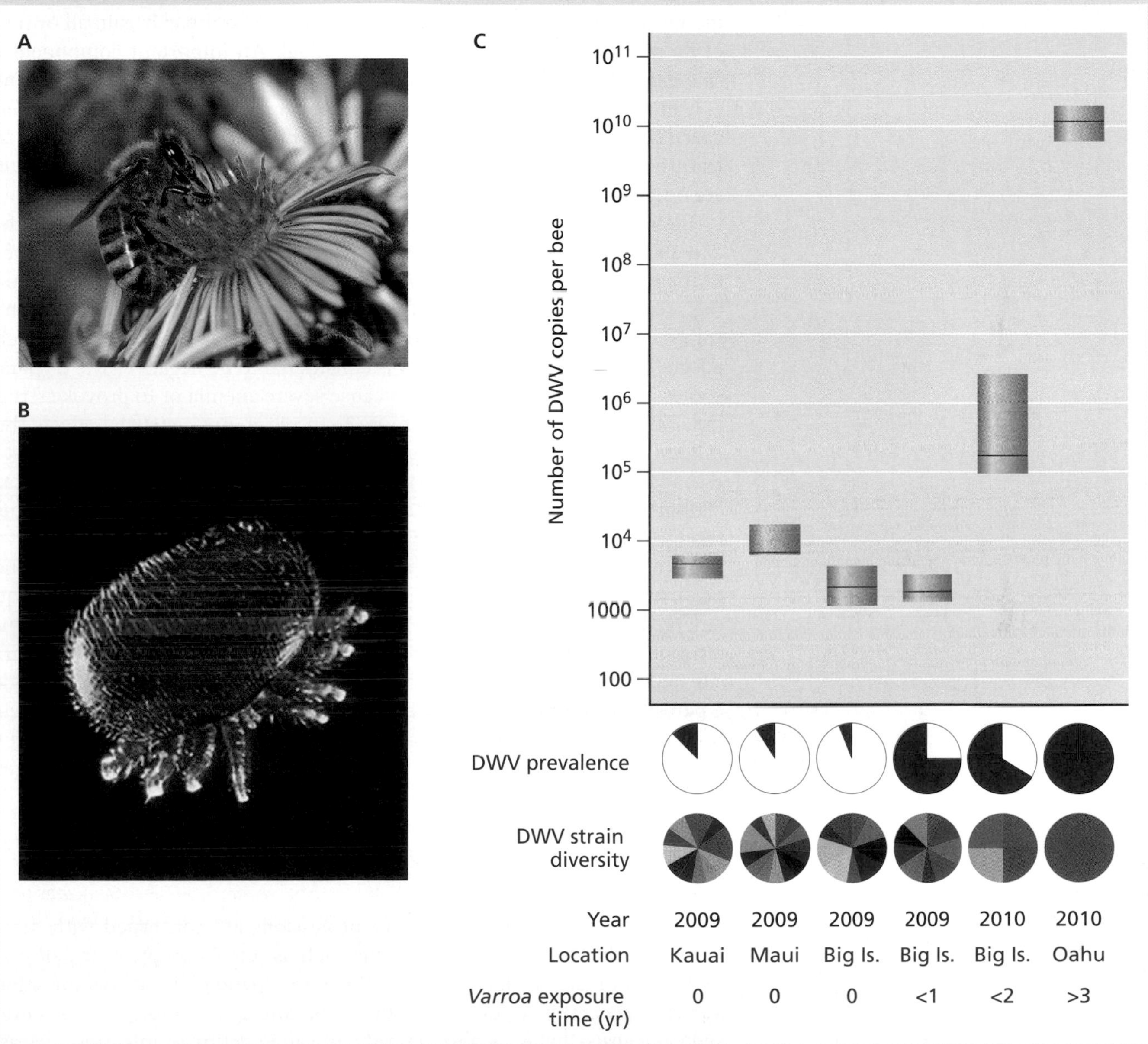

Figure 1. The decline of honeybees. (A) The honeybee *Apis mellifera*. (B) The Varroa mite, *Varroa destructor*. (C) In various Hawaiian islands, the longer the Varroa mite has been present (for times ranging from 0 to more than 3 years), the higher the viral load reported per infected honeybee (see the upper graph), the higher the prevalence of DWV (see red and white pie chart), and the lower the number of viral strains recovered (see multicolored pie charts). In Oahu where the mite has been present longest, the high viral loads and prevalence (100%) are associated with a single strain of DWV. (B, Courtesy of Hawaii Department of Agriculture. C, From Martin SJ et al [2012] *Science* 336:1304–1306. With permission from AAAS.)

people who call themselves parasitologists." This book takes a less whimsical view and describes a parasite as an organism that lives in or on another living organism (the host), obtaining from it part or all of its nutrition or needs of existence and imposing a net detrimental effect on the host. Parasites tend to be small relative to their host, and they may or may not eventually kill the host. A host is a living organism in or on which a parasite lives, that provides essential needs for the parasite and is to some extent harmed by the parasite.

This definition implies that the parasite possesses specialized attributes that enable it to survive in or on a host and that it will die if denied access to a host. Many parasites are specialized such that they can exploit only a narrow spectrum of possible hosts, but some parasites are more generalized in their host use (we return to this topic in Chapter 6). An individual of a particular life-cycle stage of a particular parasite species will ordinarily gain all or most of its nutrition from a single host individual. An important component of this definition is that the parasite accrues benefit (+) at the host's expense (–), thus making parasitism a +/– type of relationship. Parasites have been described as intimate enemies, because their interactions with their hosts feature close and prolonged physical contact, and because the interactions are often adversarial.

Biologists have long debated whether a parasite is necessarily harmful, and such terms as beneficial parasites or the goodness of parasitism appear in the parasitological literature. An example of the goodness of parasitism is the provision of useful micronutrients to a host by blood-dwelling trypanosomes. Although it is conceivable that parasites may have some beneficial effects, their net effect on the host is considered to be negative. The trypanosomes in question are also likely to cause severe anemia or to provoke strong inflammatory responses, which more than offset any beneficial effects they might have. The notion that a parasite has a net harmful effect on its host is essential to parasitism, setting it apart from other intimate symbiotic associations such as mutualism. Parasites do have useful properties or attributes though, topics discussed in Chapters 5 and 8.

As already noted, parasitism can be thought of as the mode of existence occurring between a parasite and its host. It is also one that exhibits unique emergent properties. In this case, an emergent property refers to distinct attributes or phenomena that arise when two components (such as a parasite and host) interact. In this context, some examples include the ability of a parasite to colonize a host, to establish itself within the host in the face of hazards posed by the hosts' immune system, and to achieve transmission to new hosts. Much of the content that follows describes parasitism's emergent properties.

Opinions vary on how to define some of the key aspects of parasites and their biology

Because the medical and veterinary professions are concerned with maintaining health, they tend to use terms such as infectious agent or pathogen to refer to organisms taking up residence in a patient's body. An **infectious agent** is generally considered to be an organism or suborganismal entity, such as a virus, that is capable of producing an infection or infectious disease. An **infection** is the entry and development or multiplication of an infectious agent in the body. In some cases infection results in clinical manifestations, referred to as an **infectious disease**. A **pathogen** is an infectious agent capable of causing a state of disease in a host, with **disease** broadly defined as a pathological condition with symptoms peculiar to it that set it apart from a normal body state. Disease involves an alteration in normal cell, organ, or organismal activity and is discussed at length in Chapter 5. How does our previous definition of a parasite relate to these commonly used terms, such as infectious agent or pathogen?

Different biological disciplines answer this question in different ways. For example, among microbiologists and infectious disease specialists, the term parasite is often used to refer to those organisms with the properties of parasites that are eukaryotes. **Eukaryotes** are organisms with a defined nucleus and membrane-bound organelles, such as mitochondria. Eukaryotes include unicellular protozoans, fungi, plants, and animals (this diversity is discussed in more detail in Chapter 2). Bear in mind that many eukaryotes are free-living and so are not parasites. From this vantage point, suborganismal entities, such as prions or viruses, and prokaryotes (bacteria) that infect hosts and cause them damage are not parasites. Parasites (in the restricted eukaryotic sense), viruses, and some bacteria are all considered infectious agents or pathogens by this reckoning.

Alternatively, ecologists and evolutionary biologists tend to use the word parasite more inclusively, placing more emphasis on the functional aspects of the definition (infectious, cause damage) than the taxonomic group (virus, bacteria, eukaryote) to which the parasitic entity might belong. According to this view, the term parasite could include anything from prions to viruses to organelles to some bacteria to some eukaryotes, so long as they infect their hosts and cause them harm in the process. Distinctions among the terms parasite, pathogen, and infectious agent are by no means clear-cut, and the definition of a parasite we will use in this book, if modified slightly to include such suborganismal entities as prions, viruses, and even some organelles as parasites, is more inclusive, less confusing, and more compelling because of its emphasis on the infectious and harmful nature of parasites. Consequently, be aware that the terms pathogen and parasite, as employed throughout this text imply nothing on their own about the taxonomy of the organism or suborganismal infectious agent in question. Rather, these terms reflect our broad view that any infectious agent, whether subcellular, prokaryotic, or eukaryotic, that inflicts harm on its host can justifiably be called either a pathogen or a parasite. Parasitologists tend to use the term parasite more frequently.

Let us reconsider the notion of disease, pathology, harm, or damage caused by parasites. A term often used to embody this concept of harm is **virulence**. This term, which appears repeatedly throughout this book, is also defined differently depending on the discipline. Infectious disease biologists often consider virulence to be a measure of the likelihood that an infectious agent causes disease or even fatality. Among ecologists and evolutionary biologists, virulence is generally considered and measured differently; it is the ability of a parasite to reduce its host's fitness. **Fitness** is a measure of the success of an individual in passing on its genes to future generations and is influenced by the individual's ability to survive and to reproduce. A parasite might shorten its host's life span and thereby affect the host's eventual reproductive output, or the parasite might diminish its host's reproductive output but not affect survival. If a parasite castrates or kills its host, it is considered to be especially virulent.

Note also that in some cases the term "parasitic" may be applied to organisms that are not parasitic in any real sense. For example, in the free-living marine worm *Bonellia*, the male is much smaller than the female and may even live within her body in a location where he can conveniently fertilize her eggs. Such males are sometimes identified as parasitic males. This term should probably be avoided. Although such males derive nutrition and a living place from the female, they are hardly detrimental and are required for the female to reproduce.

The residence time for a parasite in or on a host is highly variable

A key aspect of a parasite's biology is that it lives in or on its host. Without prolonged and intimate contact with a host, such that the host essentially becomes its environment and without which the parasite dies, a parasite loses its distinction from many other organisms. For example, a tapeworm such as ***Diphyllobothrium latum*** can live within the small intestine of a human host for up to 25 years. For some parasites there is no necessary requirement for a prolonged stay in or on their hosts, and they may spend much or most of their life away from a host. For example, the larvae (called **glochidia**) of bivalves of the family Unionidae must undergo an obligatory period of development within the gills, fins, or skin of fishes before they disengage from the host and develop as free-living adult bivalves. (See **Box 8.2** for more about these remarkable and endangered organisms.) Glochidia may spend only a few weeks as parasites in the gill filaments of fish and then begin a free-living life that may last a hundred years or more. Yet without their brief parasitic existence, glochidia could not develop to adulthood.

Mosquitoes and bed bugs may have only fleeting contact with their host, yet this short period of time is critical for them. Their success at obtaining a blood meal determines whether they reproduce successfully. Is a mosquito any less dependent on its host than a flea or a tick that may occupy a host's skin or a louse that typically never leaves its host? It seems that there is no defined point along such a continuum of residency that unambiguously validates an organism as being a parasite. Because organisms such as mosquitoes and bed bugs exhibit specialized behaviors for locating a host and have to overcome the behavioral and chemical defenses of the host if they are to succeed, it is difficult to exclude them from the realm of parasites. Mosquitoes, along with several other blood-feeding arthropods, which are sometimes classified as **micropredators**, are of course also relevant for our discussions because of the role they play in vectoring other organisms that are unequivocally parasitic.

There are many additional ways to categorize parasites.

Ectoparasites and Endoparasites

Given how widespread the parasitic lifestyle is, a theme developed further in Chapter 2, it is not surprising that many different kinds of parasites inhabit our world. Some such as the Varroa mite mentioned in Box 1.1, or the human body louse ***Pediculus humanus humanus***, live on the external surface of their hosts and are referred to as **ectoparasites**. They are said to cause **infestations**. In contrast, **endoparasites,** such as the adult tapeworm ***Taenia solium*** or the trematode ***Schistosoma mansoni***, live inside their host's body and are said to cause infections. As if to defy attempts at categorization, a female of the cod worm *Lernaeocera branchialis* attaches firmly to the gills of marine fish like codfish, leaving most of her body (including coiled egg strings) hanging off the external surface of the gills (**Figure 1.2**). From the anterior end of this highly modified crustacean though, a long projection extends deep into the fish and taps blood from the fish's circulatory system. Is it an ecto- or endoparasite? Several other similar examples could be noted.

Notice that, ***S. mansoni*** and ***T. solium***, both mentioned above, are examples of **helminths**, a term of convenience, with no formal taxonomic meaning, commonly used to denote a parasitic worm. Tapeworms, flukes, acanthocephalans, and parasitic nematodes are commonly called helminths (see the Rogues' Gallery for details regarding their taxonomy).

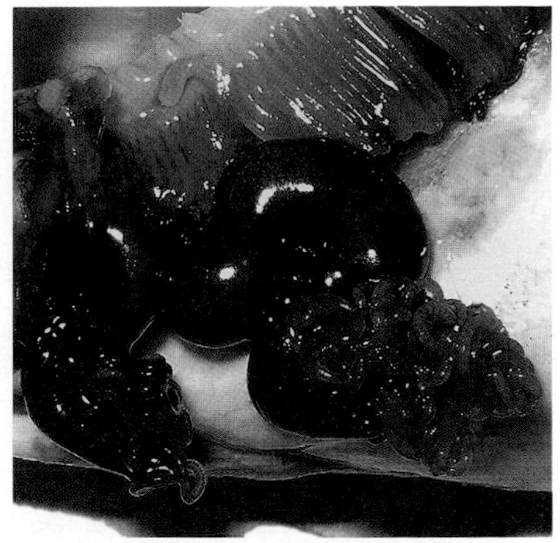

Figure 1.2 Ectoparasite or endoparasite? Two plump specimens (20–50 mm in length) of the "cod worm," *Lernaeocera branchialis*, embedded in the gill of a marine fish. Note the conspicuous coiled egg strings of this highly modified parasitic crustacean. (© Hans Hillewaert/ CC BY-SA 4.0.)

Castrators and Body Snatchers

Some parasites consistently provoke total or near-total cessation of their hosts' reproductive activities, diverting these host resources instead to their own development and reproduction. These parasites are referred to as **castrators.** An example is the larval stages of the blood fluke ***Schistosoma haematobium***. It develops in the snail *Bulinus truncatus*, which is usually castrated as a consequence.

Body snatchers are parasites that effectively take over the host, often including its behavior, such that the infected host can be viewed as an extension of the parasite's phenotype. Consequently, a snail infected with a trematode might look externally the same as an uninfected snail but would behave quite differently, favoring the continued success of the trematode larvae within at the expense of the snail itself, which has little or no future prospect of producing its own progeny. The parasite within may influence host behavior even to the extent of increasing the likelihood that another animal preys on the infected host. In the latter case, as might be anticipated, the predator is often of a species that serves as the next host in the parasite's life cycle. Further discussion of such **trophically-transmitted parasites,** parasites that are transmitted to their next host by ingestion, usually predation, and how they influence host behavior is to be found in Chapters 3, 5, and 6.

Obligatory, Facultative, and Opportunistic Parasites

An **obligatory parasite** requires a suitable host to complete its life cycle. Without such a host, obligate parasites could not exist. The large intestinal roundworm ***Ascaris lumbricoides*** must develop to adulthood in the human small intestine. Its eggs are passed in the host's feces and survive outside the host for a time, but if another human does not later ingest the eggs, the parasite could not persist. In some cases, an obligatory parasite may require a particular host simply to achieve an essential stage in development without persisting in the host to reproduce. The glochidia larvae of unionid bivalves mentioned earlier must pass through a brief but obligatory period of parasitism if they are to eventually develop to adulthood. In some cases, as with the infamous trichina worm ***Trichinella spiralis***, the parasite never leaves the body of a host: one infected animal is eaten by another that itself becomes infected in the process.

In contrast, **facultative parasites** are usually free-living but, if given the opportunity, can adopt a parasitic existence. One example is provided by *Naegleria fowleri*, normally a free-living ameba living in the muddy bottoms of aquatic environments (**Figure 1.3**). This organism is capable of colonizing and multiplying in the human brain if it is sniffed into the nasal passages of a person, say, enjoying a dip in a natural hot spring. In the latter case, the infection caused is a dead end for *N. fowleri*, as no transmission occurs either to other hosts or back to the aquatic environment.

A related concept is the **opportunistic parasite**, a parasite that takes advantage of particular circumstances to initiate an infection in a host that it normally does not infect or in which it does not normally cause disease. A person with a compromised immune system, as might occur in people infected with the human immunodeficiency virus (HIV), may become susceptible to infections with organisms such as the yeast-like fungus *Pneumocystis jirovecii*, or the apicomplexan protozoans ***Cryptosporidium parvum*** or

A

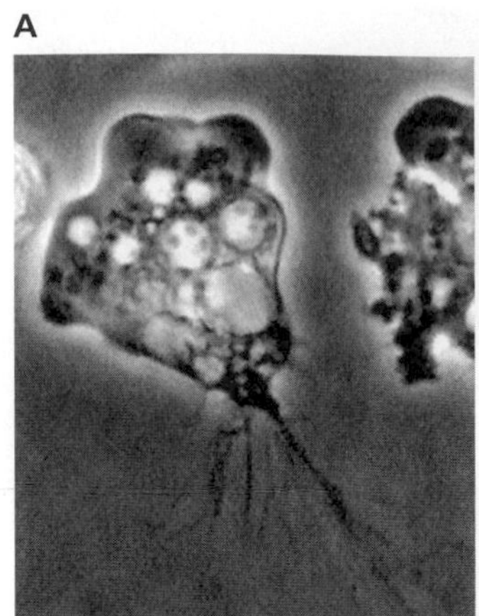

B

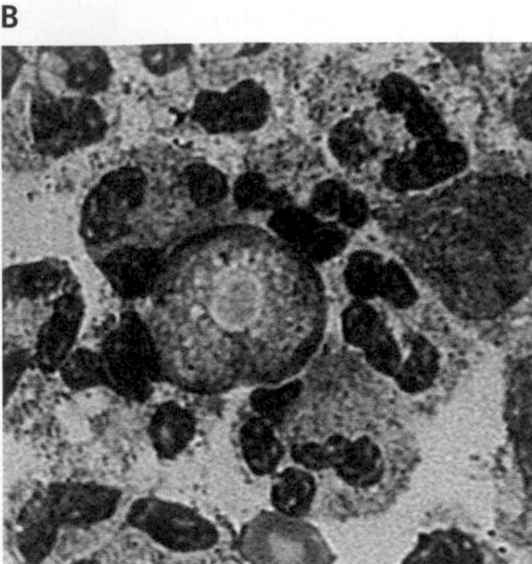

C

Figure 1.3 *Naegleria fowleri*, a facultative parasite. This ameba normally lives in soil or warm freshwater environments, often in the sediments at the bottom. (A) A free-living, feeding trophozoite. (B) A trophozoite (center) surrounded by white blood cells. The parasite has invaded the central nervous system of a person who has sniffed it into the nasal passages. From here the ameba crosses the cribiform plate to gain access to the central nervous system, where it consumes nervous tissue, causing primary amebic meningoencephalitis. This condition usually is fatal. (C) You would do well to heed the sign warning visitors to hot springs not to dip your head beneath the surface or to sniff in water from the spring. (A and B, courtesy of CDC.)

Toxoplasma gondii, all of which would normally be held in check by a person with an intact immune system. Opportunistic infections become ever more relevant in a world where many people have compromised immune systems owing to infections such as HIV or to medical procedures such as organ transplants.

Hyperparasites

The Irish author Jonathan Swift wrote:

So, naturalists observe, a flea
Has smaller fleas that on him prey;
And these have smaller still to bite 'em:
And so proceed ad infinitum.

This poem is an eloquent way of saying that parasites themselves are not immune to exploitation by other parasites, a phenomenon termed **hyperparasitism**. A **hyperparasite** is thus a parasite of a parasite. The dog flea *Ctenocephalides canis*, itself an ectoparasite of dogs, cats, and occasionally people, becomes infected with the larval stage of a tapeworm ***Dipylidium caninum***, which can thus be considered a hyperparasite. When a dog grooms and ingests a parasitized flea, the larval tapeworm develops into an adult tapeworm in the dog's intestine.

More complex examples are known in which a host insect such as the pea aphid *Acyrthosiphon pisum* harbors a nested set of parasites. Within the aphid is a larva of the parasitoid wasp, *Aphidius smithi*. This larva can in turn be infected by the larva of another parasitoid wasp, *Alloxysta victrix*. This hyperparasite can be colonized by larvae of a hyper-hyperparasite, *Asaphes californicus*, yet another parasitoid wasp. Given that *A. californicus* likely harbors a bacterium in its gut that might itself support a bacteriophage infection, a single aphid might conceivably support five nested levels of parasitism. Note the use of the term parasitoid. A **parasitoid** is an organism that spends a significant amount of its life on or within a single host, often sterilizing it, often killing it, and sometimes fully consuming it in the process. It then leaves the host and often has a free-living period of existence. As discussed later, parasitoids have many of the properties we associate with parasites, but are often more detrimental to their hosts.

Macroparasites and Microparasites

Two other terms, microparasites and macroparasites, appear in the parasitology literature. Mathematical modelers frequently use these terms, as we discuss in other parts of the text (see Chapter 6 in particular). **Microparasites** in this context are considered to be viruses, bacteria, or protozoans that establish infections in hosts for which it is difficult or impossible to quantify the actual numbers of individual infectious entities present. In cases such as this, modelers instead quantify the number of infected hosts as opposed to the number of parasites present within each host. Microparasites are small and have rapid generation times relative to their hosts.

In contrast, **macroparasites** include larger parasites such as parasitic worms or individual ectoparasites such as fleas or ticks. Because these larger parasites can be enumerated, the unit of study for the modeler is not the infected host, but the individual parasite, and models are built with the aim of predicting the size of the parasite population. Macroparasites are small and have rapid generation times relative to their hosts, but the disparities are less than between microparasites and hosts.

Parasites Included in This Book

As already noted, the approach taken in this book is to emphasize the concepts and principles relating to parasitism. Because there is much to be gained by being inclusive in our consideration of what constitutes a parasite, examples are included that involve viruses or bacteria—subjects commonly covered in virology or microbiology classes or books. Also included are descriptions of plants or fungi that are parasitic, even though these are discussed most commonly in botany or mycology courses, respectively. Such a cross-cutting approach helps to breach the artificial barriers that often arise between traditional disciplines. In deference to tradition, however, this book is devoted largely to organisms featured in most parasitology courses. Accordingly, our emphasis is on protozoans, helminths, and arthropods that infect vertebrate animals (especially humans), but examples invoking parasites infecting invertebrates or plants will also certainly be used.

1.2 HOSTS—ESSENTIAL LIFELINES FOR PARASITES

Hosts also fall into several different categories

Most parasites are obligate parasites and need a host organism to provide the basics for existence. According to Combes (1991), a host is a living environment that is "discontinuous, because individual hosts resemble islands; reproducing, because hosts produce hosts; variable when hosts evolve; and transient when hosts die." Parasites depend metabolically or physiologically on a host, a characteristic that for many lies at the heart of the concept of parasitism (for more see Chapter 3). Parasites might rely on the host for nutrition, a key digestive enzyme, or a developmental cue that enables them to mature in synchrony with their hosts. Several important terms are used to describe particular kinds of hosts relevant to a parasite's interests, and these are presented in this section.

Definitive and Intermediate Hosts

A **definitive host** is one in which a parasite achieves sexual maturity, which is usually followed by sexual reproduction within that host. This may take a familiar form in which sperm from one individual parasite fertilizes ova from another, followed by formation of a zygote, and production of progeny. Alternatively, it may be more cryptic and rare as in some protozoans, yet still result in genetic recombination. An **intermediate host** is one in which a parasite undergoes a required developmental step and may even reproduce asexually but does not reproduce sexually. As an example, for the helminth ***Fasciola hepatica***, sheep or cattle are the definitive host because it is here its adult stages are found and sexual reproduction occurs. Snails such as *Lymnaea truncatula* serve as intermediate hosts because they support a complex developmental program, including asexually reproducing larval stages. ***Fasciola hepatica*** is also an example of a parasite with an **indirect** or **heteroxenous life cycle**, meaning it lives in more than one host during its life cycle. In contrast, a parasite like ***Giardia lamblia***, the organism that causes giardiasis, has a **direct** or **monoxenous life cycle**. Actively feeding stages (trophozoites) of ***G. lamblia*** in the human small intestine eventually round up and form cysts that pass with the feces into the environment where they are infective for other people when they are ingested—no intermediate host is required. Incidentally, humans can be considered as a definitive host for ***G. lamblia*** because, even though typical sexual reproduction involving a male and female does not occur, genetic recombination does occur among individual parasites residing in the human gut.

Paratenic Hosts

In some cases, the resource provided by a host may be more in the form of a safe haven, a place in which the parasite can persist and prolong its survival, thus increasing its likelihood of transmission to a new host. Thus many parasites exploit **paratenic** or **transport hosts** in which the parasite does not undergo further or necessary development. Paratenic hosts enable the parasite to bridge a trophic gap in its life cycle, thus making transmission more probable. Chickens become infected with the nematode *Heterakis gallinarum* when they ingest the parasite's eggs on the soil. In some cases, though, earthworms eat the parasite eggs and although the eggs hatch in the earthworm, the nematode larvae that emerge from the eggs do not develop further. A sojourn in the paratenic earthworm host favors *Heterakis* in two ways, by prolonging its survival and by being packaged in a food item delectable to chickens.

The exploitation of earthworms by *Heterakis* is distinct from, yet reminiscent of, another phenomenon called **phoresy**. Phoresy is the association of one organism with another for the purpose of transport from one place to another with no physiological dependence. Many pseudoscorpions or free-living mites hitch rides on insects as shown in **Figure 1.4A**. The mites may be transported to a carcass where they disembark and may feed on eggs and larvae of flies also feeding on the carcass. A remarkable example of phoresy occurs when the female of the human bot fly *Dermatobia hominis* (Figure 1.4B) captures female mosquitoes in mid-air and oviposits on them (Figure 1.4C). When the mosquito approaches a human host to seek a blood meal, the bot fly eggs will hatch and the larva will fall to the skin of the human, either penetrate the skin or enter the hole made by the mosquito, and develop into a large larva or bot (Figure 1.4D), often under the skin of the scalp. Eventually the larva falls to the ground, pupates, and the adult bot fly emerges. By using a phoretic mosquito host, the female bot fly, which is large, scary, and conspicuous, avoids the hazards of approaching a human host and leaves it up to the mosquito to do the dirty work. Many college students have come back from spring break somewhere in Central America with a bot fly larva as a souvenir.

Reservoir Hosts

Another term frequently encountered in the parasitology literature, but one that does not have a single, settled definition, is **reservoir host**. One definition of a reservoir host is the host in which a parasite normally dwells and where it can be reliably maintained, even when not being actively transmitted, and

A

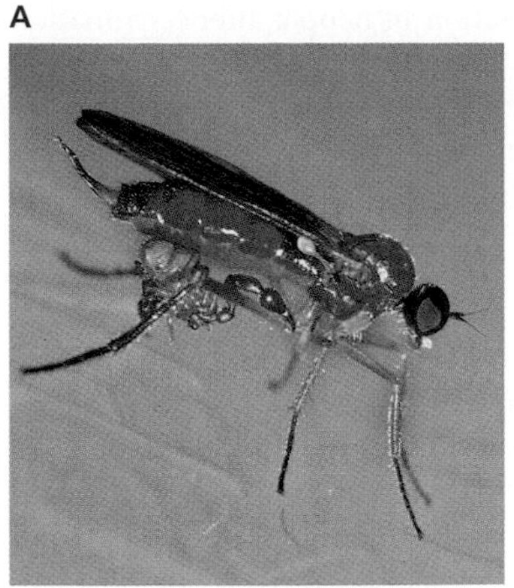

B

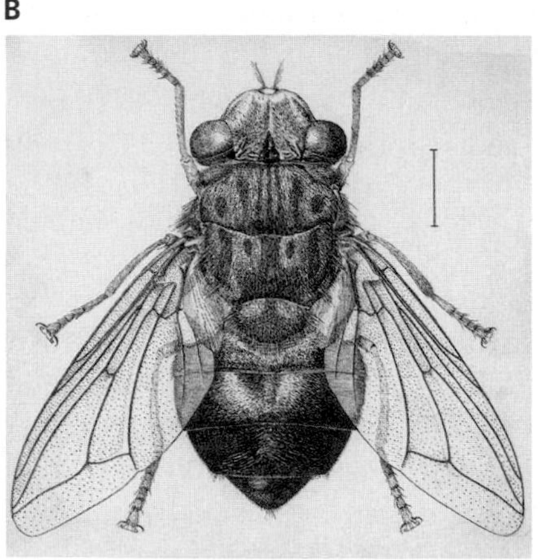

C

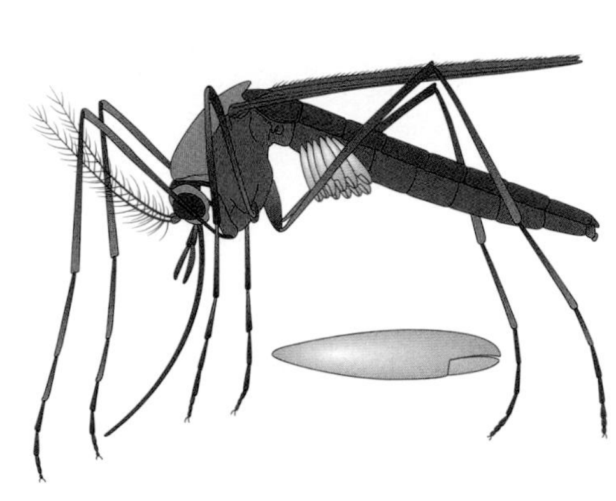

D

Figure 1.4 Examples of phoresy. (A) A phoretic pseudoscorpion *Lamprochernes* sp. attached to the leg of hybotid fly *Leptopeza flavipes.* (B) The female human bot fly *Dermatobia hominis.* (C) A cluster of *D. hominis* eggs on the under surface of the abdomen of the mosquito, which is the phoretic host in this case. (D) The large larva or bot of *D. hominis.* (A, Courtesy of Sarefo/ CC BY-SA 4.0. B and C, Courtesy of A Cushman, US Department of Agriculture. D, Courtesy of Lyle Buss, University of Florida.)

from which it can then colonize other hosts. The emphasis here is on the host that serves as the parasite's stronghold. According to this definition, canine distemper virus enjoys an abundant and successful reservoir host, namely domestic dogs, and virus originating from dogs serves as the source of infection to rare and often endangered wildlife species such as lions. Domestic dogs have also been implicated as reservoir hosts for ***Leishmania infantum***, a protozoan that can cause visceral leishmaniasis in people as well. Often, although not always, accompanying this view of a reservoir host is that it suffers little damage from the parasite. This version of the definition is often the one used when considering the emergence of a new disease in people or domestic animals. Thus Ebola virus exists in nature in its reservoir hosts, likely fruit bats, but occasionally spills over into humans. Wild macaques are the reservoir host for the malaria parasite, ***Plasmodium knowlesi***, which can also infect people.

A different, more anthropocentric definition of a reservoir host is one that serves as a source of infection and potential reinfection of humans and as a means of sustaining a parasite when it is not infecting humans. As an example, consider a parasite like ***Schistosoma mansoni*** that depends for most of its transmission on people but that can also infect wild rodents. Imagine that an active control program was able to eliminate ***S. mansoni*** in people. According to the second definition, the wild rodents could be viewed as reservoir hosts, both serving (along with snails) to sustain the parasite during the control program and as a source of eventual reinfection of people after termination of the control program. In some cases the term reservoir is even applied to a nonliving source of a pathogen. Thus the reservoir for anthrax spores may be said to be the soil. Consequently, the term reservoir host needs to be used with some care and may require some further clarification. The primary use of the term in this book will be the first one described.

Vectors

The term **vector** also deserves careful elaboration because of the significance of vectors in transmitting so many kinds of parasites. In its broadest sense, a vector is any means by which a parasite can be transmitted, whether it is wind, water, or some kind of host organism. Different categories are used if the vector is an organism.

A **mechanical vector** is one that picks up a pathogen from one host and transfers it to another, with the pathogen undergoing no development or multiplication within or on the vector. Transmission of the equine infectious anemia virus or the protozoan ***Trypanosoma evansi*** from one horse to another via the fresh blood contaminating the mouthparts of a horse fly are two examples.

A **biological vector** is one in which the parasite either develops or multiplies, or both. As an example, filarial worms like ***Wuchereria bancrofti*** undergo essential molts in their life cycle within mosquitoes but do not multiply within the mosquito. This is an example of what is known as **cyclodevelopmental transmission**, involving developmental progression but no increase in the parasite's numbers. Arboviruses such as dengue or West Nile Virus multiply greatly in numbers within mosquitoes but undergo no evident specific developmental progression. This is referred to as **propagative transmission**. Finally, in cases with malaria parasites like ***Plasmodium falciparum*** in mosquitoes, the parasite passes through essential developmental stages *and* multiplies within the mosquito, and this is referred to as **cyclopropagative transmission**. As noted above, many vectors can themselves be considered to be parasites.

Remember that vector has another common but quite distinct use in biology: that of an agent such as a plasmid or a virus used to transfer a gene or genes from one location to another. Although such vectors are important experimental tools in parasitology, vectors in this book refer primarily to organisms such as mosquitoes that transmit malaria, filariasis, or similar infections.

1.3 APPRECIATING PARASITISM'S PLACE IN NATURE

Parasitism is one of several categories of symbiotic associations

Symbiosis refers to an intimate association between organisms of two different species. Parasitism is but one category of symbiotic associations. In the manner most commonly used today, symbiosis embraces many kinds of intimate relationships, ranging from mutually beneficial (**mutualism**), to those that harm one partner at another's expense (for example, parasitism or predation), to those that clearly benefit one partner without obviously affecting another (**commensalism**).

Within the broad spectrum of symbiotic associations, many can be considered to be exploitative, in which one participant (the exploiter) benefits (+) and another is exploited (–). A parasite exploiting its host is one example. In other cases, though, the exploiter uses the exploited party, its products, or resources to avoid predators, defend territories, or reduce the cost of obtaining limited resources. The term parasite is even applied to some of these exploiters: social parasites, brood parasites, or cleptoparasites are examples. **Social parasites** invade or lay their eggs in the nest of a host organism and develop on food provided in that nest. Some birds are **brood parasites**. They surreptitiously deposit their eggs into a nest of the same or a different species of bird, with the eventual result that the foster parents will rear the progeny to fledging. A **cleptoparasite** steals or scrounges food or other resources obtained by another organism (**Figure 1.5**). In all of these associations, an organism customarily identified as a parasite clearly benefits at the expense of a host and there are often intricate and specialized behaviors or anatomical structures that facilitate the interaction. However, these kinds of exploitative associations do not involve one organism living in or on another, so we will not discuss them further. Phenomena like brood parasitism, though, can and should continue to spark debates as to what constitutes a parasite.

Figure 1.5 An example of cleptoparasitism. A cleptoparasitic fly, *Milichia patrizii*, stops an ant (*Crematogaster*). The fly grasps the ant's antennae and then taps the ant's palps, triggering an automatic regurgitation response by the ant. This enables the fly to steal food that the ant has accumulated. (From Wild AL & Brake I [2009] Field observations on *Milichia patrizii* ant-mugging flies (Diptera: Milichiidae: Milichiinae) in KwaZulu-Natal, South Africa. African Invertebrates 50: 205–212. With permission from Alex Wild.)

Parasitoids straddle the boundary between predation and parasitism

Also worthy of some consideration is the distinction between predators and parasites. Predation comes in many guises. At one extreme are conventional predators that are larger than their prey, dispatch the prey by force, and often kill and consume several distinct prey during their life. More ambiguous are predators such as spiders that might be smaller than their prey and that might consume a prey item slowly. Many different kinds of insects such as mosquitoes and bed bugs are sometimes referred to as micropredators or insidious predators. As with predators, the organism they exploit for each meal may be different. For reasons discussed above, mosquitoes and bed bugs fall more within the realm of parasites than predators, but dogmatic pigeon-holing is a hazardous business and some organisms will always defy our attempts to do so.

Now consider parasitoids, mentioned in the section on hyperparasites. A parasitoid is an organism that spends a significant amount of its life on or within a single host, often sterilizing it, often killing it, and sometimes fully consuming it in the process (**Figure 1.6A**). The parasitoid then leaves the host and often has a free-living period of existence. Parasitoids are often insects whose larvae live in or on other insects, usually resulting in the death of the parasitoid's host. Only a single host is required to complete larval development of the parasitoid and often several parasitoids may develop in the

A

B

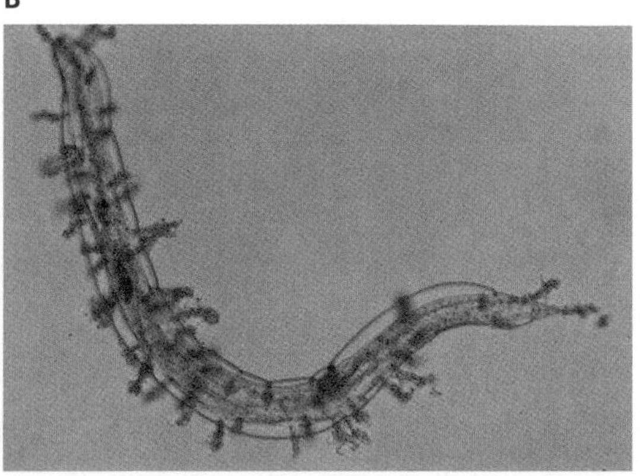

Figure 1.6. Two examples of parasitoids. (A) In this prickly situation, a female hymenopteran parasitoid wasp *Aleiodes indiscretus* oviposits in a gypsy moth (*Lymantria dispar*) caterpillar. (B) Soil-dwelling nematodes ingest the conidia of the parasitoid fungus *Harposporium anguillulae*, which get stuck in the nematode's pharynx, then germinate and grow as a mycelium within the nematode, killing it in the process. The hyphae of the fungus then penetrate the nematode's cuticle as seen here and produce conidiophores, which produce more conidia infective for other nematodes. (A, Courtesy of Scott Bauer, US Department of Agriculture. B, Courtesy of Dr Koon-Hui Wang, University of Florida.)

body of a single host. A related term is **protelean parasite**, which refers to an organism parasitic when immature but free-living as an adult.

The term parasitoid should not be reserved for insects because other unrelated organisms exhibit the same mode of existence. The fungus *Harposporium* first parasitizes and then kills soil-dwelling nematodes (Figure 1.6B). Mermithid nematodes and the horsehair worms (**Phylum Nematomorpha**; see also Box 2.3) are free-living as adults but are parasitoids in arthropods as juveniles, and the juveniles frequently, but not always, kill their hosts.

A parasitoid has much in common with a conventional parasite because it must possess all the tricks required to colonize a host, evade or subvert the host's internal defenses (see more in Chapter 4), and extract nutrition from the host without killing it, or at least not kill it until the proper time. The association is usually specific in that the parasitoid cannot indiscriminately infect any host species, and it clearly benefits the parasitoid at the host's expense. Furthermore, unlike a typical predator, the individual parasitoid is not obligated to colonize and consume another host individual to persist. For reasons like these, it is hard to exclude the fascinating aspects of parasitoid biology from the pages that follow. Note that parasitoids often directly and deliberately kill their hosts, whereas it is frequently not in the interest of a parasite to do so. However, as discussed in later chapters, some parasites do cause their host's demise. When this happens, it is often indirectly, as by increasing its host's conspicuousness or vulnerability to a predator that also is the next host in its life cycle.

Our understanding of parasitism is enhanced by an appreciation of its relationship to another ubiquitous type of symbiosis, mutualism

Whereas parasitism benefits one partner at the expense of another, mutualism benefits both interacting partners, so at first blush the two categories of symbiotic associations seem quite distinct. Before we delve into the properties of parasites in later chapters, it is important to gain a better appreciation for some of the nuances of mutualism and see how it interfaces with parasitism. As noted by Herre et al (1999), "mutualisms are ubiquitous, often ecologically dominant, and profoundly influential at all levels of biological organization."

Just as with parasitism, there is considerable debate about what comprises a mutualistic association. For example, are mutualisms obligatory or facultative and are they necessarily permanent or can they be temporary? Are they diffuse and indirect in action or highly integrated, reflective of long evolutionary accommodations among the mutualists? There is also vigorous debate about how to define and measure the benefits of mutualism and a growing appreciation that mutualisms may entail both cooperation and conflict. Indeed, mutualism has been called "reciprocal exploitation" that still derives a net benefit for both partners. Mutualisms can also exist simultaneously at multiple levels: as one example, two different kinds of bacteria living within a gutless marine oligochaete worm (*Olavius algarvensis*) enjoy a mutualistic association with one another, and both are also advantageous to their host's survival.

Parasitism or Mutualism?

Because of the difficulties of measuring net benefit and harm, it is often not clear whether a particular symbiotic association is mutualistic or parasitic,

A

B

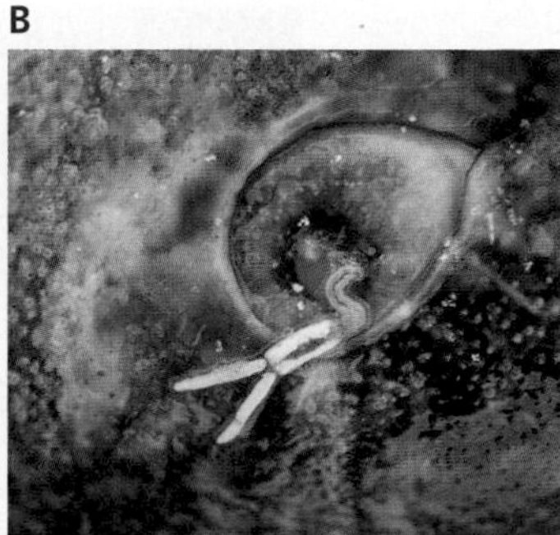

Figure 1.7. Parasitism, mutualism, or commensalism? (A) The Greenland shark (*Somniosus microcephalus*) has something dangling from its eye. (B) This is a copepod, *Ommatokoita elongata*. Is it a blatant parasite, cryptic mutualist, or surprising commensal? (A, © Ocean Images/Photoshot. B, © George W.Benz/SeaPics.com.)

or perhaps even commensalistic. Consider the case of Greenland sharks (*Somniosus microcephalus*) from Arctic waters, over 84% of which are infested with the ectoparasitic copepod *Ommatokoita elongata* (**Figure 1.7**). The copepod attaches firmly to the shark's cornea, usually one copepod per eye, and may severely impair the shark's vision. This may seem like an obvious case of parasitism, but is it? It has been speculated that the large (up to 7-cm long), yellowish-white copepods dangling from the eyes of this apparently lethargic shark species serve as lures to attract the char *Salmo alpinus*, which can then be captured and devoured by the shark. Although direct evidence for such an alluring role for the copepods is lacking, we must at least consider the possibility that this association that has obvious detrimental aspects may on balance be beneficial to the shark as well as the copepod. The thought of a copepod inhabiting the external surface of our own corneas and causing extensive scarring and visual impairment seems horrible, yet for a shark living in deep, dark waters, the essential function of the eye may be more in light detection than acute image formation. From the shark's standpoint, the infestation does not seem to be severely debilitating: after all, some of the infested sharks may be up to 5 m long in spite of having corneal scarring patterns consistent with multiple bouts of infestation. Consequently, this relationship may trend toward mutualism or commensalism, where the host is little damaged, especially from the perspective of its fitness. It serves to reinforce the notion that what seems like parasitism to us may be strongly influenced by our own preconceptions. The biological realities may differ.

Conditional Mutualisms

In some cases our judgment as to whether a particular association is mutualistic or parasitic depends entirely on the circumstances, in what may be called conditional mutualisms. The association of angiosperms (flowering seed plants) with mycorrhizal fungi is often presented as a classical mutualism, with the former providing organic carbon generated by photosynthesis and the latter providing several advantages. These advantages include the enhanced ability to uptake limiting nutrients such as phosphorous from the soil and to produce compounds that discourage plant pathogens. When soil nutrients are not limiting, however, or when a carbon-limited seedling plant is trying to establish, then the mycorrhizal fungi may either fail to help the plant or hurt it by diverting away essential organic carbon. Moreover, in some cases, fungal mycelia may become conduits that funnel nutrients away from host plants to other plants that are unambiguously parasitic (**Figure 1.8**), a phenomenon termed **mycorrhizal epiparasitism**. Tracking the subterranean connections between plants and fungi and determining the direction of net flow of resources clearly becomes a daunting task.

A

B

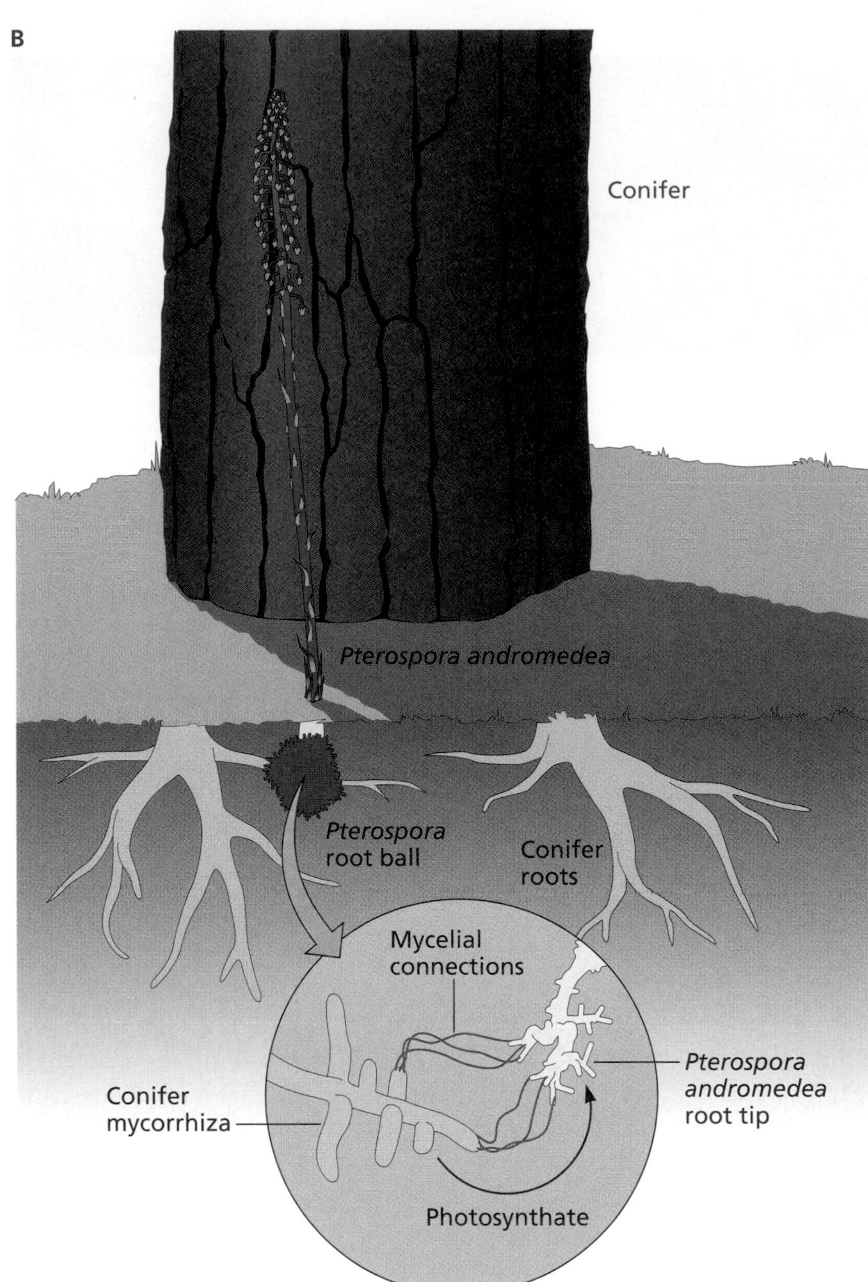

Figure 1.8. Mycorrhizal epiparasitism. *Pterospora andromedea*, also known as pine drops (A), lacks chlorophyll and (B) receives all of its organic carbon via a specific mycorrhizal fungus (*Rhizopogon*) that serves as a conduit between its roots and the roots of the host plant, often a conifer. No direct connection exists between the two plants. This is an example of mycorrhizal epiparasitism. (A, Courtesy of Walter Siegmund, CC BY-SA 3.0. B, From Cullings KW, Szaro TM & Bruns TD (1996) Nature 379:63–66. With permission from Macmillan Publishers Ltd.)

Simultaneously Parasite *and* Mutualist

It is also conceivable that a single organism can function simultaneously as mutualist and parasite. Mosquitoes feeding on rodents infected with malaria (*Plasmodium chabaudi*) had a shorter probing time than mosquitoes feeding on uninfected rodents. The enhanced ability of vectors to locate blood, possibly caused by a disruption in blood clotting resulting from *P. chabaudi*, may be a common feature of vector-borne parasite infections. At least during the interval of time that the mosquito is feeding, *P. chabaudi* could be considered as a mutualist with the mosquito vector and a parasite of the rodent host. Immediately after feeding though, the interests of *P. chabaudi* and the mosquito dramatically change. The developing malaria parasite would draw nutrients from the mosquito and exert other effects that are clearly parasitic. Here too it is worth noting that the malaria parasite is entirely dependent on its mosquito host being sufficiently healthy to survive long enough to

A

B

Figure 1.9. Some examples of cleaning symbiosis, generally considered a type of mutualism. (A) A Pacific cleaner shrimp *Lysmata amboinensis* removes parasites from the mouth of a moray eel (*Gymnothorax*). (B) A yellow-billed oxpecker (*Buphagus africanus*) gleans ticks and other parasitic arthropods from the surface of a large mammal. Examples of cleaning symbiosis carry their own ambiguities as oxpeckers feed on the blood flowing from wounds that they themselves may open, and cleaners are sometimes eaten by their clients. (A, Courtesy of Broken Inaglory/CC BY-SA 3.0. B, Steve Garvey, CC BY-SA 2.0.)

permit its development to be completed, and to be healthy enough to feed, such that the malaria parasite's progeny can be injected into another host. In other words, a delicate touch is required. We return to some of these themes in later chapters.

Parasitism and Cleaners

Last, with respect to parasitism and mutualism, the pervasiveness of the former can serve as the basis for the establishment and maintenance of the latter. This is evidenced by many types of cleaner relationships (**Figure 1.9**), in which infested hosts (clients) are relieved of a portion of their burden of ectoparasites by specialized cleaners, which clearly benefit by securing a novel source of nutrition. Upon further scrutiny, what appears at first to be an unambiguous mutualism, can be more complicated and may very much depend on the local circumstances. For example, the cleaners are often placed at obvious risk of predation and might become tempting prey for the client, particularly if it is not bothered by parasites at the time. Conversely, the cleaners might be tempted to exploit valuable resources (such as blood) from their clients. Some species mimic cleaners and overtly exploit the client, potentially requiring the client to adjust its willing acceptance of the approach of a true cleaner (see also Chapter 6). Once again, the closer we look, the more we need to exercise care with respect to interpreting the nature of intimate relationships among species.

In the following chapters, we will build on the basic framework covered here to gain a fuller appreciation of the many intrinsically fascinating aspects of the parasitic life style. This process begins in Chapter 2 by providing an overview of the diversity of parasites inhabiting our world.

Summary

Parasitism is a mode of existence in which one organism, a parasite, infects another, the host, and does some measure of harm to the host while itself deriving a benefit. Parasitism is among the most common lifestyles on earth, and thus intersects the study of biology at levels ranging from molecules and genes to entire ecosystems. The decline in honeybee populations demonstrates that parasites cause problems for species conservation and agriculture, as well as their more familiar unfavorable effects on the health and well-being of humans and our domestic animals and plants.

Not surprisingly, given the huge variety of parasitic lifestyles, there are different opinions regarding what comprises the key attributes of parasite biology. In this text, we will take an inclusive approach, with a parasite considered to be any infectious agent, whether subcellular, prokaryotic, or eukaryotic, that inflicts harm on its host. However, in keeping with traditional treatments of parasitology, the emphasis will be on parasites that are eukaryotic. Another key aspect of parasitism, namely the concept of virulence, means different things to scientists from disparate backgrounds depending on their outlook. For example, clinicians tend to define virulence as the ability to cause disease or mortality, whereas evolutionary biologists define it as the ability of a parasite to reduce host fitness (fitness meaning the success of an individual in passing on its genes to future generations).

Parasites have remarkably diverse modes of existence and can be classified by where they live, how they live, what they do to the host, how they are transmitted, and their size. Ectoparasites live on the external surface of their hosts, whereas endoparasites reside in the host's body. Castrators cause total or near-total disruption of their hosts' reproductive activities. Body snatchers take over their host, often including its behavior. Trophically-transmitted parasites are passed to their next host after ingestion by an initial host. Obligatory parasites require a suitable host to complete their life cycle; most parasites are obligate parasites. Facultative parasites are typically free-living, but they can adopt a parasitic existence. Opportunistic parasites take advantage of particular circumstances to cause infection in a host that they normally do not infect or in which they normally do not cause disease. Hyperparasites exploit other parasites. Microparasites are small and have rapid generation times as compared to their hosts and as it is difficult to quantify the number of microparasites within a host, their numbers are assessed by mathematical models. Macroparasites are larger parasites, such as worms, ticks, or fleas, and can be counted. A parasitoid is an organism that spends a significant amount of time attached to or within a single host (often sterilizing, killing, or consuming it) and after leaving the host often has a free-living period of existence.

Given that parasites are utterly dependent on their hosts, it is important to appreciate the various ways in which hosts enable parasites to fulfill their needs, and we can classify them accordingly. The definitive host is the host in which parasites undergo sexual maturation and reproduction. Parasites undergo a required developmental step, but do not reproduce sexually, in an intermediate host. Paratenic (or transport) hosts do not support parasite development but provide a place where the parasite can persist and prolong its survival. A reservoir host is one in which a parasite normally dwells and where it can be reliably maintained, even when not being actively transmitted, and from which it can then colonize other hosts, including people. A vector is a means by which a parasite can be transmitted. If vectors are organisms, they can be mechanical (used to transfer parasites from one host to another) or biological (used to provide an environment for development or multiplication).

Parasitism is a type of symbiotic association, and an awareness of the broad range of intimate symbiotic associations helps us appreciate the uniqueness of parasitism, but also points out how minor the differences can be, for example, between mutualism (where the association is beneficial to both organisms) and parasitism.

REVIEW QUESTIONS

1. Emergent properties refer to distinct attributes or phenomena that arise as a consequence of the interactions of two components (such as a parasite and host). Describe some distinctive emergent properties that characterize the interactions between hosts and parasites.

2. Definitions are elusive and it's important to realize that exceptions or complications always occur. By considering parasites that defy simple categorization, we gain a deeper appreciation of the nature of parasitism. Consider the parasite *Trichinella spiralis* referred to in Chapter 1. You may also want to look at the Rogues Gallery page referring to the life cycle of this helminth. It is said the same host individual first serves as a definitive host and later an intermediate host. How can this be?

3. Consider the parasite *Strongyloides stercoralis*, an intestinal nematode of humans often called the threadworm. This organism is also capable of free-living existence in the soil, with male and female adult worms present there that produce eggs. The eggs hatch and release larvae that undergo molts to become either larvae infective for people or that develop into free-living adults that continue the soil-based part of the cycle. Is *S. stercoralis* an obligatory parasite, a facultative parasite, or an opportunistic parasite?

4. Parasitism is such a common, pervasive phenomenon that some kinds of structures and associations among species depend on it for their existence. Provide some examples.

5. List some examples in which a mutualistic (+/+) association between two different species might readily become a parasitic or other type of (+/–) association.

6. Compare and contrast parasitoids, hyperparasites, cleptoparasites, brood parasites, and macroparasites.

References

Epe C & Kaminsky R (2013) New advancement in anthelmintic drugs in veterinary medicine. *Trends Parasitol* **29**:129–134.

Guerrant RL (1998) Why America must care about tropical medicine: threats to global health and security from tropical infectious diseases. *Am J Trop Med Hyg* **59**:3–16.

Chakraborty S & Newton AC (2011) Climate change, plant diseases and food security: an overview. *Plant Pathol* **60**:2–14.

Nichols E & Gomez A (2011) Conservation education needs more parasites. *Biol Conserv* **44**:937–941 (doi: 10.1016/j.biocon.2010.10.025).

Website of the World Federation of Parasitologists. http://www.wfpnet.org/tab_home.php

Windsor DA (1998) Most of the species on earth are parasites. *Int J Parasitol* **28**:1939–1941.

BUILDING AN UNDERSTANDING OF THE BASICS OF PARASITISM

Parasites live in or on their hosts and cause them harm

Brooks DR & McLennan DA (1993) Parascript—Parasites and the Language of Evolution, pp 5. Smithsonian Institution Press.

Goater TM, Goater CP & Esch GW (2013) Parasitism: The Diversity and Ecology of Animal Parasites. Second edition. Cambridge University Press.

Combes C (2001) Parasitism: The Ecology and Evolution of Intimate Interactions. University of Chicago Press.

Lincicome DR (1971) The goodness of parasitism: a new hypothesis. In Aspects of the Biology of Symbiosis (TC Cheng ed.), pp 139–227. University Park Press.

Roberts LS & Janovy J (2008) Gerald D. Schmidt & Larry S. Roberts' Foundations of Parasitology, 8th ed. McGraw-Hill.

Opinions vary on how to define some of the key aspects of parasites and their biology

Lehmann T (1993) Ectoparasites: direct impact on host fitness. *Parasitol Today* **9**:8–13.

Price PW (1980) Evolutionary Biology of Parasites. Princeton University Press.

Schmid-Hempel P (2011) Evolutionary Parasitology: The Integrated Study of Infections, Immunology, Ecology and Genetics. Oxford University Press.

Vollrath F (1998) Dwarf males. *Trends Ecol Evol* **13**:159–163.

The residence time for a parasite in or on a host is highly variable

Askew RR (1971) Parasitic Insects. Heinemann.

Kat PW (1984) Parasitism and the Unionacea (Bivalvia). *Biol Rev Camb Philos Soc* **59**:189–207 (doi: 10.1111/j.1469-185X.1984.tb00407.x).

Kennedy CR (1975) Ecological Animal Parasitology. Blackwell.

Leiper RT (1936) Some experiments and observations on the longevity of *Diphyllobothrium* infections. *J Helminthol* **14**:127–130.

There are many additional ways to categorize parasites

Agrios GN (2005) Plant Pathology. 5th ed. Academic Press.

Anderson RM & May RM (1991) Infectious Diseases of Humans: Dynamics and Control. Oxford University Press.

Lafferty KD & Kuris AM (2009) Parasitic castration: the evolution and ecology of body snatchers. *Trends Parasitol* **25**:564–572 (doi: 10.1016/j.pt.2009.09.003).

Sullivan DJ (1987) Insect hyperparasitism. *Annu Rev Entomol* **32**:49–70.

HOSTS—ESSENTIAL LIFELINES FOR PARASITES

Hosts too fall into several different categories

Combes C (1991) Evolution of parasite life cycles. In Parasite–Host Associations: Coexistence or Conflict? (CA Toft & A Aeschlimann eds), pp. 62–82. Oxford University Press.

Dantas-Torres F (2007) The role of dogs as reservoirs of *Leishmania* parasites, with emphasis on *Leishmania (Leishmania) infantum* and *Leishmania (Viannia) braziliensis*. *Vet Parasitol* **149**:139–146.

Packer C, Altizer S, Appel M et al (1999) Viruses of the Serengeti: patterns of infection and mortality in lions. *J Anim Ecol* **68**:1161–1178.

Smyth JD (1994) Introduction to Animal Parasitology, 3rd ed. Cambridge University Press.

Zelmer DA & Esch GW (1998) Bridging the gap: the odonate naiad as a paratenic host for *Halipegus occidualis* (Trematoda, Hemiuridae). *J Parasitol* **84**:94–96.

APPRECIATING PARASITISM'S PLACE IN NATURE

Parasitism is one of several categories of symbiotic associations

Barnard CJ (1990) Parasitic relationships. In Parasitism and Host Behavior (CJ Barnard & JM Behnke eds), pp 1–33. Taylor and Francis.

Gill FB (1995) Ornithology, 2nd ed. W.H. Freeman.

Leung TLF & Poulin R (2008) Parasitism, commensalism, and mutualism: exploring the many shades of symbioses. *Vie Milieu Paris* **58**:107–115.

Parasitoids straddle the boundary between predation and parasitism

Godfray HCJ (1994) Parasitoids—behavorial and evolutionary ecology. Princeton University Press.

Price PW (1975) Reproductive strategies of parasitoids. In Evolutionary Strategies of Parasitic Insects and Mites (PW Price ed). Plenum.

Our understanding of parasitism is enhanced by an appreciation of its relationship to another ubiquitous type of symbiosis, mutualism

Bansemer C, Grutter AS & Poulin R (2002) Geographic variation in the behaviour of the cleaner fish *Labroides dimidiatus* (Labridae). *Ethology* **108**:353–366 (doi:10.1046/j.1439-0310.2002.00777.x).

Benz GW, Borucinska JD, Lowry LF et al (2002) Ocular lesions associated with attachment of the copepod *Ommatokoita elongata* (Lernaeopodidae : Siphonostomatoida) to corneas of Pacific sleeper sharks *Somniosus pacificus* captured off Alaska in Prince William Sound. *J Parasitol* **88**:474–481 (doi:10.1645/0022-3395(2002)088[0474:OLAWAO]2.0.CO;2).

Borucinska JD, Benz GW & Whiteley HE (1998) Ocular lesions associated with attachment of the parasitic copepod *Ommatokoita elongata* (Grant) to corneas of Greenland sharks *Somniosus microcephalus* (Bloch and Schneider). *J Fish Dis* **21**:415–422.

Bronstein JL (1994) Conditional outcomes in mutualistic interactions. *Trends Ecol Evol* **9**:214–217.

Cullings KW, Szaro TM & Bruns TD (1996) Evolution of extreme specialization within a lineage of ectomycorrhizal epiparasites. *Nature* **379**:63–66.

Douglas AE (2010) The Symbiotic Habit. Princeton University Press.

Dubilier N, Mülders C, Ferdelman T et al. (2001) Endosymbiotic sulphate-reducing and sulphide-oxidizing bacteria in an oligochaete worm. *Nature* **411**:298–302.

Francke OF & Villegas-Guzman GA (2006) Symbiotic relationships between pseudoscorpions (Arachnida) and packrats (Rodentia). *J Arachnol* **34**:289–298 (doi:10.1636/04-36.1).

Grutter AS (1999) Cleaner fish really do clean. *Nature* **398**:672–673.

Herre EA, Knowlton N, Mueller UG & Rehner SA (1999) The evolution of mutualisms: exploring the paths between conflict and cooperation. *Trends Ecol Evol* **14**:49–53.

Pozo MJ, Azcon-Aguilar C, Dumas-Gaudot E & Barea JM (1999) Beta-1,3-glucanase activities in tomato roots inoculated with arbuscular mycorrhizal fungi and/or *Phytophthora parasitica* and their possible involvement in bioprotection. *Plant Sci* **141**:149–157.

Rossignol PA, Ribeiro JMC, Jungery M et al (1985) Enhanced mosquito blood-finding success on parasitemic hosts: evidence for vector–parasite mutualism. *Proc Natl Acad Sci* **82**:7725–7727.

An Overview of Parasite Diversity

2

All living species are involved in parasitism, either as parasites or as hosts.
DE MEEÛS ET AL (1998)

In This Chapter

IN CHAPTER 1 WE DESCRIBED THE BASIC FEATURES OF PARASITISM and explained how parasitism differs from other kinds of biological associations. The goal of this chapter is to present a big picture of the diversity of the world's parasite species. In this overview, we begin to develop an appreciation for the relationship of these many species to one another and to other organisms.

Such an overview is helpful in many ways. It can help us understand how different groups of parasites have diversified and adapted to their hosts. From this we gain a better appreciation for predicting which parasites might colonize new host species, have the potential to cause an emerging disease, or become invasive in new locations. This knowledge facilitates efforts to control parasitic diseases by enabling us to determine the full spectrum of parasites that might be involved and that are in need of control. The study of parasite diversity also allows us to develop more natural schemes to facilitate identification and classification of parasites. It also enables us to assess the rate at which parasite species are becoming extinct. Without a thorough inventory of parasite diversity, we may never know that certain species even existed before they are gone. An understanding of parasite diversity allows us to more fully appreciate the place of parasites in natural environments, including their distinctive role in an ecosystem. It also puts us in a better position to exploit the unique biochemical capabilities of parasites for medicinal or other purposes. Finally, investigation of parasite diversity aids us in understanding the ecological and evolutionary processes that dictate how and why parasites have diversified as they have.

Many of these topics will be revisited throughout the book, but in this chapter we first provide an overview of the immense diversity of parasites using evolutionary trees as a framework to portray this diversity. As part of this overview, we include discussion of **horizontal gene transfer** (HGT), a process in which one organism acquires genetic information from another organism without being the offspring of that organism. Thereafter, examples are provided that show how the evolutionary relationships of some enigmatic parasite groups have been revealed, while noting that others still defy resolution. Also discussed are examples of how humans have acquired some of our parasites and how we can retrace their evolutionary histories. The search for new parasite diversity is ongoing and some examples of how this search

Species names highlighted in red are included in the Rogues' Gallery, starting on page 429.

is undertaken and the diversity cataloged are described in **Box 2.1**. We also discuss some examples of improved classification schemes that are based on a thorough knowledge of parasite diversity and evolutionary relationships.

Last, although the primary emphasis of this chapter is to examine parasite diversity on a large scale, we must also remain mindful that parasite diversity can be gauged in other ways as well. Consequently, we also consider the genetic diversity inherent within parasite species and explain why this is relevant. Although it is clear that parasites have frequently arisen from free-living ancestors, the chapter concludes with a discussion of whether parasites ever give rise to free-living organisms.

BOX 2.1

An Example of the Study of Parasite Biodiversity

Suppose your task is to necropsy a monitor lizard *Varanus salvator* (Figure 1A) that was originally collected in Papua New Guinea but that recently died at the local zoo. When you examine the lungs of the animal, you discover a few striking annulated worms (Figure 1B) that upon further inspection were found to possess two pairs of stubby, clawed appendages on either side of a mouth. Beyond that, there are very few distinctive external characteristics to go on. What have you found and how do you go about confirming these findings? How would you document and reveal your findings to other scientists? What have you learned about the surprising relationships of these worms to other organisms?

A search in Google Scholar for "*Varanus salvator*," "parasite," and "lung" would quickly lead you to some appropriate references such as Riley, Spratt and Presidente (1985), where you would begin to see repeated references to a group of worms called pentastomids that are characterized by the very features of paired appendages around a mouth and an annulated body exhibited by the worms you have found. Other prominent groups of worms such as nematodes, trematodes, or tapeworms lack the appendages and, furthermore, tapeworms do not possess a mouth at all. Further perusal of the pentastome literature would reveal a worm with the specific attributes of your specimens, previously described and named as *Elenia australis* Heymons, 1932. Ideally, you would contact a pentastomid expert for verification of your identification, and your worms would eventually be sent to a museum to serve as permanently archived **voucher specimens**. Along with your specimens would go information about the date and locality where it was collected, the identity of the host and its peculiar history, where the worms were found in the host, and any photographs of the worms or necropsied host that might be useful to scientists in the future. Assuming the specimens were preserved in ethanol or some other medium suitable for molecular sequencing work, most likely a representative barcode sequence would eventually be obtained for the specimen to serve as a sequence marker to aid identification and future comparisons with other pentastomids. Your specimens might also become the subject of a complete genome sequencing project. Remarkably, these intriguing blood-sucking, lung-inhabiting worms mostly found in reptiles have been found to have as their closest relatives a group of ectoparasitic crustaceans of fish to which they bear very little superficial similarity. Pentastomids today are recognized as highly modified crustaceans and thus are relatives to lobsters, shrimp, and copepods.

Reference

Riley J, Spratt DM & Presidente PJA (1985) Pentastomids (Arthropoda) parasitic in Australian reptiles and mammals. *Aust J Zool* 33(1):39–53.

A

B

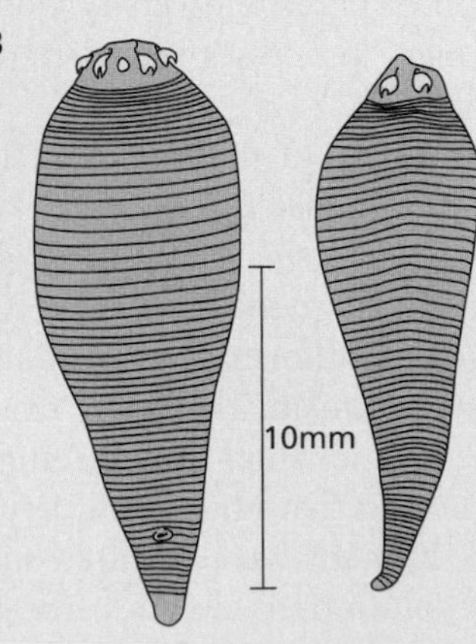

Figure 1 An example of the study of parasite diversity. (A) The water monitor *Varanus salvator*. (B) Worms from the lungs of *Varanus salvator*. (A, Courtesy of Nur Hussein, CC BY-SA 3.0. B, Riley J et al [1985] *Aust J Zool* 33(1)39-53. ©CSIRO 1985. Published by CSIRO publishing. Reproduced with permission.)

2.1 THE DIVERSITY OF PARASITE SPECIES

What constitutes a parasite species requires some explanation

Let us begin with the definition of a parasite and of a species. Recall the discussion of what constitutes a parasite in Chapter 1. For some, the term parasite applies only to eukaryotes with parasitic lifestyles. For others, there need be no kind of taxonomical constraint placed on an organism (such as a bacterium) or genetic entity (such as a virus) for it to be considered a parasite. Clearly these differing points of views will influence one's estimate of how many different kinds of parasites exist.

Many different definitions of species have been put forth over the years and all have shortcomings. We use the following widely adopted version, frequently referred to as the **biological species concept**. A **species** is a group of individuals with similar properties that are able to interbreed with one another and produce fertile offspring and that don't regularly interbreed with other species. The members of a particular parasite species may be **dioecious** (or **gonochoristic**), meaning the species is comprised of separate male and female individuals, or they may be **monoecious** (or **hermaphrodites**), having functional reproductive organs of both sexes and capable of either cross-fertilization with another individual or, possibly, self-fertilization.

Although this definition readily fits many parasite species, it is less applicable for some organisms such as prokaryotes (eubacteria and archaea) that often have high levels of genetic variability. This variability arises in part from their enormous ability to exchange genetic information by HGT with other organisms. The biological species concept also is problematic when applied to some eukaryotic parasites that may only rarely undergo sexual reproduction or that do so in unconventional or cryptic ways. As an example, for a species such as the protozoan ***Giardia lamblia***, distinct but related clonal lineages exist that are nonetheless embraced by the same species name. These lineages, however, may undergo genetic exchange when rare sexual recombination events occur, and this cohesiveness between the clones justifies the continued use of the species names we use to define them. Reproduction involving infrequent, nonconventional, or cryptic exchange of genetic information applies to other important groups of parasites such as protozoans in the genera ***Entamoeba***, ***Leishmania***, and ***Trypanosoma***.

One approach to documenting that trypanosomes do in fact undergo sex is shown in **Figure 2.1**. Experimental crosses were made between two parental stocks of ***Trypanosoma brucei***, one transfected with a gene encoding a red fluorescent pigment and one transfected with a gene encoding a green pigment. The two stocks were mixed in a single tsetse fly (***Glossina***) vector by allowing it to feed on blood containing both types of parasites. Trypanosomes were removed from the fly and expanded in numbers in mice. Some trypanosomes retrieved from the mice exhibited either parental genotype (red or green) and some revealed a yellow phenotype (a color resulting from the presence of both red and green pigments), indicative that genetic recombination had occurred between red and green parental trypanosomes. This result indicates that a named species like ***T. brucei*** is not just a collection of isolated clones but that these clones are united by sexual recombination events, even if such events do not occur as predictably or obviously as they do in other organisms.

Because of such difficulties, some have advocated as an alternative to the biological species concept using an **evolutionary species concept** for parasites. In this view, a species is considered to be a group of organisms having a single lineage with a shared evolutionary trajectory. Ambiguities arise here as

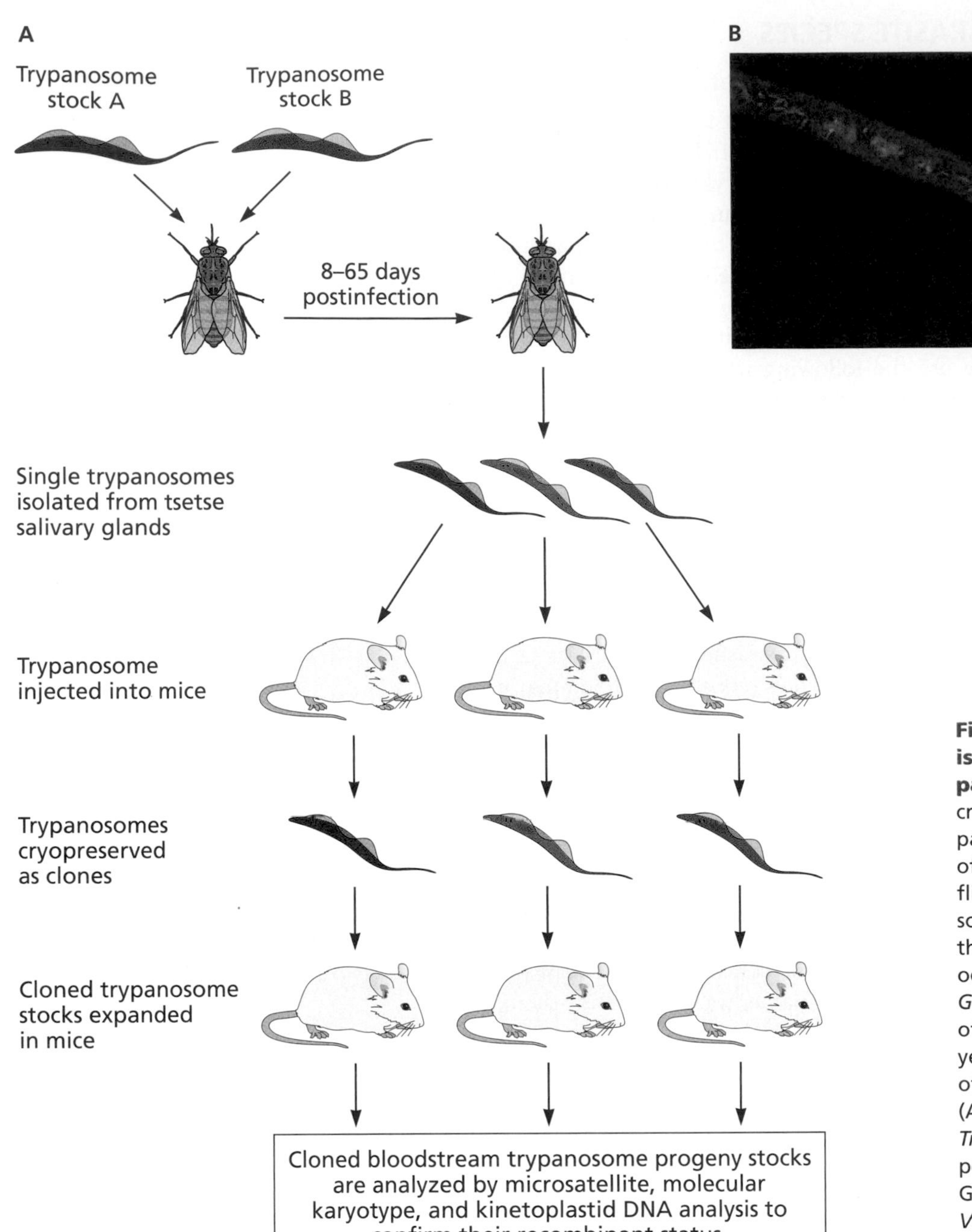

Figure 2.1 The species concept is not easily applied to some parasites. (A) Experimental crosses made between two parental stocks (red and green) of *Trypanosoma brucei* in tsetse flies (*Glossina*) leads to retrieval of some progeny (yellow), indicative that sexual reproduction has occurred. (B) A salivary gland from *Glossina* showing trypanosomes of both parental types and of a yellow type, the latter indicative of genetic recombination. (A, modified from Hide G [2008] *Trends Parasitol* 24:425–428. With permission from Elsevier. B, From Gibson W et al [2008] *Parasit Vectors* 1: Article Number: 4 [doi: 10.1186/1756-3305-1-4].)

well because how different do lineages have to be if we are to consider them as one or more species? Also, analysis of different genes might return different answers with respect to the evolutionary relationships among the organisms being considered. Ambiguities in species concepts serve to remind us that throughout this book, regardless of the species concept used, it should be considered relative to the distinctive biology of the particular parasite being studied.

Given these considerations, how many species of parasites inhabit the Earth?

A shortcoming in our understanding of life on Earth is our inability to say with confidence how many species exist on our planet. Attempts to identify and enumerate species began with Linnaeus and his publication of the *Systema Naturae* in 1735. Since then, biologists have come up with widely varying estimates; in recent years these estimates ranged from 3 million to 100 million species. There may never be a definitive answer to the question. The calculation is complicated by a lack of thorough sampling of many habitats, by the presence of cryptic species (a topic we discuss later in this chapter), and by the potential loss of species to extinction faster than we can discover, describe, and name them. Two recent estimates place the number of all species at 5 ± 3 million species, and 8.7 ± 1.3 million species. The number of species already described is reckoned at about 1.9 million species. The estimate of 8.7 million species takes the interesting approach that the higher levels of our classification schemes (phylum, class, order, and family, for example) are relatively well known and follow a predictable pattern, such that we can use these values to extrapolate the number of species.

Although it is reassuring that these estimates seem to be converging on a more defined range, great uncertainties remain, including how many of the species are actually parasites. Some groups of parasites are much more tractable than others. For instance, credible estimates of the number of helminths (tapeworms, trematodes, nematodes, and acanthocephalans) infecting vertebrates fall into the range of 75,000 to 300,000 species. The numbers of helminth species infecting invertebrates or plants are not as easy to estimate. Additional uncertainties arise in quantifying the number of parasitic protozoans, algae, plants, and fungi. We are particularly ignorant regarding the global abundance of protozoans, fungi, arthropods (especially insects and mites), and nematodes. For each of these groups, many of the species to be discovered will prove to be parasites.

The estimates given above mostly apply to eukaryotes. If we do not restrict the definition of a parasite to just the eukaryotic realm, parasites also include some of the prokaryotes, which have been systematically underrepresented in calculations of global species diversity. They are small and easily overlooked, and many cannot be easily isolated and cultured. Currently there are only about 9,500 named species of bacteria. **Metagenomics**, the characterization of genetic material recovered directly from a particular environment without the need to culture the organism present, has revolutionized our understanding of biodiversity, especially but not exclusively for prokaryotes. Using this approach, the ongoing characterization of the rich assemblage of microbial species found in humans—the human microbiome—has revealed a remarkable 500 to 1000 species of gut-inhabiting bacteria per individual. If this line of thinking is extended to all animal species, then we are presently just glimpsing the tip of the microbial iceberg. Again bearing in mind the difficulties of applying the species concept to bacteria, estimates of the true diversity of bacteria species range as high as 10^7 to 10^9 species. An unknown percentage of these many bacterial species will prove to be parasites. Add to these ranks the unnumbered hordes of distinct suborganismal entities, such as viruses and viroids, that can be considered parasites, and it becomes clear that we are still a long way from arriving at an estimate of the world's diversity of parasites. Based on such considerations, it is not hard to arrive at a conclusion reached by many, that parasitism is one of the most common modes of life on our planet. By extension from their present day ubiquity, it seems probable that every species to have ever lived has been infected by some kind of parasite. The following simple argument also supports the view of the commonness of parasite species. Consider that every free-living host species harbors at least one parasite species and that parasites themselves often have parasites. Therefore, it has to be concluded that the number of parasite species is greater than the number of host species. Caution is required though in making such broad arguments. For example, it is by no means certain that each host species harbors a unique parasite species: many parasite species can infect multiple host species. However, to counterbalance this notion, some parasites exist in still as yet unresolved complexes of cryptic species that once revealed will add further diversity to the parasite ranks. For the moment, we can state with some justification that parasitism is indeed one of the most common lifestyles on earth, but we are unable to answer with a definitive number how many parasite species exist on Earth.

Evolutionary trees are used to visualize evolutionary relationships and to display parasite diversity

Given that there are so many species of parasites, how can we possibly begin to organize them into some kind of comprehensible framework? **Taxonomy** refers to the science of identification, description, and naming of organisms. Since Linnaeus' time, biologists have labored to develop taxonomic schemes to organize biodiversity. The basic unit of such systems is the species, complete with its Latin binomial name (genus and species, such as *Homo sapiens*). Traditionally such taxonomic schemes were elaborated using morphological characteristics that enabled perceptive scientists to group them into ever more inclusive categories (species, genus, family, order, class, phylum, and kingdom).

With the publication of Darwin's *Origin of Species* came the compelling metaphor of the **tree of life** (**Figure 2.2A**). It indicates the origin of all life from common ancestry followed by diversification of lineages of life as represented by twigs on this growing tree. This view emphasizes the vertical acquisition of genetic information from one's ancestors and assumes there is a core complement of genes that is retained and reflects ancestry, with the gene encoding the RNA found in the small ribosomal subunit (SSU rRNA gene) being one. Some buds on the tree, representing successful lineages of organisms, gave rise to new branches that further diversified, whereas others withered and died out. Such a tree depicts the patterns of historical relatedness among all organisms. Given that such relationships exist, it would be optimal to have our modern taxonomic schemes follow the tree's branching patterns such that taxonomy mirrors the actual sequence of evolutionary events leading to the diversity we observe. For many years, we had no way to verify if our taxonomic schemes actually reflected evolutionary relationships. For some groups, availability of fossils exemplifying transitional forms helped to verify taxonomy, but for parasites, which are mostly soft-bodied and have left a poor fossil record (but see Box 7.5 and Figure 10.2), this is usually not an option.

Figure 2.2B modifies the concept of the Darwinian tree of life by taking into account the phenomenon of horizontal (or lateral) gene transfer (HGT), the movement of genetic information laterally between organisms that occurs without the involvement of sexual reproduction. HGT is believed to have been rampant during the early stages of cellular life and continues abundantly today, especially among bacteria and archaea, and as we discuss further, also with eukaryotic parasites. According to this view, the tree of life is really a web of life.

Enter the discipline of **phylogenetics** referring to the study of the evolutionary relationships among organisms based on molecular sequence data or morphological traits. These relationships are conveniently depicted with the use of evolutionary trees. Phylogenetics is often involved in making and evaluating hypotheses about historical patterns of descent and can be thought of

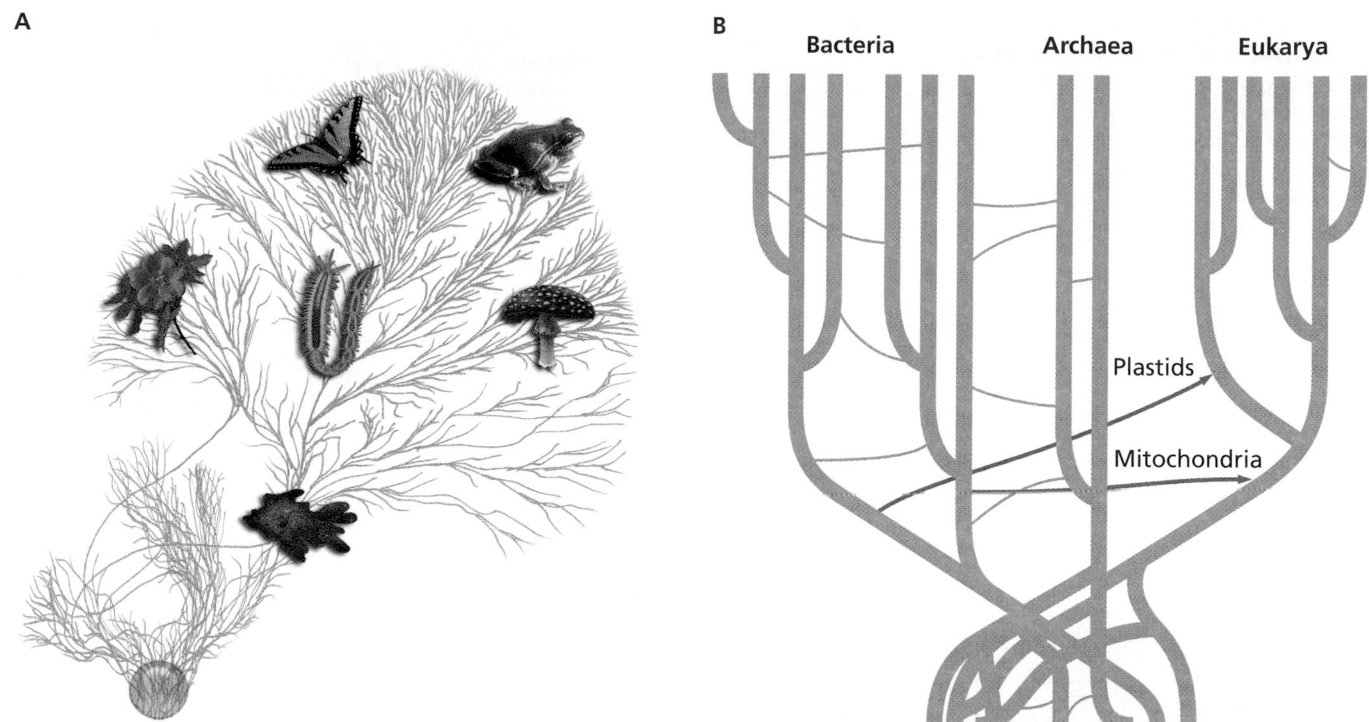

Figure 2.2 Darwin's concept of the tree of life and a modern alternative. (A) Darwin's concept of the tree of life, with life arising from common ancestry and an emphasis on vertical acquisition of genetic information. (B) This tree takes into account horizontal gene transfer (HGT). According to this view, genetic information is commonly acquired from organisms other than one's ancestors. (A, from the Tree of Life web project: http://tolweb/tree/phylogeny.html. B, modified from Smets BF & Barkay T [2005] *Nat Rev Microbiol* 3:675–678. With permission from Macmillan Publishers Ltd.)

as part of a broader subject called **systematics**, which refers to the study of the diversification of life on Earth, including the relationships among organisms over time. Evolutionary trees are constructed using algorithms that assess the degrees of similarity in DNA or RNA nucleotide sequences or in protein amino acid sequences in the organisms being compared. Many trees are also constructed based on morphological characters or a combination of morphology and sequence data. The optimal ways to construct trees remains a topic of vigorous and ongoing debate, with entire professional societies and journals devoted to the topic. It is important to understand how to interpret what such trees are saying, and the website associated with this book has a tutorial covering some of the basic features of interpreting trees.

Modern biologists portray relationships among organisms using such trees, and for good reason: trees provide a compelling way to order our thinking about the biodiversity that follows our best efforts to reconstruct the evolutionary trajectories taken. Evolutionary trees based on molecular sequence data have revolutionized our view of relationships among organisms. However, until we invent a machine that enables us to go back in time and directly witness how groups of organisms actually diversified, such trees should be considered as hypotheses of relationships. For each of the trees shown in this chapter, many alternative versions could have been provided and each of those represented will become outmoded in time. Each is viewed as a relatively conservative statement of the current understanding of the relationships for the group in question. Someday the relationships among all organisms will be based on comparisons of complete genome sequences, a trend that is already under way.

Efforts are well underway to reveal the overall tree of life

There is a steady, ongoing effort to reveal the overall topology of the tree of life, and one representation of the resulting effort is shown in **Figure 2.3**. For such an all-inclusive effort, it has been essential to choose a target gene that

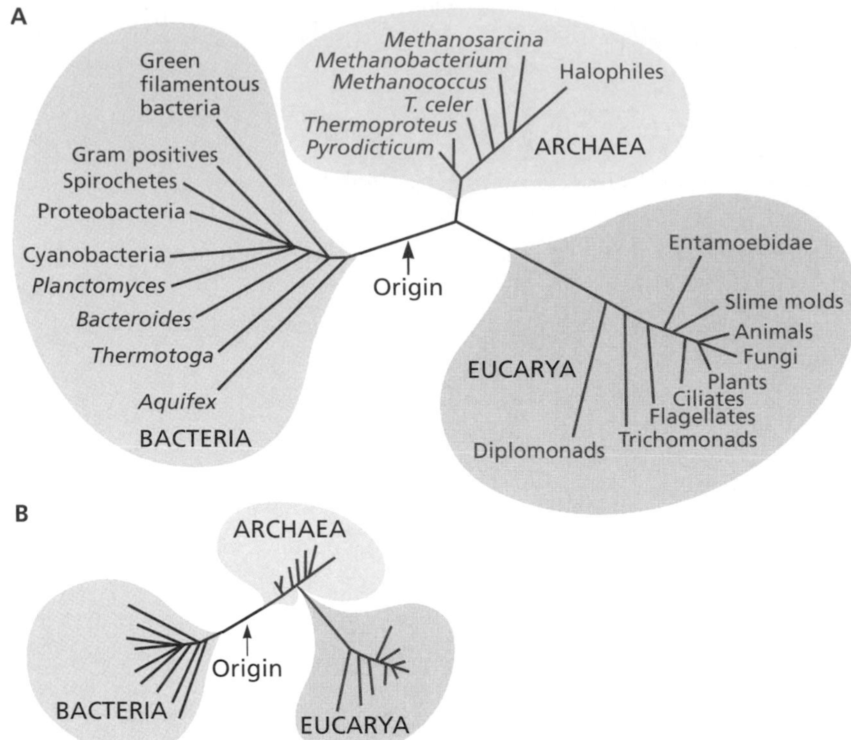

Figure 2.3 The tree of life, based on comparisons of the small ribosomal subunit gene. (A) Notice the origin (root) of the tree and that three major domains of life are demarcated: Bacteria, Archaea and Eucarya. (B) A recent challenge to the three domain concept, arguing that the origins of Eucarya lie within the Archaea, suggesting there are really two major domains, Bacteria and Archaea/Eucarya. (B, modified from Williams TA et al [2013] *Nature* 504:231–236. With permission from Macmillan Publishers Ltd.)

is universally present in all organisms, a so-called part of the core genome, and that is relatively conserved such that sequence comparisons incorporating organisms as disparate as bacteria and tapeworms can be accommodated in the same analysis. Such a gene is the one encoding the RNA that is a structural component of the small subunit of the ribosome (the small subunit rRNA gene, or SSU rDNA). Although this single gene does not contain sufficient information to enable us to resolve fully the tree of life, it nonetheless has proven to be an excellent backbone for this effort. Using this approach, the pioneering efforts of Carl Woese and colleagues in the 1970s and 1980s revealed three major domains of life: the Bacteria, the Archaea, and the Eukarya. Providing a root for such a universal tree is problematic, since virtually all living organisms are included in the ingroup to be analyzed, thus leaving no group of organisms against which to compare them (there is no outgroup). The tree in Figure 2.3 has been provided a root by analyzing other genes believed to have undergone duplication in a universal common ancestor. By studying the differences among such duplicates, it was concluded that the root lies along the branch leading to the Bacteria. The position of the root on this tree suggests that the Archaea and Eukarya had an early common history separate from the Bacteria, accounting for some of the basic biochemical similarities between the two domains.

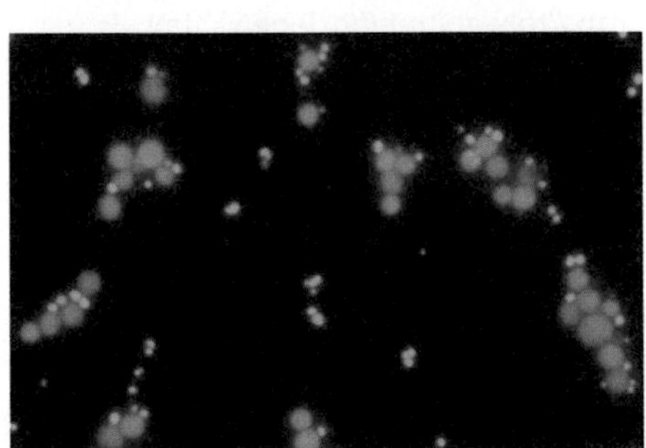

Figure 2.4 Parasitism is apparently rare in the Archaea. The only known parasitic archaea, *Nanoarchaeum equitans*, appears as the tiny bright spheres attached to the surfaces of bigger, fuzzier archaea host cells, *Ignicoccus* sp., which are less intensely stained. The tiny *N. equitans*, only 400 nm in diameter, has one of the smallest genomes known for any organism, consisting of only 490,000 base pairs. (Courtesy of Harald Huber, University of Regensburg.)

As presently known, the adoption of parasitism has not been a conspicuous feature in the evolution of the Archaea. One species of Archaea, *Nanoarchaeum equitans*, has been discovered that parasitizes another member of the Archaea (**Figure 2.4**). Both host and parasite live in scalding hot, sulfur-rich water. However, most members of the Archaea are nonparasitic. The underlying nature of relationships of archaeal species with one another and with their hosts are variable and poorly known, so parasitism may prove to be more prominent than we currently know. Archaea inhabit extreme environments, such as hot springs or saline lakes, are commonly found in aquatic habitats, or in some cases are part of the gut flora of animals, including humans. Although archaeal species are generally not parasitic, an increasing number of examples indicates that they have contributed genes by horizontal gene transfer to bacteria that are overtly parasitic. For the two remaining domains, the Bacteria and Eukarya, as we discuss in the sections that follow, adoption of parasitism has figured prominently.

Horizontal gene transfer (HGT) has been pervasive throughout the evolution of life

Availability of genome sequence information for a broad variety of organisms has made it clear that in addition to inheriting genetic information vertically from one's ancestors, genetic information can move laterally between organisms, without the involvement of sexual reproduction. This process is known as HGT, horizontal (or lateral) gene transfer (see Figure 2.2). Because of HGT, some have argued that the essential concept of the tree of life and divergence from common ancestry has little relevance for bacteria and archaea. It has been further argued that instead of the diversity of all life being derived from common ancestry, that multiple genetic and environmental circumstances acting over a long time have created the diversity of life. Also quite ambiguous with respect to the tree of life are the viruses, which probably infect the cells of all types of organisms and are very adept at moving genetic information from one host cell to another. Standard versions of the tree of life do not include viruses, yet viruses have had a profound impact on all phases of organismal life. In addition to their parasitic role and their involvement in causing disease, viruses regularly provide their host organisms with novel genetic material. A thorough discussion of viruses is beyond the bounds of this book, and the reader is referred to virology textbooks for more discussion of these inherently parasitic, important, and fascinating entities.

Many bacteria are parasites

Bacteria are normally covered in microbiology courses, and for that reason their biology is not emphasized in this book. As discussed in Chapter 1, microbiologists typically refer to bacteria that cause disease as pathogens. Nevertheless, parasitism is a phenomenon that cuts across the boundaries of traditional disciplines, and it would be an oversight in this synopsis of parasite diversity to ignore bacteria completely. Consequently we discuss them briefly here.

Currently about 30 major groups of bacteria called phyla (singular, phylum) are recognized, and this number will certainly increase. Of 16 relatively well-known bacterial lineages, 11 contain at least some parasitic representatives. Among them are bacteria causing many prominent human diseases including tuberculosis, cholera, plague, syphilis, anthrax, and leprosy. Some prominent bacterial lineages such as *Chlamydia* and *Rickettsia* consist exclusively of intracellular parasites. Other groups, such as the Spirochaetes, have prominent parasitic representatives but also include many free-living species. Several bacterial lineages are predominantly free-living but contain a few parasitic representatives. It is clear that bacteria have readily adopted parasitism on several occasions. It is also clear that HGT has played an important role in the history of parasitism in bacteria. The first evidence for this process was the documentation of the transfer of drug resistance or virulence genes on plasmids from one bacterium to another unrelated bacterial species.

Among the bacteria are several intriguing and lesser-known parasites. *Bdellovibrio* is a parasitoid that bores into and parasitizes the bodies of other

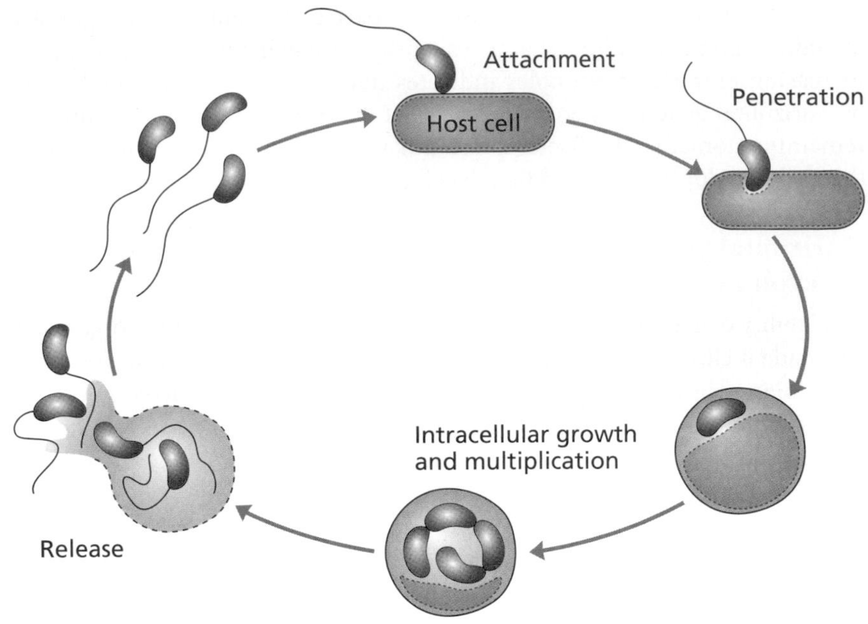

Figure 2.5 The parasitoid bacterium *Bdellovibrio*. *Bdellovibrio* attaches to, penetrates, and multiplies within its host cell (also a bacterium), eventually destroying it. It is then released and starts the cycle again.

bacteria (**Figure 2.5**). Other bacteria colonize unicellular eukaryotes and either castrate them (*Holospora*) or effectively addict them to their presence (*Caedibacter*). In some cases, unicellular eukaryotes like freshwater amebas serve as reservoir hosts for bacteria like *Legionella pneumophila* that may also parasitize people causing ailments such as Legionnaires' disease. Plants too suffer from bacterial infections as do invertebrates such as corals that are weakened by stressors like climate change and are thus more vulnerable to infections.

Eukaryotes are a very diverse group that includes many different kinds of parasites

The Eukarya, or eukaryotes, are characterized by the possession of membrane-bound structures within their cells, the most important and distinctive of these being the nucleus. As shown in **Figure 2.6**, efforts to reconstruct the history of the eukaryotes suggest there was a single common origin for eukaryotes, a group that today embraces most of all formally described and named species. Many eukaryotes are unicellular, microscopic, and motile. They have traditionally been referred to as **protozoans** or **protists**, terms that today carry no formal taxonomic meaning but are still commonly used for convenience. Eukaryotes are distinguished from the other two domains of life because they include multicellular organisms, such as some algae, plants, fungi, and animals.

Efforts to unravel the relationships among the many recognized lineages of eukaryotes are ongoing, and Figure 2.6 represents one such effort. Included in the figure are the main lineages into which several major groups of eukaryotic parasites fall. Note that this tree does not resolve relationships among all the major lineages, and it retains several **polytomies**—nodes in the tree that are not completely resolved to dichotomies. For example, within the Excavata, there are three unresolved branches stemming from one node. A polytomy can result from an inadequate amount of sequence data to enable full resolution of the tree, in which case it is called a **soft polytomy**. These are particularly likely to occur at deeper (older) nodes in the tree. In contrast,

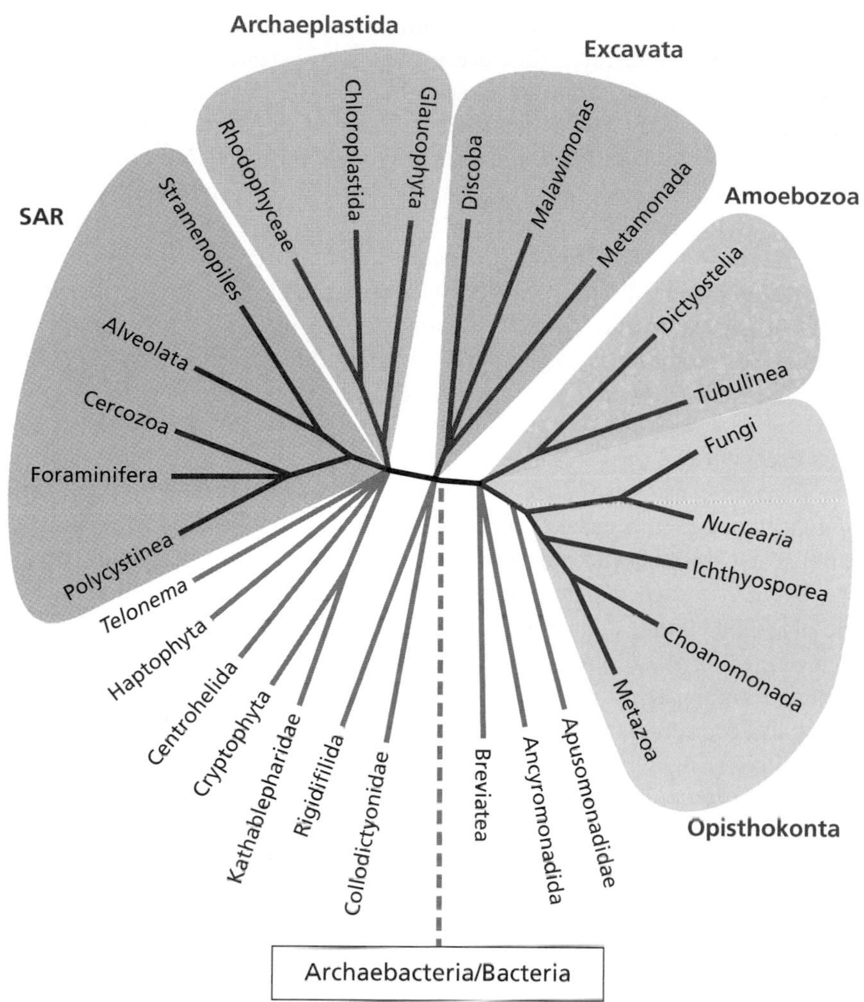

Figure 2.6 One portrayal of the relationships among the major groups of eukaryotes. Note that this is but one of several hypotheses offered for eukaryotic relationships. (Modified from Adl SM, Simpson AGB, Lane CE et al [2012] *J Eukaryot Microbiol* 59:429–493. With permission from John Wiley and Sons.)

the lineages involved may have diversified from one another over a relatively short period of time and are consequently difficult to resolve on a tree, in which case it is called a **hard polytomy**. These major eukaryotic lineages likely diverged over 1 billion years ago, which understandably makes it hard to retrace fully their history. One pattern again emerging from a perusal of the overall eukaryotic tree is that parasitism has arisen independently in several distinct lineages, a pattern that becomes more apparent when we consider the following independent lineages, each of which includes several parasitic groups.

The Stramenopila

Stramenopiles are part of the SAR (Stramenopila-Alveolata-Rhizaria) lineage shown in Figure 2.6. Stramenopiles are also called heterokonts because their motile stages produce flagella of two different shapes. Members of this group, dominated by diatoms, golden algae, and brown algae, are largely free-living, but some parasitic lineages occur, most prominent among them the water molds or oomycetes (see also Figure 2.7). The Hyphochytriales, Labyrinthules, and Thraustochyrtrids are small groups of stramenopiles that are parasites of algae or marine plants.

Additionally, two organisms that can colonize humans, *Blastocystis hominis* and *Chilomastix mesnili,* are stramenopiles. Several species of *Blastocystis* occur, and they are among the most common unicellular eukaryotes inhabiting the human intestine. They are probably transferred between humans and domestic animals and may cause mild illness in immunocompromised individuals. As with many parasitic eukaryotes inhabiting anaerobic or microaerophilic portions of the gut, *Blastocystis* does not possess conventional mitochondria capable of aerobic respiration. It has a mitochondrion-like organelle that seems intermediate in function between a mitochondrion and a **hydrogenosome**. The latter is an organelle that generates ATP from pyruvate, while giving off hydrogen (H_2) as a by-product. *Chilomastix mesnili* is found in the cecum and colon of humans, other primates, and pigs, and is an example of a **commensal** organism, one that neither harms nor benefits its host but itself clearly benefits from making a living in the host.

The Alveolata

The Alveolata, also part of the SAR lineage, is a relatively well-defined clade of eukaryotes of immense importance because it contains many prominent groups of parasites. The name indicates their possession of alveoli, which are flattened vesicles lying beneath the plasma membrane. Included among the alveolates is the Apicomplexa, a huge, nearly exclusively parasitic lineage of great medical significance. We discuss them later in Section 2.1. Also included among the alveolates are the ciliates and dinoflagellates, some of which are prominent parasites. Ciliates typically have two nuclei and many short cilia arranged in rows. The one ciliate known to be parasitic in humans is ***Balantidium coli***, which causes balantidiasis, a zoonosis transmitted by the fecal–oral route, with pigs serving as reservoirs. Another familiar example of a parasitic ciliate, one frequently posing a problem for aquarists, is *Ichthyophthirius multifiliis,* better known as ich, an ectoparasite that causes infestations of white spots to appear on the skin of freshwater fishes (see also Figure 3.11).

Although dinoflagellates are mostly free-living, mostly marine planktonic photosynthetic organisms, the group shows a predilection to form symbiotic associations. Zooxanthellae are mutualistic dinoflagellates that play an important role in formation of coral reefs, and some like *Oodinium* and *Pfiesteria* are parasitic. One species of the latter genus, *Pfiesteria piscicida*, once referred to as the "cell from hell", is instructive in that it has been implicated in producing toxins causing both fish kills and human illness, including memory loss. This provoked "*Pfiesteria* hysteria", now largely alleviated by further work, showing that it does not pose an obvious human health hazard. See Figure 2.26 for an example of dinoflagellates that are parasitoids of other dinoflagellates. Parasites of the genus *Perkinsus*, significant pathogens of marine bivalves including oysters, are also alveolates related to dinoflagellates.

The Rhizaria

The third prominent group within the SAR lineage is the Rhizaria (comprised of Cercozoa, Foraminifera, and Polycystinea shown in Figure 2.6), mostly unicellular eukaryotes that produce pseudopods. Most, such as the foraminiferans and radiolarians, are free-living. Among the Rhizaria are the Phytomyxea (also called Plasmodiophorids), which are parasites developing within the cells of plants, often causing the infected tissue to form a scab or gall. Another group of rhizarian parasites are the spore-forming Ascetosporea, which are usually parasites of marine invertebrates. One example is *Haplosporidium nelsoni*, which causes a disease known as MSX (short for multinucleated sphere unknown) that has caused crashes in commercially valuable oyster populations.

The Archaeplastida

Included here are red and green algae, glaucophytes, and the land plants. These organisms are characterized by the possession of a chloroplast, and most possess a cell wall. Among the plants are several species that are parasitic. Some red algae are also parasitic, often on other red algae.

The Excavata

Members of this prominent group of unicellular eukaryotes possess a mitochondrion, but often in a highly modified form. Most have flagella. Many prominent parasites are excavates: ***Trypanosoma***, ***Leishmania***, *Naegleria*, ***Giardia***, ***Histomonas***, ***Trichomonas***, and *Dientamoeba*. Many members of the Excavata are included in the Rogues' Gallery and are discussed elsewhere in this book. Consider ***Giardia lamblia***, the causative agent of giardiasis, first seen by van Leeuwenhoek using his microscope to examine his own stools in 1681. We have since learned that ***Giardia*** does not possess typical mitochondria, and it was thought that this organism might have diverged from eukaryotic stock before the ancestral eukaryote had acquired the mitochondrion by **primary endosymbiosis**. Primary endosymbiosis in this context refers to the acquisition of a bacterium (probably an alphaproteobacterium) by an ancestral protoeukaryote, with the metabolically versatile bacterium thereafter serving as the mitochondrion. Although ***Giardia*** is still considered to be an early diverging eukaryote, we know today that it possesses a reduced version of the mitochondrion called a mitosome. **Mitosomes** are double-membrane structures like mitochondria and are almost certainly derived from them, but they lack mitochondrial DNA. They are incapable of aerobic respiration, in keeping with the limited oxygen environment in which ***Giardia*** lives, but they can still produce ATP. A related group of excavates, including the genus ***Trichomonas***, also has modified mitochondria that in this case are considered to be true hydrogenosomes (they produce H_2 as a by-product), as described above.

Other excavates, such as ***Trypanosoma*** and ***Leishmania***, are unusual in possessing a single mitochondrion that contains a **kinetoplast**. The kinetoplast contains a network of concatenated circular DNA molecules (assembled like the chain mail in armor), some of which are maxicircles that encode in a peculiar encrypted fashion the usual mitochondrial gene products. Many minicircles are also present and encode guide RNAs, which are used to decode the encrypted maxicircles. Guide RNAs either insert or delete uridine residues in maxicircle transcripts to accomplish this. It is not clear why kinetoplastids use this unusual RNA editing process. It may have been derived from genes transferred horizontally from viruses. Whatever the origin or purpose, it is clear that disabling RNA editing is lethal for kinetoplastids. Kinetoplastids are also unusual for sequestering the enzymes of glycolysis within distinct, membrane-bound **glycosomes**.

The Amoebozoa

Amoebozoans are unicellular eukaryotes that move by formation of blunt pseudopods. Within this group are prominent parasites such as ***Entamoeba histolytica***, which causes amebic dysentery, and others such as ***Endolimax***, *Acanthamoeba*, and *Balamuthia*. ***E. histolytica*** is yet another example of an anaerobic or microaerophilic parasitic eukaryote with mitosomes. The mitosomes of ***Entamoeba*** appear to be among the most reduced of all endosymbiont-derived organelles; their function is still uncertain because they do not appear to participate in energy metabolism. Their abundance within ***E. histolytica*** trophozoites suggests that they have an essential yet still enigmatic role to play.

The Opisthokonta

This is the final major lineage of eukaryotes of note from the standpoint of parasitology. Many opisthokonts possess posteriorly directed flagella that propel them. It includes the Fungi, and a relatively closely related group, the Animalia, both with many parasitic representatives and both discussed below. The study of opisthokonts points out that we still have much to learn about the variety of parasitic organisms in nature, as exemplified by the Mesomycetozoea (or Ichthyosporea), a lineage of mostly parasitic organisms that did not come to light until 1996. These organisms are discussed further in Section 2.3.

HGT has also played a role in the evolution of eukaryotic parasites

To what extent has HGT played a role in evolution of eukaryotes, including parasites? In general, eukaryotic cells are believed to be less prone to HGT, in part because they often maintain separate somatic and germinal cells, with the latter more difficult for foreign DNA to invade. Although it is hard to detect true cases of HGT, it has been considered to be particularly likely to occur successfully in eukaryotes that are parasitic, and it may even have been the key factor enabling a particular lineage to adopt parasitism. Bacteria and viruses are both likely sources of genes for eukaryotic parasites: they are ubiquitous; they are metabolically very diverse, such that they have novel capabilities to provide; they are frequently found in the same local environments as eukaryotes; and they are often ingested by eukaryotes. Once acquired, the genes could be further shuttled serially among eukaryotes. The genomes of some of the eukaryotes mentioned above including *Blastocystis*, ***Trypanosoma***, ***Trichomonas***, and ***Entamoeba*** show evidence of repeated and considerable HGT from prokaryotes and viruses, especially those sharing environments with them.

Another example of HGT involving representatives of two distantly related eukaryotic kingdoms is the transfer of multiple genes from fungi to oomycetes (**Figure 2.7**). Oomycetes (see stramenopiles) comprise a distinct lineage of fungus-like microorganisms that are often called water molds, even though most species infect terrestrial plants. Unlike fungi with cell walls made of chitin, the cell walls of oomycetes are made of cellulose. Like fungi, they produce filamentous structures for absorption of nutrients. HGT in this case is believed to have involved several genes, including those giving oomycetes the ability to take up soluble nutrients, a phenomenon called osmotrophy. Also transferred were genes encoding products able to resist host immune responses, considered essential for the ability of oomycetes to parasitize plants. One oomycete benefiting from HGT is *Phytophthora infestans*, the organism responsible for late potato blight that caused the Great Potato Famine in Ireland from 1845 to 1852. A million Irish citizens died during the famine and another million emigrated from Ireland. Late potato blight is still a formidable problem today. Several examples of HGT between parasites and their hosts have also been documented (for instance in parasitic plants). It is probable that the importance of HGT in parasite evolution will prove to be much greater than presently appreciated.

A

B

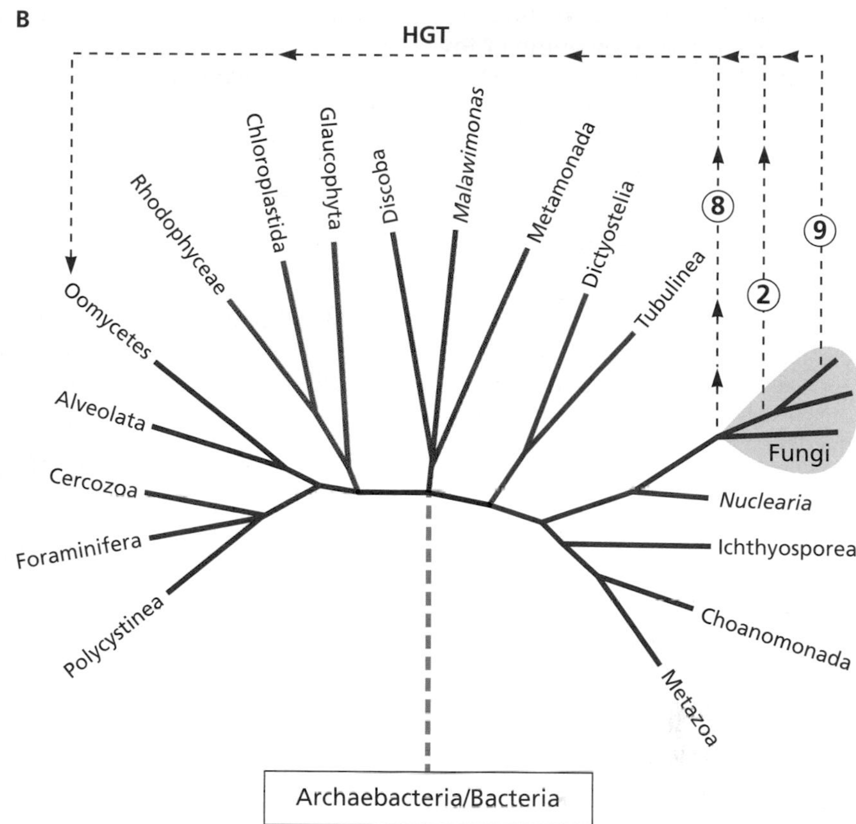

Figure 2.7 *Phytophthora infestans*, an oomycete responsible for the infamous late potato blight, has engaged in horizontal gene transfer (HGT). (A) Note the shrunken appearance of the infected potato and its rotten interior. (B) Shown on this tree are eight, two, and nine different instances in which genetic material was likely transferred by HGT (dashed lines) from fungi at different points in their diversification to members of the oomycete (including *Phytophthora*) lineage. (A, courtesy of US Department of Agriculture. B, based on Richards TA et al [2011] *Proc Natl Acad Sci USA* 108:15,258–15,263.)

The Apicomplexa is a huge, important, nearly exclusive parasitic group of organisms

The largest and arguably most important of all groups of parasites is the phylum Apicomplexa, part of the Alveolate lineage shown in Figure 2.6. Some of the most deadly pathogens of humans, such as the malaria parasites, and of domestic animals, such as the coccidians, are apicomplexans. With the exception of one small group of free-living predators, apicomplexans are otherwise exclusively intracellular parasites. They make use of an **apical complex** (**Figure 2.8A**) for which they are named to recognize, attach to, and penetrate host cells. The apical complex is composed of a **conoid**, a set of microtubules arranged in a spiral configuration, **rhoptries,** which are secretory in function, and one or more **polar rings**. The complex also may contain more slender convoluted secretory structures called **micronemes** that connect to the rhoptries.

Most apicomplexans are also notable for possessing a circular plastid genome contained within a structure called the **apicoplast**. It is likely the apicoplast was originally acquired through a process called **secondary endosymbiosis** (Figure 2.8B), believed to have occurred as follows. An algal cell containing a chloroplast was taken up into the endomembrane system of an ancestral protist. (The algal cell contained a chloroplast that is a modified cyanobacterium that was itself taken up by the ancestor of the host algal cell by **primary endosymbiosis.**) Once the protist ingested the alga, genes were transferred from both the nucleus and chloroplast of the alga to the protist nucleus. The chloroplast (the original cyanobacterium) persists and is called the apicoplast, and the protist cell in which it is found, is called an

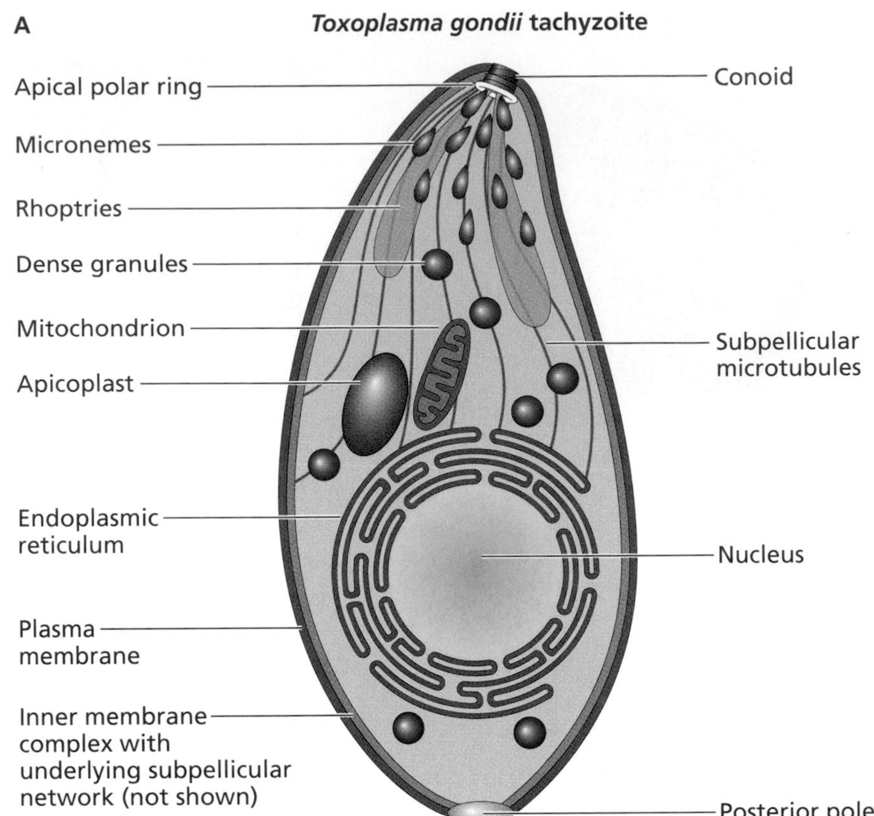

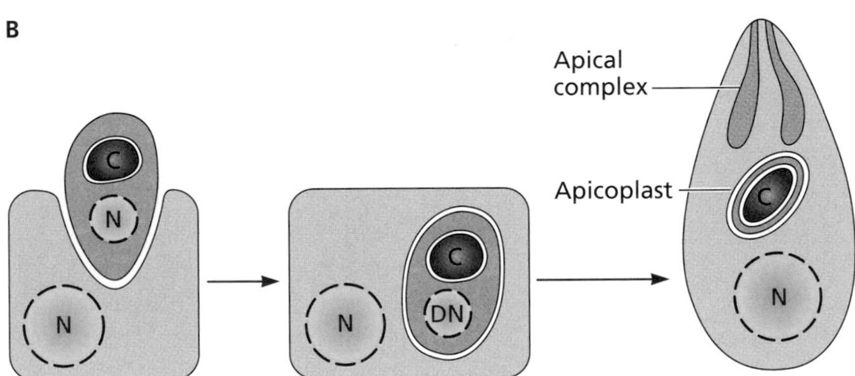

Figure 2.8 Characteristic features of apicomplexans and the origin of the apicoplast. (A) A typical apicomplexan, the tachyzoite of *Toxoplasma gondii*, shows details of the apical complex, a structure well named for indeed it is complex. Note also the presence of an apicoplast. (B) A possible sequence of events showing the origin of the apicoplast. Left, an algal cell (blue) with a chloroplast (C, red) is taken up by an ancestral protist host (tan). In the middle, the algal cell has been engulfed, and its nucleus is diminished in size (DN). Note the persistence of the chloroplast. On the right, note that the algal cell has become much smaller, lost its nucleus, and its chloroplast persists. The chloroplast is surrounded by four membranes and is called an apicoplast. The host cell is now an apicomplexan. C, chloroplast; N, nucleus; DN, diminished algal nucleus. (A, modified from Baum et al [2006] *Nat Rev Microbiol* 4:621–628. With permission from Macmillan Publishers Ltd. B, modified from Sheiner L & Striepen B [2013] *Biochim Biophys Acta* 1833(2)352–359. With permission from Elsevier.)

apicomplexan. The apicoplast lies within a four-membrane structure in the host cell. The membranes are derived from the inner chloroplast membrane of cyanobacterial origin, the outer chloroplast membrane is of cyanobacterial origin, the periplastid membrane is of algal plasma membrane origin, and the outer membrane is from the apicomplexan endomembrane system.

Although genes involved in photosynthesis have been deleted from the original plastid chromosome and apicomplexans are not capable of photosynthesis, the plastid does retain functional genes. If the apicoplast is disabled, the apicomplexan may be killed or be unable to penetrate new host cells. The apicoplast is an appealing target for development of new drugs that would selectively target apicomplexans and leave the host (lacking plastid genomes) unaffected (see Chapter 9).

It has been suggested that all free-living animal species harbor at least one apicomplexan parasite. A remarkable and largely unappreciated degree of biodiversity exists even within a single apicomplexan genus. For example, there are already about 2000 described species of ***Eimeria***, with many more species awaiting description. Similarly, enumeration of apicomplexans called gregarines, which are parasites of insects and other invertebrates, has only just begun and 1650 species have already been described. Because gregarines are found in speciose host groups such as beetles, this number will almost certainly increase dramatically. As many as 10,000 species of *Haemoproteus* and ***Plasmodium*** may eventually be enumerated.

Traditional taxonomic schemes have recognized four groups of apicomplexans: the coccidians, gregarines, haemosporidians, and the piroplasmids. Most of the apicomplexans have a conoid in their apical complex (see Figure 2.8A), but the haemosporidians (***Plasmodium*** sp.) and the piroplasms (such as ***Theileria*** and ***Babesia***) do not. The phylogenetic relationships among these groups are actively being pursued, far from definitively resolved, and **Figure 2.9** presents one recent hypothesis of relationships. Note that *Perkinsus* and *Colpodella* are outgroups to root the tree (see the website associated with this book). An outgroup represents an organism that is believed to be a close relative of the group being analyzed (in this case the apicomplexa), but does not fall within that group. Several apicomplexan genera of medical significance (for example, ***Cryptosporidium***, ***Toxoplasma***, ***Plasmodium***) will be mentioned often throughout this book.

Many well-known parasites belong to familiar groups of multicellular organisms

Also particularly worthy of discussion in this overview of parasite diversity are several groups of parasites that are multicellular (see Figure 2.6), found among the red algae, plants, fungi, and animals.

The Parasitic Rhodophytes (Red Algae)

Red algae are photosynthetic, macroscopic algae. They contain chloroplasts and distinctive red pigments. Although red algae hardly spring to mind as iconic parasites, at present about 8% of the 66 described genera of red algae (in the florideophyte lineage) include parasites. All together, there are about 116 known species of parasitic red algae. Parasitism is believed to have arisen

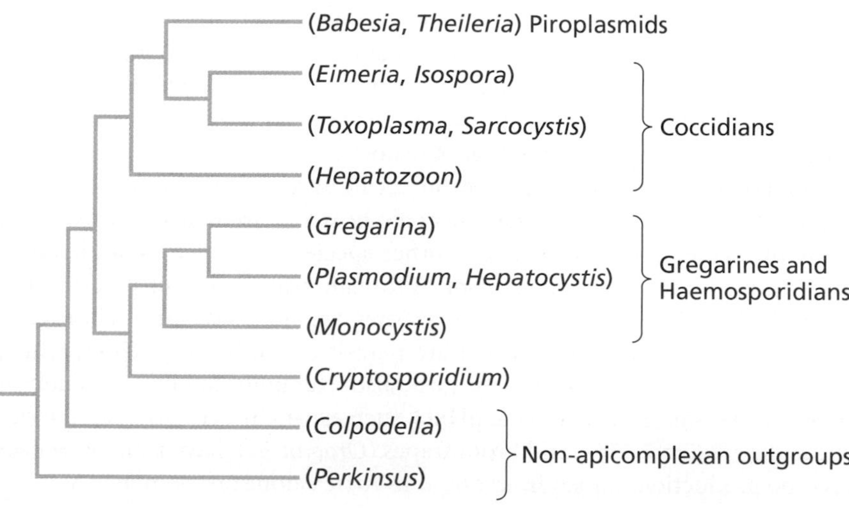

Figure 2.9 A phylogenetic tree of the Apicomplexa based on full-length sequences of the SSU rRNA gene. Included on this tree are several important parasites including *Babesia*, *Cryptosporidium*, *Eimeria*, *Plasmodium*, *Sarcocystis*, *Toxoplasma*, and *Theileria*. *Perkinsus* and *Colpodella* form the outgroups. This particular tree does not hypothesize a close relationship between haemosporidians and piroplasmids. (Redrawn from Morrison DA [2009] *Trends Parasitol* 25:375–382. With permission from Elsevier.)

independently over 100 times in this group. Red algae are distinctive in their mode of parasitism. They are able to form fusions among adjacent cells, the fusions being called secondary pit connections, that provide an avenue through which organelles such as the nucleus, mitochondria, and plastids from a parasitic cell can gain access to the cytoplasm of a host cell. Once the parasite organelles are present, they proceed to divide and spread via secondary pit connections, taking over the host's body (**Figure 2.10A**). Another peculiar feature of parasitism in red algae is that most parasites share a recent common ancestry with an extant, free-living red alga species that they specifically parasitize. Such parasites are called **adelphoparasites** (adelphose being a Greek term for kin). Other red algal parasites are called **alloparasites** and typically have broader host ranges (but the hosts are always other red algae) that include more distantly related host species in other genera or families (Figure 2.10B). One idea awaiting further study is that adelphoparasites represent an early phase in parasitism followed by adoption of more distantly related hosts as characteristic of alloparasites. Molecular studies have shown that many red algal parasites exist in a continuum between these two poles, suggesting some revision in emphasis on use of these two terms may be needed. As discussed further in Chapter 7, red algae provide a compelling model to examine the origins of parasitism from free-living ancestors.

Parasitic Plants

Plants are multicellular, photosynthetic, autotrophic eukaryotes that are members of the kingdom Plantae. They produce cell walls containing cellulose. Included are seed plants, ferns, mosses, liverworts, and hornworts. Among the seed plants are the flowering plants, or **angiosperms**. Over 4400 species in total, or about 1% of all flowering plants, are parasitic. Parasitism is estimated to have arisen 12 to 13 separate times during the evolution of angiosperms (**Figure 2.11**), with parasitic representatives known in about 30 different families.

Parasitic angiosperms invade only the roots or shoots of other land plants with their specialized invasive roots called **haustoria**, apparently having no capacity to become invasive within the bodies of other major groups of multicellular organisms. Some plants do however parasitize mycorrhizal fungi (see Figure 1.8). Formation of haustoria is an essential feature of plant parasitism, and it is through these structures that parasitic plants absorb their nutrition from their hosts (**Figure 2.12**).

Plant parasites are either **facultative parasites** and are able to live autotrophically without the need to parasitize a host, or they are **obligatory parasites** and require a suitable host to complete their life cycles. Parasitic plants are also classified according to their ability to engage in photosynthesis: some are **hemiparasites** and are still capable of photosynthesis, whereas others are **holoparasites,** which are incapable of photosynthesis and must obtain all their energy through their haustoria.

Some parasitic plant species can infect hundreds of different host plant species, whereas others have very specific host requirements. For example, some mistletoes can parasitize only other species of mistletoes, a phenomenon termed **obligate epiparasitism**. Although reminiscent of parasitic red algae described above, obligate epiparasites are not necessarily close relatives of their hosts. Mistletoes that are parasites of other mistletoes, which are themselves parasites of a plant host, also exemplify the phenomenon of **hyperparasitism**. Some parasitic plants, such as witchweeds (*Striga*), dodder (*Cuscuta*) (**Figure 2.13A**), and broomrapes (*Orobanche*), have a major impact on crop production. Losses from *Striga* in Africa alone are estimated at up to

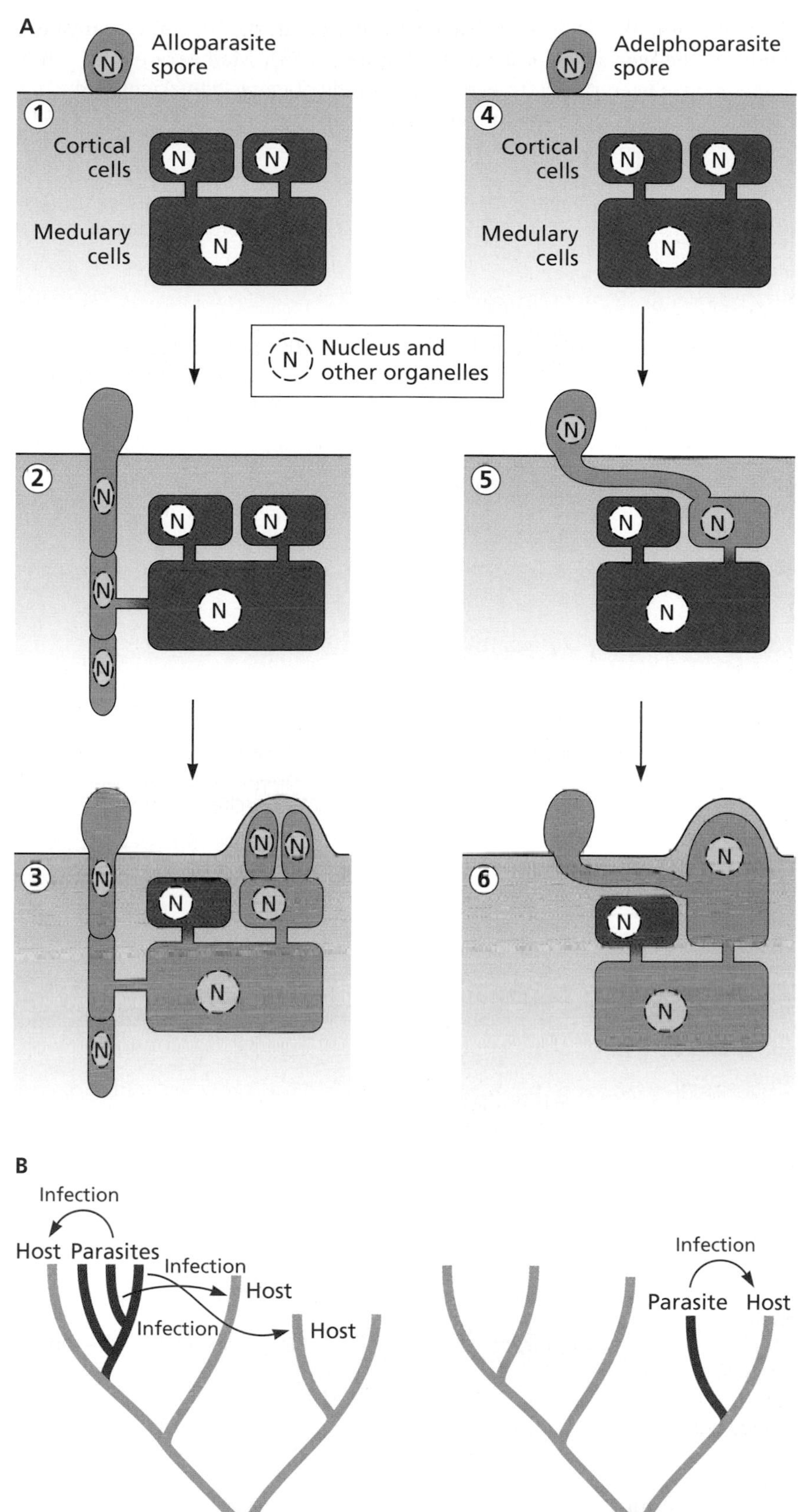

Figure 2.10 The peculiar nature of parasitism in red algae. (A) Host red algal cells are shown in red with their nuclei (N) in white. Parasite cells are shown in yellow with their nuclei (N) in gray. Germination and development for an alloparasite (*Choreocolax polysiphoniae*) is shown in panels 1–3 and for an adelphoparasite (*Gracilariophyla oryzoides*) in panels 4–6. Note how the nucleus and other organelles of the parasite spread through host cells via connections, basically taking over the host cells in the process. (B) Note the close relationship between the adelphoparasite (brown) and the host it infects (red arrow). In alloparasites, infection (red arrows) of more distant relatives can occur. (B, from Blouin NA & Lane CE [2012] *BioEssays* 34: 226–235. With permission from John Wiley and Sons.)

$7 billion annually. One remarkable parasitic plant, the "queen of parasites", produces the world's biggest flower (Figure 2.13B). Members of the genus *Rafflesia* have been shown to express hundreds of genes likely acquired from their host plants, thus providing an example of host to parasite HGT.

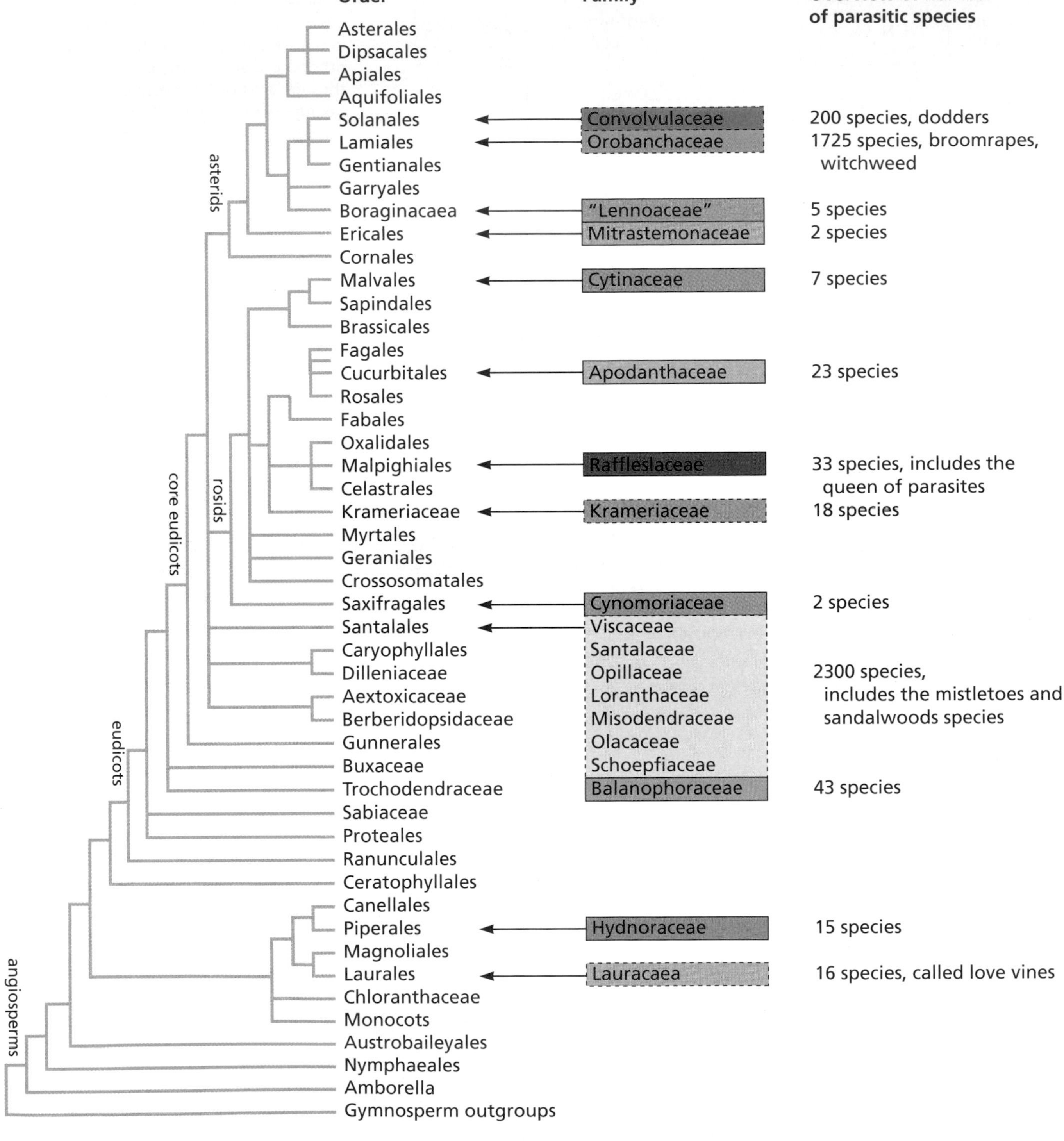

Figure 2.11 An overview of parasitism in plants. This tree provides an overview of the diversity of flowering plants and indicates the lineages in which parasitism has occurred (in red) and the approximate number of parasitic species for each. Those groups enclosed by a dashed box are hemiparasites and those enclosed by a box with solid lines holoparasites. Convolvulaceae and Orobanchaceae include both. See http://www.parasiticplants.siu.edu/ for an excellent overview of parasitic plants.

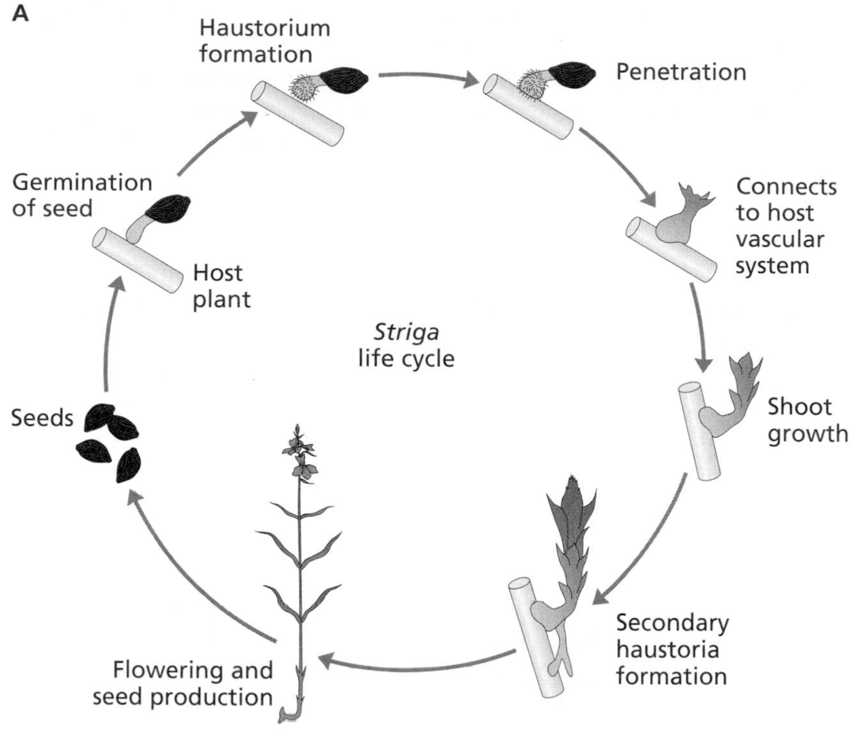

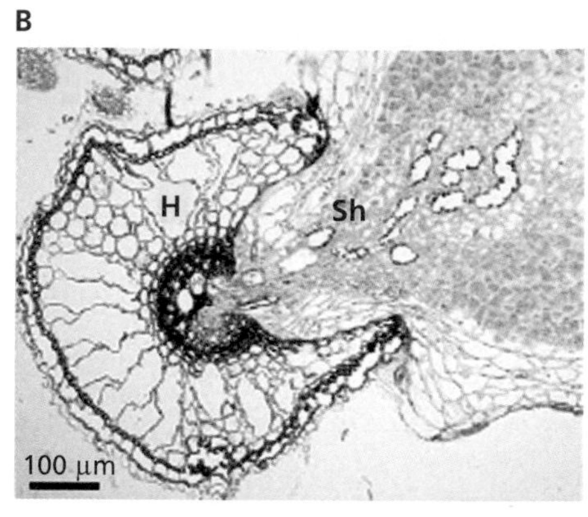

Figure 2.12 Life cycle and host colonization for a typical parasitic plant. (A) The life cycle of witchweed (*Striga* spp), (B) Cross-section of *Striga hermonthica* (Sh) parasitizing a rice root (H). (A, modified from IITA Striga manual (1997). B, from Yoshida S & Shirasu K [2012] *Curr Opin Plant Biol* 15:708–713. With permission from Elsevier.)

Figure 2.13 Some remarkable parasitic plants. (A) A species of dodder (*Cuscuta*) enveloping an acacia tree. Some species of *Cuscuta* have small amounts of chlorophyll and can engage In photosynthesis and are classified as hemiparasites, whereas others are totally dependent on their host for nutrition and are holoparasites. (B) A flower of the parasitic plant, *Rafflesia arnoldsi*, the world's largest at 100 cm in diameter. The flower smells like rotting flesh and thereby attracts flies that serve as pollinators. This remarkable plant, sometimes called the "queen of parasites", lacks leaves, stems, and roots. It produces an invasive haustorium that colonizes vines of the genus *Tetrastigma*. (A, courtesy of Khalid Mahmood, CC BY-SA 3.0. B, courtesy of Ma Suska, CC BY-SA 2.0.)

Parasitic Fungi

The kingdom Fungi includes organisms ranging from single-celled chytrids to complex, multicellular mushrooms. Fungal parasites are usually discussed in mycology courses and will not be emphasized in this book. **Mycology** is the branch of biology that deals with the study of fungi. However, parasitism is such a pervasive feature of fungal biology, and fungi are so frequently implicated as the cause of emerging diseases or of extirpations of endangered species, it would be remiss not to include them in this overview of parasite diversity. Examples of the effects of parasitic fungi appear throughout the book.

Fungi are heterotrophic with an **osmotrophic** (absorptive) mode of acquisition of nutrients. They usually possess a thallus (body) composed of branching filaments called **hyphae** that together can form a densely branched network called a **mycelium**. The hyphae grow by apical extension and have polymers of *N*-acetyl glucosamine (**chitin**) in their walls. It is estimated that two-thirds of the known ~100,000 species of fungi enter into some form of intimate association with another living organism, and many different kinds of fungi parasitize a huge variety of organisms including other fungi, plants, and many different kinds of animals, including humans. A common estimate of the number of fungal species in the world is 1.5 million species.

Although the phylogenetic relationships among fungi are far from settled, there has been considerable progress in resolving relationships and identifying component groups. **Figure 2.14**, representing one hypothesis of relationships, provides an overview of some of the major parasitic groups.

The chytrids used to be regularly excluded from the fungi but are now recognized as being among the earliest diverging members of this kingdom. Chytrids are mostly aquatic forms, but there are representatives that parasitize animals or plants. One chytrid, *Batrachochytrium dendrobatidis*, can infect over 500 species of amphibians on all continents where amphibians are found. It has contributed to nearly half of all amphibian species being in decline and has already caused the extinction of some frog species (see Chapter 8 and Box 8.1). The microsporidians, long considered to be protists because of their unicellular spores, are now recognized as fungi; all are parasites of animals. These host-specific parasites colonize all groups of animals and are probably grossly underrepresented by the approximately 1500 species thus far described. Among them are species of *Nosema* implicated in declines of bee populations. The Ascomycota is a vast group of fungi that includes several important plant parasites. These include the causative agents of Dutch elm disease (*Ophiostoma ulmi*),

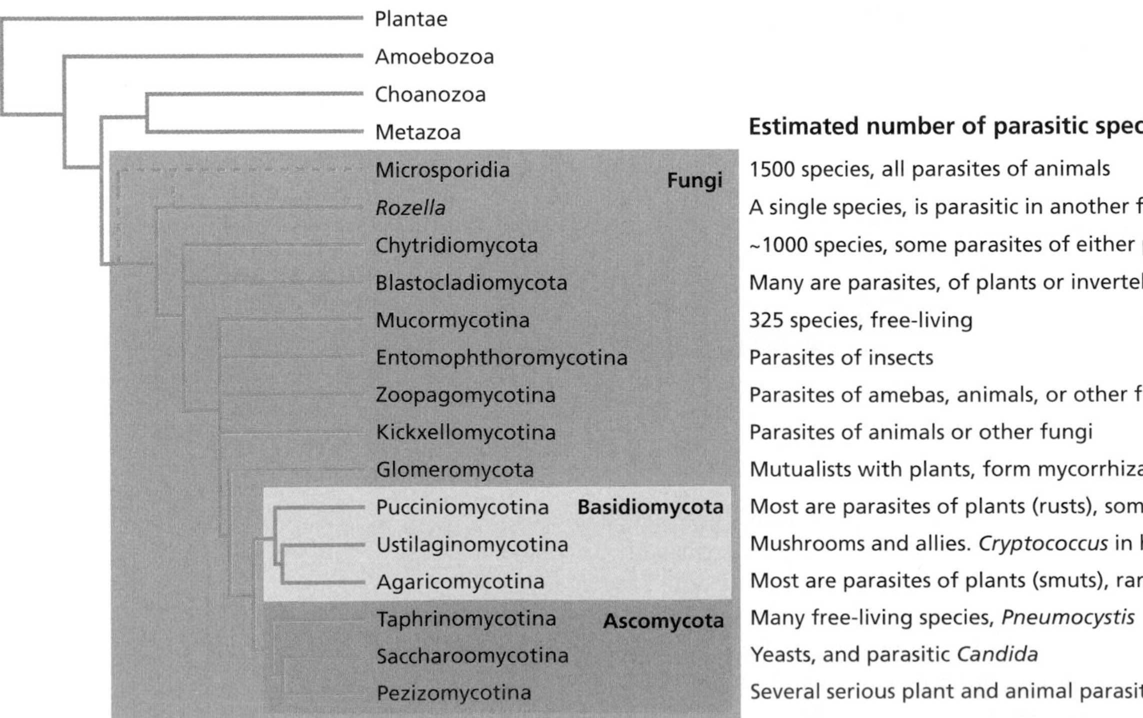

Figure 2.14 Parasitism is ubiquitous among the Fungi. An overview of the kingdom Fungi is shown in the yellow box, and two major lineages, the Basidiomycota (blue box) and Ascomycota (green box) are indicated. It is noteworthy that parasitism has originated in most fungal lineages and that the variety of hosts for fungi is huge. Note that for the Microsporidia two different possible origins are indicated by the dashed lines. (Modified from Stajich JE et al [2009] *Curr Biol* 19:R840–R845. With permission from Elsevier.)

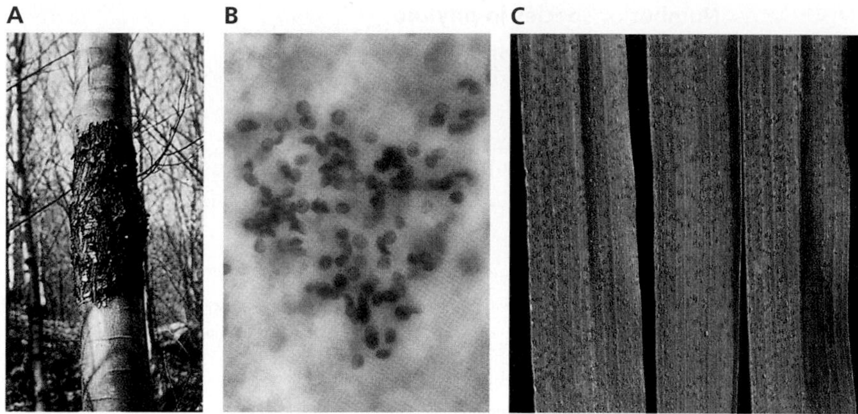

Figure 2.15 Examples of the impact and diversity of fungal parasites. (A) A lesion (canker) on a chestnut tree (*Castanea*) afflicted by chestnut blight (*Cryphonectria parasitica*). (B) Cysts of *Pneumocystis jirovecii*, an ascomycete fungus, from the lungs of a patient with pneumonia. (C) The wheat leaf rust (*Puccinia triticina*), a basidiomycete fungus that infects wheat and other related food grain plants. (A, courtesy of Daniel Rigling. B, courtesy of Pulmonary Pathology, CC BY-SA 2.0. C, courtesy of James Kolmer, US Department of Agriculture.)

chestnut blight (*Cryphonectria parasitica*) (**Figure 2.15A**), and the recently problematic ash dieback (*Chalara fraxinea*). Chestnut blight devastated the American chestnut tree (*Castanea dentata*) across the North American continent in the early 1900s. The ascomycete *Pneumocystis jirovecii* is a frequent cause of pneumonia (Figure 2.15B). *Pneumocystis jirovecii* used to be called *P. carinii*, the latter a name now reserved for a species found in animals. Fungal infections such as this one are often **opportunistic** because they do not cause disease in healthy hosts but can become problematic in unhealthy individuals. This parasite can cause fatal infections in immunocompromised individuals. The parasite is often found in those who have concurrent infection with HIV or who are taking immunosuppressive drugs following organ transplants. Another ascomycete that has achieved unwanted prominence is *Pseudogymnoascus destructans*, the fungus responsible for white-nose syndrome currently decimating some species of North American bats (see Box 8.2). The Basidiomycetes includes the familiar mushrooms, stinkhorns, and puffballs, as well as several substantial parasites of plants such as the rusts (Figure 2.15C), smuts, and wood-rotting fungi. *Puccinia triticina* and its relatives are of great concern because of their potential for devastating wheat and other food crops around the world.

Parasitic Animals

Members of the kingdom Animalia (animals, or metazoans) are unique among the world's organisms for the development in most of integrated nervous and muscular systems that give them unprecedented mobility and responsiveness to environmental circumstances. Animals are multicellular heterotrophs that usually acquire their energy from ingestion of organic compounds, although several parasitic groups acquire nutrients by absorption across their body walls. Many of the world's most familiar and medically significant parasites are found among the animals. Parasitism has arisen independently on at least 60 occasions in animals, in both major and minor lineages. Some lineages of animals are exclusively parasitic, some have a mixture of free-living and parasitic species, and some as best we know are without parasitic representatives (**Figure 2.16**).

Establishing the relative phylogenetic positions of the approximately 36 phyla of animals in the overall tree of animal evolution is a subject of ongoing intensive study. As is true with most groups of organisms, nucleotide sequence data have dramatically improved our understanding of the relationships among animals. This is particularly true for parasites because often their morphological features have been greatly modified, rendering their origins and relationships obscure (see also Figure 7.26). A conservative view of

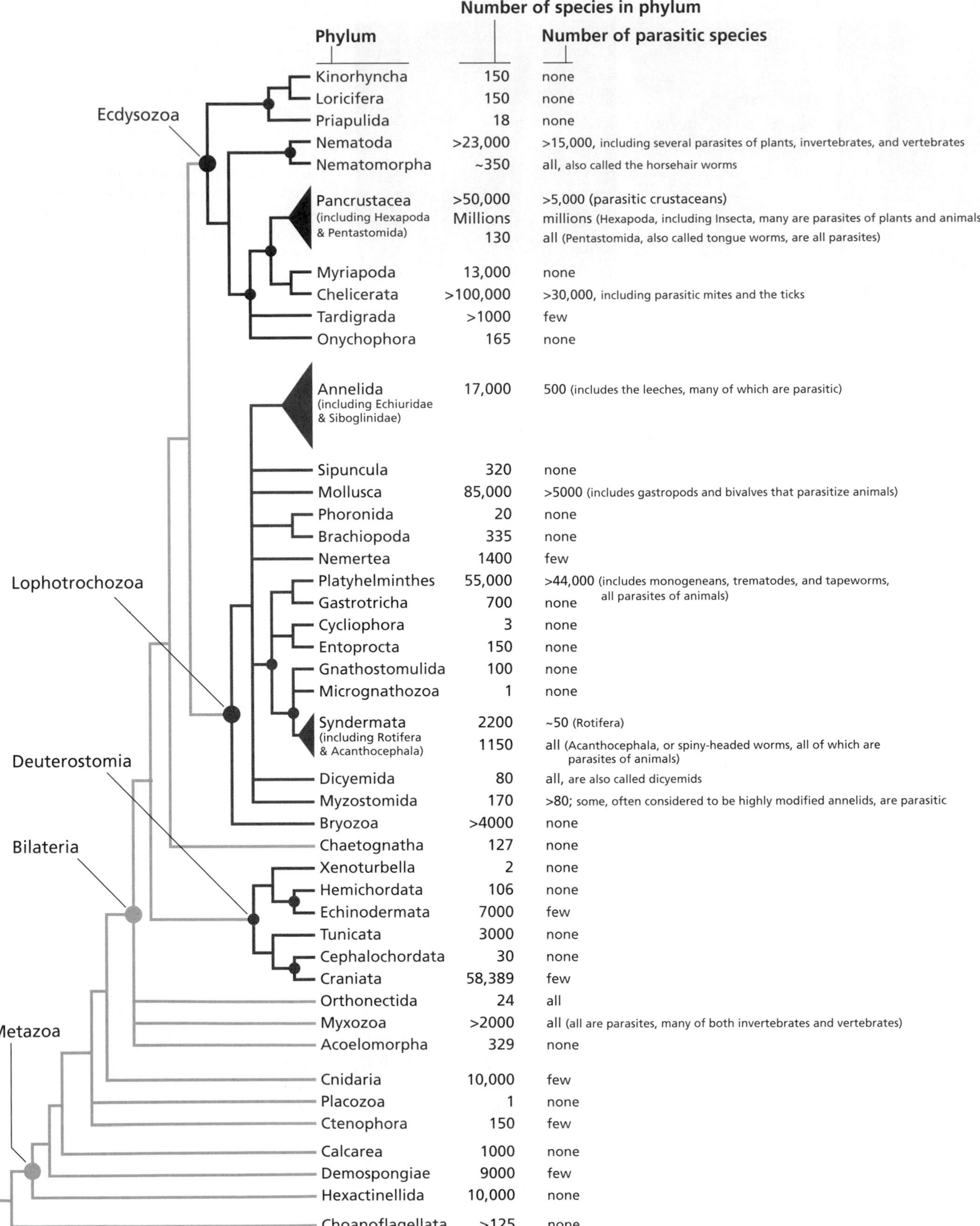

Figure 2.16 An overview of animal (metazoan) phylogeny with an emphasis on parasitic animals. This view of animal relationships is based on SSU rDNA. The dots on the tree indicate branching points delineating major groups. To the right of the tree is indicated the number of species in each phylum, followed by the number of parasitic species in that phylum.

animal relationships based on SSU rDNA provides many of the salient features (Figure 2.16). Included to the right of the hypothesized relationships are approximations of the numbers of species present in each lineage, followed by the estimated number of parasitic species.

Most trees depicting animal relationships recognize the close relationships between mesomycetozoeans and choanoflagellates with animals, and they use them as outgroups to root the animal tree. Ichthyosporeans, as noted above, are a poorly known group of (mostly parasitic) opithokonts that group near to both the Fungi and Animalia. Choanoflagellates are free-living unicellular or colonial eukaryotes that have a distinctive collarlike structure surrounding a flagellum. They resemble the choanocyte cells of sponges, which also have a collar and associated flagellum. The sponges are usually, but not always, identified as the earliest diverging animal lineage, followed by other phyla that lack bilateral symmetry (pre-bilaterians).

Among those animals with bilateral symmetry, in addition to a number of smaller phyla whose affinities are debated, three major lineages of animals are recognized, as indicated by the colors in Figure 2.16: the Deuterostomia, Lophotrochozoa, and the Ecdysozoa. Molecular studies have been particularly important in defining this last major lineage, one united by the common property of molting (**ecdysis**) of an external cuticle. The Guinea worm, ***Dracunculus medinensis*** (**Figure 2.17A**), a member of the phylum Nematoda that includes many parasitic representatives, is an example of an animal that undergoes molting (Ecdysozoa). This species is now on the verge of eradication owing to control measures to break its life cycle by filtering drinking water and preventing ingestion of infected copepods (see Chapter 9). Also included among the molting phyla is a huge group of organisms, the **Arthropoda**. Included prominently in the Arthropoda are the insects. The Arthropoda include several other groups with many parasite species such as ***Sarcoptes scabiei***, a mite that causes intense itching when it tunnels through human skin (Figure 2.17B). More specifically, it is a chelicerate arthropod meaning it has pincerlike chelicerae instead of jaws. Many species of mites

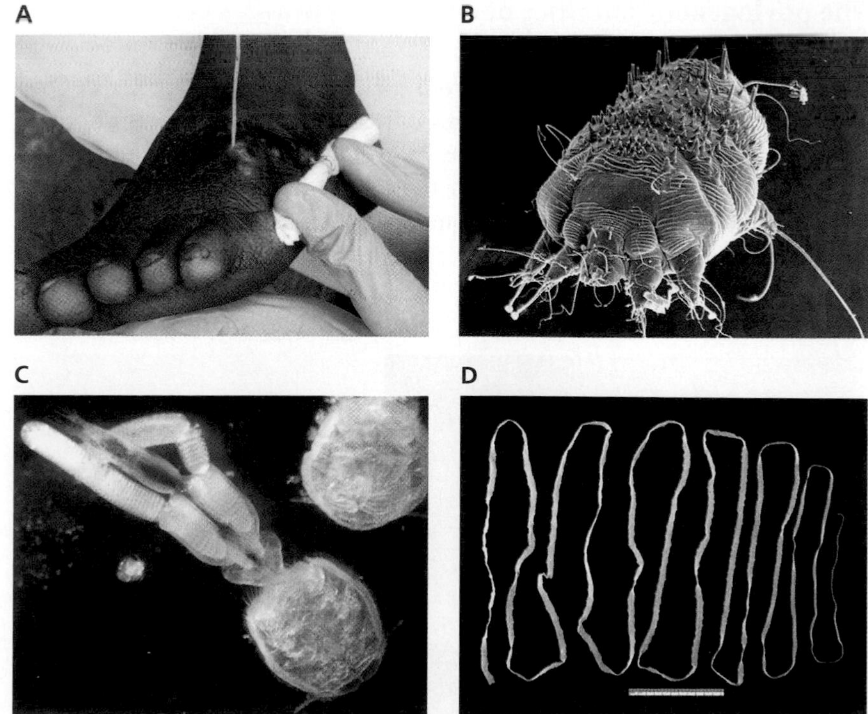

Figure 2.17 Examples of some prominent groups of animal parasites. (A) The Guinea worm *Dracunculus medinensis*, a nematode, extruding from a victim's foot. (B) *Sarcoptes scabiei*, a parasitic mite that burrows through the skin of its human host. (C) A parasitic copepod, also a member of the phylum Arthropoda. This female copepod burrows into the gills and mouth of its host, a shark, and produces two large ovisacs. (D) *Taenia saginata*, the beef tapeworm. (A, ©The Carter Center, L. Gubb. B, courtesy of Louis De Vois. C, with permission from Kelly Weinersmith. D, courtesy of the CDC.)

are parasitic. Another prominent group of arthropods are the crustaceans, many of which are parasitic, often becoming highly modified in form in the process of becoming parasitic (Figure 2.17C). Prominent among the Lophotrochozoa are members of the phylum **Platyhelminthes**, most of which are parasitic. One spectacular example is ***Taenia saginata***, a tapeworm that routinely grows to 4 to 12 meters in length, living in the small intestine of its human host (Figure 2.17D).

Figure 2.16 reveals that much of the uncertainty with respect to quantifying the number of eukaryotic species in general, and of animal parasites in particular, lies within the vast phylum **Arthropoda**. The phylum is dominated by insects of which there are an estimated one million named species. However, there are many more insect species to be described and we do not know if this number will prove to be in the millions or thousands of new species. The variety of parasitic lifestyles among the insects alone is immense (**Figure 2.18**), including organisms as diverse as gall-making parasites of plants, hundreds of thousands of wasp species that are parasitoids undergoing their larval development in other insects and invertebrates, ectoparasites such as fleas and lice, and additional blood-feeding insects such as mosquitoes, black flies, and kissing bugs that also are frequently implicated in transmission of viruses or other disease-causing organisms.

Further discussion of how parasites have both provided remarkable insights into animal relationships and confounded the evolutionary tree builders are discussed in Section 2.2.

2.2 INSIGHTS INTO PARASITISM FROM THE STUDY OF DIVERSITY

The quest to reveal the full measure of parasite diversity and to understand the relationships of parasites to one another and to other organisms has lead to many novel insights about parasitism. In this section, we provide some examples.

The phylogenetic affinities of enigmatic parasites can be revealed

Box 2.1 described the pentastomes from the lungs of a monitor lizard. Over the years, scientists have noted features of pentastomes reminiscent of those found in a number of animal groups—tardigrades, mites, onychophorans, annelids, and myriapods—but given their lack of obvious similarity with other animals, pentastomes were frequently accorded the status of a separate

A

B

Figure 2.18 Examples of parasitoids and parasites among a huge lineage containing many parasites, the Insecta. (A) A parasitoid braconid wasp *Peristenus digoneutis* in the process of ovipositing in the body of its host, the tarnished plant bug (*Lygus*), the latter a serious pest of fruit and vegetable plants. (B) The flea *Xenopsylla cheopis*, a blood-feeding ectoparasitic insect that also serves as a vector for *Yersinia pestis*, the bacterium that causes plague. (A, courtesy of US Department of Agriculture. B, Courtesy of CDC.)

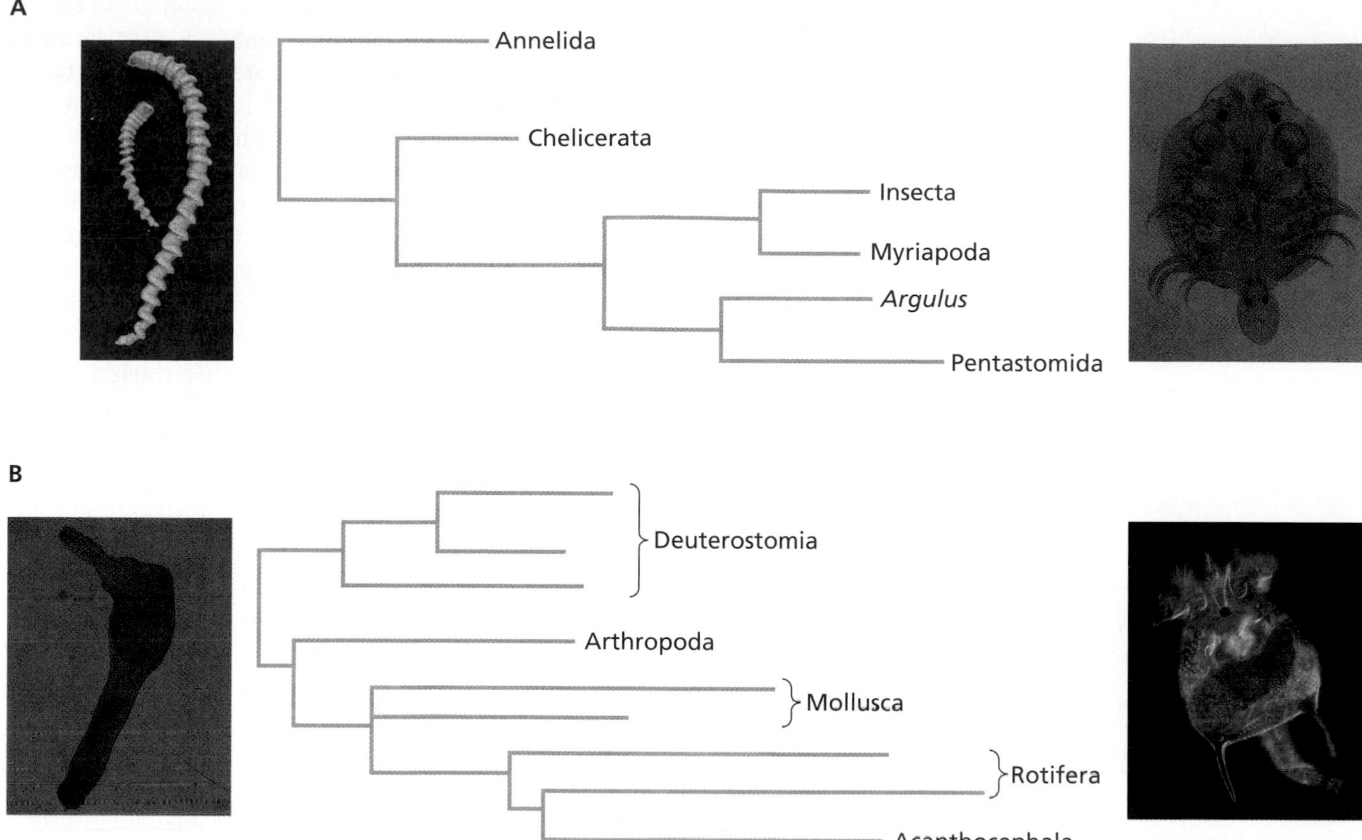

Figure 2.19 Two examples of perplexing parasites with surprising relatives. (A) On the left is a pentastome shown by the tree portraying the results of molecular phylogenetic work (center) to be related to branchiuran crustaceans such as *Argulus*, the fish louse shown on the right. (B) Acanthocephalans or thorny-headed worms, an exclusively parasitic group (left), were hypothesized in analysis (center) to be related to rotifers (right), predominantly a free-living group. (A, (from left to right) from Tappe D & Büttner DW [2009] *PLoS Negl Trop Dis* 3(2):e320. From Abele L et al [1989] *Mol Biol Evol* 6: 685–691; with permission from L Abele. Courtesy of H Yokoyama, University of Tokyo. B, (from left to right) from Garey JR et al [1998] *Hydrobiologia* 387:83–91; with permission from Springer Science and Business Media. Courtesy of Frank Fox, http://www.mikro-foto.de, CC BY-SA 3.0.)

phylum, the Pentastomida. Then it was noted that details of the structure of the spermatozoa of pentastomes shared surprising similarities with those of another parasite, but one totally dissimilar in appearance and habitat, the fish louse *Argulus* (**Figure 2.19A**). *Argulus* is a Pancrustacean (see Figure 2.16) and is typically found living in the gill chambers or on the external surfaces of fish. Following this lead, the SSU rDNA sequences of pentastomes and *Argulus* were examined, along with those of several other possible invertebrate relatives. The results were conclusive in supporting the relationships suggested by sperm morphology. This result has since been affirmed by studies of other reference sequences as well. Pentastomids are now generally considered to be nearly unrecognizably modified members of the Pancrustacea, having as relatives not only fish lice, but crabs and copepods as well, organisms we far more intuitively unite as crustaceans.

This example teaches us that painstaking anatomical work can reveal relationships otherwise not at all foreseen based on general appearance. Also, sequence data can provide an objective independent assessment of relationships. Pentastomes, like many parasites, are under selective pressure to accommodate to radically different kinds of habitats such that the normal anatomical features of the group to which they belong become obscured. Keep in mind that although the pentastome–*Argulus* relationship now seems to be well supported, it remains a hypothesis awaiting further testing. After all, exactly how was a presumptive ancestral lineage of fish ectoparasites transformed into the enigmatic endoparasitic pentastomes of reptiles, crocodilians, and mammals that we see today?

The second example in Figure 2.19B reveals a surprising relationship between an exclusively parasitic group, the **acanthocephalans** or thorny-headed worms, and the mostly free-living phylum Rotifera (the rotifers or wheel animals). **Acanthocephalan** adults use their spiny proboscis to embed in the intestinal wall of a vertebrate host, with their body, which lacks a gut, hanging into the intestinal lumen. Their larval stages are usually encysted in arthropods. Thorny-headed worms used to be placed in their own phylum, the Acanthocephala, but now they are often considered to be highly modified rotifers. Yet another similar example in which molecular phylogenetic studies have revealed previously cryptic relationships is that of the exclusively parasitic, unicellular spore-forming **microsporidians**, now generally appreciated to be members of the kingdom Fungi and not unicellular protozoans as formerly thought (see also Figure 2.14).

Some groups of parasites remain "persistent problematica"

There are several additional examples in which insights from molecular phylogenetics have been slow to fully clarify relationships for some baffling parasites. One example is provided by myzostomids (**Figure 2.20**), an obscure group of worms, many species of which are parasites or commensals of echinoderms, particularly crinoids or sea lilies. Marks found on fossil crinoids dating to the Carboniferous Period are similar to those made by modern myzostomids, indicative of an association with crinoids that is at least 300 million years old. Because of their possession of paired lateral appendages (parapodia) that bear bristles (setae), they are frequently categorized as aberrant annelid worms (phylum Annelida). This classification is further borne out by mitochondrial gene sequences that support annelid relationships for myzostomids. However, sequences from genes encoding ribosomal proteins suggest an affinity with a group of phyla including the **Platyhelminthes**, or flatworms. In this instance, different sets of genes yielded results that were not mutually supportive indicating that additional work is needed to resolve their status. Further study of large sets of genes (up to 989 different sequences) or of small noncoding microRNAs support the annelid hypothesis, but the study's authors point out that the flood of sequence data resulting from new sequencing technology need to be interpreted with care. For example, we must be sure that **orthologous genes** (genes in different species that originated by vertical descent from a single gene of the last common ancestor) are actually the ones being compared.

Another group that has yielded conflicting phylogenetic signals is the **Myxozoa**. Myxozoans are an exclusively parasitic group of animals inhabiting fish and either annelid worms or bryozoans. Some like ***Myxobolus cerebralis*** cause prominent fish diseases (**Figure 2.21**). Myxozoan genes often show rapid rates of evolutionary change, leading to long branches on evolutionary trees that can attract one another (called long-branch attraction), confounding tree-building efforts. Although it is now clear that myxozoans

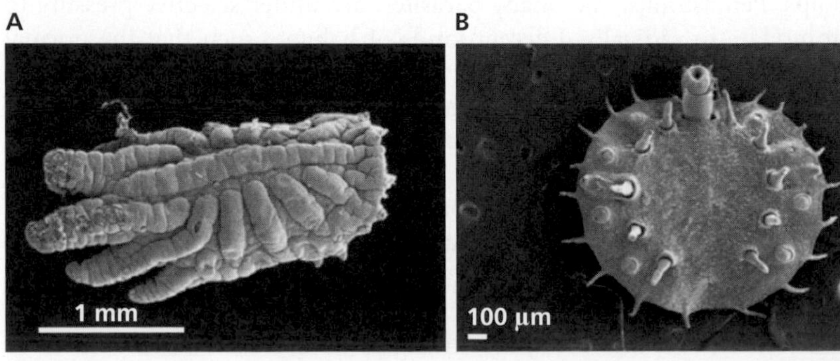

Figure 2.20 The mysterious myzostomids as an example of parasites with uncertain phylogenetic affinities. Two examples of myzostomids, *Myzostoma fissum* and *Myzostoma cirriferum*, both ectoparasites of crinoids. (From Lanterbecq D et al [2009] *Invertebr Biol* 128:283-301. With permission from John Wiley and Sons.)

A

B

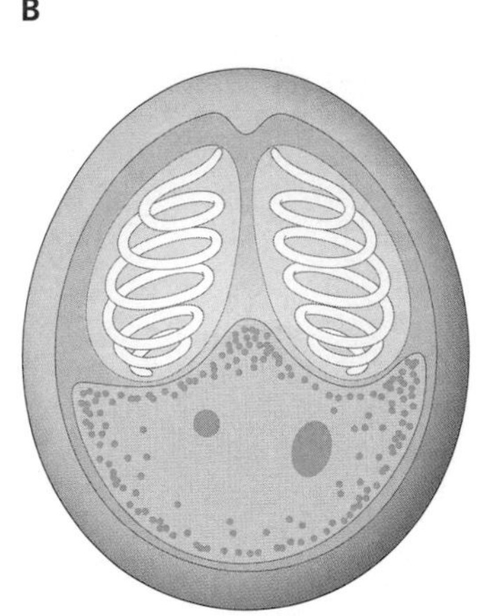

C

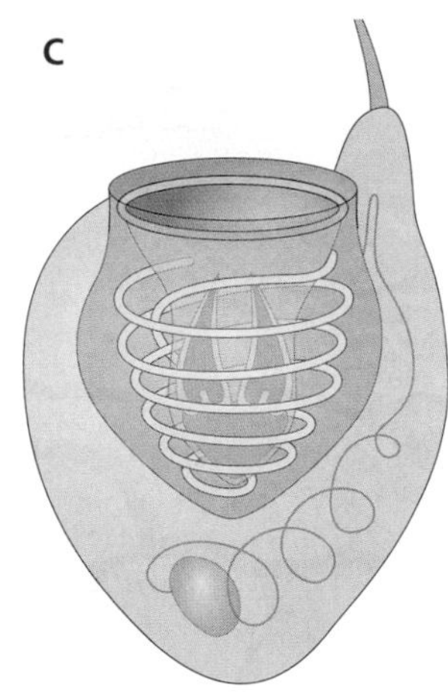

Figure 2.21 Myxozoans, exemplified by *Myxobolus cerebralis*, the organism that causes whirling disease in salmonid fishes. (A) Trout suffering from whirling disease infection. Note the misshapen spine and blackened tail. (B) Example of a myxospore showing coiled polar filaments within. These are extruded and provide attachment to the host cell to be infected. (C) An example of a cnidarian nematocyst showing the coiled tube within, which is believed to be homologous to the polar filaments of the myxospore. (A, from US Department of Agriculture. B, Eiras JC, Malta JCO, Varella AMB & Pavanelli GC [2005] *Mem Inst Oswaldo Cruz* 100:245–247; with permission from JC Eiras. C, courtesy of Ivy Livingstone.)

are morphologically simplified animals and not protozoans as once thought, myxozoans have in some cases been considered as early diverging bilaterians or as aberrant cnidarians and thus related to animals having radial symmetry, such as jellyfish and corals. The latter interpretation is supported by the similarity between cnidarian nematocysts and the polar capsules of myxozoan spores and by sequence analysis of genes found in both groups that are responsible for encoding nematocyst proteins.

If the connection between cnidarians and myxozoans seems improbable, it is worth noting that the oocytes of sturgeons are often infected with a bizarre nematocyst-bearing cnidarian called *Polypodium* that even produces medusa-like forms in the freshwater stage of its life cycle. Perhaps an organism similar to *Polypodium* gave rise to myxozoans.

Studies of parasite diversity reveal how particular parasites came to infect humans

An advantage of having a solid understanding of the phylogenetic relationships for both host and associated parasite lineages is that we gain insights into when and how hosts acquired their parasites. This is certainly also true for human parasites, as indicated in the following two examples. The first pertains to the sucking lice (Anoplura) that we harbor. All sucking lice are blood-feeding ectoparasites of mammals. Humans are unusual as compared to our nearest relatives for harboring lice representing two different species, each of a different genus, ***Pediculus humanus*** and ***Pthirus pubis***. The former species is of particular note for serving as a vector of the bacterium causing epidemic typhus (*Rickettsia prowazekii*) and other pathogens. In contrast, chimpanzees and gorillas each harbor one sucking louse species (*Pediculus schaeffi* and *Pthirus gorillae*, respectively). A number of phylogenetic studies have ascertained both the pattern of relationships among primates, including apes, and among their sucking lice (**Figure 2.22A**). Such studies have also estimated the time of divergence based on the amount of sequence change occurring for both lice and primates (**Box 2.2**). For example, the divergence time between ***Pediculus*** and ***Pthirus*** was estimated to occur 13 million years ago.

Figure 2.22B shows one hypothesis that explains how humans acquired their two louse species. Thick tan lines show the phylogeny for humans, chimpanzees, gorillas, and Old World monkeys. The relationships among lice are shown by the thin blue lines (solid and dashed). This scenario is of a parasite duplication that occurred about 13 million years ago (estimated from degree of sequence divergence) leading to ***Pediculus*** (solid lines) and ***Pthirus***

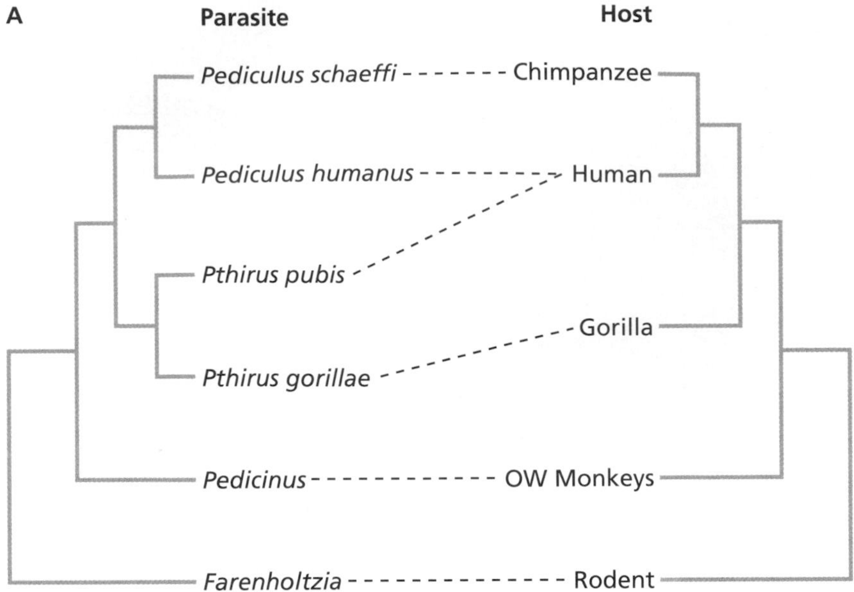

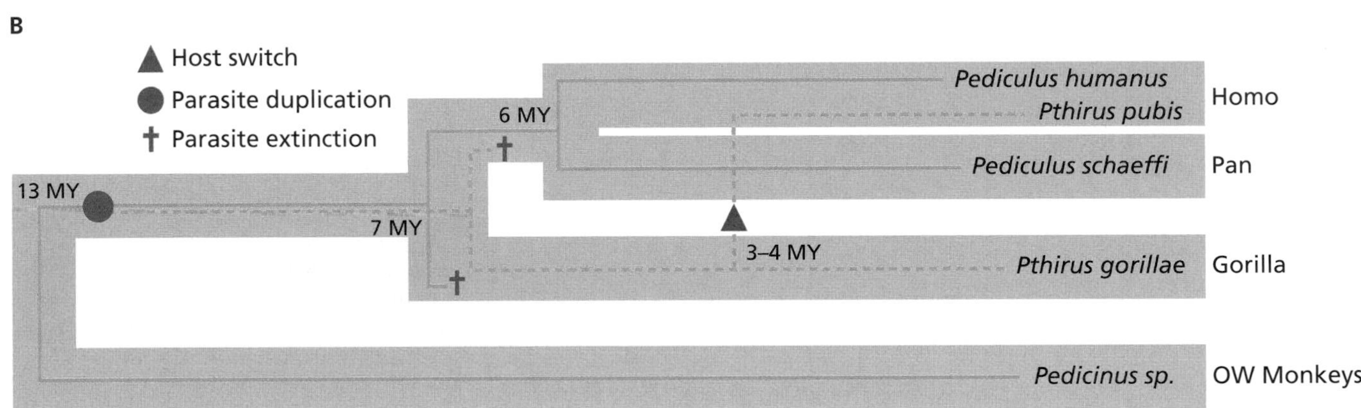

Figure 2.22 The origins of humans and sucking lice. (A) Phylogenetic trees for both primates (emphasizing humans and our closest relatives) and their anopluran sucking lice. (B) This tree provides a hypothesis for how humans acquired their two louse species. (Pair of lice lost or parasites regained: the evolutionary history of anthropoid primate lice. *BMC Biol* 5:1–11.) (Adapted from Reed et al [2007] *BMC Biol* 5:1–11.)

(dashed lines). An extinction (represented as a cross) occurred in each louse lineage. Both humans and chimpanzees acquired ***Pediculus*** from a common ancestor 5–6 million years ago: the estimated time of divergence between chimpanzees and humans and between our respective ***Pediculus*** species match. In contrast, the estimated divergence time between our ***P. pubis*** and the gorilla's *P. gorillae* occurred only 3–4 million years ago, a much shorter time than the time of the last common ancestor between gorillas and our human lineage (7 million years ago). It thus seems likely that we acquired ***Pthirus*** lice by a host switch from gorillas (represented by the vertical arrow within the ***Pthirus*** lineage). Such a switch may have been favored by humans sharing habitats with gorillas or when humans preyed on gorillas allowing lice the opportunity to move onto a new host. According to this parsimonious scenario for humans, which is by no means the only possible interpretation, ***P. humanus*** is an example of an **heirloom parasite** (one acquired from our ancestors), whereas ***P. pubis*** is a **souvenir parasite**, acquired via a host shift along the way. Recent studies suggest the divergence times between humans and our closest relatives may be longer than suggested here, so stay tuned for alternative ideas about our relationships with our lice.

The second example involves one of our most deadly parasites, ***Plasmodium falciparum***. How did we acquire it? ***P. falciparum***, an apicomplexan, is the

BOX 2.2

The Molecular Clock Hypothesis

The essential idea behind the molecular clock hypothesis is that DNA sequences change by mutation at a constant rate, such that the degree of divergence in a particular gene sequence from, for example, two related species (or between members of different genera or families) could be used to date the time when they diverged. Although the clock hypothesis is somewhat controversial and the rate of nucleotide substitution change varies among different groups of organisms, it is still widely used. It is best applied to groups where the rate of change can be validated by reference to the fossil record or to a particular geological event known to have separated the two species. In such a case, a time of divergence can be related to a distinct geological period of known age. Often for groups like parasites such calibration using fossils or geological events is not possible.

most widespread and lethal of the four species commonly implicated in human malaria. The considerable pathogenicity of ***P. falciparum*** has left its mark on human evolution. For example, it has favored the persistence of genes encoding alternative forms of hemoglobin that enhance survival in the face of heavy malaria transmission (see Chapter 7). One possibility for its acquisition is that, just as with ***Pediculus*** lice, the last common ancestor of humans and chimpanzees harbored a malaria parasite that subsequently diversified into ***P. falciparum*** in humans and *P. reichenowi* in chimpanzees. However, it was shown that the amount of genetic diversity in ***P. falciparum*** was far less than in *P. reichenowi* from chimpanzees, suggesting that humans may have acquired a parasite similar to the ***P. falciparum*** parasite from chimpanzees long after their split from a common ancestor. Although attractive, this idea has since been superseded by data suggesting that gorillas from western Africa are the source of ***P. falciparum***. The tree in **Figure 2.23** shows the diversity inherent in ***Plasmodium*** parasites of African apes. This diversity was revealed by extensive sampling of ***Plasmodium*** DNA acquired and subsequently amplified from gorilla, chimpanzee, and bonobo fecal samples (malaria parasites normally inhabit the blood, but some of their genetic material ends up in the feces of infected animals). All isolates of the human parasite ***P. falciparum*** exhibit only modest diversity in comparison and fall within one lineage (G1), indicating their closest relatives are from gorillas. The time of the hypothesized transfer of ***Plasmodium*** from gorillas to humans is not known.

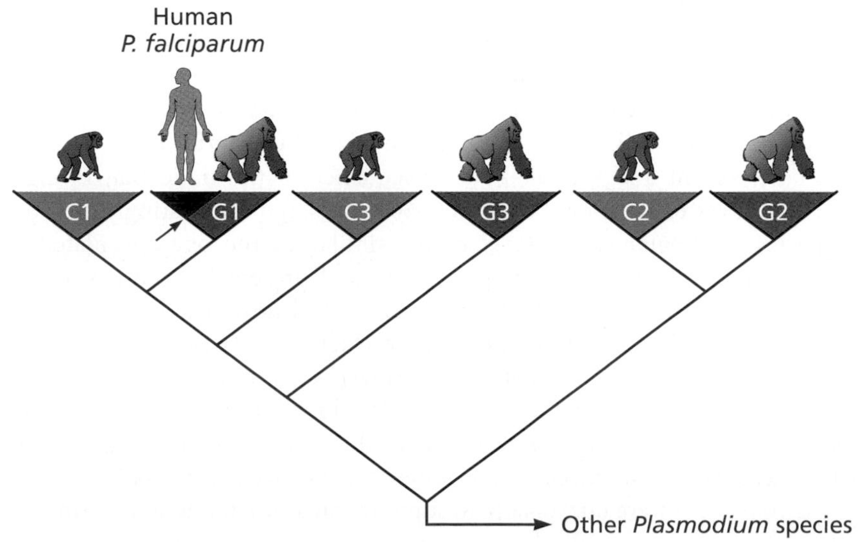

Figure 2.23 The origins of one of the most dangerous species of human parasite, *Plasmodium falciparum*. Isolates obtained from chimpanzees are shown in blue and those from gorillas from western Africa in red. All isolates of the human parasite, *Plasmodium falciparum*, are shown in black. (From Holmes EC [2010] *Nature* 467:404–405. With permission from Macmillan Publishers Ltd.)

In addition to providing a dramatic insight into the acquisition of one of our most feared parasites, this story also demonstrates how application of an innovative new technique can broaden the range of samples available, and provide a better overall picture of the evolutionary history of a parasite, in this case ***P. falciparum***. Is this the final word on the origins of ***P. falciparum***? Studies of fossil apes constantly expand our knowledge of the complexities of our evolutionary history and more specimens of ***P. falciparum*** and other malaria parasites will become available for analysis. Keep in mind as well that great ape populations are now so reduced that they may provide a misleading gauge of how frequent human–great ape contacts may have been in the past.

Studies of diversity can help reconstruct the historical biogeography of parasites

Another reason to pursue studies of parasites based on an understanding of their overall diversity and relationships is to gain insight into the historical biogeography of the parasite in question. **Historical biogeography** refers to the study of how historical aspects of geology, ecology, or climate may have influenced the past and present distributions of species. How did the parasite come to be where it is today? What can historical biogeography tell us about where it might someday go? One example is provided by the human parasite ***Schistosoma mansoni***, a causative agent of intestinal schistosomiasis. This species today infects about 90 million people, mostly in Africa but also in South America. In its adult stage, this parasite inhabits the veins around the intestine of humans, whereas its larval stages undergo obligatory development in freshwater snails of the genus *Biomphalaria*. The geographic range of ***S. mansoni*** is dictated by the presence of compatible *Biomphalaria* snails. Phylogenetic studies of ***Schistosoma*** suggest this group has diversified in both Asia and Africa (**Figure 2.24**).

Surprisingly, *Biomphalaria* snails originated and first diversified in South America, raising the question, why is ***S. mansoni*** so common in Africa today? Phylogenetic studies suggest that a South American *Biomphalaria* snail likely experienced a transoceanic colonization event in Africa sometime in the last 2 million years (Figure 2.24), long after the continents had drifted far apart. The long-distance colonization of Africa was possible because *Biomphalaria* snails are hermaphrodites (they can self-fertilize so only one snail is needed to start a population) and because they are often caught in the feathers or on the feet of birds able to fly across oceans. Once in Africa, *Biomphalaria* diversified and provided a new option for resident ***Schistosoma*** species, one of which must have switched into this newly available snail and evolved into a separate species that we know today as ***S. mansoni***.

So how did this parasite then come to colonize South America? Several lines of evidence indicate the colonization of South America by ***S. mansoni*** was fairly recent and was likely a consequence of the trade that brought infected slaves to the New World 400–500 years ago. Interestingly, because a species of *Biomphalaria* (*B. glabrata*) similar to the one speculated to have given rise to the African species was still present in South America, ***S. mansoni*** fortuitously found the necessary snail hosts needed to support its life cycle and to enable its persistence (Figure 2.24). A related species, ***Schistosoma haematobium,*** that causes urinary schistosomiasis in Africa was also almost certainly brought to the New World in slaves but did not gain a foothold there. Because its snail hosts (in the genus *Bulinus*) originated in Africa and never colonized South America, the necessary snail hosts for ***S. haematobium*** were unavailable to support transmission in South America

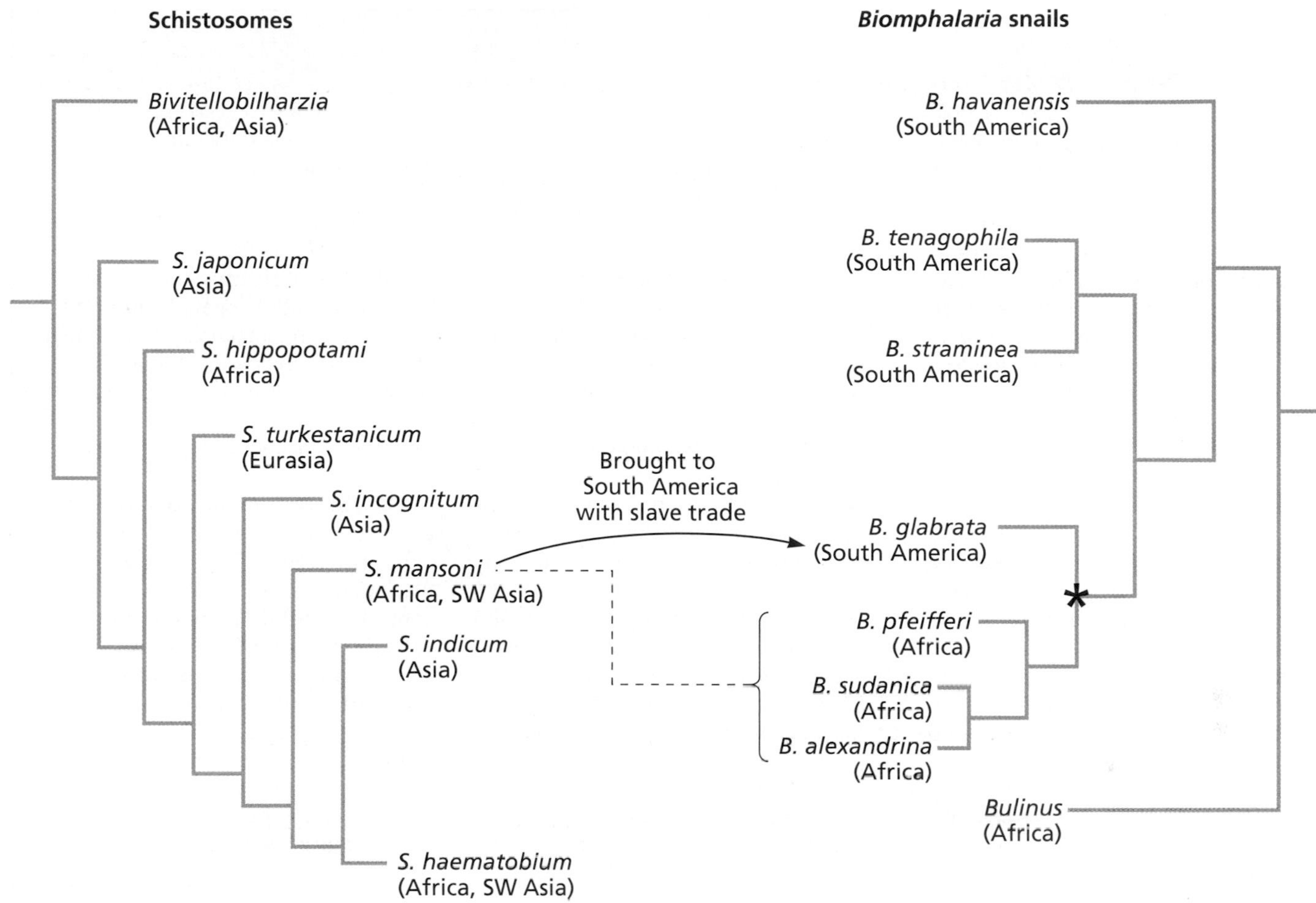

Figure 2.24 Phylogenetic studies help us reveal the historical biogeography of *Schistosoma mansoni*. Shown on the left is a phylogeny for *Schistosoma*, primarily an African and Asian genus. The distribution of *S. mansoni* is mostly African. Shown on the right is a phylogeny for *Biomphalaria* snails. Note that the *Biomphalaria* originated in South America but that Africa was later colonized (asterisk). The dashed line indicates the species of *Biomphalaria* with which *S. mansoni* evolved. Note that *S. mansoni* was later brought to South America (arrow) and was able to colonize *B. glabrata* there. Although *S. haematobium* was also brought to South America, it was unsuccessful because its snail host *Bulinus* is confined to Africa and Asia.

at the time of the slaves' arrival . This example indicates that ***S. mansoni*** is a relatively newly acquired parasite of the human lineage. It reveals the dependence of parasites on key hosts such as snails and explains why this parasite so readily spread to the New World, whereas close relatives did not.

2.3 THE ONGOING QUEST TO REVEAL AND UNDERSTAND PARASITE DIVERSITY

All who have attempted to quantify life's diversity agree on one thing—there are many species yet to find and describe. This is certainly true for parasites, which because of their small size or their hidden habitats in other organisms often go unnoticed. Furthermore, many host species have limited or remote distributions, making sampling difficult. There are vigorous efforts underway to fill in the gaps (**Box 2.3**). To go along with traditional collecting methods and descriptions of morphology are new approaches to achieve this worthy but elusive goal, some of which are highlighted in this section.

DNA barcoding is one way to catalog parasite diversity

Sequence data are used with ever greater frequency as a tool to discover new parasites and reveal their relationships. The genes selected for study are diverse. There has also been a growing trend to document for many to all eukaryotic organisms the sequence of a particular gene that can serve as a convenient species-specific marker. Just as a supermarket uses distinctive

BOX 2.3

Finding New Parasite Diversity: The Nematomorpha as a Model

One of the most exciting aspects of science is to discover something never before seen, and the ongoing efforts by parasitologists to find new species of parasites unknown to science is an example of this pursuit. How are such discoveries made, how is the new species characterized, and what is the benefit of this information? One good model that demonstrates how these questions are answered is that of a relatively small, exclusively parasitic phylum of worms called the Nematomorpha, or nematomorphs. These also are known as horsehair worms or gordiaceans. The long, sinuous nematomorph adults are free-living, often being found in knot-like aggregations of copulating groups along the margins of streams or ponds (see Figure 1A). The name gordiaceans comes from the knot of rope from the city of Gordius that was famously cut by Alexander the Great.

Nematomorph larvae are often ingested by invertebrates like aquatic insect larvae, such as midges and mayflies, where they undergo initial essential developmental steps. Eventually these early nematomorph larval stages are ingested by arthropods such as crickets, where they grow and proceed through a single molt, just before completing their larval development. The mature adults then make a dramatic exit from their host's body (Figure 1B). Being parasitic as larvae but free-living as adults, they are examples of protelean parasites. Nematomorphs can have profound effects on their hosts, such as castration. Although these effects may be temporary, they nonetheless can provoke behavioral changes believed to promote nematomorph transmission. The extent to which developing nematomorphs fill the body cavities of their hosts without actually killing them (Figure 1C) is a testament to the ability of parasites to manipulate their host's anatomy and physiology to their benefit.

Currently there are about 350 species of nematomorphs known to science, and experts estimate there are as many as 2000 species in the world. Using a detailed understanding of the known nematomorphs as a starting point, nematomorph specialists have embarked on a series of collections of new specimens that are often found as adult worms in aquatic habitats. Some of the new specimens are prepared for scanning electron microscopy so external features can be accurately described (Figure 1D). Some are extracted to acquire DNA, so that diagnostic sequences like the *CO1* barcode region or other signature genes can be obtained and checked against other known nematomorph sequences. If distinctive, an evolutionary tree placing the new worm into a broader context is constructed. Typically an exhaustive description of a new species will be made and submitted to a peer-reviewed scientific journal, and then voucher specimens will be deposited in an appropriate museum. By having a more complete picture of the diversity of nematomorphs, we will be able to clarify how parasitism evolved in this group, how

barcodes to identify its products, the goal of DNA barcoding is to use the sequence of the mitochondrial cytochrome *c* oxidase 1 gene (*CO1*) as such a marker. This is possible because the *CO1* gene is widely represented in eukaryotic genomes and, at least for animals, a 648-base-pair stretch of the gene is sufficiently variable among species to provide a distinctive reference point. In general, mitochondrial genes are prone to higher rates of mutation than genes contained in the nucleus. The higher mutation rate may result from their proximity to reactive oxygen species that are produced in the mitochondria during respiration and that are capable of causing damage. The variability of *CO1* contrasts with the SSU rRNA gene discussed earlier that has worked effectively for building a tree of life incorporating very disparate organisms but that is too invariant to serve as a species marker.

Although the actual specimen associated with a particular barcode sequence must still be identified by traditional means, often with the involvement of an expert for that group, that particular sequence and identification are then linked and made available in a barcode database. Specimens collected later can then be readily identified if they match the barcode sequence of the known species. Even if a database match is not available, the *CO1* sequence for an unknown specimen is a powerful tool to assist later with a sound identification.

Such molecular markers are also very useful tools for parasitologists interested in revealing the diversity of otherwise cryptic parasite species present in a host or in elaborating the unknown life cycle of a new parasite.

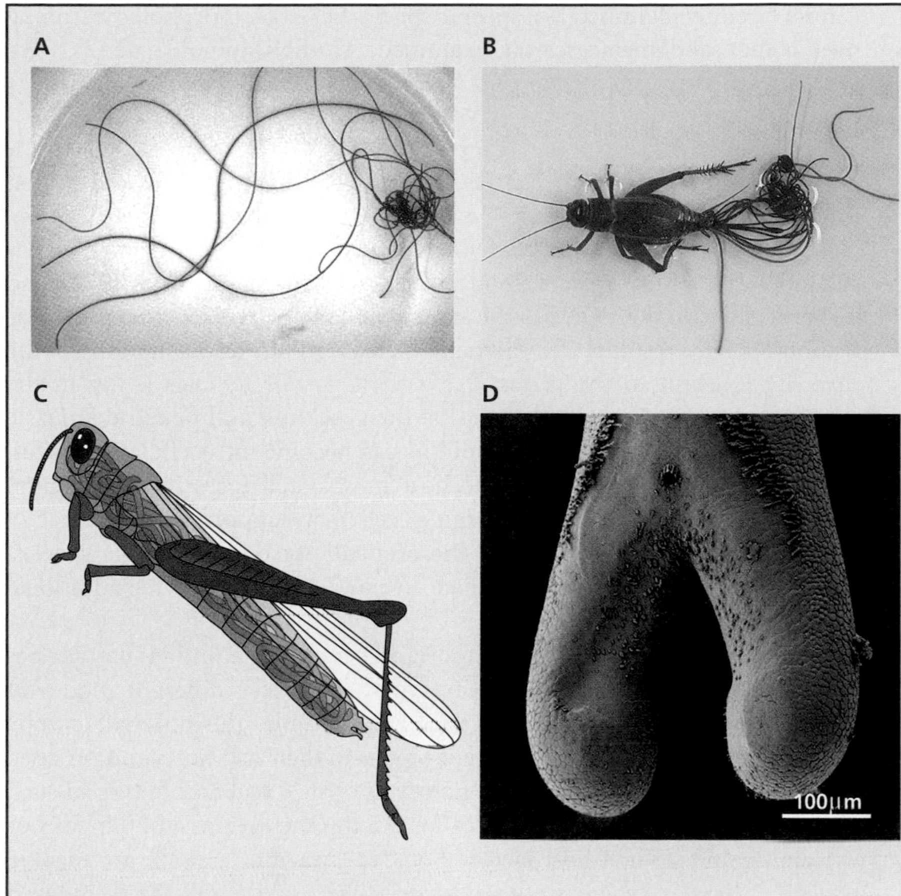

frequently horsehair worms have shifted into new lineages of hosts, and how nematomorphs have spread around the world. We might eventually understand why they are particularly common in arid climates and how their diversity might serve as a way to understand the impact of climate change.

Reference

Hanelt B, Thomas F & Schmidt-Rhaesa A (2005) Biology of the phylum Nematomorpha. *Adv Parasitol* 59:243–305.

Figure 1 Discovering new parasite diversity as exemplified by nematomorphs. **(A) A tangle of nematomorph adults, each about 25 inches in length. (B) Adults of *Paragordius varius* exiting from a cricket host, 28 days after initial infection. (C) Note the extent to which the body of the insect host is occupied by the developing nematomorph worm within. (D) Scanning electron micrograph showing the forked posterior end of a male nematomorph (*Gordionus*), with a cloacal opening. (A, B, courtesy of Ben Hanelt, University of New Mexico. C, courtesy of Rebecca Strich. D, from Begay A, Schmidt-Rhaesa A, Bolek M & Hanelt B [2012] *Zootaxa* 3406:30–38**

For example, suppose that larval stages of an unknown parasite collected from a snail are barcoded and added to the barcode database. Years later, possibly in a very different location, another researcher might collect an adult parasite from the intestine of a bird that is identical in barcode sequence to the larvae originally acquired from the snail. In this way, an important connection can be made to illuminate for the first time the life cycle of the parasite in question. Because minor variations within barcode sequences also occur among individuals within a species, barcoding also provides a means to monitor intraspecific diversity within a parasite species, even those from different locations.

Some parasites exist in complexes of cryptic species

One problem in quantifying parasite diversity that has come to light with the advent of molecular systematics methods is the surprising extent to which **cryptic species** have been reported, particularly so for helminths such as digenetic trematodes (digeneans, or flukes). Cryptic species are closely related organisms that are morphologically similar and, on these grounds, are described as a single species. Yet within the group are clusters of individuals that are sufficiently distinct genetically from other clusters to be accorded the status of distinct species.

One striking example of cryptic diversity is provided by digeneans that infect fish in the St. Lawrence River. The fish are infected with a life-cycle stage (metacercariae) of a group of digeneans (diplostomoids). These digeneans cannot be easily identified as a specific species based on morphology. Among over a thousand metacercariae examined, 47 diplostomoid species were detected based on distinctive *CO1* barcode sequences, representing a huge increase in previously unknown diversity for this group of digeneans.

The likelihood of finding complexes of cryptic species is greater for those parasite groups that lack a sufficient number of easily recognized morphological characters to enable precise species identification. Nematodes of the genus ***Trichinella***, responsible for causing trichinellosis (**Figure 2.25**), also provide a good example of how improved breadth and depth of collections coupled to the application of biochemical and molecular techniques have resulted in a dramatic increase in our understanding of the diversity inherent in the genus. ***Trichinella spiralis*** females living in the small intestine give birth to larvae that then seek out and penetrate muscle cells, demonstrating that even helminths can become intracellular parasites. Under the parasite's influence, the host muscle cell loses its contractility and becomes a highly modified nurse cell that supports the survival of the nematode larva within. From the originally recognized single species, ***T. spiralis***, there are now nine named species, with at least three additional genotypes awaiting further study.

Revealing such diversity is important because the individual species involved in cryptic species complexes may have very different modes of transmission or patterns of host use. For example, the different cryptic diplostomoid species inhabit different organs in their fish hosts and revealed patterns of host specificity not previously known. A full accounting of such diversity could give important clues for tracing the origins of outbreaks or explaining shifts to new host species. Some species of ***Trichinella*** are adapted to Arctic climates, whereas others thrive in the tropics, and the recognized species differ in host preference with some more likely to infect domestic swine, for example, than others. Similar considerations apply to the arthropod vectors that transmit disease. An example is the malaria-transmitting ***Anopheles*** mosquitoes, which often occur in complexes of cryptic species. *Anopheles culicifacies*, one of the major malaria vectors on the Indian subcontinent, is actually a complex of five cryptic species. The individual species differ in their geographic distribution, competence for hosting malaria, seasonal abundances, and response to insecticides, in addition to differences in the host species they prefer to bite. Our ability to fully understand malaria epidemiology is greatly enhanced if the cryptic species and their distinctive attributes are known.

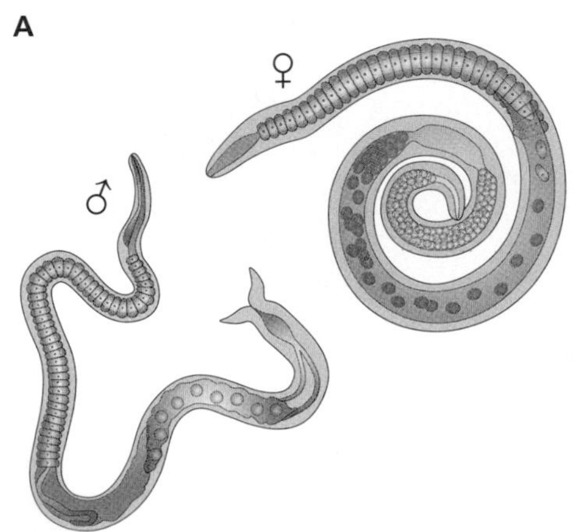

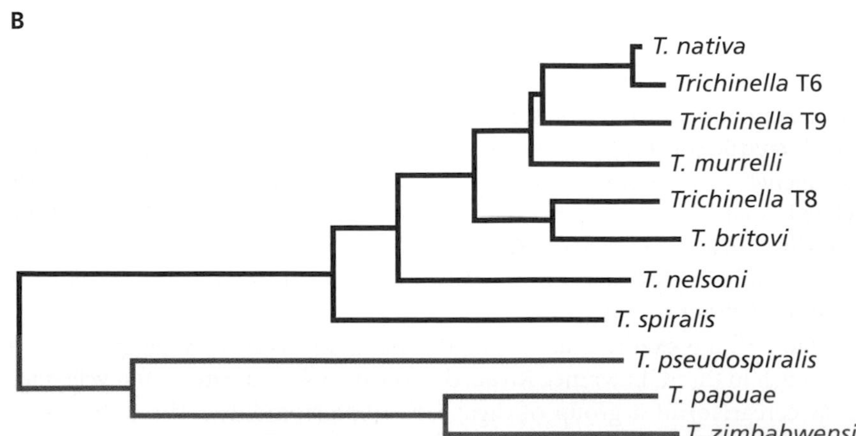

Figure 2.25 The hidden diversity of *Trichinella* species. (A) Adult worms of *Trichinella spiralis*, the nematode responsible for causing trichinellosis. (B) A phylogenetic hypothesis for *Trichinella*, based on variation in mitochondrial large subunit and cytochrome c oxidase 1 genes, with species known to be encapsulated in host cells shown in red and nonencapsulated species shown in green. (A, courtesy of Ivy Livingstone. B, From Zarlenga DS et al [2006] *Proc Natl Acad Sci USA* 103:7354–7359. Copyright (2006) National Academy of Sciences, USA.)

Whole lineages of unapparent parasites may escape our attention

The application of molecular systematics has also brought to light formerly unappreciated new lineages of parasites that we did not know existed until the component species were united by their related nucleotide sequences. An example is provided by the elusive clade of poorly known opisthokonts referred to as Mesomycetozoeans (or Ichthyosporeans) (see Figure 2.6). This lineage is made up mostly of parasites of fish that are united by similarities in SSU rDNA sequences. Because the members of this group do not share obviously diagnostic morphological features, the ichthyosporeans might never have been recognized as a distinct group. Indeed, a lively debate regarding their phylogenetic position continues, fueled by the ongoing discovery of similar new species. Why should we care? One enigmatic parasite, *Rhinosporidum seeberi*, frequently implicated in infections of the nasal mucosa in humans and other mammals, is now known to be an ichthyosporean. Other ichthyosporeans of medical or veterinary significance also may come to light.

Metagenomics provides a new way to reveal parasite diversity

In marine habitats, toxin-producing dinoflagellates (recall the Alveolata in Figure 2.6) often form harmful blooms that can cause illness and death, even in humans who come in contact with them in these environments (**Figure 2.26**). Studies sequencing 18S rRNA genes obtained directly from seawater revealed a large number of distinct sequences for other dinoflagellates (Order Syndiniales) that are parasitoids of the species involved in blooms. The parasitoids infect the dinoflagellate cells that cause the bloom, reproducing within and eventually killing them. The parasitoids become so abundant they can quickly control blooms, thus providing a compelling example of parasitism affecting entire ecosystems and potentially influencing nutrient cycles on global scales. They also exemplify how a beneficial ecosystem service, namely dispelling a harmful bloom of dinoflagellates, can result from a parasitic association.

It is particularly remarkable that use of culture-independent metagenomics methods has revealed an otherwise unappreciated diversity and abundance of these parasitoids, many of which interact with their hosts in a species-specific manner. The parasitoids contribute to control of bloom-forming dinoflagellates year after year, prompting their designation as "serial parasitic killers". Although metagenomics approaches such as this one are just beginning to be applied to the search for parasite diversity, they have also uncovered unsuspected diversity of parasites similar to *Perkinsus* in freshwater, the first records of such parasites in other than marine habitats.

A

B

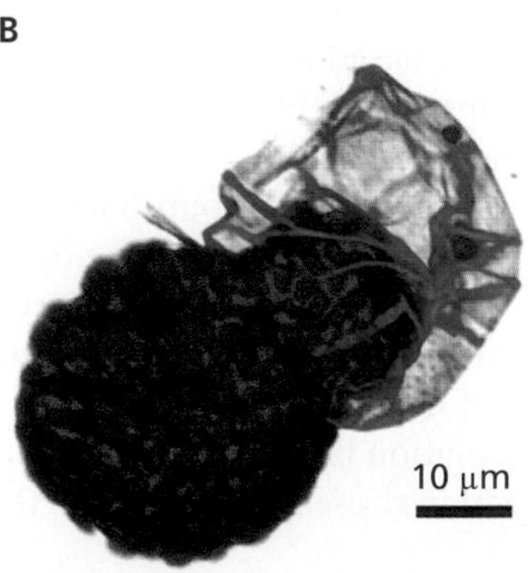

Figure 2.26 Control of toxic dinoflagellate blooms (red tides) by parasitoid dinoflagellates. (A) A dinoflagellate bloom with pigments from the dinoflagellates imparting a red color to the water. (B) *Amoebophyra* sp., a parasitoid of the bloom-causing *Alexandrium minutum*. The parasitoid is stained green and red, and the shell or theca of the host cell is blue. A host cell has lysed and the multinucleated "beehive" parasitoid within is emerging (red stain indicates individual nuclei). Eventually it will fragment to release many uninucleate dinospores that will infect other host cells. (A, from trooperworld.wikidot.com/crimson-tide, CC BY-SA 3.0. B, from Chambouvet A et al [2008] *Science* 322:1254–1257. With permission from AAAS.)

Just as metagenomics has revolutionized our understanding of bacterial diversity, a similar impact may await with respect to eukaryotic diversity, including eukaryotic parasite diversity.

Studies of parasite diversity help provide a better foundation for taxonomy

One of the goals of the study of parasite diversity is to reveal the evolutionary relationships among parasite groups, such that existing taxonomic schemes for classification can be brought into agreement with these relationships. Examples such as the acanthocephalan–rotifer and pentastomid–crustacean connections have already been discussed.

One example of the reconciliation of phylogenetic relationships with taxonomy is provided by the schistosomes infecting mammals. One genus *Orientobilharzia*, consisting of a few species infecting Eurasian ruminants, was established that differs from ***Schistosoma*** (found in many mammals from Asia and Africa) primarily on the basis of the number of testes present in male worms, 37–80 and <10, respectively (**Figure 2.27a**). This is a clear morphological difference that makes separation of the two genera straightforward. However, several molecular phylogenetic studies consistently identified *Orientobilharzia* as nested within a larger clade that otherwise contained only ***Schistosoma*** species. At the same time, worms in other described schistosome genera fall into distinct clades (Figure 2.27b). Given that other morphological characters unite *Orientobilharzia* and ***Schistosoma***, it would seem in this case that the number of testes simply is not a characteristic that reflects major evolutionary changes. As a consequence of such molecular studies, the generic name *Orientobilharzia* has since been relegated to the status of a junior synonym of ***Schistosoma***, and *Orientobilharzia turkestanicum* for example, is now most properly known as *Schistosoma turkestanicum*.

Other examples of the disparity between existing taxonomy and the revelations of molecular phylogenetics are more glaring. For example, in the phylum **Myxozoa**, spore structure is a major feature differentiating taxonomic groups. The existing taxonomic scheme does not match the relationships derived from phylogenetic studies of either SSU rRNA or elongation factor 2 (*EF-2*) genes. This disparity is probably because relatively few characters are available to describe the myxozoan spore and because some spore configurations are just more successful than others. The result is that species that differ genetically have converged to the same spore morphology; that is, they exhibit homoplasy. **Homoplasy** refers to similarities that result from convergent evolution rather than common ancestry.

Another similar example is provided by monogenean flukes of the family Capsalidae. Flukes of this family live on the external surfaces of marine fishes (see **monogeneans**). The existing taxonomic scheme recognizing ~180 species is based on relatively few morphological characters. A phylogenetic study based on three genes suggested that of the four subfamilies within the Capsalidae, three were not **monophyletic**. A monophyletic group is one that includes all the taxa derived from the most recent hypothetical common ancestor of that group. Such a related group of organisms is said to be monophyletic or to exhibit monophyly. Because the taxonomy of the Capsalidae did not reflect natural related evolutionary groups, it was concluded that some of the morphological traits used to order the taxonomic scheme exhibited homoplasy. Homoplasy is likely to be a common outcome in parasite evolution because parasites often experience similar hosts and microhabitats within those hosts that favor the evolution of similar traits.

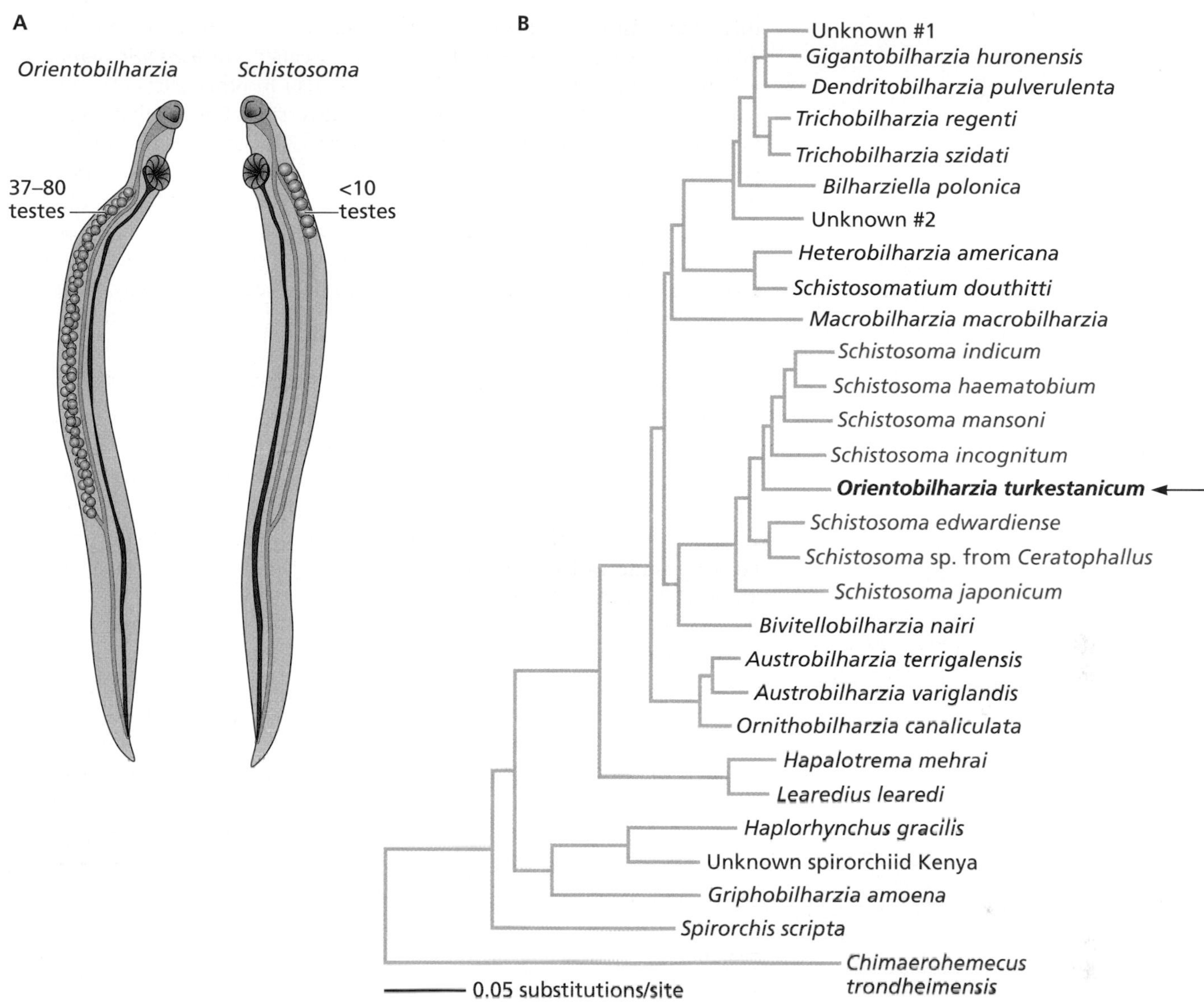

Figure 2.27 Improving taxonomy through a better fundamental understanding of parasite diversity. (A) Male schistosomes of the genus *Orientobilharzia* have 37–80 testes, whereas males of *Schistosoma* have fewer than 10. (B) On this tree, *Orientobilharzia turkestanicum* (black arrow) nests within a group containing several species of *Schistosoma*. (B, Loker ES & Brant SV [2006] *Trends Parasitol* 22:521–528. With permission from Elsevier.)

It should by no means be concluded that morphological characters are inevitably misleading in attempting to define parasite diversity. Our taxonomic hierarchies have for decades been constructed using morphological criteria and many have proven to be robust. Morphological characters often provide convenient ways to identify species and can be used in conjunction with results from molecular studies to identify traits, called **synapomorphies**, that indeed are shared by members of a monophyletic group and the group's immediate common ancestor. Our best understanding of parasite diversity will continue to derive from information from as many sources as possible, including both morphology and molecules.

2.4 OTHER WAYS TO CONSIDER PARASITE DIVERSITY

The discussion above focuses on named parasite species as one metric to define, measure, and study parasite diversity. There are at least two other forms of parasite diversity to consider as well, both of which contribute enormously to the properties that we associate with parasites and that prove to be relevant with respect to understanding the nature of parasites, including their transmission, virulence, treatment, and control. The first is the genetic diversity inherent within each species, so-called **intraspecific variation,** which is addressed in the next section and also in other parts of the book, particularly Chapter 7. The second is the diversity of genes that constitute the genetic repertoire of each parasite species. The latter topic is one which is steadily being clarified in the form of completed genome sequences for an increasing number and variety of parasite species. This topic is addressed in Chapter 3. Throughout this chapter we have discussed multiple origins of parasitism from free-living organisms. Does it ever happen that parasites give rise to free-living organisms? We discuss this question briefly at the end of this chapter.

Diversity within parasite species is extensive and important

No species consists of individuals that are completely uniform genetically. Even in species that routinely engage in self-fertilization or asexual reproduction, genetic variants nonetheless occur thanks to the ongoing process of mutation or other genetic modification that might arise. It is this intraspecific genetic variation that forms the substrate upon which natural selection acts, favoring some variants over others. For example, individual parasites with a variant gene that confers resistance to a particular drug might be favored during a control program testing the use of that drug. In Chapter 7, we revisit the importance of genetic variation and population structure within parasite species when we discuss the evolution of parasites. For now the mission is to show that diversity exists within parasite species and that this variation is consequential and needs to be considered.

First we introduce several terms that are applied to variant forms within a species. The term **isolate** describes a sample of a parasite species derived from a particular host at a particular time. **Strain** refers to an intraspecific group of parasites that differs from other such groups in one or more traits, including traits that might be relevant to control or treatment. **Subspecies** is used to identify a distinctive group of organisms within a species that may occupy a particular region and that can interbreed with other subspecies. In this case, however, the organisms typically do not interbreed because of their isolation or another reason. Subspecies are given a formal name, such as ***Trypanosoma brucei rhodesiense***, with *rhodesiense* being the subspecies name. Also, whereas studies to discriminate among species rely on less variable genetic markers such as SSU rDNA or cytochrome oxidase (see Section 2.3), studies of intraspecific diversity rely on more variable markers, such as other genes or sequences associated with the mitochondrial genome, or **microsatellite markers**. Microsatellite markers are short repeat sequences (usually 2 to 6 nucleotides long) that undergo rapid change in the number of times they are duplicated, providing a good way to discriminate among individual parasites.

One important example of how variation within a parasite species matters is illustrated by ***Trypanosoma brucei***, which is commonly recognized to consist of three subspecies, ***T. brucei brucei***, ***T. brucei rhodesiense***, and ***T. brucei gambiense***. These subspecies can be thought of as intraspecific variants with distinctive host ranges, diseases caused, and geographic distributions (**Table 2.1**).

One of the mysteries presented by this complex of subspecies is why epidemics of the subspecies that afflict humans only appear at certain times and often in the same defined locations. There is some evidence that the acquisition of a particular gene, the *SRA* gene (serum resistance-associated gene),

Table 2.1 Characteristics of the subspecies of *Trypanosoma brucei*.

Species/Subspecies	Host Range	Disease	Disease Profile in Humans	Distribution
Trypanosoma brucei brucei	Wild and domestic animals	Nagana	—	Tropical Africa
Trypanosoma brucei rhodesiense	Humans, wild and domestic animals	Rhodesian sleeping sickness	Acute	Eastern and Southern Africa
Trypanosoma brucei gambiense	Humans, primarily. Wild and domestic animals	Gambian sleeping sickness	Chronic	Western and Central Africa

From Hide G & Tait A (2009) Molecular epidemiology of African sleeping sickness. *Parasitology* 136:1491–1500. With permission from Cambridge University Press.

by a relatively rare recombination event (recall Figure 2.1) transforms the nonhuman parasite ***T. b. brucei*** into an organism able to survive in and infect people. Presence of the protein encoded by this gene confers resistance to lysis of trypanosomes when placed in human serum. The *SRA* gene happens to be present in all isolates of human-infective ***T. brucei rhodesiense*** and thus seems to serve as a marker of the presence of a trypanosome capable of infecting people and potentially able to cause an epidemic. Lest we become too confident in thinking we now hold the secret for how, where, and why all human sleeping sickness epidemics might occur, *SRA* does not serve as a marker to identify the human-infective ***T. brucei gambiense*** isolates more commonly found in West and Central Africa. Nonetheless, this example serves to remind us that differences in genetic composition among representatives of what is still defined as a single species, ***T. brucei***, can have major implications for human and animal health.

Having some appreciation for the extent of variation within a parasite species gives us a valuable way to both interpret the evolutionary history of the parasite and to understand how this variation might influence eventual control efforts. For example, when comparing human isolates of ***Plasmodium falciparum*** to those of ***Plasmodium vivax,*** the latter species has much more genetic variability. This pattern of variation suggests that ***P. vivax*** has had a longer and more stable association with people than ***P. falciparum***. The greater variability within ***P. vivax*** will require different approaches for control and eventual eradication than envisaged for ***P. falciparum***.

Another example is the Old World screwworm *Chrysomya bezziana*, a fly that causes myiasis in livestock. **Myiasis** is an infestation of tissues of a live host by larvae of flies that then feed on the living tissue, causing the host great damage. We now know this species exhibits distinct lineages that tend to occupy particular locations. This is a fundamental piece of information to have prior to the contemplated use of the sterile-male technique to control these harmful parasitic flies. This technique, discussed further in Chapter 9, basically relies on deliberate propagation, sterilization, and distribution of sterile male flies to disrupt the fly's life cycle. If multiple lineages with high aggregate diversity occurred in the same area, the sterile male technique would potentially require propagation of a broader variety of males, thereby greatly increasing the resources needed for the control program to succeed in that area.

In an example related to potential use of vaccines for control, different isolates of the nematode *Trichostrongylus colubriformis* have different forms of an immunodominant carbohydrate surface antigen that is a promising potential vaccine target (**Figure 2.28**). This forces us to take into account the intraspecific variation inherent in the target species with respect to developing a vaccine that can be broadly effective.

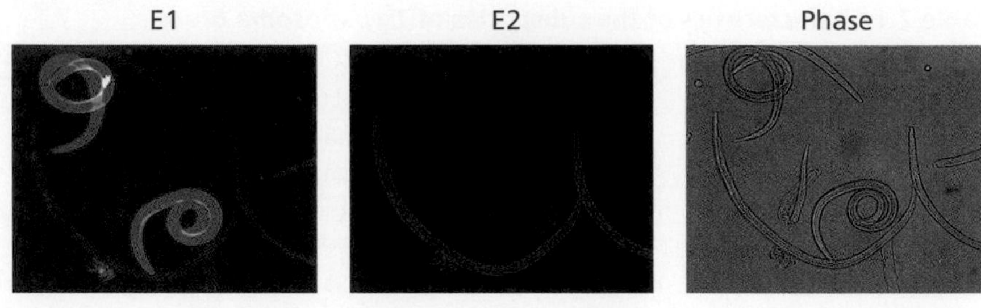

Figure 2.28 A demonstration of intraspecific variability on the surface of larvae of the nematode *Trichostrongylus colubriformis*. A sample of L3 larvae (all worms present are revealed by the phase contrast image at right) was stained with two different antibodies (E1 and E2), each labeled with a different fluorochrome (green or red), and each recognizing a different carbohydrate antigen. Note how different subsets of worms were stained by each antibody. (From Maass DR, Harrison GB, Warwick N et al [2009] *PLoS Pathog* 5:e1000597.)

Studies of intraspecific variation, as exemplified by investigations of tapeworms of the genus ***Echinococcus***, can provide new insights into the evolutionary history of parasites. Both distinctive North American and Asian variants of ***E. multilocularis*** have been discovered on St. Lawrence Island, which lies between Alaska and Siberia in the Bering Sea. This discovery provides supportive evidence that the island was part of the Beringian land bridge that once connected the two continents and enabled variants from both continents to co-occur there.

We also gain deeper insights into the nature and consequences of parasitism (including important topics such as virulence) when we have an appreciation for the variation occurring within parasite species. Variants of ***Trypanosoma cruzi*** found in the blood of infected people differ from those isolated from cardiac muscle severely damaged by this parasite, suggesting some variants have a greater predilection to attack heart muscle cells than others. Also, strains of ***Toxoplasma gondii*** found in people in the Northern Hemisphere are much more likely to be clonally propagated by carnivory of domestic herbivores, whereas other strains found in South America show much more evidence of this parasite undergoing regular sexual recombination and in inhabiting wild animals. This latter point is of considerable relevance to conservation biologists interested in protecting endangered species. Furthermore, such strains could someday cause emergent disease in people.

Finally, the symbionts harbored by parasites may also be variable among individuals. The wasp *Cotesia sesamiae* is considered to be a generalist parasitoid because it has a broad spectrum of infectivity for different insect species. However, the host species colonized depends on the particular variant of immunosuppressive polydnavirus carried by the wasp (see also Chapter 4). Different *C. sesamiae* wasps with different host preferences harbor distinctive polydnaviruses. The particular form of polydnavirus present is also influenced by symbiotic bacteria (*Wolbachia*) harbored by the wasps. Thus the variability in insect host species infectivity shown by the wasp is influenced by the version of the polydnavirus present, which is in turn influenced by the presence or absence of *Wolbachia*.

Do parasites give rise to free-living organisms?

The survey of parasite diversity in Section 2.1 emphasizes the concept that parasitism has been derived multiple times from free-living progenitors. But what about the opposite possibility? Traditionally, it has been thought that the move to parasitism is a one-way trip because organisms lose genes and structures required to return to a free-living state and they acquire adaptations that specialize them to a life of parasitism. This is in essence what is called **Dollo's law**. Indeed, the discussion of genomes in Chapter 3 highlights that genomic reduction is a frequent but by no means inevitable consequence of parasitism. Also, in groups where adoption of parasitism is recent and not yet been accompanied by extensive adaptation to parasitism, reversals might be expected.

Examples have come to light in which apparent reversions to free-living existence within groups that are parasitic have occurred. Such reversions are particularly noteworthy in mites. For example, *Mitonyssoides stercoralis* is a predatory mite living on other mites that feed on bat guano. It belongs to a family of mites (Macronyssidae) that is otherwise parasitic, and it is closely related to a species that is a parasite of bats. A plausible scenario is that some individuals of the parasitic species adopted predatory feeding habits when they fell off or left their host. Although examples like this one require us to consider that it is possible for parasites to adopt a free-living lifestyle, such examples seem to be rare by comparison to the many adoptions of parasitic lifestyles by free-living organisms (see also a discussion of the origins of parasitism in Chapter 7).

Summary

An enormous diversity of parasites inhabits our world, and there is an ongoing quest to reveal the full extent of this diversity. The unit of diversity studies is often the species and although a conventional species definition (populations of similar and interbreeding individuals) has been used for parasites, we have seen that this definition often does not describe many parasite groups in which sexual reproduction is cryptic, rare, or absent. It would be very satisfying to say with confidence how many parasite species are present on Earth, but the answer is still elusive as some geographic areas, habitat types, and groups of host species remain poorly sampled. We are getting closer to an answer for some parasites, such as the number of helminth species that infect vertebrates, but we are still in the dark for many other groups. Despite these uncertainties, a strong case can be made for the existence of more species of parasites than species of free-living organisms.

Phylogenetic trees provide a compelling way to order our thinking about parasite diversity. Although such trees are designed to trace the vertical patterns of inheritance from ancestors to progeny, we have also seen that horizontal gene transfer has undoubtedly been a factor in parasite evolution and may well prove in time to be even more important than we currently realize.

Among the unicellular eukaryotes are several prominent parasite groups that show peculiar modifications to their parasitic lifestyles, and none are more prominent than the apicomplexans, some of which cause important diseases such as malaria or toxoplasmosis. The apicomplexans harbor a huge, still largely unrevealed reservoir of diversity. Many multicellular eukaryote groups include parasites, such as the red algae, which engage in surprising and insidious strategies of infection, and plants, which have frequently adopted parasitism, even to the extent of losing their ability to engage in photosynthesis. The Fungi comprise a huge group of poorly characterized organisms, many of which are parasitic in plants or animals; fungi are also increasingly implicated in causing emerging diseases. Parasitic animals are relatively well known for their effect on the health of plants, animals, and people and it is interesting to note that parasitism has arisen at least 60 different times among animals. We have also seen that sometimes parasites give rise to free-living organisms, though it seems to be rare.

Studies of parasite diversity have helped clarify many (but not yet all) enigmatic relationships among disparate organisms that have long mystified biologists. The origins and biogeographic distributions of human-infective parasites have been revealed by modern diversity studies, which have also helped to provide convenient benchmarks (such as DNA barcodes) to categorize species. Modern molecular studies have in many cases revealed that what we once thought was a single parasite species is in fact a complex of species, and have enabled biologists to identify entire new lineages of parasites that were previously missed. Metagenomics approaches have also revealed how surprisingly common and diverse some parasite groups are. Finally, modern studies of parasite diversity have also provided fresh insights into the considerable and biologically relevant diversity that is inherent within a single parasite species.

REVIEW QUESTIONS

1. Why are studies of parasite diversity important and useful?

2. The biological species concept is difficult to apply to some parasite groups. Which ones and why?

3. How many species of parasites are there, and what groups or factors make the answer particularly hard to quantify?

4. How do the patterns of inheritance implied by Darwin's tree of life compare to those emerging from the concept of horizontal gene transfer, and how are parasites affected?

5. Many of the parasitic protozoans have unusual cell biology that features peculiar organelles. Provide some examples.

6. Why do you suppose there have been many origins of parasitic species from free-living ancestors, but relatively few examples of parasites evolving to become free-living?

7. What are metagenomics studies? Give some examples where they are relevant to the study of parasites in general.

8. What is a monophyletic group and how does the process of homoplasy obscure relationships?

9. Differentiate between an isolate, a strain, and a subspecies. Use the example of *Trypanosoma brucei* to show why intraspecific diversity matters.

References

GENERAL REFERENCES

Opening quote

De Meeûs T, Michalakis Y & Renaud F (1998) Santa Rosalia revisited: or why are there so many kinds of parasites in "the garden of earthly delights?" *Parasitol Today* **14**:10–13.

THE DIVERSITY OF PARASITE SPECIES

Given these considerations, how many species of parasites inhabit the Earth?

Dobson A, Lafferty KD, Kuris AM et al (2008) Homage to Linnaeus: how many parasites? How many hosts? *Proc Natl Acad Sci USA* **105**:11,482–11,489.

Mora C, Tittensor DP, Adl S et al (2011) How many species are there on earth and in the ocean? *PLoS Biol* **9(8)**:e1001127.

Qin JJ, Li RQ, Raes, J et al (2010) A human gut microbial gene catalogue established by metagenomic sequencing. *Nature* **464**(7285):59–65.

Wilson EO (1999) The Diversity of Life. W.W. Norton.

Windsor DA (1998) Most of the species on earth are parasites. *Int J Parasitol* **28**:1939–1941.

Evolutionary trees are used to visualize evolutionary relationships and to display parasite diversity

Gregory TR (2008) Understanding evolutionary trees. *Evolution (N Y)* **1(2)**:121–137.

Moissl-Eichinger C & Huber H (2011) Archaeal symbionts and parasites. *Curr Opin Microbiol* **14**:364–370.

Efforts are well underway to reveal the overall tree of life

Tree of Life web project. http://tolweb.org/tree/

Pace NR (2009) Mapping the tree of life: progress and prospects. *Microbiol Mol Biol Rev* **73**:565–576.

Waters E, Hohn MJ, Ahel I et al (2003) The genome of *Nanoarchaeum equitans*: Insights into early archaeal evolution and derived parasitism. *Proc Natl Acad Sci USA* **100**:12,984–12,988.

Horizontal gene transfer (HGT) has been pervasive throughout the evolution of life

Doolittle WF (2009) The practice of classification and the theory of evolution, and what the demise of Charles Darwin's tree of life hypothesis means for both of them. *Philos Trans R Soc Lond B Biol Sci* **364**:2221–2228.

Keeling PJ (2009) Functional and ecological impacts of horizontal gene transfer in eukaryotes. *Curr Opin Genet Dev* **19**:613–619.

Many bacteria are parasites

The bacterium *Bdellovibrio*. http://people.sju.edu/~jtudor/undergrads/docs/tudorweb/tudorresearch.html

Eukaryotes are a very diverse group that includes many different kinds of parasites

Aguilera P, Barry T & Tovar J (2008) *Entamoeba histolytica* mitosomes: organelles in search of a function. *Exp Parasitol* **118**:10–16.

de la Casa-Esperon E (2012) Horizontal transfer and the evolution of host–pathogen interactions. *Int J Evol Biol* (Article ID 679045).

Denoeud F, Roussel M, Noel B et al (2011) Genome sequence of the stramenopile *Blastocystis*, a human anaerobic parasite. *Genome Biol* **12**:R29.

Stuart KD, Schnaufer A, Ernst NL et al (2005) Complex management: RNA editing in trypanosomes. *Trends Biochem Sci* **30**:97–105.

HGT has also played a role in the evolution of eukaryotic parasites

Danchin EGJ, Rosso MN, Vieira P et al (2010) Multiple lateral gene transfers and duplications have promoted plant parasitism ability in nematodes. *Proc Natl Acad Sci USA* **107**:17,651–17,656.

Richards TA, Dacks JB, Jenkinson JM et al (2006) Evolution of filamentous plant pathogens: gene exchange across eukaryotic kingdoms. *Curr Biol* **16**:1857–1864.

The Apicomplexa is a huge, important, nearly exclusively parasitic group of organisms

Gubbels MJ & Duraisingh MT (2012) Evolution of apicomplexan secretory organelles. *Int J Parasitol* **42**:1071–1081.

Levine ND (1988) The Protozoan Phylum Apicomplexa, Taylor and Francis.

Morrison DA (2009) Evolution of the Apicomplexa: where are we now? *Trends Parasitol* **25**:375–382.

Rueckert SI & Leander BS (2008). Gregarina. Tree of Life web project http://tolweb.org/Gregarina/124806

Many well-known parasites belong to familiar groups of multicellular organisms

Halanych KM (2004) The new view of animal phylogeny. *Annu Rev Ecol Evol Syst* **35**:229–256.

Hawksworth DL (2001) The magnitude of fungal diversity: the 1.5 million species estimate revisited. *Mycol Res* **105**:1422–1432.

Fisher MC, Henk DA, Briggs CJ et al (2012) Emerging fungal threats to animal, plant and ecosystem health. *Nature* **484**:186–194.

Jeffries P & Young TWK (1994) Interfungal Parasitic Relationships. CAB International.

Musselman LJ & Press MC (1995) Introduction to parasitic plants. In Parasitic Plants (Press MC & Graves JD eds), pp. 1–13. Chapman and Hall.

Pechenik JA (2014) Biology of the Invertebrates. 7th ed. McGraw-Hill.

Westwood JH, Yoder JI, Timko MP et al (2010) The evolution of parasitism in plants. *Trends Plant Sci* **15**:227–235.

Xi Z, Bradley RK, Wurdack KJ et al (2012) Horizontal transfer of expressed genes in a parasitic flowering plant. *BMC Genomics* **13**:227.

Zuccarello GC, Moon D, & Goff LJ (2004) A phylogenetic study of parasitic genera placed in the family Choreocolacaceae (Rhodophyta). *J Phycol* **40**:937–945.

INSIGHTS INTO PARASITISM FROM THE STUDY OF DIVERSITY

The phylogenetic affinities of enigmatic parasites can be revealed

Fischer WM & Palmer JD (2005) Evidence from small-subunit ribosomal RNA sequences for a fungal origin of Microsporidia. *Mol Phylogenet Evol* **36**:606–622.

Some groups of parasites remain "persistent problematica"

Bleidorn C, Podsiadlowski L, Zhong M, et al (2009) On the phylogenetic position of Myzostomida: can 77 genes get it wrong? *BMC Evol Biol* **9**:Article 150.

Evans NM, Holder MT, Barbeitos MS et al (2010) The phylogenetic position of Myxozoa: exploring conflicting signals in phylogenomic and ribosomal data sets. *Mol Biol Evol* **27**:2733–2746.

Helm C, Bernhart SH, Siederdissen CHZ et al (2012) Deep sequencing of small RNAs confirms an annelid affinity of Myzostomida. *Mol Phylogenet Evol* **64**:198–203.

Holland JW, Okamura B, Hartikainen H et al (2011) A novel minicollagen gene links cnidarians and myxozoans. *Proc R Soc Lond B Biol Sci* **278**:546–553.

Jenner RA & Littlewood DTJ (2009) Invertebrate Problematica: kinds, causes, and solutions. In Animal Evolution: Genomes, Fossils, and Trees (Telford MJ, Littlewood DTJ, eds), pp 107–126. Oxford University Press.

Siddall ME, Martin DS, Bridge D et al (1995) The demise of a phylum of protists: phylogeny of myxozoa and other parasitic cnidarians. *J Parasitol* **81**:961–967.

Struck TH, Paul C, Hill N et al (2011) Phylogenomic analyses unravel annelid evolution. *Nature* **471**:95–98.

Studies of diversity can help reconstruct the historical biogeography of parasites

Campbell G, Jones CJ, Lockyer AE et al (2000) Molecular evidence supports an African affinity of the neotropical freshwater gastropod, *Biomphalaria glabrata*, Say 1818, an intermediate host for *Schistosoma mansoni*. *Proc R Soc Lond B Biol Sci* **267**:2351–2358.

THE ONGOING QUEST TO REVEAL AND UNDERSTAND PARASITE DIVERSITY

DNA barcoding is one way to catalog parasite diversity

Barcode of Life http://www.barcodeoflife.org/content/about/what-dna-barcoding

Some parasites exist in complexes of cryptic species

Barik TK, Sahu B & Swain V (2009) A review on *Anopheles culicifacies*: from bionomics to control with special reference to Indian subcontinent. *Acta Trop* **109**:87–97.

Locke SA, Mclaughlin JD & Marcogliese DJ (2010) DNA barcodes show cryptic diversity and a potential physiological basis for host specificity among Diplostomoidea (Platyhelminthes: Digenea) parasitizing freshwater fishes in the St. Lawrence River, Canada. *Mol Ecol* **19**:2813–2827.

Monis PT, Caccio SM & Thompson RCA (2009) Variation in *Giardia*: towards a taxonomic revision of the genus. *Trends Parasitol* **25**:93–100.

Perez-Ponce de Leon G & Nadler SA (2010) What we don't recognize can hurt us: a plea for awareness about cryptic species. *J Parasitol* **96**:453–464.

Pozio E. & Zarlenga DS (2013) New pieces of the *Trichinella* puzzle. *Int. J. Parasitol.* **43:**983–997.

Whole lineages of unapparent parasites may escape our attention

Ragan MA, Goggin CL, Cawthorn RJ et al (1996) A novel clade of protistan parasites near the animal–fungal divergence. *Proc Natl Acad Sci USA* **93**:11,907–11,912.

Metagenomics provides a new way to reveal parasite diversity

Brate J, Logares R, Berney C et al (2010) Freshwater Perkinsea and marine–freshwater colonizations revealed by pyrosequencing and phylogeny of environmental rDNA. *Isme J* **4**:1144–1153.

Studies of parasite diversity help provide a better foundation for taxonomy

Aldhoun JA & Littlewood DTJ (2012) *Orientobilharzia* Dutt & Srivastava, 1955 (Trematoda: Schistosomatidae), a junior synonym of *Schistosoma* Weinland, 1858. *Syst Parasitol* **82**:81–88.

Fiala I & Bartosova P (2010) History of myxozoan character evolution on the basis of rDNA and EF-2 data. *BMC Evol Biol* **10**:228.

Lockyer AE, Olson PD, Ostergaard P et al (2003) The phylogeny of the Schistosomatidae based on three genes with emphasis on the interrelationships of *Schistosoma* Weinland, 1858. *Parasitology* **126**:203–224.

Perkins EM, Donnellan SC, Bertozzi T et al (2009) Looks can deceive: molecular phylogeny of a family of flatworm ectoparasites (Monogenea: Capsalidae) does not reflect current morphological classification. *Mol Phylogenet Evol* **52**:705–714.

Snyder SD & Loker ES (2000) Evolutionary relationships among the Schistosomatidae (Platyhelminthes: Digenea) and an Asian origin for *Schistosoma*. *J Parasitol* **86**:283–288.

OTHER WAYS TO CONSIDER PARASITE DIVERSITY

Diversity within parasite species is extensive and important

Branca A, Le Ru BP, Vavre F et al (2011) Intraspecific specialization of the generalist parasitoid *Cotesia sesamiae* revealed by polyDNAvirus polymorphism and associated with different *Wolbachia* infection. *Mol Ecol* **20**:959–971.

Casulli A, Interisano M, Sreter T et al (2012) Genetic variability of *Echinococcus granulosus* sensu stricto in Europe inferred by mitochondrial DNA sequences. *Infect Genet Evol* **12**:377–383.

Colinet D, Schmitz A, Cazes D et al (2010) The origin of intraspecific variation of virulence in an eukaryotic immune suppressive parasite. *PLoS Pathogens* **6**:e1001206.

Lymbery AJ & Thompson RCA (2012) The molecular epidemiology of parasite infections: tools and applications. *Mol Biochem Parasitol* **181**:102–116.

Nakao M, Xiao N, Okamoto M et al (2009) Geographic pattern of genetic variation in the fox tapeworm *Echinococcus multilocularis*. *Parasitol Int* **58**:384–389.

Neafsey DE, Galinsky KJ, Rays HY et al (2012) The malaria parasite *Plasmodium vivax* exhibits greater genetic diversity than *Plasmodium falciparum*. *Nat Genet* **44**:1046–1050.

Ramirez JD, Guhl F, Rendon LM et al (2010) Chagas cardiomyopathy manifestations and *Trypanosoma cruzi* genotypes circulating in chronic Chagasic patients. *PLoS Negl Trop Dis* **4**:e899.

Su C, Khan A, Zhou P et al (2012) Globally diverse *Toxoplasma gondii* isolates comprise six major clades originating from a small number of distinct ancestral lineages. *Proc Natl Acad Sci USA* **109**:5844–5849.

Wardhana AH, Hall MJR, Mahamdallie SS et al (2012) Phylogenetics of the Old World screwworm fly and its significance for planning control and monitoring invasions in Asia. *Int J Parasitol* **42**:729–738.

Wendte JM, Gibson AK & Grigg ME (2011) Population genetics of *Toxoplasma gondii*: new perspectives from parasite genotypes in wildlife. *Vet Parasitol* **182**:96–111.

Do parasites give rise to free-living organisms?

Cruickshank RH & Paterson AM (2006) The great escape: do parasites break Dollo's law? *Trends Parasitol* **22**:509–515.

Radovsky FJ & Krantz GW (1998) New genus and species of predaceous mite in the parasitic family Macronyssidae (Acari: Mesostigmata). *J Med Entomol* **35**:527–537.

The Parasite's Way of Life

Only kings, presidents, editors, and people with tapeworms have the right to use the editorial "we."

Mark Twain

In This Chapter

When stripped down to the bare essentials, the bottom line for all living things is the same: successful reproduction and propagation. Those that contribute to the next generation persist. Those that do not will perish. To achieve this fundamental goal, all organisms must surmount similar problems. They all must find an appropriate environment in which they can develop properly. They must obtain nutrients and they must successfully cope with various abiotic factors. Furthermore, no organism lives in a biological vacuum. At the very least, it is necessary to find a mate or to use an alternative means of reproduction in which a mate is not essential. Other biotic interactions, including predator–prey relationships, competition, and various symbioses, are challenges in any environment. However for parasites, the host *is* the environment. Thus, for parasites, the need to find and subsequently inhabit an appropriate environment takes on greater meaning.

As noted in Chapter 1, from the parasite's perspective, hosts can be thought of as islands—relatively small and unevenly dispersed habitats that can only be reached by traversing "oceans" of intervening environment. But the island metaphor only goes so far. Genuine islands do not move. Nor do they die after at most several decades. Furthermore, unlike more conventional habitats, such as coral reefs or temperate forests, living hosts actively defend themselves against parasitic colonization and over time they evolve to lessen the impact of at least the most common parasites to which they are exposed. Reaching a new host is just the beginning. Once it arrives, the parasite must gain entry. Next it must navigate the labyrinth of the host's anatomy, locate its required site within the host, find a mate, and possibly outcompete other parasites, all while coping with often hostile conditions as well as active host defenses.

The manner in which a given parasite species overcomes these hurdles is reflected in its life cycle. In this chapter, we provide an overview of basic life-cycle patterns. In doing so, we will consider some of the colorful history of life-cycle elucidation. We will then tease apart the various components of the life cycle, considering it as a series of problems that a parasite must solve. The astonishing diversity that we observe among parasitic organisms, described in Chapter 2, is in many ways a consequence of the ingenious, devious, and sometimes perplexing solutions, that different parasitic taxa have evolved.

Species names highlighted in red are included in the Rogues' Gallery, starting on page 429.

3.1 A HISTORICAL PERSPECTIVE OF THE PARASITE LIFE CYCLE

A particular parasite's **life cycle** is one of its most characteristic biological features. The life cycle can be defined as the sequence of developmental events that occurs over the course of an organism's lifetime. For parasites, at least some of these events occur in or on the host. Many parasites require different hosts for different developmental events. In such cases, these hosts are colonized in a specific sequence.

Figure 3.1 A historical reference to *Dracunculus medinensis*, the Guinea worm. Asclepius, the Greek god of medicine and healing hold his staff (the staff of Asclepius). The snake-entwined staff is believed to come from the age-old practice of guinea worm removal, in which the long female worm is slowly wrapped on a stick. The staff of Asclepius (or sometimes the staff of Caduceus, entwined with two snakes) has come to symbolize modern medicine. (Courtesy of Her Dax, CC BY-SA 3.0.)

Early medical and natural history studies gave rise to an understanding of parasite life cycles

Because of their relatively large size, certain helminths were the earliest infectious agents recognized in humans and animals (**Figure 3.1**). Ancient Egyptian medical chronicles, including the Ebers papyrus dated at about 1500 BC, describe nematodes, possibly ***Ascaris*** and the hookworm ***Ancylostoma duodenale***, that commonly infected humans. Around 400 BC, Hippocrates described seedlike structures in human feces that were most likely tapeworm reproductive segments. In one of the earliest discussions of habitat site selection in the host, in approximately 180 AD, the Greek physician Galen described nematodes in the human intestine that prefer the distal end of the colon near the anus (probably pinworms). Other, larger worms preferred the upper portion of the small intestine (probably ***Ascaris***).

The fall of Rome, starting in approximately 400 AD, ushered in the Dark Ages, during which parasitology, like other sciences, largely regressed. Although Chinese and Arabic texts from the Middle Ages contain some references to helminths, real advancement in our understanding of parasites would not occur until the eighteenth century. In 1735, when Carl Linnaeus published his *Systema Naturae*, he described various parasitic helminths, including ***Ascaris***, pinworms, guinea worm, liver flukes, and the pork tapeworm. Yet even as the list of worms known to inhabit humans steadily grew, there was little or no understanding of the complex life cycles or developmental changes that are typical of many parasites.

In the 1830s, a Danish biologist, Johann Steenstrup, suggested that all the various organisms that he found inhabiting freshwater snails and sheep livers might be different developmental stages of the same organism that somehow were able to move from sheep to snails and back to sheep. These stages are now recognized as the sporocysts, rediae, cercariae, and adults of the liver fluke ***Fasciola hepatica***. Steenstrup also suspected that the cystlike structures that were commonly found in pork were a developmental stage of an adult worm of some sort.

Acting on Steenstrup's hunch, a German physician, Friedrich Kuchenmeister, decided to investigate. He had already observed that these cysts, which he called bladder worms, looked something like small tapeworms. Furthermore, similar bladder worms were typically found in pigs and other animals such as rodents that might be described as prey. Tapeworms, on the other hand, were usually seen in dogs, cats, and other predators. Might the bladder worms be immature tapeworms that develop into adults when a predator eats an infected prey animal? In 1851, he decided to find out by feeding the cystlike bladder worms extracted from rabbit muscle tissue to foxes. After he found adult tapeworms in the foxes, he conducted other experiments in which he fed tapeworm segments released in the feces of a dog to a sheep. When he dissected the sheep several weeks later, Kuchenmeister found cysts in the animal's brain. Finally, he showed that his human patients, whom he knew were often infected with tapeworms called ***Taenia solium***, became infected after eating pork that contained cysts. In a particularly illustrative (some might say gruesome) experiment, Kuchenmeister obtained cysts from raw pork. He then mixed these cysts into food that was destined to be eaten by a condemned murderer. Several days later the prisoner was executed. When Kuchenmeister inspected the dead man's intestine, he found young adult ***T. solium***.

As the idea that specific parasites could appear radically different at different times of their development gained acceptance, other insights into parasite transmission followed, setting the stage for the elucidation of complete parasite life cycles. In 1860, Fredrich Zenker showed that humans also became infected with ***Trichinella spiralis*** when they consumed infected, undercooked pork. Shortly thereafter, the Italian Giovanni Grassi demonstrated that transmission could occur for some parasites via the ingestion of eggs. Specifically, he consumed ***Ascaris lumbricoides*** eggs and later found eggs in his own feces.

Such self-infection was instrumental in demonstrating that certain other helminth parasites could gain access to their vertebrate host by a different route—directly through the skin. Since 1845, for example, hookworms have been recognized as a causative agent of human disease. The manner by which they infect humans, however, was not discovered until 1896. When the German Arthur Looss attempted to infect guinea pigs by feeding them hookworm larvae, he accidentally spilled some of these larvae onto his hand. When his skin became inflamed, Looss speculated that the hookworm larvae had penetrated directly through the skin on his hand. A few weeks later, he found hookworm eggs in his feces. He described the species as ***Ancylostoma duodenale***. Looss evidently thought highly of his self-infection technique. Fresh from his discovery regarding hookworms, Looss purposely put *Strongyloides stercoralis* on his skin. Seven weeks later he found larvae in his feces, demonstrating that this nematode also gained access to humans via dermal penetration. See **Box 3.1** for other parasitologists who went this extra mile in pursuit of their science.

Mosquito transmission was first demonstrated for filarial worms

Few diseases are as disfiguring as elephantiasis, which is caused by infection with nematodes in the genera ***Wuchereria*** and ***Brugia***. They are commonly known as **filarial worms** (from the Latin for threadlike). Because of the obvious and often extensive deformity observed in limbs and genitals, even ancient civilizations were familiar with the condition (**Figure 3.2**). The term elephantiasis comes from the ancient Greeks, who described those suffering from the condition as elephantlike. Elephantiasis is notable for another important reason; it is the first parasitic condition shown to be transmitted by infected arthropods.

Adult worms inhabit lymphatic vessels. The characteristic symptoms of elephantiasis result from obstruction of the vessels and subsequent inflammation. Following mating, females release large numbers of larvae called **microfilariae**. In 1863, French surgeon Jean-Nicolas Demarquay was the first to observe microfilariae in accumulated lymphatic fluid. Several years passed, however, before the Scottish physician Timothy Lewis made the link between microfilariae and elephantiasis. In 1876, the Australian Joseph Bancroft first described the adult worm, which is now known as ***Wuchereria bancrofti*** in his honor.

Patrick Manson, another Scottish physician, made a critical discovery regarding the life cycle of filarial worms when he investigated their mode of

Figure 3.2 Elephantiasis in an Egyptian pharaoh. Mentuhotep II was Pharaoh for over 50 years, from approximately 2046 BC to 1995 BC. The swollen legs carved in this statue suggest elephantiasis, caused by filarial worm infection. Blockage of the lymphatic vessels by adult worms causes fluid buildup and tissue distention. (Courtesy of Anastasia Grishchenko.)

BOX 3.1

The Grand Tradition

Parasitologists often display a lack of squeamishness when it comes to infecting themselves in pursuit of knowledge. Indeed, according to the famed parasitologist and author Asa Chandler, the good parasitologist has been infected with one or more of the organisms he or she studies. To many parasitologists, such infection represents a badge of honor, a way to earn their stripes, so to speak. A surprising number have gone as far as to participate in the grand tradition of self-infection.

The list of parasite life cycles that have been elucidated at least in part by parasitologists willing to be part of their own experiments is long, and our few examples only scratch the surface. Self-infection or the infection of others has also been carried out for several other reasons. Some were no doubt motivated by simple curiosity or perhaps other more obscure motives. Others may have felt like Clark Read, who, after attempting to infect himself with the rat tapeworm, *Hymenolepis diminuta*, explained that, "It seemed like the thing to do in light of all the rats I have infected with this parasite."

In other instances, the act may have been purely practical. In 1922, for example, Claude Barlow consumed over 100 metacercariae of the digenetic trematode *Fasciolopsis buski* to better understand how clinical symptoms developed in humans. Asa Chandler, quoted above, clearly did not simply talk the talk. Chandler wanted to know if the various species of tapeworms in the genus *Diphyllobothrium* might actually be all the same species. *Diphyllobothrium* spp. use mammals as the definitive host. A wide variety of mammals become infected when they consume freshwater fish, which serve as the second intermediate host. Species designations were based on structural differences in the various cestodes, but Chandler suspected that the differing morphologies were simply a consequence of which mammal the parasites infected. He predicted that if these so-called species all went through the same host (himself) the morphological difference would vanish. His willingness to serve as his own experimental subject allowed him to show that he was correct. In 1968 Dale Little swallowed infective eggs of the nematode *Trichuris trichiura*. The eggs were especially large, and Little wished to know if egg size was a genetic trait in these helminths. When the eggs he eventually passed all came out just as large as the ones that he consumed, he got his answer. Egg size was indeed an inheritable trait.

Other scientists have infected themselves for more pedestrian reasons: to avoid bureaucratic red tape or difficulty of obtaining permits to import parasites into the country. In the 1920s, for instance, Hans Vogel infected himself with *Fasciolopsis buski*, just to import these intestinal trematodes quietly into the United States. Likewise, Claude Barlow, infected himself with *Schistosoma haematobium* cercariae in Egypt, so that there would be no trouble with customs officials. Barlow wanted to see if *S. haematobium* was capable of infecting species of North American snails. As a bonus, his infection provided a continuous source of material for teaching, and he made a number of new observations, such as the ability of *S. haematobium* to release its eggs in seminal fluid. More recently, in the 1980s, Robin Overstreet imported *Diphyllobothrium latum* from Finland by carrying the contraband tapeworms in his intestine. Harvey Blankespoor wanted to know what happened to avian schistosomes when they penetrated human rather than bird skin. As he explained it, he carried out his experiment on himself to avoid any possibility of a lawsuit.

Then there was the Canadian student who evidently was not fond of his roommates. He allegedly slipped infective *Ascaris lumbricoides* eggs into their food. Two of the apparent victims almost died due to an allergic reaction. Although the district attorney contemplated a charge of attempted murder, charges were eventually dropped as all the evidence was circumstantial. None of the roommates passed an adult worm, meaning that no "murder weapon" could be found.

References

Mayberry LF (1996) The infectious nature of parasitology. *J Parasitol* 82:855–864.

Kean BH, Mott KE & Russel AJ (1978) Tropical Medicine and Parasitology Classic Investigations, vols I and II. Cornell University Press.

Grove DI (1990) A History of Human Helminthology. CAB International.

transmission. Manson already knew that arthropod transmission of parasitic worms was a possibility, and he was aware of earlier work demonstrating that humans became infected with Guinea worm when they inadvertently consumed infected copepods in drinking water. He suspected that filarial worms were also transmitted through the consumption of an arthropod vector, but when he found microfilariae in the blood of dogs and humans, he hypothesized that blood-feeding insects might be involved. To find out, he allowed mosquitoes to feed on the blood of a human volunteer who was afflicted with elephantiasis. He subsequently found larval worms in these mosquitoes in 1878. Yet Manson believed that larval parasites passed from the infected mosquito into water, where they might be accidently consumed by a human. It was Manson's assistant, George Low, who, in 1900, finally identified the actual mode of transmission, the bite of an infected mosquito, when he found larval filarial worms in the insects' mouthparts.

Arthropod transmission for filarial worms suggested that other diseases may be similarly transmitted

The discovery that arthropods were involved in transmission of filarial worms had enormous implications for parasitology and microbiology. As it became apparent that parasites might be spread by a previously unsuspected means, the mode of transmission for many infectious agents became clear. In 1892, for example, two Americans, Theobald Smith and Frederick Kilbourne found that redwater fever, a serious disease of cattle, was transmitted by ticks. An apicomplexan in the family Babesiidae, ***Babesia bigemina***, causes redwater fever. Kilbourne and Smith's study was the first to demonstrate that a protozoan pathogen could develop in and be transmitted by an arthropod. Moreover, once the mode of transmission was revealed, means of control became apparent. One simply had to kill the involved tick, ***Rhipicephalus annulatus***, or prevent them from feeding on cattle. Similarly, in 1898, once the Yellow Fever Commission in Cuba established that ***Aedes aegypti*** mosquitoes transmitted yellow fever, incidence of the disease fell rapidly as mosquito control measures were introduced (**Figure 3.3**).

No historical discussion of arthropod-borne parasites, of course, would be complete without considering malaria, at or near the top of anyone's list of important human diseases. Unlike some of the helminths previously discussed, whose large size made them obvious even to ancient health practitioners, there was no real understanding of diseases such as malaria or of bacterial pathogenesis until the acceptance of the germ theory of disease in the late nineteenth century.

Through much of recorded history, at least as far back as ancient Rome in 200 BC, the severe fevers induced by malaria were thought to be due to the unhealthful vapors that emanated from wetlands. Indeed the very name malaria is a literal translation of bad air in Italian. Our understanding of this

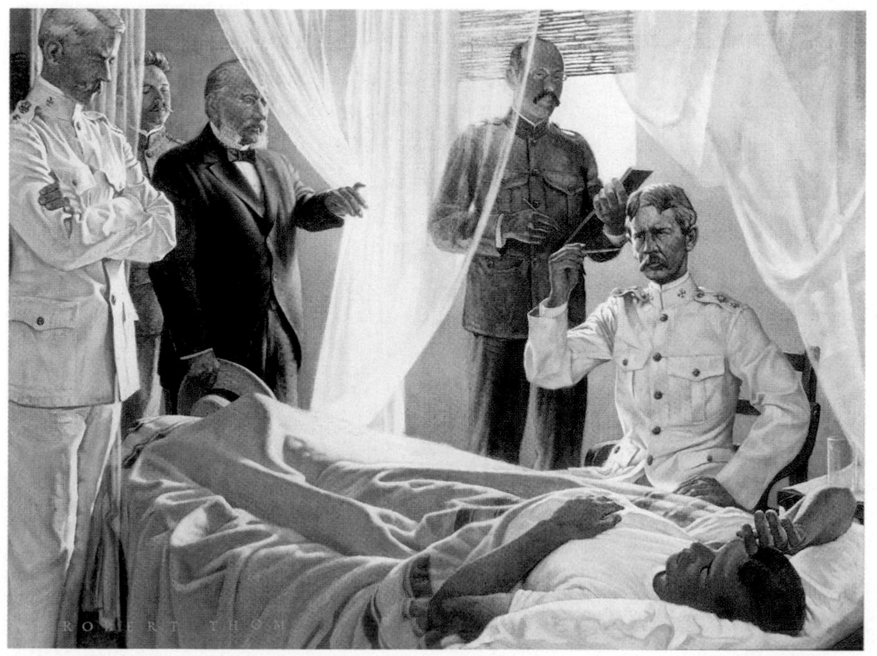

Figure 3.3 Walter Reed and the Yellow Fever Commission. In 1881 the Cuban physician Carlos Finley proposed that yellow fever might be transmitted by mosquitoes. In 1898, with the outbreak of the Spanish-American War, the US Army became especially interested in yellow fever. The Yellow Fever Commission, headed by Walter Reed (seated), was established in Cuba to determine the exact source of the disease. As part of the study, mosquitoes reared in the laboratory were allowed to feed on infected patients. These same mosquitoes were later permitted to feed on human volunteers, including Commission members James Carroll and Jesse Lazear. Both of these individuals developed yellow fever. Lazear died as the result of his infection. (Painting by Robert Thom, The Conquest of Yellow Fever. University of Michigan Museum of Art, Collection of the University of Michigan Health System, Gift of Pfizer Inc.)

disease began in 1880 when Charles Laveran, a physician with the French Foreign Legion in Algeria, found the parasite in a blood sample taken from a soldier suffering from malaria. He quickly concluded that the organisms he had found were the cause of the disease. Following his groundbreaking work on mosquito transmission of filarial worms, Patrick Manson first hypothesized mosquito transmission for malaria as well.

Manson explained his ideas to Ronald Ross, a young English physician based in India who had returned to England on leave in 1894. He convinced Ross to study mosquito transmission upon his return to India in 1895. Ross devoted 12 years to investigating malaria transmission, painstakingly dissecting thousands of mosquitoes, searching for the telltale developmental stages of the parasite. Finally in 1897, he demonstrated that avian malaria, caused by *Plasmodium relictum*, was transmitted by mosquitoes in the genus ***Culex*** (**Figure 3.4**). A year later, in 1898, a group of Italian scientists showed conclusively that the important human pathogen, ***Plasmodium falciparum***, was transmitted by mosquitoes in the genus ***Anopheles***. These researchers included Giovanni Grassi, the same individual who intentionally consumed ***Ascaris*** eggs, as described above. The discovery that mosquitoes were involved in the transmission of two other important causative agents of human malaria, ***P. malariae*** and ***P. vivax***, soon followed.

It was not until 1947 that the complete ***Plasmodium*** life cycle, including those stages that occur in the human host, were fully elucidated. By that time, it was assumed that the **sporozoite**, the developmental stage found in the salivary glands and mouthparts of infected mosquitoes, was infective to humans. Oddly, sporozoites could not be detected in the blood of recently infected people, and no blood stages were observed for over a week. Where did the sporozoites go? English scientists Henry Shortt and Cyril Garnham got the answer when they infected rhesus monkeys with sporozoites taken from the salivary glands of infected mosquitoes. One week later, they found what they called preerythrocytic stages in the monkey's liver cells. It then became evident that the sporozoites first infected liver cells, where they underwent asexual reproduction, before leaving the liver and entering red blood cells.

3.2 AN OVERVIEW OF PARASITE LIFE CYCLES

Having described a few of the historical breakthroughs that sparked our understanding of parasitic life cycles, we turn to a closer look at the life cycles themselves. All living things have a life cycle, which can be defined as the

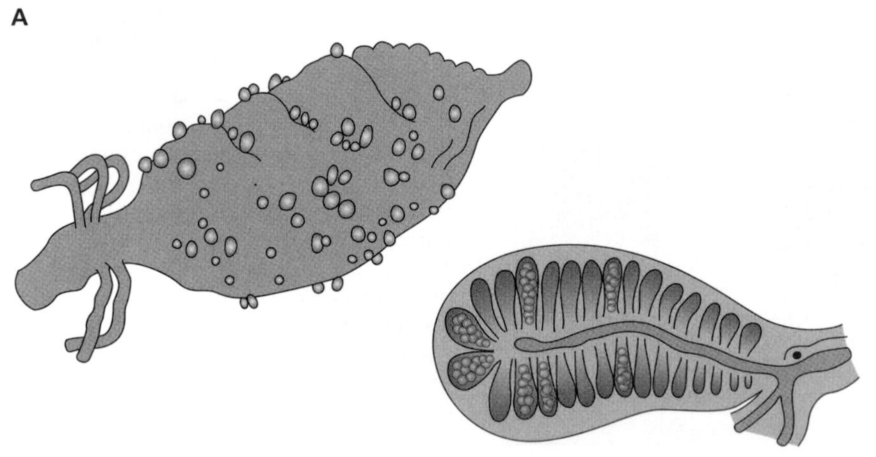

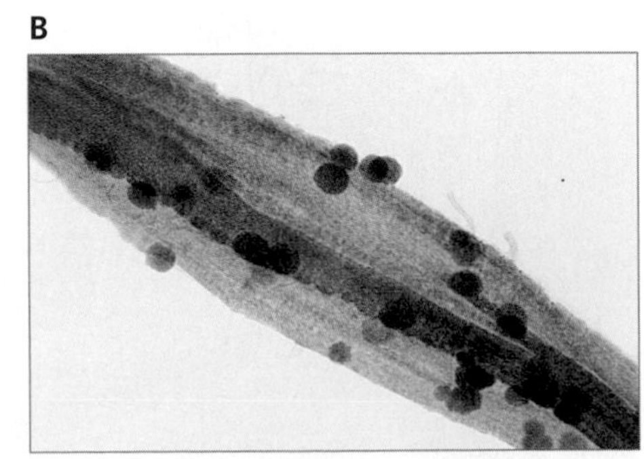

Figure 3.4 Mosquito transmission of malaria. (A) Ronald Ross's drawings of *Plasmodium* oocysts in the digestive tract of a mosquito. (B) A microscopic image of *Plasmodium* oocysts, also on the mosquito gut wall. (B, Courtesy of SJ Upton, Kansas State University.)

developmental events, including growth and reproduction, that occur in the course of an organism's life. Parasitic life cycles are distinguished by the need for host organisms in which to accomplish at least some of these events. Perhaps the simplest way to consider the various forms the parasitic life cycle can take is to distinguish between direct and indirect life cycles.

Parasites with direct life cycles use only a single host

Parasites with **direct life cycles** require only a single host to complete their development and reproduce. Such life cycles are also described as **monoxenous**. Several of the parasites discussed in the historical overview in Section 3.1 have direct life cycles. These include the nematode ***Ascaris lumbricoides***, which can complete its entire life cycle in humans. Other examples are the pinworm (*Enterobius vermicularis*), the hookworm ***Ancylostoma duodenale***, as well as *Strongyloides stercoralis*. These later two examples are noteworthy in that they also have obligatory, free-living larval stages. Many protozoa also have direct life cycles. ***Giardia lamblia*** and ***Entamoeba histolytica*** are two examples.

Parasites with direct life cycles are notable because their propagules can reinfect the same host in which sexual reproduction occurred. For example, individuals infected with adult ***A. lumbricoides*** who pass eggs in their feces may inadvertently consume these eggs, resulting in multiple generations of the same parasite within the same individual host. Although direct cycles are relatively simple, these life cycles are not necessarily boring or conventional. Figure 3.5 shows an example demonstrating that when it comes to the bizarre and curious, parasites can set the bar high.

Two or more hosts are necessary for those parasites with indirect life cycles

An **indirect life cycle**, also called a **heteroxenous** life cycle, implies that two or more different hosts are required. As discussed in Chapter 1, sexual reproduction occurs in the *definitive host*, whereas larval development takes place in the *intermediate host*. Depending on the parasite, larval development may be accompanied by asexual reproduction and in many cases there is more than one intermediate host. We also learned in Chapter 1 that some heteroxenous parasites may employ *paratenic hosts*—nonessential hosts that nevertheless increase the likelihood of transmission to the definitive host. Recall that no development or reproduction occurs in such hosts.

Figure 3.5 The direct life cycle. Parasites that use only a single host have a direct or monoxenous life cycle. (A) *Cymothoa exigua* is a parasitic crustacean in the order Isopoda. An adult female is seen here in the host's mouth. (B) The *C. exigua* life cycle. Motile larvae are released into the environment. Larvae develop to the point where they can attach to the gills of an appropriate host. Here they undergo a process known as *protandrous hermaphroditism*. They first develop into males and upon further maturation into females. Mating is believed to take place on the gills of the host. The now gravid female makes her way to the fish's mouth, where she attaches to the base of the tongue and begins to feed on blood in the tongue. The female will remain in this position for the rest of her life. As the tongue atrophies from a lack of blood, the isopod actually takes the place of the tongue, remaining firmly attached via her rear thoracic appendages. Oddly, the fish is able to use the isopod as if it were a normal tongue. This is the only recognized example of a parasite actually replacing a host's body part. As the female's eggs hatch, giving rise to a new generation of larvae, the life cycle is completed. (A, With permission from Oddy Central.)

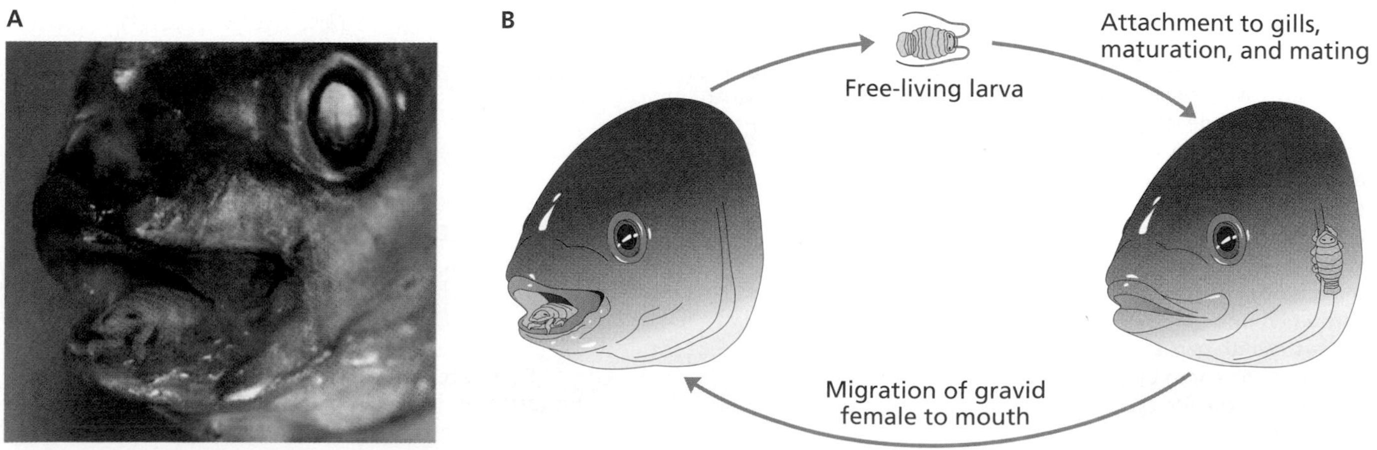

Several of the organisms discussed in the historical overview, including ***Fasciola hepatica***, ***Taenia solium***, ***Wuchereria bancrofti***, and ***Plasmodium falciparum***, have indirect life cycles. Members of the phylum **Acanthocephala** provide an additional example of a relatively straightforward two-host indirect life cycle (**Figure 3.6**). These parasitic worms inhabit and sexually reproduce in the gut of vertebrates such as fish, birds, and reptiles (see Figure 2.19B, which shows the characteristic proboscis that they use to penetrate and attach to the host's gut wall). The vertebrate consequently serves as the definitive host.

The digenetic trematode *Alaria americana* provides us with a more complex (possibly the most complex) multihost life cycle (**Figure 3.7**). It is most common in North America, typically in temperate regions near water. Carnivorous mammals, such as foxes, coyotes, or skunks, serve as definitive hosts. Adult worms are **monoecious**—they possess both male and female reproductive organs and are capable of producing both eggs and sperm. Adults mature in the small intestine and fertilized eggs are released with the feces. Such eggs must reach fresh water, where they hatch in roughly two weeks, releasing a ciliated **miracidium** (plural **miracidia**). This initial, usually free-swimming larval stage is typical of all digenetic trematodes. In the case of *Alaria*, as in most other trematodes, a miracidium seeks specific types of aquatic snails, which it penetrates upon contact. It then sheds its ciliated epithelium, enters the snail's renal veins, and develops into a saclike form called a **mother sporocyst**. The sporocyst reproduces asexually to produce numerous **daughter sporocysts**, which migrate to the snail's digestive gland. After a period of prolonged development of as long as several months, daughter sporocysts give rise to the next life-cycle stage, called **cercariae** (singular **cercaria**). This juvenile, tail-bearing stage exits the snail and swims to the surface. It may hang upside down, periodically sinking, whereupon it again swims to the surface. If nearby water currents mechanically stimulate the cercaria, the larval trematode will swim in the direction of the current, seeking its second intermediate host, a tadpole. Using protease-containing

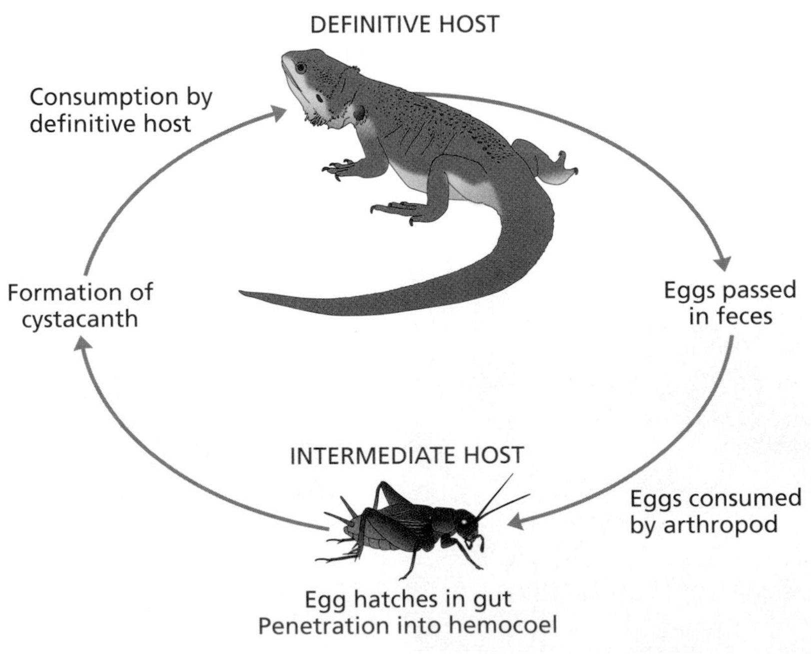

Figure 3.6 A generalized acanthocephalan life cycle. Certain details of the life cycle vary according to parasite species. In all cases, however, the definitive host is a vertebrate. Following fertilization, the female releases eggs, which are passed in the feces. The eggs may then be consumed by the intermediate host, almost always a specific type of arthropod. The egg hatches in the gut of the arthropod and the acanthor that emerges penetrates the gut, entering the body cavity, called the *hemocoel*. The larva then develops into a cyst-like stage called a *cystacanth*. If the infected arthropod is subsequently eaten by the definitive host, the life cycle is completed, as the cystacanth excysts in the digestive system, where it develops into the adult. In many cases, the infected arthropod is first consumed by a paratenic host, in which the cystacanth persists. The paratenic host must later be eaten by the definitive host to complete the life cycle.

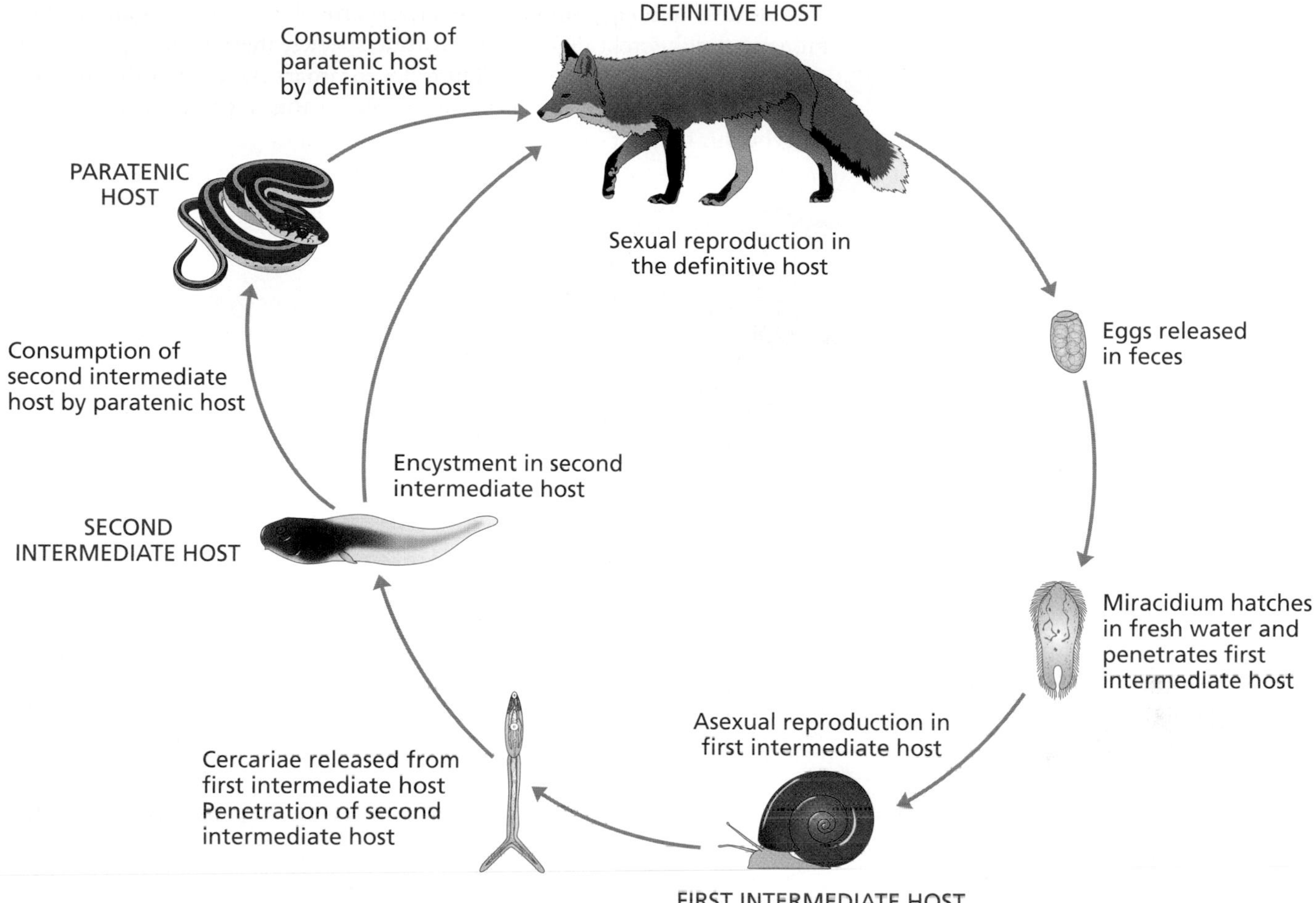

Figure 3.7 The life cycle of *Alaria americana*. The complex indirect life cycle of this digenetic trematode may include up to four hosts. A canid may either consume an infected tadpole, or the tadpole may be consumed by a reptile, which serves as a paratenic host. In this case, the life cycle is completed when a canid consumes the reptile.

penetration glands in its body, the cercaria penetrates through the skin of the tadpole, loses its tail, and enters the juvenile amphibian's body cavity. It then develops into a **mesocercaria**, the infective stage for the next host in the life cycle. Should the tadpole undergo metamorphosis, the mesocercaria will remain infective in the now adult amphibian.

Next, one of two things happens. Occasionally, the definitive host consumes an infected tadpole. If so, the mesocercariae are liberated during digestion and subsequently penetrate the intestinal epithelium, migrating to the lungs. After more than a month, the larval trematode, now called a **diplostomulum metacercaria**, moves up the trachea. After reaching the throat, it is swallowed, which returns it to the intestine. Back in the intestine, following sexual maturation in several weeks, the life cycle is completed.

However, canids or other appropriate definitive hosts eat amphibians only rarely. *Alaria americana* more commonly relies on an alternative Plan B to complete its life cycle. If a snake or lizard consumes an infected tadpole or frog, the reptile will serve as a paratenic host, harboring the still infective trematode larva, until the reptile is finally consumed by a mammal. As an added twist to this complex life cycle, once they reach their mammalian host, the larval stage of the parasite can move from an infected female mammal to her nursing young through breast milk. Transmission from mother to offspring, either in milk or to a developing fetus is termed **vertical transmission**, to be discussed below.

Note that although similar in many regards, life cycle variation is common among different digenetic trematode species; there is no typical single pattern of larval development. **Figure 3.8** summarizes some of the diversity on a common basic theme that is observed in this subclass of the phylum **Platyhelminthes**.

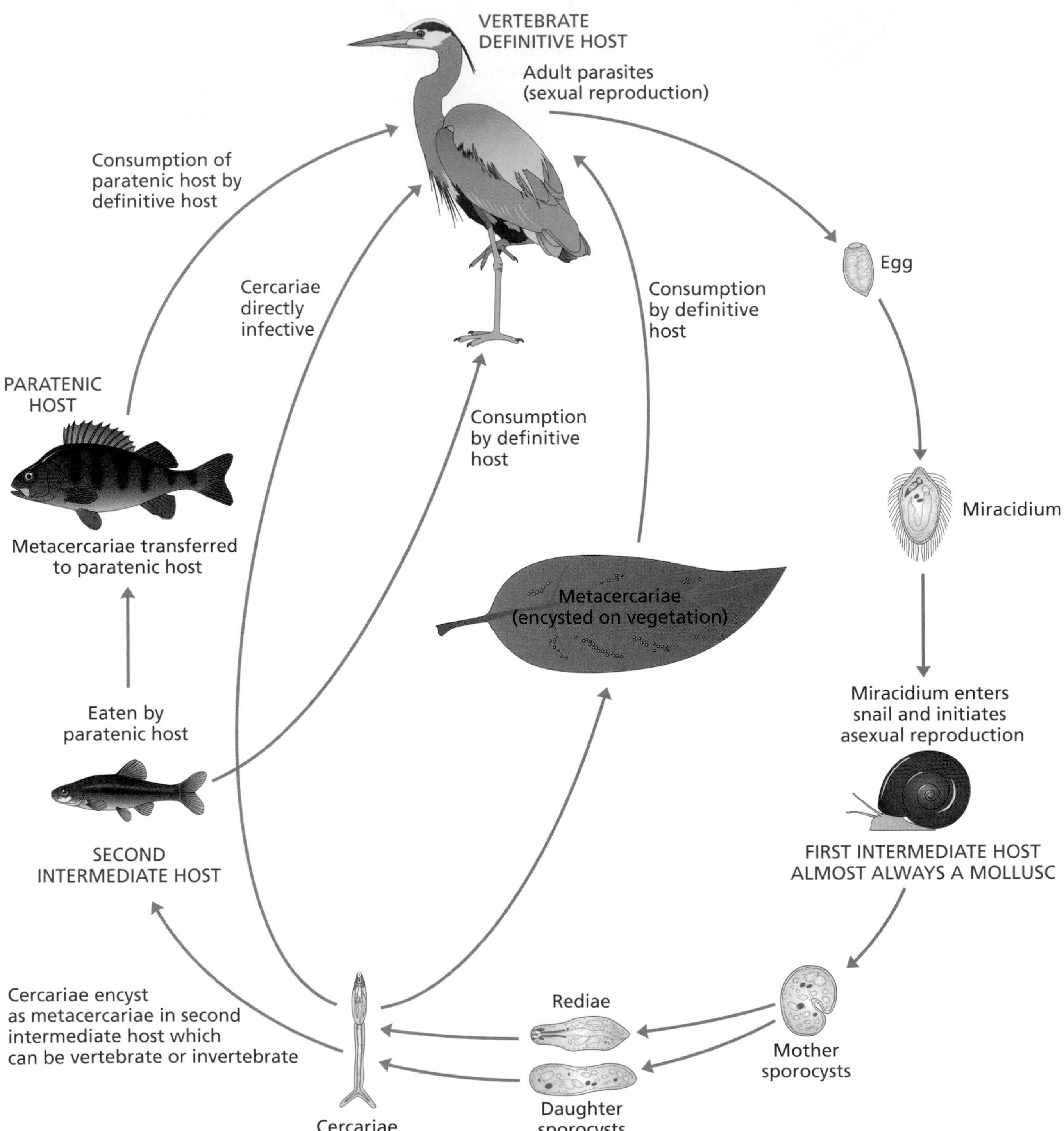

Figure 3.8 Diversity of life cycles in the digenetic trematodes. Digenetic trematodes show remarkable diversity in their life cycles , and no single species takes all the paths shown here.. Note that initially mother sporocysts produce either daughter sporocysts or rediae, but never both. If a paratenic host is used in the life cycle, no additional development occurs within the paratenic host. Not all possible life cycle variations are shown here. For instance, in the first intermediate host, there may be multiple generations of daughter sporocysts or rediae.

3.3 THE PARASITE'S TO DO LIST

Each stage of a parasite's life cycle comes with hurdles that the parasite must negotiate if it is to achieve its overriding objective of reproduction. To be successful a parasite must:

- Achieve transmission
- Enter its host
- Migrate to an appropriate target tissue or site within the host
- Maintain its position within the host
- Find a mate
- Successfully reproduce in and release progeny from the host
- Undergo essential developmental changes within the host
- Cope with ambient physiological conditions within the host
- Evade destruction by the host immune system, at least until reproduction is complete

As you might expect, different parasites use a wide range of strategies to overcome these various obstacles and not all requirements are identical for all parasites. Ectoparasites, for example, by definition do not enter their host. For some hermaphroditic parasites or for those lacking sexual reproduction, there is no absolute requirement to find a mate. We will now examine individually each life-cycle component and its associated problems for the parasite. Immune evasion will be addressed in Chapter 4.

Effective transmission is essential for all parasites

Transmission, the passage of a parasite to a host, is the most fundamental imperative for any parasite. All other life cycle concerns are moot if the parasite cannot reach a required host in the first place.

All organisms, whether free living or parasitic, must disperse from their site of origin to reduce crowding, intraspecific competition, and overuse of resources. However, for the parasite simple dispersal is not enough; unless parasites achieve effective transmission, subsequent development and, ultimately, reproduction do not occur. Parasites employ a wide variety of strategies to achieve transmission. The manner in which a parasite moves from host to host can be defined as its **mode of transmission**.

Fecal–Oral Transmission

Intestinal protozoa, as well as many intestinal helminths, rely on **fecal–oral** transmission to complete at least part of their life cycle. Such transmission suggests that propagules, the stage responsible for achieving transmission, are released in the feces. They may then contaminate food or water, which may be consumed by the subsequent host. Often the propagule is a relatively metabolically inactive egg or cyst, which can withstand long periods of time in the external environment. Its consumption by the next host is largely fortuitous, and the parasite expends little if any energy as it simply waits to be eaten.

The flagellated protozoan ***Giardia lamblia*** provides a typical example (**Figure 3.9**). Like many intestinal protozoa, they live in their host's small intestine as an actively feeding and reproducing stage called the **trophozoite**. Reproduction by binary fusion results in new generations of trophozoites. However, under specific physiological conditions within the host the trophozoite may transform into a nonreplicating and dormant **cyst**. For ***G. lamblia***, dehydration of feces entering the colon has been suggested as an important

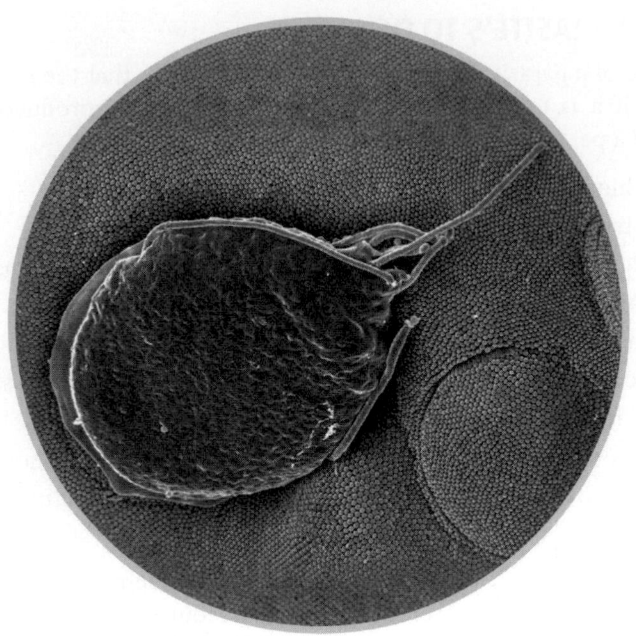

Figure 3.9 *Giardia lamblia*. In this electron micrograph of the trophozoite stage, both the ventral adhesive disk and flagella are clearly visible. Note the circular mark on the intestinal lining left by the adhesive disk. (Courtesy of CDC, Stan Erlandsen.)

stimulus for encystation. *In vitro* cyst formation has also been induced by a lack of cholesterol, which triggers the formation of cyst wall materials that are observed in the encysting trophozoite's endomembrane system. Cyst formation can also be induced experimentally by mimicking the pH and bile concentration that occurs first on the surface of the intestinal epithelium beneath the mucous layer and subsequently in the intestinal lumen. During cyst formation, the flagella and adhesive disk of the trophozoite are lost and several cyst-wall proteins are excreted via exocytosis. The cyst wall, which also incorporates various secreted carbohydrates, forms as the proteins cross-link. Once passed with the feces, cysts can persist for at least a few months under appropriate conditions of humidity and temperature until a suitable host inadvertently consumes them.

Parasites that depend on fecal–oral transmission do not necessarily rely entirely on luck. In many cases, the eggs or cysts released in the feces have specific adaptations to enhance the likelihood of transmission. Cestodes, for instance, have indirect life cycles, and for many tapeworms, the adult host passes eggs in its feces that are subsequently consumed by the intermediate host. In some tapeworms using small aquatic crustaceans called ostracods as hosts, the tapeworm eggs appear similar to the aquatic algae on which the ostracod feeds, tricking the intermediate host into consuming them. Other tapeworms use copepods as intermediate hosts, and their eggs either sink or float, depending on whether or not the specific copepod in question is benthic or pelagic.

Trophic Transmission

The fecal–oral transmission described for the cestode parasites is only half of the story. Once the intermediate host is infected, transmission back to the definitive host is achieved when the definitive host consumes the intermediate host. That is, the parasite takes advantage of the established predator–prey relationship that exists between the definitive and intermediate hosts.

Such transmission, fecal–oral alternating with trophic, is typical of cestodes as well as many other parasites with indirect life cycles. ***Hymenolepis diminuta***, the rat tapeworm, provides a representative example (**Figure 3.10**). Various insects that serve as intermediate hosts may consume eggs passed in

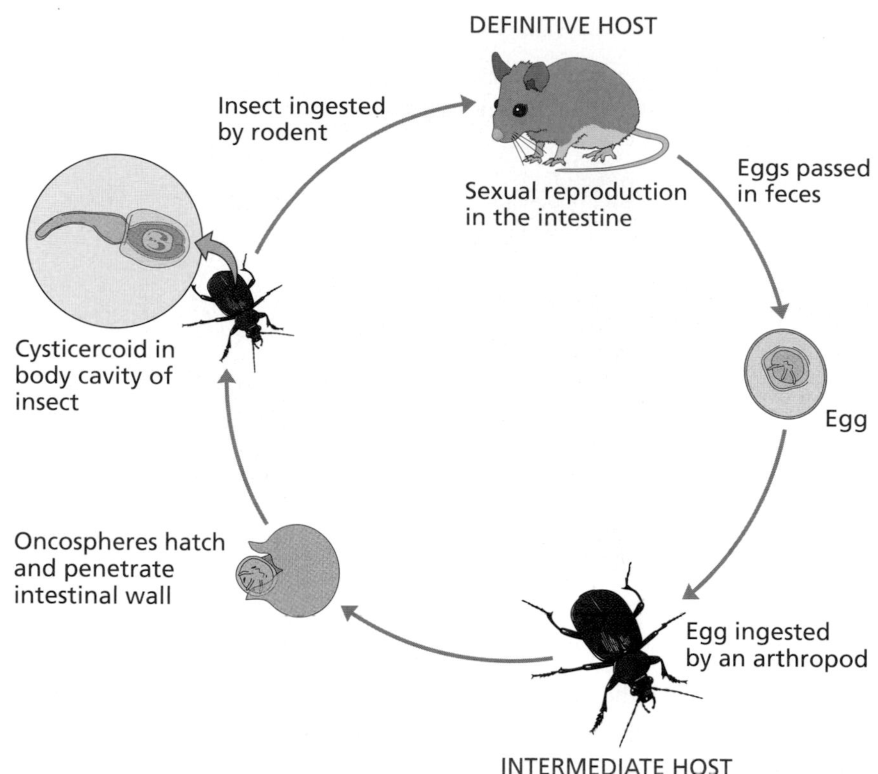

Figure 3.10 *Hymenolepis diminuta*, the rat tapeworm. The life cycle of this cestode relies on both fecal–oral transmission (for transmission from the definitive to the intermediate host) and trophic transmission (for transmission from the intermediate host back to the definitive host). Humans occasionally are infected when they inadvertently consume an infected insect that is serving as an intermediate host. (Courtesy of CDC.)

a rat's feces. The eggs hatch in the insect's digestive system, releasing a larval stage known as the **oncosphere**. The oncosphere penetrates the insect's gut wall and enters the body cavity, where it develops into a **cysticercoid**. Transmission back to the definitive host occurs when a rat consumes the infected insect. Although human infections are not common, they can occasionally occur if a human inadvertently consumes an infected insect. This is most likely to occur when a flour beetle (genus *Tribolium*) is acting as an intermediate host, as such beetles are not unusual in products such as flour or cereals.

A reliance on trophic transmission suggests that the parasite is relatively passive and that it does little or nothing other than wait to reach its next host. A fascinating aspect of transmission biology, however, is the way in which many trophically transmitted parasites actually manipulate the behavior of their hosts in a manner that increases the likelihood that a specific host will be consumed by the next host in the life cycle. This topic will be examined more fully in Chapters 5 and 6.

Direct Penetration

Some parasites take a more proactive approach to their transmission; they actively seek out their host and once they find it, they bore their way in. An example familiar to many tropical fish hobbyists is *Ichthyophthirius multifiliis*, a parasitic ciliate that causes the condition known as ich or whitespot (**Figure 3.11**). The condition is easily recognized when small white nodules, looking like grains of salt, appear on the fish's body, fins, or gills.

The life cycle of *I. multifiliis* is unusual for a parasitic protozoan in that reproduction does not occur in or on the host but via asexual binary fission in cystlike structures found on the bottom of a lake, pond, or aquarium (see Figure 3.11). Between 50 and a few thousand new trophozoites are produced in each cyst and are subsequently released into the surrounding water. These free-swimming trophozoites are also called **tomites**. Once a tomite contacts

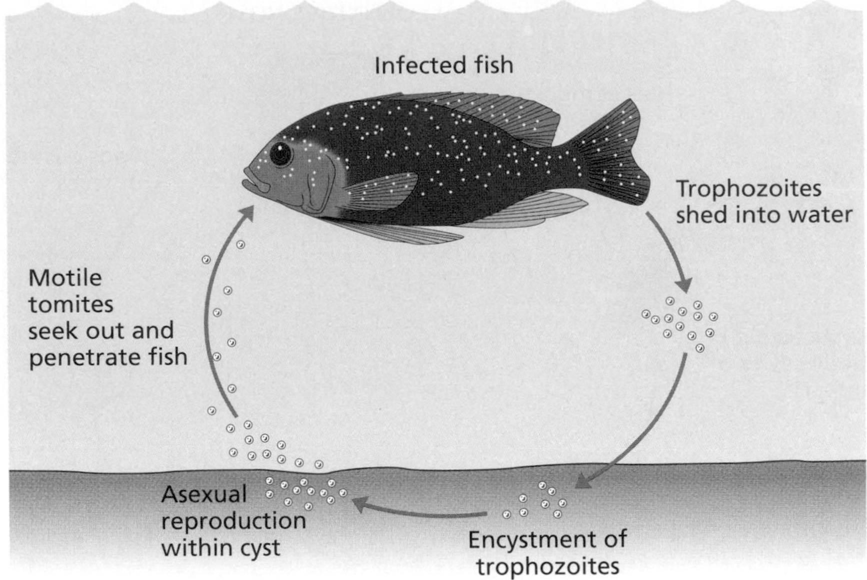

Figure 3.11 The life cycle of *Ichthyophthirus multifiliis*. Trophozoites released into the water form cysts in the bottom sediments, where they undergo extensive asexual reproduction. Free-swimming tomites released back into the water seek out a fish. Upon contact they penetrate the epithelium and transform into trophozoites. The resulting inflammation and tissue damage forms the characteristic white pustule, and the parasite feeds on nutrients released by dying host cells. The pustule eventually ruptures, releasing motile trophozoites back into the water, whereupon they return to the substrate.

a fish, it penetrates the epithelium. The resulting inflammation and tissue necrosis result in the typical white pustule. This parasitic stage lasts only a few days, during which the parasite feeds on nutrients released by dying epithelial cells. When the pustules finally rupture, motile trophozoites are again released back into the aquatic environment, returning to the substrate where they once again encyst.

Unlike those parasites relying on fecal–oral transmission, during which resistant cysts or eggs simply tough it out until they are consumed, for parasites relying on direct penetration, the clock is ticking. Survival of motile and metabolically active penetration stages in the external environment is usually measured in hours. A typical miracidium, for example, as previously seen in the *A. americana* life cycle has only a few hours in which to locate and penetrate an appropriate snail before it succumbs to the rigors of the external environment, is eaten, or depletes its energy reserves (see Figure 3.7). It is not surprising, therefore, that digenetic trematodes and other parasites with metabolically active external stages have evolved strategies to increase their likelihood of finding an appropriate host rapidly. Cercariae, for instance, are unable to locate hosts over distances more than just a few centimeters. To compensate for this deficiency, they often position themselves in that part of the water column where they are most likely to encounter a host. To illustrate, consider the digenetic trematode *Cardiocephalus longicollus*. Like most such parasites, it uses a molluscan first intermediate host. The cercariae that emerge from the marine, benthic-dwelling mollusc must find an appropriate second intermediate host—a fish such as a bream or porgie. Consequently, the cercariae swim from the bottom to an intermediate depth, where the suitable fish are most likely to be found. Once there, they allow themselves to fall back through the water column, repeating this vertical migration until the cercariae either contact a fish or die. The cercariae of other species, which use surface-dwelling fish, not surprisingly, rise to the surface and remain there.

Most cercariae also time their emergence from the molluscan host to coincide with the activity pattern of their subsequent host, a phenomenon known as **periodicity**. An excellent example is provided by digenetic trematodes in the genus *Schistosoma*, which use a freshwater snail as an intermediate host and a mammal as the definitive host. In the case of *S. bovis*, the definitive host is a cow or other ungulate. In *S. rodhaini*, rodents are the definitive host. *Schistosoma bovis* cercariae emerge in the early morning, coinciding with the time during which ungulates are most likely to visit a water source. Rats and other rodents, on the other hand, are nocturnal. *S. rodhaini* cercariae capitalize on this behavior by emerging shortly after dusk.

Vector Transmission

As discussed in both Chapter 1 and in the historical overview, many parasites rely on vectors to achieve transmission. Parasites typically move between vector and host as the vector takes a blood meal. Most vectors are arthropods, and the roster of parasites (eukaryotic, prokaryotic, or viral), which rely on arthropods to achieve transmission is long. The ***Anopheles*** mosquitoes involved in ***Plasmodium*** transmission may be the best-known example. **Mosquitoes** also serve as vectors for the transmission of those nematodes in the genera ***Wuchereria*** and ***Brugia*** that cause filariasis. Other well-known examples include tsetse flies (genus ***Glossina***), which transmit sleeping sickness (caused by kinetoplast protozoans in the ***Trypanosoma brucei*** species complex), cone-nosed or **kissing bugs,** which transmit Chagas' disease (caused by ***Trypanosoma cruzi***), and **sand flies**, which transmit leishmaniasis. Other parasites such as apicomplexans in the genus ***Babesia***, rely on noninsect arthropods, such as **ticks** in the genus ***Ixodes*** or ***Rhipicephalus***.

Not all blood-feeding arthropods can act as vectors. An ingested parasite must not only be capable of survival in the vector's digestive tract, but it must also frequently be able to penetrate the vector's gut wall; reproduce, develop, or both within the vector; migrate to the vector's salivary glands; and achieve transmission when the vector takes its next blood meal. All of these tasks must be accomplished within the milieu of the vector's body and in the face of immune defense designed to impede parasite survival. It is therefore not surprising that various parasites often adapt to the selective pressures imposed by a particular vector, just as they adapt to their other specific definitive or intermediate hosts. The term **vector competence** describes the ability of a particular vector type to become infected with a specific pathogen, to transmit a specific pathogen, or both. Vector competence is assessed in the laboratory by allowing uninfected potential vectors to feed on animals infected with a particular pathogen. These putative agents of transmission are subsequently allowed to feed on uninfected animals. Should the second group of animals become infected, vector competence for that pathogen is demonstrated.

In addition to vector competence, various other factors influence the ability of a given vector to transmit a parasite. Even the most competent vector may be unimportant in disease transmission, for instance, if it is not abundant or if it rarely feeds on a particular host. These extrinsic factors are termed **vector capacity**. Some of the other characteristics that contribute to vector capacity include:

- Regular feeding on the host in question
- Feeding for an extended period of time and taking a relatively large blood meal
- A life span long enough to allow the parasite time to reach its infective stage, to migrate to the salivary glands, and to replicate sufficiently to allow the vector to infect the next host
- Abundance
- Good dispersal ability

Thus, competent vectors differ in terms of their vector capacity. A perfect vector would have all of these characteristics. Vectors with high capacity may serve as **primary vectors**—vectors able to maintain a parasite transmission cycle in the absence of other vectors. Competent vectors with low capacity may act as **secondary vectors** because they may be able to transmit a parasite but unable to maintain the transmission cycle in the absence of the primary vector.

Let us examine just one characteristic of vector capacity. Consider *Anopheles freeborni*, a species that is known to be competent for several ***Plasmodium*** spp. Because it is common in California, *A. freeborni* has long been considered as the historical primary malaria vector in California. However, it has been shown that under typical ambient conditions in California, ***Plasmodium*** requires about 12 days in the arthropod vector before it is infective to mammals. An adult *A. freeborni* rarely lives long enough for this to occur. In fact, should an adult become infected with its first blood meal on day 3 postemergence, the probability of that mosquito living long enough to transmit ***Plasmodium*** is approximately 0.0072. We will return to this topic in Chapter 6 with respect to modeling of parasite abundance. **Box 3.2** describes another example of an arthropod that is less able to transmit pathogens then one might guess.

Many vector-borne parasites also come equipped with various adaptations designed to increase the odds of transmission. For instance, adult ***Wuchereria bancrofti*** females release microfilariae into the lymphatic fluid of the human host. From there, they enter the circulatory system. During the night, when the mosquito vector is most likely to feed, these microfilariae migrate to peripheral blood vessels near the skin, well positioned to be ingested along with a blood meal. The cues the microfilariae use are probably either changes in body temperature or in blood gas composition that occur at night, when the human host is sleeping. Where filarial worms are transmitted by day-biting mosquitoes, this periodicity is reversed. Such periodicity may also be seasonal rather than only daily. The causative agent of canine heartworm *Dirofilaria immitis* also releases microfilariae into the blood of an infected dog. These larval nematodes, infective to the mosquito vector, increasingly localize in surface blood vessels in the spring just as mosquito numbers begin to increase.

Another example is provided by sand flies that when infected with ***Leishmania*** parasites feed more frequently. Likewise, trypanosome-infected tsetse flies seem to become more active and to increase their feeding frequency, increasing the likelihood of transmission. Additionally, it may be no accident that vector-borne parasites such as trypanosomes are often among the most virulent. It has been suggested, for instance, that the extreme lethargy of humans infected with trypanosomes is actually a strategy on the part of the parasite; a comatose victim of sleeping sickness is in no condition to shoo away a tsetse fly in search of a meal. Thus, enhanced virulence may be viewed as an adaptation to increase the likelihood of transmission. In Chapter 7 the relationship between virulence and transmission will be more fully explored.

Essential developmental changes or reproductive events do not occur in all vectors. In **mechanical vector transmission**, the parasite merely adheres to the vector's body or mouthparts and essentially hitches a ride without cost to the vector. ***Trypanosoma evansi***, discussed in Chapter 1 and in **Box 3.3** provides an interesting example.

BOX 3.2
Return of the Bed Bugs

Bed bugs are back in the news. The most common of these small, blood-feeding insects (order Hemiptera) is *Cimex lectularius* (Figure 1). The name bed bug comes from their preferred habitat: dark, enclosed areas where people sleep. Mainly nocturnal, they emerge from cracks in the wall or mattresses to feed unnoticed on their human hosts.

When feeding, a bed bug penetrates the host's skin with two tubelike proboscises. The first proboscis injects saliva containing a mixture of anticoagulant and anesthetic. The other is used to obtain the blood meal, which requires approximately five to ten minutes. After feeding, the insect returns to its hiding place. Humans may never see their tormenters, whose only calling card might be their small, granular, reddish-black feces that are released following engorgement. Severe blistering and itching may occur at the bite site, although in some individuals there are no visible symptoms (Figure 1).

Bed bugs have plagued humans for millennia. In more developed countries, the insects were largely eradicated by the 1950s thanks mainly to the use of effective insecticides. Recently, however, bed bugs have been mounting a comeback. Severe infestations have been reported in many large North American cities (Figure 2). Across Europe as well, bed bugs have become an unpleasant part of life. Exactly why bed bugs are resurging is unclear. Resistance to insecticides, increased travel and a simple neglect of bed bug control have all been suggested as potential factors.

Because they feed repeatedly, it has often been speculated that bed bugs should serve as effective biological vectors. Over the years, various researchers have proposed that bed bugs might transmit a variety of diseases, including filariasis, leishmaniasis, and Chagas' disease. More recently the vector competence of beg bugs to transmit the human immunodeficiency virus (HIV) and the hepatitis B virus (HBV) has been evaluated. In no case has transmission of any infectious organism been demonstrated. Regarding filariasis, for example, when bed bugs were allowed to feed on humans infected with *Wuchereria bancrofti*, although microfilariae were found in the insects, they uniformly died before developing into infective L3 larvae. HIV was detected in bed bugs up to eight days following their experimental engorgement with blood containing high concentrations of virus. Yet the virus failed to replicate and no virus was isolated from insect saliva or feces. Similarly, attempts to transmit HBV from infected chimpanzees to uninfected and susceptible chimpanzees using bed bugs were unsuccessful.

In spite of recent high profile court cases in which attorneys representing plantiffs who were bitten by bed bugs in upscale hotels argued that their clients were at risk for various vector-borne diseases, there is not a single study demonstrating their ability to serve as vectors. Exactly why this is the case remains unclear. Remember that although nobody wishes to "let the bed bugs bite," we can "sleep tight" in the knowledge that at least they're not infecting us with deadly parasites.

References

Goddard J & deShazo R (2009) Bed bugs (*Cimex lectularius*) and clinical consequences of their bites. *JAMA* 301:1359–1366.

Goddard J (2003) Bed bugs bounce back: but do they transmit disease? *Infect Med* 20:473–474.

Reinhardt K & Siva-Jothy MT (2007) Biology of the bed bugs (Cimicidae). *Annu Rev Entomol* 52:351–374.

A

B

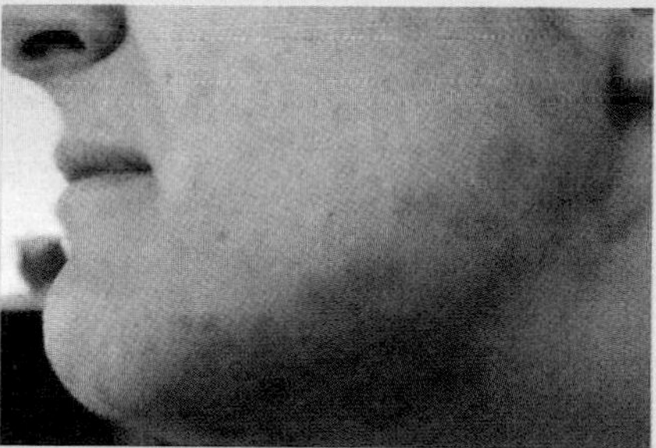

Figure 1 Bed bugs. (A) *Cimex lectularius*, the common bed bug. (B) Blistering on the face, caused by inflammation in response to bed bug feeding. Individuals vary considerably in the severity of these blisters. (A, courtesy of CDC.)

Figure 2 Bed bugs in the Big Apple. New York City, as well as other cities across North America and Europe, have been. plagued by an increase in bed bugs. (Courtesy of Rick Moser.)

Sexual transmission

Eukaryotic parasites may also be transmitted via sexual contact. *Trypanosoma equiperdum* (see Box 3.3) is an example. Another example is provided by ***Tritrichomonas foetus***, a flagellated protozoan parasite that can cause serious disease in cattle. The parasite generally resides in the foreskin of bulls, in which symptoms are generally unapparent. When transferred to the female

BOX 3.3

Trypanosomiasis for the Veterinarian

In any discussion of trypanosomiasis, the big three subspecies of ***Trypanosoma brucei*** (***T.b. brucei***, ***T.b. gambiense***, and ***T.b. rhodesiense***) as well as *T. cruzi*, grab most of our attention—and for good reason. These organisms collectively account for most of the misery that livestock and humans suffer at the hands of trypanosomes. Specifically, ***T.b. gambiense*** and ***T.b. rhodesiense*** cause African sleeping sickness in humans, and ***T.b. brucei*** causes the serious disease **Nagana** in livestock such as cattle and sheep introduced into Africa. Chagas' disease in Latin America is caused by *T. cruzi*.

Three other trypanosome species deserve mention because of their veterinary importance (Table 1). They also illustrate that when it comes to transmission, trypanosomes show considerable diversity. *Trypanosoma equinum*, for instance, causes mal de caderas (hip disease) in horses. As the name implies, infected horses have trouble walking and may eventually be unable to stand. Severe weight loss is another symptom. Mal de caderas is most prevalent in central South America. Unlike the species mentioned above that rely on biological vector transmission, *T. equinum* is principally transmitted mechanically by various flies (see Table 1). Horse flies in the genus *Tabanus*, notoriously vicious biters, are probably the most important mechanical vector, infecting new hosts with contaminated mouthparts.

Trypanosoma equiperdum is of concern to horse owners throughout Africa, Asia, Mexico, and parts of Europe. It causes a disease called dourine in horses, mules, and donkeys. Remarkably, transmission occurs sexually; no arthropod vector has ever been identified (see Table 1). Infected equines ultimately develop paralysis that starts with the neck and face and eventually spreads across the entire body. Unless treated, dourine is generally fatal. It has been successfully eradicated in Western Europe and the United States. All horses imported into the US must be tested for this disease before they can be admitted.

Trypanosoma evansi is an additional parasite of veterinary concern. Although it looks the same as members of the ***T. brucei*** complex, its life cycle differs. Like *T. equinum*, transmission usually occurs mechanically via biting, tabinid flies (see Table 1). Vampire bats have also been implicated as vectors in South America, and it is believed that ***T. evansi*** can also be transmitted vertically via the milk of lactating female mammals.

The disease known as surra in Asia and murrina in Latin America affects a variety of mammals. The name surra comes from the Marathi (a language of western and central India) for the sound of heavy breathing through the nostrils. In addition to labored respiration, symptoms in infected animals include fever, weakness, and lethargy, weight loss, and anemia. The acute blood-borne infection, if untreated, is almost always fatal in horses and other equids, elephants, dogs, cats, deer, and camelids. In certain other ruminants such as cattle and buffalo, the disease is less severe and may enter a chronic state. Suramin is often an effective treatment in horses, camels and dogs. Melarsomine can be used to treat camels.

Table 1 Diversity of transmission in *Trypanosoma*. Several members of the genus *Trypanosoma* (class Kinetoplasta) are important pathogens for both humans and domestic animals. The variety of transmission modes used by these organisms is highlighted here. This list of trypanosomes is incomplete; numerous other trypanosomes, generally of lesser medical or veterinary importance infect mammals, birds, reptiles, amphibians, and fish.

Species	Mode of Transmission	Disease	Affected Mammals
T. brucei brucei	Biological vector: tsetse flies	Nagana	Domestic livestock
T.b. rhodesiense	Biological vector: tsetse flies	Sleeping sickness	Humans, livestock, wildlife
T.b. gambiense	Biological vector: tsetse flies	Sleeping sickness	Humans
T. equinum	Mechanical vector: horse flies	Mal de caderas	Equines
T. equiperdum	Sexual transmission	Dourine	Equines
T. evansi	Mechanical vector:horse flies, vampire bats Vertical transmission: milk of lactating females	Surra (murrina)	Equines, elephants, dogs, cats, deer, camelids

reproductive tract, however, it can cause uterine infection (pyometra), infertility, and spontaneous abortion. In the past, this parasite imposed a large financial burden on the cattle industry. In recent years, the number of cases has declined as a result of the use of artificial insemination. Remarkably, however, it is now considered to be an emerging pathogen in cats, where, unlike cattle, it inhabits the intestinal tract and is excreted in the feces.

An additional example is ***Pthirus pubis***. Commonly called crabs, these organisms are not crustaceans at all, but sucking lice—insects in the suborder Anoplura. Crabs have prominent, grasping appendages reminiscent of crab claws, with which they hold tightly to their host's body hair (**Figure 3.12**). ***P. pubis*** only infects humans. The related *P. gorilla* infects gorillas (see Figure 2.22). Although it can be found in armpits, eyebrows, and other hairy body parts, it is most commonly found in the external genital region, where it feeds on the host's blood. The primary symptom is itching, which is sometimes severe. In some cases the skin takes on a characteristic bluish coloration. Transmission occurs through close contact between individuals, most commonly sexual intercourse. The lice adhere closely to their host and cannot survive for long away from the human body. Following mating the female cements her eggs, called nits, to the host's pubic hair. Infection in children has a possible forensic angle; although certainly not conclusive, it at least raises the suspicion of sexual abuse.

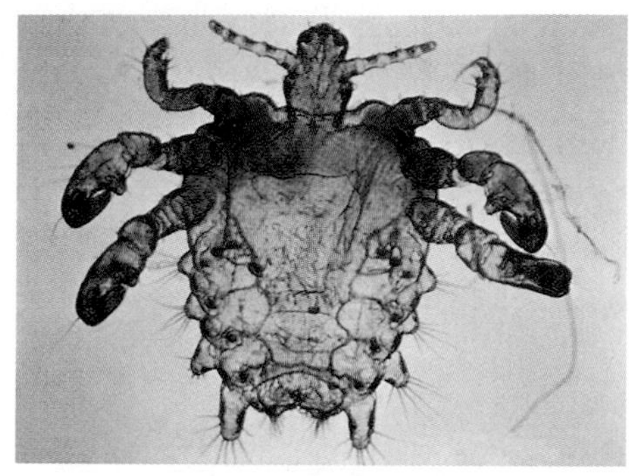

Figure 3.12 *Pthirus pubis*. A micrograph of a male pubic louse. Note the prominent appendages enabling the louse to hold tightly onto its host's pubic hair. *Pthirus pubis* is also known by a number of colorful names including crab louse (owing to its crablike appearance), cooties, and le papillon d'amour. Another human ectoparasitic louse, the head louse *Pediculus humanus capitis*, has a more elongated body. (Photo courtesy of CDC.)

Vertical Transmission

Vertical transmission refers to the transmission of parasites from mother to offspring, either across the placenta or through breast milk in the case of mammals or via infected gametes. *Toxocara canis* for example is a common nematode parasite of dogs. Puppies are frequently infected before birth as a consequence of transplacental transmission. Earlier in this chapter, we mentioned that *Alaria americana* can be transmitted through breast milk. In Box 3.3 we learned that ***Trypanosoma evansi*** may also be transmitted vertically.

Babesia bigemina, the causative agent of redwater fever in cattle, relies on ticks in the genus ***Rhipicephalus*** as vectors. The parasite, however, can be passed from an infected female tick to her eggs, a type of vertical transmission known as **transovarian transmission**. Thus, newly hatched larval ticks are born already infected and can pass on the parasite to cattle when they take their first blood meal, making ***Rhipicephalus*** somewhat of a rarity among arthropod vectors. Typically, most blood-feeding arthropods must feed on at least two different hosts to serve as vectors—first to acquire an infection and then to pass the infection on to a new host. Ticks, depending on the species, use from one to three hosts and can consequently be categorized as one-host, two-host, or three-host ticks (**Figure 3.13**). One-host ticks are generally not involved in parasite transmission. But some ***Rhipicephalus*** spp. are one-host ticks, thereby breaking this rule. The newly hatched and already infected larval ticks transmit the parasite when they feed on their first (and only) host.

High reproductive rates are common in many parasite life cycles

A relatively high investment in reproduction is common among living things for which the likelihood that any single progeny will survive is low. Many plants and sessile aquatic animals, for instance, rely on **broadcast reproduction**. Propagules are released into the environment to be dispersed by wind or water currents. Most fail to survive. High fecundity allows these organisms to compensate for the high mortality incurred by their progeny.

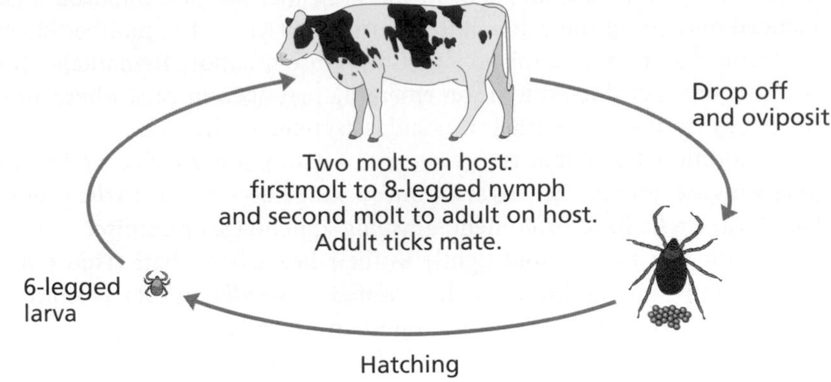

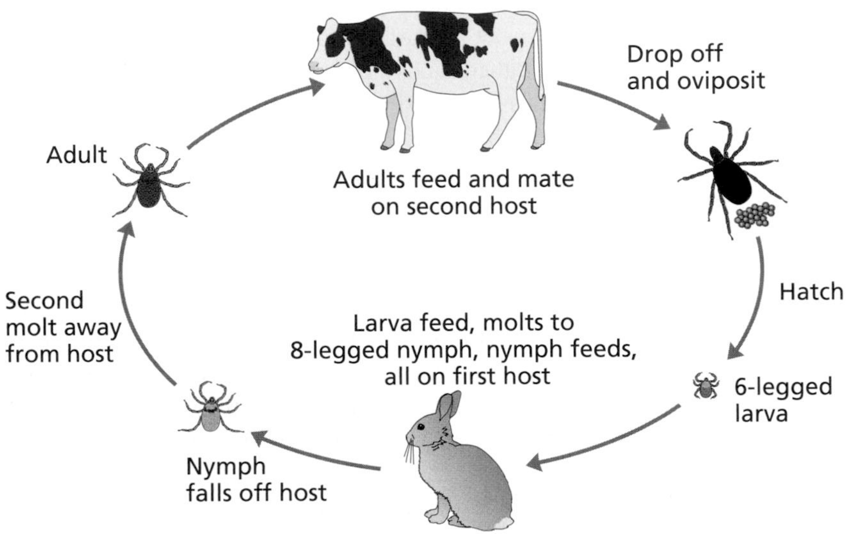

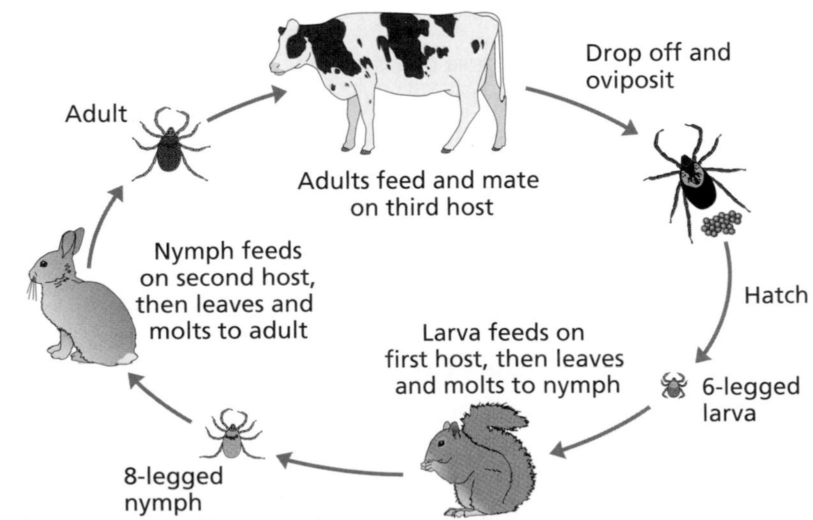

Figure 3.13 Life cycles of hard ticks. Hard ticks (family Ixodidae) can be distinguished based on the number of hosts they use over the course of their life cycle. (A) In one-host ticks, adults, larvae, and nymphs all take a single blood meal on the same host. They are generally not involved in pathogen transmission. *Rhipicephalus (Boophilus) annulatus*, which transmits *Babesia bigemina*, is an exception, owing to the transovarian transmission of the parasite. (B) *Hyalomma marginatum*, an example of a two-host tick, can also transmit *B. bigemina* as well as some viruses. (C) *Dermacentor andersoni*, an example of a three-host tick, is involved in tick paralysis, an unusual disease caused by the tick itself, rather than an infectious organism. A neurotoxin produced in the tick's salivary gland causes the disease.

Many parasites also produce large numbers of relatively small progeny to counterbalance the low probability of success of any single egg or larva. Parasites, like other organisms, must contend with various potentially lethal abiotic and biotic factors. Parasites, however, must cope with the additional problem of transmission. The high host specificity of many parasites further complicates the problem of finding a suitable host environment (see Chapter 6).

The egg-laying capacity of some helminths is truly astonishing. It is estimated that a mature female ***Ascaris lumbricoides*** can produce up to 200,000 eggs a day over the course of her six-to-twelve month life span. Even more impressive are some cestodes. The pork tapeworm, ***Taenia solium***, can produce up to 300,000 eggs daily, whereas ***T. saginata***, the beef tapeworm, produces as many as 700,000 eggs each day. Top honors, however, may go to *Polygonoporus giganticus*, a tapeworm of whales. These behemoths often achieve lengths of close to 40 meters. Each worm is composed of up to 45,000 segments, and we can only speculate about the vast number of eggs released each day. Cestodes have the added advantage over certain other dioecious helminths, such as most nematodes and acanthocephalans, in which sexes are separate. Cestodes, on the other hand, are usually monoecious. Each segment, called a **proglottid**, is equipped with both testes and ovaries. In some species, such as *P. gigantigus*, there are even multiple sets per proglottid. Mating can occur between two proglottids in the same individual or between two different tapeworms. Cestodes do not encounter the problem of finding a mate, a prerequisite for reproduction in dioecious organisms.

Most digenetic trematodes, which like cestodes are in the phylum **Platyhelminthes**, are also monoecious, although most are preferential outcrossers. Compared to cestodes, however, their egg production can appear somewhat paltry. ***Schistosoma haematobium*** adult females (schistosomes, unlike most trematodes, are dioecious), for example, can do no better than one or two hundred eggs daily. The other two principal schistosome species that infect humans, ***S. mansoni*** and ***S. japonicum***, produce about 300 and between 1500 and 3500 eggs, respectively, each day. Certain other trematodes are more fecund. The liver fluke ***Fasciola hepatica*** is capable of producing over 20,000 eggs per day. However, some of these trematodes live for many years, resulting in a very high cumulative egg production. Also, as previously mentioned, they are capable of substantial asexual reproduction in their intermediate hosts. Some cestodes, most notably those in the genus ***Echinococcus***, also reproduce asexually in the intermediate host, but this mode of reproduction is most pronounced in digenetic trematodes. Indeed the name digenean refers to the two generations, both sexual and asexual, that are characteristic of these trematodes.

In digenean trematodes this characteristic asexual reproduction is known as **polyembryony**, the development of a single embryo into numerous others. Following infection of the molluscan host by the miracidium, the offspring (cercariae) eventually dervied from the miracidium are all clonal; except for mutations that may have occurred during cell division they are genetically identical. Consequently, polyembryony differs from other forms of asexual reproduction such as budding or binary fission in which the adult gives rise to progeny that are genetically identical to the adult.

Parasitic protozoans also are capable of impressive feats of reproduction. Asexual reproduction usually via binary fission is a common feature of the parasitic protozoan life cycle. ***Entamoeba histolytica***, for instance, is the primary agent responsible for amebic dysentery in humans. This strictly human parasite has a direct life cycle. Transmission is via the fecal–oral route. The infective stage (the cyst) contains four trophozoites formed by two rounds of cell division. Following ingestion, the cyst passes through the stomach to the lower portion of the small intestine where excystation and trophozoite release occurs. These trophozoites migrate to the large intestine where they feed on cellular debris and bacteria and undergo multiple rounds of binary fission. Under certain conditions, most of which remain to be characterized, trophozoites encyst and are released with the feces.

Kinetoplastids are flagellated protozoans, which include among their ranks some of the most important human and animal parasites. Trypanosomes are responsible for both human and animal trypanosomiasis, whereas ***Leishmania*** spp. cause either cutaneous or visceral leishmaniasis. These parasites have indirect life cycles and use vector transmission. In all of them, there is extensive asexual reproduction via binary fission in both the vertebrate host and the arthropod vector. ***Leishmania*** spp. are transmitted by sand flies in the genera ***Phlebotomus*** (in Africa and Asia) and *Lutzomyia* (in the Americas). In the vector, replication of a flagellated stage known as a promastigote occurs in the sand fly gut. Promastigotes are ultimately transferred to the vertebrate host as the sand fly feeds. Within the vertebrate, promastigotes are ingested by phagocytic cells such as macrophages, where they develop into amastigotes. Amastigotes are also capable of extensive binary fission within the phagocytic cell. As the phagocytic cell releases the newly formed amastigotes, they may be taken up by a different cell, resulting in a new round of replication. Ultimately a sand fly may ingest an infected phagocytic cell as it feeds, completing the cycle.

Both sexual and asexual reproduction are used by apicomplexans such as *Toxoplasma gondii*

The apicomplexan life cycle typically includes both sexual and asexual reproduction (**Figure 3.14**). As an example, we will consider ***Toxoplasma gondii***, the causative agent of toxoplasmosis. Only cats and other felids can serve as the definitive host. The intermediate hosts are rodents, birds, or other small vertebrates on which a cat may prey.

Cats acquire an infection when they consume a prey animal infected with tissue cysts (**Figure 3.15**). Each cyst contains up to a thousand or more **bradyzoites**. This slowly growing, encysted life-cycle stage (brady, from the Greek for slow) is responsible for chronic toxoplasmosis. Bradyzoites are

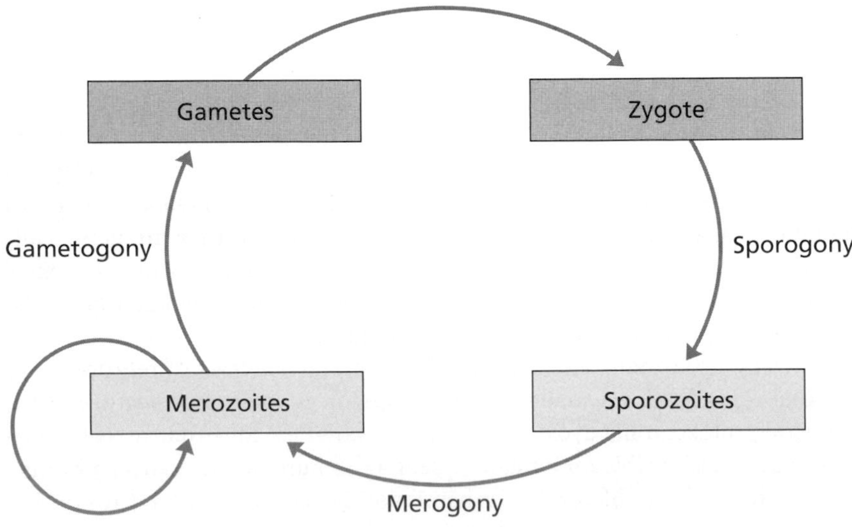

Figure 3.14 Generalized life cycle of apicomplexan parasites. Apicomplexans have a complex life cycle that alternates between sexual reproduction, during which gametes are formed, and asexual reproduction (sporogony and merogony). Merozoites and sporozoites are invasive forms resulting from merogony and sporogony, respectively. (From Wiser MF [2011] Protozoa and Human Disease 1ed. Garland Science.)

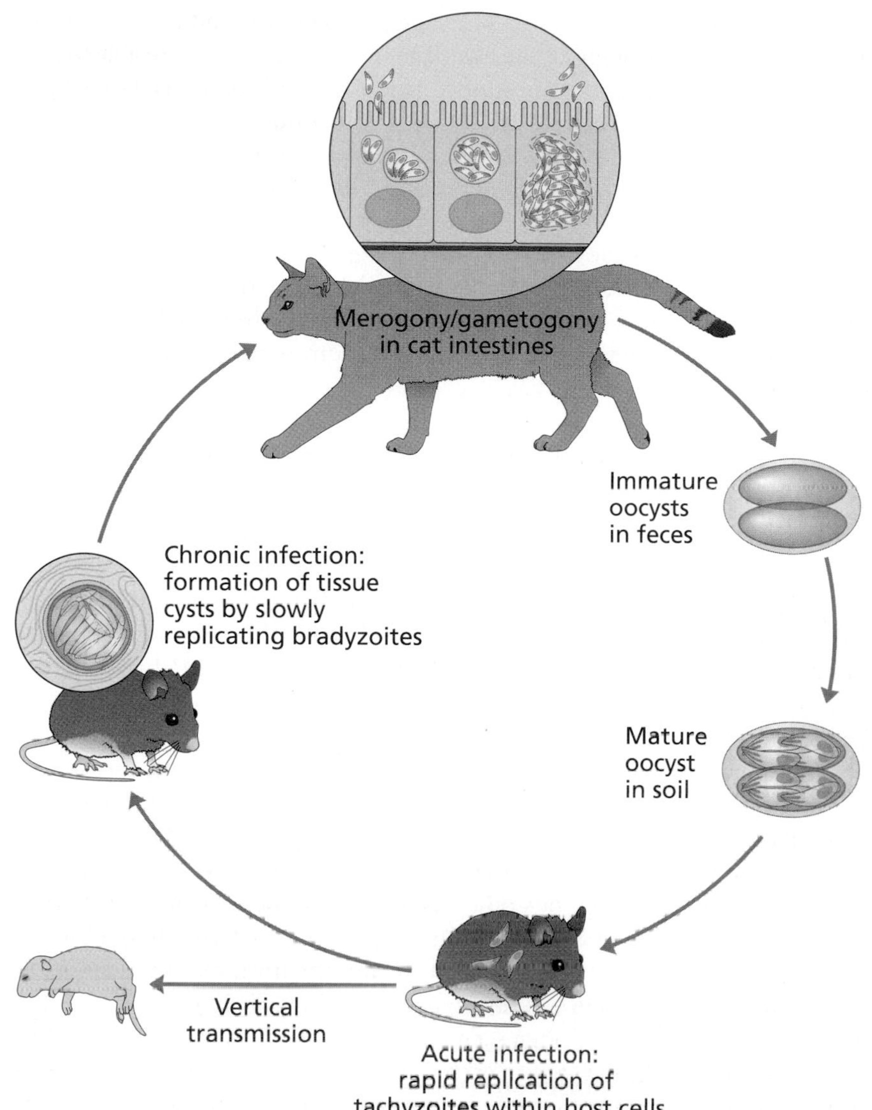

Figure 3.15 Life cycle of *Toxoplasma gondii*. *Toxoplasma gondii* uses both trophic and fecal–oral transmission. Cats, the definitive host, become infected by consuming a small animal harboring tissue cysts that serves as an intermediate host. Following an intestinal infection, first asexual reproduction (merogony) and then sexual reproduction (gametogony) take place. Fertilization following gametogony gives rise to oocysts, which are released in the feces. Oocysts become infective following sporogony. Ingestion of infective oocysts by an animal results in an initial acute infection, characterized by the production of intracellular tachyzoites and followed by the formation of tissue cysts by bradyzoites. Vertical transmission from an infected female intermediate host to her offspring can also occur. See the Rogues Gallery for additional details of the *T. gondii* life cycle. (From Wiser MF [2011] Protozoa and Human Disease 1ed. Garland Science.)

released in the cat's intestine, whereupon they invade intestinal epithelial cells. Here they undergo a form of asexual reproduction called **merogony**. Unlike binary fission, merogony is a type of multiple fission in which the nucleus and other organelles of the bradyzoite divide several times before cell division, resulting in the simultaneous production of many **merozoites**. The merozoites are released from infected cells, whereupon they invade new cells. They may then commence a new cycle of merogony or they may enter **gametogony**, the sexual stage of the life cycle during which gametes are produced. As in many other apicomplexans, two types of gametes, the smaller **microgamete** and the larger **macrogamete**, are produced. After they are released from epithelial cells, microgametes enter a different host cell containing a macrogamete. Fertilization occurs and the newly formed zygote gives rise to the cystlike **oocyst**, which is passed with the cat's feces. A second round of asexual multiple fission known as **sporogony** takes place within the oocyst. After a few days, the now mature oocyst contains two sporocysts, each containing four **sporozoites**, which are the infective stage for the intermediate host.

Once a rodent or bird ingests a mature oocyst, the released sporozoites invade intestinal epithelial cells where they undergo additional merogony.

The newly formed merozoites then invade macrophages and dendritic cells, which carry them throughout the host. The invading parasites are able to survive inside the host phagocytic cells because as they invade the cell they form a protective membranous structure called a **parasitophorous vacuole**. While inside the parasitophorous vacuole, a rapid internal budding process gives rise to **tachyzoites** (tachy, from the Greek for fast), so named because of their rapid rate of reproduction. The macrophage eventually bursts, releasing the tachyzoites, which invade new cells and initiate new rounds of asexual reproduction. Eventually the rate of replication slows, giving rise to the slowly replicating bradyzoites. Cells infected with bradyzoites become encapsulated, giving rise to the tissue cysts, which are most commonly found in the brain and muscle tissue.

Parasites may use strategies other than high fecundity to achieve transmission

Returning to the theme of high fecundity and parasitism, a caveat must be added: the relationship between parasitism and high fecundity is not always as straightforward as one might assume. Like all living things, parasites must balance many competing metabolic demands, and the observed fecundity for any parasitic species represents a trade-off with other life history characteristics. For example, it has been found that parasitic copepods that use invertebrates as hosts produce far fewer eggs than those that parasitize fish. This difference is thought to reflect the relative ease of locating and attaching to a sedentary, benthic invertebrate as opposed to a highly mobile and active fish. Those relatively few eggs produced by species using invertebrates as hosts tend to be larger, reflecting the higher investment in each individual reproductive event. Numerous other factors might likewise act to influence the optimal fecundity of a parasite. These factors include but are not limited to adult size, life span of both the parasite and the host, and the host range of the parasite (**Figure 3.16**). In those species that engage in both asexual and sexual reproduction, there often seems to be a compromise of sorts regarding the amount of energy devoted to each of these two processes. Some cestodes in the genus ***Taenia***, for example, reproduce asexually in the intermediate host, whereas others do not. In those that do, adult worms tend to be smaller. They also have shorter life spans relative to exclusively sexually reproducing species. Likewise, in mammalian schistosomes, some of the species with the highest rate of egg production in the definitive host produce the lowest number of cercariae per miracidium in the intermediate host.

Many factors can complicate an understanding of parasite transmission

Obviously transmission is a complex affair. Various factors, including those discussed in the previous section, influence the metabolic commitment that a specific parasite makes to reproduction and the likelihood of successful transmission to the next host. And that's just the beginning. Many other variables further complicate transmission and make simple statements about its efficiency problematic.

Co-infection with other parasites is one such variable. Mice infected with ***Plasmodium***, for instance, have more gametocytes in their blood when they are co-infected with digenetic trematodes, compared to when these trematodes are absent. Mosquitoes that feed on such mice ingest more gametocytes. These mosquitoes consequently produce more sporozoites that are subsequently passed on to the rodent host in greater numbers. Other parasites

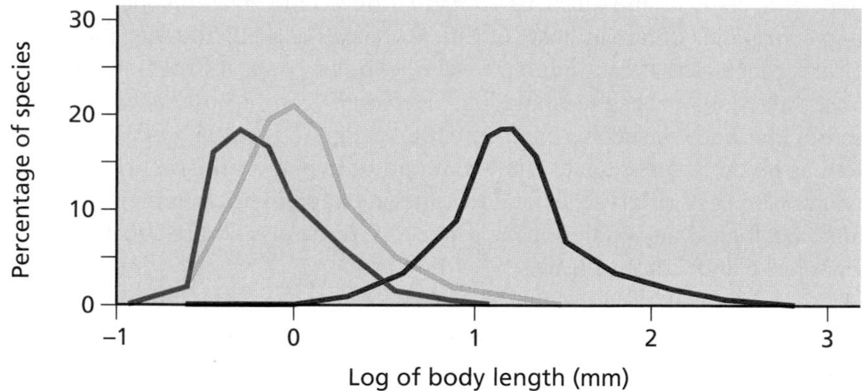

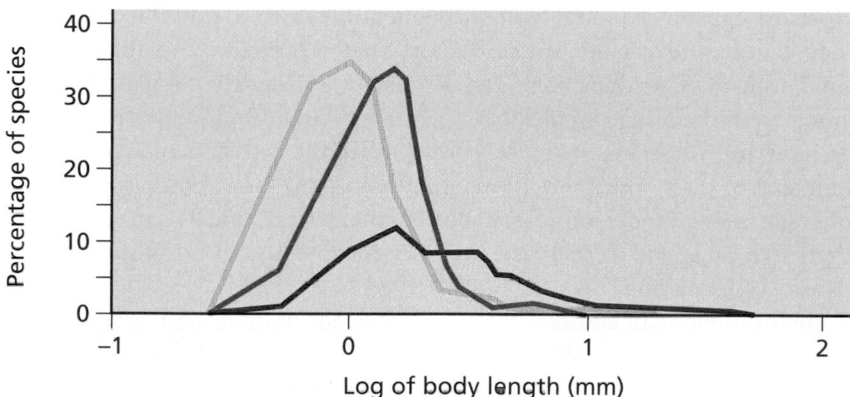

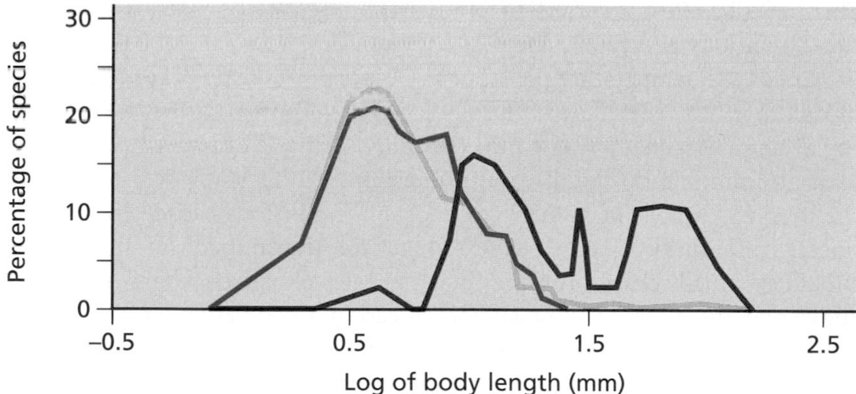

Figure 3.16 Frequency distribution of body sizes. Body-size distributions of free-living organisms are shown in yellow. Those parasitic on invertebrates are shown in blue, and those parasitizing vertebrates are shown in red. (A) Nematodes. (B) Copepods. (C) Isopods. Although there is much overlap, parasites of vertebrates tend to be the largest in all three taxa. Invertebrates, which are generally smaller than vertebrates, harbor somewhat smaller parasites. Free-living organisms are generally not larger than their parasitic relatives. (Adapted from Poulin R [1995] *Parasitol Today* 11:342–345. With permission from Elsevier.)

that use more than one mode of transmission might vary their transmission strategy based on specific environmental variables. The host's sex, for instance influences transmission for the gregarine *Ascogregarina culicus*, which infects ***Aedes aegypti*** mosquitoes. The infective stage of the parasite, the oocyst, infects mosquito larvae. If a specific larva is female, the oocysts are retained during pupation, giving rise to an infected adult female. Oocysts are released either as the female oviposits in an aquatic habitat or when the infected mosquito dies at a breeding site. Male mosquitoes, of course, do not oviposit, and they are far less likely to visit and die at aquatic breeding sites. Infected male larvae generally die when they pupate, releasing oocysts as a consequence. Not only does this example indicate that the parasite detects the sex of its host and alters its transmission accordingly but that the parasite's virulence differs in hosts of different sexes as well. There is a long list of other factors that can influence the likelihood of successful transmission once infectious propagules have been released from a host. These include abiotic parameters such as environmental temperature, pH, and salinity, as well as biotic factors such as the likelihood of hyperparasitism, of predation on parasite eggs or larvae, or of encountering and entering an inappropriate host. Ecological and evolutionary aspects of transmission are considered in Chapters 6 and 7, respectively.

Mathematical models provide a useful tool to predict transmission rates

Because transmission can be so complex, the use of mathematical models in parasite transmission studies has great appeal. Models provide a tool with which to explain complex biological phenomena in a simplified manner. They can also reveal the critical factors that are required to fully understand transmission dynamics. The simplicity of models compared to the almost overwhelming complexity of genuine biological processes is both the strength and weakness. It is a tall order indeed to reduce a biological phenomenon to a set of mathematical equations. Yet if the various assumptions inherent in any model are reasonable, mathematical models can have great predictive value and as such can provide valuable insight into the nature of parasite transmission.

For example, climatic conditions, specifically rainfall and temperature, affect both the survival and development of ***Anopheles*** mosquitoes and the ***Plasmodium*** species that they transmit. A model incorporating rainfall and temperature patterns in sub-Saharan Africa can predict how these climatic parameters affect the ability of mosquito vectors to transmit ***P. falciparum***. This has allowed researchers to predict the intensity of malaria transmission in various geographic locations (**Figure 3.17**). Specifically, the impact of rainfall and temperature on $\mathbf{R_0}$, or the **basic reproductive rate** of infected mosquitoes, was considered. The R_0 value corresponds to the mean number of new cases that a single infected individual will cause in a population with no immunity to the disease in question. In this study, R_0 represented the average number of human ***P. falciparum*** infections caused by a single infected mosquito. When R_0 is less than one, transmission of any disease should eventually cease, provided infection rates remain constant. As R_0 rises above one, the disease will be able to spread in a population, and the larger the R_0 value becomes, the more intense the transmission. The researchers found that the ability of ***Anopheles*** mosquitoes to transmit malaria (that is, their vector capacity) could be predicted based on rainfall and temperature. Vector capacity was then used to develop a model that could predict R_0, and consequently the likelihood of malaria transmission in any geographic location. It was suggested that such a model could be used to determine how transmission of malaria and other diseases might change in response to global climatic change. We will return to the topic of climate change in Chapters 8 and 10.

Mathematical models have repeatedly demonstrated their value in predicting the effectiveness of various intervention strategies to disrupt transmission. ***Schistosoma japonicum***, for instance, is unusual in that, unlike other important human schistosomes, this species uses a wide variety of mammals as definitive hosts. By assessing the number of eggs released in

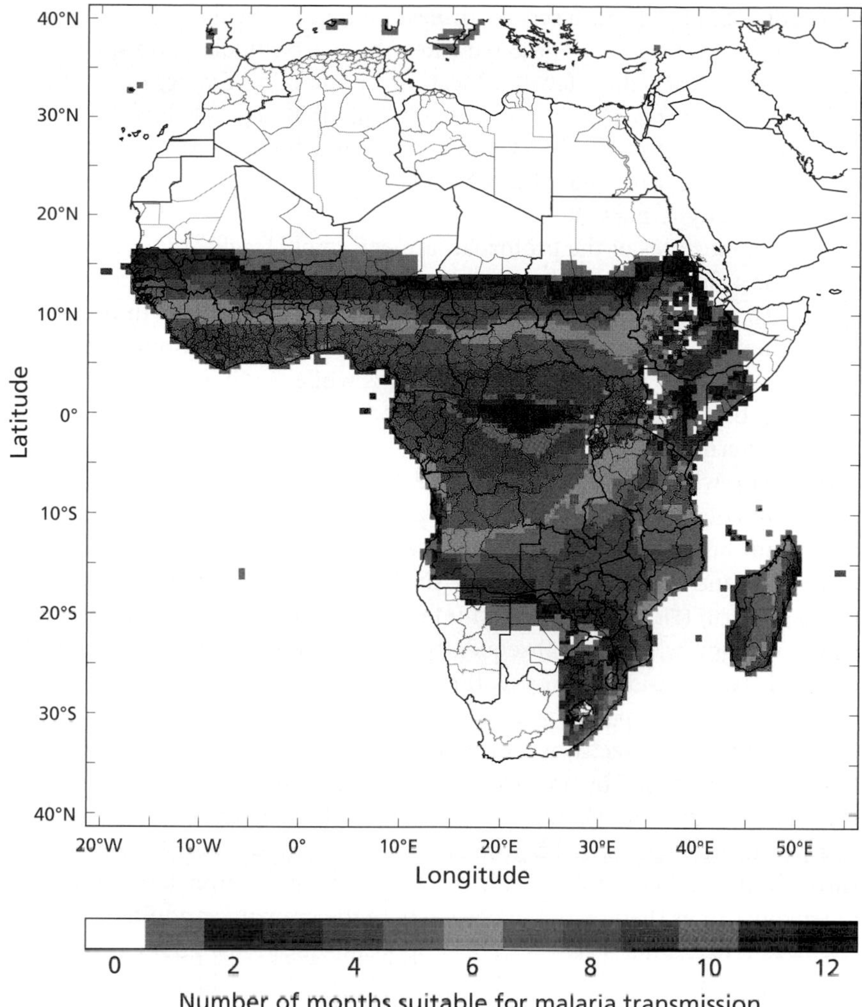

Figure 3.17 Malaria transmission based on climatic conditions. The map shows the number of months each year that are suitable for *Plasmodium falciparum* transmission, based on a model that incorporates rainfall, mean humidity, and mean temperature. Suitability is defined as annual precipitation greater than 80 mm, mean temperature between 18°C and 32°C, and relative humidity greater than 60 percent. Such conditions are conducive for the survival, development, and reproduction of both the parasite and the mosquito vector. (From Grover-Kopec EK, Blumenthal MB, Ceccato P et al [2006] *Malar J* 5:38.)

the feces of various mammalian hosts, including humans, researchers developed a transmission model that assessed the roles of different mammalian hosts in ***S. japonicum*** transmission in the Philippines. The model predicted that larger animals such as water buffaloes and dogs played a relatively slight role in transmission to humans. Smaller mammals, including rats, were of much greater importance. This finding suggests that control efforts aimed at reducing infection in water buffaloes or dogs will have little impact on overall infection rates in humans. Rat control, however, could pay large dividends in reducing human disease prevalence and intensity. We will revisit the topic of modeling parasite transmission in Chapter 6.

Many parasites must migrate to specific sites or tissues within the host

Reaching and entering an appropriate host is just the beginning. Once they have gained entry, many parasites make elaborate and often bewildering migrations within the host in an effort to arrive at a particular location. Such **site specificity** is a common feature in many parasitic infections. ***Leishmania*** spp. are found only in macrophages within their vertebrate host, whereas ***Plasmodium*** is restricted to either hepatocytes or erythrocytes in mammals, depending on the developmental stage. Poultry raised in backyards are

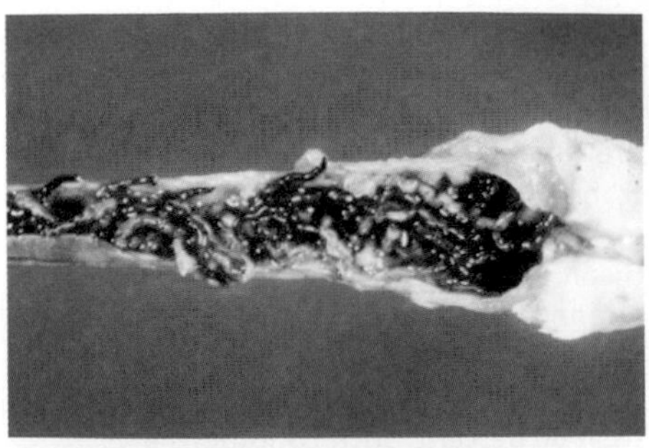

Figure 3.18 *Syngamus trachea*, the gapeworm. *S. trachea* adults in the trachea of a ring-necked pheasant that died of gapeworm. To prevent gapeworm infection in domestic birds, it is a good idea to keep them from feeding on newly plowed fields where earthworms (the paratenic hosts) are easily available. Earthworm populations can also be reduced with soil treatments containing compounds such as ethylene dibromide. Infected birds can be successfully treated with a variety of anthelmintics, such as ivermectin. (© 2012 Jakowski RM. From Veterinary Gross Pathology Image Collection. http://ocw.tufts.edu/Content/72/imagegallery/1362314/1368464/1373596. With permission from author and publisher.)

susceptible to blackhead, a serious disease caused by the flagellate protozoan ***Histomonas meleagridis***. These parasites show a particular tropism for the cecum. They reach their favored location by infecting the eggs of a different parasite, *Heterakis gallinarum*, a nematode that also inhabits the cecum of ground-feeding poultry, such as chickens and turkeys. As first described in Chapter 1, as a parasite of a parasite, ***H. meleagridis*** is consequently an example of a **hyperparasite**.

The trophozoites of the protozoan are extremely fragile and only remain viable for a few hours if released into the environment. After infecting the eggs of *H. gallinarum*, however, they are passed with the feces into the soil, where they can remain infective for up to two years. Birds become infected as they inadvertently ingest the nematode eggs while feeding. Once the nematode eggs hatch, the *H. gallinarum* larvae release ***H. meleagridis*** in the bird's cecum. Alternatively, as noted in Chapter 1, the eggs may be consumed by an earthworm, which serves as a paratenic host that might eventually be consumed by a bird. A related nematode of poultry, *Syngamus trachea*, infects not the cecum but the trachea. Here they cause respiratory symptoms resulting in gasping and coughing, giving rise to the common name for the parasite: the gapeworm (**Figure 3.18**). As an infected bird struggles to breathe, it may cough up eggs, which are subsequently swallowed. Larval worms within the egg molt twice inside the egg as they transit through the digestive system. Eggs are ultimately passed with the feces. As with *H. gallinarum*, eggs can either be ingested directly by a bird or they may reach a bird indirectly, after first being consumed by an earthworm. Once the bird consumes the eggs, they hatch in the gut, whereupon the infective larvae penetrate the gut wall. The larvae are then carried by the blood to the lungs, where they penetrate through the capillaries into alveoli. A final climb up the bronchiole and bronchi finally brings them to the trachea, where they remain as adults.

These few examples raise several important questions. First and perhaps easiest to answer is, why are many parasites so site specific? From the parasite's perspective, any host offers hundreds or even thousands of habitats that may be colonized. Each of these potential habitats differs in pH, nutrient availability, oxygen concentration, and other factors. All parasites must cope with the host immune response, and the manner in which a host responds differs in different sites within the body. Considering all the variables that confront any parasite in the host environment, site specificity becomes easily understandable in terms of adaptation to a site-specific range of conditions. It is an unusual parasite indeed that can successfully survive and reproduce anywhere within the host environment. Complicating such adaptation to specific conditions are the other parasites with which any given species must contend. Although it has proven difficult to demonstrate that competition within a host takes place (to be discussed more fully in Chapter 6), it seems likely that at least in heavy infections, interspecific competition further enforces site selectivity, resulting in parasite habitat specialists able to outcompete other species and successfully reproduce in their preferred location.

This is not to say that all parasites are restricted to highly specific niches within the host. The tissue cyst stages of tapeworms in the genera ***Taenia*** and ***Echinococcus***, for instance, may be found in various tissues or organs. Even then, however, such cysts tend to be in a highly predictable, relatively short list of potential sites within the intermediate host.

A second question is, how do parasites manage to find their preferred location? The ability to home in on such a site may seem less problematic for ectoparasites, many of which take site specificity to something of an extreme. Certain lice on birds, for example, only attach to feathers with barbs of a specific diameter (**Figure 3.19**). The monogenean trematodes that parasitize fish in some cases limit themselves to a particular gill filament. In such cases, specificity is often a consequence of morphological adaptations for attachment, which permit strong adherence only at particular sites.

Internal parasites have a more complex task. Navigation within an environment that is dark, churning with digested food or flowing blood, and full of molecules such as antibodies designed to impede progress is far from trivial. The host environment, however, although complicated, is predictable. Because of anatomic similarity among members of the same host species and because of homeostatic mechanisms that maintain relatively constant chemical conditions in specific regions of the body, parasites are presented with a predictable set of navigational cues. Over evolutionary time, parasites have evolved to recognize such cues and alter their migration within the host accordingly.

It has been shown that those parasites undergoing complex migrations within their hosts actually key in on a very small number of the environmental stimuli to which they are exposed. That is, such parasites live in a relatively limited sensory world and respond to only a few of the environmental signals that they encounter. Furthermore, unlike many animals that navigate by following a concentration gradient of airborne or waterborne chemicals, parasites are far less likely to rely on chemical gradients. In a tightly constrained environment, such as an intestine or blood vessel, such gradients cannot be established because fluid within the space is continually bounced back and forth as it encounters anatomical barriers. Consequently, animal parasites exhibit genetically programmed stereotypic behaviors when they encounter a particular threshold level of a given stimulus. Such behaviors help ensure that they respond in an appropriate manner.

Figure 3.19 Site specificity in feather lice. *Columbicola columbae* are lice specializing on rock pigeons. Like other feather lice, they are usually closely adapted to their avian host, with appendages that match the diameter of the individual feather barbs. By specifically attaching to barbs of a particular diameter, the lice become much less vulnerable to being removed by the bird when it preens its feathers with its beak. (Courtesy of Jack Scott, University of Alberta.)

As an example, consider the liver fluke ***Fasciola hepatica***, and its movement from the intestine to the liver. When a mammal consumes the infective metacercaria, contact with low stomach pH initiates excystment. Once the partially excysted larva reaches the duodenum, it responds to the presence of bile with violent spasms that release it from the cyst. Only bile causes such a response. These same spasms in tandem with secreted lytic enzymes allow the fluke to penetrate the intestinal lining into the peritoneal cavity. Within the cavity, the nonfeeding larva adheres via its anterior and posterior suckers to the endothelium lining the cavity. It subsequently begins to move along the peritoneal endothelium until it eventually reaches the liver. Interestingly, it does not necessarily navigate directly toward the liver and no sophisticated orientation is required. Because the peritoneal cavity is like the inside of a sphere, movement in any direction will eventually bring the fluke into contact with the liver, which sits in the anterior portion of the cavity, directly beneath the diaphragm. Once it finds the liver parenchyma, it begins to feed upon liver tissue.

Less motile protozoan parasites often find their target tissue as a consequence of an affinity between certain molecules on the parasite surface and other molecules found on the membrane of the target tissue. We saw earlier in this chapter, for example, that the initial round of replication by ***Toxoplasma gondii*** in gut epithelial cells of the intermediate host is followed by invasion of phagocytic cells. These phagocytic cells leave the intestine and effectively disseminate the parasite throughout the host's body. Like other apicomplexans, ***T. gondii*** has adhesion proteins associated with its plasma membrane. The attachment between a ***T. gondii*** merozoite and a phagocytic cell occurs when the parasite adhesion protein, called mic2, binds to ICAM-1 on the surface of the phagocytic cell. The parasite then penetrates the phagocytic cell using **gliding motility**, a mode of locomotion unique to the apicomplexans. This type of movement occurs only on substrates and in the absence of locomotory organelles and involves the interaction of the parasite surface proteins with actinomyosin found beneath the parasite cell membrane in the cytoskeleton. In effect, the myosin motor proteins moving along actin filaments pull the apicomplexan along the substrate. In this case, such movement is involved in actual host-cell penetration.

Phagocytic cells are able to cross epithelia by squeezing between epithelial cells in a process called **diapedisis**. ***T. gondii***, while in its host's cells, thus exploits normal cell-trafficking pathways in the host to cross cellular barriers, including the placenta and the blood–brain barrier. A similar process involving a close association between host and parasite surface proteins allows ***Plasmodium*** merozoites to enter erythrocytes.

The evolution of complex migration within a host is not always clear

Another question relates to the evolution of the complex migration observed in some parasites. The tissue migration of many intestinal nematodes with direct life cycles, for example, appears to defy logic. ***Ascaris suum*** is illustrative (**Figure 3.20**). This species is a common intestinal roundworm of swine and is similar to the human parasite ***A. lumbricoides***. As in all nematodes, there are characteristic life-cycle stages. Fertile eggs are produced by the adult female. A first stage larva, called an L1 larva, develops within the egg. The larva undergoes four molts that are followed by periods of growth giving rise to L2, L3, and L4 larvae and ultimately to an adult. The larval stage that emerges from the egg may be the L1, L2, or L3 larva, depending on species. For many nematode species, L3 is the infective stage for the definitive host. Alternatively, for nematodes in the genus ***Ascaris***, L2 larvae are infective. Note that some researchers prefer to call immature nematodes juveniles instead of larvae and thus refer to the L1 to L4 stages as J1 to J4.

Following the ingestion of eggs containing L2 larvae, the larvae emerge in the intestine and embark on a seemingly byzantine trek that starts when they penetrate the intestinal epithelium. They then enter the circulatory system,

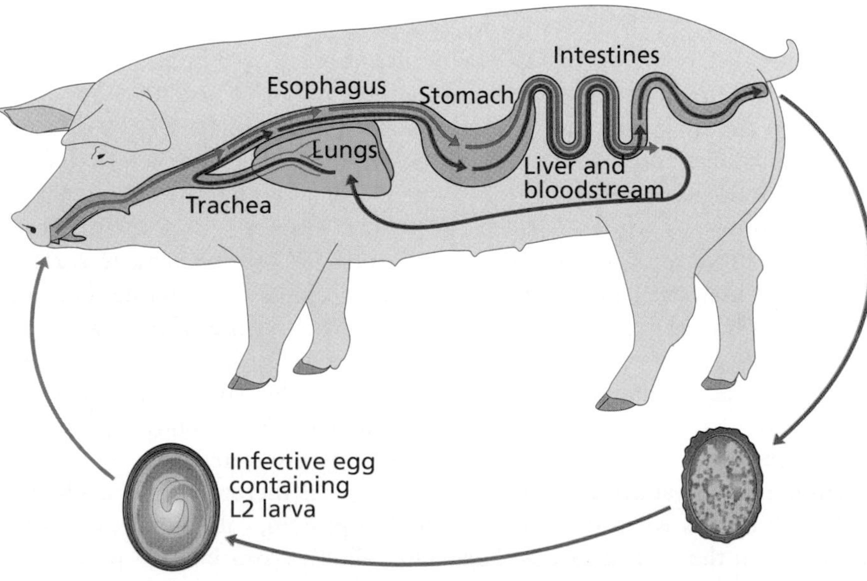

Figure 3.20 Internal migration of *Ascaris suum* within the pig. *Ascaris suum* a complex migration within its host. Infection occurs when the pig consumes infective eggs in which the larvae have developed into infective second stage (L2) larvae. The eggs hatch in the intestine, whereupon the L2 larvae penetrate the intestinal epithelium. After entering the circulatory system, they pass through the liver, eventually reaching the lungs. After penetrating through the capillaries into the alveoli, larvae first molt into L3 and subsequently into L4 larvae. These L4 larvae migrate up the trachea. They then return to the intestine via the esophagus. Having returned to the intestine, L4 larvae undergo a final molt into the adult form. Adults remain in the intestine, where reproduction occurs.

where they are transported to the lungs. Arriving in the lungs, they break into the alveoli, migrate up the trachea, and return to the intestine via the esophagus, when they are swallowed. Why do the larvae leave the intestine, if their only goal is to return there? Why do emerging larvae not simply remain in the intestine, maturing to adulthood, without migrating through the tissues?

One possibility is that ***A. suum*** evolved from a species that had an indirect life cycle and tissue migration in the ancestral intermediate host was required for proper larval development. According to this hypothesis, the developmental requirement for tissue migration was retained as ***A. suum*** evolved a direct life cycle. Alternatively, it has been observed that nematodes such as ***A. suum*** that undergo extensive tissue migration grow faster and are larger than nonmigrating nematodes. Because larger nematodes are more fecund, the maintenance of tissue migration over evolutionary time may be a selectively advantageous life-history strategy.

Finally, before we leave the topic of parasite migration, it is worth noting that such migration is often unsuccessful; migrating stages often become lost and many are killed as a consequence of host immune responses. Aberrant migration is particularly common when infective stages enter an inappropriate host, in which the anatomy, the necessary navigational cues, or both are sufficiently different from those found in the normal host. A well-known example involves the infection of humans with hookworms that normally infect dogs or other animals. When encountering their normal host, hookworm larvae penetrate through the skin and subsequently migrate through the circulatory system and lungs, eventually reaching the intestine, where they mature into adults. Consider, however, *Ancylostoma braziliense*, which typically infects dogs. When infective *A. braziliense* larvae contact human skin, they may still penetrate the outermost epithelial layer. They cannot, however, move through the deeper basal layer. At this point they may simply wander aimlessly through the skin, leaving a trail of inflammation in their wake. The inflamed, itchy skin condition is called **cutaneous larval migrans** (**Figure 3.21**).

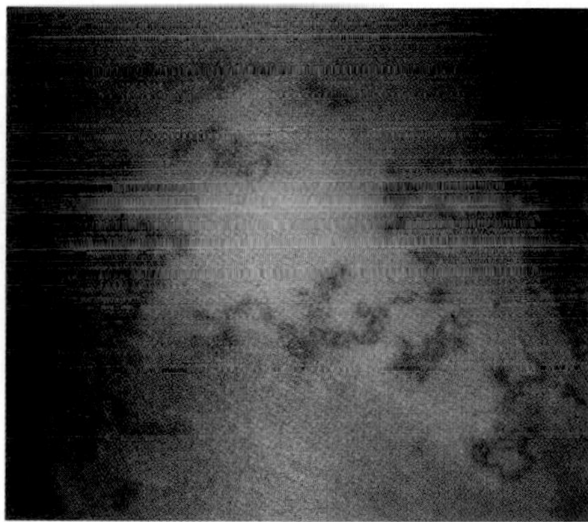

Figure 3.21 Cutaneous larval migrans. When infective L3 larvae of canine or feline hookworms such as *Ancylostoma braziliense* penetrate human skin, they are unable to reach the circulatory system. As the larvae migrate aimlessly, they provoke a telltale trail of inflammation. Older homes often have crawl spaces underneath them, where pets may often defecate. Because plumbers often had to enter crawl spaces to work on piping, they were frequently at elevated risk for this condition, which in consequence is sometimes referred to as plumber's itch. An alternative common name is creeping eruption.

Parasites are adapted to maintain their position on or within the host

The host environment is not particularly conducive to a stationary existence. Parasites inhabiting the digestive tract must contend with powerful peristalsis. Those found in the circulatory system must cope with the surge of rushing blood. Ectoparasites are confronted with environmental factors such as wind or rain or host behaviors such as grooming that make remaining in place more problematic.

We have already mentioned some of the strategies used by various parasites to overcome this obstacle. Gut parasites may invade the intestinal epithelium, as we saw in the case of ***Toxoplasma gondii***, or like ***Giardia lamblia***, they may adhere with adhesive disks (see Figure 3.9). Acanthocephalans and tapeworms rely on their proboscis or scolex, respectively, whereas many intestinal nematodes utilize powerful longitudinal muscles to maintain their location via vigorous sinusoidal motion. Trematodes generally come equipped with powerful ventral suckers or **acetabula** that assist in adherence. Male and female schistosomes form pairs with the female lying within the gynecophoral canal of the male. A male with it's more powerful musculature and with the aid of it's oral sucker and acetabulum moves the pair upstream into smaller venules where the pair becomes lodged. Some ***Plasmodium*** species, on the other hand, once they have adhered to and entered erythrocytes, enjoy a moveable feast of sorts, circulating through the host in the safety of their cellular shuttle. ***P. falciparum***, on the other hand, the most pathogenic of the human malaria parasites, takes a different tack. Once inside erythrocytes, this species encodes a protein called *Plasmodium falciparum* erythrocyte membrane protein 1 (PfEMP1) that protrudes from the surface of infected cells. This protein is involved in the formation of knobs, which bind the infected cells to capillary walls and consequently interfere with blood flow. Such interference is an important factor in the severe pathology caused by ***P. falciparum***.

Ectoparasites also employ a wide range of adaptations to maintain their position in a specific site. Of particular interest are the parasitic copepods, some of which appear to have stretched morphological plasticity to the limit in terms of their specific adaptations for adherence. For instance in Chapter 1 we discussed *Ommatokoita elongata*, the females of which imbed themselves permanently in the cornea of Greenland and Pacific sleeper sharks (see Figure 1.7). Another noteworthy parasitic copepod is *Pennella balaenopterae*, the colossus of the copepod world. Over 30 cm in length, these copepods along with other members of the family Penellidae have the distinction of being the only copepods to parasitize whales. Like other members of its family, *P. balaenopterae* has lost all traces of external segmentation, and its locomotory appendages are largely vestigial in adults. Females of this species, however, more than compensate for these anatomical losses with the development of two prominent anterior anchoring appendages. Each of the two processes expands and bifurcates, allowing the female to anchor herself permanently into the whale's blubber.

Finding a mate is a requirement for many sexually reproducing parasites

The site specificity we have been discussing, as well as the adaptations that permit it, serve another purpose; for sexually reproducing parasites it facilitates finding a mate. This, of course, is not a problem for asexual or nonreproducing life-cycle stages. Monoecious species are often preferential outcrossers, but for those that readily self-fertilize, finding a mate is of reduced importance. However, for dioecious sexually reproducing parasites, finding a member of the opposite sex is less problematic when both sexes are homing in on the same target tissue. In this sense, parasites conform to the **Allee effect**. This population biology concept notes that in small populations, individual fitness will increase as population density increases. It has even been proposed that for dioecious parasites with indirect life cycles, trophic transmission evolved as an efficient way for parasites to meet a sexual partner, as it serves to concentrate parasites at a single location.

Once they have arrived at their site of choice, evidence suggests that many parasites rely on chemoattractants to bring them into physical contact. In echinostomes, for example, lipophilic chemoattractant factors have been characterized. In ***Schistosoma mansoni***, surface receptors containing *N*-acetyl-D-galactosamine mediate chemoreception in females. Curiously, the surface molecules characteristic of human blood type A also contain *N*-acetyl-D-galactosamine. Anti-A antibodies, circulating in the blood of individuals with type B or O blood may cross-react with receptors on female schistosomes, interfering with chemoattraction. Consequently, reproduction and therefore egg production is lower in individuals with types B or O blood relative to individuals with type A. This difference may explain the observation that morbidity due to ***S. mansoni*** infection is generally higher in individuals with type A blood relative to those with types B or O.

Parasite genomes reflect their adaptations to a parasitic lifestyle

Thus far we have discussed some of the adaptations that allow various parasites to complete their life cycles and successfully reproduce. Yet we have only scratched the surface. Parasites must, for instance, absorb or otherwise acquire nutrients. Many are adapted to live in anoxic environments or are able to resist the action of damaging digestive enzymes or bile in a host's digestive system. We will discuss many of these adaptations throughout the text. In Chapter 4, for example, we will examine some of the remarkable ways that parasites subvert, manipulate, or evade the host immune system. Here, however, we consider how many such adaptations are mirrored in a particular parasite's genome.

Recent advances in genomics have greatly facilitated our ability to investigate the size, content, and organization of a parasite's genome and to compare it to free-living forms. As new genomes become available, progress in our understanding of how genomes reflect lifestyle will no doubt accelerate. Certain trends are already well established. Prokaryotic intracellular parasites have among the smallest genomes of any cellular life form. *Rickettsia prowazekii*, for example, has approximately 830 protein-encoding genes, whereas *Mycoplasma genitalium* has only about 470. *Escherichia coli*, on the other hand, a fairly typical free-living species, has about 4200 protein-encoding genes. This trend toward genome reduction reflects a loss in biosynthetic and metabolic pathways as the parasite becomes increasingly dependent on the host. It has also been suggested that the reduced gene redundancy observed in many intracellular bacteria results from the relative stability of the host environment. In relatively unstable environments genetic redundancy provides the ability to adapt to changing conditions. In predictable environments this genetic flexibility is less of a concern. In other words, if it never rains, there is no need to carry an umbrella.

Genome reduction is characteristic of many eukaryotic parasites as well. ***Cryptosporidium parvum***, an apicomplexan, and *Gracilariophyla oryzoides*, a red alga parasitic on other red algae, are both thought to have undergone such reduction. The human body louse ***Pediculus humanus humanus***, with a genome of only 100 megabases, has the smallest known insect genome to date, a phenomenon attributed to its relatively homogeneous environment.

Microsporidia are especially notable. Some microsporidians with fewer than 2000 protein encoding genes have smaller genomes than typical free-living bacteria. *Encephalitozoon intestinalis*, with only 2.25 million base pairs, has one of the smallest genomes of any known eukaryote. *Nosema ceranae*, implicated as a possible cause of colony collapse syndrome of honeybees (see Box 1.2), does not encode enzymes required for the Krebs cycle or other pathways involved in oxidative metabolism. Likewise, many of the pathways involved in biosynthetic processes such as lipid synthesis are lacking. *Enterocytozoon bieneusi* parasitizes swine and is also found in AIDS patients. This species appears to have taken dependence on the host to the limit, lacking even those genes encoding enzymes required for glycolysis. With no fully functional means to generate ATP from biological molecules, *E. bieneusi* relies entirely on the host, importing ATP via cell-surface transport proteins.

The relationship between parasitism and genome size is not always clear

The trend between gene reduction and parasitism is far from absolute (**Table 3.1**). Consider ***Trichomonas vaginalis***, the causative agent of trichomoniasis. This flagellate has a genome of about 160 megabases, approximately

Table 3.1 Diversity of genome size. Note that both the largest (*T. vaginalis*) and smallest (*P. falciparum*) numbers of protein-encoding genes belong to parasites, highlighting the difficulty involved in developing generalities between genome size and parasitism. Other nonparasitic organisms are provided for comparison.

Species and Common Name	Estimated Total Size of Genome (bp)	Estimated Number of Protein-Encoding Genes*
Saccharomyces cerevisiae (unicellular budding yeast)	12 million	6000
Trichomonas vaginalis	160 million	60,000
Plasmodium falciparum (unicellular malaria parasite)	23 million	5000
Caenorhabditis elegans (nematode)	95.5 million	18,000
Drosophila melanogaster (fruit fly)	170 million	14,000
Arabidopsis thaliana (mustard; thale cress)	125 million	25,000
Oryza sativa (rice)	470 million	51,000
Gallus gallus (chicken)	1 billion	20,000–23,000
Canis familiaris (domestic dog)	2.4 billion	19,000
Mus musculus (laboratory mouse)	2.5 billion	30,000
Homo sapiens (human)	2.9 billion	20,000–25,000

Source: Pray L (2008) *Nature Education* **1(1)**. With permission from Macmillan Publishers Ltd.

ten times larger than expected. Up to two-thirds of the genome consists of repetitive elements and transposons, suggestive of a massive and evolutionarily recent genome expansion.

Exactly why this organism has undergone such an unprecedented genomic expansion remains unclear. Many of the amplified gene families are involved in pathogenesis and in the phagocytosis and digestion of host proteins. It has been suggested that this amplification may reflect adaptations to the urogenital environment. Most of the amplification observed in ***T. vaginalis*** is lacking in its closely related sister taxon, ***T. tenax***, which inhabits the mouth. Thus, it is at least possible that ***T. vaginalis*** underwent its massive expansion during its transition to the urogenetial habitat and after it diverged from ***T. tenax***.

The genome expansion of ***T. vaginalis*** is certainly unusual in its extent, but the amplification of certain gene families is common in many eukaryotic parasites, even when their overall genome size is small in comparison to free-living forms. ***Giardia lamblia***, for example, although it has a relatively compact genome, has over two dozen recently duplicated genes that code for variant specific surface proteins (VSPs). These proteins are thought to play an essential role in the alteration of surface antigens or **antigenic variation**, which allows the parasite to evade host immune responses. The topic of antigenic variation will be explored in detail in Chapter 4. Likewise, an analysis of the genome of ***Plasmodium falciparum*** suggests that with only 5300 genes this parasite has a small genome compared to free-living eukaryotic microorganisms. However, there has been a substantial expansion in the number of genes devoted to host–parasite interaction and immune evasion.

Furthermore, when they invade erythrocytes, ***Plasmodium*** merozoites form a parasitophorous vacuole similar to that described previously for ***Toxoplasma gondii***. The parasitophorous vacuole helps to protect the parasite from toxic iron-rich components of hemoglobin, but it also forms a barrier to the import of essential nutrients. Consequently, the parasite encodes a large number of proteins involved in the transport of nutrients across both the erythrocyte and the parasitophorous vacuole membrane. The genes encoding these proteins are believed to reflect a large genetic expansion. Because many of these genes are common in both ***P. falciparum*** and ***P. vivax***, it is possible that the radiation of these genes occurred in a common ancestor before these two species diverged. Likewise, the genomes of ***Schistosoma mansoni*** and ***S. japonicum*** show clear evidence of both genetic gains and losses compared to free-living platyhelminths. Because the host provides most fatty acids and steroids, the number of genes involved in lipid metabolism has been reduced. Both schistosome species, however, show expansions in those genes coding for proteases, such as those involved in penetration of either the definitive or the intermediate host. They also both have large numbers of genes involved with detection of chemical signals, light, or temperature—not particularly surprising for an organism that must navigate through often murky freshwater, as well as through the bodies of its mammalian and molluscan hosts.

Propagules are released through a portal of exit

Following parasite reproduction within a host, eggs or larvae must often be released to the external environment or must pass directly to the next host in the life cycle. The anatomical structure through which propagules move is called the **portal of exit**. For parasites of the gastrointestinal tract, the usual portal of exit is the anus. Vector-borne parasites rely on the feeding behavior of an arthropod both for host ingress and egress. Genital contact serves this same dual function for sexually transmitted parasites such as ***Trichomonas vaginalis***. It is not always true, however, that the portal of exit is part of the same anatomical structure or even the same organ system as the one through which the parasite enters a host. Hookworms, for instance, infect their definitive host when larvae burrow through the skin. Following an extensive migration, they ultimately arrive in the intestine where they mature sexually. Fertile eggs exit the definitive host with the feces. In the schistosomes, cercariae enter the definitive host by direct penetration of the skin. Eggs, however, are released through the feces or the urine, depending on species (**Figure 3.22**).

Parasites undergo complex developmental changes in response to environmental cues

The developmental changes that occur in all living things during their lives are especially noticeable in those organisms that use very different habitats at different points in their life cycle. For instance, striking changes in both body plan and physiology are typical of insects with aquatic larvae and terrestrial adults. Many parasites also undergo dramatic developmental changes as they move from one habitat or host to the next. Recall, for instance, the complex life cycle of *Alaria americana* discussed earlier in this chapter. Although we know a great deal about precisely what developmental changes transpire during a parasite's life cycle, we know comparatively little about how these changes occur. Many questions remain regarding the signals used by parasites to cue their developmental changes, how such signals are detected and transduced, how genes are differentially expressed in different stages of development, and the consequences of this differential expression. Yet some trends are evident and, as fields such as transcriptome analysis continue to advance, pieces of this puzzle are beginning to fall into place.

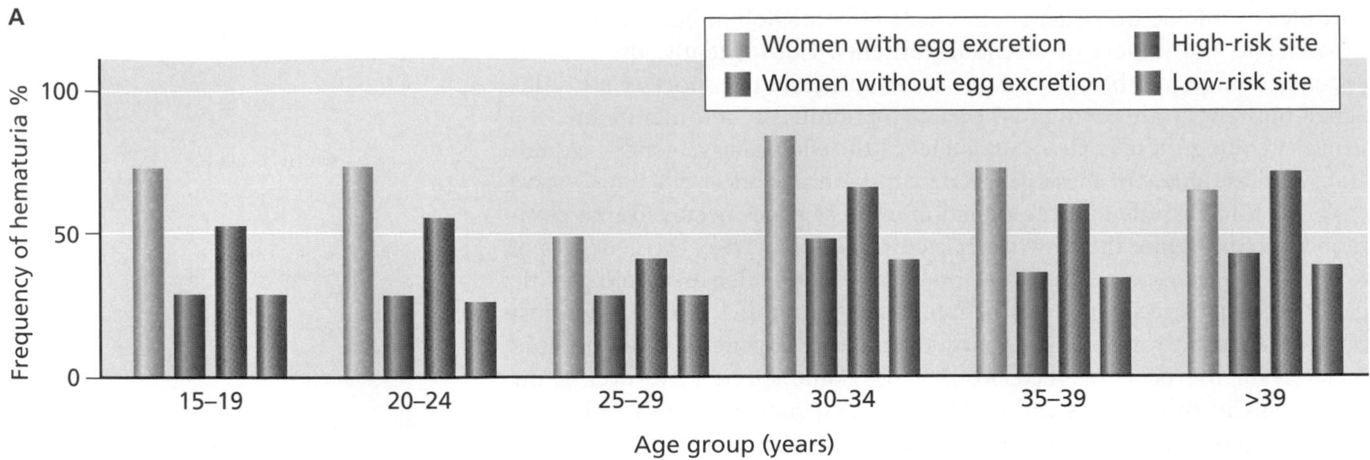

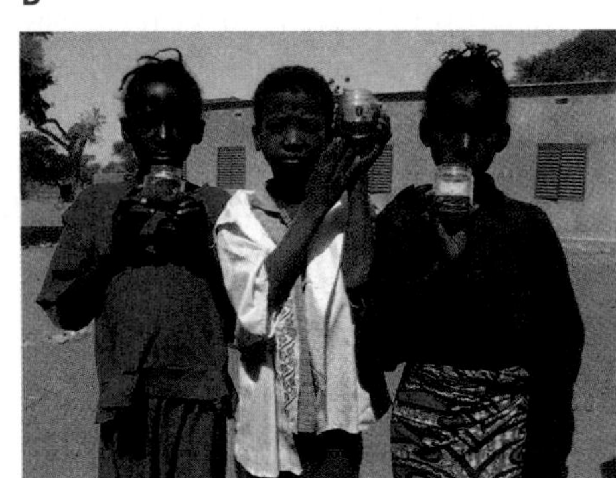

Figure 3.22 Portal of exit for *Schistosoma haematobium*. Although the infective stage, the cercaria, enters a human by direct penetration of the skin, the portal of exit for *S. haematobium* is the urine. Bloody urine is a common sign of *S. haematobium* infection, caused by damage to the circulatory epithelium during transit of eggs into the bladder. (A) Frequency of bloody urine (hematuria) in women either passing or not passing eggs in their urine, or in all women from either high-risk or low-risk areas for urinary schistosomiasis in Tanzania. Hematuria in women not passing eggs may be due to menstruation or urinary tract infection. (B) Children in Niger holding urine samples indicative of heavy infections. (A, Poggensee G, Krantz I, Kiweku I et al [2000] *Bull World Health Organ* 78(4):542–548. With permission from WHO. B, Courtesy of Juerg Utzinger, University of Basel.)

One obvious trend is that, not surprisingly, most developmental changes occur either in anticipation of transmission or upon arrival in the body of a new host. These, of course, correspond to those times in the life cycle of any parasite when the environment is either about to change or has suddenly changed most radically. It is also becoming increasingly clear that parasites actively monitor their environment and, in doing so, are able to regulate their developmental changes to best insure either successful transmission or survival in the new host.

We would expect that the dramatic changes occurring in parasites as they transition through life-cycle stages are accompanied by substantial alteration of gene expression and protein synthesis. In the last several years this expectation has been documented for a wide range of parasites. ***Leishmania major***, an important cause of cutaneous leishmaniasis, provides a protozoan example. Microarray analysis indicates that approximately 9% of whole-genome expression is altered as the parasite moves from sand fly vector to mammalian host. The functions of many of these differentially expressed genes remains unknown. Others have been characterized as genes involved in metabolism, cellular organization, and transport. Likewise, many stage-specific genes have been identified in ***Trypanosoma brucei gambiense***. Approximately 22% of the trypanosome genome is expressed differentially in the procyclic form found in the tsetse fly vector as opposed to the bloodstream form found in the mammalian host. At least some differentially regulated genes that are up-regulated in the bloodstream form are involved in the production of the parasite's major surface proteins and thus play a role in their ability to alter their surface antigens, enabling them to avoid an adaptive immune response in the mammalian host. Other protozoan parasites for which stage-specific gene expression has been demonstrated include apicomplexans in the genus ***Plasmodium*** and ***Toxoplasma gondii***.

Differential gene expression in different life-cycle stages appears to be a common phenomenon among helminth parasites as well. Schistosomes, for example, undergo dramatic changes in morphology and cellular composition in the course of their life cycle. These changes are accompanied by extensive alterations in gene expression patterns (**Figure 3.23**). Genes that are up- or down-regulated in a stage-specific manner often have functions related to nutrient acquisition, calcium signaling, lipid metabolism, and subversion of the host immune response. In ***Clonorchis sinensis***, metacercariae express high levels of genes coding for cytoskeletal proteins, relative to adult worms. In both the hookworm ***Ancylostoma caninum*** and the strongyloid nematode *Strongyloides ratti*, transition from the free-living to the parasitic state is accompanied by significant transcriptome change. In ***A. caninum*** specifically, infective L3 larvae were found to up-regulate over 600 genes, including many that encoded proteins related to pathogenesis. Other up-regulated genes included proteases, which are thought to be involved in both morphological restructuring and host penetration.

It has also become apparent that parasite development does not occur on a strict timetable; parasites do not simply charge ahead and undergo developmental changes at a defined time, no matter the circumstances. On the contrary, parasites interact with a complex array of signals of both host and parasite origin. These signals indicate when the time is right to morph into the next developmental stage. ***Plasmodium*** spp., for example, alter their production of transmissible gametocytes in response to host stress hormones or as a consequence of competition with other blood-borne parasites. The mammalian bloodstream form of ***Trypanosoma brucei*** undergoes a change from a long slender form to a shorter form known as stumpy in preparation for transmission back to the tsetse fly vector. Development into this transmissible form is controlled by the parasites themselves, which monitor their density in the blood. In a process very similar to the quorum sensing observed in many bacteria, the slender forms release a signal called stumpy induction factor (SIF). Although SIF remains uncharacterized, it appears

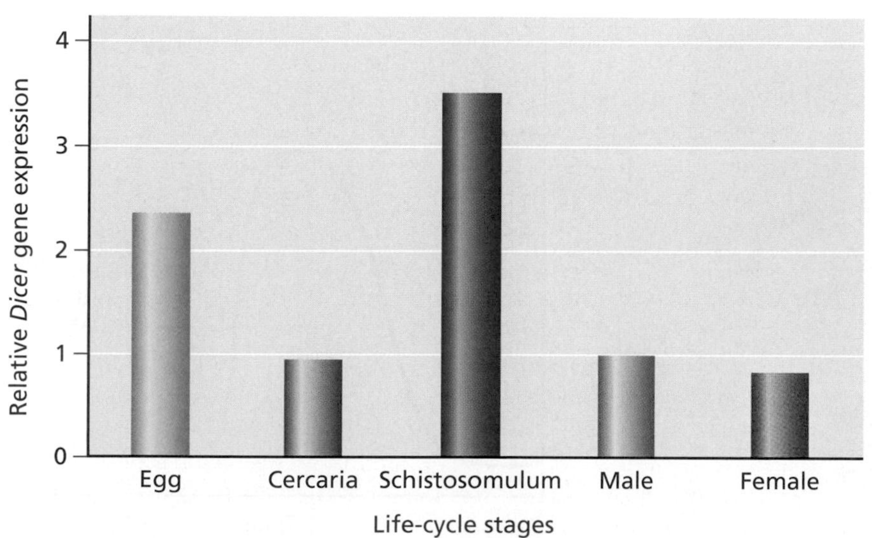

Figure 3.23 Differential gene expression at different life-cycle stages for *Schistosoma mansoni*. Expression of the *Dicer* gene varies widely in different life-cycle stages of *S. mansoni*. The dicer protein is an endoribonuclease that cleaves double stranded RNA into short double-stranded RNA fragments called small interfering RNA (siRNA). The high level of *Dicer* expression during the schistosomulum stage is thought to control the activation of transposons that may be more prone to occur during the schistosomulum stage because of the accelerated larval cell differentiation and growth. This idea does not, however, explain the relatively high levels seen in eggs. The expression of *Dicer* in all life stages examined suggests that the protein provides an important function for the parasite throughout its life, likely involving nucleic acid metabolism and gene regulation. (Adapted from Krautz-Peterson G & Skelly PJ [2008] *Exp Parasitol* 118:122–128. With permission from Elsevier.)

that when slender forms are exposed to enough SIF, indicating that the population of parasites in the blood is sufficiently high, the signal to develop into the stumpy form passes a threshold and development proceeds (**Figure 3.24**). Metazoans too, seem to adjust their life cycle in response to such environmental cues. Filarial worms, for instance, increase their production of microfilariae in response to a more aggressive host response.

Parasitic nematodes with free-living larval stages often arrest development of those stages when external environmental conditions are not compatible with larval survival. Such developmental arrest, called **hypobiosis**, makes sense from an adaptive perspective, as development in parasites is usually irreversible; once a parasite undergoes a developmental change, there is no going back. Consequently, the parasite must regulate development carefully to prevent premature development that may not allow survival or transmission. Hypobiosis may either be induced or ended as specific environmental cues are detected.

Both hookworms and strongyloids are well-known examples. The stimuli that induce such arrest vary according to species, but they are generally assumed to include factors such as temperature, soil moisture, changing photoperiod, or some combination of these and other abiotic factors. Many nematodes are likewise able to arrest their development inside the host when conditions warrant. The microfilariae of filarial worms, for instance, may circulate in the blood for months without further development until their arthropod vector consumes them. It has more recently become clear that developmental arrest may occur in protozoans as well. In ***T. brucei***, for example, progression to the transmissible stumpy form described above is repressed by a tyrosine phosphatase, which maintains the parasites poised for developmental change once the appropriate signal is received.

However, once transmission is successful and the parasite reaches a new host, rapid development is often imperative. Conditions within the newly acquired host may differ radically from those in the previous environment, necessitating speedy change. One common strategy seems to be to hold premade mRNA molecules ready for translation as soon as needed. In ***Plasmodium***, for instance, mRNA is held quiescent in gametocytes inside

A

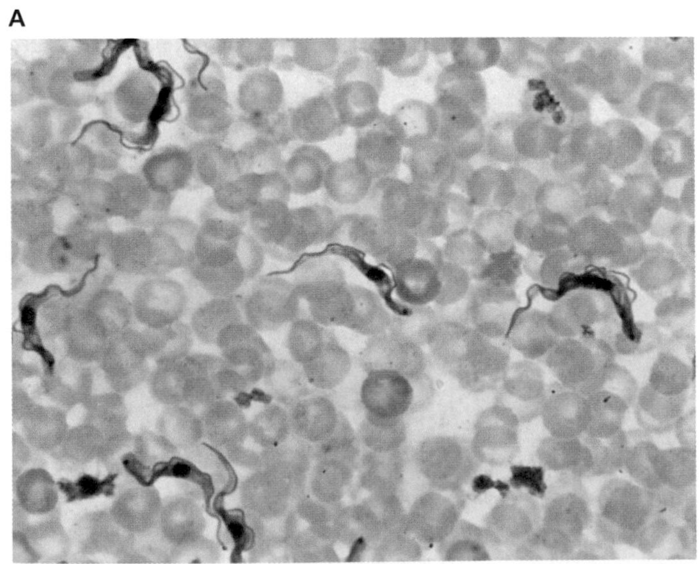

B

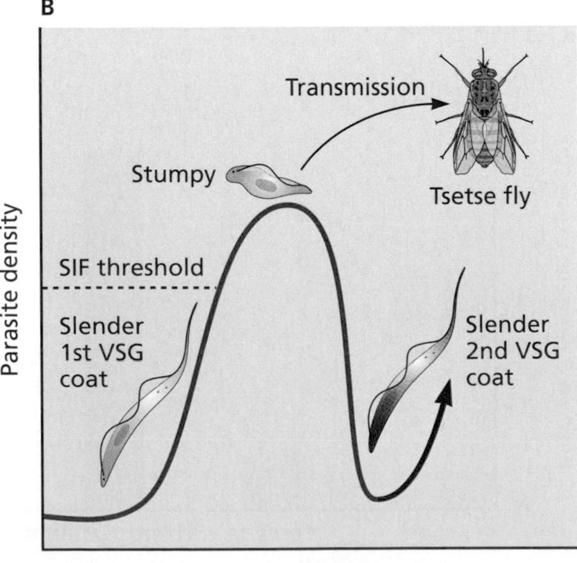

Figure 3.24 Trypanosome developmental stages. (A) Blood smear from an animal infected with *Trypanosoma brucei*. Slender forms of the parasite are visible. The stumpy form, as the name suggests, has a somewhat shorter, thicker morphology. (B) The progression from slender to stumpy form is in response to stumpy induction factor (SIF), released by slender forms. As slender forms replicate in the blood, parasitemia rises, as does the concentration of SIF, inducing some parasites to differentiate into nonreplicating stumpy forms that are transmissible to the tsetse fly vector, once a threshold in the blood concentration of SIF is reached. As slender forms differentiate into stumpy forms, most of them are cleared by the host immune response, which recognizes a surface antigen on the parasite called VSG. This leads to a crash in parasitemia. However, some slender forms are able to switch the form of VSG that they display on their surface, and these parasites are not recognized immunologically. This allows a second wave of parasitemia, as parasite numbers once again increase. (A, courtesy of CDC, Mae Melvin. B, Adapted from Pollitt LC, MacGregor P, Matthews K & Reece SE [2011] *Trends Parasitol* 27:197–202. With permission from Elsevier.)

cytoplasmic granules called P granules. These mRNA transcripts will be translated following ingestion of gametocytes by the mosquito vector. The signals that activate translation are poorly understood, but they may involve the sudden drop of temperature within the mosquito vector, the sudden rise in pH within the gut, and the presence of xanthurenic acid, a component in the mosquito pigment pathway. Whatever the exact mechanism, it allows the parasite to remain poised for action as soon as transmission to a new host is achieved. A recent comprehensive analysis of protein kinases in the rodent parasite *P. berghei* identified 66 such enzymes as components of signal transduction cascades regulating development. Twenty-three of these kinases had redundant functions related to asexual development within erythrocytes. Others were required for the gametogenesis prior to transmission from rodent to mosquito host and formation of sporozoites, the infective stage for the mammalian host. Because kinases are largely conserved among ***Plasmodium*** species, it is likely that these identified enzymes are involved in regulation of development and stage-specific gene expression in other malaria parasites as well.

Epigenetic phenomena and co-opting of host signaling molecules may be important in parasite development

Epigenetics refers to heritable changes in phenotype or gene expression caused by mechanisms other than an alteration in the DNA sequence. DNA methylation in eukaryotes, which can silence genes without changing those genes' nucleotide sequences, is a familiar example.

An additional trend that is becoming increasingly clear is the importance of epigenetic processes in parasite development. In the transition of one parasite developmental stage to the next, histone acetylation in particular, which also results in a rapid change in a cell's transcription profile, may be particularly important. Such histone modification has already been shown to play a key role in the stage conversion of apicomplexans such as ***Plasmodium*** and ***Toxoplasma***.

Of particular interest is the manner in which some parasites co-opt host hormones or immunomodulatory signals to regulate their own development. Insulin, epidermal growth factor, and fibroblast growth factor all have positive impacts on the growth and development of the cestode ***Echinococcus multilocularis***. Certain **cytokines** (small proteins involved in intercellular communication) stimulate the development of protoscoleces within the hydatid cysts. The involved tyrosine kinase receptors are all structurally analogous to those found in the mammalian host. In schistosomes, it has been shown that activation of host monocytes via cytokines also stimulates parasite development (**Figure 3.25**).

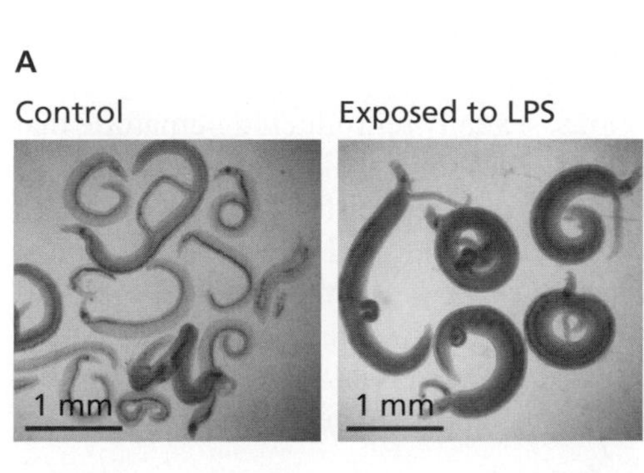

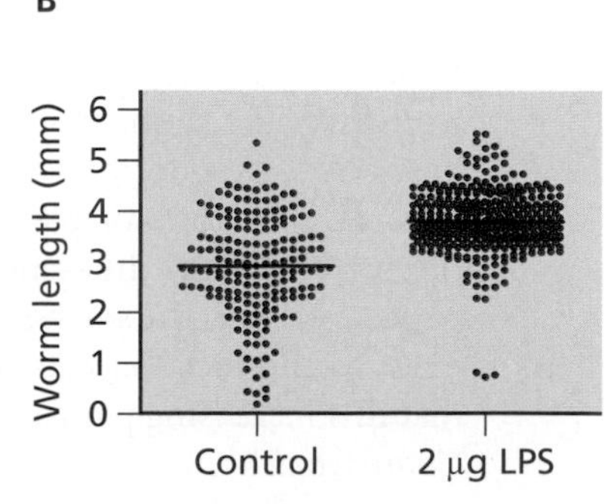

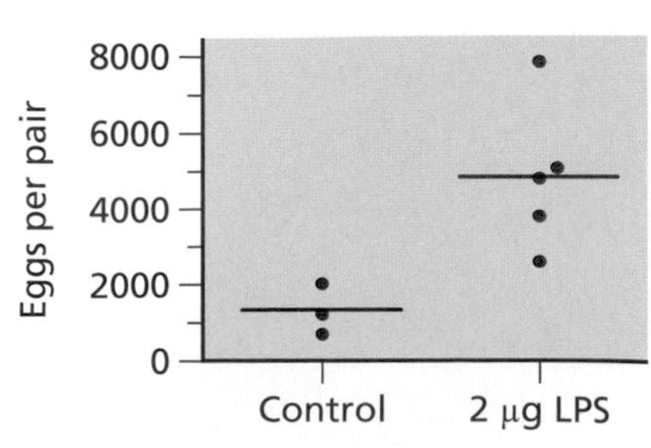

Figure 3.25 Monocyte activation and schistosome development. Development of *Schistosoma mansoni* in mice depends at least in part on the activation of host monocytes. (A) Microscopic images of adult schistosomes isolated from mice 6-weeks post infection. The worms to the right were from mice stimulated with lipopolysaccharide (LPS), which is a powerful activator of monocytes. Worms in the presence of activated monocytes were larger and more robust compared to control worms, seen in the photo to the left. Adults recovered at 6 weeks following *in vivo* stimulation with LPS were both (B) longer and (C) produced more eggs, than controls in which *in vivo* monocytes were not exposed to LPS. (A, Lamb EW, Walls CD, Pesce JT et al [2010] *PLoS Pathog* 6:E1000892.)

Clearly, many questions remain regarding the signal transduction mechanisms that underlie parasite development. As we continue to tease apart the various aspects of such mechanisms, we can expect a more coherent picture of parasite development to emerge. Not only will such a picture enhance our knowledge of development in general, but it opens the door to new and innovative strategies to control parasitic diseases by interfering with specific required events that occur in the course of a parasite's life cycle.

Summary

In general, the problems confronting parasites are similar to those of all living things, but the parasitic lifestyle requires adaptation to the unique host environment, which differs considerably from more familiar and conventional habitats. The manner in which parasites have adapted to the host environment and evolved in response to the overriding need to be transmitted between hosts has resulted in various life-cycle patterns. Understanding these life cycles provides insight into numerous other aspects of the parasitic mode of existence.

As the problems that any parasite must overcome are essentially the same, there are basic patterns to parasite life cycles, but there is seemingly no end to the number of ways different parasite species modify these patterns to suit their specific needs. Parasites transmitted via the fecal–oral route need eggs or larvae that are capable of temporary survival in the external environment, their infective stages must be able to cope with the low pH encountered in the stomach, and the life-cycle stages inhabiting the intestine have to withstand powerful digestive enzymes and peristalsis. Other parasites, using other modes of transmission, encounter somewhat different environmental conditions that necessitate different solutions.

Many of the generalities historically accepted about parasites are not necessarily true and sometimes the adaptations observed in a particular parasite seem at first glance to be counterintuitive. For example, contrary to the commonly accepted belief, parasites are not always more fecund relative to their free-living relatives. Likewise, although it is true that many parasites display genome reduction over evolutionary time, examples of genome expansion in parasites or the extensive duplication of certain gene families remind us that simple statements about what constitutes a parasite rarely stand up to close inspection.

Like all living things, parasites must juggle a large number of competing selection pressures and the result is frequently a compromise between these competing demands. The variety of selection pressures acting on parasites insure that they evolve along a wide variety of trajectories, just as free-living organisms do.

REVIEW QUESTIONS

1. *Gnathostoma procyonis* is a diecious, sexually reproducing nematode that parasitizes the stomach of raccoons. Raccoons, which serve as definitive hosts, become infected when they consume either fish serving as second intermediate hosts or paratenic hosts (other vertebrates that consumed the infected fish). Imagine a raccoon that is lightly infected with this parasite early in the spring. As spring progresses, the intensity of infection rises, as the raccoon eats additional infected fish. How do you predict that this increasing parasite intensity will impact the reproduction rate of *G. procyonis*?

2. Discuss the "hosts as islands" analogy. In what ways is it appropriate and in which ways is it inappropriate?

3. You are studying a previously undescribed apicomplexan parasite, about which almost nothing is currently known. In the course of your investigations you observe that the life cycle of this parasite includes a metabolically inactive "cyst-like" stage. Which mode or modes of transmission can you likely rule out, based on this observation?

4. Parasites relying on vector-borne transmission are often among the most pathogenic. Can you think of a mode of transmission, for which selection might work to lessen pathogenicity?

5. The R value, related to the basic reproduction number (R_0) but taking into account the immune status of the population, describes the average number of new cases in susceptible individuals derived from one infected individual. If members of the potential host population are immune to the infectious agent in question, they cannot become infected, and this causes R to change as the proportion of individuals that are immune changes. In the following cases would R be expected to rise or fall:
   a) If a sudden influx of new potential host individuals who have not previously been exposed to the parasite in question entered the population?
   b) If a number of hosts became infected with the parasite, and subsequently became immune to later infections with the same parasite?
   c) If a number of potential hosts were vaccinated against this particular parasite?
   d) If a number of new babies are born?

6. *Onchocerca lupi* is an emerging parasite of veterinary importance. Like the related *O. volvulus*, it can result in vision loss for the definitive host. Whereas *O. volvulus* is a human parasite, *O. lupi* appears to affect mainly dogs and other canines. The vector of this parasite has yet to be identified. Suppose that you wished to identify potential vectors, and found that a species of biting fly was at least occasionally infected with larval stages of *O. lupi*. What would be your next step, if you wished to further investigate the role of this fly as a potential *O. lupi* vector?

References

A HISTORICAL PERSPECTIVE OF THE PARASITE LIFE CYCLE

Early medical and natural history studies gave rise to an understanding of parasite life cycles

Cox FEG (2002) A history of human parasitology. *Clin Microbiol Rev* **15**:595–612.

Arthropod transmission for filarial worms suggested that other diseases may be similarly transmitted

Cox FEG (2010) The history of the discovery of the malaria parasites and their vectors. *Parasit Vectors* **3**:5 (doi 10.1186/1756-3305-3-5).

AN OVERVIEW OF PARASITE LIFE CYCLES

Parasites with direct life cycles use only a single host

Nikander S & Saari S (2006) An SEM study of the reindeer sinus worm (*Linguatula arctica*). *Rangifer* **26**:15–24.

Two or more hosts are necessary for those parasites with indirect life cycles

Zelmer E, Wetzel J & Esch GW (1999) The role of habitat in structuring *Halipegus occidualis* metapopulations in the green frog. *J Parasitol* **85**:19–24.

Shoop WL & Corkum KC (1984) Transmammary infection of newborns by larval trematodes. *Science* **223**:1082–1083.

THE PARASITE'S TO DO LIST

Effective transmission is essential for all parasites

Jarecka L (1961) Morphological adaptations of tapeworm eggs and their importance in life cycles. *Acta Parasitol* **9**:409–426.

Henni L, Molyneux DH, Livesey L & Galun R (1980) Feeding behavior of tsetse flies infected with salivarian trypanosomes. *Nature* **283**:383–385.

McHugh CP (1989) Ecology of a semi-isolated population of adult *Anopheles freeborni*: abundance, trophic status, parity, survivorship, gonotrophic cycle length, and host selection. *Am J Trop Hyg* **41**:169–176.

Combes C (2004) Parasitism: The Ecology and Evolution of Intimate Interactions. University of Chicago Press.

Lauwaet T, Davids BJ, Reiner DS & Gillin FD (2007) Encystation of *Giardia lamblia*: a model for other parasites. *Curr Opin Microbiol* **10**:554–559.

Little TJ, Shuker DM, Colegrave N et al (2010) The coevolution of virulence: tolerance in perspective. *PLoS Pathog* **6(9)**: e1001006. (doi:10.1371/journal.ppat.1001006).

High reproductive rates are common in many parasite life cycles

Loker ES (1983) A comparative study of the life-histories of some mammalian schistosomes. *Parasitology* **87**:343–369.

Moore J (1981) Asexual reproduction and environmental predictability in cestodes (Cyclophyllidea: Taeniidae). *Evolution (N Y)* **35**:723–741.

Poulin R (1995) Clutch size and egg size in free-living and parasitic copepods: a comparative analysis. *Evolution (N Y)* **49**:325–336.

Poulin R (1995) Evolution of parasite life history traits: myth and reality. *Parasitol Today* **11**:342–345.

Abrous M, Rondelaud D & Dreyfuss G (2000) Cercarial productivity of redial generations in single-miracidium infections of *Lymnaea truncatula* with *Paramphistomum daubneyi* or *Fasciola hepatica. J. Helminthol* **74**:1–5.

Many factors can complicate an understanding of parasite transmission

Poulin R & Lefebvre F (2006) Alternative life-history and transmission strategies in a parasite: first come, first served? *Parasitology* **132**:135–141.

Noland GS, Graczyk TK, Fried B & Nirbhay K (2007) Enhanced malaria parasite transmission from helminth co-infected mice. *Am J Trop Med Hyg* **76**:1052–1056.

Thieltges DW, Jensen KT & Poulin R (2008) The role of biotic factors in the transmission of free-living endohelminth stages. *Parasitology* **135**:407–426.

Fellous S & Koella JC (2009) Different transmission strategies of a parasite in male and female hosts. *J Evol Biol* **22**:582–588.

Mathematical models provide a useful tool to predict transmission rates

Craig MH, Snow RW & Le Sueur D (1999) A climate-based distribution model of malaria transmission in sub-Saharan Africa. *Parasitol Today* **15**:105–111.

Riley S, Carabin H, Bélisle P et al (2008) Multihost transmission dynamics of *Schistosoma japonicum* in Samar province, the Philippines. *PLoS Med* **5(1)**:e18. (doi:10.1371/journal.pmed.0050018).

Many parasites must migrate to specific sites or tissues within the host

Sukhdeo MVH & Sukhdeo SC (2002) Fixed behaviors and migration in parasitic flatworms. *Int J Parasitol* **32**:329–342.

Henrik L, Hitziger, N, Dellacasa I et al (2006) Induction of dendritic cell migration upon *Toxoplasma gondii* infection potentiates parasite dissemination. *Cell Microbiol* **8**:1611–1623.

Moreira CK, Templeton TJ, Lavazec C et al (2008) The *Plasmodium* TRAP/MIC2 family member, TRAP-like protein (TLP), is involved in tissue traversal by sporozoites. *Cell Microbiol* **10**:1505–1516. (doi:10.1111/j.1462-5822.2008.01143.x).

The manner in which complex migration within the host evolved is not always clear

Read AF & Sharping A (1995) The evolution of tissue migration by parasitic nematode larvae. *Parasitology* **111**:359–371 (doi:10.1017/S0031182000081919).

Finding a mate is a requirement for many sexually reproducing parasites

Kanev I, Sterner M, Radeu V et al (2000) An overview of the biology of echinostomes. In Echinostomes as Experimental Models for Biological Research (Fried B & Graczyk TK eds), pp 1–29. Kluwer.

Brown SP, Renaud F, Gueagan JF et al (2001) Evolution of trophic transmission in parasites: the need to reach a mating place? *J Evol Biol* **14**:815–820.

Haseeb AMA, Thors C, Linder E et al (2008) *Schistosoma mansoni*: Chemoreception through *n*-acetyl-D-galactosamine-containing receptors in females offers insight into increased severity of schistosomiasis in individuals with blood group A. *Exp Parasitol* **119**:67–73.

Parasite genomes reflect their adaptations to a parasitic lifestyle

Lehker MW & Alderete JF (1999) Resolution of six chromosomes of *Trichomonas vaginalis* and conservation of size and number among isolates. *J Parasitol* **85**:976–9 (doi:10.2307/3285842).

Keeling PJ (2004) Reduction and compaction in the genome of the apicomplexan parasite *Cryptosporidium parvum*. *Dev Cell* **6**:614–616.

Pain A, Renauld H, Berriman M, et al (2005) Genome of the host cell transforming parasites *Theileria annulata* compared with *Theileria parva*. *Science* **309**:131–133 (doi:10.1126/science.1110418).

Sargeant TJ, Marti M, Caler E et al (2006) Lineage-specific expansion of proteins exported to erythrocytes in malaria parasites. *Genome Biol* **7**:R12 (doi:10.1186/gb-2006-7-2-12).

Carlton JM, Hirt JP, Silva RC et al (2007) Draft genome sequence of the sexually transmitted pathogen *Trichomonas vaginalis*. *Science* **315**:207–212 (doi:10.1126/science.1132894).

Berriman M, Hass BJ, LoVerde PT et al (2009) The genome of the blood fluke *Schistosoma mansoni*. *Nature* **460**:352–358 (doi:10.1038/nature 08160).

Wang L, Zhang J, Zhou Y et al (2009) The *Schistosoma japonicum* genome reveals features of host–parasite interplay. *Nature* **460**:345–351 (doi:10.1038/nature08140).

Cai H, Gu J & Wang Y (2010) Core genome components and lineage-specific expansions in *Plasmodium* malaria parasites. *BMC Genomics* **11**, Suppl3:513 (doi:10.1186/1471-2164-11-53-513).

Hancock L, Goff L & Lane C (2010) Red algae lose key mitochondrial genes in response to becoming parasitic. *Genome Biol Evol* **2**:897–910.

Kirkness EF, Hass BJ, Sun W et al (2010) Genome of the human body louse, *Pediculus humanus humanus*. *Proc Natl Acad Sci* **107**:12,168–12,173.

Sun J, Jiang H, Flores R et al (2010) Gene duplication in the genome of parasitic *Giardia lamblia*. *BMC Evol Biol* **10**:49 (doi:10.1186/1471-2148-10-49).

Mendonca AG, Alves RJ & Pereira-Leal JB (2011) Loss of genetic redundancy in a reductive genome. *PLoS Comput Biol* **7**:e1001082 (doi:10.1371/journal.pcbi.1001082).

Parasites undergo complex developmental changes in response to environmental cues

Vassella E, Reuner B, Yutzy B et al (1997) Differentiation of African trypanosomes is controlled by a density sensing mechanism, which signals cell cycle arrest via the cAMP pathway. *J Cell Sci* **110**:2661–2671.

Billker O, Lindo V, Panico M et al (1998) Identification of xanthurenic acid as the putative inducer of malaria development in the mosquito. *Nature* **392**:289–292 (doi:10.1038/32667).

Radke JR, Behnke MS, Mackey AJ et al (2005) The transcriptome of *Toxoplasma gondii*. *BMC Biol* **3**:26 (doi:10.1186/1741-7007-3-26).

Szoor B, Wilson J, McElhinney H et al (2006) Protein tyrosine phosphatase *Tb*PTP1: a molecular switch controlling life cycle differentiation in trypanosomes. *J Cell Biol* **175**:293–303 (doi:10.1083/jcb.200605090).

Brehm K & Spilotis M (2008) The influence of host hormones and cytokines on *Echinococcus multilocularis* signaling and development. *Parasite* **15**:286–290.

Datu BJD, Gasser RB, Nagaraj SH et al (2008) Transcriptional changes In the hookworm *Ancylostoma caninum* during the transition from a free-living to a parasitic larva. *PLoS Negl Trop Dis* **2(1)**:e130 (doi:10.1371/journal.pntd.0000130).

Rochette A, Raymond F, Ubeda JM et al (2008) Genome-wide gene expression profiling analysis of *Leishmania major* and *L. infantum* developmental stages reveals substantial differences between the two species. *BMC Genomics* **9**:255 (doi:10.1186/1471-2164-9-255).

Bougdour A, Maubon D, Baldacci P et al (2009) Drug inhibition of HDAC3 and epigenetic control of differentiation in Apicomplexa parasites. *J Exp Med* **206**:953–966.

Cho PY, Kim TI, Whang SM & Hong SJ (2009) Gene expression profile of *Clonorchis sinesis* metacercariae. *Parasitol Res* **102**:277–282.

Gobert GN, Moertel L, Brindley PJ & McManus DP (2009) Developmental gene expression profiles of the human pathogen *Schistosoma japonicum*. *BMC Genomics* **10**:128 (doi:10.1186/1471-2164-10-128).

Reece SE, Drew DR & Gardner A (2009) Sex ratio adjustment and kin discrimination in malaria parasites. *Nature* **453**:609–614 (doi:10.1038/nature06954).

Babayan SA, Read AF, Lawrence RA et al (2010) Filarial parasites develop faster and reproduce earlier in response to host immune effectors that determine filarial life expectancy. *PLoS Biol* **8**:e1000525 (doi:10.1371/journal.pbio.1000525).

Lamb EW, Walls CD, Pesce JT et al (2010) Blood fluke exploitation of non-cognate CD4+ T cell help to facilitate parasite development. *PLoS Pathog* **6**:e1000892 (doi:10.1371/journal.ppat.1000892).

Mair GR, Lasonder E, Garver LS et al (2010) Universal features of post-transcriptional gene regulation are critical for *Plasmodium* zygote development. *PLoS Pathog* **6**:e1000767 (doi:10.137/journal.ppat.1000767).

Tewari R, Straschil U, Bateman A et al (2010) The systematic functional analysis of *Plasmodium* protein kinases identifies essential regulators of mosquito transmission. *Cell Host Microbe* **8**:377 (doi:10.1016/j.chom.2010.09.006).

Veitch NJ, Johnson PCD, Trivedi U et al (2010) Digital gene expression analysis of two life cycle stages of the human-infective parasite, *Trypanosoma brucei gambiense* reveals differentially expressed clusters of co-regulated genes. *BMC Genomics* **11**:124 (doi:10.1186/1471-2164-11-124).

Host Defense and Parasite Evasion

4

Parasitology has done more for immunology than immunology has ever done for parasitology.

Franz von Lichtenberg

In This Chapter

In 430 BC, at the height of the Peloponnesian War, a plague of unknown origin struck Athens. Between one-third and two-thirds of the city's population died, giving Sparta the edge it needed to ultimately defeat its Athenian foe (**Figure 4.1**). What little we know of this epidemic, we owe to the Athenian historian Thucydides, who chronicled the disease's symptoms and its high mortality rate.

Thucydides described symptoms that included rapid onset, violent retching, and extreme thirst. Explosive diarrhea ensued, with pustules eventually forming on the skin. Typhus, smallpox, and measles have all been suggested as possible causes. Alternatively, a disease that no longer exists may have caused the plague. Thucydides himself contracted and recovered from the disease. His recovery, he realized, rendered him resistant to a subsequent infection. He also noted that others who recovered from the illness were those best suited to care for the sick owing to this same resistance to reinfection.

Thucydides provides us with what may be the first description of **immunological memory** in Western literature. Immunological memory is a hallmark of vertebrate immunity. It is the ability of the immune system to respond more rapidly and effectively to specific pathogens upon second or subsequent encounters. It explains why when vertebrates recover from a particular illness, they are unlikely to suffer from the same illness again, at least for a period of time.

In this chapter, we investigate the **immune system**, the complex assemblage of mechanisms that constitute the host's defenses against infection. The immune system explains why hosts can persist and thrive in spite of their relentless exposure to viral, prokaryotic, and eukaryotic parasites. Indeed, in most cases disease is the exception rather than the rule. And when it does occur, disease often resolves on its own. But not always. Host defense is imperfect at best. Hosts regularly succumb to disease or remain infected for prolonged periods of time. Furthermore, in many cases the overt symptoms of disease that are observed in any infection have more to do with host responses than they do with direct effects of the parasite.

All organisms are assailed by parasites, and the need for immunity is not restricted to vertebrates. Accordingly, to provide some evolutionary perspective on immunity, we first briefly highlight a few of the remarkable

Species names highlighted in red are included in the Rogues' Gallery, starting on page 429.

Figure 4.1 The plague of Athens. Before the initial outbreak, the population of Athens was estimated to have been 315,000. Of these, somewhere between 79,000 to 105,000 Athenians succumbed to the mysterious illness, including Pericles, the Athenean leader. In his history of the plague, Thucydides wrote, "The bodies of dying men lay one upon another, and half-dead creatures reeled about in the streets. The catastrophe became so overwhelming that men cared nothing for any rule of religion or law." (Painting by Michiel Sweerts, Plague in an Ancient City. LACMA.)

defense capacities of prokaryotes in their ongoing battles with viruses. We next discuss some of the immune responses of plants, which must also contend with a broad a range of parasites. We then examine immunity in invertebrate animals and proceed to vertebrate host defense and the special problems posed by eukaryotic parasites. We conclude the chapter by scrutinizing the insidious ploys that parasites use to evade or manipulate host defenses. As we contemplate this broad palette of biological interactions and seek general principles, we must also realize that even within particular host lineages, there is no one stereotypic solution to achieving protection from parasite attack. Parasites with different evolutionary histories and with highly variable life cycles have selected for marvelously diverse defense systems, the ongoing study of which continues to reveal unforeseen complexities and surprises.

A complete overview of basic immunity is beyond the scope of this text. Our focus here is **immunoparasitology**—the immunology governing the relationship between parasites and their hosts. Yet it is imperative that anyone attempting to comprehend this relationship be well grounded in the basic functioning of immune processes. Courses in immunology will provide this essential background. Alternatively, for those lacking this background or for those who would like a brief refresher on basic vertebrate immunology, a full review is provided on the website associated with this book. Readers are recommended to examine this material before proceeding.

4.1 AN EVOLUTIONARY PERSPECTIVE ON ANTI-PARASITIC IMMUNE RESPONSES

Prokaryotes have developed remarkable immune innovations during their billions of years encountering parasites

Prokaryotes have systems to protect them from their own parasites that range from transposons to plasmids to viruses to even other bacteria. These systems have developed over the approximately 3.5 billion years of prokaryote life on our planet. In seeking a sound fundamental understanding of the immunobiology of host–parasite interactions, remember that immune systems, even with a form of immunological memory, were bacterial innovations that took place long before eukaryotes came on the scene.

One noteworthy defense strategy is the restriction modification system in bacteria. A typical system couples an endonuclease with a modification enzyme. The endonuclease recognizes specific sequences in phage DNA and cleaves the DNA at these sites, thus disabling the phage. The modification enzyme, often a methyltransferase, adds methyl group to the host bacterium's DNA, thereby protecting it from endonuclease attack. This is a system of self–nonself recognition that can selectively disable nonself entities such as viruses. In response, phages have been selected to avoid using nucleotide sequences targeted by restriction enzymes. They may also incorporate unusual or modified nucleotides at target sites, or they may protect the sites by cloaking them in proteins. Alternatively, they may hijack host methytransferases to methylate and thus protect their own target sequences. Clearly, the classic elements of an arms race exist within this system.

Another amazing system possessed by 40%–70% of bacteria and 90% of archaea to promote their defense from viruses and foreign plasmids is **CRISPR** (Figure 4.2). This acronym stands for clustered regularly interspaced short palindromic repeats. It refers to a particular region of bacteria DNA in which are found short segments of DNA (21–72 bp long) called spacers. These DNA segments are derived from various phages that have been taken up, with each spacer flanked by two of the characteristic conserved short repeat sequences that first caught the attention of microbiologists. These phage sequences, consequently, represent an archive of previous exposure events for the bacterium. The process of acquiring and storing these snippets of exotic DNA has been likened to immunization. This information can later be deployed to attack, in a sequence-specific way, the DNA of a comparable phage, should the bacterium, or its progeny, encounter it again. This phage-specific response happens by transcribing the short segments of stored phage sequence, producing transcripts called crRNA. These transcripts interact with a system of host enzymes (the Cas system) that are effectively guided by the crRNA to the comparable sequence on the newly invading phage, where they then attack and disable it. The system has also been referred to as an acquired immune system because it stores information acquired from previous encounters with phage. It has also been described as a type of memory because it retains the information from the previous encounters, and, furthermore, it is passed on in heritable form to its progeny. The presence of such elaborate systems of antiviral defense long before the appearance of vertebrates humbles us once again. The CRISPR system has also been considered to exemplify **Lamarckian inheritance**, in which a trait acquired in an organism's lifetime can be passed on to its progeny.

Many kinds of parasites compromise the health of plants so it is important to know how plants defend themselves

Plants contend with parasites as diverse as viroids, viruses, bacteria, protozoans, oomycetes, fungi, other plants, nematodes, and insects. Knowing how plants respond to and protect themselves from this diversity of disease agents not only sheds fresh light on the nature of parasite–host interactions, but is also of huge practical importance because of our obvious dependency on plants (for food). For example, the recent emergence and spread of a new variant of the fungus responsible for wheat stem rust *Puccinia graminis* (**Figure 4.3**) poses grave concerns as many of the widespread cultivars of wheat (*Triticum* sp.) are vulnerable to this new rust variant. By understanding

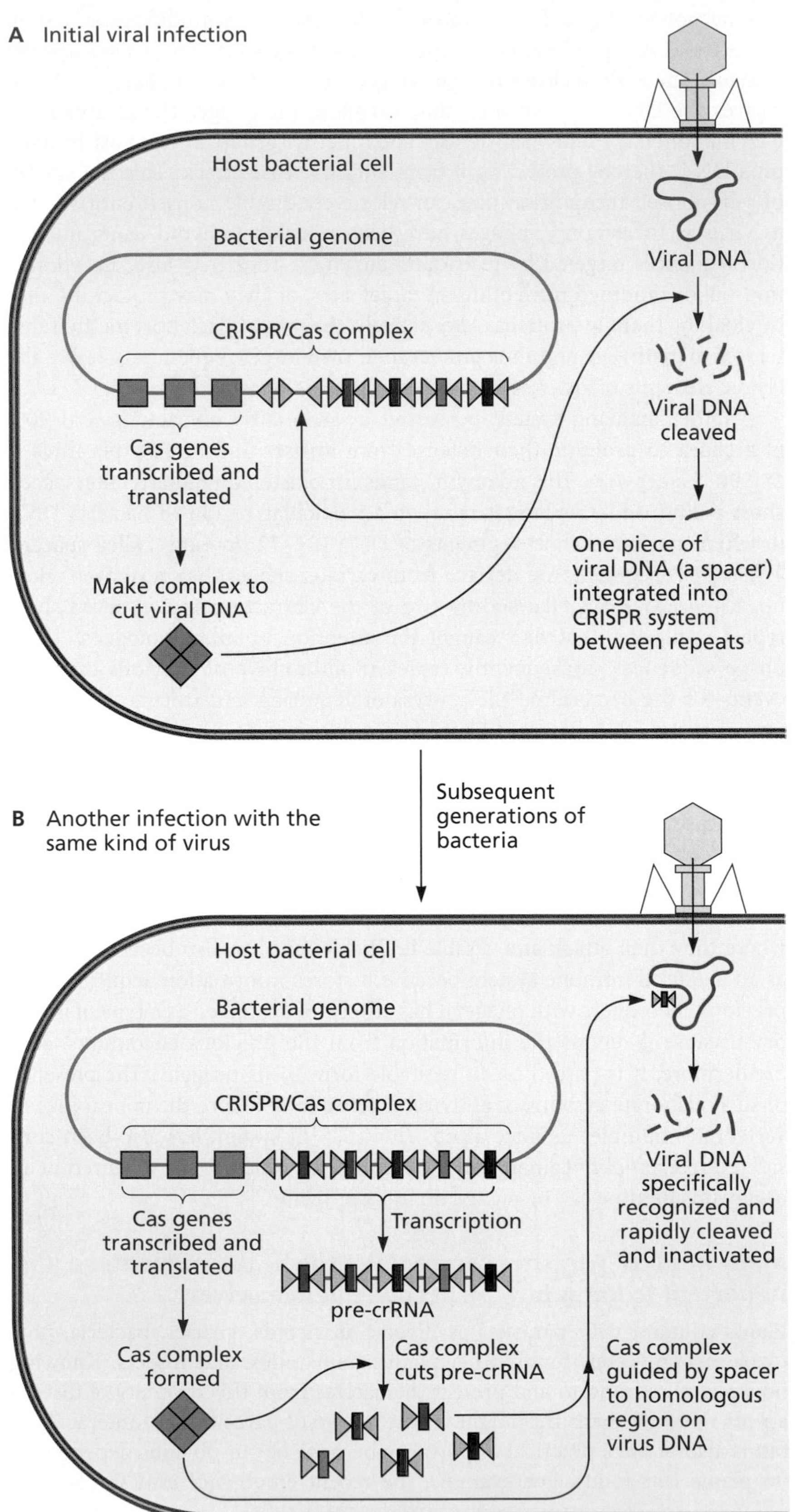

Figure 4.2 An overview of the CRISPR/Cas system. (A) An immunization process. Successful immunization may occur if the host cell is infected with an incompetent virus that is unable to kill the cell, or with a plasmid. Exogenous DNA from the virus or plasmid infects the bacterium and is cleaved by the Cas complex. A piece of the exogenous DNA called a novel spacer is inserted into the bacterium's CRISPR array, bracketed by repeat units at one end of the CRISPR locus. (B) The immunity process. When the bacterium or its descendents are again infected with exogenous DNA from the same type of virus or plasmid, the set of CRISPR repeat-spacer units is transcribed into a pre-crRNA that is then cleaved by the Cas complex into several small crRNAs. If one of the crRNAs has a sequence that matches the invading virus or plasmid DNA, in conjunction with the Cas complex, it will align with the matching sequence in the invading DNA. The Cas complex will then cut and disable the invading DNA. Repeats are represented as triangles and spacers as rectangles.

the fundamentals of plant immunology, we have a much better chance to wisely manage our food plants and protect them from parasites.

Although plants lack specialized immune cells, they still can mount effective, long-term responses to parasites

Plants lack specialized mobile immune cells and a circulatory system to distribute them, but they have many defenses at their disposal. These include both **constitutive defenses** that are constant and enduring in their expression and **inducible defense** systems that can lead to structural reinforcements at the site of infection or to specialized responses, such as the hypersensitive response or the systemic acquired resistance response. An inducible defense response is one that can be activated above a baseline level in response to exposure to a parasite. This response is usually triggered by reception of a signal of infection, followed by activation of an intracellular signaling pathway that culminates in increased transcription of immune-related genes.

Figure 4.3 *Puccinia graminis*, the wheat stem rust. Rust-colored lesions appear on the stems. Infection reduces grain production, breaks stems, and may cause the death of the plant. (Courtesy of US Department of Agriculture.)

Plants express **pattern recognition receptors (PRR)** on their cell surfaces that recognize and bind common and conserved molecules associated with particular group of pathogens known as **pathogen-associated molecular patterns (PAMPs)**. If a particular PRR is engaged by a PAMP, then PAMP-triggered immunity is activated. PAMP-triggered immunity is also called pattern-triggered immunity, either of which may be referred to as **PTI**. PTI involves activation by fairly nonspecific PAMPs of an intracellular signaling pathway that can lead to relatively generic responses such as strengthening of cell walls to prevent infection or production of lytic enzymes or antimicrobial compounds.

In some cases, plant parasites inoculate effector proteins into the cytoplasm of the plant cell. These effectors often have the ability to degrade host cell molecules that are key hubs in the plant's signaling pathways. As such, key signaling molecules are degraded and the plant's defense response is effectively short-circuited. However, the plant immune system is very sensitive to the presence of effector proteins and it actively protects its signaling pathway molecules by producing another category of defense molecule, the **NB-LRRs (nucleotide-binding, leucine-rich repeat proteins)**. These are involved in activation of signal pathway molecules that lead to downstream immune responses that are collectively referred to as **effector-triggered immunity (ETI)**.

NB-LRRs are products of R or resistance genes, first discovered in the 1950s and described in classic studies of the interactions between flax (*Linum usitatissimum*) and flax rust fungus (*Melampsora lini*). Based on these studies, the gene-for-gene hypothesis was proposed, which postulated that the inheritance of resistance in flax and the rust's ability to cause infection are controlled by pairs of matched genes in the two organisms. The product of a particular R gene (an NB-LRR, also called an R protein) interacts specifically with the product of a flax-rust gene (an effector). Further study has revealed that R genes are often polymorphic and that, in a host population, several different alleles may exist: they encode NB-LRRs with different recognition capabilities. Parasites too possess many different variant alleles, each able to produce a different effector protein. Furthermore, we also now know that many different R genes exist in a plant species: for example, *Arabidopsis* possesses 150 and rice about 500. It is considered that these different R genes may be involved in controlling different parasites and that in general ETI is more specific than PTI.

In some cases, R proteins bind the parasite-produced effector and directly disable it. In other cases, the R proteins are involved in guarding the target host-cell molecule. In this case, if the parasite effector binds to the target molecule and cleaves it, the R protein can recognize the disabled target as modified self, and then triggers a response. In this way, a single type of guarding R protein can potentially protect against several different effectors. The **guard concept** in plant immunology is reminiscent of what has been called the **danger hypothesis** for mammalian immunity. The basic idea of the danger hypothesis is that certain immune system cells respond to danger signals, such as components released from injured cells (analogous to a cleaved target molecule in the case of plants). These serve as alarm signals to warn the immune system there is a problem requiring a response.

If an ETI response is aroused, then a **hypersensitive response** (HR) follows. The HR is associated with programmed cell death of the plant's cells in a localized area at the site of infection, which is called a lesion. In addition to HR, in response to local infection, plants also mount a slower, but longer lasting and more widespread response called **SAR**, or **systemic acquired resistance**. SAR can provide long-lasting protection not only to the parasite that initiated the original attack, but to others as well. Study of plant parasites has shown that parasites can target several different steps of both HR and SAR for disruption.

Many nematode species are specialized to parasitize plants

Worth a special mention among the many groups of parasites encountered by plants are nematodes, most of which parasitize the roots. Some are ectoparasitic, whereas others colonize plant tissues and are either mobile or sedentary within the plant. Remarkably, sedentary nematodes often trick the plant into producing feeding structures, which in some cases consist of multinucleate giant plant cells. The nematode must be able to complete feeding on such structures within an approximately six-week time frame and must interfere with host defenses long enough to conclude its development. Plant-parasitic nematodes typically have a prominent needlelike stylet (**Figure 4.4**) used to tease apart adjacent host cells or to pierce and deliver a variety of molecules into host cells. Several of these injected factors are indeed known to be immunosuppressive. The nematodes produce glutathione-S-transferases to detoxify plant toxins such as terpenoids. Nematodes also produce antioxidant proteins such as glutathione peroxidase to detoxify the reactive oxygen species produced by plants. Interestingly, the latter molecule is also produced prominently by nematodes parasitic in animals, ostensibly for the same

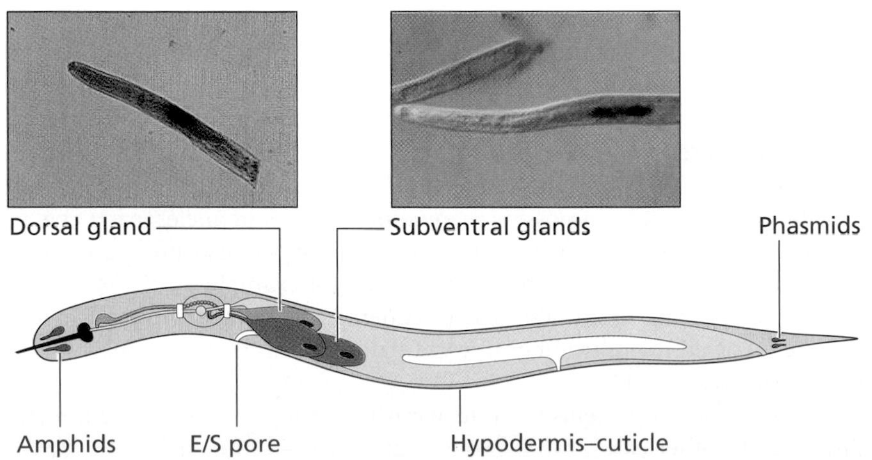

Figure 4.4 Plant-parasitic nematodes. Schematic diagram of a typical plant-parasitic nematode showing its most important secretory organs, which include amphids (flanking the stylet), dorsal glands, subventral glands, and the hypodermis–cuticle complex. These organs secrete factors involved in countering plant defenses. The site of production of these factors can be visualized using *in situ* hybridization in which a labeled complementary DNA or RNA strand is used as a probe to localize a specific messenger RNA (corresponding to the protein factor) in the tissue of the nematode where the mRNA is being expressed. Immunosupressive factors are localized to both the dorsal and subventral glands. (Figure adapted from Haegeman et al [2012] *Gene* 492:19–31. With permission from Elsevier. Photos left to right from Jones et al [2009] *Mol Plant Pathol* 10:815–828 and Jones et al [2003] *Mol Plant Pathol* 4:43–50. With permission from John Wiley and Sons.)

purpose. Plant nematodes also produce effectors (such as Gp-RBP-1 produced by *Globodera pallida*, the potato cyst nematode) that are recognized by corresponding R proteins (GPA2). Nematodes may also interfere with production of salicylic acid and other signaling pathways involved in defense, such as those producing jasmonate, another known plant defense-related hormone.

Although there are many additional fascinating topics deserving of attention, we can only touch on the basics of plant immunity in this discussion. We know that plants are routinely infected with fungal symbionts to which they have made special mutualistic accommodations and that may play an important role in assisting plants to defend against bacteria or even rival fungi. Last, there is intriguing evidence to suggest that plants respond to stress, including parasite exposure, by increasing recombination rates of their genetic material. Such recombination could potentially lead to somatic diversification of their R genes by unequal crossing over or gene conversion. This could conceivably provide plants (which are often very long-lived with slow generation times) with a much more flexible way to respond to fast evolving plant parasites. Furthermore, these rearranged R genes can be transferred to the plant's progeny, potentially conferring on them increased resistance to infection.

Invertebrates have distinctive and diverse innate immune systems

All but one of the 36 or so major animal lineages or phyla are composed exclusively of invertebrates. Even our own phylum, the Chordata, contains numerous invertebrate representatives. Some of the general rules of thumb regarding invertebrate immunity are that invertebrates possess the ability to distinguish self from nonself and that they are capable of both constitutive and inducible defenses. Furthermore, their defense responses generally involve both humoral and cellular components and usually employ limited repertoires of defense molecules. Consequently, the degree of specificity and the immunological memory shown by invertebrate defense responses are limited when compared to the responses of jawed vertebrates. It is frequently stated that invertebrates exhibit **innate immunity** characterized by a relative rapid response time and deployment of basic categories of PRRs that recognize PAMPs and that can be up-regulated upon exposure to parasites. In contrast, jawed vertebrates like ourselves also possess **adaptive immunity** that works in conjunction with the innate immune system. Adaptive immunity is the predominant work of the lymphoid system and, though slower in its deployment than innate immunity, is capable of exquisitely specific detection of a broad spectrum of **antigens**. An antigen is any molecule that is recognized immunologically in a highly specific manner, inducing a targeted immune response. Vertebrate animals owe their ability to respond to so many antigens to the existence of large and diverse lymphocyte populations, that are able to recognize many different antigen types when first presented. Vertebrate defense systems are discussed at length later in this chapter.

Another part of the conventional wisdom for invertebrates is that many of the building blocks of innate immunity have been highly conserved over evolutionary time and that there has been an undeniable tendency to arrive at similar defense solutions in phyla as distinct as cnidarians and chordates. However, lest we draw an erroneous conclusion that all invertebrates have similar immune systems, consider that these animals have been evolving and diversifying for a long time—about 800 million years—usually under very different ecological circumstances. In recent years, particularly with the advent of genomics research, we have gained a much greater understanding of invertebrate immune capacities, revealing a number of surprises in the process. In this discussion, we highlight some of the surprising aspects of the interactions between some invertebrates and the parasites they host, with an emphasis on those involved in transmission of medically important parasites. We are getting closer to the point at which some of this knowledge can actually be put to work to help us fight certain infectious diseases, and this too is described. See **Box 4.1** for a consideration of a simple, yet often overlooked question; do animal parasites themselves have immune systems?

Invertebrates, including vectors and intermediate hosts, mount immune responses to contend with their parasites

Let us consider immunoparasitology from the perspective of the invertebrate host. As we have noted, invertebrates are generally considered to have limited repertoires of PRRs. How then can invertebrates, some of which live longer than we do, keep up with the pressures posed by their parasites, which evolve relatively quickly? One answer to this question is that invertebrates may not be quite so constrained in their ability to produce diverse PRR repertoires as first thought. Consider the example of fibrinogen-related proteins (or FREPs) produced by snails such as *Biomphalaria glabrata*, one of the most important intermediate hosts for the trematode ***Schistosoma mansoni***. Although we are most familiar with fibrinogen molecules functioning in blood coagulation in vertebrates, in invertebrates fibrinogen-related molecules seem to be more involved in nonself recognition. Snail defense cells called **hemocytes** produce FREPs. FREPs are able to agglutinate parasite

BOX 4.1

Do Parasitic Helminths Have Immune Systems?

It might seem nonsensical to ask about helminth immune systems. After all, helminths are parasites, and immunity exists to protect hosts from parasites. However, helminths are routinely exposed to viruses, bacteria, and eukaryotic parasites in both the free-living and parasitic phases of their life cycles, and indeed helminths do suffer from infections. Cestode larvae are infected with viruses, and *Salmonella* bacteria attach to the surface tegument and are found in the gut of adult schistosomes (Figure 1). Some trematodes, such as *Nanophyetus salmincola*, are routinely infected with bacteria, such as *Neorickettsia helminthoeca*, that they can pass on to their vertebrate hosts, in some instances with fatal consequences.

Caenorhabditis elegans, the intensively studied free-living model nematode is infected naturally with at least four different parasites: the gram-negative bacterium *Microbacterium nematophilum*, the fungus *Drechmeria coniospora*, the microsporidian parasite *Nematocida parisii*, and a nodavirus-like Orsay virus (Figure 2). *C. elegans* reminds us that for any organism, if you look hard enough, distinctive parasites will be found. Furthermore, these relatively newly discovered parasites will prove to be very useful tools to supplement the powerful genetic toolkit available to reveal mechanistic processes in *C. elegans* that will shed light on the origins and functions of our own immune systems. This model system will eventually answer such questions as: How do helminths keep their bodies from routinely being overrun by viruses, bacteria, or other parasites? What do we know at this still early stage about parasitic helminths in particular?

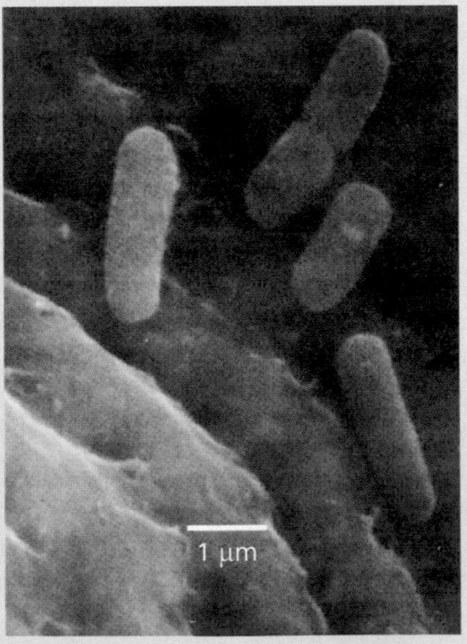

Figure 1 A parasite of a parasitic helminth. Scanning electron micrograph showing rod-shaped *Salmonella typhimurium* LT2 attached to the tegument of *Schistosoma mansoni*. (From LoVerde P et al [1980] *J Infect Dis* 141:177–185. With permission from Oxford University Press.)

Helminths including nematodes, trematodes, and cestodes lack specialized defense cells such as hemocytes and, at least for nematodes, are dependent on the secretion of antimicrobial peptides to counter their parasites. The signaling pathways involved are distinctive and culminate in the production of unique categories of antimicrobial peptides such as caenopores that cause leaks to form in bacterial membranes. As compared to the free-living bacteriovore *C. elegans*, the genome of the root knot plant parasitic nematode *Meloidogyne incognita* carries fewer immune genes. Genes homologous to those of most of the innate immune signaling pathways of *C. elegans* are found in *M. incognita*, but immune effectors such as lysozyme, C-type lectins, and chitinases were much less abundant in the parasite. Entire classes of immune effector genes were lacking in *M. incognita*, including both antibacterial and antifungal genes. It was speculated that *M. incognita* is sheltered from a variety of biotic and abiotic stresses as a consequence of its privileged life embedded in roots of plants.

For the filarial worm ***Brugia malayi***, the innate immune system is largely comparable to that of *C. elegans*, having thioester proteins, scavenger receptors, C-type lectins, and galectins. However, antimicrobial peptides were conspicuously lacking in ***B. malayi***, suggesting either that filarial worms lack this effector arm altogether or use their own distinctive small peptides. It is noteworthy that another parasitic nematode, ***Ascaris suum***, does have antimicrobial peptides.

Another way to view the defense responses of parasitic nematodes is that they are actually geared largely toward countering the aggressive attacks from the immune systems of their hosts. For ***B. malayi***, these hosts include both the mosquito vector and the mammalian definitive host. This nematode produces several antioxidants that may protect its surface from attack by oxygen radicals in either kind of host. Similarly ***A. suum*** produces a number of compounds believed to inhibit host B lymphocytes, factors that mimic host immune components, anti-inflammatory factors, and peptides that may mask the parasite's antigens by mimicking host molecules.

References

Abad P, Gouzy J & Aury JM (2008) Genome sequence of the metazoan plant-parasitic nematode *Meloidogyne incognita*. *Nat Biotechnol* 26:909–915 (doi:10.1038/nbt.1482).

Barnhill AE, Novozhilova E, Day TA et al (2011) *Schistosoma*-associated *Salmonella* resist antibiotics via specific fimbrial attachments to the flatworm. *Parasit Vectors* 4:123 (doi:10.1186/1756-3305-4-123).

Ghedin E, Wang S, Spiro D et al (2007) Draft genome of the filarial nematode parasite *Brugia malayi*. *Science* 317:1756–1760 (doi:10.1126/science.1145406).

Headley SA, Scorpio DG, Vidotto O et al (2011) *Neorickettsia helminthoeca* and salmon poisoning disease: A review. *Vet J* 187:165–173 (doi:10.1016/j.tvjl.2009.11.019).

Jex AR, Liu S, Li B et al (2011) *Ascaris suum* draft genome. *Nature* 479:529–U257 (doi:10.1038/nature10553).

LoVerde P, Amento C, Higashi G (1980) Parasite-parasite interaction of *Salmonella typhimurium* and *Schistosoma*. *J Infect Dis* 141:177–185.

Marsh EK & May RC (2012) *Caenorhabditis elegans*, a model organism for investigating immunity. *Appl Environ Microbiol* 78:2075–2081 (doi:10.1128/AEM.07486-11).

Ottens H & Dickerson G (1972) Studies on the effects of bacteria on experimental schistosome infections in animals. *Trans R Soc Trop Med Hyg* 66:85–107.

Mueller JF & Strano AJ (1974) *Sparganum-proliferum*, a sparganum infected with a virus. *J Parasitol* 60:15–19 (doi: 10.2307/3278671).

Tarr D & Ellen K (2012) Distribution and characteristics of ABFs, cecropins, nemapores, and lysozymes in nematodes. *Dev Comp Immunol* 36:502–520 (doi: 10.1016/j.dci.2011.09.007).

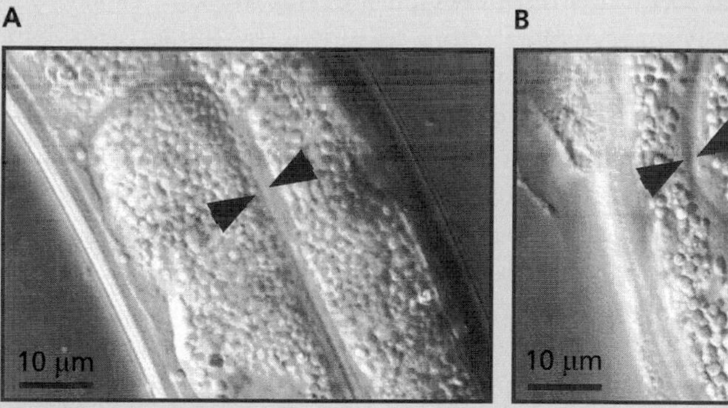

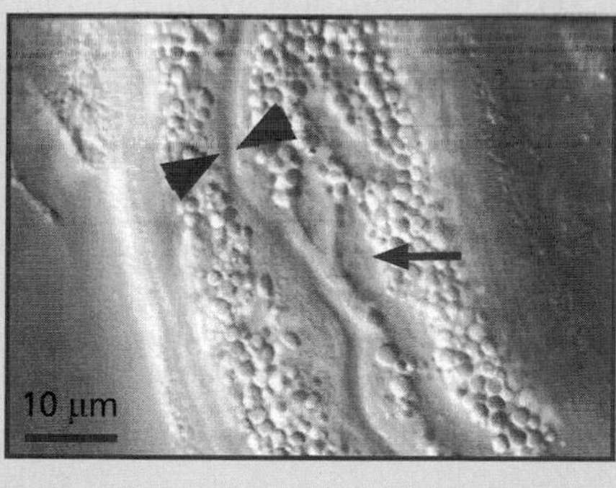

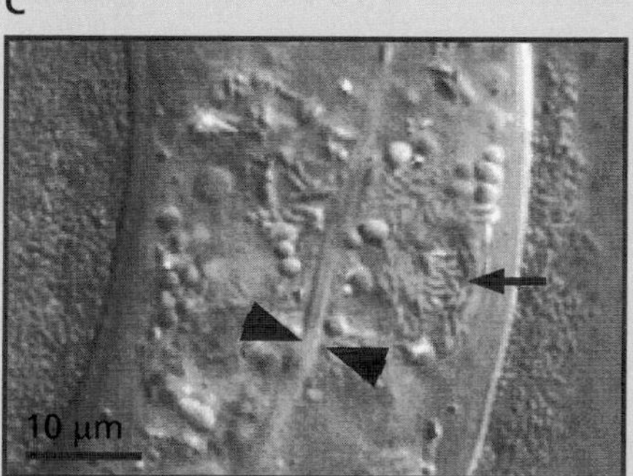

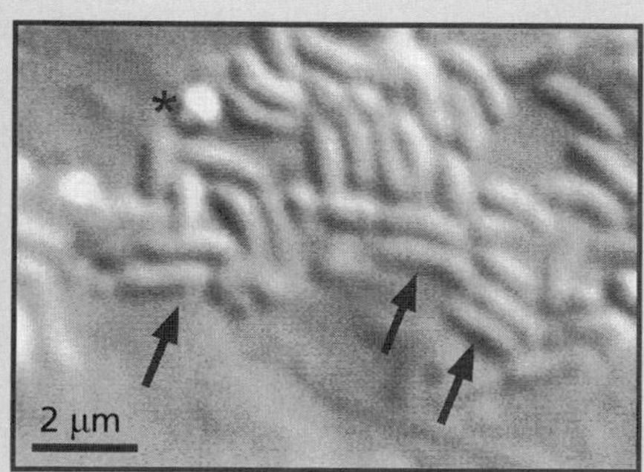

Figure 2 A microsporidian parasite of the worm, *Caenorhabditis elegans*. (A) Intestine of uninfected *C. elegans* surrounded with cells having normal gut granules. Arrowheads indicate intestinal lumen in A–C. (B) Infection by the microsporidian *Nematocida parisii* causes displacement (arrow) of gut granules. (C) Small rod-shaped microsporidians (arrow) in intestine. (D) Higher magnification of rod-shaped microsporidians. (From Troemel ER et al [2008] *PLoS Biol* 6(12): e309. doi:10.1371/journal.pbio.0060309.)

antigens, and they can act as **opsonins**, molecules that bind to a target and enhance phagocytosis. Consequently, binding of an opsonin to a parasite surface increases the likelihood of subsequent phagocytosis or encapsulation of the parasite. Because they are organized in tandem arrays, FREP-encoding genes can engage in gene conversion, with the result that FREPs can be diversified during the course of the snail's ontogeny (an example of somatic diversification). As a consequence, the hemocytes within a given snail do not necessarily all express the same FREP proteins. There is some evidence to suggest that diverse FREPs interact with highly variable polymorphic mucin molecules produced by life-cycle stages of ***S. mansoni*** that infect snails. Furthermore, suppressed production of one FREP, FREP3, results in partially diminished levels of resistance to trematodes in *B. glabrata*. Consequently, outcomes of interactions between snails and schistosomes could be governed by interaction between parasite- and snail-produced molecules that are more specific and varied than generally conceived. Thus the standard interpretation of invertebrate immune systems as possessing but limited sets of PRRs may require some modification. Some form of diversified defense molecules have been found in several invertebrate groups such as crustaceans, snails, bivalves, insects, echinoderms, tunicates, and amphioxus.

Our standard notions regarding invertebrate immunobiology also need to be reconsidered when evaluating the ability of some invertebrates to mount specific heightened responses upon secondary exposure to the same antigen (**Figure 4.5**), a response that might be called **innate immune memory**. This response differs somewhat from the long-term and exquisitely specific form of immunological memory exhibited by mammals, which is carried out by specialized populations of immune system cells. In invertebrates, there is as yet no evidence for the existence of populations of persistent immune cells that are sequestered and await a secondary exposure to the same antigen. Rather, it appears that some immune cells can be boosted in activity such that when they are exposed again to the same antigen, they can respond more rapidly and effectively and with some specificity.

Such heightened secondary responses may be relevant with respect to parasite transmission by invertebrate vectors, as shown by studies of ***Anopheles gambiae*** mosquitoes exposed to ookinetes of ***Plasmodium***. When ookinetes penetrate the mosquito gut wall, they enable gut bacteria to contact damaged gut epithelial cells. This sensitization is believed to stimulate a robust, long-lived, and enhanced antibacterial response that also plays a role in reducing survival of ***Plasmodium*** ookinetes in later exposures. The priming exposure to ookinetes results in increased production and activity of granulocyte cells in the mosquito's hemolymph, which are more effective in damaging ookinetes in later exposures.

Invertebrates also adopt distinctive behaviors to supplement their anti-parasite immune responses

As we will discuss in Chapter 6, immune responses are costly to mount, and heavy investment in defenses can limit the resources available for other key functions such as reproduction. This general constraint would certainly apply to invertebrates too. Consequently, it is not surprising that animals resort to particular behaviors to limit their exposure to parasites or, in the context here, to supplement their immune responses. One example is provided by the larvae of *Drosophila melanogaster*, which consume yeast developing in rotting fruit. The larvae are susceptible to attack by parasitoid wasps (species of *Leptopilina*), the females of which inject their eggs into the fly larvae.

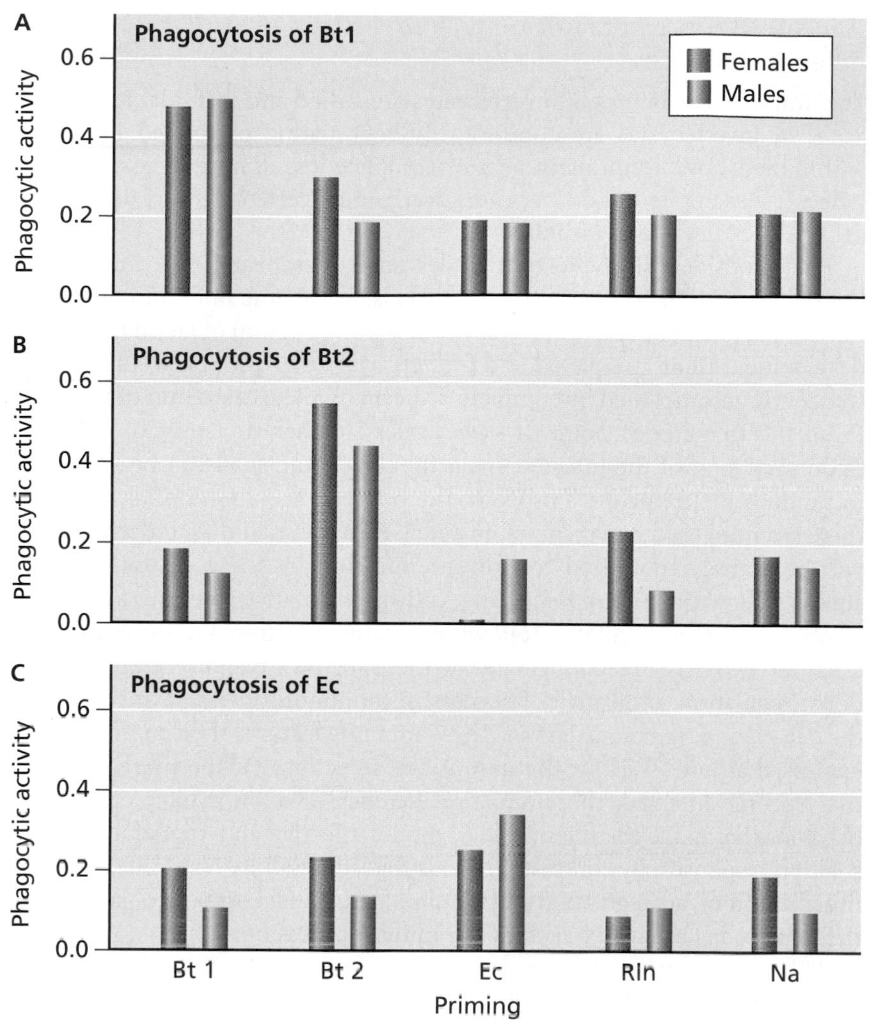

Figure 4.5 Innate immune memory. Shown for females (orange bars) and males (blue bars) of the woodlouse *Porcellio scaber* (Crustacea: Isopoda) is phagocytic activity (vertical axis is hemocytes phagocytosing/total hemocytes) for bacteria. Woodlice were first primed (horizontal axis) by exposure to a particular strain or species of bacteria. Bacteria used were Bt1 (*Bacillus thuringiensis* strain 1), Bt2 (*B. thuringiensis* strain 2); Ec (*Escherichia coli*), or Rin (for Ringer's saline). Na refers to animals that were not exposed to a priming dose and that serve as controls. The three panels (top to bottom) indicate woodlice that were then challenged two weeks later with Bt1, Bt2 or Ec. Note that phagocytosis rates upon secondary exposure were always highest in animals that had been primed with the same (homologous) bacterium. This trend was significant for the Bt1 and Bt2 priming experiments and trended the same way for the Ec experiment. Specificity was evident even at the level of strains of bacteria used. (From Roth O & Kurtz J [2009] *Dev Comp Immunol* 33:1151–1155. With permission from Elsevier.)

The developing wasp larvae then consume and eventually kill the fly larvae. Upon wasp attack, fly larvae seek out food containing ethanol, which proves to be an advantageous drinking habit. Wasp larvae are more sensitive to the effects of ethanol than the fly larvae, and ethanol supplementation in the larval fly diet reduces wasp infection success, increases killing of larval wasp larvae, and promotes survivorship of larval flies without the fly larvae having to mount typical defense responses.

The parasite avoidance behaviors of social insects such as ants and bees are legendary, perhaps explaining why they seem to have lost immune genes as compared to other insects. One example of how behavior can facilitate a more effective colonywide defense from parasites is provided by the observation that uninfected ants (*Lasius neglectus*) frequently rub against colony members infected with the fungus *Metarhizium anisopliae*, in the process becoming exposed to a low level of infection. The exposure stimulates up-regulation of a set of immune genes that have the overall effect of boosting antifungal immunity specifically. Interestingly, the ants do not show evidence of boosted antibacterial responses. By virtue of this controlled exposure to the fungal pathogen, ants are usually not killed and exhibit an enhanced ability to inhibit fungal growth. This process has been referred to as **social immunization**.

Parasites suppress, manipulate, and destroy invertebrate defense responses

The immune capabilities of invertebrates are varied and formidable, yet parasites of invertebrates are common, often initiating prolonged infections within their hosts resulting in near or complete loss of host fitness. We provide a few examples of how parasites overcome invertebrate host defenses to achieve such devastating results.

First, consider digenetic trematodes such as schistosomes undergoing their obligatory larval development in snails. In general, once an infection is initiated, it takes about a month for the complex program of larval trematode development to be completed and for cercariae to be produced. Infection is frequently accompanied by complete or near complete castration of the snail. From the trematode's point of view, unless further transmission depends upon predation of the infected snail, the longer the snail can be kept alive to produce the trematode's progeny, the better. Thus, upon initial infection, there is a need to first overcome the defenses of the snail that understandably will be actively mobilized to prevent infection by these castrating parasites. As shown in **Figure 4.6**, many known or putative immune features of snails are first up-regulated following exposure to trematodes, but within a day or two post-infection many of these features begin to trend toward down-regulation, indicative of successful immunosuppression mediated by the developing trematode larvae. These immune features then continue to be expressed at low levels for the duration of infection. Furthermore, in snails that become infected, the circulating hemocytes often exhibit changes in behavior that make them less able to mount effective anti-trematode encapsulation responses. By contrast, in snails that successfully resist infection, the initial trend of up-regulation of immune features is maintained, suggesting the trematode has failed to achieve immunosuppression.

Although parasitized snails essentially have their bodies taken over by the parasites, given the relatively long time course of infection (from weeks to years), it is not in the parasite's interest to have its host so immunocompromised as to succumb to any opportunistic pathogen. Hemocytes from

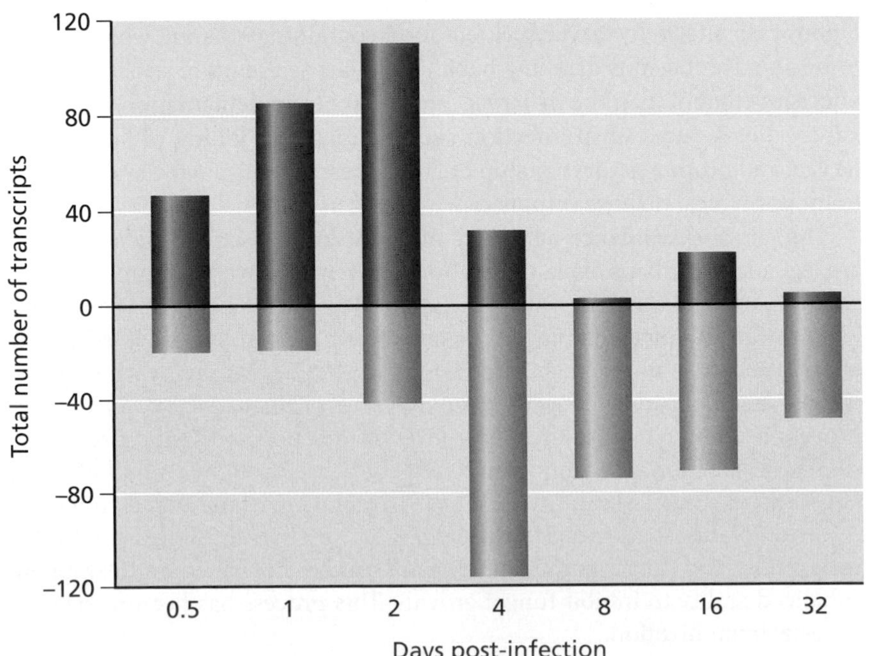

Figure 4.6 Infection and altered gene expression. The graph shows the total number of differentially expressed transcripts (up-regulated in purple and down-regulated in blue) in *Biomphalaria glabrata* after infection with *Schistosoma mansoni*. Snails infected with *S. mansoni* were collected at 0.5, 1, 2, 4, 8, 16, and 32 days post-infection and analyzed using a *B. glabrata* microarray targeting immune- and stress-associated transcripts. Each time point represents data collected from 15 individual snails, pooled into three experimental replicates. Transcripts expressed at the cutoff for significance of 5% false positive rate and ±1.0 $\log_2$ in all experimental groups were considered to be differentially expressed. (From Hanington PC et al [2010] *Int J Parasitol* 40:819–831. With permission from Elsevier.)

infected snails retain the ability to phagocytose bacteria or encapsulate nematode larvae and, as described in Chapter 6, trematode larvae themselves may actively contribute to the defense of their host. The host defense may come in the form of overt attack by trematode rediae on the larvae of newly invading trematodes or by secretion of soluble factors that suppress growth of other trematode species. Although the extent to which trematode larvae or other parasites might contribute to the defense of their hosts is largely unknown and worthy of study, as we have noted before, trematode infections can be remarkably persistent with some snails living for over a decade with trematode infections. This is particularly remarkable because a succession of different trematode species may occupy the same snail over such long intervals of infection.

Some parasites rely on symbiotic partners to subvert the immune responses of their invertebrate hosts

There are other equally fascinating examples of complex relationships between parasites and their invertebrate host immune systems. Consider wasps belonging to two huge hymenopteran families, the Ichneumonidae and the Bracondidae, that routinely oviposit in other insects (**Figure 4.7**). Their larvae then proceed to grow within the insect host, consuming it from within and eventually killing it—these wasps are parasitoids (see also Figures 1.6A). They rely on a remarkable symbiotic association, one that is best described as mutualistic, with particular kinds of viruses called polydnaviruses to achieve infection of their hosts. Polydnaviruses are found nowhere else but in these two parasitoid wasp families. The viruses are so named because their genome

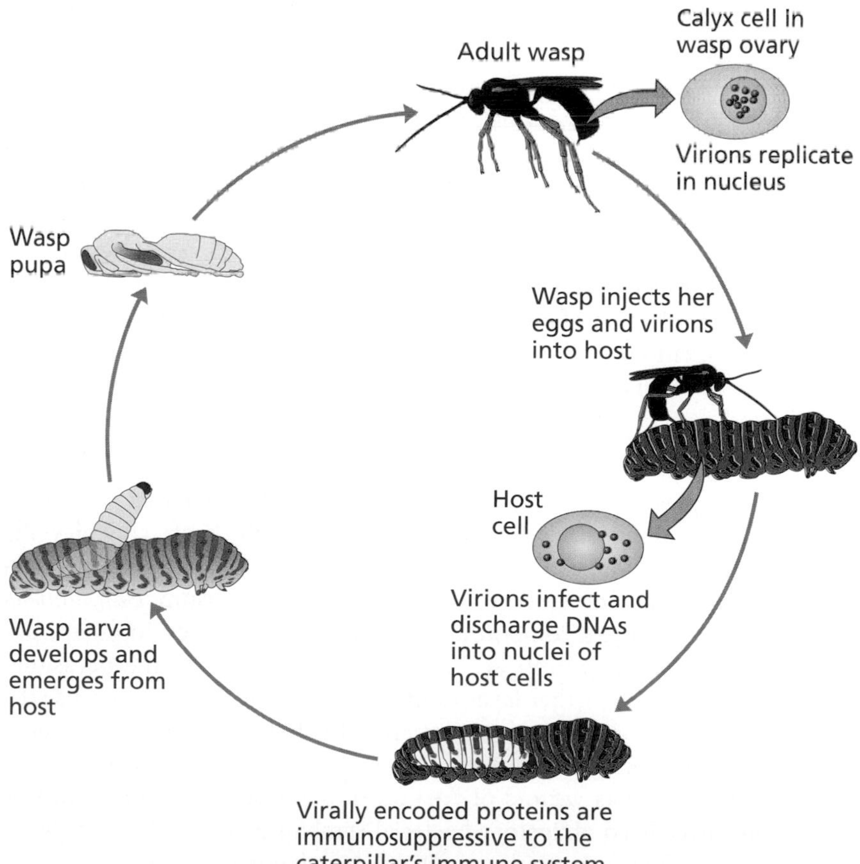

Figure 4.7 Life cycle of a bracovirus (BV). The BV genome is integrated into the genome of a braconid wasp and in this state is referred to as a provirus. The virus replicates to produce virions only in the nuclei of calyx cells in the ovary of an adult female wasp. The wasp injects virions plus one or more of her eggs containing the proviral genome into the host insect, a caterpillar. Each wasp egg hatches and releases a wasp larva that feeds on the host. Virions infect and discharge their DNAs into host cell nuclei, which then rapidly integrate into the genome of the host cell. Viral virulence genes are then transcribed in host cells over the time required for the wasp larvae to complete their development. Virally encoded proteins are immunosuppressive to the caterpillar's immune system. Upon completing development, the wasp larva emerges from the host caterpillar to pupate; the caterpillar dies. Each germ line and somatic cell of the newly produced wasp contains the BV genome in its proviral form. (Modified from Burke GR & Strand MR [2012] *Insects* 3:91–119.)

consists of several separate pieces of double-stranded DNA. These peculiar viruses are stably integrated into the genome of the wasp as proviruses, so they are regularly propagated from generation to generation when the wasp reproduces. Actual virus particles (virions) are produced only in one specific location, the calyx cells in the ovary of the female wasp. They then accumulate in large numbers in her reproductive tract. The virions are injected into the host insect along with the wasp's egg. The virions then infect hemocytes, fat bodies, and other tissues of the host. The virus enters host cell nuclei but does not integrate, nor is it able to replicate there. Nonetheless, viral genes (called virulence genes) are quickly expressed, and many of the proteins involved have the explicit task of disabling the host's immune response. For example, the ability of host hemocytes to bind to wasp eggs is disabled, host hemocyte signaling cascades are disrupted, and the phenoloxidase enzyme is inhibited, thereby preventing the host from melanizing and killing the wasp larvae. In particular, virally encoded proteins Ank-H4 and Ank-N5 bind strongly to Relish in the host's IMD (Immune Deficiency) pathway and strongly inhibit its processing, thereby preventing its translocation to the nucleus and activation of synthesis of antimicrobial peptides. Polydnavirus gene products also alter host insect development in ways to favor success of the wasp's larvae. Even though the virus cannot multiply in the wasp's host insect, they nonetheless make it possible for the wasp's larvae to successfully develop, thereby ensuring their own survival within the wasp's genome.

Another example of a collaborative, mutualistic arrangement to achieve infection of an invertebrate host is provided by entomopathogenic nematodes such as *Steinernema* and *Heterorhabditis* (see also Box 7.2). Infective *Steinernema* larvae actively search for, find, and infect insect hosts. Once in the insect, they release from special areas of their intestine mutualistic bacteria (*Xenorhabdus*) that then proliferate rapidly in the host insect's hemocoel, producing a number of factors that damage the host's hemocytes. The bacteria also inhibit expression of host-produced antimicrobial peptides, such as cecropin, and inhibit prophenoloxidase and thus melanization. Furthermore, the bacteria produce antimicrobial factors that prevent the growth of opportunistic bacteria and release enzymes that degrade molecules produced by the host insect, providing a nutrient soup that favors growth of their associated nematodes. Consequently, both *Xenorhabdus* and *Steinernema* proliferate in the host, which is soon killed. Eventually thousands of larval nematodes leave the host insect, each carrying an inoculum of these specialized bacteria to facilitate infection of a hapless new host. Because of their efficiency in killing insects, entomopathogenic nematodes have been used widely as biological control agents.

Some invertebrates enlist symbionts to aid in their defense

The two examples noted above are mutualistic partnerships that infect and overcome invertebrate host defenses, but invertebrate hosts can play the symbiont game too, enlisting them to augment their defenses against parasite attack. By their possession of the symbiotic bacterium *Hamiltonella defensa*, which produces compounds that suppress wasp development, pea aphids (*Acyrthosiphon pisum*) are protected from attack by the parasitoid wasp *Aphidius ervi*. A further layer of intrigue exists in this case because the factors that prevent wasp development are actually toxins encoded in genes found in a bacteriophage that infects *H. defensa*. Tsetse flies such as ***Glossina morsitans*** that serve as vectors of African trypanosomes harbor a number of different specialized obligatory bacterial symbionts, some that can play a protective role. One such symbiont, *Wigglesworthia glossinidia* (**Figure 4.8**),

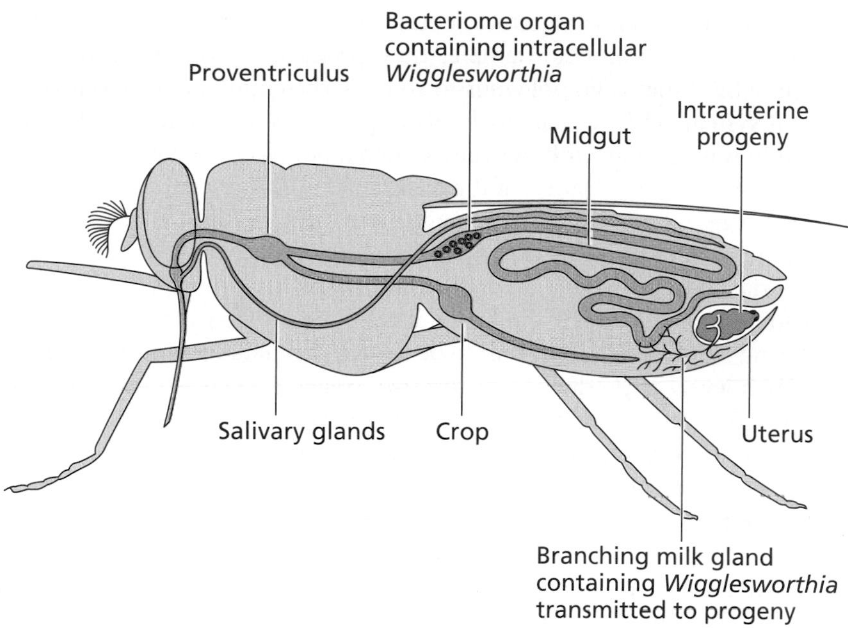

Figure 4.8 Tsetse flies and their symbionts. A female tsetse fly (*Glossina*) is shown. Its obligate mutualist symbiont, *Wigglesworthia glossinidia,* develops intracellularly in the bacteriome near the midgut of the fly. During their intra-development, trypanosomes reside nearby in the midgut and then in the salivary glands of the fly. *Wigglesworthia glossinidia* is transmitted to the intrauterine progeny in the milk produced by the branching milk glands of the female fly. (Modified from Weiss B & Aksoy S [2011] *Trends Parasitol* 27:514–522. With permission from Elsevier.)

provides vitamins that are not otherwise available in the exclusively blood-derived diet of tsetse flies. In addition, if flies are denied access to *W. glossinidia* during their development, they are both sterile and immunocompromised. They are vulnerable to infections with *Escherichia coli* and exhibit reduced expression of genes that encode antimicrobial peptides (cecropin and attacin), other hemocyte-derived proteins (thioester-containing proteins 2 and 4 and prophenoloxidase), and signal-mediating molecules. Furthermore, they have reduced hemocyte populations, and they become more susceptible to trypanosome infections. This evidence indicates that *W. glossinidia* must be present if the tsetse immune system is to develop properly. Studies of the human **microbiome** (the collection of microorganisms that normally reside in or on the body) reveal a similar trend: full immune maturation depends on the presence and stimulation provided by human microbial symbionts. In the case of the tsetse flies, there are also some indications that other species of bacterial symbionts can also provide nutrients such as phenylalanine needed for trypanosome development and so may actually favor development of trypanosome infections in some species of ***Glossina***.

Researchers hope to manipulate invertebrate immune systems to achieve parasite control

A major research effort in recent years has been to increase our understanding of the immune systems of prominent vectors such as mosquitoes. Coupled with improved understanding of the evolutionary biology of vector–parasite relationships (see Chapters 7 and 10), we could potentially exploit this knowledge to increase mosquito refractoriness to major parasites such as arboviruses (for example, dengue virus) or ***Plasmodium*** that they vector. It is envisioned that parasite-refractory mosquitoes could then be introduced into natural populations or otherwise be encouraged to proliferate, such that the ability of mosquito populations to support and transmit dengue or malaria infections to humans would be reduced. By diminishing transmission, this approach could complement chemotherapy or vaccine development to achieve control that is more sustainable over the long haul. Are we making progress with such an endeavor?

We now know that mosquitoes mount diverse defenses against parasites such as ***Plasmodium***. For example, they produce a thioester-containing protein (TEP1) that is homologous to some of the components of our complement system. TEP1 covalently binds to parasite surfaces, leading to their demise in ways that are still not fully understood. If TEP1 expression levels are diminished by application of the technique of RNA interference (RNAi), then ***Anopheles*** mosquitoes become more susceptible to ***Plasmodium*** infection. Furthermore, in at least some systems, mosquitoes susceptible to malaria infection have a different TEP1 allele than mosquitoes that are refractory to infection. TEP1 interacts with two leucine-rich repeat proteins, LRM1 and APLC1, that also seem to be required to make mosquitoes refractory. High levels of production of reactive oxygen species (ROS) and nitric oxide (NO) are also associated with refractoriness.

If allelic versions of genes associated with refractoriness could be identified, appropriately engineered, and successfully introduced into natural populations, then the hoped for effects of diminished disease transmission might be achieved. However, it seems unlikely these allelic versions would spread easily and naturally through the mosquito population and achieve high frequency on their own. That is, if they are so advantageous to the mosquito, why hasn't natural selection already favored their spread? Several possible explanations exist. One explanation for why traits conferring refractoriness have not spread is that the natural prevalence rate of malaria infection in mosquito populations is usually low, suggesting ***Plasmodium*** per se may not be a major driver of mosquito evolution. A second explanation is that ***Plasmodium*** infections are not acutely pathogenic to mosquitoes, which makes sense from the parasite's point of view because infection depends on the mosquito being reasonably intact such that it can deliver it to and inject it into a new host. Consequently, there is reduced impetus for mosquitoes to develop strong anti-***Plasmodium*** responses. A third explanation is that there may be physiological and energetic costs associated with being refractory. For example, the reproductive output or longevity of mosquitoes could be reduced if resistance is energetically costly. All these factors may conspire against the success of getting anti-***Plasmodium*** genes to spread in the mosquito population.

Therefore, special mechanisms may need to be employed. For example, selfish genetic elements such as homing endonuclease genes could be used to drive the genes conferring resistance into natural populations. These selfish elements encode endonucleases that cleave a particular site in the genome and then effectively insert the genes encoding the endonuclease (and potentially engineered to contain anti-***Plasmodium*** factors, too) into the cleaved site. In this way, they continually spread to other homologous sites that lack the endonuclease genes, inevitably becoming more common in the process. Some evidence suggests these drive mechanisms will work. If further work shows conclusively that we can successfully drive genetic elements conferring resistance into mosquito populations, then we also need to be sure that engineered ***Plasmodium***-refractory mosquitoes are not also more refractory to all other mosquito pathogens as well; it would not be a good outcome to release super disease-resistant mosquitoes that then might become even more abundant than they already are. Also, this approach could quickly select for variants of ***Plasmodium*** that can overcome the engineered resistance genes. Although these may prove to be insurmountable problems, it can be argued we have nonetheless learned a great deal about mosquito biology, and this knowledge has paid off in other ways. For example, dengue virus vectors such as ***Aedes aegypti*** have been successfully infected with *Wolbachia* bacterial symbionts (see Chapters 2, 7, and 9 for more discussion of *Wolbachia*). These bacteria are close relatives of *Wolbachia* from *Drosophila* that are known to skew sex ratios, kill male flies, and favor fertilization only by males that are also infected. In the case of *Wolbachia*-bearing ***A. aegypti***, they can spread upon release into natural populations and replace *Wolbachia*-free ***A. aegypti***. Furthermore, the bacterial infection is pathogenic enough to kill many dengue-infected mosquitoes before they can transmit the infection to another person. A similar goal—to reduce longevity of the infected vector below the developmental time required by the vectored parasite—could also be used to prevent ***Plasmodium*** infections from achieving sporozoite production (see Chapters 6 and 9). A similar approach could also be pursued for larval schistosomes developing in snails, which also take a long time to develop relative to the life spans of their snail hosts. That is, a way forward may be to exploit our knowledge of vector or intermediate host immunology just enough to prevent transmission from occurring.

4.2 AN OVERVIEW OF VERTEBRATE DEFENSE

As noted earlier, an understanding of immunoparasitology requires a solid grounding in basic immunological processes. Yet here, before we specifically address the manner in which vertebrate immune systems and eukaryotic parasites interact, we provide a brief overview of the overall strategy of vertebrate immune defense.

As we discussed in Chapter 3, reaching a host is just the beginning if a parasite is to successfully colonize that host, reproduce, or undergo required developmental changes. Vertebrates, like the other hosts we have just examined, come equipped with diverse defenses known as **barriers to entry** that stop most parasites before they can reach their target tissue. Intact skin, for example, is a formidable mechanical barricade, through which very little can penetrate (schistosome cercariae and hookworm and strongyloid L3 larvae are exceptions). Indeed, effective barriers to entry explain why most interactions between host and parasite do not result in actual colonization and infection. The attempted colonization is nipped in the bud before it begins.

However, barriers can be compromised and parasites have evolved to overcome even intact barriers in specific targeted hosts. When parasites inevitably circumvent these barriers in individual hosts, the host's defensive strategy switches from keeping invaders out to destroying the invaders (**Figure 4.9**). The vertebrate immune response to successful invasion can conveniently be divided into two parts. Initially, components of the innate immune system recognize and respond to the infection. The innate immune system represents those immune components that are immediately activated and responsive to pathogens in a relatively nonspecific manner. Phagocytosis, tissue inflammation, and fever all represent innate responses to infection.

Sometimes innate immunity alone is enough to contain and destroy invading parasites. In such cases, innate responses abate following parasite elimination. In other situations, innate defenses are unable to quickly eradicate the parasite. In such circumstances, elements of the innate immune system activate the adaptive immune system. The adaptive immune response is the highly targeted counterattack on specific pathogens. Through the activation of specific cell types and effector molecules, the adaptive immune system can recognize and respond to an essentially limitless number of pathogens. Unlike an innate response, an adaptive response to a particular pathogen becomes faster and stronger with each subsequent exposure. This is an aspect of the immunological memory that characterizes adaptive immunity, as first described by Thucydides.

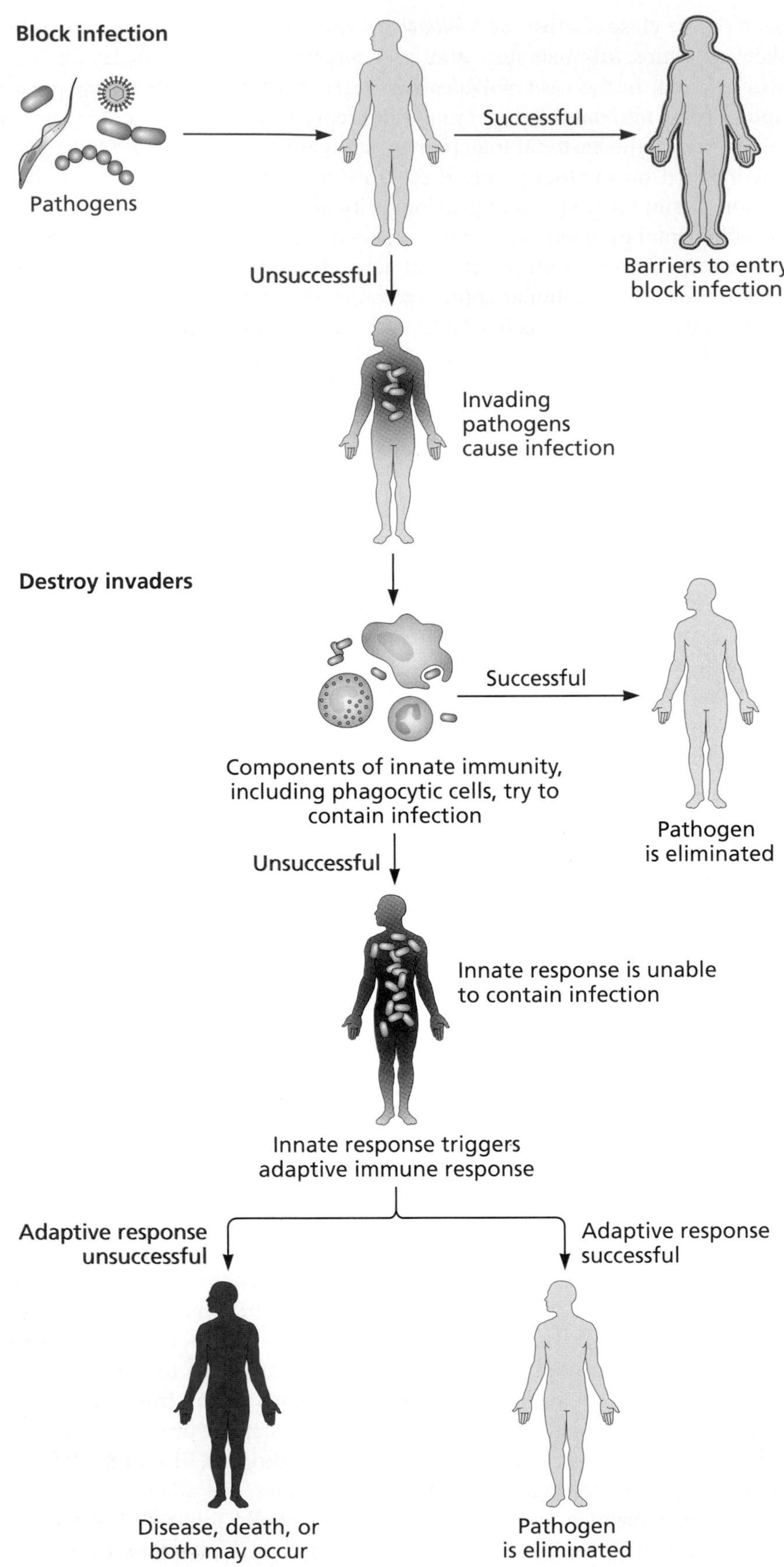

Figure 4.9 The strategy of immune defense. Initially, the host attempts to prevent infection using various barriers to entry. Should these barriers fail, an immune response is initiated. First, induced innate responses are activated, in an attempt to eliminate the pathogen. If innate responses alone are insufficient, elements of innate immunity activate adaptive immunity, which mobilizes a highly specific, targeted attack on the pathogen. (From Hofkin BV (2011) Living in a Microbial World 1ed. Garland Science.)

If the parasite is eliminated as a result of an adaptive immune response, health is generally restored and there are often no long-term consequences (see Figure 4.9). On the other hand, if the parasite is too virulent or if it successfully evades adaptive immune defenses, the host may die. These alternative scenarios are typical of acute infections, such as those caused by ***Plasmodium falciparum***. Other infections result in chronic disease, in which the immune response may contain but not eliminate the pathogen. Chronically infected hosts may continue to display certain symptoms of disease indefinitely, and the battle between host and parasite may be best described as a stalemate. Chronic infections are often the result of sophisticated immune evasion or manipulation on the part of the parasite. Chronic infections are the rule for many parasitic infections in vertebrate hosts. Infections with schistosomes and intestinal nematodes, as well as with many protozoa including *Giardia* and *Entamoeba*, are a few such examples. The ability of so many parasites to establish chronic infections indicates that the immune response to parasites is imperfect. Indeed, although most parasites stimulate strong immune responses, many such responses do not appear to be protective. Consider hookworms, for example. If infected individuals are treated with anthelmintics, they often become reinfected, demonstrating that long-term, protective immunity has not been established.

4.3 IMMUNE RESPONSES TO EUKARYOTIC PARASITES

We now turn to the vertebrate immune response as it occurs specifically against protozoa and helminths. The nature of that response will, of course, vary considerably based on the parasite in question. Intestinal parasites, for example, elicit different responses than those inhabiting the blood or burrowed within various tissues.

There certainly is no guarantee that any immune response will successfully eliminate the parasite. Indeed, many parasites are noteworthy for their ability to establish long-term chronic infections, even in the face of a robust immune response. As master immunologists in their own right, numerous parasites have evolved sophisticated mechanisms to hide from, suppress, or skew the immune response in such a way that their persistence is assured. Such persistence gives them the time they require to mature, reproduce, and achieve successful transmission to new hosts.

We will first consider immune responses to protozoa. Subsequently, we examine anti-helminthic immunology. Our goal is not to provide an encyclopedic description of immunity to every important parasite. Rather, we will look for general patterns in the way the vertebrate immune system recognizes and responds to these two very broad groups of eukaryotic parasites.

Recognition of PAMPS initiates the immune response to protozoa

Vertebrates, of course, are vulnerable to a wide assortment of protozoan infections. It is difficult to provide a comprehensive overview of the immune response to protozoa because these organisms are so variable in terms of life cycle, tissue tropism within the host, and other aspects of their basic biology. To cite just one example, protozoa may be predominantly intracellular or extracellular parasites. The specific type of immune response that might be expected to confer maximum resistance varies accordingly. Nevertheless, some general trends are increasingly evident.

When infected with a protozoan, a host's innate immune system must rapidly detect and respond to the parasite. Failure to do so results in the parasite overwhelming the host, with disease and death as likely outcomes. In this way, the response to protozoa is no different from the response to other pathogens. Until very recently, however, it has been unclear how the initial detection was achieved. The discovery of Toll-like receptors (TLRs) in the mid-1990s and the manner in which they initiate responses to bacterial and viral pathogen-associated molecular patterns (PAMPs), suggested the possibility that protozoa might be recognized in a similar way. Recently it has become clear that protozoa do indeed bear PAMPs that bind specific TLRs, initiating signal transduction and cytokine expression. Not only is such PAMP engagement critical for the development of a protective innate response, but also for the orchestration of an appropriate adaptive response. Many such protozoan PAMPs from various parasites are now known to bind specific mammalian TLRs (**Table 4.1**).

The most compelling evidence regarding the importance of TLR recognition of protozoan PAMPs comes from studies of mice lacking MyD88, a principal adaptor molecule and signal transducer for several TLR signaling pathways. Mice lacking MyD88, for example, are extremely susceptible to infection by ***Toxoplasma gondii*** and ***Trypanosoma cruzi***. In both cases there is unusually high parasitemia, greatly reduced production of inflammatory cytokines, and enhanced morbidity that generally leads to rapid death. Yet in some instances, even if mice lack a specific TLR, a PAMP that ordinarily binds this TLR may still elicit a response. This finding suggests that there is at least some redundancy in TLR response to protozoa and that no single TLR is essential for immune protection.

TLRs are found on the surfaces of macrophages and dendritic cells. Among the cytokines that are frequently up-regulated in response to TLR engagement is IL-12. One of the effector functions of this cytokine is the activation of **natural killer (NK) cells**. NK cells are lymphocytes that lack the specificity of T or B cells. They play an especially important role in the early response to many pathogens, prior to the full activation of an adaptive

Table 4.1 Protozoan PAMPs and TLR binding. The PAMPs listed are known to bind various TLRs. Many of these PAMPs are the membrane anchor regions of glycosyl phosphatidylinositol (GPI). These anchors may have somewhat different structures in different protozoa, or in different life-cycle stages. Thus these anchors do not all bind the same TLR.

PAMP	Protozoan	Expression stage	TLR
GPI anchors	*L. major*	Promastigote	TLR-2
	L. donovani	Promastigote	TLR-2
	T. cruzi	Trypomastigote	TLR-2
		Epimastigote	TLR-4
	T. brucei	Trypomastigote	TLR-2?
	P. falciparum	Merozoite	TLR-2, TLR-4
	T. gondii	Tachyzoite	TLR-2, TLR-4
Unmethylated CpG DNA	*T. brucei*	All stages	TLR-9
	T. cruzi	All stages	TLR-9
	P. falciparum	Merozoites	TLR-9
Profilin-like proteins	*T. gondii*	Tachyzoites	TLR-11

(Adapted from Gazzinelli RT & Denkers EY [2006] *Nat Rev Immunol* 6:895–906. With permission from Macmillan Publishers Ltd.)

response. Once activated, they serve as an early source of IFN-γ, which acts to potentiate a Th-1 response. IFN-γ also serves to activate phagocytes, and it increases their ability to kill ingested parasites through the production of reactive oxygen and nitrogen molecules (**Figure 4.10**). Such activation is crucial to early containment of a protozoan infection prior to the initiation of an adaptive response.

TLR engagement may also result in the expression of IL-10 and TGFβ, which counteract and suppress the potent inflammatory response to protozoa that might otherwise cause pathological and even life-threatening damage to host organs and tissue. Thus, the cytokine production by antigen presenting cells in response to TLR binding is an attempt to walk a fine line between sufficient but not excessive response. This delicate balancing act may explain a consistent feature not only of protozoan infections but of eukaryotic parasitic infections in general; complete expulsion or killing of the parasite is only rarely achieved. The costs of such complete immunity may simply be too high in terms of damage to host tissues and energy expenditure; the benefits of eradicating all parasites are exceeded by these costs. The solution is for the host to tolerate a certain level of parasitemia so long as the associated pathology is not too severe. Tolerance is discussed in Chapter 6.

Immune responses to protozoa include both humoral and cell mediated components

Like other pathogens, protozoan parasites stimulate both humoral and cell-mediated adaptive responses. In general, certain types of responses are consistent with immune protection and increased resistance, whereas others are more likely to result in increased morbidity and higher parasitemia.

A particularly striking example is provided by leishmaniasis. The factors that determine the outcome of infection with ***Leishmania*** are complex and depend in large part on the species of parasite involved. Host factors also

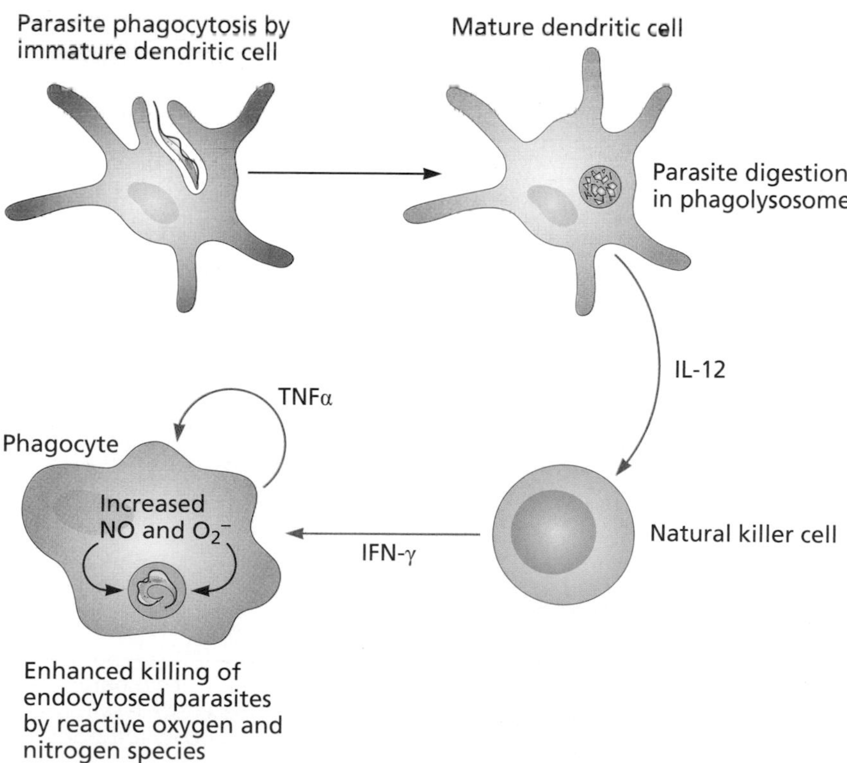

Figure 4.10 Activation of phagocytic cells by natural killer cells. Dendritic cells that have endocytosed a parasite release IL-12. One of the functions of this cytokine is the activation of natural killer cells. Activated NK cells release IFN-γ, which activates phagocytes, enhancing their ability to kill ingested parasites by stimulating them to produce increased levels of reactive oxygen and nitrogen species. The phagocyte's release of self-stimulating TNFα enhances this response. NK cells are a particularly important source of IFN-γ early in an immune response. Later in the response, Th-1 cells release large amounts of this interferon.

play an important part in disease progression. ***L. mexicana*** and ***L. tropica***, for instance generally cause a relatively benign form of the disease called **localized cutaneous leishmaniasis** (**Figure 4.11**). In this case, infected macrophages do not move far beyond the initial site of inoculation and the lesions that form are typically self-healing. However, on occasion the disease can progress to a far more serious **disseminated cutaneous leishmaniasis**. This condition is characterized by numerous disfiguring lesions over a wide area that are caused by the dissemination of infected macrophages. In patients who respond with a primarily cell mediated, Th-1 response, localized cutaneous leishmaniasis is the rule. Activated Th-1 cells release abundant IFN-γ, which enhances the ability of macrophages to kill parasites, effectively terminating the infection, and allowing healing to commence. There is also increased activation of cytotoxic T cells, which are able to identify and kill infected cells. Few antibodies are observed. In individuals in whom the response is primarily humoral, although antibody titers are high, there is little or no effect on this intracellular parasite. Consequently, the parasites continue to survive and replicate and, as infected macrophages disseminate away from the original site of infection, the gross tissue damage associated with the disseminated form of the disease is observed.

Both cell-mediated and humoral immunity come in to play in response to ***T. gondii***, ***Cryptosporidium***, and other apicomplexans. Although such parasites provoke a strong humoral response, antibodies appear to play little role in the primary response to infection with apicomplexans. Yet extensive experimental evidence indicates the importance of humoral immunity

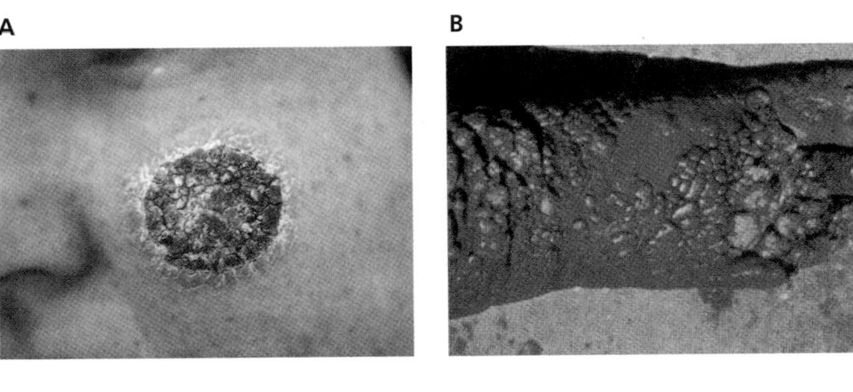

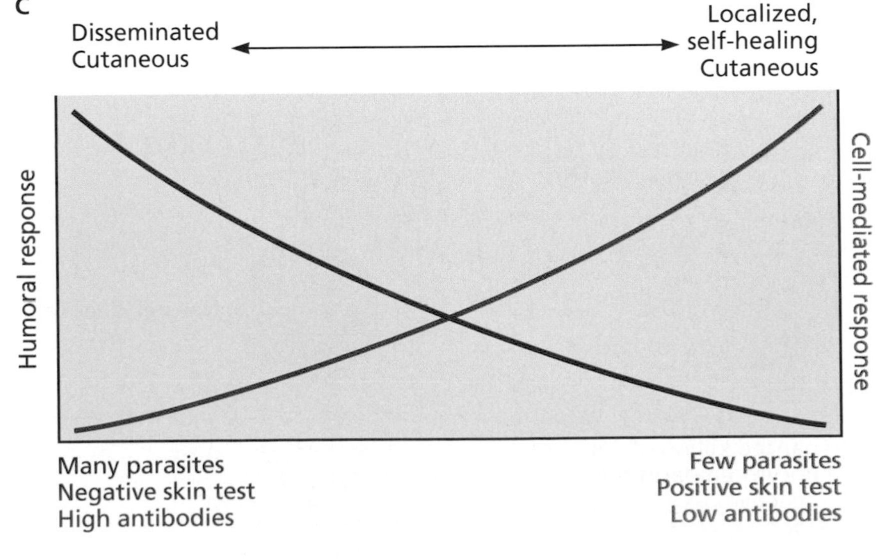

Figure 4.11 Localized and disseminated cutaneous leishmaniasis. Examples of (A) localized and (B) disseminated cutaneous leishmaniasis. Leishmaniasis presents a wide array of clinical pictures, and these are only two examples. (C) Much of the variation in the course of the disease depends upon the nature of the host immune response. In those individuals responding with a primarily cell-mediated Th-1 response, the lesions are self-healing and few parasites can be recovered from the lesion. Antibody levels are low. The skin test used to detect a type-1 response is positive. When the adaptive response is primarily humoral, cutaneous leishmaniasis is much more likely to disseminate. Many parasites can be present in the lesions, and antibody levels, largely ineffective against this intracellular parasite, are high. The negative skin test indicates the inactivity of Th-1 cells. (A, Aytekin S, Ertem M, Yağdiran O & Aytekin N [2006] *Dermatol Online J* 12(3):14. With permission the author. B, with permission from Dr Khalil Redah Al-Yousifi.)

in subsequent memory responses. IgA on the intestinal mucosa has been seen to protect mice from challenge infections with *T. gondii*, for example. Chickens can be protected from challenge infections with ***Eimeria*** by passively infusing them with anti-***Eimeria*** IgG.

It is generally assumed that an adaptive humoral response is more protective against exclusively extracellular protozoa such as ***Giardia lamblia***, which inhabits the intestinal lumen. Yet in some cases, experimental results have been equivocal and counterintuitive. For example, secretory IgA is often stated to be of prime importance in defense against ***G. lamblia***. Yet, while humans develop a strong IgA response to infection, the importance of this response remains unclear. Human patients with B-cell defects such as Bruton agammaglobulinemia, in which affected individuals produce no antibodies, are more prone to chronic giardiasis. Patients with selective IgA deficiency, however, show no such tendency toward chronicity. These data suggest that, although antibodies do contribute to host defense, IgA in particular is not critical and other alternative host defense mechanisms are involved. In murine models, mice lacking the **poly-IgA receptor** necessary to transport IgA across the intestinal epithelium and into the lumen were still able to control ***G. lamblia*** infections, although infections with the rodent parasite *Giardia muris* had an increased tendency to become chronic. The mechanism by which mice are able to eradicate the human pathogen remains unknown.

Data for ***Trichomonas vaginalis*** are similarly ambiguous. A comparison of asymptomatic and symptomatic mice infected with this parasite found that the inflammatory Th-1-type cytokines were actually elevated in asymptomatic animals. Titers of IgA were similar in both groups. A study in chronically infected symptomatic and asymptomatic women yielded similar results.

These examples underscore the complexity, interconnectivity, and redundancy of the vertebrate immune system. The diversity of the protozoan parasites themselves thwarts any attempt to identify single immune pathways or processes that are uniformly invoked in a similar manner. Complicating the issue still further are the varied ways that protozoa use to subvert host immune responses, a topic we will address more fully later in this chapter.

Protective immunity to malaria develops as a consequence of repeated exposure

Because of their overriding impact on human health, ***Plasmodium*** species and the immune responses they generate deserve special consideration. In highly endemic areas, infected mosquitoes repeatedly bite almost everyone. The resulting immunological memory means that in most instances only the initial exposure is likely to cause severe disease. Consequently, the most severe cases occur in young children, infected for the first time. However, the immunity that develops is not sterilizing—it does not eradicate all parasites in an infected individual. It is also of short duration; if a person from a highly endemic area leaves that area for a year or two and then returns, he or she is once again vulnerable to serious disease. This phenomenon, in which regular and repeated infection is required to prevent the onset of more serious, debilitating disease, is known as **premunition** (**Figure 4.12**).

It is ironic that the short-term and incomplete nature of immunity to malaria makes those living in highly endemic areas better off in some ways than those living in areas of sporadic transmission. The premunition in the former group insures that if they survive infancy, they are unlikely to suffer from the worst complications of malaria. Furthermore, it at least raises a

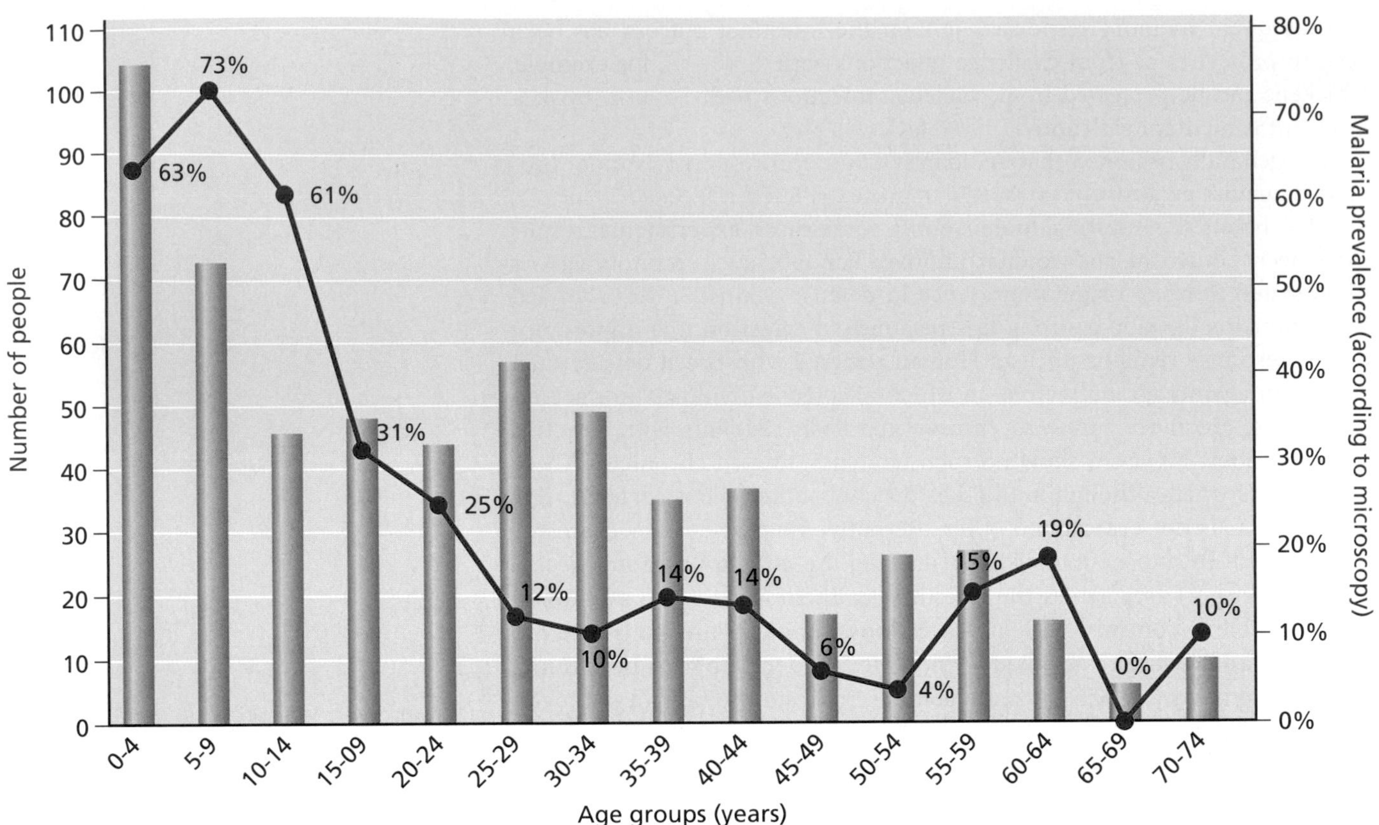

Figure 4.12 Premunition to malaria. These data were collected in a hyperendemic area of intense malaria transmission in Tanzania. Young children are far more likely to show an obvious blood parasitemia and to experience the most severe clinical cases. Premunition explains why older individuals are less likely to display parasitemia and less prone to severe complications of disease. Blue columns represent the number of tested individuals in each five-year age group. The red line represents malaria prevalence in each age group based on microscopic evidence of parasitemia. (From Laurent A et al [2010] *Malar J* 9:294.)

question about the long-term value of control programs geared at reducing human–mosquito contact, such as the use of bed nets, in areas of high transmission. Do such measures merely allow any adaptive immunity to wane, setting the stage for a severe malaria episode in the future? The premunition observed in malaria infections also casts a shadow on the prospects for an antimalaria vaccine that offers more than short-term and incomplete protection; any vaccine providing solid, long-term immunity would have to do something that natural infections do not. This obstacle for vaccine researchers will be revisited in Chapter 9.

Immune responses are generated against each stage in the *Plasmodium* life cycle

Immunity to malaria is limited in part because of its complex life cycle, during which each parasite stage expresses different antigens. As we will see shortly, even these stage-specific antigens may change during successive waves of replication. Yet the greatly reduced severity of disease in those who are regularly exposed demonstrates that human and other vertebrate hosts develop some protective immunity to ***Plasmodium*** infection. As in other immune responses, many humoral and cellular mechanisms are in play, although the precise role of these mechanisms is far from completely resolved. Yet it is clear that immune responses are mounted against each stage in the parasite life cycle, including sporozoites, and intracellular liver and erythrocyte stages. Most evidence suggests that the immunological memory that is established is directed primarily at the merozoites infecting red blood cells.

As in other infections, one of the key early events in the initiation of an immune response against ***Plasmodium*** is the engagement of PRRs on the surface of antigen-presenting cells (**Figure 4.13**). Upon activation,

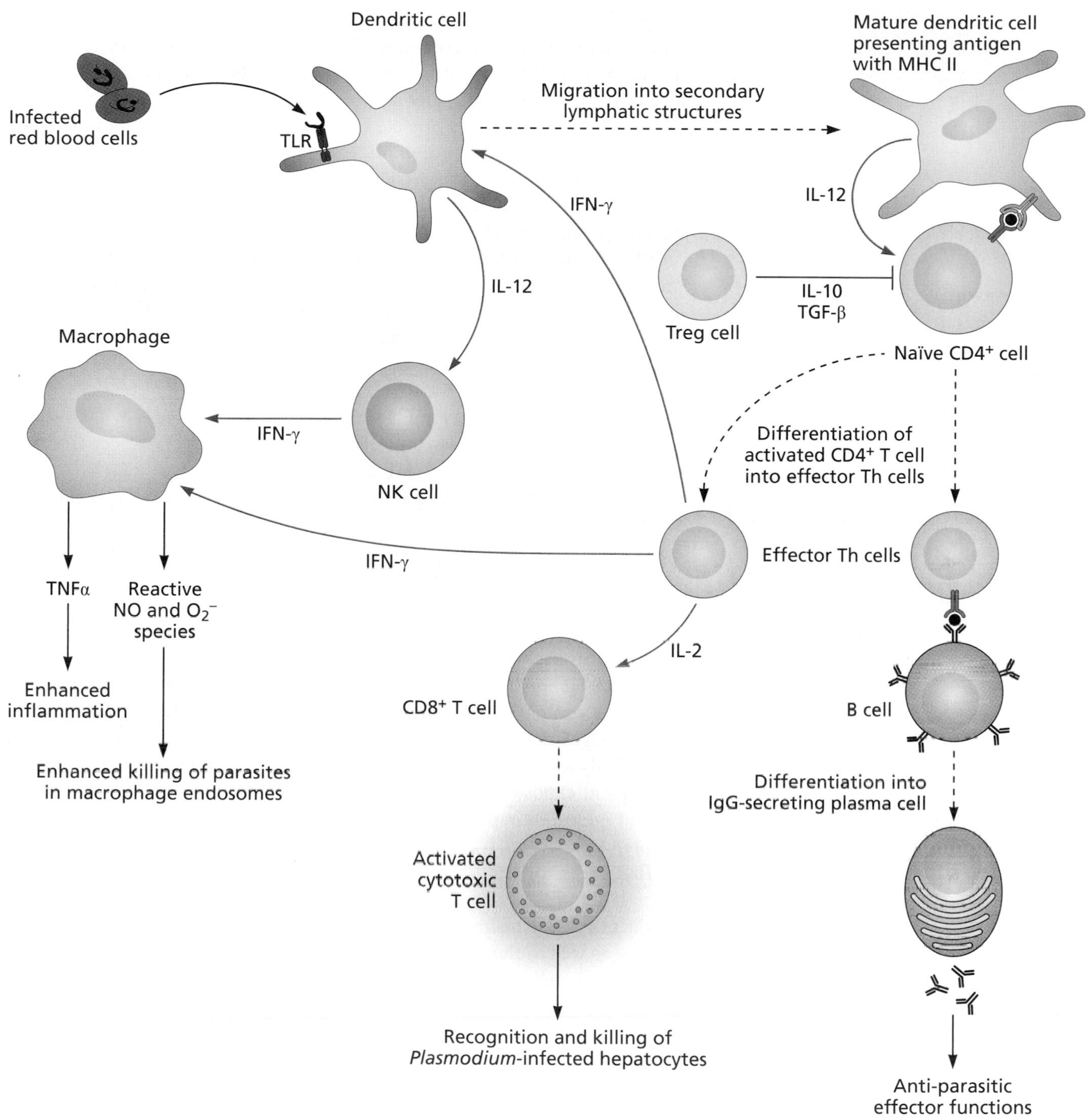

Figure 4.13 Overview of the immune response to malaria. For simplicity, only some of the major components and effector processes are shown. The immune system is alerted to the presence of *Plasmodium* when parasite PAMPs bind Toll-like receptors. Glycosyl phosphatidylinositol (GPI) is known to bind TLR-2, whereas hemozoin binds TLR-9. Dendritic cells subsequently secrete IL-12, which among its effects, activates natural killer (NK) cells. NK cells provide an early source of IFN-γ, which activates macrophages and increases their ability to produce toxic reactive oxygen and nitrogen species, with which they kill ingested parasites. Activated macrophages also secrete TNFα to enhance inflammation. Dendritic cells, meanwhile, migrate into secondary lymphatic structures, where they present antigen and continue to release IL-12, activating naïve CD4+ T cells, inducing them to differentiate into effector T cells, primarily Th-1 and Tfh cells. Effector Th-1 cells release IL-2, which activates CD8+ T cells. The now active CD8 T cells (cytotoxic T cells) are able to recognize and kill infected hepatocytes. Th-1 cells also produce IFN-γ, which continues to activate macrophages and dendritic cells. Other effector CD4+ cells activate B cells and cause antibody class switching, mainly to IgG. This immune activation is counteracted by the suppressive effects of IL-10 and TGFβ, released by Treg cells, inhibiting both inflammation and T-cell activation. (Adapted from Stevenson MM et al [2004] *Nature Rev Immunol* 4:169–180. With permission from Macmillan Publishers Ltd.)

antigen-presenting cells (APCs) secrete IL-12. Among the targets of this cytokine are natural killer (NK) cells. As discussed earlier, NK cells are important in the early response to intracellular pathogens, prior to the full activation of an adaptive response. Once activated, they provide an early source of IFN-γ, which serves to potentiate a Th-1 response. IFN-γ also activates macrophages and increases their ability to kill ingested parasites as described. NK cells have also been observed to lyse ***Plasmodium***-infected erythrocytes directly, at least *in vitro*, and they secrete chemokines that recruit other cells during a malaria infection.

NK cells are not the only lymphocytes that appear to mediate an initial innate response to ***Plasmodium***. Unlike other T cells we have discussed, **invariant natural killer T cells (iNKT cells)** bear neither CD4 nor CD8. Consequently, they do not recognize antigen presented by MHC. They have T-cell receptors (TCRs) with extremely limited diversity, and unlike TCRs on conventional T cells, which recognize only peptides, iNKT receptors seem to recognize glycolipids. Glycolipids are presented to iNKT cells by a molecule called **CD1**, which is similar to MHC in its ability to present certain antigens. The main response of iNKT cells seems to be the very rapid secretion of certain cytokines, including IL-4. Consequently, they are able to promote a Th-2 response. They also seem to activate another type of unusual lymphocyte called a **B1 B cell**. These cells are quite distinct from the conventional B cells we have discussed so far. B1 B cells have very limited receptor diversity and seem to make antibody primarily against polysaccharide antigens without the assistance of CD4$^+$ T cells. Because they do not require Th cells to become activated, B1 B cells can secrete antibody within 48 hours of an infection—well before the time required by conventional B cells to proliferate and differentiate into plasma cells. The antibody secreted by B1 B cells is primarily of the IgM class. There is no B1 B-cell memory response and the amount of antibody and the speed with which it is produced does not increase with subsequent exposure. Nevertheless, the antibody they produce may represent an important rapid and specific humoral response to ***Plasmodium*** that is independent of MHC.

Th-1 cells drive the primary adaptive response against liver-stage ***Plasmodium*** parasites (see Figure 4.13). Once they are activated, Th-1 cells produce IFN-γ, which promotes activation and enhanced killing by macrophages, as well as the proliferation and activation of CD8$^+$ cytotoxic T cells. Cytotoxic T cells can recognize and kill infected cells bearing their specific antigen complexed with MHC I (see **Box 4.2** for a discussion of MHC diversity). Infected mammalian erythrocytes, on the other hand, are not vulnerable to cytotoxic T cell-mediated killing. Mammalian erythrocytes lose their nuclei as they develop and consequently do not produce protein, including MHC, following maturation.

Accordingly, humoral immunity is the primary adaptive effector mechanism once ***Plasmodium*** has left the liver. Most of the antibody is of the IgM and IgG classes. The presence of elevated IgE levels indicates the involvement of Th-2 cells, perhaps activated in part in response to IL-4 secreted by iNKT cells as described above.

Antibodies produced in response to ***Plasmodium*** blood stages may help to reduce parasitemia in any of several ways. They may neutralize newly released merozoites before they can infect new erythrocytes or they may bind to parasite antigens found on the membranes of infected erythrocytes, causing the agglutination of infected cells. In the case of ***P. falciparum***, they may block cytoadherence in capillary beds by binding to the surface of infected cells. These infected cells, unable to sequester themselves, might then be transported to the spleen for destruction.

Helminth parasites provoke a strong Th-2 response

Although helminths generally elicit a strong immunological response, that response is often not completely effective. High rates of reinfection following treatment, the chronic nature of many helminth infections, and the very high prevalence in some endemic areas all suggest that the immune response is only partially successful in providing long-term protection. Nevertheless, evidence such as the decreased prevalence of most helminths infections with age suggests some degree of protective immunity. Additional evidence for such protection comes from those who are immunosuppressed and suffer increased susceptibility to helminth infection.

A complete understanding of immune response to helminths is complicated by the fact that, like protozoa, helminths infecting vertebrates have a wide variety of life histories. They consequently may activate immunity differently and provoke different effector mechanisms. Many helminths are intestinal parasites, some of which make complex migrations through host tissue before arriving in the intestine where they mature and reproduce. Others, such as schistosomes or filarial worms, mature sexually in the blood or other tissues. When one considers the various life cycles, tissue tropisms, and evasion mechanisms of these parasites, it is not surprising that so many elements of immune response to helminths are redundant and that various alternative mechanisms are available to elicit the same immune effector functions. In addition to being orders of magnitude larger than microparasites, helminths differ from other parasites in another important way: they do not ordinarily replicate within the vertebrate host. Infective stages not only must colonize the host, they must also reach sexual maturity and produce propagules to be transmitted to the next host, all while under immune assault. Consequently, the immune response to helminths differs in several respects to that generated against other types of parasites.

All of these factors make it difficult to develop broad generalizations regarding the immune response of vertebrates to helminths. However, although questions remain, several recent and exciting discoveries are shedding light on previously unresolved issues and have brought us much closer to a complete picture of anti-helminthic immunity.

One point upon which there is general agreement is the central role of Th-2 cells. Similar to other adaptive responses, the differentiation of naïve $CD4^+$ T cells into Th-2 cells begins with antigen presentation by antigen presenting cells (APCs) in secondary lymphatic structures. Yet it has been unclear exactly how the development of a Th-2 response is orchestrated and precisely how parasites may be cleared in either a primary or subsequent encounter.

It is known, for example, that naïve CD4 T cells differentiate into Th-2 cells in the presence of cytokines such as IL-4. Once they have fully differentiated, Th-2 cells themselves produce IL-4, which among its effects, further strengthens the type-2 response by positive feedback. The initial source of IL-4 has remained elusive, however. Both mast cells and basophils release IL-4 and have been suggested as potential early sources of this cytokine. Basophils also produce IL-25, which impedes a type-1 response by inhibiting IL-12 release from dendritic cells. Yet mice engineered to lack basophils still produced Th-2 cells and normal amounts of IgE when infected with the intestinal nematode *Nippostrongylus brasiliensis*. Mice deficient in mast cells,

BOX 4.2
MHC's Balancing Act

The Major Histocompatibility Complex (MHC) proteins are encoded by multiple loci. Humans and some other mammals have three MHC I loci, identified as *A*, *B*, and *C*. Consequently, mammals encode three types of MHC I. Cells will bear each of these variants. Alleles at each locus are co-dominant. Therefore, any human produces between three (if homozygous at all three loci) and six (if heterozygous at all three loci) forms of MHC I. MHC II is a dimeric protein, consisting of alpha and beta chains. The DP, DQ, and DR loci encode these chains. Thus, there are a total of six MHC II loci: DP-alpha and DP-beta, DQ-alpha and DQ-beta, and DR-alpha and DR-beta chain loci. If heterozygous at all loci, an individual encodes 12 types of MHC II. All 12 will be present on the surface of antigen-presenting cells.

MHC genes are also the most polymorphic genes known (Figure 1). At each MHC I locus and at all but one MHC II loci, there are dozens of alleles; in one case, there are more than 1000 alleles. The chances are excellent that any individual will be heterozygous at all or most MHC loci.

The high polymorphism and the high number of separate MHC loci (high polygenicity) raise the question, why is there so much MHC genetic diversity? To answer this question, we need to know a few additional things about MHC. First, cells present foreign antigen complexed with either MHC I or MHC II. In the absence of infection and therefore foreign antigen, MHC presents self-antigen. These self-peptides should not elicit any immune response; ordinarily any T cells that respond to self-antigen are removed in the thymus during T cell development. Called **clonal deletion**, this process is meant to prevent attacks on self (**autoimmunity**) later in life.

Second, different MHC variants present overlapping but different peptides. If comparing two MHC I *A* alleles, for example, we would find that the expressed MHC molecules for these alleles present many of the same peptides but many that are also different. Consequently, any individual can present an especially wide range of antigens to T cells. Because most individuals are heterozygous at most loci, most potential hosts express close to the maximum number of MHC types. Nevertheless, owing to high polymorphism, different individuals within a population generally express different MHC molecules. This variability partly explains why some individuals are more resistant to a particular pathogen than others. They are simply better at presenting that pathogen's antigens.

Evolutionary biologists usually attribute high MHC diversity to pathogen-driven selection. Pathogens themselves are under selection to alter their antigens in order to reduce the likelihood of a protective immune response. Hosts respond by increasing the number of MHC alleles at each locus and by increasing the number of loci. Consequently, high MHC diversity may have evolved to outmaneuver the evasive strategies of pathogens. With numerous loci and a high probability that an individual will be heterozygous at each locus, each individual can present a broad range of antigens and is able to mount a strong adaptive response against most pathogens. Common alleles are under greatest pathogen pressure because different pathogens attempt to interfere with or evade successful antigen presentation. This process drives the positive selection of uncommon alleles to which pathogens are less adapted. As pathogen pressure on the previously common alleles decreases and that on previously rare alleles increases, the frequency of these relatively uncommon alleles in the population stabilizes.

The value of high MHC polymorphism in a population is observed in small or endangered populations with especially low MHC polymorphism. In such cases, any pathogens that can evade presentation by the predominant alleles may sweep through the population, causing high morbidity and mortality. In larger populations with more MHC diversity, no single pathogen is likely to evade MHC presentation in all individuals. A recent example is provided by devil

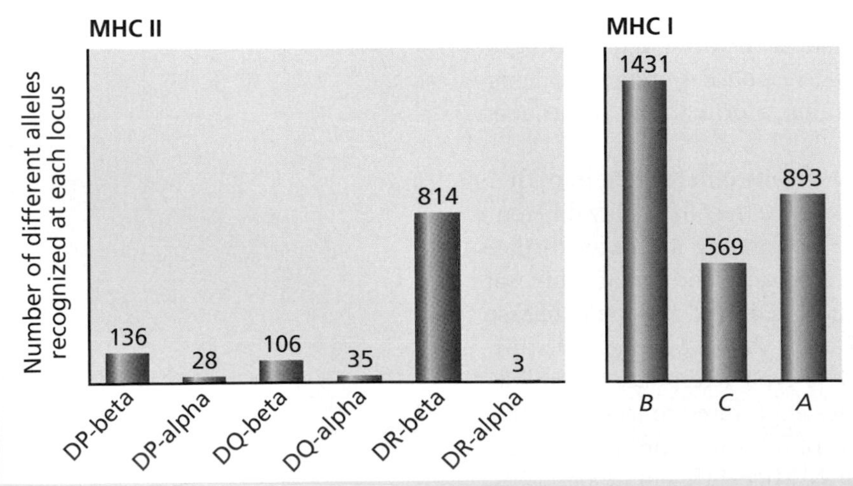

Figure 1 Polymorphism in human MHC genes. Gene loci encoding MHC, with the exception of DR-alpha, are the most polymorphic known in humans. The heights of the bars are the number of different MHC alleles currently recognized at each locus. (Adapted from Murphy K [2012] Janeway's Immunobiology 8ed. Garland Science.)

facial tumor disease, a transmissible form of cancer, currently causing high mortality in Tasmanian devils (*Sarcophilus harrisii*). These animals have unusually low levels of MHC polymorphism, and most infected individuals do not present a key tumor antigen. Consequently, adaptive immunity is limited and serious disease ensues. More evidence is provided by the fact that regular exposure to certain pathogens can select for certain MHC alleles that are especially good at presenting key antigens. A particular MHC I *B* allele, for instance, *B53*, is especially common in parts of West Africa where malaria is endemic. It is a rare allele in areas where malaria is uncommon. Individuals expressing *B53* are more likely to recover from malaria infection.

Yet if high MHC diversity is good, wouldn't even greater diversity be better? More loci would provide for even broader antigen presentation. It might make it even more likely that each individual could respond strongly to any potential pathogen. But if more MHC proteins were expressed, an even greater range of self-peptides would be presented in the thymus to developing T cells. As more and more T cells are eliminated by clonal deletion, the advantage of presenting a broader range of antigens is outweighed by the disadvantage of increased T-cell loss. It therefore appears that the current number of MHC loci is an attempt to balance the need for an increased range of antigen presentation with the need not to eliminate too many T-cell clones during thymic development. See also Chapter 7, and Figure 7.29.

References

Froeschke G & Sommer S (2012) Insights into the complex associations between MHC class II DRB polymorphism and multiple gastrointestinal parasite infestations in the striped mouse. *PLoS One* 7(2):e31820 (doi:10.1371/journal.pone.0031820).

Messaoudi I, Patino JAG, Dyall R et al (2002) Direct link between MHC polymorphism, T cell avidity and diversity in immune defense. *Science* 298:1797–1800.

Potts WK & Slev PR (1995) Pathogen-based models favoring MHC genetic diversity. *Immunol Rev* 143:181–197.

on the other hand, failed to generate a strong Th-2 response when infected with the intestinal nematodes *Heligmosomoides bakeri* or *Trichuris muris*. Furthermore, it has recently become clear that epithelial cells in the intestine also secrete the cytokines IL-25, IL-33, and thymic stromal lymphopoietin (TSLP). These so-called **alarm cytokines** or **alarmins** apparently assist in initiating a Th-2 response. Mice lacking mast cells also failed to express high levels of alarmins. When mast-cell activity was restored, alarmin expression returned to normal levels. Exactly how mast cells might stimulate the epithelium to secrete alarmins, as these results suggest, remains unclear. Taken together, these data imply that in addition to directly stimulating CD4$^+$ T cells to differentiate into Th-2 cells, mast cells also indirectly stimulate Th-2 cell development via their stimulation of epithelial cells.

Another intriguing recent development is the identification of a new class of lymphocyte called **natural helper cells** or **nuocytes**. These novel lymphocytes, which lack conventional T-cell receptors, were first discovered in lymph nodes of *N. brasiliensis*-infected mice. It has since been observed that in response to the alarmins released by epithelial cells, nuocytes release cytokines such as IL-5 and IL-13 that like IL-4 help to skew the differentiation of naïve CD4$^+$ T cells along the Th-2 pathway. As such, nuocytes are now viewed as a potentially important early source of the cytokines required to provoke Th-2 immunity.

In addition to stimulation by mast cells, epithelial cells may independently detect the need to initiate Th-2 immunity, causing them to release their alarmins. Parasite proteases used to penetrate the intestinal epithelium have been shown to stimulate TSLP release, whereas necrotic epithelial cells release IL-33.

Consequently, a picture is emerging in which at least some helminth infections are initially detected by either IL-4-secreting mast cells or the intestinal epithelium, causing the epithelium to release IL-25, IL-33, and TSLP (the alarmins) (**Figure 4.14**). This in turn helps to induce nuocytes to express the IL-5 and IL-13 necessary for Th-2 differentiation. The Th-2 cells then express high levels of IL-4 and IL-13, which activate anti-helminthic effector functions.

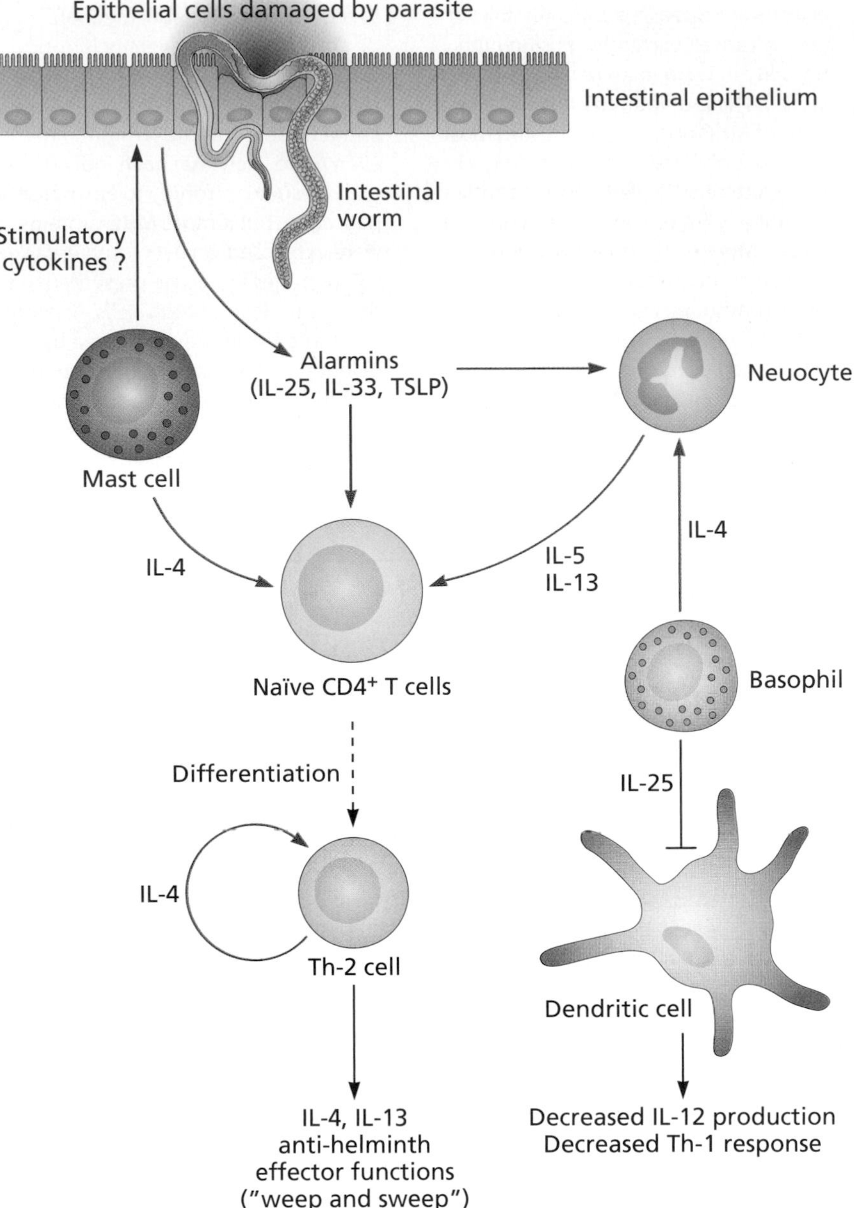

Figure 4.14 Possible mechanisms for the initiation of Th-2 immunity to intestinal helminths. Parasite damage to the intestinal epithelium, as well as factors released by mast cells, may induce the epithelium to release IL-25, IL-33, and TSLP, collectively referred to as *alarmins*, in response to helminth infection. These alarmins may then stimulate the release of IL-5 and IL-13, which in concert with IL-4 released by basophils and mast cells may cause the differentiation of naïve $CD4^+$ T cells into Th-2 cells. Once they have been activated, Th-2 cells release abundant IL-4, further stimulating the development of a type-2 response. Basophils may also inhibit the development of a Th-1 response via their release of IL-25, which inhibits Th-1 and induces IL-12 release from dendritic cells. Note that several of these proposed mechanisms are redundant, meaning that a Th-2 response may still be elicited even without the certain mechanisms as shown here.

Extensive changes to the intestinal epithelium occur in response to intestinal helminths

Our understanding of these effector functions has also been substantially revised as of late. As we have previously noted, Th-2 cells are critical for antibody class switching to IgE. Immunoglobulin, however, now appears to play only a limited role during a primary response to many helminths, especially those inhabiting the gut. Rather, it is the effect of Th-2 cytokines, IL-4 and IL-13 in particular, on the mucosal epithelium itself that appears to take center stage in a primary response to intestinal worms.

Specifically, the intestinal mucosa with receptors for IL-4 and IL-13 undergoes several important changes in response to these cytokines. First, increased smooth muscle contraction enhances peristalsis, helping to dislodge worms from the intestinal wall. Increased mucus secretion further creates difficulties for worms trying to attach to or penetrate the mucosal lining. Accelerated cell division by goblet cells in intestinal crypts (**hypertrophy**) replaces cells damaged by worms and reduces mucosal surface area, providing less available surface for worm attachment. Less surface area also means reduced nutrient absorption by the host, which causes some of the pathological symptoms observed in intestinal helminth infections. One of the recently described and potentially most important mucosal responses is the production of **resistan-like molecule β (RELMβ)**, which follows ligand engagement by epithelial cell IL-4 receptors. The exact mechanism by which RELMβ exerts its anti-helminthic activity remains unclear, but this molecule has been shown to eliminate intestinal nematodes from infected mice. Some evidence suggests that it interferes with parasite feeding on host tissues or that it may block pore-like structures with chemosensory function (for example, amphids; see Figure 4.4), consequently interfering with chemotaxis. Furthermore, IL-4 and IL-13 produced by Th-2 cells can cause the differentiation and activation of **alternatively activated macrophages**, also known as **M2 macrophages**. The macrophages we have discussed so far are the so-called **conventional macrophages**. They are activated by Th-1 cells, whereupon they release pro-inflammatory cytokines. M2 macrophages, on the other hand, produce products that enhance goblet cell division, mucosal epithelial repair, and smooth muscle contraction, contributing to the creation of microhabitats in the intestine that are inhospitable to worms. The enzyme arginase is the most important of the chemical mediators released by M2 macrophages responsible for these effector functions.

Thus, a growing body of evidence seems to indicate that the primary response to intestinal helminths has more to do with extensive tissue remodeling than it does with immunoglobulin. These processes are often collectively referred to as a "weep-and-sweep" strategy (see Figure 4.14).

An immunoglobulin response, on the other hand, may form the basis of any long-term immunological memory that is generated in response to helminths. IL-4 released by Th-2 cells induces antibody class switching to IgE. This immunoglobulin binds to Fc receptors on the surface of mast cells, basophils, and eosinophils. Th-2 cells also release IL-5, which recruits and activates eosinophils, as well as IL-3 and IL-9, which serve to recruit mast cells to the site of infection. These effector cells degranulate upon contact with specific helminths antigens, releasing toxic products that may kill the parasite. This process is known as **antibody-dependent cell-mediated cytotoxicity** (**Figure 4.15**). Furthermore, as constant, local sources of IL-4 and IL-13 in response to worm antigens, these cells may help provide a local environment appropriate for the induction of Th-2 immunity. Indeed, this may be the primary anti-helminthic activity of basophils. We earlier mentioned that mice lacking basophils were still able to generate a primary response against intestinal nematodes. These same mice, however, were unable to develop long-term IgE-mediated protection against these same parasites, suggesting an important role for basophils in anti-helminthic secondary responses.

Immunocompromised hosts are more vulnerable to parasitic infection and increased pathology

The importance of a properly functioning immune system can perhaps best be visualized when a normal immune response is compared to that which occurs in immunocompromised hosts. In such individuals, infections that are normally not serious or at least not life threatening can become especially ominous. ***Toxoplasma gondii*** infection in humans provides a

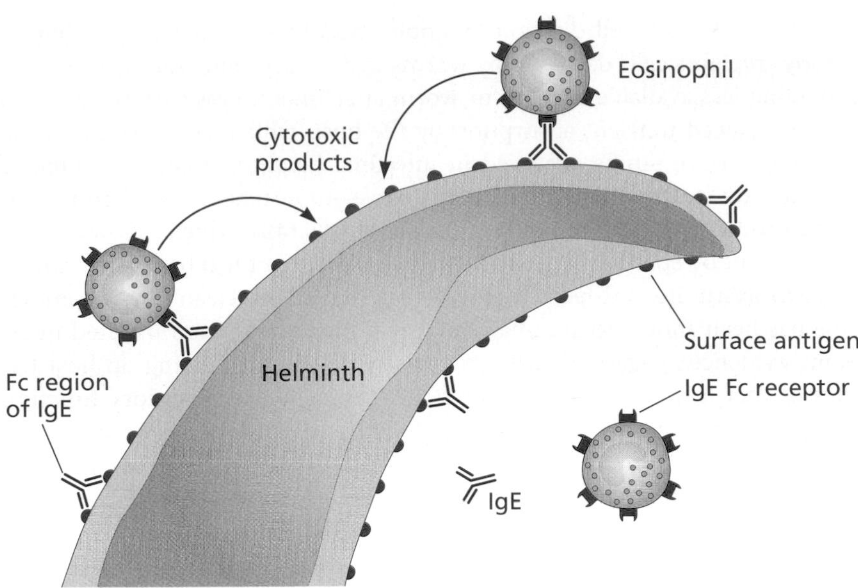

Figure 4.15 Antibody-dependent, cell-mediated cytotoxity (ADCC). IgE Fc receptors allow eosinophils to bind to IgE, which has bound surface antigens on the tegument of helminths. Eosinophils also have IgG Fc receptors for this same purpose. Upon binding, eosinophils degranulate, releasing the contents of their cytoplasmic granules. The released molecules, including major basic protein, eosinophil cationic protein, eosinophil peroxidase, and eosinophil-derived neurotoxin, are cytotoxic and thus play an important role in immune response to helminths. Other cell types, including mast cells and natural killer cells, are also able to attack helminths via ADCC.

particularly good example. Generally, no signs or symptoms of clinical disease are apparent in immunocompentent individuals. Should the host become immunocompromised, however, bradyzoites within tissue cysts may reactivate, giving rise to rapidly reproducing tachyzoites (see Chapter 3, pages 89–90). In the brain, such reactivation can result in encephalitis with severe, possibly lethal consequences. Something similar is observed in humans infected with ***Cryptosporidium***. Individuals with normally functioning immune systems generally experience a self-limiting acute bout of intestinal distress and diarrhea. AIDS patients, on the other hand, are subject to long-term chronic infection. Patients experience profuse and watery diarrhea that can last for months or years. Death is a frequent outcome if the condition is untreated. With the advent of combination retroviral therapy in the mid-1990s, however, patients may now remain symptom-free indefinitely. HIV specifically targets $CD4^+$ T cells. Suppression of the virus through the use of antiviral drugs allows the numbers of these T cells to rebound, permitting containment of ***Cryptosporidium*** infections.

Untreated individuals who are HIV^+ or other patients suffering from various forms of immunodeficiency are even at elevated risk for infection with ectoparasites such as the mite ***Sarcoptes scabiei***, the causative agent of scabies (**Figure 4.16**).

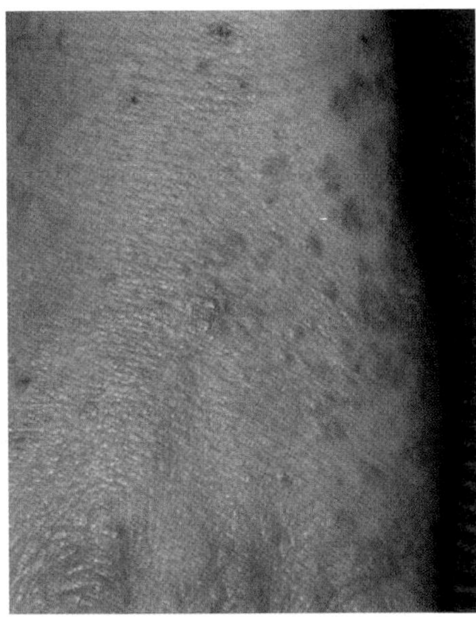

Figure 4.16 Scabies in an HIV^+ individual. Scabies is caused by infection with parasitic mites in the family Sarcoptidae. In humans, the most likely cause of infection is *Sarcoptes scabiei*. The closely related *S. scabiei canis* causes sarcoptic mange in dogs. HIV^+ individuals are at elevated risk for scabies because of their depressed numbers of $CD4^+$ T cells and, consequently, their inability to mount a proper immune response to this ectoparasite. Scabies and other infectious diseases that are more common in immunosupressed individuals are known as opportunistic infections. (Courtesy of Erin Lord, CC BY-SA 3.0.)

A final example is provided by *Strongyloides stercoralis*. Infection with this nematode is often asymptomatic in immunocompetent individuals. Eggs usually hatch in the intestine and L1 larvae are passed in the feces. In some cases, the larvae manage to molt twice before they are shed. As L3 larvae, they have reached the infective stage for their mammalian host, whereupon they can penetrate the gut wall. They then retrace their larval migration through the body, ultimately either returning to the intestine where they mature into adults or disseminating widely through the body. This process, called **autoinfection**, occurs only very early in a primary infection in immunocompetent hosts, before the onset of an adaptive response. In severely immunocompromised individuals, autoinfection can occur indefinitely, resulting in an overwhelming and potentially lethal hyperinfection.

4.4 PARASITE EVASION OF HOST DEFENSES

Successful parasites must survive long enough within their hosts to develop, reproduce, and otherwise complete their life cycle. To do so, they must avoid immune destruction, at least until transmission is complete. This fact of life for any parasite is one of the critical factors determining host range. If immune-mediated elimination is too immediate or complete, a potential host is simply off limits as far as the parasite is concerned.

Parasites do not just tough it out when it comes to dealing with their preferred host's immune system. Quite the opposite: they have evolved a wide variety of sophisticated strategies to manipulate, evade, or otherwise outwit host defenses. We have already mentioned a few of these remarkable mechanisms. The cytoadherence of ***Plasmodium falciparum***-infected erythrocytes to the endothelium in capillary beds that allows the parasite to avoid destruction in the spleen is just one example. Every aspect of the vertebrate immune system represents a potential weak link as far as parasites are concerned and here we will investigate how parasites have taken the initiative when it comes to exploiting such vulnerabilities.

Many parasites are able to evade complement-mediated innate immune responses

The complement cascade (see on-line tutorial) is an important, early innate response to infection. Yet many eukaryotic parasites avoid the effects of complement, using an impressive number of evasive tactics.

Leishmania spp. provide an excellent example. Because they are largely intracellular, these parasites have little to fear from soluble immune proteins such as complement during most of their life cycle. There is a transient period, however, following inoculation into the blood by their sand fly vectors and before their entry into target cells, when they might be assumed to be vulnerable. Yet as the promastigotes in the sand fly develop into infective metacyclic promastigotes, they undergo a change in membrane structure that renders them resistant to the effects of complement. Specifically, a modified surface lipophosphoglycan is expressed that is roughly double the length of the form found in promastigotes. Owing to this developmental change, the plasma membrane of the metacyclic promastigote is no longer vulnerable to insertion of the membrane attack complex formed by complement activation and is therefore no longer at risk of complement-mediated lysis. The membrane is further modified by the incorporation of a protease that cleaves C3b into an inactive form, preventing the subsequent synthesis of terminal, active complement proteins. As if to add insult to injury, the now inactive C3b (called iC3b) facilitates entry of the parasite into its macrophage target cell by acting as an opsonin for the host cell.

Other protozoa are similarly innovative. In the genus ***Entamoeba***, for instance, both invasive and pathogenic ***E. histolytica*** and noninvasive ***E. dispar*** activate complement via the alternative pathway. Yet although ***E. histolytica*** resists the effects of the membrane attack complex, ***E. dispar*** remains susceptible. This is thought to arise from the presence of surface receptors that bind galactose and *N*-acetyl-D-galactosamine in ***E. histolytica*** but not in ***E. dispar***. These receptors bind the C8 and C9 components of the membrane attack complex, preventing complex assembly and subsequent parasite lysis. Thus, the difference in pathogenicity between these two closely related ***Entamoeba*** species may reflect, at least in part, their different susceptibilities to complement-mediated lysis.

Helminths have also evolved novel means to avoid complement-mediated damage. ***Schistosoma mansoni***, for instance, encodes several proteins that interfere with the complement cascade. A protein called complement C2 receptor inhibitor trispanning, suppresses classical complement activation by binding and inactivating the C2 protein fragment in both humans and mice. Mammalian stages also express a type of paramyosin on their surface called SCIP-1 that binds C9, inhibiting the polymerization leading to the formation of the membrane attack complex. ***Fasciola hepatica*** inhibits alternative pathway activation by shedding its glycocalyx coat as it develops into an adult. C3b, crucial to alternative activation, can bind the glycocalyx found on larvae. Free of their glycocalyx coat, adults are not bound by C3b and consequently fail to elicit a complement response. The microfilariae of ***Onchocerca volvulus***, responsible for retinal damage leading to river blindness, bind and inactivate factor H, a necessary component in the alternative pathway.

Intracellular parasites have evolved mechanisms to avoid destruction by host cells

Invading pathogens that enter host cells are often destroyed by the low pH and proteolytic enzymes found within the acidified vesicles of the endomembrane system. This is certainly true for phagocytic cells such as macrophages, which specialize in such pathogen destruction, but it is likewise true for other nucleated cells that are invaded by intracellular pathogens. Certain intracellular protozoa, however, owe their success to their ability to avoid such destruction and thereby undermine an effective innate response. Clearly, if such intracellular parasites are able to thwart induced innate immunity in this way, many of the points made in previous sections regarding the initiation of an adaptive response become moot, allowing the parasite to establish a long-term, chronic infection. The different strategies of three representative intracellular protozoa are shown in **Figure 4.17**.

Helminths, of course, are too large to be phagocytosed. Yet they too are vulnerable to the cytotoxic effects of compounds released by innate effector

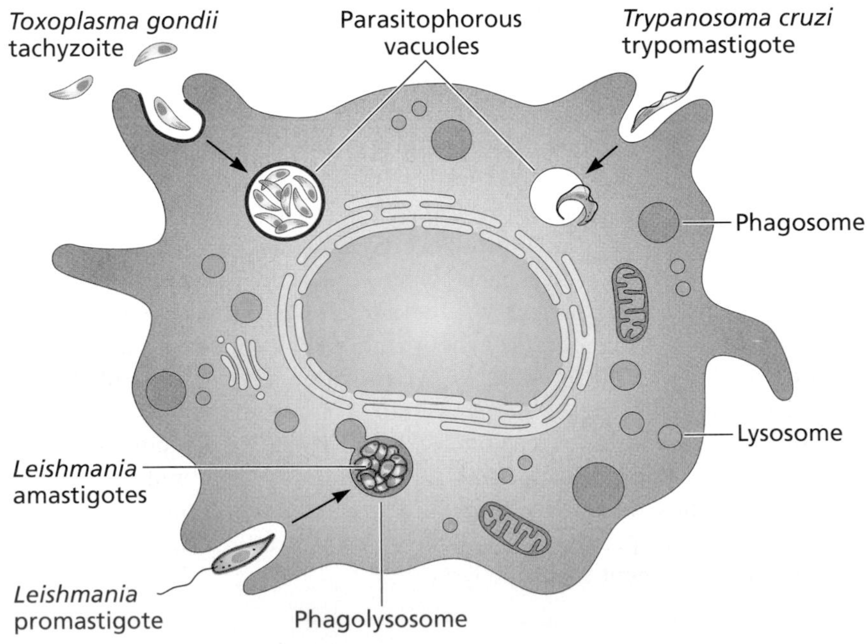

Figure 4.17 Protozoan survival within the macrophage. *Toxoplasma gondii*, *Trypanosoma cruzi*, and *Leishmania* spp. all survive within host macrophages. *T. gondii* tachyzoites actively penetrate macrophages. The vacuole that forms around the tachyzoite during penetration is the parasitophorus vacuole (blue). The parasite remodels the membrane around the parasitophorous vacuole. It lacks membrane proteins of host origin, but it contains proteins secreted by the tachyzoite. These modifications prevent vacuole acidification and fusion with lysosomes, permitting parasite survival. *T. cruzi* also actively penetrates host cells, but the resulting parasitophorous vacuole is not modified. Instead, *T. cruzi* secretes a protein called Tc-Tox, which forms a pore in the vacuole that allows the parasite to escape to the cytoplasm prior to vacuole fusion with lysosomes. *Leishmania* spp. enter cells via receptor-mediated endocytosis. *Leishmania* spp. promastigotes and amastigotes remain in the acidified phagolysosome, protected from destruction by the glycolipids and glycoproteins that they incorporate into their plasma membrane (green). (Sacks D & Sher A (2002) *Nat Immunol* 3:1041–1047. With permission from Macmillan Publishers Ltd.)

cells. The production and release of reactive oxygen species by macrophages and neutrophils, for example, is a principal means by which helminths are attacked. Not too surprisingly, many helminths defuse such attacks via abundant production of antioxidants. Schistosomes, ***Fasciola hepatica***, and filarial worms in the genus ***Brugia*** are all known to produce antioxidants such as superoxide dismutase, catalase, and various peroxidases.

Parasites may interfere with intracellular signaling pathways

In some cases, parasites inhibit an immune response by interfering with cell signaling events. A good example involves macrophages, which, as we have discussed, have an extensive oxidative metabolism that results in the production of toxic reactive oxygen molecules. In this process, NADPH oxidase transfers electrons from NADPH to O_2 to form superoxide and hydrogen peroxide. The increased oxygen consumption by the macrophage during this time is known as the **respiratory burst**. During phagocytosis of the parasite, a signal transduction cascade is initiated that results in protein kinase C (PKC) activation. PKC, in turn activates NADPH oxidase (**Figure 4.18**). PKC activation, however, is inhibited by hemozoin, a waste product of hemoglobin digestion, produced by malaria parasites. We will take a closer look at hemozoin and its effects in Chapter 5. See **Box 4.3** for another consequence of hemozoin release.

Intracellular parasites may also interfere with apoptosis, ordinarily a primary defense against such parasites. By preventing apoptosis, parasites extend the life of their host cell until their replication within the cell is complete. Many apicomplexans are particularly adept at such host cell manipulation. ***Eimeria tenella***, to cite one example, invades epithelial cells in the intestinal ceca of chickens. This serious pathogen can cause high mortality, especially in young birds. Early in an infection, the parasite causes activation of a host cell transcription factor, resulting in the increased expression of proteins that block an apoptotic response.

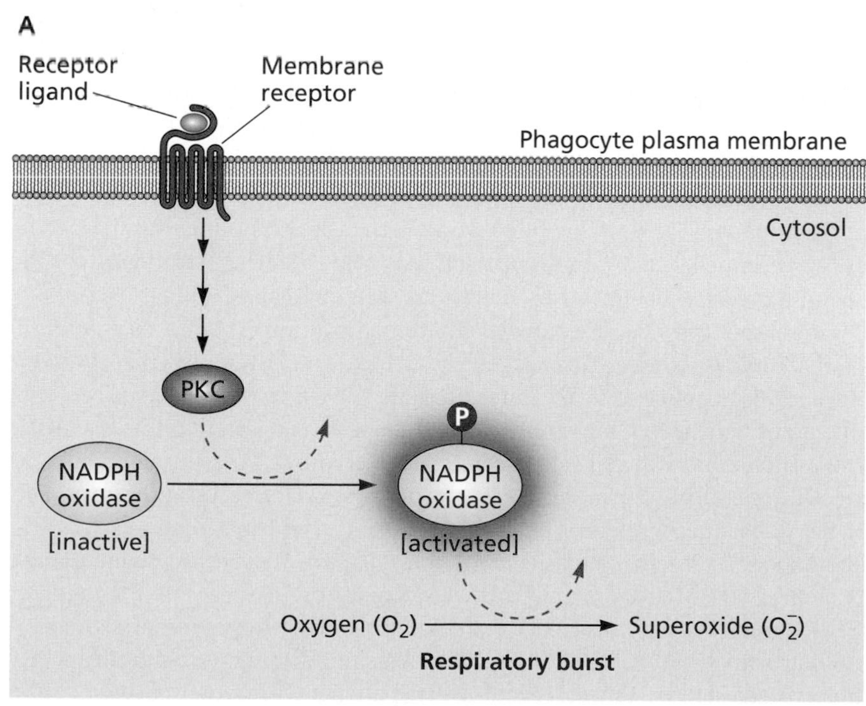

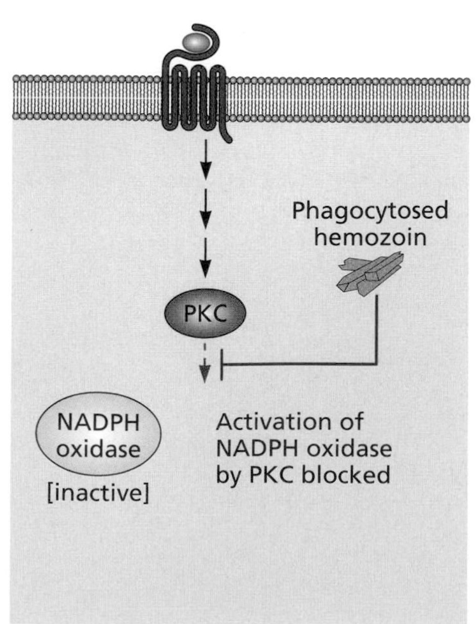

Figure 4.18 Inhibition of the respiratory burst by hemozoin. (A) In phagocytes, protein kinase C (PKC) is normally activated as part of a signal transduction cascade in response to ligand binding by a specific receptor. The now-active PKC then phosphorylates NADPH oxidase, resulting in its activation. This enzyme, in turn, converts molecular oxygen into the superoxide free radical, as part of the phagocyte's respiratory burst. (B) Hemozoin is a waste product of *Plasmodium*, generated when the parasite digests hemoglobin in infected erythrocytes. When the infected erythrocytes lyse, hemozoin is released and may be phagocytosed. Once inside the phagocyte, hemozoin can interfere with the activity of PKC. The result is a decrease in superoxide production and a less effective respiratory burst.

BOX 4.3
Malaria and Immunosupression

When the activity of phagocytes is disrupted by hemozoin, the consequences spread further than a simple evasion of host immunity by *Plasmodium*. On the contrary, the reduced ability of phagocytes to destroy ingested material is accompanied by an inhibition of MHC II presentation by antigen-presenting cells that take up hemozoin. In addition to the reduced overall immune response to *Plasmodium*, a general immunosupression may ensue that increases vulnerability to other infections as well. For example, in most individuals, infection with Epstein-Barr virus results in a latent infection in which no viral replication occurs. The infected individual remains symptom-free as the virus lies dormant in B cells. In individuals also suffering from malaria, however, the virus may become reactivated owing to the *Plasmodium*-induced immunosupression. The result is a B-cell lymphoma known as Burkitt's lymphoma (Figure 1). This same immunosupression is believed to result in lower efficacy of all vaccines observed in areas of high malaria transmission.

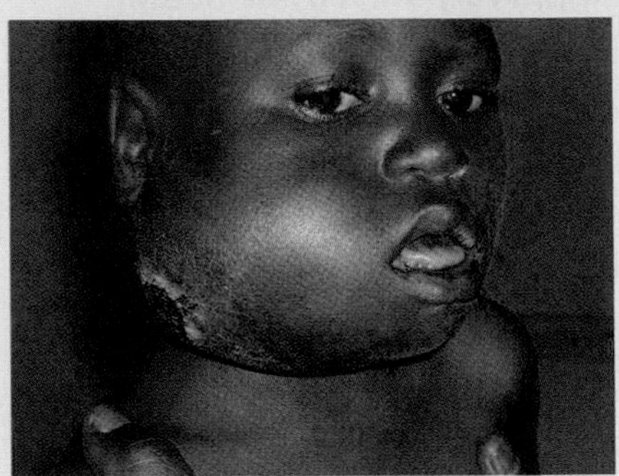

Figure 1 Burkitt's lymphoma. A B-cell malignancy caused by the Epstein-Barr virus in a seven-year-old Nigerian boy. Immunosupression due to chronic malaria infection is thought to be a contributing factor to the onset of this disease. (Courtesy of Mike Blyth, CC BY-SA 2.5.)

References

Millington OR, Di Lorenzo C, Phillips RS et al (2006) Suppression of adaptive immunity to heterologous antigens during *Plasmodium* infection through hemozoin-induced failure of dendritic cell function. *J Biol* 5:5–22.

Hisaedaa H, Yasutomob K & Himeno K (2005) Malaria: immune evasion by parasites. *Int J Biochem Cell Biol* 37(4):700–706.

Some parasites interfere with antigen presentation, resulting in an impaired immune response

Many studies demonstrate that certain parasites suppress the generation of a protective adaptive response by disrupting proper antigen presentation. As we discussed, for example, ***Toxoplasma gondii*** remains essentially isolated in the cytoplasm of macrophages and dendritic cells because the parasitophorous vacuole does not fuse with host cell vesicles (see Figure 4.17). Such isolation may prevent subsequent antigen processing and association with MHC II. In ***Trypanosoma cruzi***, it has been demonstrated that virulent strains of the parasite can down-regulate MHC II expression more effectively than less virulent strains, suggesting that such down-regulation is an important component of pathogenicity in this parasite. As such, strain-specific ability to impede MHC II presentation may be part of the explanation for the striking diversity of clinical presentations seen in Chagas' disease patients.

These examples feature parasites that normally infect APCs. ***Plasmodium falciparum***, of course, first targets hepatocytes and then erythrocytes for intracellular replication. Yet since the late 1990s, it has been known that infected erythrocytes can somehow reduce MHC II expression on dendritic cells. Such dendritic cells also secrete an altered cytokine profile, which, as we will see shortly, results in a less protective adaptive immune response. It has subsequently been found that infected erythrocytes express a parasite-encoded protein, called ***Plasmodium falciparum*** erythrocyte membrane protein 1 (PfEMP1), on their surfaces. This protein serves as a ligand for CD36, found on the surface of dendritic cells. Such binding initiates an intracellular signal that inhibits MHC II expression. The clinical significance of this is suggested by the observation that patients with either severe or mild malaria have reduced MHC II expression on dendritic cells as compared to healthy controls. Paradoxically, it has been observed that infected red blood cells with high CD36 binding affinity are more frequently observed in patients with mild rather than severe malaria. This difference may result from the reduced expression of pro-inflammatory cytokines in response to the parasite that exacerbate the symptoms of disease.

The ability to alter antigen presentation is certainly not limited to protozoa. A number of animal parasites are known to produce **cystatins**, a family of cysteine protease inhibitors with known immunoregulatory functions. Cystatins negatively affect MHC II presentation by inhibiting host proteases required for antigen processing. ***Brugia malayi*** is just one nematode that owes part of its success to cystatin production. Another example is the filarial nematode ***Onchocerca volvulus***, in which infection is characterized by a significant lack of protective immunity. Part of the explanation lies in the secretion of a cystatin known as onchocystatin. And cystatins are not limited to nematodes. The soft tick *Ornithodoros moubata*, for instance, is an important vector for African swine fever virus and the spirochete *Borrelia duttoni*. The cystatin present in its saliva, OmC2, reduces the ability of dendritic cells at the bite site to present antigen and secrete inflammatory cytokines. This prevents the acquisition of immunity to tick feeding, and it permits more efficient transmission of the tick-borne pathogens. Yet the discovery of immunosuppressive parasite cystatins comes with a silver lining. It has been proposed, for instance, that OmC2 might make a good vaccine target to protect animals against both feeding ticks and the pathogens they transmit. Furthermore, because of their immunosuppressive activity, synthesized parasite cystatins might have value for the treatment of certain autoimmune diseases, such as rheumatoid arthritis, in which symptoms are caused by an aberrant immune response.

Some parasites regularly change their surface antigens to avoid immune responses

One of the most dramatic parasite evasion strategies is the **antigenic variation** employed by African trypanosomes. The blood-borne stage in vertebrates, the trypomastigote, regularly changes the composition of its surface coat. Antibodies generated against the previous coat become ineffective. By changing its single coat protein, known as the **variant surface glycoprotein (VSG)**, to antigenically distinct forms, the parasite in effect remains one jump ahead of the immune response, which must constantly initiate new antibody responses against the new antigens (**Figure 4.19**). The result is a chronic infection, characterized by waves of increased parasitemia, each followed by a decrease in parasite numbers, as the lagging immune response struggles to catch up.

The trypanosome genome contains over a thousand copies of the VSG gene. All these genes encode a VSG surface protein with a conserved C-terminus, which anchors the protein in the plasma membrane. The VSG N-terminus, exposed to the extracellular environment and consequently to host antibodies varies from gene to gene.

Only a single VSG gene is expressed at any one time. Humoral immunity generated to the VSG protein greatly reduces parasitemia. As some of the remaining parasites switch their VSG expression, however, parasitemia rises, and the previously effective antibodies are no longer able to bind to the parasite (see Figure 4.19). This cat-and-mouse game continues in recurring waves throughout the chronic infection, giving rise to the waxing and waning of symptoms observed in infected patients. The antigenic switching appears

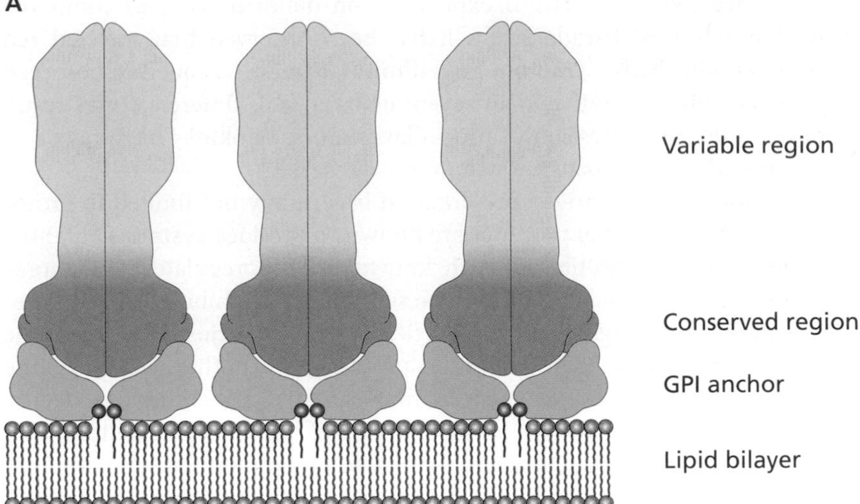

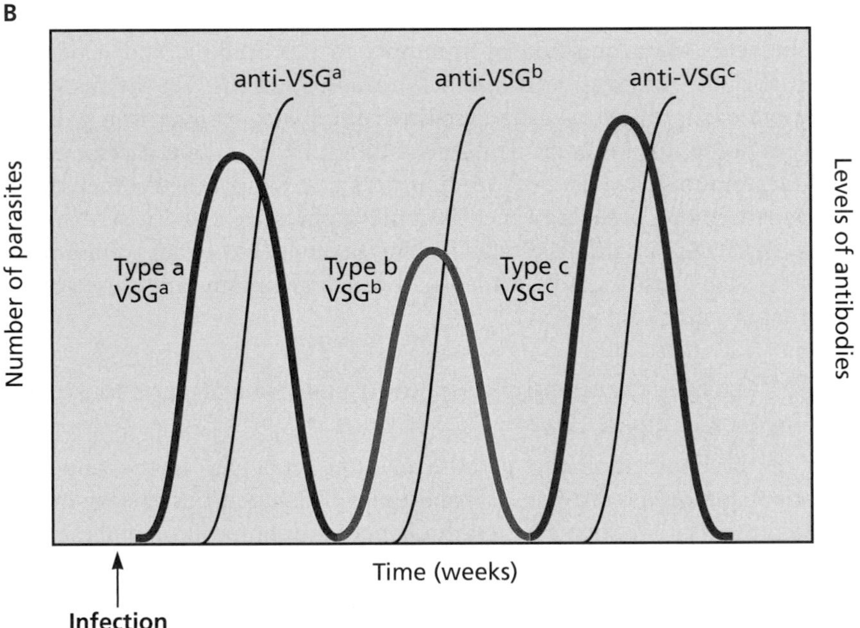

Figure 4.19 Antigenic variation in African trypanosomes. (A) A schematic representation of the trypanosome variable surface gene (VSG) protein on the surface of a trypanosome. The variable region is exposed on the outer surface of the membrane, and the conserved region is attached to the plasma membrane via a GPI anchor. (B) Over 1000 VSG genes encoding antigenically distinct VSG proteins are present in the trypanosome genome. The graph shows parasite levels and antibody response during three VSG antigenic switches (from type a (red) to type b (yellow) to type c (blue). A new antibody response must be generated to each of these phenotypically distinct VSG proteins. Once antibodies to a particular VSG are present, parasite numbers fall. A switch to a different VSG allows parasite numbers to once again increase. Symptoms of disease rise and fall with the change in parasitemia. (A, from Wiser MF [2011] Protozoa and Human Disease 1ed. Garland Science. B, from Murphy K [2012] Janeway's Immunobiology 8ed. Garland Science.)

to occur spontaneously in every 100 or so generations of the parasite rather than in response to the antibodies themselves.

Approximately three dozen VSG promoter sites are found on the subtelomeric portions of particular trypanosome chromosomes. The VSG genes themselves are found in three different locations and the manner in which antigenic switching occurs appears to vary with the location of the VSG genes in question. For those subtelomeric VSG genes located adjacent to promoters, switching is believed to rely on **allelic exclusion**, the expression of only one of various alternative alleles at a particular time. The expression site that is active at a particular time appears to be located in a particular site within the nucleus, and this site has room for only a single promoter and VSG gene at a time. Subsequently, a nuclear reorganization occurs in which the previously active promoter and adjacent VSG gene are replaced by a new promoter–VSG gene combination. The newly expressed VSG variant protein will now replace the previously expressed protein on the parasite surface.

This process appears to occur only very early in infection of the vertebrate host, before the transition from metacyclic to blood form trypomastigote.

Two additional mechanisms of allelic variation involve expression of VSG genes that are not immediately adjacent to promoters. Some of these genes remain unexpressed until a crossover event places one of them in an expression site. As the crossover occurs, the new VSG gene replaces the previously expressed VSG gene, which is then moved away from the promoter site. Finally, many VSG genes are pseuodogenes, arranged in long subtelomeric arrays (**Figure 4.20**). Because they are not adjacent to promoters, these genes cannot be directly expressed. Rather, they are first replicated and then the duplicate copy of the gene replaces the currently active VSG gene at the expression site. The pseudogene serving as a template remains in place. Thus, the pseudogenes are like a stored collection of genes that might be copied and subsequently expressed at any time. Both of these mechanisms occur following transition to the blood-form trypomastigote, with the use of pseudogenes more likely to occur later, following the establishment of the chronic infection. The triggers that account for VSG switching remain under investigation.

African trypanosomes are not alone in their use of antigenic variation to establish chronic infections. ***Giardia lamblia***, ***Trichomonas vaginalis***, ***Plasmodium falciparium***, and ***Theileria microti*** also employ this strategy, using different regulatory mechanisms. *G. lamblia*, for instance, has a repertoire of about 190 variant-specific surface protein (VSP) encoding genes. Like trypanosomes, only one VSP is found on the surface of each parasite, but in every 10 or 12 generations, there is a spontaneous switch to a different

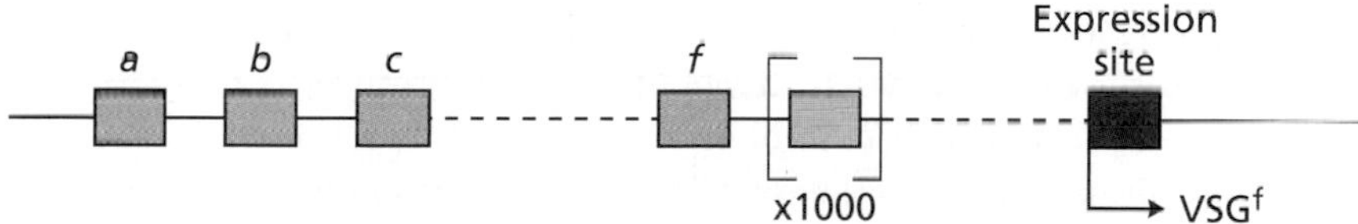

There are many inactive trypanosome pseudogenes, but only one site for expression.

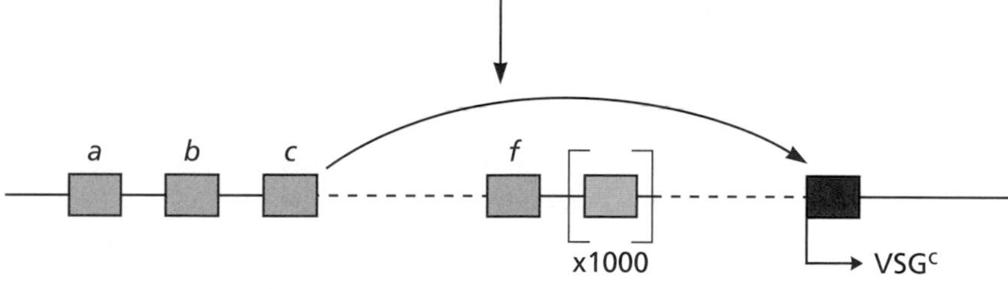

Pseudogenes are copied and transposed into the expression site.

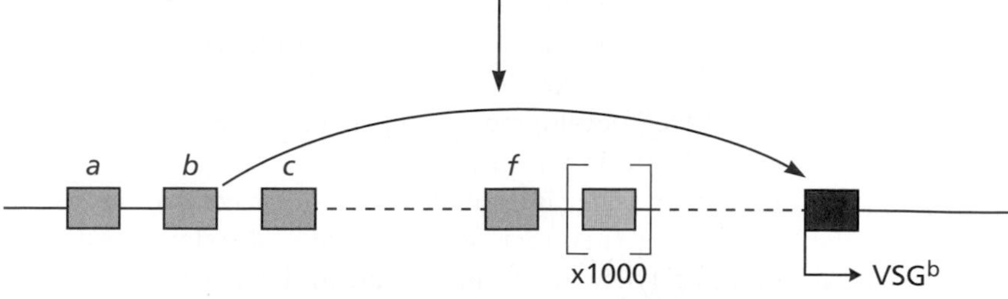

The process of pseudogene copying and transposition can occur repeatedly, allowing different VSG genes to be expressed each time.

Figure 4.20 Expression of pseudogenes as a mechanism of trypanosome antigenic variation. Inactive genes, or genes not necessarily expressed because they lack a promoter, are called pseudogenes. In African trypanosomes, many of the VSG genes are pseudogenes that are arranged in long arrays adjacent to the single promoter at the telomeric expression site. During each antigenic shift, a pseudogene is copied. It is subsequently transposed to the expression site, where it replaces the currently active VSG gene. Only a single gene, the one located at the expression site, is active at any given time. Of the three mechanisms used by trypanosomes to undergo antigenic switching, pseudogene expression is the most common.

VSP. Unlike trypanosomes, however, no gene crossover or DNA recombination appears to be involved. Instead, ***G. lamblia*** seems to transcribe several VSP genes at a time, but the detection of small antisense RNAs corresponding to silenced VSP genes indicates the involvement of RNA interference as a means of control. When components of this control such as the endoribonuclease dicer are experimentally silenced, even a single parasite expresses multiple VSPs on its surface.

Parasites frequently suppress or alter host immune responses by interfering with cell communication

Many parasites can directly intervene in the normal intercellular communication that occurs during an immune response. In doing so, they modify the response in such a way that their survival is more assured. Parasites may either suppress immunity entirely or they may shift an immune response to one that is more compatible with their survival. Parasites more vulnerable to cellular responses, for example, may skew immunity toward a less protective humoral response.

Earlier in the chapter, for instance, we discussed ***Leishmania mexicana*** and how infection with this kinetoplast can result in either a localized or disseminated form of cutaneous leishmaniasis. When the response is primarily cell mediated, the result is the less serious, localized disease. When the response is primarily humoral, the disseminated form is manifest. The parasite itself is involved in shifting the response away from the more protective Th-1 response by suppressing IL-12 release from infected macrophages. More specifically, amastigotes produce a cysteine protease that degrades NFκB, the necessary transcription factor for IL-12 synthesis (**Figure 4.21**). ***Toxoplasma gondii*** can also block IL-12 expression via a different mechanism, creating a cytokine milieu more compatible with its survival.

Helminths are also known to create immune environments more to their liking. As we previously discussed, one of the most typical features of a helminth infection is the induction of a Th-2 immune response. In many cases, it appears that the ability of helminth parasites to create a Th-2 environment explains their remarkable longevity in the host and their capacity to cause long-term chronic infections. Modification of immunity is frequently a result of various excretory products that in one way or another alter communication between cells.

Brugia malayi provides a useful example. This filarial worm, a causative agent of lymphatic filariasis, secretes an analog of the mammalian cytokine macrophage migration inhibitory factor (MIF). This molecule seems to mimic the effects of the host cytokine, helping to prevent the pro-inflammatory activation of macrophages. ***B. malayi*** also produces a TGFβ analog, which may promote Treg development, as has been found for mammalian TGFβ. A different nematode, *Heligmosomoides polygyrus*, a common intestinal parasite of rodents, also produces a TGFβ analog. When $CD4^+$ T cells are exposed to this parasite molecule *in vitro*, they up-regulate their expression of FoxP3, an important marker for Treg cells. Thus it appears that the nematode may induce Treg development, suggesting an important role in immunosuppression and parasite survival.

H. polygyrus and certain other intestinal nematodes have another useful strategy in their repertoire; they secrete the enzyme acetylcholinesterase. This enzyme degrades acetylcholine, a neurotransmitter released by motor neurons. In the intestine, acetylcholine stimulates smooth muscle contraction (intestinal peristalsis) and mucus secretion. Earlier we discussed the

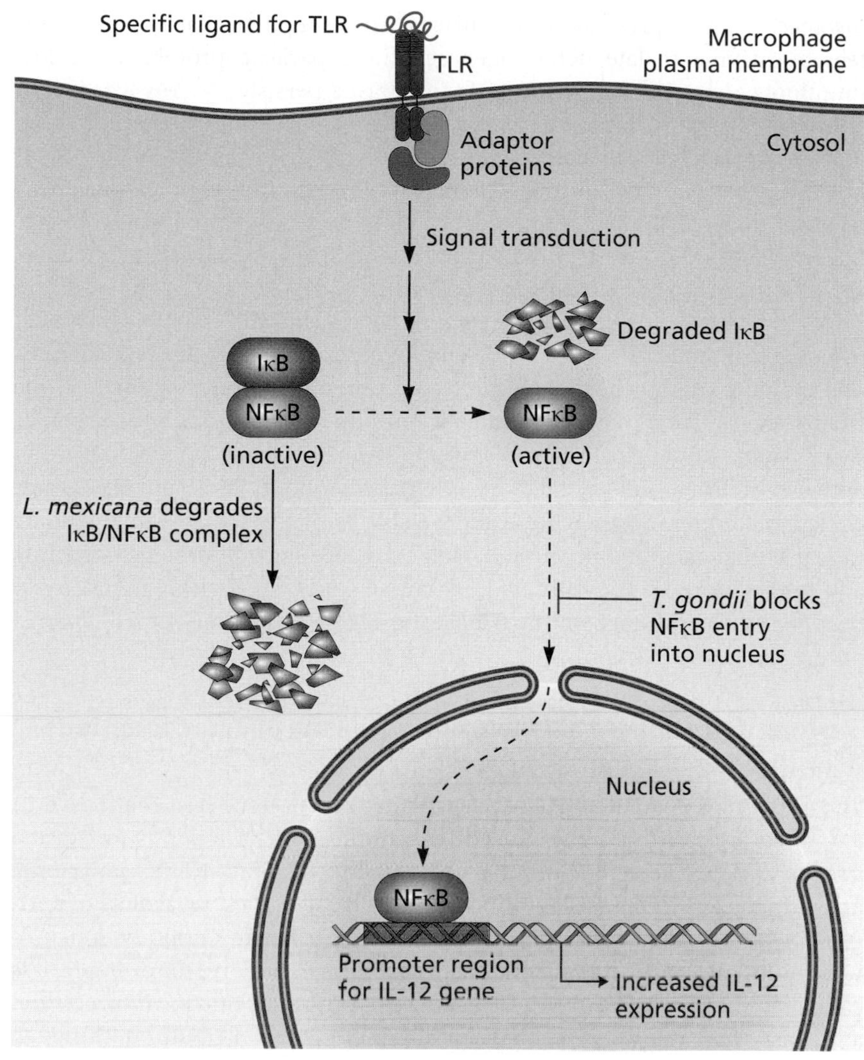

Figure 4.21 Interference of IL-12 production by protozoan parasites. IL-12 is one of the principal pro-inflammatory cytokines released by antigen-presenting cells to promote a protective, Th-1 response against intracellular parasites. The figure shows the manner in which two intracellular protozoa, *Leishmania mexicana* and *Toxoplasma gondii* suppress IL-12 expression, thus modifying the immune response and increasing the likelihood of parasite survival. IL-12 induction can be stimulated by either of these parasites when the important PAMP GPI in their plasma membranes binds to TLR-2 or TLR-4. Such binding activates a signal transduction cascade, during which the inhibitory subunit (IκB) is removed from the transcription factor NFκB. The now active NFκB enters the nucleus where it binds the promoter for the IL-12 gene. Expression of the gene ensues, resulting in the release of the cytokine. *L. mexicana* amastigotes, however, produce a cysteine protease that degrades the IκB–NFκB complex. *Toxoplasma gondii* blocks NFκB translocation to the nucleus. In either case, the transcription factor fails to bind the promoter, resulting in decreased IL-12 production.

importance of both peristalsis and mucus secretion in the clearance of gut helminths. Consequently, by interfering in this way with the host's weep-and-sweep response, these parasites are better able to ensure their persistence in the intestine.

Some parasites render themselves invisible to immune detection

The most fundamental distinction made by the immune system is that between self and nonself. During the development of lymphocytes, those cells bearing antigen receptors that recognize self-antigens are ordinarily eliminated in a process called **clonal deletion**. Lymphocytes that are permitted to survive are those that recognize foreign antigens. The process is far from perfect; autoimmune disease is basically a result of the presence of self-reactive lymphocytes that escaped clonal deletion. In general, though, we expect lymphocytes to largely ignore potentially antigenic components of our own bodies.

Certain parasites have taken advantage of this nonresponse to self. They may do so in one of two ways. Some parasites encode proteins that are similar to self-proteins of the host. That is, they have evolved to mask themselves with proteins that have structural resemblance to host molecules, thereby tricking the immune system into ignoring them. *Trichomonas vaginalis*, for instance, displays proteins on its surface with a high similarity to the vertebrate enzyme malate dehydrogenase. These parasite proteins have low immunogenicity and may contribute to parasite persistence. ***Toxocara canis*** encodes proteins analogous to the host protein CD23. The host protein is an Fc receptor for IgE, normally found on the surface of mast cells and basophils. It is unclear whether the parasite protein functions primarily as a way to make the parasite appear more hostlike or as a means to actually engage IgE Fc regions and thereby protect the parasite from the damaging effects of IgE binding via its Fab domains.

Another means of camouflage is for the parasite to simply adsorb host proteins to its surface. ***T. vaginalis*** employs this trick, but the real champions of appropriating host molecules are certainly the schistosomes. These trematodes, apparently so vulnerable to immune surveillance in the circulatory system, avoid detection by disguising themselves with host blood-group antigens, clotting inhibitory factors, low-density lipoproteins, and other molecules that adhere to their tegument. Analysis of the schistosome genome indicates that as added protection, they may also encode proteins structurally similar to those found in its host; many genes have striking analogy to host genes. The products of the trematode gene analogs might further complicate immune detection.

Various parasites are able to undermine the effector functions of antibodies

The antibodies generated against a parasite may be rendered ineffective in any of several ways. One way used by a number of protozoa is to cause a general nonspecific activation of B cells. Such activation, termed **polyclonal activation**, leads to the proliferation of many B cell clones, regardless of antigen specificity. The serum is consequently flooded with antibody, much of which is nonreactive with the parasite. In malaria infections, for example, it is estimated that only about 10% of serum antibody is specific for ***Plasmodium*** antigens. The remaining nonspecific antibodies contribute nothing to immune defense but may actually impede the ability of ***Plasmodium***-specific antibodies to bind and neutralize the parasite.

Other parasites take a less subtle approach; they simply degrade enzymatically any antibodies that manage to adhere to them, using proteases incorporated into their tegument. *Dirofilaria immitis*, the filarial worm responsible for canine heartworm, ***Schistosoma mansoni***, the liver fluke ***Fasciola hepatica***, and the protozoa ***Giardia lamblia***, ***Trichomonas vaginalis***, and ***Entamoeba histolytica*** all avail themselves of this tactic. The case of ***E. histolytica*** is particularly interesting. We mentioned earlier that this invasive and pathogenic species is capable of avoiding complement-mediated lysis, whereas the noninvasive and nonpathogenic ***E. dispar*** is not. Likewise, although ***E. histolytica*** rapidly degrades secretory IgA, ***E. dispar*** lacks this capacity. These mechanisms of immune evasion may explain in part the differences in pathogenicity between these two amebas.

Not content to merely destroy antibodies, ***E. histolytica*** has yet another way to dispose of these potentially damaging host molecules. Surface antigens on the amebas interact with elements of the cytoskeleton directly beneath the plasma membrane. When these antigens are bound and cross-linked by antibodies, cytoskeletal proteins are directly involved in transporting these antibody–antigen complexes to a specific region on the cell surface. These polarized molecular aggregates are subsequently eliminated. The removal of antibodies in this way, known as **antibody capping**, does not occur in *E. dispar*.

Summary

All living things are under assault from a dizzying array of parasites, and because of this some form of host defense is a universal feature of both prokaryotic and eukaryotic organisms. As parasites are so diverse, the various mechanisms used to limit their damage are equally numerous and sundry, and it can be difficult to develop general, underlying principles of immunity. This is true, not only when comparing very different host types, but even within a particular host lineage. Furthermore, recent discoveries such as the CRISPR system in prokaryotes, the systemic acquired resistance observed in many plants, and the concept of innate immune memory in invertebrate animals, suggest that many of the traditional paradigms of comparative immunology, such as adaptive immunity being limited to vertebrates, may require reevaluation.

Certain long-standing assumptions of vertebrate immunoparasitology are also in need of rethinking. For instance, mast cells and basophils, long described as part of the innate immune system, now seem important in the orchestration of the adaptive response to intestinal worms. As a result of findings such as this, the study of immunoparasitology has further blurred the already somewhat artificial line between innate and adaptive immunity. Host immunity is a complex, interconnected, and redundant process, many elements of which still await full elucidation.

Immune responses are energetically expensive and can cause their own pathology and must therefore be carefully regulated to achieve a delicate balance between "enough response" and "too much response". Subtle shifts in the precise nature of the response can have important consequences for the course of an infection. The pathology observed in cutaneous leishmaniasis, for instance, depends in large measure on whether the immune response is skewed toward a predominantly cell-mediated or humoral adaptive response. In many instances, parasites owe their ability to survive within the host to their own highly developed evasion strategies. The imperative of the immune system to protect the host without too much collateral damage, along with the capacity of successful parasites to avoid complete immune destruction, explains why so many parasitic infections are persistent and long-lasting.

REVIEW QUESTIONS

1. Based on recent findings, it has been suggested that even prokaryotes have an adaptive immune system. Upon what is that argument based?

2. How has the discovery and characterization of FREPs allowed us to at least partially answer an important question about the ability of invertebrates to survive constant exposure to relatively fast-evolving parasites?

3. Why is it that complete elimination of eukaryotic parasites by the immune system is rarely achieved in vertebrate hosts?

4. It has been demonstrated that the treatment of cutaneous leishmaniasis can be augmented by direct administration of IL-12 into skin lesions. Based on what you have learned about the immune response to this parasite, and the role of IL-12, explain the efficacy of IL-12 treatment.

5. In what way does the phenomenon of premunition to *Plasmodium* infection impact our thinking about the possibility of a fully protective malaria vaccine?

6. What problems might an individual with a deficiency of Natural Killer cells experience in terms of response to eukaryotic parasites?

7. We learned in this chapter that *Toxocara canis* encodes a protein that is similar in structure to a host protein called CD23. The normal function of CD23 is to act as a receptor for the Fc region of IgE molecules on the surface of mast cells and basophils. Its precise value to *T. canis* is unresolved. It may help to make the parasite appear more "hostlike". Alternatively, it may bind the Fc region of IgE antibodies that might ordinarily damage the parasite by binding it with its Fab region. Can you describe an experiment which might allow a researcher to determine which of these two possibilities, if either, is more important?

References

AN EVOLUTIONARY PERSPECTIVE ON ANTI-PARASITIC IMMUNE RESPONSES

Prokaryotes have developed remarkable immune innovations during their billions of years encountering parasites

Makarova KS, Wolf YI & Koonin EV (2013) Comparative genomics of defense systems in archaea and bacteria. *Nucleic Acids Res* **41**:4360–4377.

Horvath P & Barrangou R (2010) CRISPR/Cas, the immune system of bacteria and archaea. *Science* **327**:167–170 (doi:10.1126/science.1179555).

Many kinds of parasites compromise the health of plants so it is important to know how plants defend themselves

Agrios GN (ed) (2005) Plant Pathology, 5th ed. Academic Press.

Although plants lack specialized immune cells, they still can mount effective, long-term responses to parasites

Boyko A & Kovalchuk I (2011) Genetic and epigenetic effects of plant–pathogen interactions: an evolutionary perspective. *Mol Plant* **4**:1014–1023 (doi:10.1093/mp/ssr022).

Gust AA & Nuernberger T (2012) Plant immunology: a life or death switch. *Nature* **486**:198–199.

McDowell JM & Simon SA (2008) Molecular diversity at the plant–pathogen interface. *Dev Comp Immunol* **32**:736–744 (doi:10.1016/j.dci.2007.11.005).

Spoel S & Dong X (2012) How do plants achieve immunity? Defense without specialized immune cells. *Nat Rev Immunol* **12**:89–100.

Many nematode species are specialized to parasitize plants

Haegeman A, Mantelin S, Jones JT et al (2012) Functional roles of effectors of plant–parasitic nematodes. *Gene* **492**:19–31 (doi:10.1016/j.gene.2011.10.040).

Jones JT, Kumar A, Pylypenko LA et al (2009) Identification and functional characterization of effectors in expressed sequence tags from various life cycle stages of the potato cyst nematode *Globodera pallida. Mol Plant Pathol* **10**:815–828.

Invertebrates have distinctive and diverse innate immune systems

Lemaitre B & Hoffmann J (2007) The host defense of *Drosophila melanogaster. Annu Rev Immunol* **25**:697–743 (doi:10.1146/annurev.immunol.25.022106.141615).

Sôderhall K (2011) Invertebrate immunity. In Advances in Experimental Biology and Medicine. Landes Press, Springer Sciences

Invertebrates, including vectors and intermediate hosts, mount immune responses to contend with their parasites

Christensen BM, Li JY, Chen CC et al (2005) Melanization immune responses in mosquito vectors. *Trends Parasitol* **21**:192–199 (doi:10.1016/j.pt.2005.02.007).

Hanington PC, Forys MA & Loker ES (2012) A somatically diversified defense factor, FREP3, is a determinant of snail resistance to schistosome infection. *PLoS Negl Trop Dis* **6**:e1591 (doi:10.1371/journal.pntd.0001591).

Rodrigues J, Brayner FA, Alves LC et al (2010) Hemocyte differentiation mediates innate immune memory in *Anopheles gambiae* mosquitoes. *Science* **329**:1353–1355 (doi:10.1126/science.1190689).

Roth O & Kurtz J (2009) Phagocytosis mediates specificity in the immune defence of an invertebrate, the woodlouse *Porcellio scaber* (Crustacea: Isopoda). *Dev Comp Immunol* **33**:1151–1155 (doi:10.1016/j.dci.2009.04.005).

Invertebrates also adopt distinctive behaviors to supplement their anti-parasite immune responses

Konrad M, Vyleta ML, Theis FJ et al (2012) Social transfer of pathogenic fungus promotes active immunisation in ant colonies. *PLoS Biol* **10**:e1001300 (doi:10.1371/journal.pbio.1001300).

Milan NF, Kacsoh BZ, & Schlenke TA (2012) Alcohol consumption as self-medication against blood-borne parasites in the fruit fly. *Curr Biol* **22**:488–493 (doi:10.1016/j.cub.2012.01.045).

Parasites suppress, manipulate and destroy invertebrate defense responses

Castillo JC, Reynolds SE & Eleftherianos I (2011) Insect immune responses to nematode parasites. *Trends Parasitol* **27**:537–547 (doi:10.1016/j.pt.2011.09.001).

Hanington PC, Lun CM, Adema CM et al (2010) Time series analysis of the transcriptional responses of *Biomphalaria glabrata* throughout the course of intramolluscan development of *Schistosoma mansoni* and *Echinostoma paraensei*. *Int J Parasitol* **40**:819–831 (doi:10.1016/j.ijpara.2009.12.005).

Some parasites rely on symbiotic partners to subvert the immune responses of their invertebrate hosts

Bitra K, Suderman RJ & Strand MR (2012) Polydnavirus Ank proteins bind NF-kappa B homodimers and inhibit processing of relish source. *PLoS Pathog* **8**:e1002722 (doi:10.1371/journal.ppat.1002722).

Burke GR & Strand MR (2012) Polydnaviruses of parasitic wasps: domestication of viruses to act as gene delivery vectors. *Insects* **3**:91–119.

Some invertebrates enlist symbionts to aid in their defense

Oliver KM, Moran NA & Hunter MS (2005) Variation in resistance to parasitism in aphids is due to symbionts not host genotype. *Proc Natl Acad Sci USA* **102**:12,795–12,800 (doi:10.1073/pnas.0506131102).

Oliver KM, Noge K, Huang EM et al (2012) Parasitic wasp responses to symbiont-based defense in aphids. *BMC Biol* **10**:1–10.

Rio RVM, Symula RE, Wang J et al (2012) Insight into the transmission biology and species-specific functional capabilities of tsetse (Diptera: Glossinidae) obligate symbiont *Wigglesworthia*. *MBio* **3**:e00240 (doi:10.1128/mBio.00240-11).

Weiss B & Aksoy S (2011) Microbiome influences on insect host vector competence. *Trends Parasitol* **27**:514–522 (doi:10.1016/j.pt.2011.05.001).

We are hoping to manipulate invertebrate immune systems to achieve parasite control

Marois E (2011) The multifaceted mosquito anti-*Plasmodium* response. *Curr Opin Microbiol* **14**:429–435 (doi:10.1016/j.mib.2011.07.016).

Marshall JM & Taylor CE (2009) Malaria control with transgenic mosquitoes. *PloS Med* **6**:164–168:e1000020 (doi:10.1371/journal.pmed.1000020).

Mitri C & Vernick KD (2012) *Anopheles gambiae* pathogen susceptibility: the intersection of genetics, immunity and ecology. *Curr Opin Microbiol* **15**:285–291 (doi:10.1016/j.mib.2012.04.001).

Windbichler N, Menichelli M, Papathanos PA et al (2011) A synthetic homing endonuclease-based gene drive system in the human malaria mosquito. *Nature* **473**:212 (doi:10.1038/nature09937).

AN OVERVIEW OF VERTEBRATE DEFENSE

Murphy K (2012) Janeway's Immunology, 8th ed. Garland Science.

Bellanti JA (2012) Immunology IV. I Care Press.

IMMUNE RESPONSES TO EUKARYOTIC PARASITES

Immune response to protozoa is initiated by the recognition of PAMPS

Rodrigues MM, Oliveira AC & Bellio M (2012) The immune response to *Trypanosoma cruzi*: role of Toll-like receptors and perspectives for vaccine development. *J Parasitol* (doi:10.1155/2012/507874).

Gazzinelli RT & Denkers EY (2006) Protozoan encounters with Toll-like receptor signalling pathways: implications for host parasitism. *Nat Rev Immunol* **6**:895–906.

Immune responses to protozoa include both humoral and cell mediated components

Reithinger R et al (2007) Cutaneous leishmaniasis. *Lancet Infect Dis* **7**:581–596.

Golgher D & Gazzinelli RT (2004) Innate and acquired immunity in the pathogenesis of Chagas disease. *Autoimmunity* **37**:399–409.

Padilla AM, Bustamante JM & Tarleton RL (2009) $CD8^+$ T cells in *Trypanosoma cruzi* infection. *Curr Opin Immunol* **21**:385–390.

Frolich S, Entzeroth R & Wallach M (2012) Comparison of protective immune responses to apicomplexan parasites. *J Parasitol Res* Article ID 852591 (doi:10.1155/2012/852591).

Amezcua Vesley MCA, Bermejo DA, Montes CL et al (2012) B-cell response during protozoan parasite infections. *J Parasitol Res Article* ID 362131 (doi:10.1155/2012/362131).

Pantenburg B, Dann SM, Wang HC et al (2008) Intestinal immune response to human *Cryptosporidium* infection. *Infect Immun* **76**:23–29.

Solaymani-Mohammadi S & Singer SM (2010) The double-edged sword of immune responses in giardiasis. *Exp Parasit* **126**:292-297.

Lopes MF, Zamboni DS, Lujan HD et al (2012) Immunty to protozoan parasites. *J Parasitol Res* Article ID 250793 (doi:1155/2012/250793).

Protective immunity to malaria develops as a consequence of repeated exposure

Doolan DL, Dobaño C & Baird JK (2009) Acquired immunity to malaria. *Clin Microbiol Rev* **22(1)**:13–36.

Immune responses are generated against each stage in the *Plasmodium* life cycle

Roetynck S, Baratin M, Vivier E & Ugolini S (2006) NK cells and innate immunity to malaria. *Med Sci (Paris)* **22(8–9)**:739–744.

Stevenson MM & Riley EM (2004) Innate immunity to malaria. *Nat Rev Immunol* **4**:169-180.

Hisaedaa H, Yasutomob K & Himeno K (2005) Malaria: immune evasion by parasites. *Int J Biochem Cell Biol* **37(4)**:700–706.

Frolich S, Entzeroth R & Wallach M (2012) Comparison of protective immune responses to apicomplexan parasites. *J Parasitol Res* Article ID 852591 (doi:10.1155/2012/852591).

Leirinao P, Albuquerque SS, Corso S et al (2005) HGF/MET signalling protects *Plasmodium*-infected host cells from apoptosis. *Cell Microbiol* **7(4)**: 603–609.

Helminth parasites provoke a strong Th-2 response

Hodgkin J (2004) Dissecting worm immunity. *Nat Immunol* **5**:471–472.

Hepworth MR et al (2012) Mast cells orchestrate type-2 immunity to helminths through regulation of tissue derived cytokines. *Proc Natl Acad Sci* **109**:6644–6649.

Harris N & Gause WC (2011) To B or not to B: B cells and the Th2-type immune response to helminths. *Trends Immunol* **32**:80–88.

Saenz SA et al (2010) Innate immune cell populations function as initiators and effectors in Th2 cytokine responses. *Trends Immunol* **31**:407–413.

Neill DR, Wong SH, Bellosi A et al (2010) Nuocytes represent a new innate effector leukocyte that mediates type-2 immunity. *Nature* **464**:1367–1370.

Min B & Paul WE (2008) Basophils and type 2 immunity. *Curr Opin Hematol* **15**:59–63.

Extensive changes to the intestinal epithelium occur in response to intestinal helminths

Allen JE & Maizels RM (2011) Diversity and dialogue in immunity to helminths. *Nat Rev Immunol* **11**:375–388.

Harris NL (2011) Advances in helminth immunology: optimism for future vaccine design? *Trends Parasitol* **27**:288–293.

Ohnmact C, Schwartz C, Panzer M et al (2010) Basophils orchestrate chronic allergic dermatitis and protective immunity against helminths. *Immunity* **33**:364–374.

Finkelman FD et al (2004) Interleukin-4 and interleukin-13-mediated host protection against intestinal nematode parasites. *Immunol Rev* **201**:139–155.

Herbert DR, Yang J, Hogan SP et al (2009). Intestinal epithelial cell secretion of RELM beta protects against intestinal worm infection. *J Exp Med* **206**:2947–2957.

Anthony RM et al (2006) Memory Th-2 cells induce alternatively activated macrophages to mediate protection against nematode parasites. *Nat Med* **12**:955–960.

Immunocompromised hosts are more vulnerable to parasitic infection and increased pathology

Marcos LA, Terashima A, Dupont HL & Gotuzzo E (2008) *Strongyloides* hyperinfection syndrome: an emerging global infectious disease. *Trans R Soc Trop Med Hyg* **102(4)**:314–318.

Borkow G & Bentwich Z (2006) HIV and helminths co-infection: is deworming necessary? *Parasite Immunol* **28**:605–612.

PARASITE EVASION OF HOST DEFENSES

Sacks D & Sher A (2002) Evasion of innate immunity by parasitic protozoa. *Nat Immunol* **3**:1041–1047.

Schmid-Hempel P (2008) Parasite immune evasion: a momentous molecular war. *Trend Ecol Evol* **23**:318–326.

Schmid-Hempel P (2009) Immune defense, parasite evasion strategies and thelr relevance for macroscopic phenomena such as virulence. *Phil Trans R Soc B* **364**:85–98.

Many parasites are able to evade complement-mediated innate immune responses

Kedzierski L (2004) A leucine-rich repeat motif of *Leishmania* surface antigen 2 binds to macrophages through complement receptor 3. *J Immunol* **172**:4902–4906.

Tham WH et al (2010) Complement receptor 1 is the host erythrocyte receptor for *Plasmodium falciparum* PfRh4 invasion ligand. *Proc Natl Acad Sci* **107**:17,327–17,332.

Deng J, Gold D, Loverde PT et al (2003) Inhibition of complement membrane attach complex by *Schistosoma mansoni*. *Infect Immun* **71**:6402–6410.

Awandare GA, Spadafora C, Moch JK et al (2011) *Plasmodium falciparum* field isolates use complement receptor 1 (CR1) as a receptor for invasion of erythrocytes. *Mol Biochem Parasitol* **177**:57–60.

Meri T, Jokiranta TS, Hellwage J et al (2002) *Onchocerca volvulus* microfilariae avoid complement attack by direct binding of factor H. *J Infect Dis* **185**:1786–1793.

Campos-Rodriguez R & Jarillo-Luna A (2005) The pathogenicity of *Entamoeba histolytica* is related to the capacity of evading innate immunity. *Parasitol Immunol* 27:1-18.

Many intracellular parasites have evolved mechanisms to avoid destruction by host cells

Barbieri CL (2006) Immunology of canine leishmaniasis. *Parasite Immunol* **28**:329–337.

Sibley LD & Andrews NW (2000) Cell invasion by un-palatable parasites. *Traffic* **1**:100–106.

Rittig MG & Bogdan C (2000) *Leishmania*–host cell interaction: complexities and alternative views. *Parasitol Today* **16**:292–297.

Franco LH, Beverly SM, Zamboni DS et al (2012) Innate immune activation and subversion of mammalian functions by *Leishmania* lipophosphoglycan. *J Parasitol Res* Article ID 165126 (doi:10>1155/2012/165126).

Parasites may interfere with intracellular signaling pathways

Millington OR, Di Lorenzo C, Phillips RS et al (2006) Suppression of adaptive immunity to heterologous antigens during *Plasmodium* infection through hemozoin-induced failure of dendritic cell function. *J Biol* **5**:5 (doi:10.1186/jbiol34).

Nathan C & Shiloh MU (2000) Reactive oxygen and nitrogen intermediates in the relationship between mammalian hosts and microbial pathogens. *Proc Natl Acad Sci* **97**:8841–8848.

Xu X, Sumita K, Feng C et al (2001) Down-regulation of IL-12 P40 gene in *Plasmodium berghei*-infected mice. *J Immunol* **107**:235-241.

Dobbin CA, Smith NC, Johnson, AM et al (2002) Heat shock protein 70 is a potential virulence factor in murine Toxoplasma infection via immunomodulation of host NF-Kappa B and nitric oxide. *J Immunol* **169**:958–965.

Campos-Rodriguez R & Jarillo-Luna A (2005) The pathogenicity of *Entamoeba histolytica* is related to the capacity of evading innate immunity. *Parasite Immunol* **27**:1–18.

Some parasites interfere with antigen presentation, resulting in an impaired immune response

Luder CGK, Lang T, Beuerle B et al (1998) Down-regulation of MHC class II molecules and inability to up-regulate Class I molecules in murine macrophages after infection with *Toxoplasma gondii*. *Clin Exp Immunol* **112**:308–316.

Schonemeyer AH, Lucius R, Sonnenburg B et al (2001) Modulation of human T-cell responses and macrophage function by oncocystatin, a secreted protein of the filarial nematode *Oncocerca volvulus*. *J Immunol* **167**:3207–3215.

McKee AS, Dzierszinski F, Boes M et al (2004) Functional inactivation of immature dendritic cells by the intracellular parasite *Toxoplasma gondii*. *J Immunol* **173**:2632–2640.

Soto CD, Mirkin GA. Solana ME et al (2003) *Trypanosoma cruzi* infection modulates in vivo expression of major histocompatibility complex class II molecules on antigen presenting cells and T-cell stimulatory activity of dendritic cells in a strain-dependent manner. *Infect Immun* **71**:1194–1199.

Dey R, Khan S, Pahari S et al 2007) Functional paradox in host–pathogen interaction dictates fate of parasites. *Future Microbiol* **2**:425–437.

Maizels RM, Balic A, Gomez-Escobar N et al (2004) Helminth parasites—masters of regulation. *Immunol Rev* **201**:89–116.

Fujiwara RT et al (2009) *Necator americanus* infection: a possible cause of altered dendritic cell differentiation and eosinophil profile in chronically infected individuals. *PLoS Negl Trop Dis* **3(3)**: e399 (doi:10.1371/journal.pntd.0000399).

Lundie RJ (2011) Antigen presentation in immunity to murine malaria. *Curr Opin Immunol* **23**:119–123.

Klotz C, Zeigler T, Danilowicz-Lubert E et al (2011) Cystatins of parasitic organisms. *Adv Exp Med Biol* **712**:208–221.

Salat J, Paesen G, Rezacova P et al (2010) Crystal structure and functional characteristics of an immunomodulatory salivary cystatin from the soft tick *Ornithodoros moubata*. *Biochem J* **429**:103–112.

Some parasites regularly change their surface antigens to avoid immune responses

Prucca CG, Slavin I, Quiroga R et al (2008) Antigenic variation in *Giardia lamblia* is regulated by RNA interference. *Nature* **456(7223)**:750–754.

Recker M et al (2011) Antigenic variation in *Plasmodium falciparum* malaria involves a highly structured switching pattern. *PLoS Pathog* **7(3)**:e1001306 (doi:10:1371/journal.ppat.1001306).

Barry JD & McCulloch R (2001) Antigenic variation in trypanosomes. Enhanced phenotypic variation in a eukaryotic parasite. *Adv Parasitol* **49**:1–70.

Parasites frequently suppress or alter host immune responses by interfering with cell communication

Hewitson JP (2009) Helminth immunoregulation: The role of parasite secreted proteins in modulating host immunity. *Mol Biochem Parasitol* **167**:1–11.

Blader I & Saeij J (2009) Communication between *Toxoplasma gondii* and its host: impact on parasite growth, development, immune evasion and virulence. *APMIS* **117**:458–476.

Lambert H & Baragan A (2010) Modelling parasite dissemination: host cell subversion and immune evasion by *Toxoplasma gondii*. *Cell Microbiol* **12**:292–300.

Harnett W & Harnett MM (2006) Molecular basis of worm-induced immunomodulation. *Parasite Immunol* **28**:535–543.

Some parasites may render themselves invisible to immune detection

Newport GR & Colley DG (1993) Schistosomiasis. In Immunology and Molecular Biology of Parasitic Infections (KS Warren ed), 3rd ed. Blackwell.

Mountford AP (2005) Immunological aspects of schistosomiasis. *Parasite Immunol* **27**:243–246.

Various parasites are able to undermine the effector functions of antibodies

Soulsby EH (1987) The evasion of the immune response and immunological unresponsiveness: parasitic helminth infection. *Immunol Lett* **16**:315–320.

Tang, C, Lei J, Wang T et al (2011) Effect of CD4+ CD25+ regulatory T cells on the Immune evasion of *Schistosoma japonicum*. *Parasitol Res* **108**:477–480.

Parasite Versus Host: Pathology and Disease

My water's gone black.
BERKELEY, From the movie Out of Africa

Courtesy of Universal Studios Licensing LLC.

In This Chapter

BY NOW IT SHOULD BE CLEAR; PARASITOLOGY IS A LITTLE LIKE CHESS, consisting of a never-ending series of host and parasite moves and countermoves. In Chapter 4 we investigated important and fundamental aspects of this theme—the manner in which hosts attempt to defend themselves against parasites and the strategies employed by parasites to elude, manipulate, or otherwise frustrate host defense. As we discussed, many of the same mechanisms that are involved in host defense are also involved in **pathogenesis**—the capacity of a parasite to cause disease. If a host's immune response successfully thwarts a parasite's attempt to colonize and infect that host, disease is avoided. However, if a parasite is able to initially infect a host, even a successful immune response can often contribute to pathology. Frequently, the same immune mechanisms that are intended to protect us cause the symptoms observed during a parasitic infection.

Parasites may affect their hosts in various other ways as well. Depending on the particular parasite and host, we may observe altered metabolism, physiology, anatomy, reproductive success, or even behavioral changes in the host, and pathology can run the gamut from inapparent or very mild to serious or lethal. From an evolutionary perspective, the degree of virulence (see Chapter 1) observed in an infection is frequently a balance between the evolutionary cost of virulence to the parasite and the cost of resistance to the host. Owing to certain characteristics such as the life cycle of the parasite and the parasite's degree of specialization on the host in question, selection may act to either increase or decrease both parasite virulence and host resistance. The evolution of virulence will be examined in Chapter 7. In this chapter, we take a more immediate view of pathology, examining the various ways that parasites can affect their hosts and investigating the consequences of those effects. We will conclude by discussing a few curious instances in which it has been proposed that sometimes parasites can actually benefit their hosts, giving with one hand, so to speak, even as they take away with the other.

Having examined host defense in Chapter 4 and parasite-induced pathology in this chapter, we will come to understand the overall disease process as it occurs in parasitic infections and to appreciate how and why disease occurs in the first place.

Species names highlighted in red are included in the Rogues' Gallery, starting on page 429.

5.1 PATHOLOGY RESULTING FROM PARASITIC INFECTIONS

Disease, defined in Chapter 1, can be simply regarded as an alteration in normal cell, organ, or organismal activity. Much disease, of course, has nothing to do with infections of any kind. Here, we will limit our discussion to disease induced by parasites that in some manner compromises the normal activity of the host.

Certainly not all parasitic infections result in overt disease. In light infections or when the involved parasite is of limited virulence, the resulting pathology may be so mild as to result in a **subclinical infection**, an infection with no accompanying signs or symptoms. When parasite infects host, the likelihood of disease depends on three factors:

- The status of host defenses
- The number of parasites present
- The pathogenicity of those parasites

These three factors are closely interrelated. If a particular parasite strain or species is highly virulent, for instance, fewer parasites might cause enough pathology to result in disease (**Figure 5.1**). The number of parasites required to cause detectable signs and symptoms in an infected individual is called the **threshold of disease**. As the virulence of a particular parasite increases, fewer parasites are needed to reach this threshold. If the host is in someway immunocompromised, even fewer pathogenic parasites may cause serious illness. If we know something about these three factors in a particular host–parasite interaction, we can predict the likely outcome of that interaction. An important piece of this puzzle is the specific manner in which a particular parasite causes pathology.

Parasites can induce pathogenesis in various ways

We have had ample opportunity in earlier chapters to emphasize that when it comes to parasites, diversity is common. That diversity is also reflected in the large number of ways that parasites can make us sick. That is, the pathogenicity observed in parasitic infections is caused by sundry mechanisms that range from vitamin deficiencies caused by certain tapeworm infections, to ***Plasmodium***-induced kidney disease, to anemia resulting from infection with certain intestinal helminths.

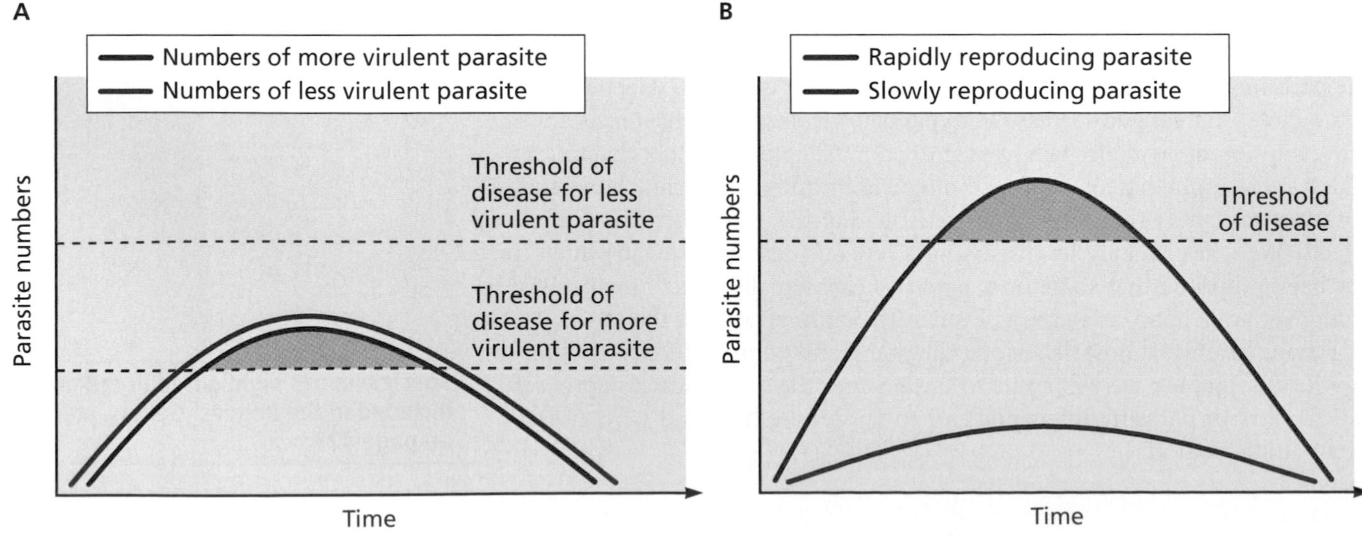

Figure 5.1 The likelihood of disease. When a parasite infects a host, there are an almost limitless number of possible outcomes. Four simplified possible outcomes of parasitic infection are shown in the graphs. Of course, many parasitic infections become chronic, and the parasites are not ultimately eliminated as in these examples. (A) Possible scenarios during an infection with a more virulent (red) and a less virulent (blue) parasite. If parasite numbers are the same for both species and if both parasites are cleared by the host immune system in the same amount of time, it is possible that only the more virulent species will cross its threshold and cause disease. Infection with the less virulent species is subclinical. (B) Two additional scenarios in which one parasite (red) reproduces much more quickly than the second (in blue). In these scenarios it is assumed that the two parasites are similarly virulent and that, once again, they are cleared immunologically in approximately the same amount of time. Only the more rapidly reproducing parasite is able to cross the threshold and cause disease before it is contained by the immune system. (From Hofkin BV [2011] Living in a Microbial World 1ed. Garland Science.)

In some cases, parasite-induced pathology is merely an unfortunate by-product of the infection that is of no obvious adaptive value to the parasite. ***Opisthorchis viverrini***, for example, commonly called the Southeast Asian liver fluke, is a digenetic trematode that as an adult inhabits the bile duct of its mammalian definitive host. Chronic infections may lead to a cancer of the bile duct known as **cholangiocarcinoma**. It is not entirely clear how the trematode predisposes an infected individual to this type of cancer. Mechanical damage caused by the fluke, inflammation in response to secreted fluke antigens, or both, may be involved in setting the stage for carcinogenesis. *Ov*-GRN-1, a secreted parasite protein with considerable sequence similarity to the mammalian growth factor granulin, may be particularly important. Provoking cancer in its host provides no benefit to the fluke. Cholangiocarcinoma is simply a side effect of an *O. viverrini* infection.

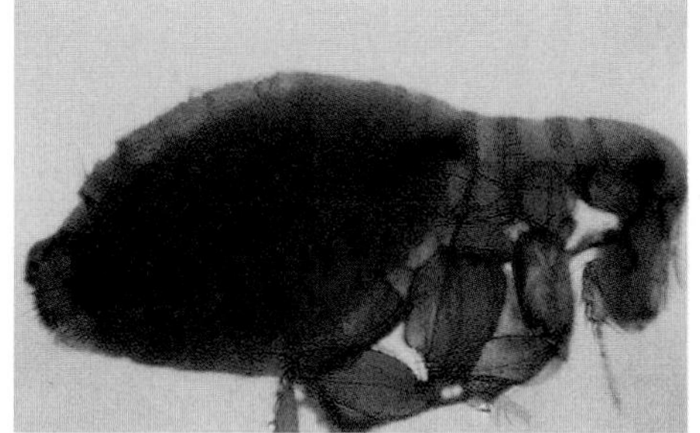

Figure 5.2 A blocked flea. This oriental rat flea, *Xenopsylla cheopsis*, is infected with the bacterium *Yersinia pestis*. The bacteria replicate within the flea's proventriculus, forming a large mass that blocks the digestive system and the flea's ability to feed. The dark red coloration from previously digested blood meals is typical of such blockage. At the anterior end of the red coloration, the blocked protoventriculus is seen in dark brown. Because they are unable to feed efficiently, blocked fleas attempt to feed repeatedly, spreading the bacteria into new hosts as they probe for blood. (Photo courtesy of CDC.)

Alternatively, pathology is sometimes best understood as an adaptive strategy on the part of the parasite to enhance its transmission or otherwise complete its life cycle. A well-known example is provided by the interaction between *Yersinia pestis*, the bacterium causing bubonic plague, and its flea vectors, one being *Xenopsylla cheopis*, the Oriental rat flea. Plague is primarily a disease of rodents. Fleas become infected when they take a blood meal from an infected animal. Humans can enter the transmission cycle when bitten by infected fleas.

When a flea feeds on an infected animal, the ingested bacteria proliferate in the flea's digestive system. Ultimately they may reproduce to such an extent that they form a mass, which partially or completely blocks the flea's proventriculus or gizzard (**Figure 5.2**). A flea that is affected in this way cannot feed properly. It consequently becomes increasingly hungry and as it attempts to feed by repeatedly probing potential hosts, it regurgitates bacteria into the bite site, infecting new vertebrate hosts. Such a flea, known as a blocked flea, is a particularly good plague vector because of its repeated attempts to feed. Clearly in this case, the pathology in the flea is in the interest of the bacteria, as transmission is substantially enhanced.

Pathology can be categorized as one of several general types

Our task of discussing the myriad ways that parasites can cause pathology is made easier because most forms of pathology can be identified as one of several types. In general, pathology from parasitic infection may be described as the result of:

- Parasite-induced trauma to cells, tissues, and organs
- Changes in cellular growth patterns
- Interference with host nutrient acquisition
- Toxins released by the parasites
- The host immune response to infection

In the following section we will discuss each of these mechanisms of pathology. At the same time, we will receive a fresh reminder of a point made in Chapter 1: in parasitology, as in so many aspects of the life sciences, natural phenomena resist our attempts at categorization and pigeon-holing. In many instances, a type of pathology in one category might just as easily be placed in different or multiple categories.

Note also that many parasites induce pathology in more than one way. Take ascariasis, the condition associated with *Ascaris lumbricoides* infection, as an example. Although many patients remain asymptomatic indefinitely, as larval worms migrate from the lumen of the intestine into the circulatory system, they may cause trauma to the epithelium resulting in peritonitis. Respiratory symptoms may occur when larvae penetrate through the capillaries into the alveoli. In a heavy infection, there may be intestinal blockage caused by a large mass of worms. Such a problem is more likely in children because fewer worms are required to block their smaller intestine. Having returned to the intestine, adult worms sometimes migrate into the bile or pancreatic duct, where they can likewise cause obstructions.

These symptoms of ascariasis can be categorized mainly as physical trauma, but inflammation in response to such trauma indicates a role for the host immune response in pathology as well. Furthermore, heavily infected individuals, especially children, can also suffer nutritional deficiency. Not only do ***A. lumbricoides*** absorb digested nutrients in the intestine intended for the host, but they also cause intestinal distress resulting in a loss of appetite and consequent reduction in food intake. Long-term infection with ***Ascaris*** and other roundworms is implicated in decreased childhood growth rate.

This chapter will not provide a comprehensive review of how all important parasites cause disease. The goal is to simply develop an understanding of the varied mechanisms involved in the disease process. Although specific examples will be provided along the way, the list of examples is far from complete. Additional examples of pathology caused by parasites of medical or veterinary importance can be found in the Rogues' Gallery at the end of the text.

Parasites can cause direct trauma to host cells, tissues, and organs

As parasites migrate, develop, feed, or reproduce, there are often damaging consequences for the host. Some of these changes are purely mechanical. The intestinal obstruction that can occur during an ***A. lumbricoides*** infection is an example. Mechanical insult is also behind some of the pathology caused by ***Taenia solium*** or ***Echinococcus*** spp. Like other tapeworms, both of these cestodes form cyst-like structures (see Section 3.3) in the tissues or organs of their intermediate hosts. These cysts can be quite large—large enough to impinge upon and damage adjacent tissue and organs. In the case of ***T. solium***, the observed symptoms and severity of disease varies with the location, number, and size of the tissue cysts (known as cysticerci). As the site of encystation is often the central nervous system, symptoms often include headaches, vision problems, dizziness, and seizures. Occasionally, pressure exerted by the cyst leads to disturbance of cerebrospinal fluid circulation, which can cause dementia.

Pathology is similar in individuals infected with ***Echinococcus granulosus***, the causative agent of hydatid disease (**Figure 5.3**). The cysts are usually larger and fewer in number than they are in a ***T. solium*** infection. Over time each cyst, called a **unilocular hydatid cyst**, develops a thick outer acellular layer and a thin inner germinal layer that produces larval stages called protoscolices via asexual reproduction.

Hydatid cysts grow exceedingly slowly; it may take a decade or more before symptoms become apparent. As with ***T. solium***, these symptoms depend on the location of the cyst, and as the cyst grows it interferes with the functions of adjacent tissues. Often, chronic pressure on a bone causes necrosis and the first sign of infection is sometimes a spontaneous fracture of the necrotic bone. The ultimate size of the cyst is in part determined by its location in the host, but when hydatid cysts are free to grow in a body cavity

Figure 5.3 Hydatid cysts in a sheep. The sheep's liver is filled with numerous hydatid cysts (white areas). The liver is the most common site for such cysts. As the cyst grows, it damages adjacent liver tissue, compromising the activity of the organ. (Photo courtesy of CDC.)

without impediment, they can be huge—ultimately bearing millions of protoscolices and over three gallons of fluid.

The tapeworms described above as well as many other helminth parasites cause tissue trauma in a second way: as a consequence of their larval migration through the host. In Chapter 3, we discussed the intricate and sometimes difficult to explain migrations that many helminths make in their hosts. We also saw that some parasites such as schistosomes penetrate directly through the skin to gain entry into their hosts. Both the tissue damage that occurs as these parasites enter or migrate through host tissues and the resulting inflammatory response contribute to the disease state. The involvement of host-mediated inflammation in these examples further highlights the difficulty of categorizing various types of pathology. Do observed symptoms result from parasite-mediated tissue damage or from the host immune response to that damage? Symptoms occur due to both of these factors, and often other factors as well.

Certain protozoa also induce pathology as a result of their direct destruction of host tissue. In Chapter 2, for instance, we discussed the severe deformations caused in fish by the myxozoan ***Myxobolus cerebralis***. ***Entamoeba histolytica*** provides another useful example. Although many infections with this ameba remain asymptomatic, some of those infected experience intestinal disorders such as abdominal distress and diarrhea. Such an infection may resolve on its own or it may develop into a long-term chronic condition. About 10% of infected individuals develop a far more invasive infection called **amebic dysentery**, in which amebas penetrate and feed on the intestinal epithelium (**Figure 5.4**). As they spread from their initial point of entry, they create characteristic, necrotic flask-shaped lesions. At this time, the ameba trophozoites are reproducing rapidly and the infected individual experiences frequent bloody bowel movements.

As bad as amebic dysentery is, things can get a lot worse. As the ulcers continue to expand, they may eventually reach the thin, underlying serous membrane or the musculature surrounding the colon. If trophozoites feeding at the edge of these ulcers penetrate the serous membrane and muscle, the result is a perforated bowel (see Figure 5.4). This condition is frequently accompanied by secondary bacterial infection of the peritoneal cavity. Penetration by infectious organisms such as amebas is a common cause of **peritonitis**, an inflammation of the peritoneal cavity. Any infection of this normally sterile cavity is a serious medical condition, requiring quick attention. With invasive ***E. histolytica***, the situation can be even more dire, as amebas enter the circulatory system and spread throughout the body. The disseminated amebas may end up in any organ, although the liver is the most frequent target. In the liver or elsewhere, feeding trophozoites form additional lesions and areas of necrosis.

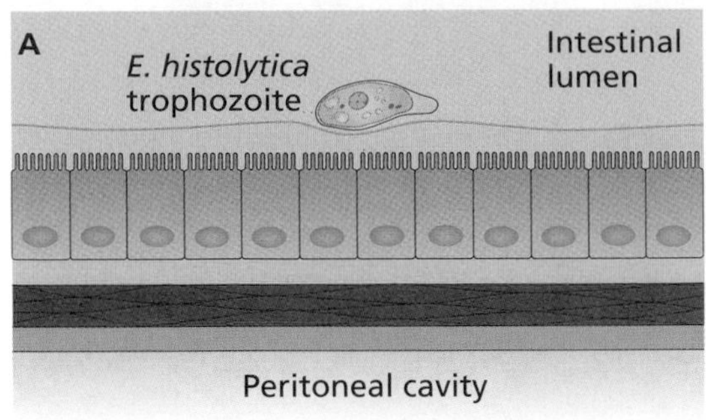

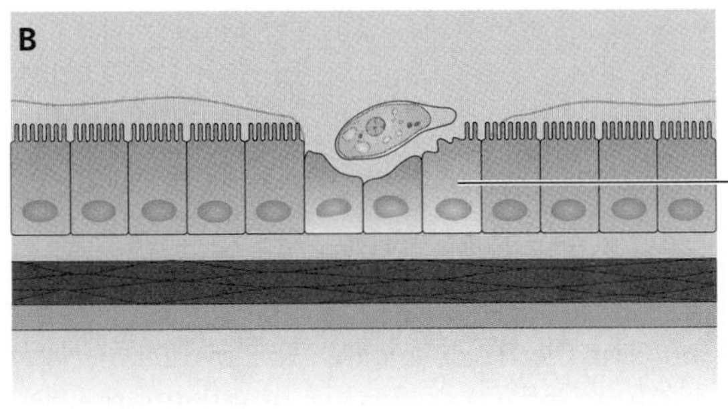

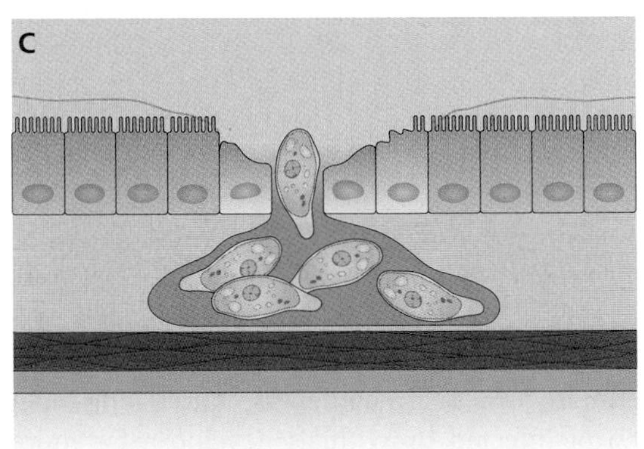

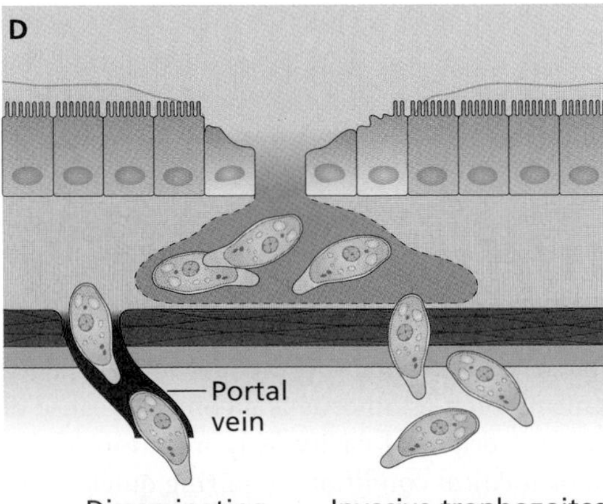

Figure 5.4 Invasiveness of *Entamoeba histolytica*. (A) Ameba trophozoites typically adhere to the mucous layer of the large intestine via surface adhesion molecules. When the mucous layer remains intact, the infection may persist in this noninvasive stage. Infected individuals may remain asymptomatic or display only mild symptoms. (B) Should the mucous layer break down, trophozoites may come in contact with epithelial cells, resulting in cell death. Cells may be killed by the activation of caspase-dependent apoptosis or by the activity of the ameba's cysteine proteases or amebapores, or both. Necrosis of the intestinal epithelium leads to invasive amebiasis. (C) As the trophozoites move into the submucosa, they produce a characteristic flask-shaped lesion. During this time, the rapidly reproducing trophozoites feed on host cells. (D) An intestinal perforation can occur as the amebas penetrate the surrounding musculature and serous layer. In such cases, symptoms may be severe and include hemorrhaging, peritonitis, and secondary bacterial infections. At this stage, if trophozoites enter the circulatory system, they can be disseminated throughout the body. Such dissemination is especially common in the liver. As amebas reaching the liver via the portal vein continue to reproduce and feed, they cause additional hepatic necrosis and associated lesions.

Mechanisms underlying the pathogenicity of *Entamoeba histolytica* remain obscure

Although we can describe what can happen during an ***E. histolytica*** infection in some detail, many questions remain about how it occurs. One way to investigate the underlying mechanisms of pathology is to compare this species with the closely related, yet nonpathogenic ***Entamoeba dispar***. These two taxa have only been recognized as distinct species since the early 1990s. Previously, they were thought to represent pathogenic and nonpathogenic forms of ***E. histolytica***.

Clearly, an important aspect of ***E. histolytica*** pathology is its ability to become invasive. ***E. dispar***, on the other hand, is noninvasive. At worst, it causes only superficial damage to the intestinal epithelium.

To become invasive, ***E. histolytica*** must first adhere to galactose or *N*-acetyl-galactosamine residues on the surface of host epithelial cells via specific **lectins** (carbohydrate-binding proteins) on the trophozoite surface. ***E. dispar*** has similar surface lectins, but only those of ***E. histolytica*** elicit an inflammatory response. This difference is thought to be caused by variation in the primary structure of the lectins in these two species. The ensuing inflammation (or lack of it) may consequently be one of the factors explaining the difference in pathogenicity between these two species, and it may set the stage for subsequent invasion of the colonic epithelium.

E. histolytica lectins may also play a direct part in host cell death. Binding of the lectin to its host-cell surface ligand results in caspase activation and subsequent apoptosis. Alternatively, trophozoites may kill host cells by directly perforating their membranes. Both ***E. histolytica*** and ***E. dispar*** produce proteins called **amebapores**, which they use to kill bacteria in their food vacuoles. Little or no direct evidence indicates that amebapores also puncture the cells of their eukaryotic hosts. Yet such perforating potential is suggested because at least for one molecule, amebapore A, the activity of this protein in ***E. dispar*** is approximately 50% of that in ***E. histolytica***. Furthermore, strains of ***E. histolytica*** that have been genetically manipulated to express low levels of amebapore A become less invasive. When amebapore A expression is completely blocked, such amebas are unable to cause liver abcesses in experimental animals. It has recently been shown that amebas may kill by ingesting distinct pieces of living cells, resulting in intracellular calcium elevation and eventual cell death. After cell killing, the amebas detach and cease ingestion.

E. histolytica trophozoites also come equipped with membrane-bound cysteine proteases that can degrade proteins found in the colonic mucus. Because such proteins generally impede access to the underlying colonic epithelium, their degradation may be an important step toward invasiveness. In general, ***E. histolytica*** produces higher levels of such proteases than ***E. dispar***.

Yet the exact role of these cysteine proteases in invasiveness has been hard to characterize. Most of the genes for cysteine proteases that have been identified in the ***E. histolytica*** genome are also found in ***E. dispar***. However, one protease unique to ***E. histolytica***, CP5, is expressed at particularly high levels. ***E. histolytica*** strains with low-CP5 expression are less invasive, and when expression in these strains is enhanced via transfection, invasiveness increases. Although these data suggest that CP5 is important in invasiveness, it is still unclear exactly why. The manner in which other cysteine proteases may increase invasiveness has remained similarly obscure.

Consequently, although several differences between these two species have been revealed, the manner in which these differences influence pathology awaits clarification. Other questions remain as well. Why, for instance, do so many ***E. histolytica*** infections remain noninvasive and even asymptomatic, whereas others can prove life-threatening? A number of factors are thought to be involved. Some of these factors, such as those discussed above related to adherence and the ability to degrade and kill host cells and tissues may be explained by genetic differences in different parasite strains. Other factors have more to do with host defenses. In the colon, for instance, the most important barrier preventing access of amebas to epithelium cells is the thick mucous layer overlying the epithelium (see Figure 5.4). Because galactose and *N*-acetyl galactosamine are abundant in the mucus, trophozoite surface lectins often bind the mucus rather than the epithelial cells themselves. In this case, the infection is noninvasive and may remain asymptomatic. Clearly, if the mucous layer is compromised in any way, the likelihood of clinical disease increases. Variables such as the makeup of the natural bacterial flora in the intestine may be important in this regard. Inflammation associated with the host immune response probably also influences the likelihood that amebas will invade the submucosa, highlighting the interplay between host and parasite factors, discussed earlier in this chapter.

Parasitic infection can alter host-cell growth patterns

Host cells are not necessarily destroyed when they are subjected to physiological or pathological insults. Before they are killed, cells may attempt to adapt in one of several ways. Parasites can serve as the trigger for this type of adaptation, resulting in altered cell and tissue growth. The pathology accompanying parasite-induced growth changes can range from fairly benign to severe.

Hypertrophy

An increase in the size of an organ or tissue due to an enlargement of its component cells is known as **hypertrophy** (**Figure 5.5**). Parasite-induced hypertrophy is a well-known phenomenon. For example, monocytes infected with ***Leishmania*** become hypertrophic. In Chagas' disease patients, cardiomyocytes undergo hypertrophy in response to ***Trypanosoma cruzi*** infection. During the hypertrophic response, there is increased expression for contractile proteins in the host cells, following by a doubling of cell size. Interleukin 1β, which is expressed rapidly upon initial establishment of a *T.* ***cruzi*** infection, appears to promote the hypertrophic response.

Hyperplasia

As opposed to hypertrophy, in which cells are increasing in size, **hyperplasia** refers to an increase in cell number (see Figure 5.5). Other than their accelerated proliferation, hyperplastic cells appear normal. In some cases, hyperplasia is a normal response to a specific stimulus. An example might be the proliferation of new skin cells to compensate for those cells lost to injury or natural senescence. In other cases, including those in which hyperplasia is induced by parasites, it may be viewed as a response to abnormal stimuli, cell irritation, or as a proliferation of cells precipitated by parasite-induced inflammation.

In some cases, hyperplasia may directly benefit the parasite responsible for the cellular proliferation. One well-documented hyperplastic response is caused by the liver fluke ***Fasciola hepatica***. When adults reach their preferred location in the bile duct, there is a proliferation of duct epithelial cells. The flukes feed upon these newly produced cells. In other instances, hyperplasia benefits the host. Many freshwater bivalves, for example, produce a parasitic larval stage called a **glochidium** (plural glochidia). These larval molluscs are released into the water, where they attach to a fish's gills

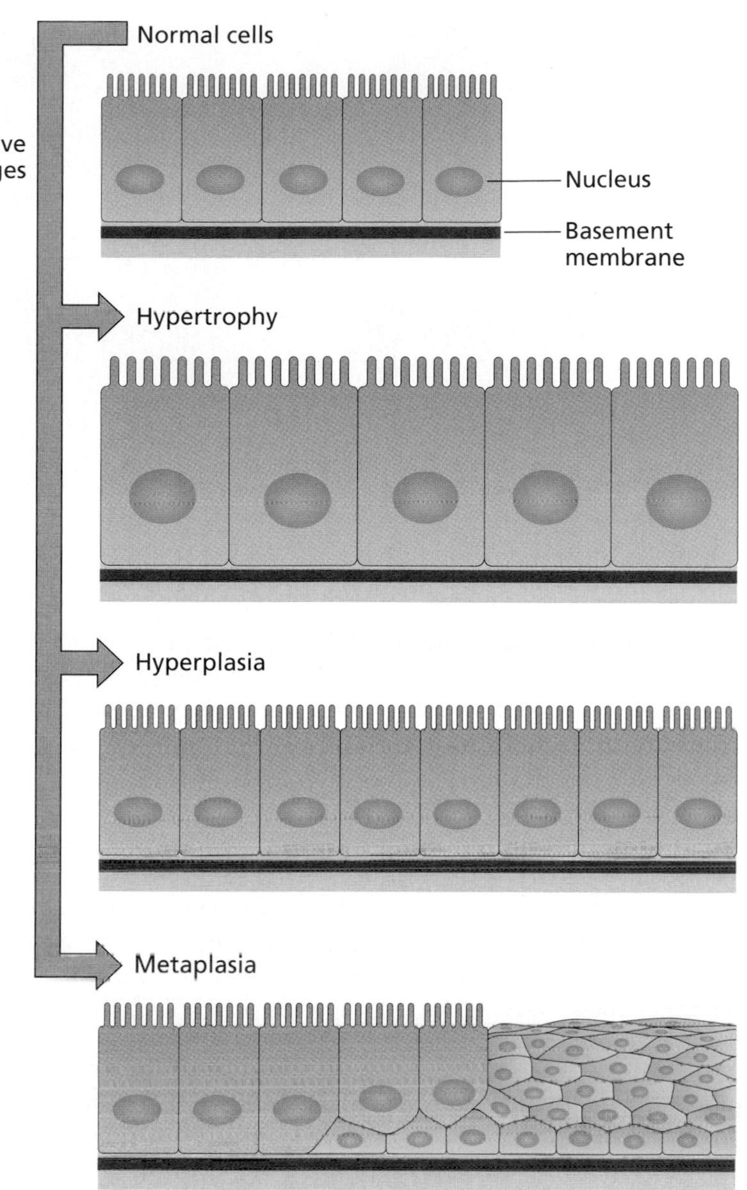

Figure 5.5 Parasite-induced changes in host cell growth. Normal cells within a particular tissue have a characteristic size and rate of replication. Various parasitic infections can alter these cell characteristics. Some parasites stimulate the cells themselves to grow larger than normal. Tissues or organs composed of such cells are likewise abnormally large. This phenomenon is called hypertrophy. Alternatively, other parasites release host cells from normal cell cycle constraints, resulting in an overproduction of cells. This process is known as hyperplasia. Metaplasia is another possible consequence of parasitic infection. Metaplasia refers to the replacement of cells of one type by cells of a different type.

using specialized hooks found on the valves (see Box 8.2). After a period of time as an ectoparasite, a glochidium undergoes metamorphosis to a typical juvenile bivalve. It then detaches from the fish and settles to the substrate. Yet some fish, such as Coho salmon (*Oncorhynchus kisutch*), are resistant to infection. As soon as glochidia attach, the cells of the gill filaments begin to replicate rapidly. Within a few days these proliferating cells are sloughed off, along with the parasites. In contrast, no such hyperplasia occurs in Chinook salmon (*O. tshawytscha*). Consequently, glochidia remain in place until their development can be completed.

A particularly striking example of hyperplasia involves the tapeworm *Spirometra mansonoides* and mice. The larval stage, which can encyst in a mouse's muscle tissues, is called the **plerocercoid** and it produces a curious substance known as **plerocercoid growth factor (PGF)**. This molecule has an affinity for mammalian growth hormone receptors, resulting in the generation of a signal that activates the cell cycle. Consequently, infected mice grow to unusually large size (**Figure 5.6**). Surprisingly, there is little or no sequence

Figure 5.6 Weight gain due to tapeworm infection. Two littermate male hamsters. The hamster on the left was infected experimentally with *Spirometra mansonoides* nine weeks previously and weighs 70 g more than its uninfected sibling to the right. *S. mansonoides* produces a compound (PGF) that mimics the effects of mammalian growth factor hormones, causing advancement of the cell cycle and accelerated mitosis. (From Mueller JF [1965] *J Parasitol* 51:523–531. With permission from Allen Press Publishing Services.)

similarity between the cestode *PGF* gene and those of mammalian hormones involved in growth. The gene does, however, show considerable similarity with genes for cysteine proteases. PGF itself has protease activity. Collagen appears to be its principal substrate. Because of its ability to digest components of the intercellular matrix, PGF may be involved in larval migration within the rodent host.

Metaplasia

Sometimes chronic irritation can cause cells to change from one differentiated cell type into another. In general, more delicate cells are replaced with more robust cells that are better able to withstand the irritant. This process, called **metaplasia** (see Figure 5.5), is usually reversible. If the initiating irritant is removed, cells revert to their original type. Metaplasia usually occurs when stem cells that are stimulated to follow a different developmental path, rather than causing a change in already differentiated cells. The immediate consequences of metaplasia are generally not serious. When it occurs in mammals, metaplasia is not a direct cause of cancer. However, metaplastic cells eventually have a higher likelihood of becoming cancerous if the abnormal stimulus is not eliminated. A well-characterized example is the replacement of pseudostratified columnar epithelial cells in the respiratory system with stratified squamous epithelial cells in response to tobacco smoke.

Certain parasites can also induce metaplasia. Monogeneans in the genus *Dactylogyrus*, for instance, adhere to the gill filaments of certain freshwater fish, which can cause soft tissue in the gills to ossify. In humans, infection with ***Schistosoma haematobium*** is associated with an increased likelihood of urinary tract metaplasia. Because infection with this digenetic trematode has also been associated with an increased risk of bladder cancer, the onset of metaplasia may be a contributing factor to tumor formation.

No discussion of metaplasia is complete without mentioning nematodes in the genus ***Trichinella***. In this case, fully differentiated cells are reprogrammed into very different cells. Several species in this genus share a bizarre life cycle in which the same mammal can serve as both definitive and intermediate host. The most important species in humans is ***T. spiralis***, usually acquired by consuming undercooked pork.

The complete ***Trichinella*** life cycle is presented in the Rogues' Gallery. Here we focus on the ability of these nematodes to dramatically alter the morphology of the muscle cells they inhabit. Adults inhabit the cytoplasm of several adjacent mucosal epithelial cells in the intestine, snaking their way through these cells like a needle and thread (**Figure 5.7**). L1 larvae travel through the circulatory system until they reach skeletal muscle. Once they have penetrated a muscle fiber cell, they alter gene expression within that cell. Expression of contractile proteins drops and mitochondria degenerate, while the nucleus enlarges and the amount of smooth endoplasmic reticulum rises. The newly modified cell secretes large amounts of collagen, which forms a protective capsule around both the cell and its nematode inhabitant. The modified cell, now called a **nurse cell**, is no longer capable of contraction, its original function. In its new guise, the sole purpose of the nurse cell is to protect and nourish its uninvited guest. Although molecules secreted by the parasite seem to be involved in the formation of nurse cells, the molecular mechanism underlying this remarkable transformation is still under investigation.

Neoplasia

If the offending stimulus causing hyperplasia or metaplasia is terminated or removed, cell growth patterns usually return to normal. In contrast, **neoplasia**, an abnormal proliferation of cells in response to a stimulus, continues even after cessation of the stimulus. If neoplasms do not invade adjacent tissues and remain localized, they are considered benign. Those that invade adjacent tissues or spread to other parts of the body are malignant cancers. There are several parasites that are associated with neoplasia in the host. Earlier in this chapter, for example, we discussed the cancer of the bile duct (cholangiocarcinoma) induced by ***Opisthorchis viverrini***.

Another example of interest to veterinarians and dog owners is *Spirocerca lupi*, a nematode commonly called the esophageal worm. With *S. lupi* we observe the strongest cause-and-effect relationship between a parasite and a particular type of cancer. *S. lupi* is widespread in tropical areas. It is also found across the southern United States. Insects serve as intermediate

A

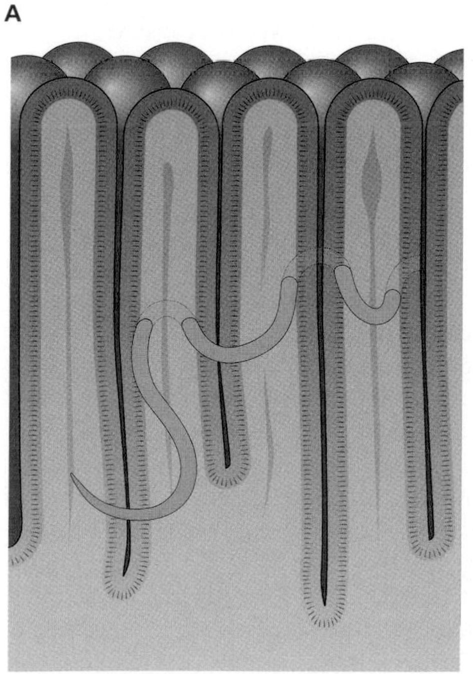

B

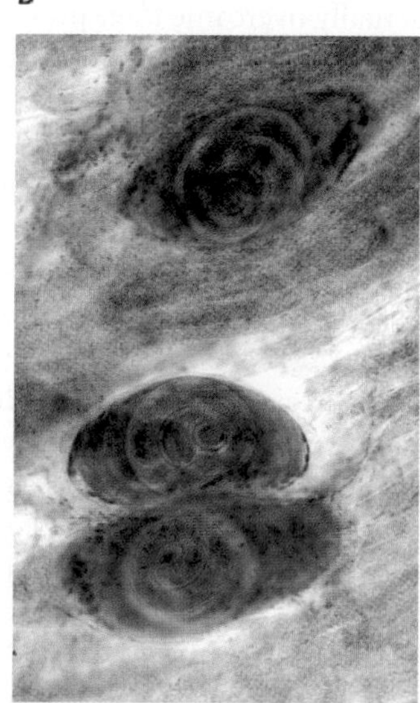

Figure 5.7 *Trichinella spiralis.* (A) Rendering of an adult worm inhabiting the cytoplasm of several adjacent intestinal epithelial cells. (B) Larvae encysted inside nurse cells within skeletal muscle. Larvae inside nurse cells are also able to stimulate angiogenesis (the formation of new blood vessels) around the nurse cell to provide nourishment. (B, Courtesy of Steve Upton, Kansas State University.)

hosts. Dogs become infected by consuming an infected insect or by eating a paratenic host such as a rodent. Larvae penetrate the dog's stomach wall and enter the circulatory system. They eventually penetrate through the aorta and subsequently migrate to the esophagus. Here the larvae enter the esophageal submucosa, where they are surrounded by a cystic nodule. Development is completed within the nodule. Eggs are passed into the esophagus, eventually exiting with the feces, where they may be consumed by an insect such as a dung beetle.

Many infected dogs exhibit no symptoms, whereas others suffer serious pathology. An odd symptom that apparently occurs as a consequence of larval migration is a type of osteoarthritis of the joints in the vertebral column called **spondylosis**.

The esophageal cysts containing the adult worms can interfere with the ability of the dog to swallow. Such dogs eventually become emaciated. These cysts can also transform into neoplasms, resulting in malignant tumors called **esophageal sarcomas**. As these tumors continue to grow, they further interfere with the ability of the dog to eat. Although the exact manner in which *S. lupi* triggers malignant neoplasia is unknown, it has recently been shown that neoplastic nodules contain a large number of activated and rapidly dividing fibroblasts compared to non-neoplastic nodules. Cells in neoplastic nodules express high levels of fibroblast growth factor and vascular epithelial growth factor. Curiously and for reasons nobody has yet adequately explained, beagles and other hound breeds appear to be especially susceptible to *S. lupi* infection and esophageal sarcoma.

Many parasites adversely affect host nutrition

In the mid-1950s, famed opera singer Maria Callas lost more than 60 pounds over several months. It was rumored at the time that Callas intentionally ingested tapeworm larvae; the tapeworms, absorbing a large portion of her digested food, would help her shed pounds. If true, Callas may not have been the first to look to tapeworms as a diet aid. In the late nineteenth and early twentieth century, tapeworms were marketed commercially as a way to help women maintain a svelte figure (see Box 6.4). It is unclear whether the diet pills actually contained tapeworm larvae as advertised or how many people actually overcame their probable disgust and swallowed the capsules.

It is not too likely that such a weight-loss strategy would produce the desired results. Most tapeworms absorb only a small fraction of the host's food and it would generally require a very heavy infection to cause rapid weight loss. Nevertheless, the logic behind the idea is sound; tapeworms and other parasites do in fact divert nutrients to themselves, sometimes with adverse consequences, including weight loss, for the host.

One tapeworm that does interfere with host nutrient acquisition is ***Diphyllobothrium latum***, the broad fish tapeworm. Fish serve as a second intermediate host for this cestode, and many fish-eating mammals, including humans, can serve as definitive hosts. The adults that develop in the intestine of the definitive host are enormous, reaching up to 10 m in length, and they can live up to 20 years. Most infected humans are asymptomatic or suffer relatively mild symptoms, which include diarrhea, weight loss, and abdominal discomfort. In some individuals, however, infection leads to a severe vitamin B_{12} deficiency. The affinity of the cestode for this vitamin is such that between 40% and 80% of the host's vitamin B_{12} intake is diverted to the parasite. Owing to vitamin B_{12} depletion, an infected individual is at risk for a condition known as megaloblastic anemia. Vitamin B_{12} is necessary for cell metabolism. If levels of this vitamin are too low, DNA synthesis is impeded. Consequently, synthesis of new red blood cells is impeded and anemia ensues. Because sufficient vitamin B_{12} is also necessary for lipid synthesis, chronic infection can also result in neuron demyelization, causing neurological symptoms. Many other intestinal helminths such as hookworms are known to affect nutrient acquisition. The effects are generally most noticeable in children, in whom infection can compromise growth and cognitive development.

Giardia lamblia is a common intestinal protozoan parasite with well-known effects on host nutrition. As they proliferate in the intestine, the trophozoites adhere to the microvilli via their ventral disc and interfere with nutrient absorption. These nutrients are lost as they are passed with the feces. Fat absorption is particularly compromised, explaining why many infected individuals pass feces that have a greasy appearance. Loss of appetite is another common symptom, further decreasing nutrient acquisition, and contributing to weight loss. *G. lamblia* is also thought to cause intestinal epithelial cells to secrete chloride ions into the intestinal lumen, resulting in a loss of both electrolytes and water. **Figure 5.8** shows another interesting component of this flagellate's pathology.

Plasmodium infections can result in host iron deficiency

Finally, we should mention that it is not strictly necessary for a parasite to inhabit the gut for it to adversely affect host nutrition. Iron deficiency, for example, is common in malaria infections. This deficiency has generally been explained as a consequence of red blood cell lysis as mature, intracellular merozoites are released. However, recent studies suggest that this view may be an oversimplification and that iron deficiency also results from competition between parasite strains.

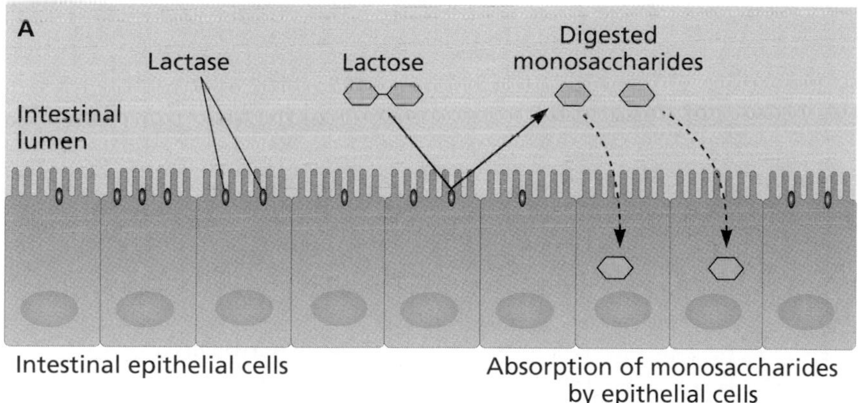

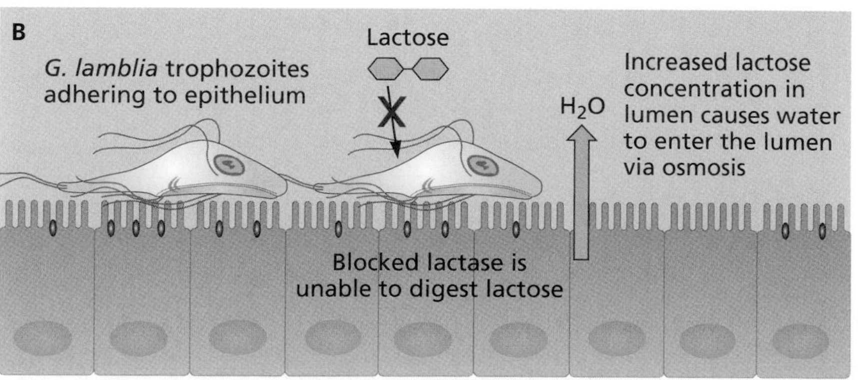

Figure 5.8 A mechanism for increased lactose intolerance and diarrhea in *G. lamblia* infections. (A) Ordinarily, lactose is digested to its component monosaccharides, glucose and galactose, in the intestinal lumen. This digestion is catalyzed by lactase associated with the plasma membrane of intestinal epithelial cells. (B) As it adheres to the epithelium, however, *G. lamblia* can interfere with this digestion, resulting in a reduced ability to digest lactose and a buildup of the disaccharide in the lumen. The increased lactose concentration alters the osmotic balance between the epithelial cells and the lumen, resulting in the movement of water into the lumen. This excess water contributes to the watery diarrhea characteristic of *Giardia* infection.

Once they reach a certain concentration in the blood, replicating merozoites have a strategy to prevent newly inoculated sporozoites from replicating in the liver. They do so by stimulating the host to release the iron regulatory hormone hepcidin. As hepcidin levels rise, iron concentration in hepatocytes declines, inhibiting the growth of liver-stage malaria parasites. That is, when it comes to parasites that must replicate in their mammalian host, it is a case of first come, first served. Interestingly, iron supplements are often prescribed in malaria cases to reverse anemia. It is now speculated that such supplements may actually open the door to hyperinfection.

Almost counterintuitively, host immunity actually conspires to make hyperinfection more likely as children age (**Figure 5.9**). Very young children, with less previous exposure to malaria, develop a high blood parasitemia. The large number of blood-stage ***Plasmodium*** parasites protects them against subsequent infections via the mechanism described. As they age, however, and are repeatedly exposed to malaria, they develop a measure of immunity. Consequently, later infections often result in lower parasitemia—low enough that the infecting parasite strain is less able to block hyperinfection.

Anemia, whether due to infection with hookworms, malaria, or other parasites, is no trivial matter. In sub-Saharan Africa alone, the prevalence of childhood anemia reaches 67%, which translates into approximately 85 million children. Childhood anemia in general is associated with reduced growth, immune function, cognitive development, and survival. Certainly,

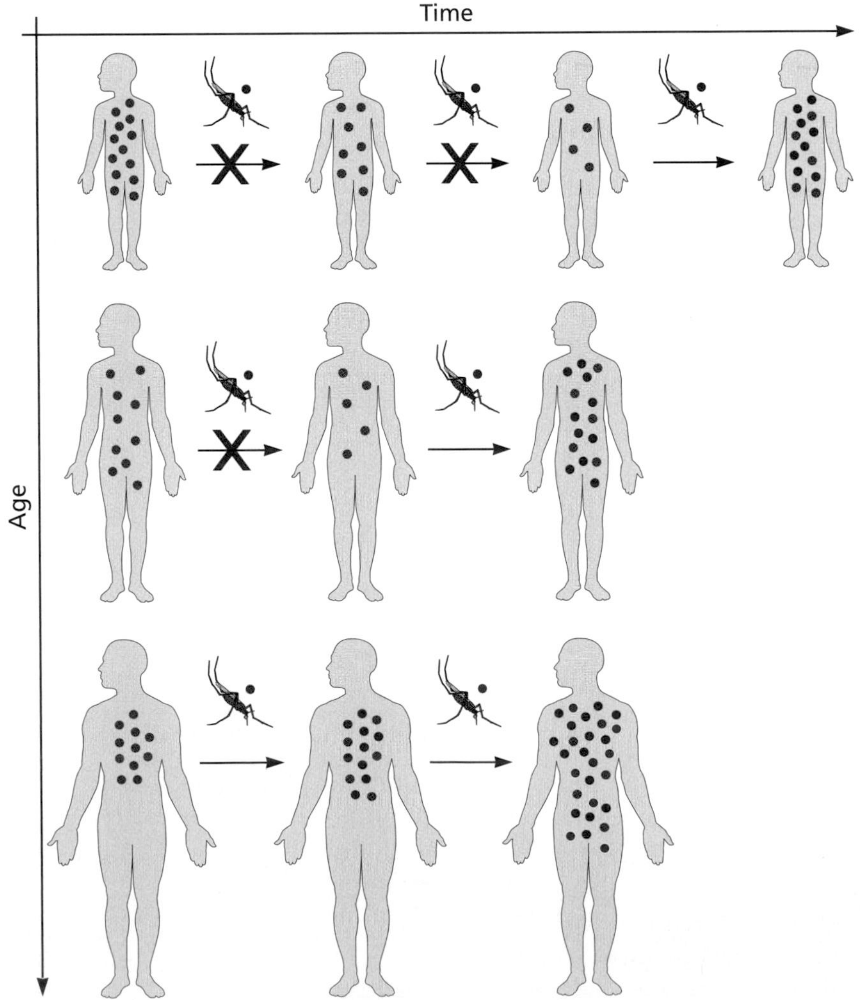

Figure 5.9 Age-related decline in protection from malaria hyperinfection. A schematic representation of the parasite-dependent protective effect over age and time. The three rows of people represent young children (top row), older children (middle row), and young adults (bottom row). The colored dots superimposed over the people (green, red, or blue) represent different genetic strains of *Plasmodium*. Malaria infections in young children (top row) are often typified by high levels of blood-stage parasitemia that protect them against hyperinfections with different genetic strains, as they prevent them from acquiring sufficient iron. These infected children only become susceptible to further infections once parasite levels fall below a critical threshold. As they grow older (middle row), because they have developed some level of immunity, older individuals lose the protective effect against superinfection sooner, and such multiclonal infections are more likely to occur. As young adults (bottom row), whose level of immunity against the initially infecting strain is even higher, the protection against superinfection is further reduced, increasing the likelihood of infection with multiple strains. (From Portugal S, Carret C, Recker M et al [2011] *Nat Med* 17:732–737. With permission from Macmillan Publishers Ltd.)

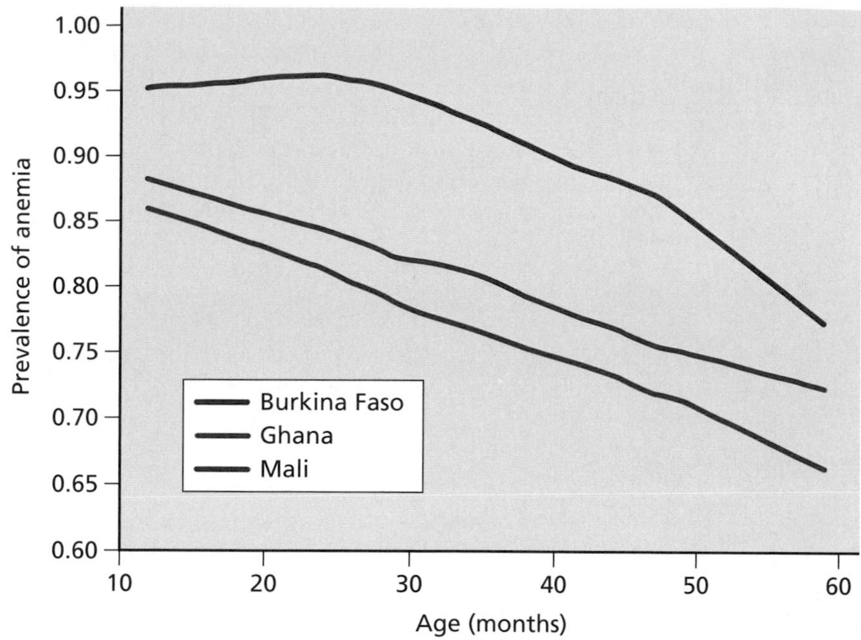

Figure 5.10 Anemia prevalence in three West African countries. Overall proportions of children with anemia for Burkina Faso, Ghana, and Mali are shown as a function of age. Anemia results largely from malnutrition or parasitic infection (mainly malaria and intestinal helminths), or both. (Adapted from Soares Magalhães RJ & Clements ACA [2011] *PLoS Med* 8(6):e1000438 [doi:10.1371/journal.pmed.1000438].)

not all childhood anemia is caused by parasitic infection. Malnutrition and even inherited hemoglobin abnormalities such as sickle cell anemia are major contributors. When parasites are involved, we have seen that they do not all cause anemia in the same way. The extent of childhood anemia in three West African countries is illustrated in **Figure 5.10**.

Toxins are a less frequent component of parasite pathology

Toxins produced by various bacteria contribute to their disease-causing capacity. Some fungal pathogens likewise cause disease, at least in part from the toxins they produce. An interesting example is *Claviceps purpurea*, a parasite of rye grass. If infected grain is used to produce bread, anyone consuming the bread may also ingest a fungal toxin called ergot. Ergot causes vasoconstriction and mental disturbances.

The direct involvement of toxins in the pathology caused by protozoan and metazoan parasites has been harder to characterize. It has been suggested over the years, for example, that secreted toxins are involved in the pathology of both New and Old World trypanosomiasis, but attempts to identify such a toxin have been equivocal. As discussed in Chapter 4, many parasites release various antigens or other molecules that stimulate a host immune response. This response may itself have pathological consequences for the host. Molecules that induce pathology by stimulating host immunity, however, are not toxins in the traditional sense; they do not target and adversely affect specific molecules, leading to direct interference with host processes. The effect of the parasite in such a case is therefore indirect and not based on any inherent toxic properties.

Hemozoin, produced as an insoluble waste of hemoglobin digestion by ***Plasmodium***, has often been cited as a true toxin (**Figure 5.11**). Ironically, although hemozoin has been described as toxic to host cells, it is produced by ***Plasmodium*** to avoid being killed by heme, an initial product in the breakdown of hemoglobin. Following infection of erythrocytes, malaria parasites take up host hemoglobin, which they digest into amino acids. The initial cleavage event, mediated by aspartic acid proteases called plasmepsins,

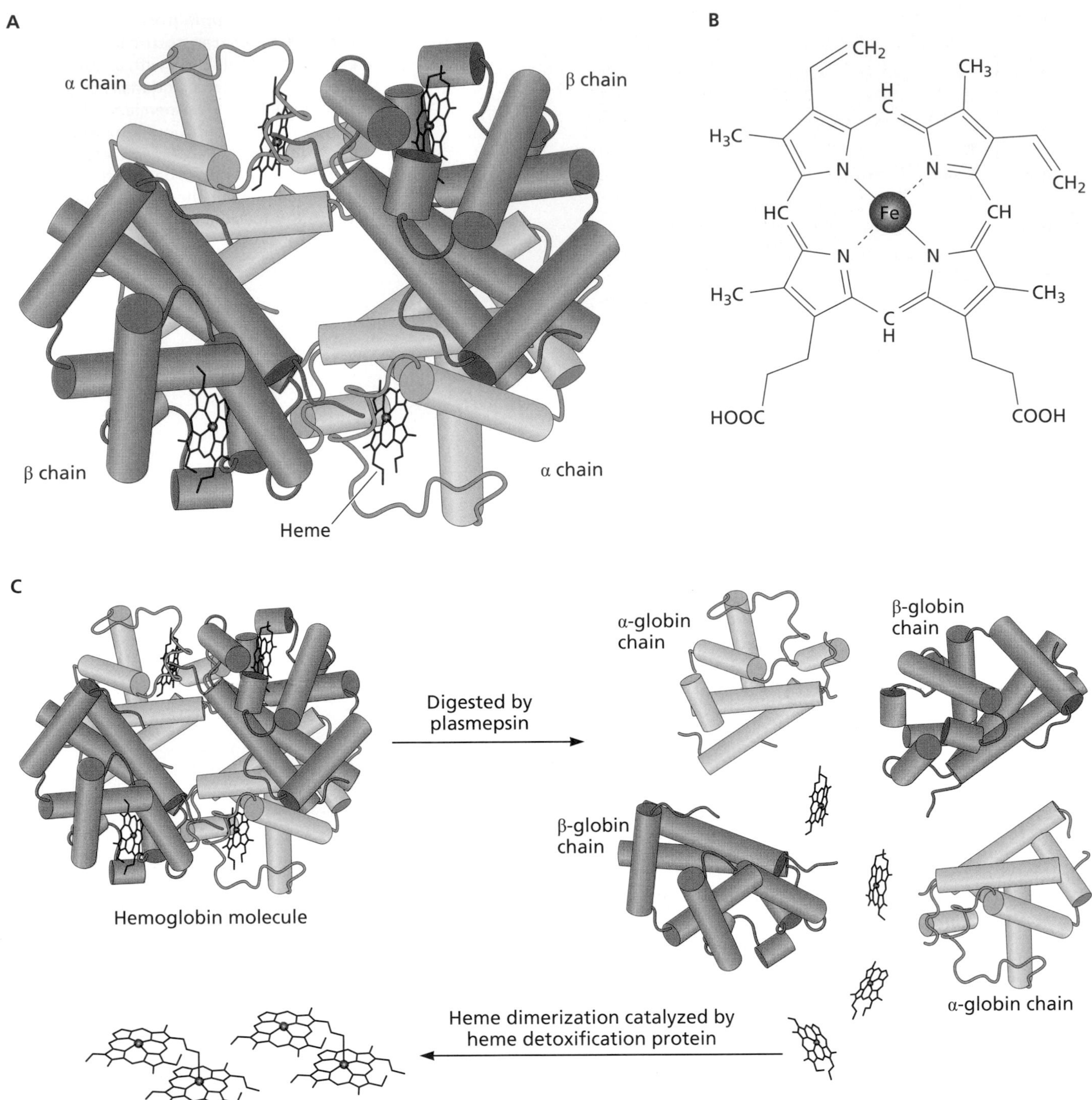

Figure 5.11 The production of hemozoin. (A) The structure of hemoglobin, showing the a- and β-globin protein chains, as well as the location of the iron-containing heme groups. (B) Heme structure. (C) A proposed mechanism for the formation of hemozoin from heme. (A, Modified from Kuriyan J, Konforti B, & Wemmer D. [2012] The Molecules of Life: Physical and Chemical Principles. Garland Science, New York.)

releases the iron-containing heme group and globin protein. The globin is then digested into its component amino acids by various proteases. Heme, on the other hand, is toxic in its free state because it can lyse membranes and inhibit the activity of certain enzymes. To eliminate this toxic substance, ***Plasmodium*** induces heme dimerization, forming molecules of hemozoin. This process appears to be catalyzed by a recently characterized molecule called **heme detoxification protein (HDP)**. When it is expressed by the parasite, HDP is first secreted into the cytosol of the infected erythrocyte. It subsequently re-enters the parasite via endocytosis and is delivered to the food vacuole, where conversion of heme to hemozoin takes place.

Hemozoin's toxic effect on host cells is observed when it is released from lysed erythrocytes. Following its release, hemozoin is phagocytosed by macrophages and other phagocytes. Hemozoin, however, cannot be digested within the endomembrane system of the phagocyte and further phagocytosis is impeded. Yet recent work suggests that even this molecule may exert much of its impact indirectly, through its stimulation of innate immunity. Specifically, some evidence suggests that hemozoin binds to ***Plasmodium*** DNA and that this hemozoin–DNA complex binds to TLR-9 (Toll-like receptor-9) found on phagocytes. The subsequent signal transduction results in an inflammatory response via an up-regulation of the cytokines IL-12 and TNFα.

Glycosylphosphatidylinositol (GPI) (see Section 4.3) is also often cited as a ***Plasmodium*** toxin. GPI is thought to bind TLR-2. In doing so, it stimulates an inflammatory response. There is also some evidence that the GPI released from ***P. falciparum*** can insert itself into the membranes of nonparasitized red blood cells. If the inserted parasite GPI is then bound by circulating anti-GPI antibodies, the cell may be destroyed. Consequently, GPI may be involved in the anemia that is often observed in individuals infected with ***P. falciparum***.

Pathology often results from immune-mediated damage to host cells and tissues

The previous discussion of the involvement of hemozoin and GPI in malaria pathology illustrates two important points. First, it highlights that pathology is often difficult to categorize. Second, it reminds us that much of the pathology observed in parasitic infections is only indirectly caused by the parasites themselves. Indeed, if one wished to rank the various mechanisms of pathology that we have been discussing, this last category, indirect damage to the host as a consequence of immune response, might be at the top of the list.

The list of parasites for which host immunity is at least part of the pathology story is long. Malaria, leishmaniasis, trypanosomiasis, toxoplasmosis, schistosomiasis, and filariasis are just a few of the diseases caused by parasites in which host immune response is part of the problem (**Box 5.1**). The underlying immunological principles that allow us to understand this type of pathology were presented in Chapter 4. The damage to the host that occurs as a result of an inappropriate immune response is called **immunopathology**.

The phenomenon of immunopathology raises the question as to why it exists in the first place. Why has natural selection not eliminated these harmful responses? The answer is actually straightforward. It's a dangerous world out there. Potential hosts simply must have a powerful array of defensive, immunological weapons on hand to protect them from the never-ending onslaught of both micro- and macroparasites. From the parasites' perspective, hosts are resources to be exploited. Without an immune system, potential hosts could never survive. In addition, it is worth repeating that evolution via natural selection never results in perfectly adapted organisms. Evolution results in a compromise. The various characteristics that we observe in any organism reflect the evolutionary trade-off that has been struck. Both immunopathology and successful immune defense represent the adaptive middle ground between competing demands that any living thing must deal with. This topic will be revisited in Chapter 6, where we explore the fascinating and recently emerging field of immunoecology, and in Chapter 7, where we address the topic of parasite evolution specifically.

The delicate equilibrium between immune protection and pathology in response to parasites is well illustrated by the manner in which mammals react to intestinal helminths, as discussed in Chapter 4. Simply put, certain aspects of the immune response, those most likely to result in parasite killing or expulsion, are also implicated in damage to and inflammation of the

BOX 5.1
Worms and *Wolbachia*

Wolbachia is a genus of gram-negative bacteria that infects a wide range of arthropods. They are obligate intracellular symbionts and although they infect many different organs and tissues, they are especially conspicuous in the ovaries and testes. Although bacteria that only infect invertebrates often receive scant attention, *Wolbachia* spp. have attracted more than their fair share, because of the odd impact they often have on their hosts. Some insects infected with *Wolbachia*, for example, are less prone to certain viral infections (see Chapter 4, pages 126–127) than those which are uninfected.

Wolbachia also infects filarial nematodes. One species in particular, *W. pipientis*, infects some of the most notorious of the pathogenic filarial worms, including *Onchocerca volvulus*, which causes onchocerciasis; *Dirofilaria immitis*, the causative agent of canine heartworm; and both *Wuchereria bancrofti* and *Brugia malayi*, the two most common causes of human filariasis or elephantiasis. In addition, the symbiotic bacteria lurking within the tissues of the worms play a sinister role in the disease process in these conditions.

Chronic inflammation contributes significantly to diseases caused by filarial worms. It is increasingly apparent, however, that the trigger for this inflammation is not the worms themselves. Rather, it is *W. pipientis*, which is released by dying worms that initiates the sequence of events leading to an inflammatory response. More specifically, the bacteria have a protein called **Wolbachia Surface Protein (WSP)** imbedded in their outer membrane. When this protein is released, it binds to and activates TLR-2 and TLR-4 on the surface of antigen-presenting dendritic cells and macrophages. This in turn activates a signal transduction cascade, which results in the up-regulation of several pro-inflammatory cytokines. Furthermore, as in most helminthic infections, the adaptive response is skewed toward a Th-2 response. Among the cytokines released by Th-2 lymphocytes is IL-4. One of the effector functions of this cytokine is to cause eosinophil degranulation. Mediators released by eosinophils contribute to the cycle of host tissue damage observed in filarial worm infection.

Consider, for example, the most devastating symptom of onchocerciasis: vision loss (Figure 1). When *O. volvulus* adults produce live microfilariae, many of these larval worms invade the eye. Live microfilariae cause little inflammation. When they die, however, *W. pipientis* is released. As these bacteria begin to die they release WSP and inflammation ensues. Because many *O. volvulus* microfilariae migrate to the surface of the cornea, inflammation of this structure (keratitis) is a frequent complication of onchocerciasis. If inflammation persists, the cornea can become scarred, resulting in sclerosing keratitis. Over time, the entire cornea becomes increasingly opaque and vision is impaired. Ultimately, complete blindness may result. Other parts of the eye, including the retina and optic nerve, may be similarly damaged by inflammation.

The recognition that bacteria underlie much of the pathology in these diseases has resulted in new treatment options. Certain antibiotics such as doxycycline are now used to eliminate *Wolbachia* from filarial worms. Without their symbionts, the worms may die, or in the case of females, become sterile. Microfilariae are especially susceptible to the effects of doxycycline. In one trial involving patients suffering from elephantiasis, an eight-week course of the drug almost completed eliminated microfilariae.

References

Taylor MJ, Makunde WH, McGarry HF et al (2005) Macrofilaricidal activity after doxycycline treatment of *Wuchereria bancrofti:* a double-blind, randomised placebo-controlled trial. *Lancet* 365 9477:2116–21.

Baldo L, Desjardins CA, Russell JA et al (2010) Accelerated microevolution in an outer membrane protein (OMP) of the intracellular bacteria *Wolbachia. BMC Evol Biol* 10:48 (doi:10.1186/1471-2148-10-48).

Tamarozzi F, Halliday A, Gentil K et al (2011) Onchocerciasis: the role of *Wolbachia* bacterial endosymbionts in parasite biology, disease pathogenesis, and treatment clinical microbiology. *Clin.Microbiol. Rev* 24: 459–468. (doi:10.1128/CMR.00057-10).

A

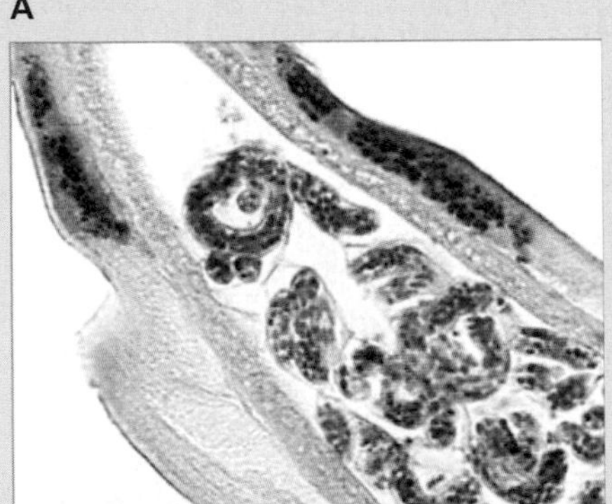

B

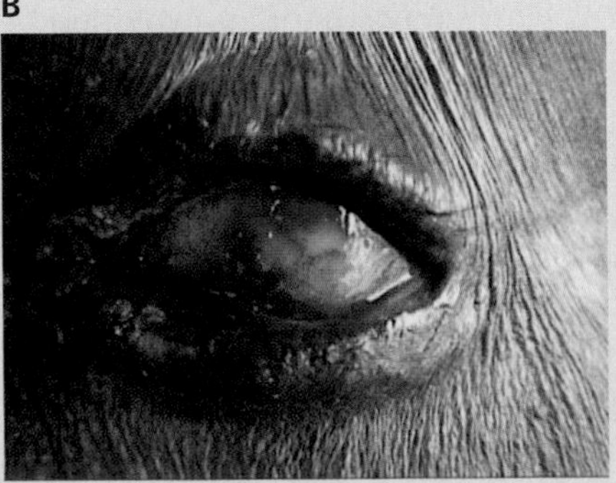

Figure 1 *Wolbachia* and river blindness. **(A) *Wolbachia* bacteria (in red) in the hypodermal cells of an adult *Onchocerca volvulus*. (B) Severe scarring of the cornea due to prolonged corneal inflammation (keratitis). As microfilariae in the eye die, symbiotic *Wolbachia* are released. The inflammation that results in pathology is largely in response to WSP (*Wolbachia* surface protein), released by dying bacteria. (A, Courtesy of Eric Pearlman, Case Western Reserve University.)**

intestine. Thus, the immune system must perform a very careful balancing act, negotiating a path between too little response and too much. Recall the extensive remodeling of the intestinal epithelium that occurs in response to intestinal helminths. Changes such as increased mucus production and rapid cell division help to dislodge helminths that have adhered to the epithelium. However, these same changes compromise the ability of the intestine to secrete digestive enzymes and absorb nutrients (see Figure 4.14). The newly produced cells are less differentiated and immature and therefore less able to carry out their normal activities. Furthermore, such accelerated cell replication is costly in terms of metabolism, and energy spent on mitosis is unavailable for other energy-dependent processes.

Immunopathology plays at least some role in just about all infectious diseases. Fever, for example, a consequence of many infections, occurs when macrophages release certain cytokines, primarily IL-1 and TNFα, in response to those infections. Like other immune responses, fever can be beneficial in that it increases the activity of phagocytes and decreases the growth rate of many microorganisms. When it is unregulated it can be uncomfortable at best and life threatening at worst.

Immunopathology is an important component of the pathology observed in malaria

Immunopathology takes many forms. An overly robust inflammatory response or one that becomes chronic in response to persistent parasite antigen is an important component in the pathology of many parasitic infections. For example, we have recently learned that the assault of ***Plasmodium*** on its host is multifaceted; red blood cell hypertrophy, induction of anemia, and the production of toxins all contribute to the pathology seen in a malaria infection. In Chapter 3 we learned how ***P. falciparium*** encodes proteins that form knobs on the surface of infected red blood cells. These protein knobs allow the red blood cells to adhere to capillaries, which interferes with blood flow. But the parasite has other tricks up its sleeve as well. Excessive inflammation from the release of pro-inflammatory cytokines is an important component of malaria pathology. These cytokines are primarily released by antigen presenting cells, following the activation of Toll-like receptors on their surfaces. This exacerbated inflammation contributes to the development of **cerebral malaria**—the most serious consequence of a malaria infection (**Figure 5.12**).

Cerebral malaria is typically described as being caused by the adherence of infected erythrocytes to the capillary endothelium, as mentioned above. As the number of adhered erythrocytes grows, the capillary may become partially or completely blocked. The result is downstream anoxia and subsequent necrosis of brain tissue. Furthermore, if these red blood cells breach the blood–brain barrier, paralysis, coma, and death are possible outcomes. Because inflammation increases the permeability of the endothelium, these most extreme consequences of malaria become more likely. Recently a clearer picture of the mechanism behind this inflammatory response has emerged. The specific TLRs involved appear to be TLR-8 and TLR-9, both of which bind to and are activated by microbial nucleic acid. In a study using a mouse model of malaria, it was demonstrated that the most severe complications of cerebral malaria could be reversed or prevented, when the mice were administered synthetic TLR-8 and TLR-9 antagonists that bound, but did not activate, these receptors. These data, of course, suggest the possibility that such antagonists may have clinical value in the treatment of human cerebral malaria as well.

A

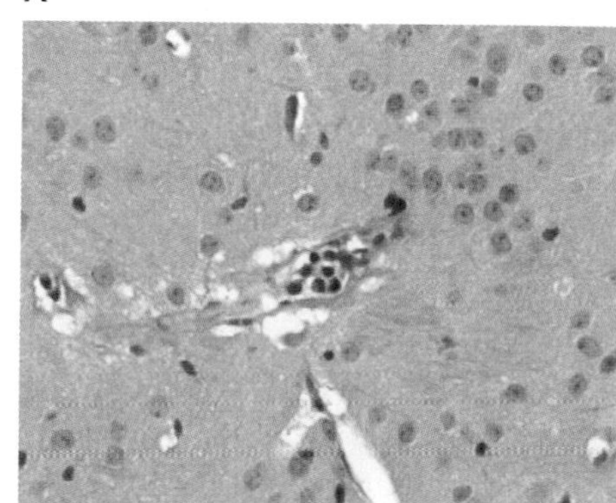

B

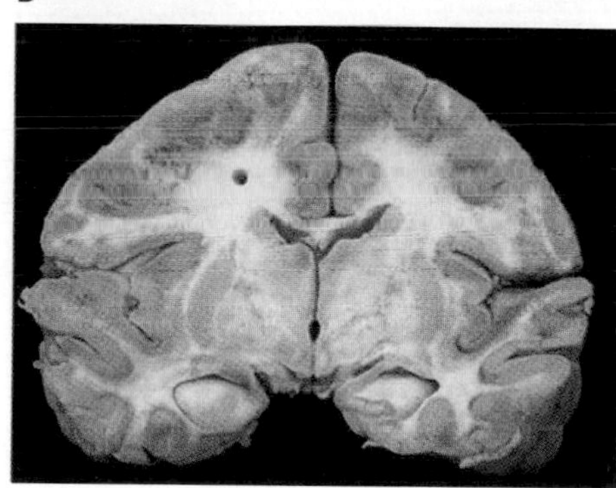

Figure 5.12 Cerebral malaria. (A) A brain section from a patient killed by cerebral malaria. The darkly stained parasite-infected red blood cells causing vessel occlusion can be observed in unstained blood vessels. (B) Human brain affected by cerebral malaria. Darker areas, primarily along the outer edge, indicate areas of tissue necrosis. (A, from Campanella GSV, Tager AM, El Khoury JK et al [2008] *Proc Natl Acad Sci* 105(12):4814-4819. Copyright (2008) National Academy of Sciences, USA. B, Milner DA, Montgomery J, Seyder KB & Rogerson SJ [2008] *Trends Parasitol* 24(12):590–595. With permission from Elsevier.)

A

B

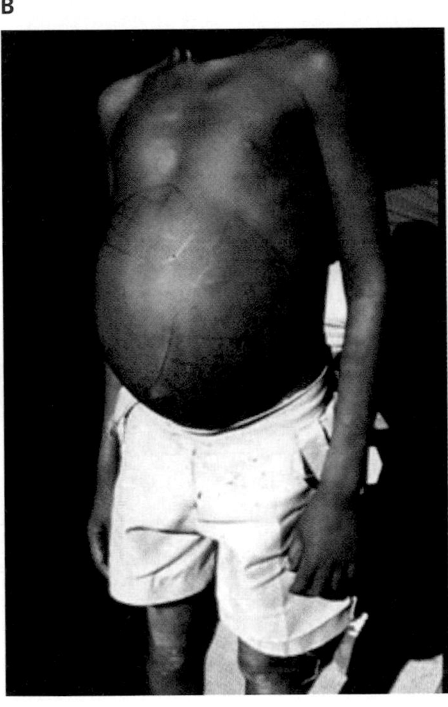

Figure 5.13 Schistosomiasis pathology. (A) A granuloma in the liver caused by an *S. mansoni* egg. The granuloma that forms around the egg consists of various cell types, including T and B cells, macrophages, epithelial cells, mast cells, fibroblasts, and eosinophils. All of these cell types play a substantial role in granuloma formation as they interact with each other, mainly through cytokines. (B) Chronic infection and fibrosis in the liver can eventually lead to portal hypertension, which results in the buildup of fluid in the abdomen known as ascites. (A and B, courtesy of Cambridge University, Pathology Department.)

Granulomas formed in response to parasite antigen are both protective and pathological

Granulomas are a common outcome in many parasitic infections. They result from a **delayed-type hypersensitivity response (DTH)**. Such a response relies on components of both innate and adaptive immunity. A DTH response begins when antigen presenting cells release cytokines that both attract and activate Th-1 cells. If these Th-1 cells recognize their specific antigen on the surface of an antigen presenting cell, they respond by releasing inflammatory cytokines, including interferon γ (IFN-γ) and tissue necrosis factor-α (TNF-α). These cytokines cause increased macrophage activation, leading to a positive feedback cycle that ultimately causes destruction of local tissue and the formation of the granuloma. The granulomas that form in response to *Mycobacterium tuberculosis* infection, called tubercles, are perhaps the textbook example of this type of pathology. An impressive roster of eukaryotic pathogens, ranging from protozoans such as ***Toxoplasma gondii*** to helminths such as cestodes and schistosomes provoke granuloma formation. In the case of schistosomes, it is the response to egg antigens rather than to the adults that triggers granuloma formation. Let us take a closer look at the immunopathology caused by these digenetic trematodes.

Three schistosome species cause the lion's share of human infections. ***Schistosoma mansoni*** and ***S. japonicum*** adults live in the mesenteric venules that drain the intestines. ***S. haematobium*** adults are mainly found in the veins of the urinary bladder. To complete transmission, eggs must either cross into the intestinal or urinary bladder lumen, but many eggs become lodged in various tissues, principally the liver (for ***S. mansoni*** or ***S. japonicum***) or the bladder (for ***S. haematobium***). The disease schistosomiasis is primarily caused by a host reaction to these trapped eggs and the granuloma formation they induce.

A granuloma represents the host's attempt to isolate an antigenic source and mitigate the damage caused by this source. In schistosomiasis, the source of antigens is the egg itself (**Figure 5.13**). Granulomas result in considerable pathology. Once granulomas have initially formed, the excess collagen and other extracellular matrix material deposited around them cause scarring. In the liver, such scarring disrupts liver function. In severe cases, it can occlude the portal veins, causing portal hypertension. Ultimately, **ascites**, the buildup of tissue fluid in the mesenteries and abdominal cavity, can result. Nevertheless, things would be much worse if granulomas did not form. Rampant inflammation in response to egg antigens would result in progressive tissue necrosis and ultimate organ failure.

The details of granuloma formation for the three principal human schistosome species vary, but certain generalizations are possible. Granuloma formation begins with the gradual buildup of macrophages, neutrophils, and eosinophils around the newly trapped egg. As the granuloma matures, epithelial precursor cells, fibrocytes, and later plasma cells and other lymphocytes form a peripheral layer around the lesion. During this stage, the egg disintegrates. Ultimately, fibrocytes, collagen, and other extracellular components become the dominant feature of the granuloma.

The initial adaptive immune response to schistosome infection largely consists of a Th-1 response directed at migrating larvae and immature adults. This response, lasting for approximately five weeks, is characterized by pro-inflammatory cytokines such as TNFα, IL-1, and IL-6. As egg laying commences, a shift to a Th-2 response occurs, in which IL-4, IL-5, IL-10, and IL-13 are the predominant cytokines. This response peaks at approximately eight weeks post-infection. The immune response is subsequently down-regulated, leading to a long-term, chronic infection.

Numerous studies in mice indicate that IL-4, IL-5, and IL-13 are necessary for granuloma formation. IL-4, as discussed, induces Th-2 proliferation. IL-5 recruits eosinophils to the site of granuloma formation. IL-13 appears to promote collagen deposition and fibrosis. Furthermore, IL-4 and IL-13 act in concert to activate alternatively activated macrophages, (AAMs) which are thought to play a role in the tissue alterations that occur as part of granuloma formation.

Meanwhile, early in granuloma formation, IL-10 mediates the shift from a primarily Th-1 to a primarily Th-2 response, suppressing the release of pro-inflammatory cytokines. Treg cells may be the primary source of the IL-10. Treg cells have also been shown to suppress Th-2 cytokine production, and therefore may also play a part in preventing an excessive and damaging Th-2 response. Not all Treg cells are the same; rather they are a diverse group of cells, many of which have poorly understood roles in suppressing immunity. The best characterized are the so-called natural regulatory T cells, which, like T-helper cells, are $CD4^+$. Precisely how they down-regulate an adaptive response is still under investigation. Abundantly clear, however, is that without this moderation, the survival of the host would be in question.

Parasites may serve as a trigger for autoimmunity

Some pathogens contribute to immunopathology in a different way: by sparking an autoimmune response. Autoimmunity, in a nutshell, refers to the loss of tolerance to self-antigens. As an immune response is generated against self-antigen, tissue damage ensues.

Various mechanisms have been proposed as an explanation for why self-tolerance is lost. The release of normally sequestered antigen from usually immunologically privileged sites may explain some autoimmune diseases. An example is sympathetic ophthalmia. Trauma to an eye can release antigens normally not encountered by the immune system, resulting in immune system priming and a subsequent attack on the eye. Inappropriately high expression of MHC on the surface of pancreatic islet cells may be involved in the development of insulin-dependent diabetes. In some cases, however, **molecular mimicry** may be to blame. In this scenario, an individual is infected with a pathogen bearing antigens that are very similar to host antigens. Lymphocytes, activated in response to the pathogen, also direct their immune assault against the similar host antigen, damaging the tissue bearing those antigens. Such molecular mimicry has been difficult to demonstrate; by the time the affected individual starts displaying symptoms of autoimmunity, the offending pathogen is usually long since destroyed. Yet for some conditions, the evidence for such molecular mimicry is strong. Rheumatic fever, for instance, is initiated by a *Streptococcus* infection. In some individuals, antibodies generated against the bacterium cross-react with antigens on heart valves.

Several eukaryotic parasites have also been at least suspected of involvement in autoimmunity. Malaria, for example, often results in the generation of autoantibodies against uninfected erythrocytes. Destruction of erythrocytes in this way contributes to ***Plasmodium***-mediated anemia. Another prime suspect is ***Trypanosoma cruzi***, the causative agent of Chagas' disease. Among the many symptoms endured by Chagas' disease patients is chronic **cardiomyopathy**—the deterioration of the heart's musculature (**Figure 5.14**). A body of evidence has grown to support the idea that this cardiomyopathy is autoimmune in origin. The delayed onset of symptoms is consistent with

A

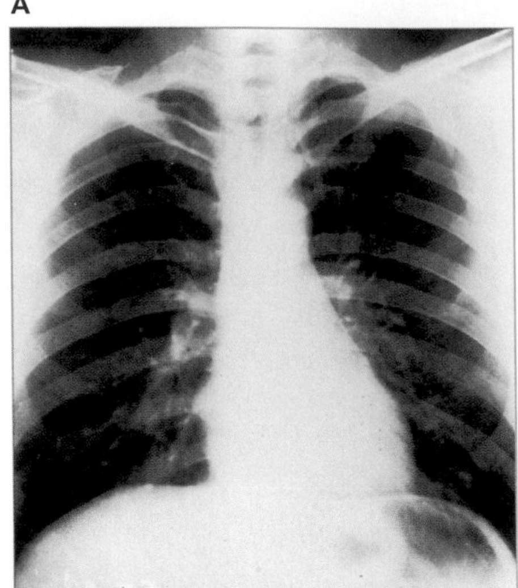

B

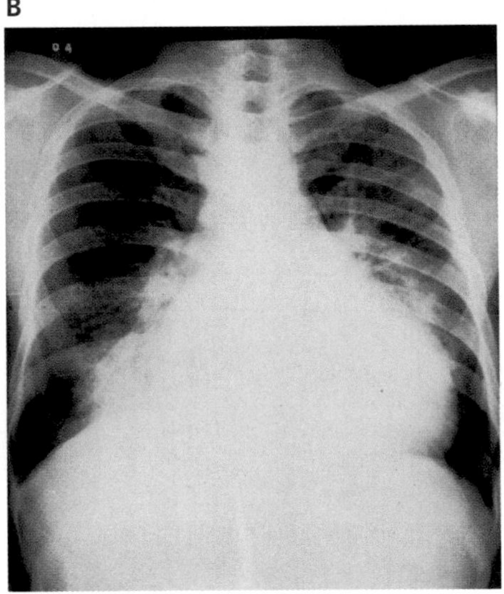

Figure 5.14 Chronic cardiomyopathy in Chagas' disease. Chest X-rays of (A) a normal individual and (B) a patient suffering from chronic Chagas' disease. Weakening of the cardiac muscle causes a large increase in the size of the ventricles, resulting in a grossly enlarged heart. Cross-reactivity of heart muscle antigens with those of *Trypanosoma cruzi* are thought to be a key component of this pathology. (Tarleton RL, Reithinger R, Urbina JA et al [2007] *PLoS Med* 4(12):e332 [doi:10.1371/journal.pmed.0040332].)

autoimmunity, as is the observation that few or no parasites are found associated with the diseased heart muscle. Anti-self antibodies, detected during a *T. cruzi* infection, add to the suspicion. Yet if *T. cruzi* does result in the loss of tolerance to self-antigens, it remains unclear how it occurs. Either molecular mimicry or the release of sequestered antigens, exposed as a result of parasite-induced tissue damage, may be involved. The "one size fits all" model has no place in parasitology. Different parasites, if they do cause autoimmunity, almost certainly do so in different ways. Later in this chapter we will cite evidence that some parasites actually reduce autoimmune responses.

Toxoplasma gondii may both contribute to and help to prevent artherosclerosis

As a final example of immunopathology, consider the recently suggested odd relationship between toxoplasmosis and atherosclerosis. It is now understood that atherosclerosis is more complex than a simple buildup of cholesterol. Clot formation and inflammation are also involved. ***Toxoplasma gondii***, the apicomplexan responsible for toxoplasmosis, plays an ambiguous role in atherosclerosis. ***T. gondii*** is dependent on host cholesterol for maximal growth and reproduction. Because it reduces host serum cholesterol, it has been suggested that infection may actually help to protect against high cholesterol levels and the subsequent buildup of fatty material along arterial walls. However, ***T. gondii*** also initially stimulates a strong Th-1-mediated inflammatory response in an infected host. This, in turn, may contribute to the development of vascular disease. It has been difficult to tease apart these two apparently opposite effects. This example, however, reminds us that the manner in which parasites and the immune response interact is complex indeed and that occasional surprising or even counterintuitive outcomes are likely. We will discuss other possible outcomes later in this chapter when we consider if and when parasite infection is ever beneficial.

5.2 PARASITES AND HOST BEHAVIOR

One of the more intriguing aspects of parasitology is the way in which parasites sometimes affect the behavior of their hosts. In Chapter 7 (Figure 7.26), for example, we will discuss the manner in which the parasitic barnacle *Sacculina carcini* alters the behavior of the crabs it infects in bizarre fashion. In this case, as in many others, the altered host behavior is thought to enhance parasite transmission. Alternatively, in Chapter 6 we will also consider how the behavior of hosts might evolve to reduce the likelihood of infection.

In a chapter focused on parasite-induced disease, why we are considering host behavior? The answer is not complicated. Pathology does not necessarily refer only to host tissue or cellular damage. Rather, it must account for all of the negative consequences of parasitic infection. Maladaptive behavior is one such consequence. Alterations in behavior can realistically be viewed as pathology of the central nervous system, which may have serious ramifications for the host.

Some parasites may modify host behavior to facilitate transmission

One of the first questions asked about host behavioral changes is whether or not the observed changes represent deliberate manipulation on the part of the parasite. Is the ability of the parasite to alter host behavior an evolved adaptation that is beneficial for the parasite or can it best be viewed as a mere side effect of infection? That is, behavioral change might be a result of the mere presence of a parasite in a certain host tissue, which coincidently has behavioral implications. Certainly, for example, if your intestine is blocked by a huge bolus of ***Ascaris***, you may not feel like eating. It is hard to envision a way that a lack of appetite is beneficial from the parasite's perspective or how natural selection on ***Ascaris*** would result in this response. Alternatively, earlier in this chapter we discussed how *Yersinia pestis* bacteria form a mass in their flea vector's protoventriculus and how this mass encourages the infected flea to attempt feeding more often, spreading the bacteria to new hosts with each feeding attempt. One can certainly conceive of how this characteristic of the bacteria would be favored by selection as it results in enhanced transmission. The topic of such selection on vector-transmitted parasites will be revisited in Chapter 7.

Parasitologists interested in such phenomena must be on guard lest they see insidious craftiness and intent on the part of parasites in every instance, even when it does not exist. To determine whether or not a given change in behavior actually represents direct manipulation by the parasite, it must be demonstrated that the behavioral change results in higher fitness for the parasite. It has been extraordinarily difficult to demonstrate this with the necessary experimental evidence. In a very few cases, the evidence is compelling. In most other examples, although "parasite mind control" seems logical and makes a riveting story, there is little supporting data.

When a host behavior change is adaptive from the parasite's perspective, it almost always is intended to increase transmission and make an infected host more vulnerable to predation (**Figure 5.15**). Thus, we expect to observe this phenomenon most commonly in parasites with indirect life cycles in which trophic transmission occurs. The topic of how parasites manipulate their hosts to facilitate transmission is most accurately viewed as an aspect of parasite ecology, and consequently, it will be more fully explored in Chapter 6. Here we will explore what is known about how parasites might actually induce pathological behavioral changes.

A

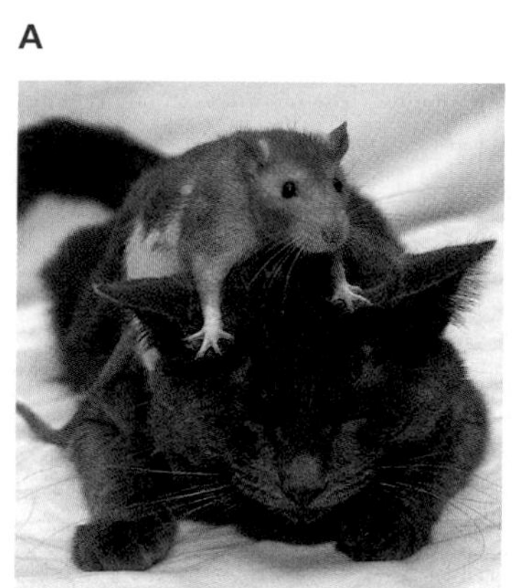

B

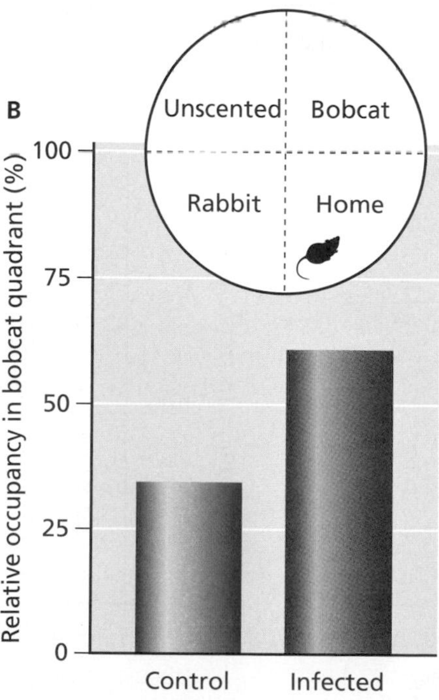

Figure 5.15 Parasite manipulation of host behavior. (A) Evidence supports the idea that when rats (the intermediate host) are infected with *T. gondii*, they lose their fear of cats (the definitive host). Because cats become infected when they eat the intermediate host, the ability to alter rat behavior in this way is thought to increase transmission of the parasite. It is thus adaptive from the parasite's point of view and represents a type of neuropathology in the rat. (B) In this experiment, experimental rats infected with *T. gondii* and uninfected control rats were placed in circular arenas. The rats were allowed to construct nests, and that quadrant of the area was designated as home. Next, one quadrant was scented with bobcat urine and another quadrant was scented with rabbit urine. The fourth quadrant was left unscented. The bar graph depicts the proportion of their time that infected and uninfected rats spent in the quadrant scented with bobcat urine. Control rats spent approximately 35% of their time in the quadrant scented with bobcat urine. Infected rats spent about 60% of their time in the quadrant scented with bobcat urine. (B, Adapted from Vyas A, Kim SK, Giacomini N et al [2007] *Proc Natl Acad Sci* 104:6642-6647. Copyright [2007] National Academy of Sciences, USA.)

The mechanisms that parasites use to alter host behavior are obscure

We know that at least some parasites alter host behavior. As we will see in Chapter 6, we also have a reasonably good understanding regarding the implications of behavioral manipulation for parasite transmission. Yet exactly how parasites influence behavioral changes in their hosts remains largely unknown.

Parasites might modify behavior by secreting chemicals that alter nervous system activity in their hosts. Alternatively, their mere presence in host tissues might interfere with neurological signaling or biochemical pathways, resulting in behavioral changes. As far as actual parasite-secreted molecules with neurological activity are concerned, details are scant. We know somewhat more, at least in a few cases, about how parasites themselves impinge on host tissues and organs to cause behavioral changes. This is especially true in the case of arthropod vectors and how infection may alter their feeding behavior. We have discussed how the bacterium *Yersinia pestis* causes infected fleas to increase their probing rate. Another well-known example involves members of the genus ***Leishmania***. Like *Y. pestis*, these flagellated protozoa interfere with the ability of their vectors (sand flies in either the genus ***Phlebotomus*** or *Lutzomyia*) to take a blood meal. They do so by secreting a viscous, gel-like substance composed of proteophosphoglycan that essentially gums up the gut and mouthparts. Consequently, the flies must bite more readily, often from many different mammals, in order to obtain sufficient blood. The parasite is thus dispersed widely and its transmission is substantially increased. Recent work has shown that this manipulation is more fine-tuned than previously understood. Secretion of the gel does not begin until the parasites have developed into metacyclic promastigotes, the infective stage for mammals (**Figure 5.16**). Thus, the parasite times its manipulation to coincide with the point when it will actually increase transmission. Earlier in the infection, interfering with its vector's feeding would be counterproductive from the parasite's perspective, as it can shorten the sand fly's life span, resulting in a decreased likelihood of transmission. The timing of the manipulation of its vector's behavior so closely with transmission certainly suggests strongly that such manipulation is specific parasite adaptation, rather than a simple side effect of infection.

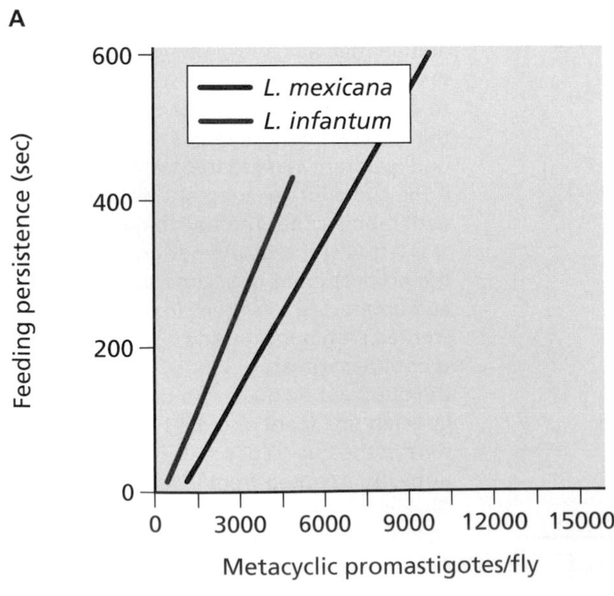

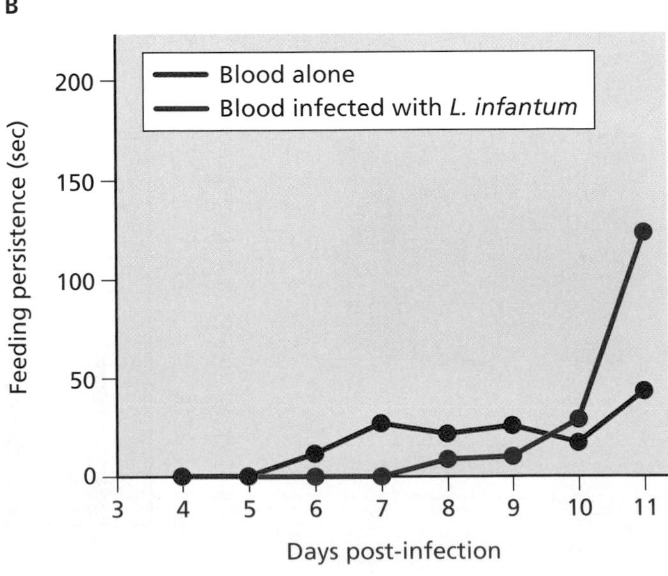

Figure 5.16 Increased sand fly feeding when infected with *Leishmania* parasites. (A) Relationship between the number of metacyclic promastigotes in sand flies (*Lutzomyia longipalpis*) and feeding persistence measured in seconds for two species of *Leishmania*. Metacyclic promastigotes are the infective stage for the vertebrate host. (B) Feeding persistence of flies previously fed on blood alone or on blood infected with *Leishmania infantum*. Note that persistence increases with the number of metacyclic promastigotes and the number of these infective stages rises rapidly after about 10 days. Thus, the increased feeding persistence corresponds to the time when there are large numbers of infective parasites that are ready to be transmitted. (Adapted from Rogers ME & Bates PA (2007) *PLoS Pathog* 3:(6):e91.)

Plasmodium uses a similar strategy in its mosquito vectors. Uninfected mosquitoes produce an enzyme called apyrase in their salivary glands. As the mosquito feeds, apyrase enters the wound, preventing blood coagulation and facilitating feeding by the mosquito. In infected mosquitoes, the infective stage for vertebrates, sporozoites, enter the salivary glands in preparation for transmission. While there, they damage the gland, such that apyrase production declines (**Figure 5.17**). As a result, when mosquitoes bearing sporozoites attempt to feed, the blood of their host quickly coagulates, reducing feeding efficiency. The mosquito must then seek a new host, and as it repeatedly attempts to feed, parasite transmission is enhanced.

Infected hosts may display unusual neurotransmitter profiles in their central nervous systems

Things become more ambiguous when parasites are believed to alter host neuroprocesses. It is assumed that parasites achieve such alterations by secreting substances that alter the concentration of host hormones and neurotransmitters in the central nervous system. Although a number of studies have demonstrated that parasitic infection can alter the host's brain chemistry, little progress has been made identifying the molecules released by parasites responsible for such alterations.

For the most part, evidence to date is indirect. Most data consist of altered neurochemistry in the brains of infected hosts relative to those of uninfected hosts. It is postulated that substances released by the parasites trigger these novel profiles of neurotransmitters and other modulators of neurological activity. For example, the aquatic amphipod *Gammarus pulex*, which is normally photophobic, shows a positive taxis to light when infected with the acanthocephalan *Pomphorhynchus laevis*. This change is believed to increase the likelihood that the amphipod, serving as an intermediate host, will be consumed by the definitive host. Infected amphipods express unusually high levels of serotonin in their central nervous system. The role of serotonin in photophilia has been demonstrated experimentally by injecting uninfected amphipods with exogenous serotonin. Amphipods treated in this manner also become attracted to light. Other hosts for which altered serotonin or dopamine profiles in the central nervous system have been correlated with behavioral changes include small mammals infected with the apicomplexan

Figure 5.17 *Plasmodium* development in the mosquito vector. All times are approximate and vary according to temperature. Inhibition of mosquito apyrase activity only occurs after about 16 days, when sporozoites (the infective stage for the vertebrate host) arrive in the salivary glands. (Modified from Riehle MA, Srinivasan P, Moreira CK & Jacobs-Lorena M [2003] *J Exp Biol* 206:3809–3816.)

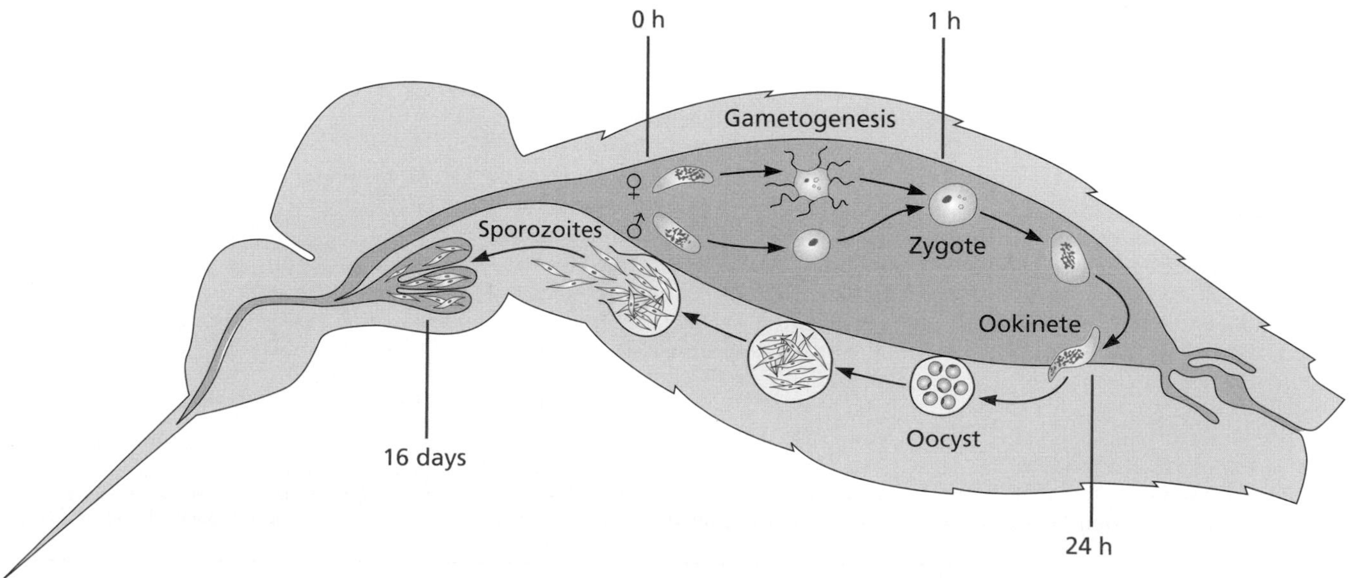

Toxoplasma gondii (**Box 5.2**; see also Figure 5.15), and killifish (*Fundulus parvipinnis*) infected with the trematode *Euhaplorchis californiensis*.

The latter case involving the small fish and the trematode is especially interesting. The killifish are the second intermediate host for *E. californiensis*. Seabirds that readily consume the fish serve as definitive hosts. Normally killifish remain away from the surface, decreasing their exposure to bird predation. Not only do killifish with encysted metacercariae in their brain swim at the surface, they also repeatedly roll over, exposing their white bellies. In effect, they advertise their presence to birds and practically scream out to be consumed (**Figure 5.18**). The phenomenon is density dependent; as the number of metacercariae in the brain increases, the higher the levels of dopamine and serotonin and the more pronounced the behavioral change.

BOX 5.2

Toxoplasma and Human Behavior

Although rodents or small birds serve as the intermediate host for *Toxoplasma gondii*, humans can also become infected through exposure to infective oocysts or by ingesting undercooked meat harboring bradyzoites. Infected humans are unlikely to be consumed by domestic cats (the definitive host); however, humans are eaten by large felid species.

Although most infected humans remain asymptomatic with latent infections, reactivation in immunocompromised individuals can result in severe disease, with serious neurological symptoms (see the Rogues' Gallery). If a pregnant woman develops an acute infection, there is a significant chance of vertical transmission to her fetus, resulting in potentially serious birth defects.

Particularly interesting is the manner in which *T. gondii* appears to manipulate its intermediate host in order to increase its likelihood of transmission (see Figure 5.15). This behavior raises an intriguing question: if infected rodents actually become less fearful of cats, might infected humans display odd behavior as well?

The answer so far is maybe. Unequivocal data are unavailable, but there is no shortage of suggestive correlations between *T. gondii* infection and certain behavioral changes in humans. Several studies, for example, report that patients with schizophrenia are more likely to have antibodies against *T. gondii* than individuals without schizophrenia. Yet other studies found no such correlation. However, it remains possible that at least some schizophrenia results in part from an otherwise asymptomatic infection, although the impact of the parasite on this psychiatric disorder appears to be modest at best. Furthermore, if such a link exists, the mechanism remains obscure. Yet latent *T. gondii* cysts produce tyrosine hydroxylase, an enzyme involved in dopamine synthesis. A breakdown in the regulation of dopamine in the brain has long been recognized as a contributing factor to schizophrenia.

Schizophrenia is just the start. Correlations between *T. gondii* infections and a startling number of other behavioral changes and psychological disorders have been revealed. Anxiety, depression, and epilepsy are just a few. It has been suggested that infected humans are less interested in new experiences and that they have slower reaction times. It has even been proposed that car accidents are 2.5 times more common in people with Rh-negative blood who have also been infected by the parasite. The author of the study believed that this tendency to rear-end fellow motorists was attributable to the slower reactions of infected people. He claimed that a million deaths a year from traffic fatalities resulted from *T. gondii* infections!

Finally, there is the alleged disparity regarding how infection differs in males and females. *T. gondii* seems to bring out the worst in men, making them less likely to follow rules and increasing their jealousy. Women, on the other hand, are said to become more nurturing and caring, thanks to the parasite. Perhaps not surprising in light of their alleged loutish behavior, infected men are claimed to be less attractive to women, whereas infected women are more attractive to men. Although the results are not conclusive, some evidence suggests that mice size up potential mates in a similar way.

We really need more than correlations. Experiments in which human behavior is assessed before and after infection are probably necessary before an impact on behavior can be conclusively demonstrated. Of course, that experiment isn't going to happen; so at least for now, a link between this apicomplexan and oddball behavior will remain more curiosity than fact. Naturally, it's only a matter of time before some crafty lawyer employs the *Toxoplasma* defense. "Ladies and gentlemen of the jury, the parasite made my client do it!"

References

Henriquez SA, Brett R, Alexander J et al (2009) Neuropsychiatric disease and *Toxoplasma gondii* infection. *Neuroimmunomodulation* 16(2):122-33 (doi: 10.1159/000180267).

Flegr J (2007) Effects of *Toxoplasma* on human behaviour. *Schizophr Bull* 33(3):757–760 (doi:10.1093/schbul/sbl074).

Flegr J, Klose J, Novotná M et al (2009) Increased incidence of traffic accidents in *Toxoplasma*-infected military drivers and protective effect the RhD molecule revealed by a large-scale prospective cohort study. *BMC Infect Dis* 9:72 (doi:10.1186/1471-2334-9-72).

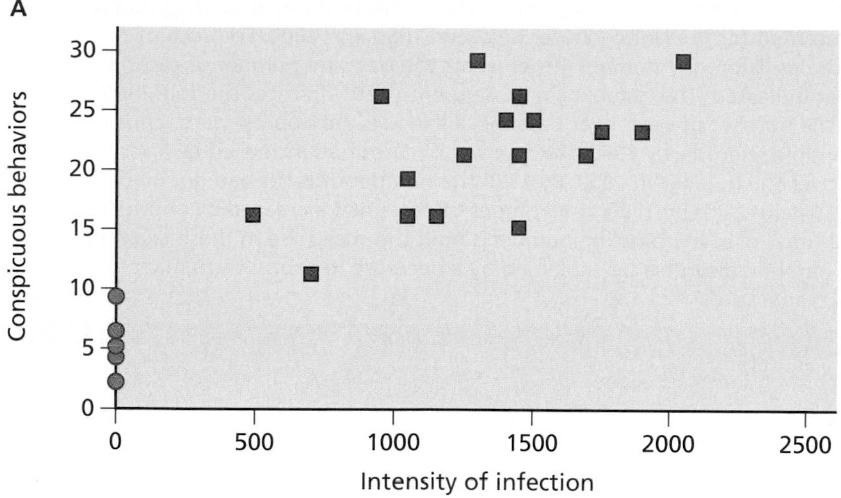

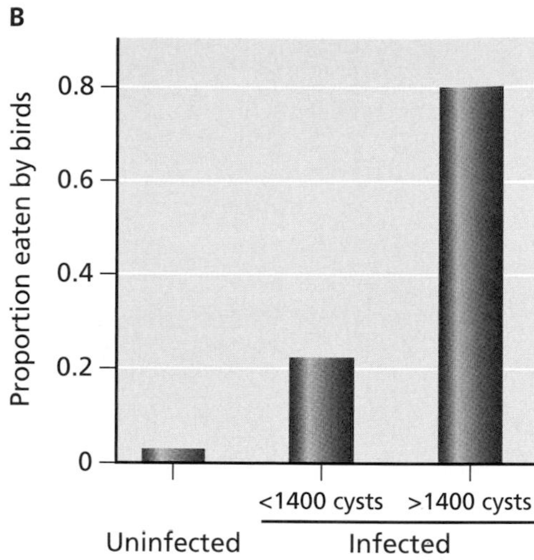

Figure 5.18 Parasite-induced behavior modification in killifish. The trematode *Euhaplorchis californiensis* modifies the behavior of its second intermediate host, the killifish *Fundulus parvipinnis*, in such a way that the fish are more likely to be consumed by seabirds, serving as the definitive host. (A) The frequency of conspicuous behaviors such as belly flashing and surfacing displayed by fish in a 30-minute observation period. The intensity of infection is measured by the number of *E. californiensis* metacercariae in each fish's brain. Green circles represent uninfected control fish. Red squares represent infected fish. (B) A comparison of the proportion of uninfected and infected fish consumed by birds during experimental field trials. (Adapted from Lafferty KD & Morris AK (1996) *Ecology* 77:1390–1397. With permission from Ecological Society of America.)

Few if any molecules secreted by parasites have been unambiguously linked to behavioral changes of the type described above. It has been suggested that, because altered protein synthesis as a result of parasitic infection may reflect altered gene expression, parasites might be affecting such expression in their hosts directly. Because processes such as DNA methylation and histone acetylation regulate eukaryotic gene expression (see Section 3.3), perhaps parasites produce substances that manipulate these processes. Although a body of indirect evidence suggests that this may indeed be the case, conclusive evidence remains lacking.

5.3 PARASITE-MEDIATED AMELIORATION OF PATHOLOGY

From the host's perspective, the discussion so far in this chapter and throughout the text, casts parasites in a less than favorable light. Parasites use hosts to advance their own agenda, and the results are rarely compatible with the host's self-interest. In this chapter we have focused on the diseases that hosts suffer as a result of parasitic infection, and we have even observed that in some cases hosts become zombie-like drones, mindlessly acting in a way that is beneficial to the parasite and causing harm or death to the host.

Nevertheless, in a few situations parasitic infection may provide unexpected benefits to the host. The best-documented examples are those in which parasites lessen the pathology caused by misdirected or unregulated host immune responses. More specifically, some parasites reduce the severity of autoimmunity or allergic response.

Beginning in the early 1980s, investigators noticed that in parts of the world where parasitic infection was reduced, allergies and autoimmunity seemed to increase. On some South Pacific islands, to cite just one early observation, a campaign to eradicate filariasis was followed by an unusual increase in anaphylaxis caused by shellfish consumption. The inverse correlation between parasitic infection and immune-related conditions was most notable in developed parts of the world, where incidence of conditions such as asthma have soared in recent decades (**Table 5.1**). Over the years, as more controlled studies have been conducted, it increasingly appears that such observations are not necessarily coincidental. This has led to the development of the **hygiene hypothesis** to explain the phenomenon.

Table 5.1 Allergies and autoimmunity in the twentieth century. Data are presented for the United States between 1950 and 2000. For each of the four diseases listed, the number 1 represents the baseline number of cases in the year indicated. The numbers in subsequent years indicates the fold-increase in the number of cases over baseline. All values are approximate. Thus, for multiple sclerosis, by 1960, the number of cases had increased by approximately ⅓ over the number in 1950. By 1970 the number of cases had doubled over the 1950 value, and by 1990, the number of cases had increased by approximately 3.3 times over the baseline number. Could the sharp rise in these examples of immune disorders be explained by a decrease in exposure to micro and macroparasites?

	YEAR					
	1950	1960	1970	1980	1990	2000
Multiple sclerosis	1	×1.3	×2.0	×2.3	×3.3	–
Crohn's disease	1	×1.6	×1.6	×2.3	×3.0	×4.0
Type 1 diabetes	–	–	–	×1.5	×2.3	×2.75
Asthma	–	–	–	1	×2.3	×3.0

Parasitic infection may be required for proper immune system development

According to the hygiene hypothesis, young children who live in places where sanitation and hygiene are poor are exposed to a wide range of bacteria and other parasites. Exposure to these organisms, with which humans have evolved for millennia, is crucial for proper immune system development. As sanitation improves, however, such exposure declines, and the result is an increase in immune disregulation and a concomitant rise in allergies and autoimmunity.

The hygiene hypothesis in its current form was first proposed in 1989, when it was observed that allergic disorders were less common in the younger siblings of large families. It was then suggested that with additional older siblings, younger family members are exposed to more infectious agents and that such exposure somehow protected them from hay fever and other allergies. A recent refinement of the hygiene hypothesis is the **old friends hypothesis**. According to this hypothesis, some infectious agents do more than simply fine-tune immunological development of the host; they actually encode certain gene products necessary for a properly functioning immune system. In other words, some genes have been entrusted to "old friends": microbes with which we have coevolved and that were historically always present in our intestine. Without these symbionts, their genes upon which we rely are missing. The result is an increased likelihood of conditions such as inflammatory bowel disease. It was reported in 2008, for instance, that a molecule produced by the common intestinal bacterium *Bacteroides fragilis* called polysaccharide A primes the immune system in such a way that inflammation of the intestinal lining is greatly reduced. Without *B. fragilis* as part of the normal flora and consequently without exposure to polysaccharide A, the stage is set for inflammatory bowel disease later in life.

Some helminth parasites may also contribute to normal immune regulation. Studies with animal models have indicated that infection with various helminths can ameliorate specific immunological conditions. It is likely that in at least some cases, they do the same for humans.

Certain intestinal helminths may reduce the host inflammatory response

How exactly do they do it? What is the mechanism underlying the proposed beneficial effects of some intestinal helminths? As we learned in Chapter 4, parasites modulate their host's immune response in diverse ways. They do so not to benefit their hosts but to increase their likelihood of survival in the face of host immune defense. In some cases, though, the capacity of the parasite to dampen an immune response pays dividends for the host.

Such immune modulation may occur at any of the many points at which a worm intersects with its host immune system. As we discussed earlier, in most intestinal helminth infections, the immune response is largely a Th-2 response. A Th-1 response, which is associated with increased inflammation, is reduced. As discussed in detail in Chapter 4, helminth excretory products are involved in the shift to Th-2 immunity and the suppression of inflammation.

Another important means by which parasites may lessen inflammation is through their engagement of alternatively activated macrophages (AAMs) (**Figure 5.19**). As we also described in Chapter 4, AAMs are also activated by Th-2 activating cytokines such IL-4 and IL-13. This is as opposed to classically activated macrophages, which are activated by IFN-γ, TNF, and other cytokines consistent with an inflammatory Th-1 response. AAMs help to polarize the immune response further toward a Th-2 profile and away from a Th-1 response by releasing IL-10, transforming growth factor β (TGFβ),

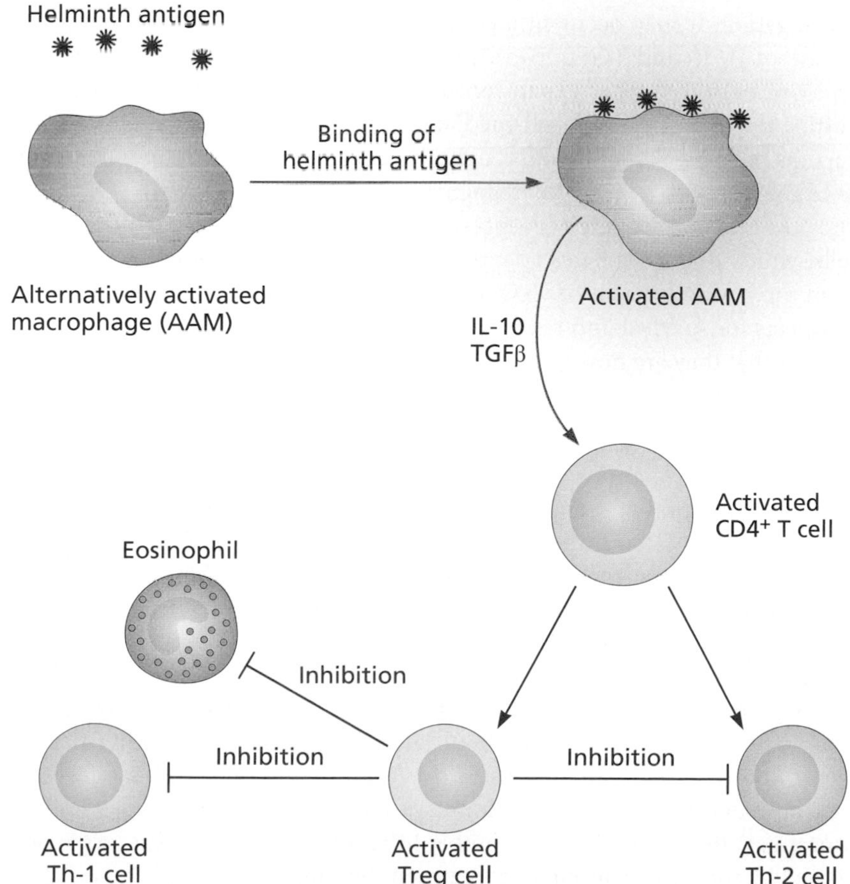

Figure 5.19 A possible mechanism for helminth immune system down-regulation. In response to helminth antigens, alternatively activated macrophages (AAMs) undergo activation. These cells then activate naïve CD4+ T cells via the cytokines IL-10 and transforming growth factor β (TGFβ). Activated CD4+ cells cause the now activated T cells to skew heavily toward a Th-2 response. IL-10 and TGFβ also generate large numbers of Treg cells, which in turn, inhibit both Th-1 and Th-2 cells, damping both cell-mediated and humoral immunity. Treg cells also inhibit eosinophils, which are then less likely to degranulate, further reducing inflammation.

and IL-1 decoy receptor molecules. They also produce a molecule called programmed death ligand 1 that induces **anergy** (an inability to become activated) in T cells. The result is reduced inflammatory damage to tissues. Mouse strains unable to develop high AAM levels, for instance, suffer from increased inflammation of the bowel when infected with the cestode *Taenia crassiceps*. Ordinarily, the inflammatory response is held in check, at least in part by IL-10 secreted by AAMs. Similarly, in murine models of schistosomiasis, when AAMs are lacking, egg-mediated liver pathology never resolves and infected mice generally die.

Intestinal helminth infection results in activation of regulatory T cells

Clearly, a shift to a Th-2 adaptive immune profile cannot be the whole story. If it were, the frequently observed inverse correlation between intestinal infection and allergies would be puzzling and counterintuitive. Many types of allergies, including hay fever and asthma, are mediated by IgE. A shift to a Th-2 response might therefore be expected to increase rather than decrease these allergic conditions. Even though some helminth infections are associated with a higher incidence of allergy (***Ascaris*** and asthma is an example), in many cases, the relationship between parasitic infection and less severe allergies appears to hold up.

At least part of the explanation for this apparent paradox involves the activation of regulatory T cells (Treg cells). As we discussed in Chapter 4 (Section 4.3), Treg cells are involved in the suppression of adaptive immunity. In addition to their role in dampening excessive immune responses, Treg cells also help to maintain tolerance to self-antigens, thus preventing autoimmunity. Precisely how they down-regulate an adaptive response is still under investigation. It may occur in a contact-dependent manner or through the release of IL-10 and TGFβ. Nevertheless, however they accomplish it, Treg cells prevent the further expansion of both Th-1 and Th-2 cells, effectively putting the brakes on both cell-mediated and humoral adaptive immunity.

Thus the parasitic modulation of both autoimmunity and allergies appears to begin in part when AAMs recognize parasite antigens (see Figure 5.19). The now initiated cascade of events ultimately results in the activation of Treg cells, which down-regulates adaptive immunity. This of course is beneficial from the parasite's point of view. Reduced immunity increases the parasite's prospects for survival and reproduction. Hosts can take some comfort from the fact that they are now less prone to wheezing, runny noses or inflammatory bowel disease.

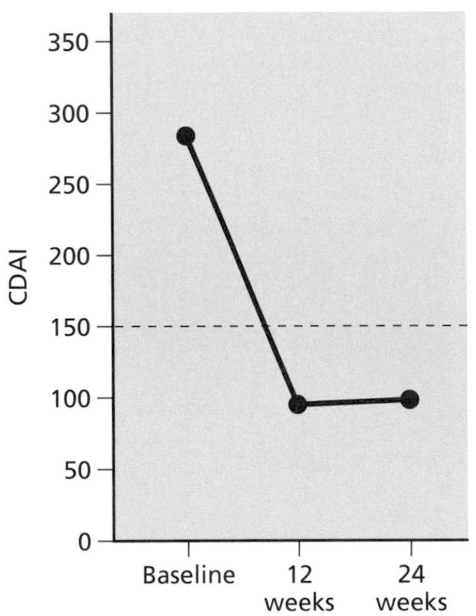

Figure 5.20 *Trichuris suis* therapy for Crohn's disease. Twenty nine patients between the ages of 18 and 72 with Crohn's disease participated in this study. The severity of Crohn's disease was assessed by determining each volunteer's Crohn's disease activity index (CDAI). A CDAI score of 220–450 is diagnostic of Crohn's disease. Each patient ingested 2500 live *T. suis* ova every three weeks for 24 weeks. Their CDAI was assessed every six weeks over the course of the study. A CDAI value of <150 indicates that disease is in remission. The graph shows the mean change in CDAI over the 24-week study period. (Adapted from Summers RW, Elliott DE, Urban JF et al [2005] *Gut* 54:87–90. With permission from BMJ Publishing Group Ltd.)

Intestinal helminths can be administered therapeutically

The apparent link between infection with certain parasites and a reduction in the severity of certain medical conditions has given rise to what seems oxymoronic: **helminth therapy**. This is the deliberate infection of patients with specific intestinal nematodes to alleviate symptoms of certain clinical conditions. To date, most research and treatment has focused on the hookworm ***Necator americanus*** and especially on the pig whipworm, *Trichuris suis* (**Figure 5.20**). Both of these species have shown promise in the treatment of a wide range of autoimmune and allergic conditions, including inflammatory bowel disease, multiple sclerosis, asthma, eczema, and hay fever. In fact, although not exactly commonplace and although perhaps not for the squeamish, the use of *T. suis* to combat inflammatory bowel disorders is now viewed by some as an acceptable treatment option.

The kinder, gentler side of parasitic infections may extend beyond allergies and inflammation of the intestine. It has recently been shown that an influx of eosinophils into fat tissue may help to prevent insulin resistance and consequently lessen the likelihood of type-2 diabetes. Eosinophils are granulocytes that develop from the myeloid stem cell line, and they play an especially important role in combating helminth parasites. Among the many chemical mediators that they produce and secrete is IL-4, which in turn activates AAMs. When AAMs enter fat tissue, they help to maintain glucose homeostasis in the blood. The researchers showed that they could induce insulin resistance and glucose imbalance by feeding mice a high-fat diet. Yet when the mice were later infected with the nematode *Nippostrongylus brasiliensis*, glucose sensitivity and homeostasis improved. This response presumably arose from the parasite-induced influx of eosinophils and subsequent activation of AAMs. The researchers even speculated that eosinophils may have evolved to help maintain metabolic homeostasis in the face of chronic infection with intestinal parasites.

Summary

Parasites affect their hosts in a variety of complex and sometimes unexpected ways. In general, the manner in which a particular parasite causes disease in its host can be assigned to one of five basic categories. Parasites may directly damage host tissues as a consequence of their feeding, attachment, growth, or migration. Some parasites alter cell growth patterns in ways that can interfere with normal host activities. Other parasites produce toxic compounds, whereas many others interfere in various ways with host nutrient acquisition. However, probably the most important pathological mechanism is an indirect one: the stimulation of an aberrant or misdirected immune response. In many, or even most, cases, the negative effects of parasitic infection involve two or more mechanisms that fall into multiple categories.

In some cases, the pathology caused by parasites is merely an unfortunate consequence of infection of no adaptive value to the parasite. In other cases, the pathology in some manner contributes to the success of the parasite. These include the difficult-to-prove but nonetheless intriguing instances in which parasites are thought to manipulate the behavior of their hosts in an effort to facilitate transmission. Often these manipulations are thought to help insure that a parasitized host will be preyed upon by the next host in the parasite life cycle.

There are even some surprising cases where parasites, through no benevolence of their own, lessen the pathology associated with certain medical conditions such as allergies, thus demonstrating once again that host–parasite interactions are rarely if ever as straightforward as they may at first seem.

REVIEW QUESTIONS

1. In Figure 5.1a, parasites with different thresholds of disease are illustrated. Two factors that may influence virulence, and therefore whether the threshold is relatively high or low, are the mode of transmission utilized by the parasite as well as its infective dose (the average number of infectious parasites to which hosts must be exposed, to infect 50% of those hosts). In general, as the ease of transmission goes up or as the infective dose goes down, the threshold decreases. Explain these observations.

2. Although the pathology caused by *Entamoeba histolytica* was discussed under the heading of "Trauma to host cells, tissues, or organs", it might have been discussed under headings for different types of pathology as well. Under which other headings might its discussion also have been appropriate, and why?

3. Consider Figure 5.9. If the discussion of how malaria parasites prevent hyperinfection by stimulating the host to produce hepcidin is confirmed, how might Figure 5.9 be different if all of the individuals portrayed in the figure were on iron supplements?

4. After reading about how hemozoin contributes to the pathology of *Plasmodium*, do you think that this molecule merits its traditional status as a true parasite toxin?

5. In an experiment, mice that are FoxP3$^{-/-}$ are infected with S*chistosoma mansoni*. FoxP3 is an important transcription factor expressed in Treg cells. Thus, a Treg response is absent in these mice. Predict the likely outcome of infection in these mice, and explain this outcome, based on the lack of a Treg response.

6. The hydatid cyst that forms in the intermediate host of *Echinococcus granulosis* can be filled with more than several liters of liquid and can contain millions of larval stages called protoscoleces. When a hydatid cyst is removed from a human, the cyst is often frozen with liquid nitrogen prior to its removal. Explain this necessary precaution, in the light of what you have learned about immunopathology.

7. Suppose that a study is published in which it is shown that the cardiomyopathy associated with Chagas' disease is worse in those who have previously had a heart attack. What does this suggest about the possibility that Chagas'-associated cardiomyopathy is due at least in part to autoimmunity and what does it suggest about the manner in which such autoimmunity is initiated?

8. In our discussion of how parasites might alter host behavior as a way to increase the likelihood of trophic transmission, we considered the example of Euhaplorchis californiensis, and the manner in which it might alter the behavior of killifish, making these fish more prone to predation by seabirds serving as definitive hosts. Why haven't the birds evolved to avoid killifish that are acting erratically and thus avoid becoming infected themselves? Can you think of a reason why this has not occurred?

References

GENERAL REFERENCES

Combes C (1997) Fitness of parasites: pathology and selection. *Int J Parasitol* **27**:1–10 (doi:10.1016/S0020-7519(96)00168-3).

Gutierrez Y (2000) Diagnostic Pathology of Parasitic Infections. Oxford University Press.

PATHOLOGY RESULTING FROM PARASITIC INFECTIONS

Parasites can induce pathogenesis in various ways

Gage KL & Kosoy MY (2005) Natural history of plague: perspectives from more than a century of research. *Annu Rev Entomol* **50**:505–528 (doi:10.1146/annurev.ento.50.071803.130337).

Young ND, Campbell BE, Hall RS et al (2010) Unlocking the transcriptomes of two carcinogenic parasites, *Clonorchis sinensis* and *Opisthorchis viverrini*. *PLoS Negl Trop Dis* **4(6)**:e719 (doi:10.1371/journal.pntd.0000719).

Sripa B, Brindley PJ, Mulvenna J. et al (2012) The tumorigenic liver fluke *Opisthorchis viverrini*—multiple pathways to cancer. *Trends Parasitol* **28**:395–407.

Pathology can be categorized as one of several general types

Hadju V, Stephenson LS, Abadi K et al (1996) Improvements in appetite and growth in helminth-infected schoolboys three and seven weeks after a single dose of pyrantel pamoate. *Parasitology* **113**:497–504.

Kightlinger LK, Seed RJ & Kightlinger MB (1996). *Ascaris lumbricoides* aggregation in relation to child growth status, delayed cutaneous hypersensitivity, and plant anthelmintic use in Madagascar. *J Parasitol* **82**:25–33.

Williams-Blangero S, VandeBerg JL, Subedi J et al (2002) Genes on chromosomes 1 and 13 have significant effects on *Ascaris* infection. *Proc Natl Acad Sci* **99**:5533–5538 (doi:10.1073/pnas.082115999).

Parasites can cause direct trauma to host cells, tissues, and organs

Diamond LS & Clark CG (1993) A redescription of *Entamoeba histolytica* Schaudinn, 1903 (emended Walker, 1911) separating it from *Entamoeba dispar* Brumpt, 1925. *J Eukaryot Microbiol* **40**:340–344.

Mechanisms underlying the pathogenicity of *Entamoeba histolytica* remain obscure

Bruchhaus I, Loftus BJ, Hall N & Tannich E (2003) The intestinal protozoan parasite *Entamoeba histolytica* contains 20 cysteine protease genes, of which only a small subset is expressed during in vitro culture. *Eukaryot Cell* **2**:501–509.

Leippe M, Bruhn H, Hecht O & Grotzinger J (2005) Ancient weapons: the three-dimensional structure of amoebapore A. *Trends Parasitol* **21**:5–7.

Ralston KS, Solga MD, Mackey-Lawrence NM et al (2014) Trogocytosis by *Entamoeba histolytica* contributes to cell killing and tissue invasion. *Nature* **508**: 526-530.

Sharma M, Vohra H & Bhasin D (2005) Enhanced proinflammatory chemokine/cytokine response triggered by pathogenic *Entamoeba histolytica*: basis of invasive disease. *Parasitology* **131**:783–796.

Parasitic infection can alter host cell growth patterns

Dvir E & Clift SJ (2010) Evaluation of selected growth factor expression in canine spirocercosis (*Spirocerca lupi*)-associated non-neoplastic nodules and sarcomas. *Vet Parasitol* **174**:257–266 (doi:10.1016/j.vetpar.2010.08.032).

Fustish CA & Millemann RE (1978) Glochidiosis of salmonid fishes II. Comparison of tissue response of Coho and Chinook salmon to experimental infection with *Margaritifera margaritifera* (L) (Pelecypoda: Margaritanidae). *J Parasitol* **64**:155–157.

Govett PD, Rotstein DS & Lewbart GA (2004) Gill metaplasia in a goldfish, *Carassius auratus auratus* (L.). *J Fish Dis* **27**:419–423.

Hodder SL, Mahmoud AF, Sorenson K et al (2000) Predisposition to urinary tract epithelial metaplasia in *Schistosoma haematobium* infection. *Am J Trop Med Hyg* **63**:133–138.

Pauley GB & Becker CD (1968) *Aspidogaster conchiola* in mollusks of the Columbia River system, with comment on the host's pathological response. *J Parasitol* **54**:917–920.

Petersen CA & Burleigh BA (2003) Role for Interleukin-1β in *Trypanosoma cruzi*-induced cardiomyocyte hypertrophy. *Infect Immun* **71**:4441–4447 (doi:10.1128/IAI.71.8.4441-4447.2003).

Phares K (1996) An unusual host–parasite relationship: The growth hormone-like factor from plerocercoids of spirometrid tapeworms. *Int J Parasitol* **26**:575–588.

Many parasites adversely affect host nutrition

Coop RL & Holmes PH (1996) Nutrition and parasite interaction. *Int J Parasitol* **8–9**:951–962.

Nyberg W, Grasbeck R, Saarni M et al (1961) Serum vitamin B12 levels and incidence of tapeworm anemia in a population heavily infected with *Diphyllobothrium latum*. *Am J Clin Nutr* **9**:606–612.

Stekelee RW (2003) Pregnancy, nutrition and parasitic diseases. *J Nutr* **133**:16,615–16,675.

Stephenson LS, Latham MC & Ottesen EA (2000) Malnutrition and parasitic helminth infections. *Parasitology* **121**:23–28.

Plasmodium infections can result in host iron deficiency

Portugal S, Carret C, Recker M et al (2011) Host-mediated regulation of superinfection in malaria. *Nat Med* **17**:732–737 (doi:10.1038/nm.2368).

Magalhães RJ & Clements AC (2011) Mapping the risk of anaemia in preschool-age children: the contribution of malnutrition, malaria and helminth infections in West Africa. *PLoS Med* **8(6)**:e1000438 (doi:10.1371/journal.pmed.1000438).

World Health Organization (2008) Worldwide prevalence of anaemia 1993–2005: WHO global database on anaemia. Geneva: World Health Organization. http://whqlibdoc.who.int/publications/2008/9789241596657_eng

Toxins are a less frequent component of parasite pathology

Brattig NW, Kowalsky K, Liu X et al (2008) *Plasmodium falciparum* glycosylphosphatidylinositol toxin interacts with the membrane of non-parasitized red blood cells: a putative mechanism contributing to malaria anemia. *Microbes and Infection* **10**:885–891.

Jani D, Nagarkatti R, Beatty W et al (2008) HDP—A novel heme detoxification protein from the malaria parasite. *PLoS Pathog* **4(4)**:e1000053 (doi:10.1371/journal.ppat.1000053).

Parroche P, Lauw FN, Goutagny N et al (2007) Malaria hemozoin is immunologically inert but radically enhances innate responses by presenting malaria DNA to Toll-like receptor 9. *Proc Natl Acad Sci* **104**:1919–1924.

Turrini F, Schwarzer FE & Arese P (1993) The involvement of hemozoin toxicity in depression of cellular immunity. *Parasitol Today* **9**:297–300.

Pathology often results from immune-mediated damage to host cells and tissues

Long GH & Boots M (2011) How can immunopathology shape the evolution of parasite virulence? *Trends Parasitol* **27**:300–305.

Immunopathology is an important component of the pathology observed in malaria

Franklin BS, Ishizaka ST, Lamphier M et al (2011) Therapeutical targeting of nucleic acid-sensing Toll-like receptors prevents experimental cerebral malaria. *Proc Natl Acad Sci* **108**:3689–3694 (doi:10.1073/pnas.1015406108).

Parasites may serve as a trigger for autoimmunity

Girones N & Fresno M (2003) Etiology of Chagas' disease myocarditis: autoimmunity, parasite persistence, or both? *Trends Parasitol* **19**:19–22.

Toxoplasma gondii may both contribute to and help to prevent atherosclerosis

Portugal LR, Fernandes LR & Alvarez-Leite JI (2009) Host cholesterol and inflammation as common key regulators of toxoplasmosis and atherosclerosis development. *Exp Rev Anti-Infect Ther* **7**:807-819.

PARASITES AND HOST BEHAVIOR

Some parasites may modify host behavior to facilitate transmission

Moore J (2002) Parasites and the Behavior of Animals. Oxford University Press.

The mechanisms that parasites use to alter host behavior are obscure

Lefèvre T, Adamo SA, Biron DG et al (2009) Invasion of the body snatchers: the diversity and evolution of manipulative strategies in host–parasite interactions. *Adv Parasitol* **68**:45–83.

Rogers ME & Bates PA (2007) *Leishmania* manipulation of sand fly feeding behavior results in enhanced transmission. *PLoS Pathog* **3(6):e91**. (doi:10.1371/journal.ppat.030091).

Poulin R (2010) Parasite manipulation of host behavior: an update and frequently asked questions. In Advances in the Study of Behavior, vol. 41, pp. 151–186. Academic Press.

Infected hosts may display unusual neurotransmitter profiles in their central nervous systems

Lafferty KD & Kimo A (1996) Altered behavior of parasitized killifish increases susceptibility to predation by bird final hosts. *Ecology* **77**:1390–1397.

Øverli Ø, Páll M, Borg M et al[Q34] (2001) Effects of *Schistocephalus solidus* infection on brain monoaminergic activity in female three-spined sticklebacks *Gasterosteus aculeatus*. *Proc R Soc Lond B* **268**:1411–1415.

Shaw JC, Korzan, WJ, Carpenter RE et al (2009) Parasite manipulation of brain monoamines in California killifish (*Fundulus parvipinnis*) by the trematode *Euhaplorchis californiensis*. *Proc R Soc Lond B Biol Sci* **276**:1137–1146.

Tain L, Perrot-Minnot MJ & Cézilly F (2007) Differential influence of *Pomphorhynchus laevis* (Acanthocephala) on brain serotonergic activity in two congeneric host species. *Biol Lett* **3**:68–71.

Webster JP, Lamberton PHL, Donnelly CA et al (2006) Parasites as causative agents of human affective disorders? The impact of anti-psychotic, mood-stabilizer and anti-parasite medication on *Toxoplasma gondii*'s ability to alter host behaviour. *Proc R Soc Lond B Biol Sci* **273**:1023–1030.

PARASITE-MEDIATED AMELIORATION OF PATHOLOGY

Microbial and parasitic infection may be required for proper immune system development

Mazmanian SK, Round JL & Kasper DL (2008) A microbial symbiosis factor prevents intestinal inflammatory disease. *Nature* **453**:620–625.

Strachan DP (1989) Hay fever, hygiene and household size. *Br Med J* **299**:1259–1260.

Zaccone P, Burton OT & Cook A (2007) Interplay of parasite-driven immune responses and autoimmunity. *Trends Parasitol* **24**:35–42.

Certain intestinal helminths may reduce the host inflammatory response

Dixon H, Blanchard C, Deschoolmeester ML et al (2006) The role of Th-2 cytokines, chemokines and parasite products in eosinophil recruitment to the gastrointestinal mucosa during helminth infection. *Eur J Immunol* **36**:1753–1763.

Maizels RM & Yazdanbakhsh M (2003) Immune regulation by helminth parasites: cellular and molecular mechanisms. *Nat Rev Immunol* **3**:733–744.

Intestinal helminth infection results in activation of regulatory T cells

Wu D, Molofsky AB, Liang HE et al (2011) Eosinophils sustain adipose alternatively activated macrophages associated with glucose homeostasis. *Science* **332**:243–247.

Wilson M, Maizels R, Capron M et al (2006) Regulatory T cells induced by parasites and the modulation of allergic responses. *Chem Immunol Allergy* **90**:176–195 (doi:10.1159/000088892).

Intestinal helminths can be administered therapeutically

Mortimer K, Brown A, Feary J et al (2006) Dose-ranging study for trials of therapeutic infection with *Necator americanus* in humans. *Am J Trop Med Hyg* **75**:914–920.

Garg SK, Croft AM & Bage P (2014). Helminth therapy for induction of remission of inflammatory bowel disease. *Cochrane Database Syst Rev* doi: 10.1002/14651858.CD0094.pub2.

The Ecology of Parasitism

6

The last word in ignorance is the man who says of an animal or plant: "What good is it?"

Aldo Leopold, *Round River: From the Journals of Aldo Leopold*

In This Chapter

Like all organisms, parasites are enmeshed in complex webs of interactions within both their abiotic and biotic environments. Surprisingly, even though parasitism is among the most common lifestyle on Earth, parasites have often been relegated to a back seat in ecological studies. As we will see, in recent years the study of parasite ecology has taken on greater significance and has generated increased interest among biologists. For instance, the prominent role that parasites play in ecosystems has been revealed in recent studies highlighted in this chapter and they have helped to reinvigorate the study of parasite ecology. Not only do parasites influence interactions among their hosts, but they also have their own distinctive relationships with their environments and with one another. Parasite ecology has proven to be a fertile field of study, yielding surprising new insights that require us to reappraise our view of the natural world.

In this chapter, we will first examine the environments in which parasites are found, touching on the important topics of habitat preference and host specificity. Next to be considered will be parasite populations. As we will see, they differ from populations of more familiar host organisms. We will then discuss how different species of parasites interact with one another and determine if consistent patterns can be observed in communities of parasites within hosts. The role of parasites in larger scale processes will next be considered. What role do parasites play in food webs and in ecosystems? Can patterns of parasite abundance on global scales be discerned? We will then examine some of the effects parasites have on the ecology of their hosts. Finally, we will cover some new approaches to the study of parasite ecology and follow this with a discussion of the science of epidemiology, explaining how this important discipline interfaces with studies of the ecology of parasitism.

6.1 DEFINING THE HABITATS OF PARASITES

As with any discipline, ecology has its own terminology (**Box 6.1**). We begin by discussing parasite niches, a topic also touched on in Chapter 3 with respect to the site specificity shown by many parasites. Defining an organism's niche is challenging because so many different variables (microhabitat

Species names highlighted in red are included in the Rogues' Gallery, starting on page 429.

BOX 6.1

Some Basic Ecology Terms

Ecology is the study of the distribution, abundance, and diversity of organisms, including study of the underlying processes that dictate the patterns observed. Ecology relates organisms to both abiotic (nonliving) and living components of the environments in which they are found. It is a discipline that pays particular attention to the flow of resources—materials and energy—through biological scales ranging from the individual, to populations of particular species, to communities comprised of multiple species, to ecosystems, to broader scale biomes, and finally to the Earth as a whole. A population is composed of all the individuals of a particular species that live in the same geographic area. A community is a group of interacting species that share the same environment. The concepts of both the population and the community require further qualification when applied to parasites. An ecosystem consists of all the organisms living in a particular area, and it includes the abiotic components of the environment with which the organisms interact (for example, a lake and its catchment basin). A biome refers to a particular category of ecosystem, such as a grassland, that shares similar biotic, climatic, and geographic features and that is widespread across the Earth's surface. Every species has a characteristic ecological niche that is defined by its particular habitat and function (how it gains resources and interacts with other organisms) within an ecosystem. Parasites are no exception.

preference, temperature, salinity, pH, food sources, need for micronutrients, to name just a few) contribute to the definition. With respect to their habitats, parasites differ from free-living organisms because, at least for part of their life cycle, their habitat is in or on another living organism. There are three aspects of parasite habitats worthy of attention: parasites sequentially experience multiple environments; they often have very specific microhabitat preferences within hosts; and they often live within very limited subsets of potential hosts (they are often host specific).

Parasites occupy multiple habitats in succession

As emphasized in Chapter 3, parasites pass through different biotic and abiotic environments sequentially and predictably during their life cycles. This passage is particularly remarkable for parasites with complex life cycles such as digenetic trematodes. Such parasites regularly inhabit both vertebrate and invertebrate hosts and typically have free-living life cycle stages that must find new hosts to complete development. Moreover, as with malaria parasites in mosquitoes, they might occupy ecologically distinct microhabitats (midgut, hemolymph, and salivary glands) within the same host individual at different phases of their life cycle. Even parasites with relatively simple, direct life cycles such as ***Entamoeba histolytica*** must be able to survive in both external and host environments. In this discussion of parasite ecology, keep in mind that parasites have complex relationships with all of their biotic and abiotic environments and adverse conditions in any of their required habitats may cause their death. All these environments need to be considered in fully defining the niche for a particular parasite species. All present potential opportunities for controlling a parasite, if that is a consideration.

Parasites have microhabitat preferences and occupy specific sites within their hosts

Also highlighted in Chapter 3 was the idea that parasites occupy distinctive microhabitats within their hosts and exhibit an amazing ability to locate and then remain within their preferred microhabitats. Human parasites demonstrate this tendency (**Figure 6.1**). Recall the example of ***Giardia lamblia*** and its use of an adhesive disk to colonize the epithelium of the human small

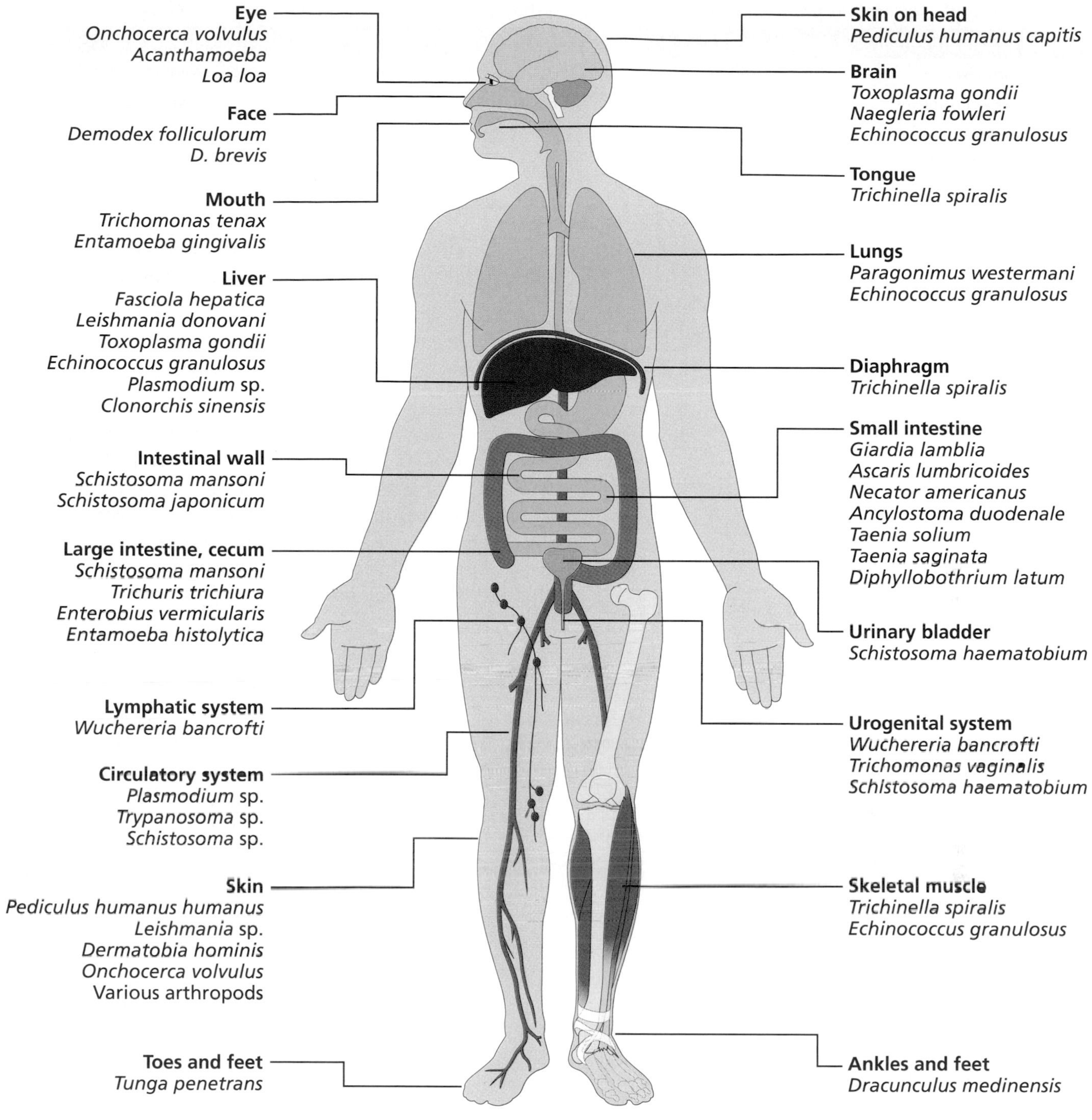

Figure 6.1 An overview of some specific habitats occupied by parasites of humans. This figure identifies the characteristic habitat preferences of many species.

intestine discussed in Chapter 3. Likewise, the whipworm ***Trichuris trichiura*** consistently inhabits the cecum and large intestine, in contrast to ***Ascaris lumbricoides***, which is typically found in the upper small intestine.

Although it is clear that parasites have distinctive habitat preferences, it is usually not at all clear how natural selection has operated to shape these preferences. **Table 6.1** lists a few examples of parasites with distinctive habitat preferences. Included with each are some possible underlying factors that might have contributed to these habitat preferences. Note that a wide variety of factors could be involved and often multiple factors are postulated. In addition, we typically do not know the exact nutrient needs or preferences of parasites, and we might be overlooking important information regarding the

Table 6.1 Examples of habitat preferences of parasites within their hosts and proposed benefits of the habitats

Parasite or Parasite Group	Habitat Used	Factors Likely Involved in Habitat Selection
Plasmodium vivax	Within new RBCs with Duffy antigen	Antigen an essential receptor; avoids immunity
Leishmania amastigotes	Within phagolysosomes of macrophage	Avoids humoral immunity; resists lytic enzymes
Adult tapeworms or flukes	In small intestine of host	High nutrient availability; ready egress of eggs
Trichinella spiralis	Within skeletal muscle cells	Stable habitat; avoids immunity; eaten by next host
Schistosoma japonicum	Mesenteric veins of hepatic portal system	High nutrient availability; finds mates
Schistosoma haematobium	Veins around urinary bladder	Avoids competition; ready egress of eggs in urine
Diplostomoid eye flukes	Within eyes of fish	Avoids immunity; increases chance of predation
Monogeneans in fish	Specific gill arches	Fit to host structure; finding mates; avoid competition
Lice in birds or mammals	Specific size feathers or hairs	Ensures attachment; avoids grooming; finds mates
Ectoparasites of marine fish	In mouth or folds in skin	Avoids activities of cleaning fish or shrimp

factors that shape their habitat preferences. Furthermore, this work needs to consider the phylogenetic history of the particular parasite considered.

Not surprisingly, nutrition is of paramount importance in influencing microhabitat choice. In addition to the example of ***G. lamblia*** mentioned in Chapter 3, consider too the tapeworm ***Hymenolepis diminuta***, which lives in the small intestine of rats and occasionally humans. This example is distinctive because the tapeworm undergoes a daily circadian migration, moving first to the anterior part of the small intestine as the host feeds (**Figure 6.2**).

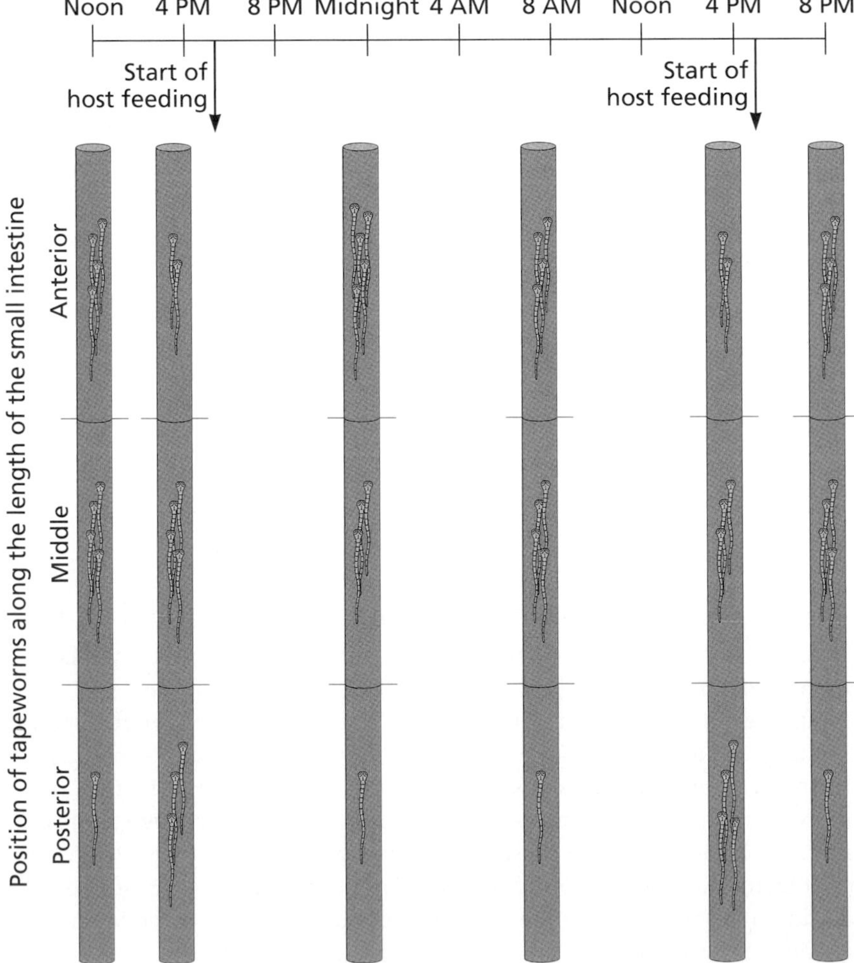

Figure 6.2 Parasites can change their locations depending on the circumstances. Shown on the vertical axis is the position of scolices (a scolex is the anterior, holdfast organ) of tapeworms "*Hymenolepis diminuta*" recovered in different parts of the small intestine of rats, at different times of the day. Anterior refers to the first 10 inches of the small intestine, middle to the second 10 inches, and posterior to the remainder of the small intestine. Rats were fed at night, from 5:30 pm to 7:30 am. Note the consistent circadian pattern of migration, where the tapeworms move to the anterior portion of the small intestine shortly after the host feeds. (Modified from Read CP & Kilejian AZ [1969] *J. Parasitol* 55:574–578.)

As the host's food passes from the stomach into the small intestine, the tapeworms move posteriorly with the food, likely as a way to maintain high rates of nutrient absorption (especially glucose) derived from the food. They pass to the posterior portion of the small intestine, by which time most nutrients have been extracted. If the feeding regime of the rats is reversed and they are fed during the day, the timing of the daily tapeworm migration is also reversed, suggesting the worm's behavior is tied to host feeding. This pattern is further supported by the observation that withholding food leads to delays in the anterior migration of the worm. The parasite is probably using cues such as hormonal changes in the gut associated with feeding to stimulate its migration.

In recent years, biologists have gained a vastly improved understanding of the composition, complexity, and function of the gut flora (the bacteria living in the intestine) of humans and other animals. Our gut contains about 10^{13} bacteria, and it would certainly not be surprising to think that they influence the biology of eukaryotic organisms such as protozoan and helminth parasites that share the gut with them. The nematode *Trichuris muris* lives in the cecum and large intestine of mice. Hatching of the characteristic eggs of this species depends both on elevated temperatures found in the host's gut and on the establishment of physical contact with the polar plugs found on each end of the egg by microbes living in the gut (**Figure 6.3**). The investigators who first observed this speculated that contact of bacterial fimbriae with the polar plugs signals the nematode larva within the egg to emerge. Treatment of mice with antibiotics to reduce the expected helper effect of bacteria on egg hatching did indeed have the expected effect of diminishing the number of *T. muris* worms able to establish as compared to mice not receiving antibiotics. To the list of factors influencing microhabitat preferences we must also add interactions with other gut symbionts.

We conclude this section by noting that although the critical factors that dictate parasite microhabitat preference are unknown in many cases, they are evidently very strong. For example, regardless of whether it is retrieved from a bird or a mammal, the trematode *Zygocotyle lunata* is consistently found in

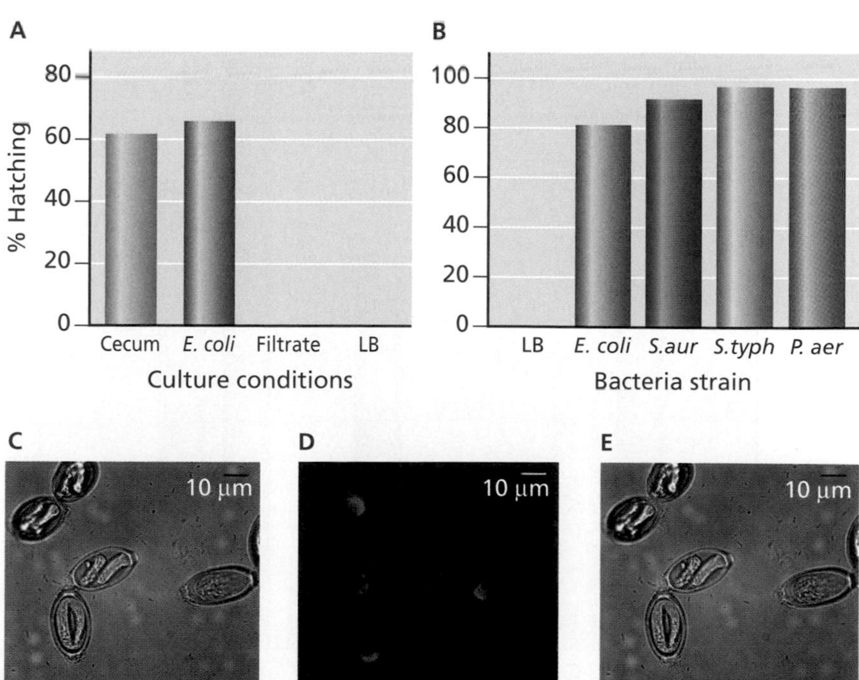

Figure 6.3 Parasites may be influenced by symbionts sharing their environments. (A) Exposure of eggs of *Trichuris muris* to either their host's cecal contents or to cultures of the bacterium *Escherichia coli* promoted hatching, whereas eggs did not hatch when exposed to bacterial filtrates or to bacterial culture medium (LB). (B) Culture of eggs with four different bacteria (*E. coli*, *Staphylococcus aureus*, *Salmonella typhimurium*, or *Pseudomonas aeruginosa*) promoted hatching, whereas exposure to LB did not. (C–E) *T. muris* eggs were cultured with GFP-expressing bacteria (they glow green under fluorescence microscopy). (C) Regular bright-field image. (D) Fluorescence image. (E) Combined image. (From Hayes KS, Bancroft AL, Goldrick M et al [2010] *Science* 5984:1391-1394. With permission from AAAS.)

the cecum. Here, site specificity seems to override host specificity, the next topic on our agenda.

Host specificity is one of parasitism's most distinctive properties

The topic of host specificity was discussed in Chapter 3 and is revisited here in light of its central relevance to the ecology of parasitism. Parasites of a given species do not occur in all available microhabitats within their hosts, nor do they occur in all available host species. That is, most species exhibit host specificity. Host specificity refers to the widely noted pattern that most parasite species are able to develop and succeed in only a limited subset of available host species (**Figure 6.4**). Some parasite species can develop in only a single host species. For example, in nature, the beef tapeworm ***Taenia saginata*** develops as an adult only in the small intestine of humans. Although its host range could possibly be extended by experimental infection of other primates, this species in its adult stage in nature is a parasite specific to humans. As another example, 67% of all louse species are found on a single host species. Such parasites are often referred to as being specialists.

Note, however, that many parasite species use multiple host species. These include most of the parasites infecting humans and domestic livestock (see also Chapter 7 and Figure 7.20). Furthermore, a few species are able to successfully colonize surprisingly broad ranges of hosts and are referred to as generalists. For example, tachyzoites of the apicomplexan ***Toxoplasma gondii***, the causative agent of toxoplasmosis, are able to infect numerous nucleated cell types in nearly all warm-blooded animals, mammal and bird alike. Also, for parasites with complex life cycles, the specificity exhibited by different life cycle stages can be variable. For ***T. gondii***, although its tachyzoites are extreme generalists, the sexual stages of its life cycle occur only in epithelial cells in the small intestine of domestic and wild cats. Whereas the adult worms of ***Schistosoma japonicum*** can be found in at least 40 species of

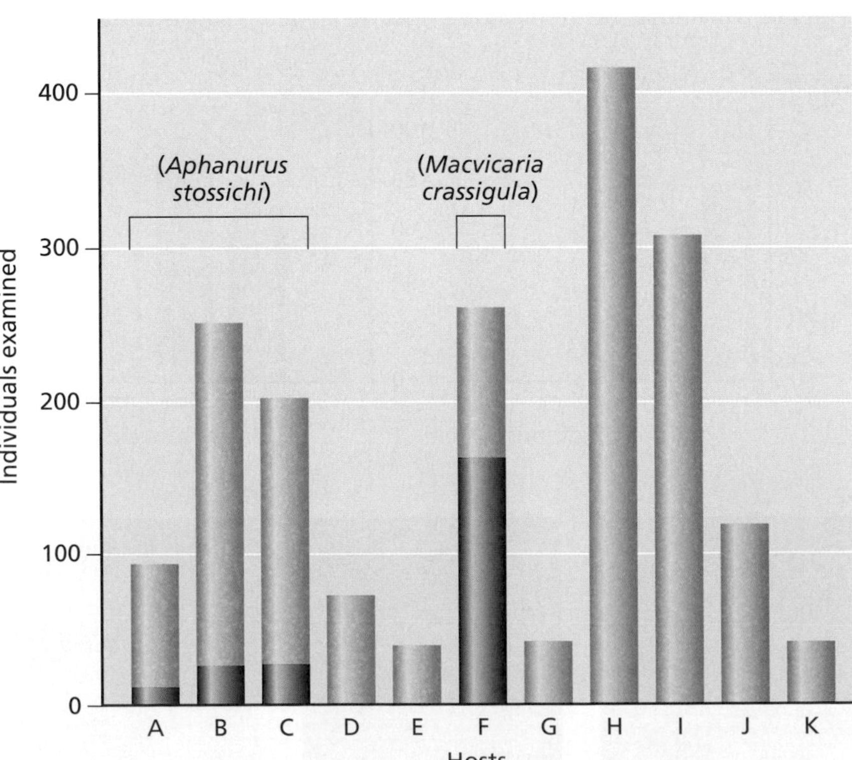

Figure 6.4 Some parasites show greater host specificity than others. An example of parasite host specificity from related fishes inhabiting the same geographic area, the eastern coast of the Mediterranean Sea. Different species of teleost fishes (A–K) are shown on the horizontal axis, and the vertical axis indicates the number of each fish species examined. The trematode *Aphanurus stossichi* is found in three of the 11 host species (see black portions of bars) and is less specific than the trematode *Macvicaria crassigula*, which infects only one of the 11 host species available to it. (Adapted from Combes C [2001] Parasitism. University of Chicago Press.)

mammals ranging from rodents to people, the larval stages occur only in snails of the genus *Oncomelania*.

Keep in mind that studies of host specificity are potentially complicated by several factors. First, the lack of precise identifications of either the hosts or parasites may distort our knowledge of the extent of host specificity. As noted in Chapter 2, both parasites (such as the diplostomoid trematodes noted in Table 6.1) and hosts (such as mosquitoes transmitting malaria) may exist in complexes of cryptic species that we have fully failed to appreciate and identify. With the application of molecular tools, we are getting better at discriminating differences among such species. More often than not, we have learned that parasite–host associations are more specific than previously realized. Furthermore, we may lack sufficient sampling for a particular parasite species such that our list of host species is incomplete. In addition, we must also acknowledge that in some cases a particular parasite species clearly achieves a much greater abundance in numbers or biomass in one host species than another in which it is minimally yet persistently represented. As the relative abundance of host species varies considerably, this too must be taken into account when assessing the importance of various host species for a parasite's persistence. Often experimental infections of particular host species can help reveal the parasite's inherent specificity, but undertaking such infections poses many practical difficulties.

Encounter and compatibility filters determine the range of host species used by a parasite

Two general types of filters have been envisioned to account for the range of host species colonized by a particular parasite species: the **encounter filter** and the **compatibility filter** (**Figure 6.5**). The encounter filter is shut if the parasite and host live in different environments or are active during different seasons and thus never come in contact or if behavioral preferences of one or the other organism prevent contact. The compatibility filter is shut when the parasite is able to initiate infection but ultimately fails, either because the host

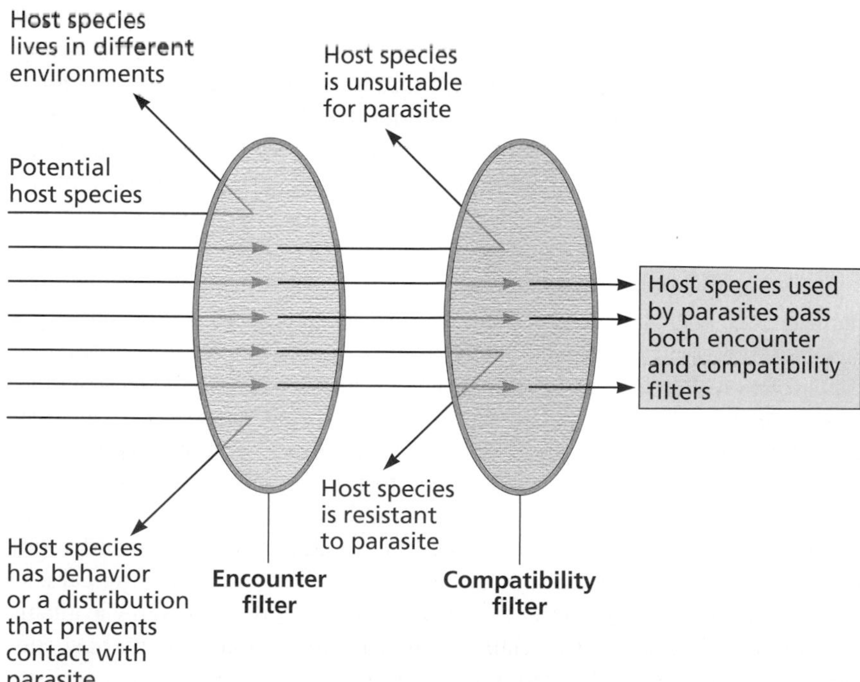

Figure 6.5 Encounter and compatibility filters. Many potential host species for a particular hypothetical parasite species are shown on the left, with each host species represented by an arrow. For some of these host species, the encounter filter prevents them from serving as hosts for the parasite. A compatibility filter also prevents other potential host species from serving as hosts. As shown on the right, fewer host species have passed both types of filters and actually serve as hosts for the parasite.

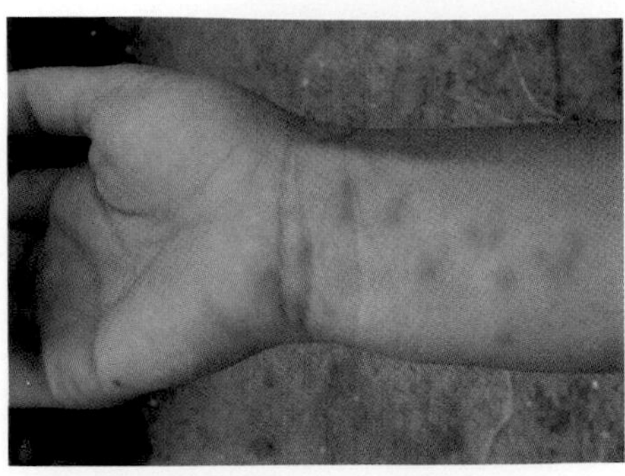

Figure 6.6 An example of a closed compatibility filter. While collecting freshwater snails, the skin of this individual (by the way, one of the world's experts on *Trichobilharzia*) was penetrated by cercariae of a schistosome, *Trichobilharzia* sp., that happened to be in the water. In this case, a hypersensitivity defense mechanism has been activated that has killed the cercariae in the skin. By contrast, in the appropriate avian host, the cercariae would survive and develop into adult worms following skin penetration. (Courtesy of SV Brant, University of New Mexico.)

does not provide essential habitats or resources or because the host mounts an active defense response that prevents the infection from succeeding. Neither of these filters works on an all-or-none principle; the host spectrum for a parasite is defined by those hosts for which both the encounter and compatibility filters are sufficiently open to enable infection, development, and subsequent production of progeny.

One example of application of the filter concept is provided by the monogenean parasite *Diplozoon gracile*, which parasitizes the gills of fishes in streams in the south of France. This species is readily recovered from four species of fish including *Barbus meridionalis*, but never from a fifth species, *Barbus barbus*, even though it is a close relative of *B. meridionalis*. Experimental studies show that *B. barbus* can be successfully infected, suggesting the lack of natural infections does not arise from the compatibility filter. However, the free-swimming oncomiracidia of *D. gracile* remain near the edges of habitats where all the fish occur except for *B. barbus*, which is found in the deeper central waters of larger streams. For *D. gracile*, the encounter filter for *B. barbus* is closed.

An example of a compatibility filter is provided by schistosomes of the genus *Trichobilharzia*, individual species of which normally develop in particular species of avian hosts. Cercariae of *Trichobilharzia* species shed from snails in aquatic habitats can and do penetrate human skin (**Figure 6.6**). However, the cercariae provoke pronounced immune reactions in the skin of people and are typically killed without further migration. Subsequent exposure to cercariae results in an even more rapid and stronger immune response. In this case, the encounter filter is frequently open, but the compatibility filter is closed, preventing successful infection. Another similar example of cutaneous larval migrans was shown in Figure 3.21.

The origins and consequences of host specificity are debated

Several hypotheses have been proposed to account for observed patterns in host specificity noted among different parasite groups. For example, it has been argued that host specificity is to a large extent determined by the phylogenetic history of the parasites, such that closely related parasites should have similar spectra of hosts. In accordance with this idea, sister species of fleas share more species of hosts than expected by chance. Similarly, helminths of the same genus (**congeners**) share more hosts than expected by chance, although this is more the case for trematodes than for other helminth groups. That phylogenetic history does not fully predict the potential host range of a given parasite indicates that multiple factors such as the complexity of the parasite's life cycle, ecological factors, and historical peculiarities are also important in determining patterns of specificity.

It has also been argued that evolutionary interactions between host and parasite lead the parasite to become more specialized and that such specialization is typically irreversible. However, there is evidence contrary to this assertion. For feather lice of doves (see also Chapter 7), ancestral species were in fact shown to be strictly host specific (specialists), whereas generalists repeatedly evolved from their specialist ancestors. Competition with congeneric louse species is believed to explain why some louse species became generalists. By acquiring new hosts, they could minimize the extent of competition, which would be favored by selection. Availability and abundance of host species have also been suggested as factors influencing parasite specificity, the argument being that hosts that are more predictably available or abundant should favor specialists, whereas parasites colonizing host species prone to crashes or limited in population size might become generalists.

It is often reasoned that if a parasite specializes on one host, it will lose its ability to infect other hosts. By developing specific tactics for evading the immune response of one host species, a parasite may, for instance, acquire greater susceptibility to attack by the different immune systems of other host species. That is, there is a trade-off: success in one host may predispose parasites to failure in another host species. Such a possibility may account for the observation that generalist parasites often achieve lower abundance in or on hosts than specialist parasites. For example, comparing 13 ***Plasmodium*** species that differ with respect to the number of vertebrate host species they use, peak **parasitemia** (the number of infected red blood cells per unit volume of blood) in hosts was higher for specialist parasites than generalists (**Figure 6.7**). However, others who have studied species of ***Plasmodium*** and *Haemoproteus* inhabiting a number of bird species have come to a different conclusion. They found that parasite species able to be transmitted across a broad host range were also the most prevalent parasites within their compatible hosts. They argued that parasites with a broad host range have higher overall encounter rates, thus potentially compensating for reduced performance in each host species. In this latter case, the jack-of-all-trades is also still a master in at least some of its host species. It was also argued that because birds, particularly migratory species, are exposed to such a wide range of avian blood parasites, they may have developed a generalist immune system to protect them from diverse rather than specific parasites. To sum up, it is fair to say that parasitologists lack an overarching theory to fully explain patterns of host specificity across many parasite groups.

Underlying mechanisms dictating specificity are also often not known

One of the most conspicuous shortcomings in our understanding of host specificity and of parasite–host interactions in general is our lack of a precise understanding of the molecular basis of specificity. Studies of viruses and bacteria are beginning to identify particular molecules that influence host species range. For example, the ability of the mutualistic bacterium *Vibrio fischeri* to colonize the light organ of the squid *Euprymna scolopes* has been

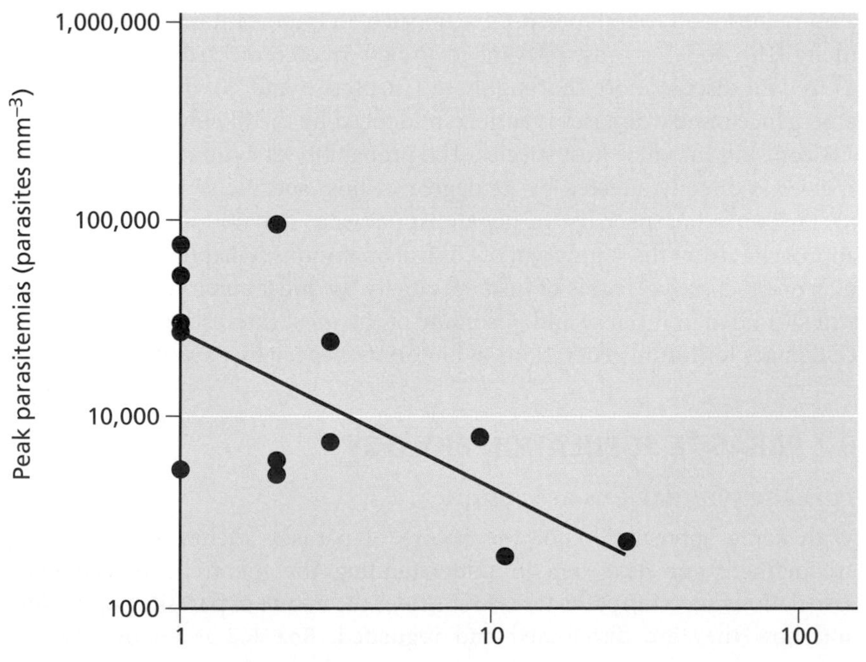

Figure 6.7 Generalist parasites often have lower abundance in their hosts than specialist parasites. Shown for 13 different *Plasmodium* species are the degree of specialization (indicated on the horizontal axis by the number of host species used) and peak parasitemia recorded on the vertical axis. More specialized species produced higher parasitemias ($r_{phyl} = -0.71$, $n = 13$, $P < 0.01$). (Adapted from Garamszegi LZ [2006] *Ecol Lett* 9:933–940. With permission from John Wiley and Sons.)

pinpointed to the presence of a regulatory gene (*kinase RcsS*). The product of this gene activates synthesis of a polysaccharide that mediates biofilm formation and favors colonization of the light organ. If the *RcsS* gene is expressed in *V. fischeri* strains that normally infect only fish, the bacteria acquire the ability to colonize *E. scolopes*.

Parasitologists have made considerable progress in identifying molecules likely to govern host species range in medically important parasites, such as for species of ***Plasmodium***. For example, sporozoites of malaria parasites are covered with the circumsporozoite protein (CSP) and this is known to play an important docking role with host hepatocytes (see also Chapter 9 for its role in malaria vaccine development). Furthermore, CSP is known to vary among ***Plasmodium*** species. If recombinant forms of the ***P. falciparum*** version of CSP are present, they can block infection of human liver cells with the sporozoites of this species. By contrast, recombinant forms of the CSP from nonhuman malaria species have no such blocking effect. Insofar as infection of hepatocytes initiates infection in the vertebrate host, CSP would seem to play a role in determining the host species ranges for ***Plasmodium*** species. Host specificity of malaria parasites is also influenced by interactions between merozoites and the glycoproteins found on host erythrocytes. Whereas ***P. falciparum*** readily penetrates human erythrocytes, it is relatively unsuccessful in colonizing chimpanzee erythrocytes because the latter have different glycoproteins on their surfaces.

In the face of such specificity, how are the tachyzoites of ***T. gondii*** able to infect such a broad spectrum of host cells? Does the parasite use a ubiquitously expressed host cell receptor, does it insert its own receptor onto the host cell, or does it have multiple mechanisms of cell entry? Although we do not yet have a definitive answer, molecular parasitologists have learned a great deal about host-cell penetration in this species. The parasite engages in an active, multi-step penetration process, with each step increasing the parasite's grip on the cell. Moreover, we are beginning to zero in on particular genes, such as the rhoptry kinase gene *ROP18*, the products of which seem to be capable of broadly inactivating intracellular host defenses. We return to this topic in Chapter 10.

As has been often noted in this book, the details of the biology of particular parasites and their hosts must be carefully considered before we can gain a more informed overall perspective on such an important topic as host specificity. This topic is most relevant in today's interconnected world because, as we will discuss more thoroughly in Chapters 8 and 10, the probability of emergence of new diseases is largely influenced by the likelihood that a parasite can shift into new host species. The probability of extinction of a parasite species is directly affected by its degree of host specificity (Chapters 7 and 8). Furthermore, specialist vs. generalist parasites may require very different approaches from the standpoint of control operations (Chapter 9). The molecular or biochemical basis of host specificity for most eukaryotic parasites is still shrouded in mystery and poses one of the most interesting and daunting challenges for future generations of biologists, especially parasitologists.

6.2 PARASITE POPULATION BIOLOGY

Parasite populations are complex

With some appreciation for the nature of parasite niches and how they are defined, our next step in understanding the hierarchy of ecological associations affecting parasites is to consider the nature of parasite populations and how they are distributed and regulated. **Box 6.2** is an overview of

BOX 6.2
Some Population Biology Basics for Parasites

The relevant parameters for any population include the current size of the population, *N*, and the instantaneous rate of growth for the species involved. This latter parameter, *r* (also called "little *r*"), is influenced both by the birth rate and the death rate for the species involved. Immigration and emigration of individuals also contribute to the size of the population at any given point in time. Although a detailed account of the mathematics of population growth is beyond the scope of this book, suffice it to say for now that population sizes are often very dynamic and subject to change. Populations are influenced by factors that act in either a **density-independent** or a **density-dependent** fashion. In the former case, the action of the factor involved, such as environmental temperature, is not influenced by the density of the population. In the latter case, the strength of action of a particular density-dependent factor, such as food supply, is often increased as the density of the population increases. Consequently, density-dependent factors have the potential to regulate population size. For example, with respect to food supply, as the population increases, per capita food supply diminishes. Starvation of some individuals becomes more likely, and population size may therefore decline. With a smaller population consuming the food source, the per capita food supply would be expected to increase, thus favoring an increase in population size. In this way, the population might be regulated around a particular mean size. This mean size is referred to as the **carrying capacity** of the environment for that population.

For parasites, one prominent aspect of their population biology is how their numbers are distributed among different individual hosts (Figure 1). A group of parasites of a single species occupying a single host individual is called an **infrapopulation**. A **component population** sums all the parasites of a given species from all the infrapopulations found within a particular host species from a particular ecosystem. There is yet one more level. A **suprapopulation** refers to all the individuals of a given parasite species summed across all the component populations from all different host species in an ecosystem. For the sake of completeness, the free-living stages of that particular parasite species are also considered to be part of the suprapopulation. These definitions embrace only a particular ecosystem, and many parasite species can be found in multiple ecosystems.

Understandably, quantifying and following suprapopulations is a daunting task, so most studies of parasite populations focus on infrapopulations or component populations. Also, quantification of individuals in populations of macroparasites, such as helminths and arthropods, is frequently undertaken because they are big enough to be readily seen and often exist in small enough numbers to feasibly count. By contrast, it is by no means routine to enumerate the population size of specific microparasites, such as protozoans. Consequently, studies of the population biology of microparasites focus on enumeration of infected hosts rather than the number of parasites in each host. In general, most studies of parasite populations and of parasite ecology currently focus on macroparasites.

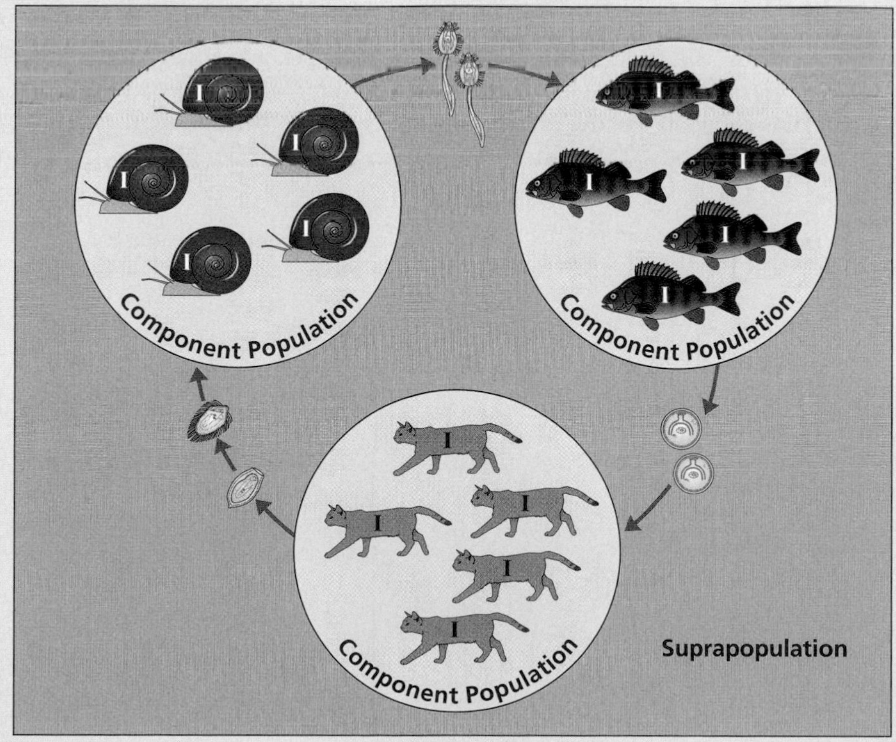

Figure 1 The hierarchical organization of parasite populations into infrapopulations, component populations, and the suprapopulation. This hypothetical parasite has a three-host life cycle involving snails, fish, and cats. For snails, each parasite infrapopulation (the population of our hypothetical parasite occurring within an infected snail) is indicated by an I. The component population sums all the infrapopulations from infected snail and is delimited by a circle. Both infrapopulations and component populations are also shown for fish and cats. Also shown are the free-living stages that connect the hosts. The arrows indicate transmission from one type of host to another. The large box circumscribing the entire figure represents the suprapopulation.

population biology, particularly as it applies to parasites. Because parasites exist in subdivided groups within hosts, the box presents some definitions particular to them.

With the structure of parasite populations noted in Box 6.2 in mind, consider how the infrapopulation size might vary among hosts for a particular parasite species. One possibility is parasites are distributed at random among different host individuals (**Figure 6.8A**). In this case, the mean infrapopulation size is equal to the variance in infrapopulation size among hosts. Another possibility is that parasites are distributed uniformly among their hosts, in which case the mean population size is greater than the variance (Figure 6.8B). The third possibility is that parasites show aggregated distributions among their individual hosts (Figure 6.8C), with some hosts having more and some having fewer parasites than expected at random. In this case, the ratio of the variance to the mean is greater than 1.

Parasites often show aggregated (overdispersed) distributions in their hosts

With respect to metazoan parasite infrapopulations, one of the striking patterns to emerge from years of study is that they are almost always aggregated in their distribution. Such an aggregated distribution has in the parasitology literature often been called an overdispersed distribution. An aggregated distribution, as shown in the graph in Figure 6.8C, is frequently described by the negative binomial probability distribution. This distribution is described by two parameters, the mean, μ, and k, a measure of aggregation. When k is low (usually <1) the parasite distribution is highly aggregated, and when k is large (>8), the parasites are essentially randomly distributed among their hosts and their distribution looks more like the graph shown in Figure 6.8A.

So pervasive is this trend toward aggregated infrapopulations that some parasitologists have concluded that occurrence in an aggregated distribution should be a part of the very definition of parasitism. Others have noted it assumes the status of a "law" of parasite ecology, although exceptions do occur. Not only are parasite distributions aggregated, but the degree of aggregation is also fairly predictable. For example, as shown in **Figure 6.9**, the points are scattered closely about the fitted regression line, and the mean explains about 90% of the variability in the variance. That is, aggregation varies within bounds. This pattern makes sense biologically because if certain

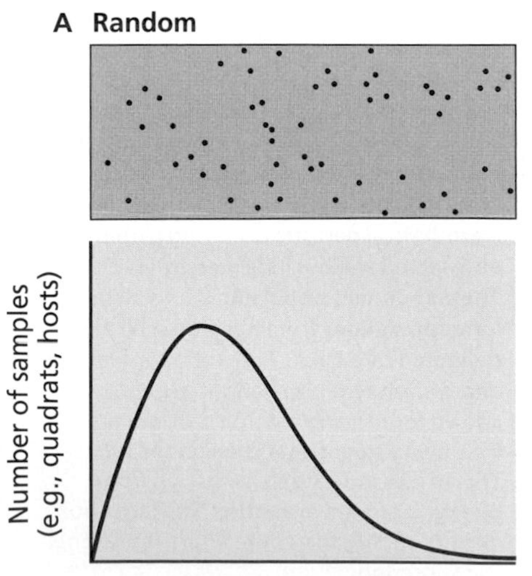

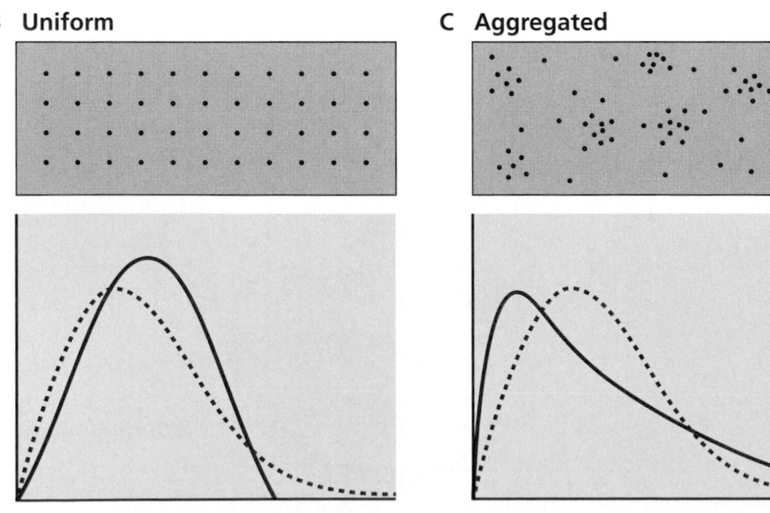

Figure 6.8 Three different patterns of distribution in space for populations. Populations may be (A) random, (B) uniform, or (C) aggregated in distribution. The graph below each pattern represents the frequency curve for the distribution (the dashed line shown in the uniform and aggregated distributions is just the random frequency distribution for comparison). (Adapted from Bush AO, Fernandez JC, Esch GW and Seed JR [2001] Parasitism: The Diversity and Ecology of Animal Parasites. Cambridge University Press. With permission from Cambridge University Press.)

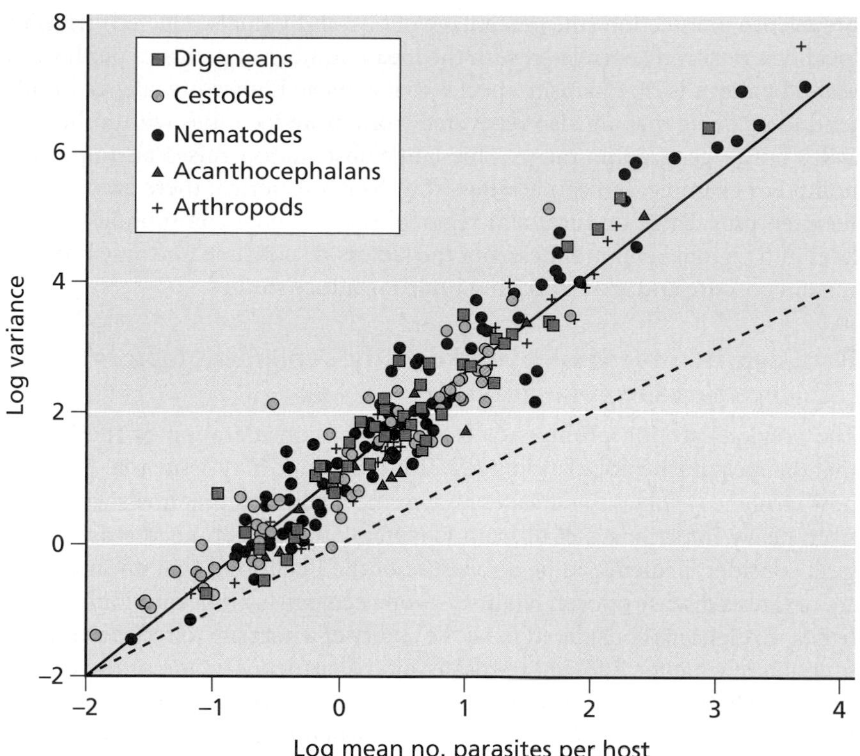

Figure 6.9 A law of parasite ecology? This graph shows the relationship ($r^2 = 0.87$) between the variance and the mean across 269 natural populations of metazoan parasites (digeneans, cestodes, nematodes, acanthocephalans, and arthropods). Notice that the slope of the relationship is greater than one, indicative of aggregated distributions. The dashed line with a slope of one, which is expected of a random distribution, is shown for comparison. (Adapted from Poulin R [2007] *Parasitology* 134:763–776. With permission from Cambridge University Press.)

individual hosts harbored too many parasites, they might succumb to parasite-induced mortality. If the number of parasites per host was too low, then the parasites might fail to find mates with whom they can reproduce.

Both parasite- and host-related factors can play a role in generating aggregated distributions. Some parasites produce progeny that remain grouped together within a particular host, such that if that host is ingested, the next host would in one event acquire a large number of parasites. Imagine a dog consuming viscera of a sheep containing a hydatid cyst of the cestode ***Echinococcus granulosus***. Hydatid cysts are distinctive for producing many protoscolices by asexual reproduction. In this way, that dog would come to harbor an enormous number of ***E. granulosus*** adults, whereas another dog that did not happen to ingest parts of the sheep contaminated with a hydatid cyst would have none. Owing to its direct life cycle and hermaphroditic mode of reproduction, a single monogenean colonizing a fish may soon build a massive infestation on that same fish.

Aggregated parasite distributions are more frequently accounted for by host-related factors. Genetic differences in innate resistance, differences in immune system competence, or differences in behavior (recall the compatibility and encounter filters, respectively, described above) may dictate that some individuals acquire more parasites than others. Also, immune competence can vary with the season. During the mating season, the high androgen levels typical of males can impair immune function (see the handicap hypothesis in Chapter 7). Potential male hosts may consequently become more susceptible, acquiring greater parasite burdens. The mobility of hosts and the extent to which they engage in grooming could also explain variability among hosts in parasite burdens. Likewise social status can influence parasite burden, in some cases high status resulting in lower parasite burdens, and in other cases, higher burdens of parasitism.

Other patterns with respect to parasite populations are also commonly reported. One is that when comparing different parasite species, the

prevalence of infection (the percentage of host individuals infected) within a locality is positively correlated with the mean number of parasites per host. A second pattern is that parasite species that achieve higher prevalence locally tend to be those that are also recovered from more localities within the parasite's entire geographic range. This latter observation raises an important point. For example, in fish parasites from North America, there are linkages between prevalence on local and regional scales. Local commonness translates into regional commonness, but the factors responsible for these linkages remain obscure and are an exciting topic for future studies.

Both density-independent and density-dependent factors influence parasite population size

The previous section prompts us to a broader consideration of the factors that dictate parasite population sizes. This is an important issue that bears on how virulence evolves in parasites and on the likelihood that hosts will suffer from heavy infections. Recall from Chapter 1 that infectious disease biologists consider virulence to be a measure of the likelihood that an infectious agent causes disease or even fatality. Among ecologists and evolutionary biologists, virulence is considered to be the ability of a parasite to reduce its host's fitness (see Chapter 7 for more discussion of virulence). One of the central issues is whether parasite populations are controlled by density-independent factors, such as temperature or rainfall, or whether density-dependent factors, such as competition with conspecifics or with other parasite species, play a prominent role. In the former case, it is likely that parasite population size will fluctuate considerably because density-independent factors are often unpredictable. In the latter case, with the operation of density-dependent mechanisms, it is likely that parasite populations fluctuate much less extensively. In this instance, the populations are regulated, implying they are stable or vary in size in a predictable way. In our discussion, we will refer to regulation of infrapopulation size; the factors regulating suprapopulation size are much harder to ascertain.

Given that parasite infrapopulations are aggregated, we would expect density-dependent factors to be triggered earliest and to act most strongly in larger infrapopulations. Aggregation in general is considered to be a factor that would dictate that density-dependent mechanisms come into play. What might such mechanisms be?

There are three categories of density-dependent regulation: **decision-dependent regulation**; **host-death dependent regulation**; and **competition-dependent regulation**. In the first case, the parasite avoids a particular host if the host is already infected. Such behavior is well known among parasitoid wasps. An ovipositing female wasp often has the ability to sense if parasitoid larvae have already infected a particular host. If so, she will move on and find another parasite-free host in which to oviposit. In some cases, trematode miracidia prefer to penetrate uninfected snails. This behavior makes sense as snails harboring a trematode infection are typically already packed with sporocysts or rediae and therefore provide a nutrient-depleted and potentially hazardous environment for a newly arrived miracidium.

A second means of density-dependent regulation of parasite infrapopulations is for the most heavily infected hosts to succumb to their infections and die; this type of regulation is known as host death-dependent regulation. In one experiment, fruit flies (*Drosophila putrida* and *D. neotestacea*) that had been infected in the laboratory with the nematode *Howardula aoronymphium* were released into the field. It was noted that the proportion of released flies retrieved that were infected steadily declined with time (**Figure 6.10A**). This

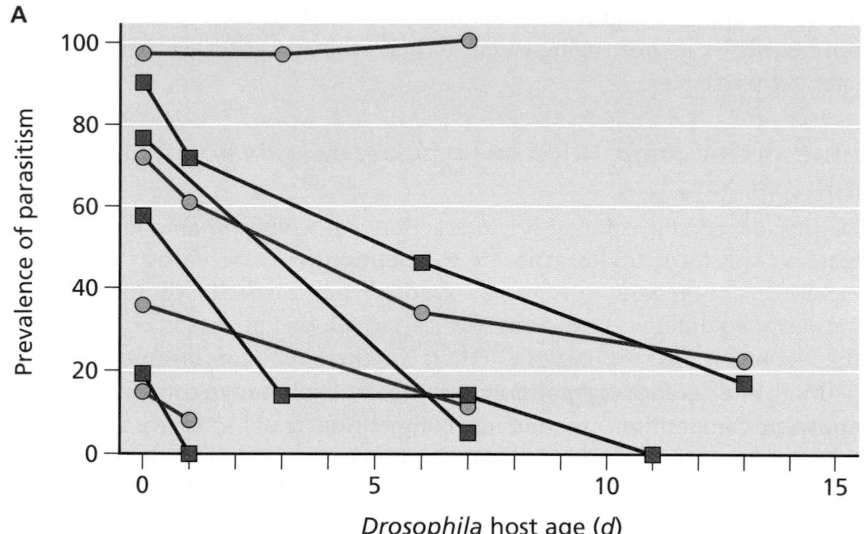

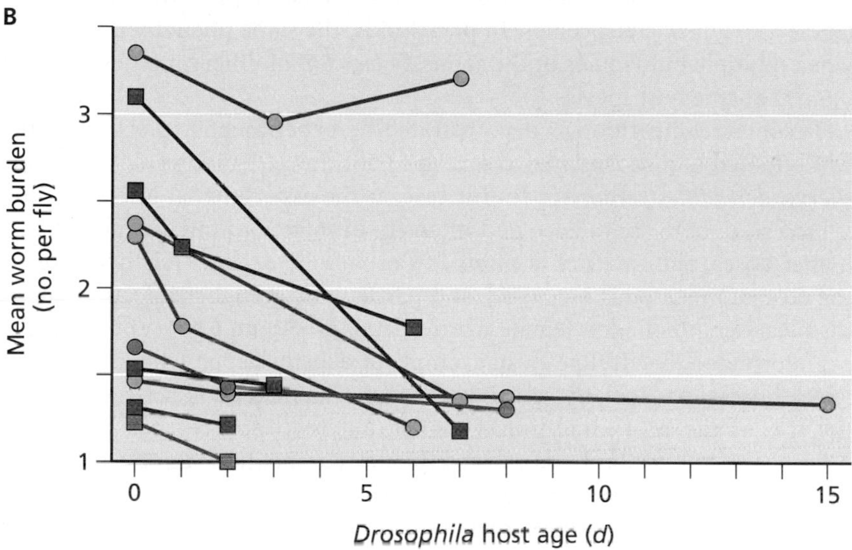

Figure 6.10 Parasites can impose death-dependent regulation on their hosts. (A) Declining prevalence of infection (percentage of flies captured that were infected) with time of *Drosophila putrida* or *D. neotestacea* that had been infected in the laboratory with the nematode *Howardula aoronymphium* and then released into the field. Each line represents a different trial. (B) The mean worm burden of the flies also generally declined with time, suggesting that the individuals infected most heavily were most affected by the parasites. (From Jaenike J, Benway H & Stevens G [1995] *Ecology* 76:383–391. With permission from Ecological Society of America.)

pattern is indicative of parasite-associated mortality, a possibility supported by other experiments indicating the flies had not simply lost their infections. Furthermore, the negative effect of parasites on host survival increased with worm burden (Figure 6.10B), prompting the conclusion that parasite-induced host mortality can be a mechanism for density-dependent regulation of these parasite populations. At higher levels of infection of the water flea *Daphnia magna* with the fungus *Metschnikowiella biscuspidata*, the host was killed without producing transmission stages. The researchers involved concluded that in this system, parasite population growth is not only regulated by within-host, density-dependent processes but also by the increased mortality of intensely infected hosts.

It makes intuitive sense that the most heavily infected hosts would be more likely to die. Red grouse in Scotland suffer higher mortality as their dose of infection with the cecal nematode *Trichostrongylus tenuis* increases. The higher the parasite burden, the more likely they are to be found by predators, because heavy infection causes the birds to defecate more frequently. This creates a scent more detectable by predators. Furthermore, being weakened by parasites, the most heavily infected birds have to forage more for food and become more conspicuous to predators. Note that the cecal nematode has a direct life cycle with grouse as the definitive host. Therefore *T. tenuis* does not benefit by its host being eaten, unlike other parasites we will consider later in this chapter.

Intraspecific competition can regulate parasite populations in different ways

Density-dependent competitive interactions are common among parasites. Here we will focus on **intraspecific competition** (as between two individual tapeworms of the same species), as opposed to **interspecific competition** (as between two different tapeworm species), which will be discussed further in the section on infracommunities. There are three types of intraspecific competition: **interference competition**; **resource** or **exploitative competition**; and **apparent competition**. Interference competition refers to active aggression between interacting individuals. One example is provided by the trematode *Echinostoma paraensei*, which in the snail *Biomphalaria glabrata* produces a mother redia with a large sucker able to attack and damage or kill later-arriving parasites of the same species but of different genotypes. Other trematode larvae also attack conspecifics. In parasitoids, the same phenomenon occurs when multiple individuals of the same species but of different parentage end up in the same host species.

Resource competition is demonstrated by experimental infection of the white-footed mouse *Peromyscus leucopis* with the gastrointestinal nematode *Pterygodermatites peromysci*. In this case, as the experimental infection dose is increased, both prevalence and intensity of infection initially increase. At higher doses, both metrics level off and eventually decline. Furthermore, as the dose of infection is increased, and particularly at higher infection doses, the mean length of adult female worms declines (**Figure 6.11**). The researchers interpreted this decline as an example of density-dependent competition for resources, particularly space, because this nematode is large relative to the size of the host's intestine. Competition for resources has frequently been suggested as the basis for interclonal competition among strains of ***Plasmodium*** within the same vertebrate host. Interclonal competition is predicted to favor clones of higher virulence (virulence in this specific example referring to faster reproductive rates) in most circumstances (see also the "short-sighted" hypothesis in Chapter 7).

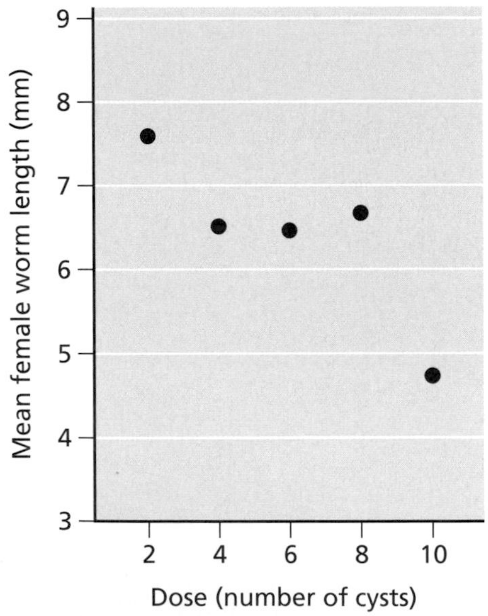

Other helminths, such as tapeworms, ascarids, and schistosomes, also show a negative relationship between the number of worms present in an infrapopulation and their size or weight. This is known as the crowding effect. Competition for nutrients, the involvement of uncharacterized interfering crowding factors, and the induction of a heightened immune response have all been suggested as explanations of the effect.

In a study of competition between two different clones of ***Trypanosoma brucei*** in mice (**Figure 6.12A**), exposure to a virulent clone (with red fluorescent protein marker) resulted in the shortest periods of mouse survival, whereas exposure to an avirulent clone (with green marker) resulted in much longer periods of survival. Exposure to both strains simultaneously significantly increased mouse survival times by 15%, suggesting the avirulent clone was competing with the virulent clone. When either clone was grown in mice alone (Figure 6.12B), the density achieved by the virulent clone was much higher than for the avirulent clone. When they were grown together though, each exerted a significant negative effect on the other, such that the total population size of trypanosomes was less than in the mice given the virulent strain alone. The basis of the competition was not determined but could have been mediated by production of harmful compounds by one

Figure 6.11 Competition for resources occurs among parasites. The relationship between inoculation dose (number of *Pterygodermatites peromysci* cysts) and mean length (mm) of adult *P. peromysci* females recovered. Worms were significantly smaller at the highest exposure rate compared with the lower dose levels. (Adapted from Luong LT, Vigliotti BA, Hudson PJ [2011] *Int J Parasitol* 41:505–511. With permission from Elsevier.)

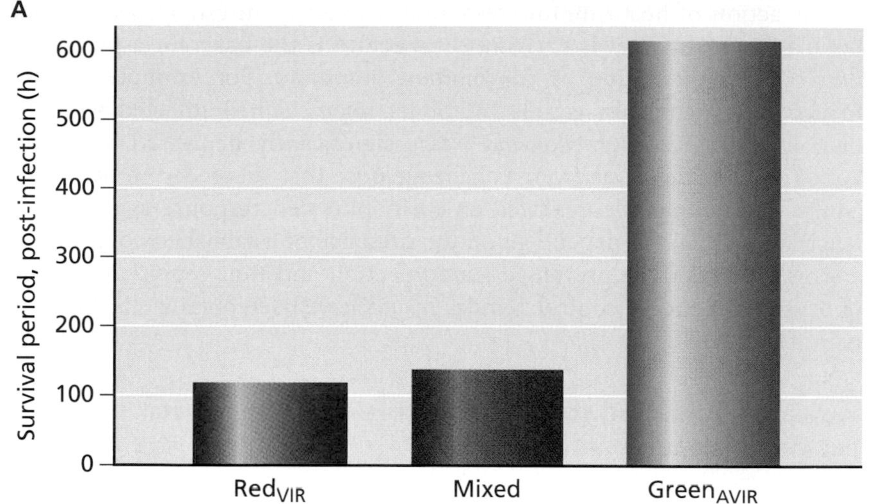

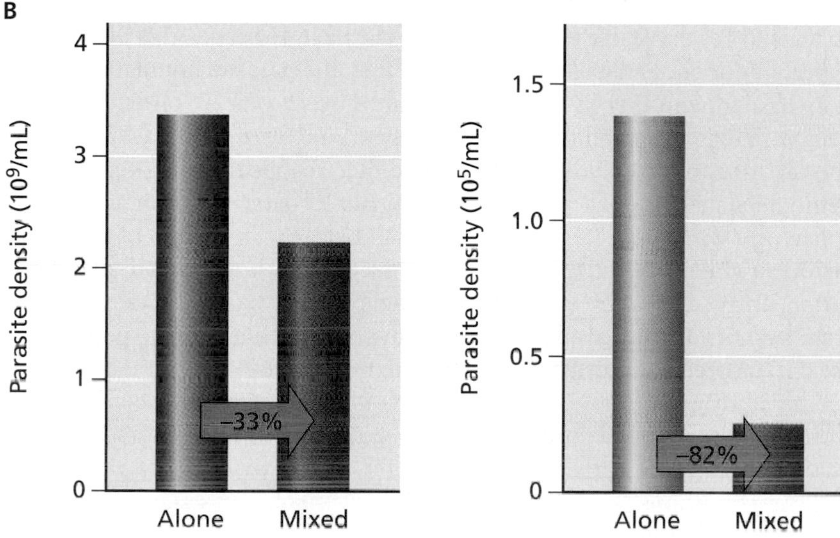

Figure 6.12 Intraspecific competition between two clones (each with a distinctive fluorescent label) of *Trypanosoma brucei* in mice. (A) The vertical axis represents the survival period in hours of mice exposed to infection. Mice exposed to the virulent (red) clone had short survival times compared to mice exposed to the avirulent (green) clone. Although the difference in mean survival times was not great (about 15%), mice exposed to both strains simultaneously had significantly increased survival times compared to mice exposed just to the virulent strain ($p = 0.015$). (B) The parasite density achieved by the virulent clone in mice was reduced by 33% when in a mixed infection with the avirulent clone. The density for the avirulent strain was reduced by 82% when grown in a mixed infection with the virulent clone. Note the different scales on the vertical axis for these two graphs. (From Balmer O, Stearns SC, Schotzau A et al [2009] *Ecology* 90:3367–3378. With permission from Ecological Society of America.)

clone (interference competition) or by apparent competition mediated by the host's immune system.

Apparent competition refers to an indirect negative interaction mediated by a third party. For example, in the face of drug pressure, selection might favor a slower growing clone able to resist the drug over an otherwise faster growing clone sensitive to the drug. Often the third party is the host's immune response. The basic idea is that as the density of parasites (and their antigens) within a host increases, the likelihood that the host's immune response will be provoked is increased and the response engendered then limits the parasite's density. Again consider the *Peromyscus–Pterygodermatites* system described above. Mice that had been exposed to infection were treated to remove their initial infections and then reinfected. The heavier the initial infection the mice had had, the lower the prevalence and intensity of infection the mice developed upon secondary exposure. The decline was correlated with an increase in abundance of anti-parasite IgG antibody found in the blood of the mice. The acquired immunity acted as a density-dependent constraint. That is, in the case of reinfections, what might have been viewed as resource-based competition was actually apparent competition from the host's immune response.

The action of host immune responses caused by an established parasite population against newly arriving conspecifics is the basis for a phenomenon called **premunition** or **concomitant immunity**. For example, relative to uninfected controls, gerbils (*Meriones unguiculatus*) infected with the filarial worm *Acanthocheilonema viteae* significantly depressed the number of conspecific larvae from a challenge dose that succeeded in becoming adults. Infected gerbils mounted an eosinophil-rich response against challenge larvae that was dependent on the presence of established adult female worms. The response prevented superinfection and thus represents a form of density-dependent control. Similar results have been obtained for malaria parasites (see Chapter 4).

Parasite population studies often require a long-term perspective and detailed sampling

Intraspecific competitive interactions clearly occur among parasites, but whether they regulate natural parasite populations remains uncertain. Populations of larvae of the tapeworm *Ligula intestinalis* in fish point out the uncertainties. If enumerated only at a certain time of the year, the population appears to be very stable from one year to the next. However, this illusion disappears upon more detailed sampling. In fact, the parasite population grows, is extirpated, and then is reintroduced anew every year by birds. Instead of being a paragon of stability, *Ligula* might be characterized instead as a "tramp species", one that regularly disappears and then reappears. Long-term studies with adequate sampling allow parasitologists to determine, for instance, if apparently stable parasite populations result from the operation of regulatory factors or if density-independent abiotic factors, such as rainfall, pH, desiccation, and temperature, just happen to be extremely stable. Until more data from such studies are available, the null hypothesis that parasite populations are unregulated and unstable may be difficult to reject.

The study of parasite infrapopulations is challenging and benefits both from laboratory-based studies that expose mechanisms of interactions and field studies, particularly long-term studies, that provide more variable yet realistic conditions than the laboratory experiments. At stake is the opportunity to gain a deeper understanding of what controls parasite populations, including species important in health or conservation. In the next section, we will discuss how and when parasite populations may be constrained through interactions with parasites of different species.

6.3 PARASITE COMMUNITIES

Parasite community ecology is devoted to understanding patterns and associated underlying processes that determine how many parasite species occur within a particular host individual or population. Why do some hosts have more kinds of parasites than others? Also, when these different parasites occur together, do they compete (or in some other way interact with one another) or are they effectively isolationist and noninteractive? Are there patterns that can be seen time and again regardless of the host species that allow us to identify rules for how parasite communities are assembled or are parasite communities assembled randomly? Do communities of parasites change with time and place or do they tend to be stable and predictable? The study of parasite community ecology is one valuable approach to help us to gain a better overall grasp of biodiversity and how it is partitioned, and it enables us to better appreciate the interconnectedness of species in nature. As with populations, communities of parasites require additional definitions, mostly because multiple parasite species are found both within particular host individuals and among the individuals in a host population (**Box 6.3**).

One practical reason to study parasite communities is to appreciate the implications of targeting one member of the community in a control program. If a species targeted by the control program competes with other parasite species, removing this competitor might favor the remaining species and then cause new problems. Conversely, you might consider deliberately infecting livestock with a benign parasite if you knew it to be a potent competitor against another species that was virulent for the host (see the example of the trypanosomes mentioned above). If the presence of one parasite species somehow facilitated the existence of another harmful species that was difficult to control, then a better approach might be to target the facilitating species in a control program. Community ecology can also help us understand the outcome of invasions of new parasite species. An invasive gut parasite might be more likely to succeed if it colonizes a host species that lacks a complex and highly interactive community of native gut parasites (see Chapter 8 for more discussion of invasive species).

Whereas almost all parasites exhibit a predictable aggregated pattern of abundance among infrapopulations, there is little about parasite community ecology —at least at present—that is predictable. This lack of predictability by no means implies that the study of parasite communities has been fruitless or that deep underlying principles will not someday come to light. Rather, the complexity arises because parasite community ecology considers multiple parasite species within a host species. Each parasite species may vary in abundance both spatially and temporally, all against the backdrop of a physical environment that exhibits both predictable and unpredictable variations in temperature, rainfall, and many other factors. When viewed in this light, it is perhaps less surprising that the characteristics of parasite communities

BOX 6.3

Getting a Handle on Parasite Communities

An **infracommunity** consists of all the parasite infrapopulations found in a particular host individual. The life span of the particular host supporting the community limits the longevity of any infracommunity. Being relatively short-lived, infracommunities are assembled by processes occurring on ecological time scales. A **component community** consists of all the parasite infrapopulations associated with a particular host population, often restricted to a particular area. A component community is consequently a much longer-lived entity, lasting for the life span of the host population it inhabits. As such, processes occurring on longer evolutionary time scales dictate its composition.

Another term frequently used in community ecology studies of parasites is the **guild**, which refers to all the different parasite species within a particular host species that have a similar niche and that acquire their resources in a similar way. A parasite ecologist might, for example, refer to the guild of macroparasites found in the gut of a particular fish species, with cestodes, nematodes, trematodes, and acanthocephalans as members. One could more precisely define the guild in this host to include just those parasite species with a particular feeding style: for example, the guild of macroparasites that absorb their nutrition (including the cestodes and acanthocephalans). Members of a guild need not be close phylogenetic relatives, and parasite guilds can be defined in different ways even for a particular host species. Members of a guild might compete or otherwise interact strongly with one another by virtue of their similar needs.

Community ecology often refers to **core** and **satellite species**. The former would be regionally common and locally abundant species that consistently occur together in a particular host species, whereas the latter refers to species that are regionally uncommon and locally rare, so may play less of a role in the structuring of parasite communities. Finally, the terms interactive and isolationist are sometimes used to describe parasite communities. An **interactive community** is one in which the different species influence one another's position or abundance (say, by competition) in the host, whereas an **isolationist community** is one in which the species present do not directly or indirectly influence one another's position, presence, or abundance.

seem to depend strongly on local circumstances or on the particular year; they usually do not show predictable large-scale patterns that repeat exactly year after year. For this reason, most parasite communities can be described as being contingent upon locality or year or upon host species.

The richness of parasite communities varies among host species for reasons that are still debated

A fundamental aspect of the infracommunity of parasites is how many parasite species it contains (its species richness). Species richness, in turn, is a function of the number of parasite species found in the species to which the host belongs (namely, the host species' parasite fauna). The richness of the parasite fauna for any host species is determined by such factors as the host's phylogenetic history and its ecology. For example, as we mentioned in Chapter 2, some parasite species are heirlooms because they are acquired from one's ancestors. This list can be further augmented by parasites acquired from other hosts as souvenirs, as by host switching (see also Chapter 7). Ecological overlaps might even lead to transfers of parasites among unrelated hosts. Attributes of the host such as body size, longevity, mobility, variety and amount of food consumed, immune competence, size of its geographic range, population density or size, aquatic vs. terrestrial habits, and latitudinal distribution are all likely to influence parasite species richness.

Widely accepted generalizations have yet to emerge regarding the determinants of parasite species richness. For example, one study found that among fish, reptiles, birds, and mammals, aquatic birds have the richest and fish the poorest parasite communities. This pattern was related to birds' high mobility. Birds are thus likely to encounter a broader range of parasites than fish (see also Figure 7.4). Also the breadth of diet and complexity of the host's physiology and gut morphology (greater for homeotherms) have been considered to be important factors in determining parasite species richness. The role of ecological factors such as use of aquatic habitats (**Figure 6.13**) was also considered to have a greater role than the phylogenetic background

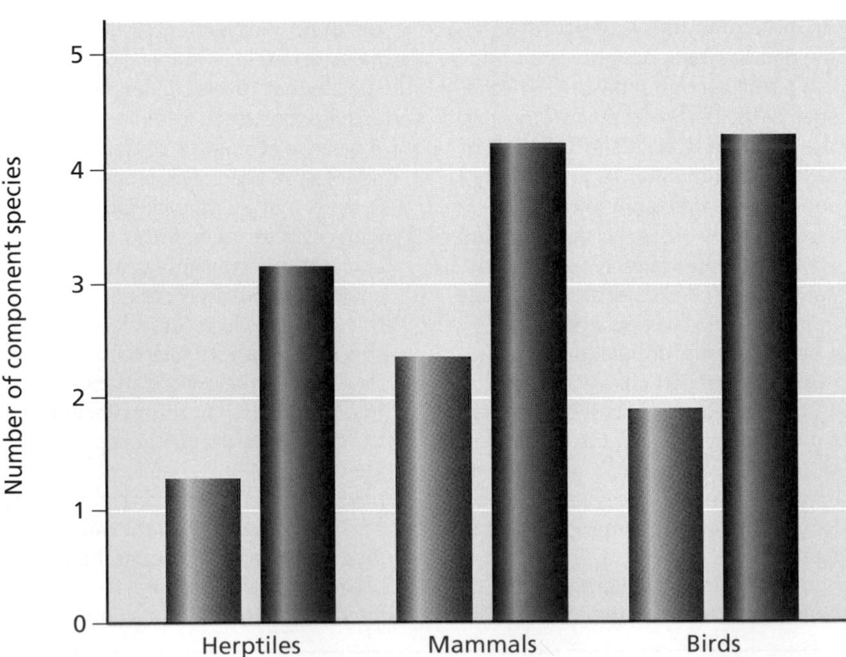

Figure 6.13 Aquatic hosts have more parasite species than terrestrial hosts. The number of parasite species for terrestrial (orange bars) or aquatic (purple bars) representatives of herptiles (amphibians and reptiles), mammals, or birds. (Modified from Bush AO, Aho JM, Kennedy CR [1990] *Evol Ecol* 4:1–20. With permission from Springer Science and Business Media.)

with respect to determining the parasite species richness of a particular host species. Some have argued that these patterns in parasite species are instead strongly influenced by the phylogenetic history of the hosts, and if phylogenetic relatedness is controlled for, then the differences in species richness among endothermic and ectothermic vertebrates or among aquatic and terrestrial birds are not supported. Investigation of the factors that dictate the richness of a host species' parasite fauna is a lively field of inquiry.

What about species richness of infracommunities (the number of parasites of all species found in a single host individual)? Are all the species found in that host species' parasite fauna, or even in the local component community, found in a single infracommunity? Or is there an upper limit to how many parasite species can be harbored in one host individual? Not surprisingly, it is probably always the case that the parasite species richness in an infracommunity is less than in that host species' parasite fauna or in the corresponding component community. In some cases, as in gastrointestinal parasites of birds and mammals, there is a linear increase in infracommunity species richness as the species richness of the component community increases. In other cases, as in gastrointestinal helminths in eels, there is a pronounced leveling off in infracommunity species richness with increasing component community richness. This latter trend suggests that in at least some cases infracommunities become saturated with parasite species.

Most studies suggest parasite communities are stochastic in nature

Another aspect of interest for parasite communities is whether there are predictable assemblages of species within infracommunities or if they are random or stochastic in their composition. In general, most, but not all, studies conclude with the observation that parasite communities are non-equilibrial in nature (not saturated with species), that local contingencies predominate, and that there is little repeatability in species richness and composition across space and time. It is possible that once more communities have been studied or that more realistic approaches to characterizing communities are devised (for example, using the biomass of each parasite species as opposed to simply counting the number of individuals), that patterns will emerge.

Against this general backdrop, note that in some communities either negative or positive associations might occur among parasite species more often than expected by chance alone. Although the overall significance of such interactions in dictating large-scale or long-term patterns in community structure can be questioned, they are intrinsically interesting and have certainly generated a great deal of interest among parasitologists, as discussed in the next section.

Parasite species within infracommunities engage in negative and positive interactions with one another

With respect to negative associations, interspecific competition has often been recorded within parasite infracommunities. The classic work of Holmes provides a conceptual framework that shows how interspecific competition among parasite species can shape community structure. One possibility is that when two species compete, one will actively exclude the other (**competitive exclusion**). Another possibility is that one or both of the competing species will react to the presence of the other and then occupy a different segment of its habitat (and so diminish overlap) than either would if alone (**interactive site segregation**). By contrast, **selective site segregation** refers to the situation in which two potential competitors do not overlap but occupy different parts of the habitat. Selective site segregation has been referred to as the "ghost of competition past". The basic idea is that over an evolutionary time frame, one marked by frequent adverse interactions between the two species, each has evolved to avoid overlap. Competition might be manifested in either **numerical** or **functional responses**, the former referring to a reduction in numbers of individuals owing to the competition and the latter to a shift in the niche occupied by the parasite.

Studies of rats infected with adults of the cestode ***Hymenolepis diminuta***, the acanthocephalan *Moniliformis dubius*, or both indicated that when present alone, each species preferred to inhabit the anterior third of the small intestine. When rats were infected with both parasites simultaneously, ***H. diminuta*** occupied a more posterior position in the gut, whereas *M. dubius* retained its original position. This finding suggested the acanthocephalan was outcompeting the tapeworm for their shared preferred attachment site. Although the presence of one species did not influence the number of worms of the other species that could colonize the gut, it did affect the size of the worms recovered. As expected based on the previous results, ***H. diminuta*** was more affected by the presence of *M. dubius* than *M. dubius* was by ***H. diminuta***. It was concluded that this represented an example of interactive site segregation arising from competition for nutrients in the gut. Although this was a laboratory study, broadly similar results have been noted for parasite species interacting in naturally infected hosts. Note that interference competition may play a role here too, as some factors such as cGMP (cyclic guanosine monophosphate) released from tapeworms have been shown to have adverse effects on growth of conspecifics and might also adversely affect other gut helminths.

As noted previously, the interactions among different parasite species within an individual host can be extensive and complex, such that removal of one targeted species could have an unforeseen effect on either the remaining parasites or the host. Apparent competition mediated by a third partner (namely the host's immune system) can also occur between species in infracommunities, just as it can among individual parasites of the same species within parasite infrapopulations. In the system shown in **Figure 6.14**, Parasite 1 has a beneficial (positive) effect on the other three parasites present because it has a general immunosuppressive effect on the host. Note that Parasite 2 has a beneficial impact on Parasite 1, but a negative impact on Parasite 4. In this case, Parasite 2 might support the general immunosuppression mediated by Parasite 1 and thereby favor that parasite, but there may be something caused by the host's response to Parasite 2 that also negatively affects Parasite 4 (perhaps 2 and 4 share a cross-reactive antigen). For Parasites 2 and 3, imagine that both have the same vector, and the vector might commonly be infected simultaneously with both parasites. In this case, because one bite might deliver two parasites, there would be a strong positive association between them in the host. Note though in this case that what is normally a beneficial impact of Parasite 2 on 3 might turn negative in some cases, as for example, during a chronic infection. This example serves to highlight that the interactions between species in a host are dynamic and may lead to different outcomes depending on the circumstances. Finally, note that Parasites 3 and 4 have negative impacts on one another. This situation might be because both parasites colonize the same specific habitat that thereby becomes limiting, or both parasites share strongly cross-reacting antigens, such that each elicits a host response that affects the other.

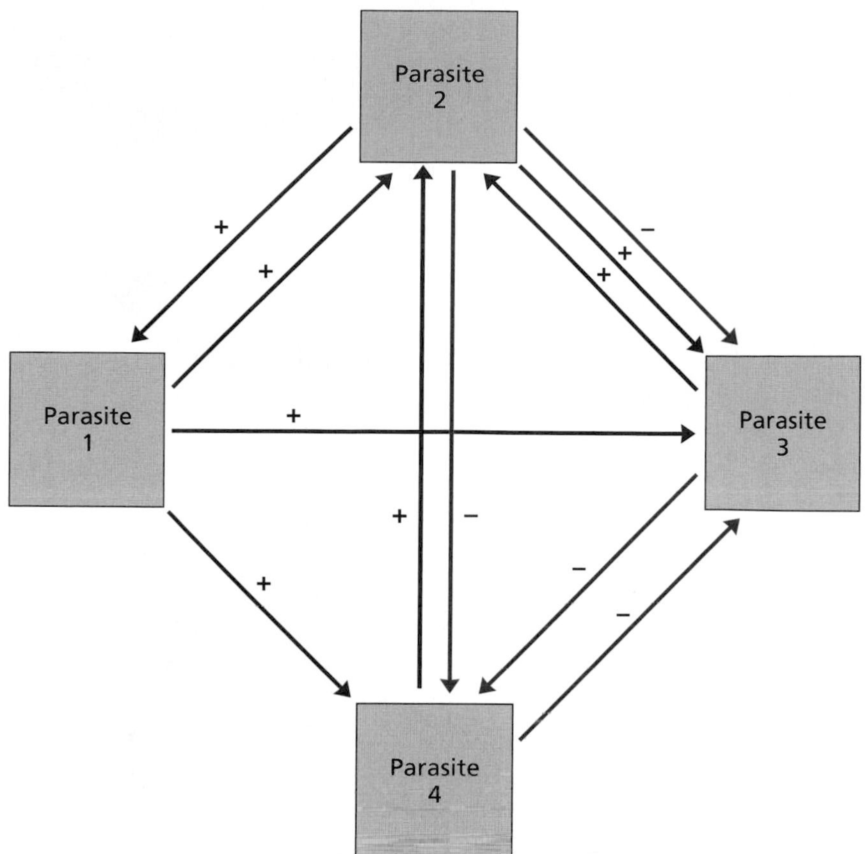

Figure 6.14 Different parasite species interact in complex ways within hosts. In this hypothetical example, four different parasite species occupy a particular host species, and the impact of one parasite on another is diagramed as being either beneficial (red line) or harmful (blue line).

In a long-term study of interactions among gastrointestinal helminths in the European rabbit *Oryctolagus cuniculus*, a complex interaction network among five species present was revealed. In most cases the underlying basis for positive or negative associations among the helminths in this system are unknown. However, the researchers concluded that acquired immunity generated against the nematode *Trichostrongylus retortaeformis* (living in the small intestine) interferes with development of the stomach worm *Graphidium strigosum*. Interestingly, the latter species can modulate host immunity, so if present in the host first, it can actually have a beneficial effect on *T. retortaeformis* by preventing the host from developing immunity. Many of the other interactions between helminth species were also of a positive nature in this system.

An excellent example of how negative interactions—in this case mediated primarily by predatory attacks and secondarily by resource competition—can strongly influence parasite community structure is provided by the guild of 17 species of larval digenetic trematodes, which inhabit the California horn snail *Cerithidea californica* in estuaries along the California coast. Interspecific interactions among trematode larvae within snails would be expected to be intense because the larvae undergo asexual reproduction in the snail and come to occupy much of the volume of the snail's body. In this system, and in similar situations elsewhere, the rediae (intramolluscan larvae with a sucker around the mouth and a gut) produced by some digenean species will attack, kill, and consume the larvae of other species when found in the same individual snail (**Figure 6.15**).

Interestingly, for particular combinations of two trematode species, one species will predictably displace the other, and an overall dominance

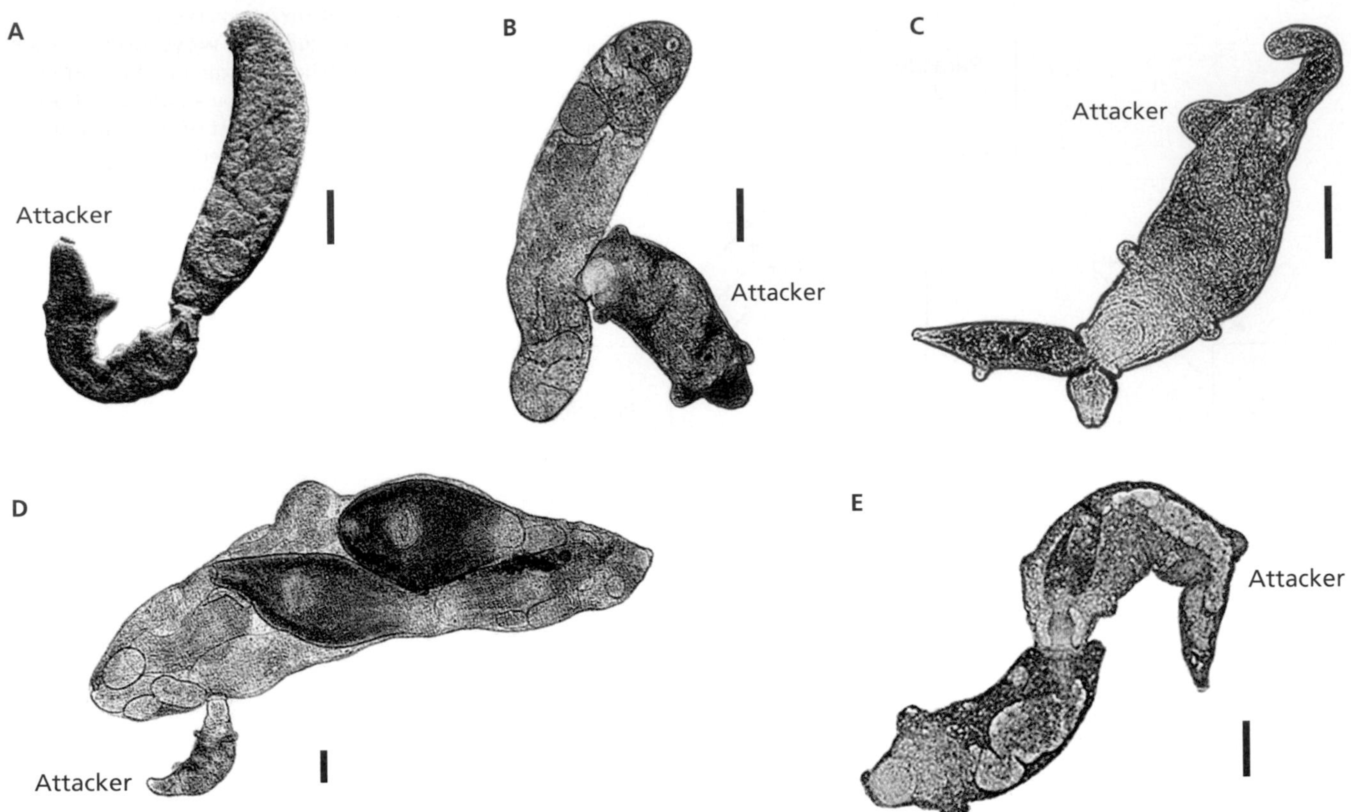

Figure 6.15 One parasite species may prey upon another when competing for a limiting resource. Several examples of rediae of *Himasthla* sp. B attacking rediae of other species. (A) Attacking redia of *Euhaplorchis californiensis*, the posterior end of which is visible in the attacker's pharynx. (B) Another attack on an *E. californiensis* redia. Note presence of eyespots of cercariae of prey in the gut of the attacker. (C, D) Attacking rediae of *Parorchis acanthus*. Note this species can mount attacks of its own and is competitively dominant to *Himasthla* (see Figure 6.16). (E) Attacking a conspecific redia from another colony (from another snail), with the posterior end of the redia being consumed. Scale bar = 0.2mm. (From Hechinger RF, Wood AC, Kuris AM et al [2011] *Proc R Soc Lond B Biol Sci* 278: 656–665. With permission from Royal Society Publishing.)

hierarchy can be established among the species involved (**Figure 6.16**). As shown in the Figure 6.16A, parasite species 2 will predictably replace 1, parasite 3 will replace 2, and parasite 4 will replace 3, but once 4 is established, it can not be replaced by 1, 2, or 3. Also, as shown in Figure 6.16B, an individual snail might suffer a succession of infections with dominant species replacing subordinate ones. In this manner, a hierarchy like the one shown in Figure 6.16C can be established that reliably predicts which species will dominate others. Species 7 is the top dog in the hypothetical hierarchy shown, followed by 6, etc. Thus in reference to the system described in Figure 6.15, *Himasthla* sp. B defeats *Echinoparyphium* sp., but it is predictably consumed and displaced by *Parorchis acanthus*. The net result is that when infracommunity structure is examined, it is far from random. Given the overall abundance of trematodes in this system, many fewer snails harbor more than one trematode species than expected by chance.

In addition to some of the interactions among helminths of the European rabbit noted above, another example of a positive association between parasites is provided by infection of mice with the gastrointestinal nematode *Heligmosomoides polygyrus* and the malaria parasite *Plasmodium chabaudi*. Nematode infection strongly modulates several aspects of the acquired immune response of mice to *P. chabaudi*, resulting in significantly increased peak malaria parasitemias and heightened mortality from malaria. Prior infection with *H. polygyrus* also interfered with vaccine-induced protection of mice against *P. chabaudi*, a result that warrants consideration when children in the tropics, who often bear a heavy burden of helminth infection, are eventually given malaria vaccines.

The manner in which parasites are acquired can also account for positive associations between parasites. If a particular species of intermediate host becomes infected with the larval stages of several parasites, then it stands

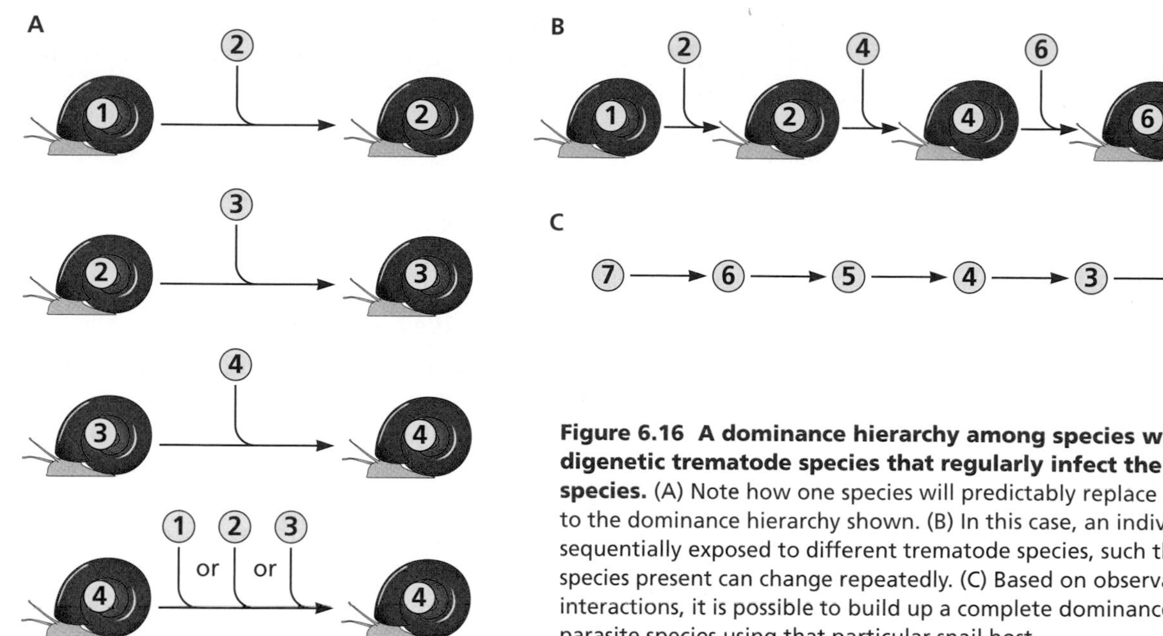

Figure 6.16 A dominance hierarchy among species within a guild of digenetic trematode species that regularly infect the same host snail species. (A) Note how one species will predictably replace another according to the dominance hierarchy shown. (B) In this case, an individual snail might be sequentially exposed to different trematode species, such that the dominant species present can change repeatedly. (C) Based on observations of pairwise interactions, it is possible to build up a complete dominance hierarchy for all the parasite species using that particular snail host.

to reason that ingestion of such a host would increase the probability that the definitive host will acquire more than one parasite species at the same time. Parasite infracommunities of the Lesser Scaup (*Aythya affinis*) were examined and positive associations were found between two cestodes and one acanthocephalan, the larval stages of which are harbored by the amphipod *Hyalella*. Similarly, three cestodes and one acanthocephalan are acquired by eating the amphipod *Gammarus*. Examples of positive or negative associations are intriguing, but it is also relevant to note that in some cases, different parasite species in the same host may simply have no influence on one another.

Generalizable patterns are also elusive in component communities of parasites

A component parasite community, by virtue of its association with a host population, is a longer-lived entity than an infracommunity, and perhaps more generalizable aspects of parasite communities can be detected at this level. As with infracommunities, the species richness of a single component community is usually lower than recorded for the entire parasite fauna of a particular host species. Also, as a general rule it might be expected that heterogeneity among different parasite component communities would be lower for highly mobile host species. Such species experience higher rates of interpopulation exchange, meaning parasite exchange is also more likely. We might expect that the larger the host population, or the greater the size and species diversity of the habitat occupied by a particular host population, the greater the richness of parasite component communities. Another vital ingredient to building component community size is the time the host population has been in place. A newly created habitat occupied by the host may lack overall species diversity, thus denying parasites access to intermediate hosts needed to support their transmission (see also Chapter 8). Also influencing parasite component communities is the distance between host populations, with a reduction in similarity expected as a function of the distance between host populations. The role of chance introductions of parasites or the history of a particular host population (such as whether the population has been stable or undergone bottlenecks) can also strongly affect the size and composition of parasite component communities.

There is evidence supporting the role of many of these factors. However, contingencies based on location or host species as yet preclude parasitologists from framing general patterns in the structure of parasite component communities. As Kennedy (2009) concludes, "The search for patterns in space and time must continue."

Human parasites have a distinctive community ecology

Humans harbor an impressive list of parasite species accumulated across the huge geographic range they inhabit. A total of 1415 species of infectious entities or organisms, including 217 viruses and prions, 538 bacteria, 307 fungi, 66 protists, and 287 helminths, were recorded as parasites of humans, this over a decade ago. This already impressive number is further augmented by inclusion of parasitic arthropods and by the ongoing discovery of new parasites, particularly viruses. New viruses of humans continue to be found at a rate of >2 per year. Most of the parasites that humans harbor were not acquired from our ancestors but have been acquired from the many kinds of domestic and wild animals closely associated with us over the years. Only about 15% of our parasites are strictly specific to humans. Many human parasites are shared with domestic animals (see also Figure 7.20), and the longer the period of domestication, the more parasites are shared (**Figure 6.17**).

In a world where over 40 million people have HIV/AIDS, about a third of the world's population is infected with *Mycobacterium tuberculosis*, malaria strikes over 200 million people per year, and overall prevalence of soil-borne helminths such as ***Ascaris*** and ***Trichuris*** is about 25% (up to 90% in some regions), many co-infections occur. Co-infections are especially common in underdeveloped parts of the tropics. Are these and other parasite species noninteractive when they co-occur, or does the presence of one either positively or negatively affect the other?

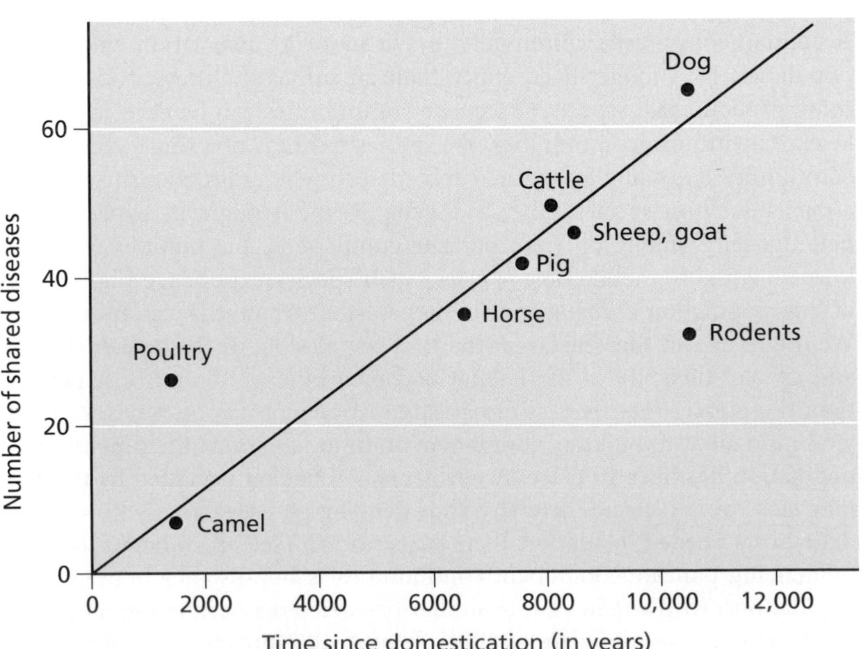

Figure 6.17 The relationship between the time since domestication and the number of parasites shared with humans. (Perrin P, Herbreteau V, Hugot J-P, Morand S et al [2010] In The Biogeography of Host–Parasite Interactions [Morand S & Krasnov BR eds], pp. 41–57. Oxford University Press. With permission from Oxford University Press.)

Of course a tragic hallmark of HIV/AIDS that became apparent early on is its immunosuppressive effects, mediated by depletion of $CD4^{+}$ T lymphocytes. Immunosuppression facilitates opportunistic infections with other viruses, bacteria (including *M. tuberculosis*), fungi, protozoans (including malaria), and helminths. Conversely, it has also been speculated that chronic helminth infections, by virtue of skewing the immune system toward a Th-2 cytokine profile and up-regulating CCR5 (a co-receptor used by HIV to enter cells), might greatly favor progression of both HIV and TB and undermine eventual attempts to vaccinate against them. Damage to the female urogenital system by the helminth ***Schistosoma haematobium*** may facilitate heterosexual transmission of HIV and infection with ***Schistosoma mansoni*** may favor reactivation of latent tuberculosis. Infections with ***Trichomonas vaginalis*** are associated with long-term infections with human papilloma virus (HPV) and increased risk of acquiring HIV. Furthermore, ***T. vaginalis*** itself is frequently colonized by *Mycoplasma hominis*, an intracellular bacterium that can also colonize urogenital mucous membranes. This relationship seems to be mutually beneficial and ***T. vaginalis*** serves as a Trojan horse for transmission of *M. hominis* into a new human host. The close association of these two organisms is reminiscent of the suspected exploitation of the eggs of the human pinworm (*Enterobius vermicularis*) by the diarrhea-causing protozoan *Dientamoeba fragilis* for its transmission. These kinds of dependencies of course result in the two parasites involved being positively associated in the human host.

With respect to malaria, one meta-analysis indicates a trend for both ***Ascaris lumbricoides*** and ***S. haematobium*** to have a protective effect against the pathogenesis and incidence of malaria, whereas ***S. mansoni*** and hookworms have a worsening effect. Spatial studies suggest the filarial worm ***Wuchereria bancrofti*** and ***Plasmodium*** spp. are negatively associated, in some cases owing to within-host, immune-mediated competition (see apparent competition above). As an example of another negative association, ***Ascaris lumbricoides*** and ***S. mansoni*** interact antagonistically, likely because both stimulate a general anti-helminthic response.

These kinds of studies indicate that the unraveling of the complex interactions among human parasites has begun, but that parasitologists are still far from a comprehensive understanding of the factors dictating the nature of the associations. There is particularly a long way to go before we can hope to fully understand the implications for intervention programs. These interventions may yield surprising outcomes if the target parasite interacts strongly with other parasites that are themselves not the subject of control operations.

6.4 THE ROLE OF PARASITES IN FOOD WEBS AND ECOSYSTEMS

The studies thus far discussed emphasize the size and composition of parasite communities and the interactions of parasites with one another. Yet to be discussed are the kinds of roles parasites play in the web of interactions among all species. Do parasites significantly influence the overall network of interactions among all species in an ecosystem or affect how energy flows through ecosystems? Or, on these grander, ecosystem-level scales, do parasites shrink to insignificance? Historically, because parasite biomass has been presumed to be small relative to other components of the ecosystem, there has been a tendency to believe that parasites are insignificant with respect to overall ecosystem structure or to energy flow through ecosystems.

However, parasites warrant a much closer look for several reasons. First, parasites can influence the size of host populations and alter host competitive ability. Second, they can significantly alter the probability that a predator eats a particular individual host. Also, both free-living stages and parasites within hosts can in some cases be significant prey items. Furthermore, parasites may have a direct bearing on the species richness of the community by functioning as general drivers of biodiversity.

One approach to better understanding the role of parasites in ecosystems is to include them in the construction of food webs. A **food web** is an explicit topology or map that represents how energy flows in an ecosystem (who eats whom). Food webs are typically divided into: 1) primary producers (such as plants) that convert solar energy into organic biomass; 2) primary consumers (or herbivores) that consume the primary producers; and 3) secondary consumers (or predators) that consume herbivores. Additionally, smaller predators may be consumed by larger predators, and so on. Food chains are characterized by how many species are present, the length of the trophic cascade that exists within the food web, and the strength of the connections (interactions) among species. A food web can be characterized by the magnitude of energy transfer indicated by various links in the food web.

One study of a salt marsh in California showed that inclusion of parasites in the food web dramatically increases the connectedness of the food web by 93% (**Figure 6.18**). A total of 78% of all food web links involved parasitism. There were 615 links involving parasites and hosts. There were also 910 links involving parasite stages being consumed by predators, as contrasted to 505 links involving more conventional predator–prey interactions. This finding supports the idea that parasitism is more common than predation as a consumer lifestyle. The increased connectedness noted is a feature suggested to greatly increase food web stability. Ecosystems are often portrayed as trophic biomass pyramids with predators occupying the top level of the pyramid. However, if it is considered that these apex predators often harbor parasites that consume resources from the predator, should these parasites be depicted at an even higher level on the pyramid?

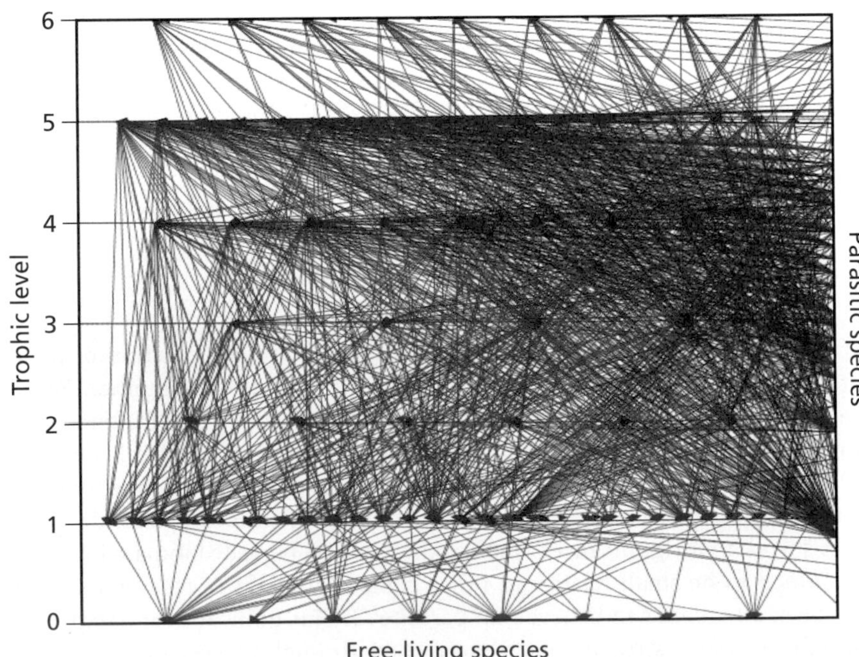

Figure 6.18 A food web for a California salt marsh that includes parasites. Each line represents the connection between a consumer species and a consumed species. Along the horizontal axis are free-living species with trophic level increasing on the left vertical axis, from basal level (0) to top predator (6). Parasite species are arranged along the right vertical axis. Links between predator and prey are shown by blue lines and parasite–host links by red lines. Note how the inclusion of parasites in the food web greatly alters the overall topology of the web. (From Lafferty KD, Hechinger RF, Shaw JC, et al [2006] In Disease Ecology Community Structure and Pathogen Dynamics [S Collinge & C Ray eds], pp. 119–134. Oxford University Press. With permission from Oxford University Press.)

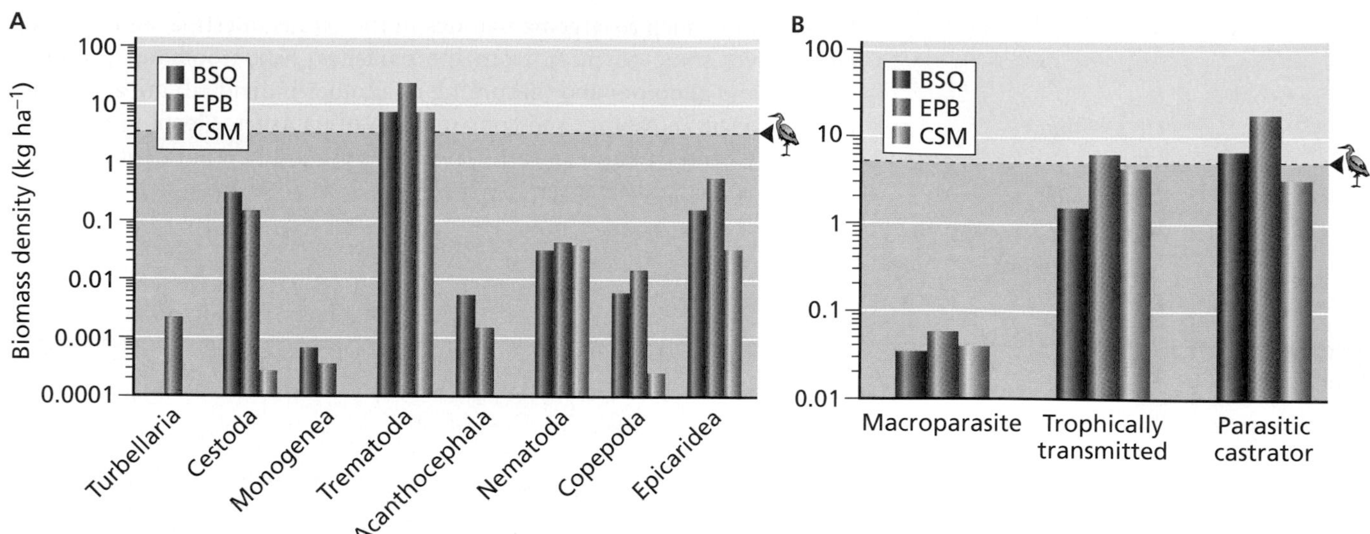

Figure 6.19 Parasites can account for large proportions of the biomass in ecosystems. Estimates of parasite biomass from three Pacific coast estuaries (BSQ, EPB, and CSM), showing parasites grouped either by major taxa (A) or by functional groups (B). The arrow on the right side of each graph compares the biomass in each habitat contributed by birds (in winter). Note that some parasite groups exceed birds in their contributed biomass. (From Kuris AM, Hechinger RF, Shaw JC et al [2008] *Nature* 454:515–518. With permission from Macmillan Publishers Ltd.)

The small size of most parasites tends to bias thinking toward the notion that they must be insignificant with respect to energy flow within ecosystems. Once again though, consider a study undertaken to quantify the biomass of all organisms present in three estuarine ecosystems along the California coast (**Figure 6.19**). It was discovered that parasites exceeded predators in biomass and that the annual biomass of cercariae produced by larval digenetic trematodes (the same system as described above with respect to negative interactions among parasites) was greater than the biomass of birds in the system. The researchers concluded that parasites significantly influence food web structure and that, by influencing the likelihood of predation of some hosts by others, they increase the strength of some links in the food web. They concluded by noting that the traditional approach of ignoring the role of parasites in ecosystem functioning may lead ecologists to misrepresent how ecosystems truly function. Results of ecosystem studies of grassland plants in Minnesota had comparable results. The biomass of fungal plant parasites was greater than herbivore biomass and the biomass of grass was better predicted by the abundance of the fungi than the herbivores. The impact of parasites on food web structure might be particularly telling if the host affected was a **keystone species** (one that is interconnected with many other species in the ecosystem and if removed causes significant changes in the system). Elimination of a keystone species, *Diadema* sea urchins, on the coral reefs of the Caribbean by a bacterial parasite had the effect of causing a major change in ecosystem structure. The lack of algae-consuming urchins enabled algae to become abundant, overgrowing coral reefs in the process.

Parasites can be a food source for other organisms

A realization emerging from the study of food webs is how commonly parasites serve as prey items for other species. The idea of parasites as prey has significance in several contexts: as a way to better understand how food webs are constructed and how energy flows through ecosystems; as a way in which parasite abundance and diversity are regulated; as an underappreciated way to control parasites of medical or veterinary importance; and as a way to understand why invasive parasites succeed or fail. Parasites also provide a very distinctive page in the annals of human gastronomy as well, as described in **Box 6.4**.

Parasites can become prey in various ways. **Concomitant predation** occurs when a predator, such as a lion eating a zebra, also consumes the parasites living in prey, such as large nematodes in the zebra's intestine, and the predator derives some nutrition from the parasites. **Autogrooming** (grooming oneself) and **allogrooming** (grooming of another individual) are also activities that result in capture and consumption of parasites. **Cleaning symbiosis** (see also Chapter 1 and Figure 1.9), which is especially prominent in marine habitats, is mediated by over 130 species of crustaceans and fish and results

BOX 6.4

Parasites as Accidental and Deliberate Human Food Sources

By this stage of the game, it should come as no surprise that we acquire some of our natural burden of parasites by unintentional ingestion of food containing encysted parasite life-cycle stages. For example, we can acquire the tapeworm *Taenia solium*, the nematode *Trichinella spiralis*, or the apicomplexan *Toxoplasma gondii* through just the ingestion of poorly cooked pork. More surprising, though, is that we also have a grand tradition of deliberately eating parasites, as beautifully laid out in authoritative revlews by Mayberry (1996) and Overstreet (2003) (see also Box 3.1). Given that we have some instinctive feel that ingestion of parasites would be harmful and lead to infection, why on earth do we do this? It turns out there are several reasons: (1) for perceived benefits for weight loss (Hollywood "slimming pills" of the 1930s contained cysticerci of *T. solium*; recall the discussion of Maria Callas in chapter 5); (2) for traditional and more modern attempts to treat disease (swallowing ground-up *Ascaris* worms as a vermifuge, ingestion of live lice to treat jaundice, or as noted in Chapter 5, ingestion of *Trichuris* eggs to treat Crohn's disease); (3) to shock others and so gain social status (ingestion of live fish with bellies swollen with parasites), as with US college students; (4) for scientific purposes (when parasitologists infect themselves to complete an unknown life cycle or as a source for experimental or teaching material); (5) while grooming during social interactions (catching, cracking and consuming live head or body lice); and (6) perhaps most importantly, as a coveted source of nutrition. Examples include bot fly larvae from reindeer, parasitic crustaceans embedded in the blubber of whales, adult tapeworms from the gut of possums in New Guinea (Figure 1A), and "liver butterflies", "little livers", or "little flapjacks", which are *Fascioloides magna* flukes recovered from deer livers, in North America (Figure 1B). In most cases when parasites are deliberately eaten as delicacies or to provide nutrition, the parasites involved are not infective to humans nor do they cause any overt harm.

If you happen to eat a particularly delectable piece of pork, bear in mind that pork infected with the cysticerci of *T. solium* (called measley pork because of the presence of "measles" or cysticerci), has been claimed to be more tasty than pork from uninfected animals. In this latter example, infection with an adult of *T. solium* is emphatically not a desired outcome as it can lead to autoinfection and cysticercosis, a potentially life-threatening condition. The old admonition to thoroughly cook your pork remains excellent advice to this day.

A

B

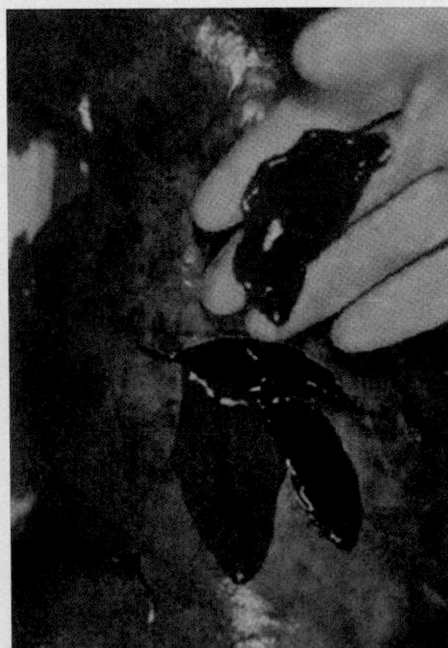

Figure 1 Parasites as human food. (A) Consumption by a Papua New Guinea native of an adult anoplocephalid tapeworm *Bertiella esculenta* from the intestine of the coppery ringtail possum *Pseudocheirus cupreus*. (B) An adult of the fluke *Fascioloides magna* recovered from the liver of a deer. (A, permission pending. B, Courtesy of Michigan Department of Natural Resources.)

References

Flannery TF (1998) Throwim Way Leg. Atlantic Monthly Press.

Mayberry LF (1996) The infectious nature of parasitology. *J Parasitol* 82:856–864 .

Overstreet RM (2003) Flavor buds and other delights. *J Parasitol* 89:1093–1107 (doi:10.1645/GE-236).

in consumption of significant numbers of ectoparasites. Some cleaners such as the cleaner wrasse *Labroides dimidiatus* are highly specialized to feed on only ectoparasites, and so are essentially totally dependent on parasites for food. One study estimated that two-thirds of the gnathiid isopods infesting reef fishes are consumed by cleaners, including up to 1200 ectoparasites per cleaner fish per day. Cleaners such as oxpecker birds (*Buphagus*) also consume ectoparasites on terrestrial hosts.

The free-living stages of parasites, such as swimming miracidia or cercariae of trematodes, oncomiracidia of monogeneans, or coracidia of tapeworms, are consumed by a huge variety of aquatic vertebrate and invertebrate predators (**Figure 6.20**). Parasite eggs or larvae in fecal samples are also consumed and killed by organisms as diverse as earthworms, dung beetles, snare-forming fungi, or even feces-consuming pigs or dogs. The magnitude of this effect is largely unknown but is likely to be considerable, simply because so many of these parasite life-cycle stages are produced. For example, it is estimated there are 5 grams per m^2 of cercariae produced by the infected snails living in a California salt marsh, enough to support an estimated 2%–3% of the annual food needs of the fish living there.

Yet another way that parasites are consumed as prey is by other parasites. This situation is commonly found with hymenopteran wasp parasitoids, where one species of parasitoid colonizes a host and then the female of another species of parasitoid injects her progeny into the body of the first parasitoid, creating a situation of hyperparasitism (see Chapter 1). Eventually the second parasitoid consumes the first, with the original host usually being killed upon emergence of the successful parasitoid. Some parasitoids are obligatory in their need to insert their progeny into the body of other developing parasitoids. In some cases, a true Russian dolls scenario can result in which a host may contain a primary parasitoid and nested within it are secondary, tertiary, and even quaternary parasitoids (see the example of pea aphids in Chapter 1). Another example is provided by larval trematodes developing in snails, when the rediae of one species will attack and consume the larvae of another (see Figure 6.15). The prolific ability of larval trematodes to multiply within their snail hosts means that there would be considerable impetus to detect and destroy a new, incoming parasite before it could initiate its own asexual reproductive program. Both the parasitoid and larval trematode examples also illustrate **intraguild predation**, when one member of a guild attacks and consumes another.

Much work remains to be done to assess the relative importance of parasites as prey items in the functioning of ecosystems. Because of the large

A

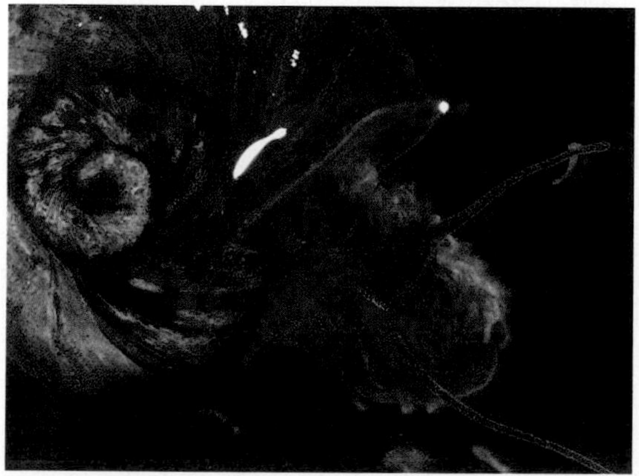

B

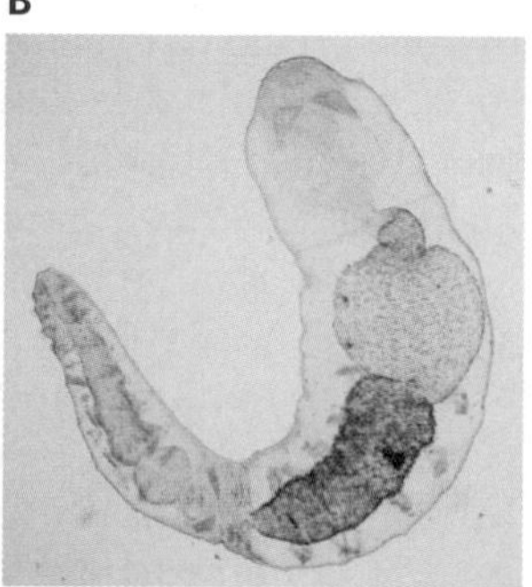

Figure 6.20 Parasites as prey. (A) Shown is the head-foot of a freshwater snail (*Helisoma* sp.) covered with oligochaete worms of the genus *Chaetogaster*, which prey upon miracidia of trematodes attempting to penetrate the snail or upon cercariae that emerge from an infected snail. (B) An oligochaete containing an ingested trematode cercaria. (© P Johnson & S Orlofeske. From Johnson PTJ, Dobson A, Lafferty KD et al [2010] *Trends Ecol Evol* 25:362–371. With permission from P Johnson.)

diversity of parasites present in a typical ecosystem, they are believed to contribute to food web stability. Moreover, revealing the complexity of such linkages has caused us to re-examine why some parasites remain uncommon, whereas others emerge to pose new disease threats. It is now acknowledged that the transmission of a particular parasite may be reduced as a result of the complexity of the environment in which it is found (including the presence of potential predators). This reduction in transmission is called the **dilution effect**. Simplification of environments with a concomitant reduction in food web complexity may then enable some parasites to become more common, a topic we return to in Chapter 10.

6.5 GLOBAL PATTERNS IN PARASITE DIVERSITY

Our discussion of the ecology of parasitism would be incomplete without some consideration of parasite distribution on scales even broader than ecosystems, as on a global scale. If patterns are observed, what underlying factors act to dictate global patterns of parasite distribution? One of the most salient features of the global distributions of free-living organisms is that as you proceed to lower and lower latitudes (toward the equator), species diversity of many groups of organisms increases. Do parasites show a similar pattern, and if so, why might this be so?

One study of the latitudinal distributions of 229 human parasites, including viruses, bacteria, fungi, protozoa, and helminths, concluded that the species richness of these parasites was inversely correlated with latitude. Just as with free-living organisms, human parasite diversity was highest in the tropics. This study also showed that the distribution of these human parasites was nested. That is, those relatively few species found in more temperate latitudes are also found in tropical latitudes, but not vice versa.

What are the underlying factors responsible for such a latitudinal pattern? Of the many factors examined in this particular study, surprisingly the one that best explained the pattern was not mean annual temperature or precipitation but rather the maximum annual range of precipitation. This finding suggests that human parasites and their vectors tend to be well adapted to regions having more contrasted wet and dry seasons. The pattern was especially true for parasite species that intimately experience environmental conditions, such as those with free-living life-cycle stages or that require vectors. Another important conclusion from this study was that when parasites are included, the gradient in diversity seen for all species is even stronger than noted just for free-living organisms. Here a note of caution is warranted as several other factors also vary with latitude including socioeconomic conditions, access to medical care and, importantly, the overall diversity of other animal species that might serve as vectors or alternative hosts for the parasites.

What about global patterns in the abundance of nonhuman parasites? Here, and perhaps not at all surprisingly, there is a growing appreciation that the best predictor of parasite species richness per area, is the richness in host species available to exploit. Hence, greater overall parasite diversity might be expected in the tropics. Additionally, the general harshness of conditions in higher latitudes is expected to disfavor survival of life-cycle stages such as helminth eggs or protozoan cysts that are exposed to the environment. These are general factors that might support higher overall diversity, on a unit area basis, of all kinds of parasites in the tropics. However, no strong latitudinal patterns on a per unit area basis have been found for the relatively well-studied helminths or fleas of mammals, suggesting that, as is often the case, we need to examine the biology of particular parasite groups more specifically or that other factors we have yet to grasp play a critical role. With respect to marine parasites, again we lack sufficient data, especially from the deep sea or from the myriads of small animals living in the sediments beneath the ocean's surface, to make sweeping generalizations about large-scale patterns in parasite species richness. Studies of both monogeneans (ectoparasites) and digenans (endoparasites) of marine fish show that their diversity peaks in the tropics. However, when the number of host species available to infect is taken into account, the increase in parasite species per host species approaching the equator was modest for monogeneans and nonexistent for digeneans. This observation suggests that the increase in parasite diversity seen in tropical seas was mostly a result of the increase in diversity of host species to infect. Such cross-latitudinal studies of richness of parasite species and host species are complicated by the fact that the taxonomic composition of both parasites and hosts change with latitude, making comparisons difficult. It is only fair to say that such studies are in their infancy and that more basic survey information is needed to inform such studies.

6.6 PARASITE EFFECTS ON HOST ECOLOGY

The discussion of the ecology of parasitism has to this point emphasized the ecology of parasites per se. In this section, the focus shifts to the effect of parasites on the ecology of their hosts. The definition of parasitism has at its core the notion that parasites do harm to their hosts, so it is hardly surprising that there are many examples of the effect of parasites on host growth, survivorship, fecundity, distributions, etc., all important components of the host's ecology. In addition, the parasites rarely exert their adverse effects in isolation: they usually act in combination with other stressors such as limited food or crowding to exert their negative effects on hosts. As just one example, a study of tadpoles of the gray tree frog *Hyla versicolor* showed that the negative impact of the trematode *Telorchis* sp. on mass at metamorphosis and on survivorship was only manifested when habitats occupied by the tadpoles were also drying out (**Figure 6.21**).

In this discussion we will examine a subset of possible parasite effects on host ecology, choosing topics that are of particular interest or importance, or that have attracted considerable attention. These topics provoke a variety of questions. How do parasites affect their host's behavior, potentially altering their energy budgets or competitive abilities or the increasing likelihood they migrate to new locations or are consumed by predators? Can parasites regulate the size of host populations? How are host communities affected by the involvement of parasites? As we shall see, a growing body of evidence suggests that parasites, either directly or indirectly, profoundly influence the interactions of their hosts with their wider environment. Understanding how parasites affect host ecology becomes an important consideration if the host in question is an agricultural pest or an endangered species. It is also important to understand how parasites influence the ecology of the host species we are most invested in, namely *Homo sapiens*.

Hosts try both to avoid infection and to actively remove parasites if they do become infected

As you might imagine, there should be considerable advantages to hosts if they could take measures to avoid acquiring parasites in the first place. They would sidestep the possibility of getting sick, the suffering caused by the altered behavior that makes them vulnerable to predation, or the need to mount a costly immune response. An ounce of prevention is indeed worth a

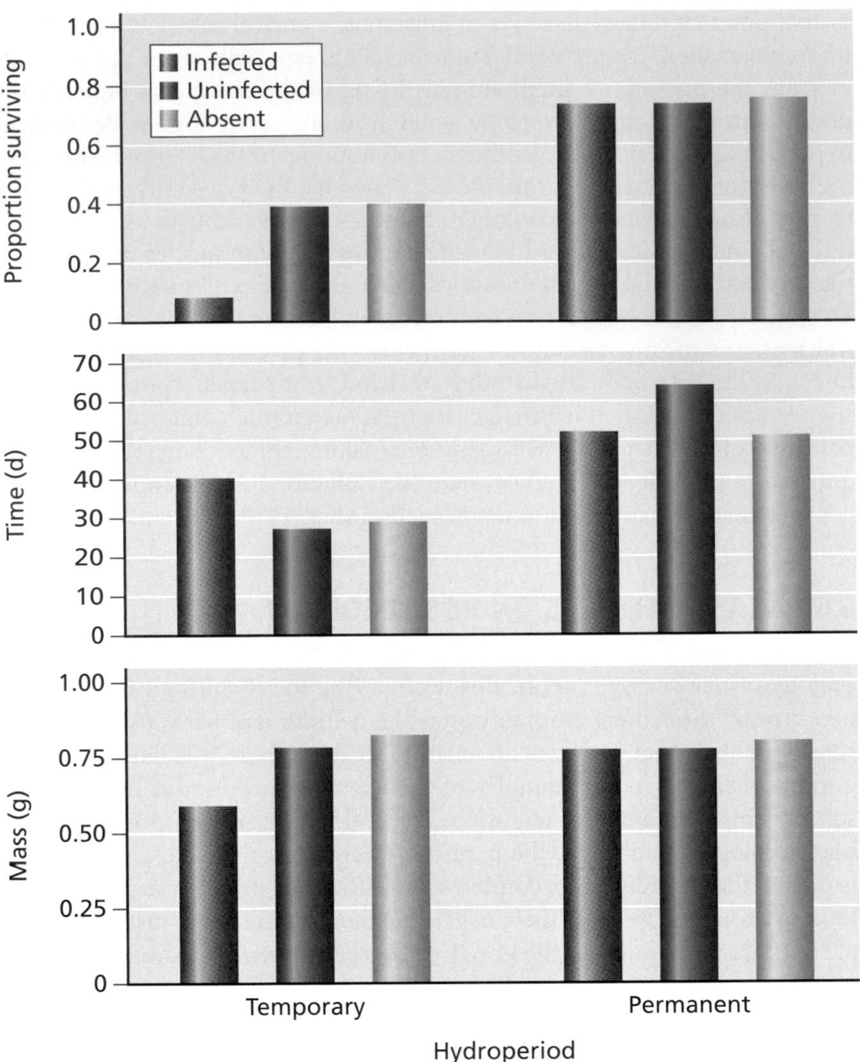

Figure 6.21 Parasites often interact with other stressors. The effects of the trematode *Telorchis* sp. on tadpoles of *Hyla versicolor* in artificial ponds that were temporary and subject to drying or that were permanent. Tadpoles were held in ponds that were infected (contained infected snails shedding cercariae of *Telorchis*, such that tadpoles become infected), in ponds that were uninfected (contained uninfected snails), or in ponds in which snails were absent. The effects of parasites on survival to metamorphosis, time taken to reach metamorphosis, and size at metamorphosis were measured. (Kiesecker JM & Skelly DK [2001] *Ecology* 82:1956–1963. With permission from Ecological Society of America.)

pound of cure. There are many examples to suggest hosts implement means to minimize the extent to which the encounter filter with respect to them is open. Many of these examples seem to reflect instinctive behaviors shaped by natural selection that has favored those individuals that avoid situations placing them at risk to acquiring infections. Trade-offs have no doubt also shaped these avoidance behaviors. For example, parasite avoidance behaviors might increase the likelihood of predation by making the host less vigilant or more conspicuous. In general, to be considered as real anti-parasite strategies, the parasites involved should have negative fitness consequences for the host, and hosts that engage in avoidance behavior should have higher fitness than those that don't.

Much of the supportive literature is anecdotal, but more and more studies document active parasite avoidance. Many animals ranging from lobsters to tadpoles to chimpanzees to humans have aversions to interacting with individuals who are obviously sick with infectious diseases. Fish shoal (group together) to minimize exposure to, and penetration by, trematode cercariae that can eventually encyst in their eyes and cause blindness. In this case the fish can actually discriminate the harmful cercariae in the water and shoal in response, lessening their subsequent parasite burden compared to unshoaled fish. Some animals such as horses and llamas have preferred defecation middens that are believed to be a means to circumscribe parasite contamination.

Tree-dwelling primates avoid branches contaminated with feces, and both domestic and wild ungulates preferably avoid grazing on habitat patches contaminated with feces from conspecifics. The latter behaviors minimize the risk of acquiring parasite infective stages such as cysts, eggs, or larvae. Areas that harbor abundant ectoparasites such as ticks may also be avoided, even if they are otherwise excellent foraging habitats. It is not hard to imagine that responses such as these could have cascading influences on other ecological interactions at larger scales, such as altering patterns of habitat use (see also parasite impacts on host migration).

Mere avoidance behavior is one thing, but other examples suggest a far more proactive response by hosts to prevent parasitism. Chimpanzees make sleeping platforms nightly in particular kinds of trees with aromatic leaves that help to minimize their exposure to night-biting vectors (such platforms also have other advantages, including predator avoidance). Birds have long been known to build nests that incorporate plants that exude volatile chemicals with properties that repel arthropods. Whether such plants actually are effective in deterring nest ectoparasites or other biting vectors remains an active area of investigation and alternative hypotheses have been entertained.

What if you are an urban-dwelling bird where such plant luxuries might not be available? Populations of house sparrows (*Passer domesticus*) and house finches (*Carpodacus mexicanus*) in Mexico City incorporate cigarette butt material in their nests and the more such material is present, the lower is the mite ectoparasite load suffered by the nest (**Figure 6.22**). It is interesting that nicotine from tobacco has been used deliberately to repel herbivorous arthropods and control poultry ectoparasites. Such studies are surely intriguing and have the effect of stimulating many more studies. For instance, what are the exact parasites being deterred, do the birds deliberately seek and collect the butts, can the butts be shown to deter parasites under controlled conditions, and does this behavior actually improve the fitness of these birds relative to ones not using butts?

If a host has already succumbed to infection, it may engage in grooming to remove ectoparasites (see parasites as prey above), a behavior that can occupy a significant proportion of its time. Birds spend about 10% of their time in maintenance behavior, most of which is grooming, and this behavior will increase if infestations are high. Energy costs associated with anti-parasite grooming can be considerable. One study estimated bats to experience a

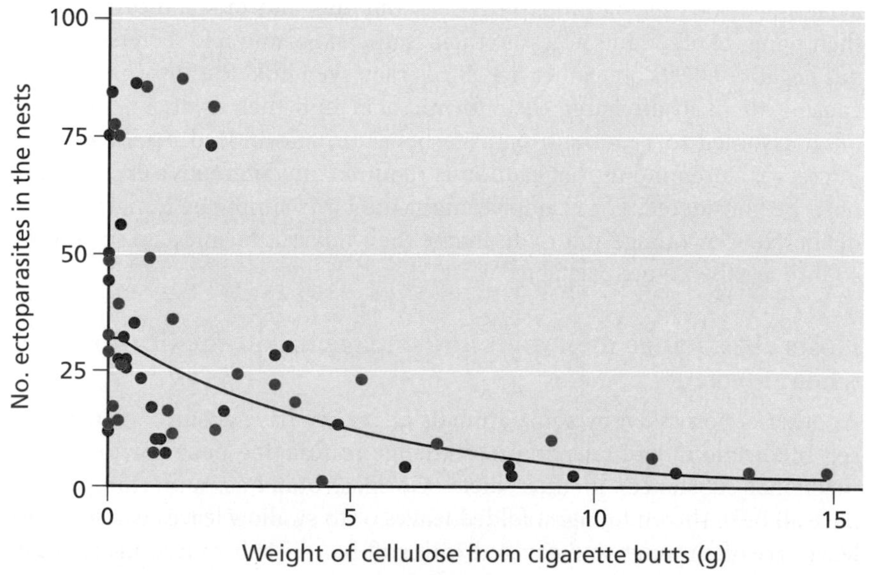

Figure 6.22 The influence of cigarette butts on nest ectoparasite burden. Shown for both house finches, *Carpodacus mexicanus* (blue circles), and house sparrows, *Passer domesticus* (red circles), is a negative relationship between the amount of cigarette butt material contained in the nest and parasite load. (From Suarez-Rodriguez M, Lopez-Rull I, Macias-Garcia C et al [2012] *Biol Lett* 9:20120931 [doi:10.1098/rsbl.2012.0931]. With permission from the Royal Society.)

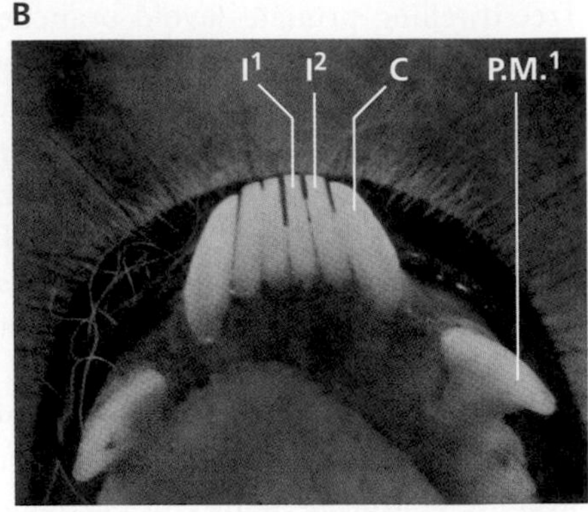

Figure 6.23 Grooming. (A) Japanese macaques engaged in allogrooming. (B) A toothcomb on the lower jaw of a lemur. Other animals including hyraxes, colugos, and some antelopes have similar modifications of their lower incisors for grooming. In addition to the incisors, I^1 and I^2, the canine teeth, C, have also been modified to form part of the toothcomb. P.M., pre-molars. (A, courtesy of Noneotuho, CC BY-SA 3.0. B, courtesy of Alex Dunkel, CC BY 3.0.)

0.5% increase in overall metabolism per ectoparasitic mite harbored. Anti-parasite grooming is an extensive field of study unto itself, and we provide only a few highlights. The close correspondence of ectoparasite holdfast structures to the feathers or hairs in preferred sites on host surfaces are legendary and serve to remind us of the impact grooming has had on parasite evolution (see Figure 3.19). Grooming is so pervasive among many animals, especially primates, that it has become fully integrated into the social fabric of societies, as a way to build friendships and alliances, to calm tensions, or to secure mates (**Figure 6.23A**). As we have noted above, the need to remove parasites also forms the basis of interspecific mutualistic interactions between cleaner fish, shrimp, or some birds and parasite-infested hosts.

Examination of hosts with defective grooming ability indicates the importance of grooming. Impala (*Aepyceros melampus*) prevented from normal grooming by wearing a neck harness harbored 20 times more female ticks than individuals wearing control harnesses that did not interfere with grooming. Also, the extent to which ectoparasites congregate on areas of a host's body not easily reached by grooming activities is another indication of grooming's obvious deterrent effect. The elaboration of particular devices or behaviors to enable parasite removal is noteworthy. Among them is the toothcomb (Figure 6.23B), a modification of the lower incisors of lemurs and some antelopes to comb the pelage to remove ectoparasites. Animals also engage in dust or mud baths to discourage their parasites, although the extent to which this is effective or might actually encourage infestations, or is done just to provide relief to itchy skin, is not well known. In one striking avian behavior called anting, birds seek out ants and in some cases spread their wings to allow ants access to their feathers. Formic acid secreted by ants has negative effects on feather lice. Birds may even hold the ants in their bills causing them to discharge their formic acid into their feathers. Although often assumed to be an anti-parasite behavior, and indeed it may be, anting serves to remind us that caution is required and alternative explanations need be considered. For example, might the birds simply be hungry and so deliberately encourage ants to discharge their noxious formic acid secretions, so they are then more palatable?

Hosts also change their diets and engage in self-medication when infected

Another action taken by some animals after they have acquired parasites is self-medication, here referring to a change in foraging behavior to include medicinal substances in their diets. Gorillas, bonobos, and chimpanzees have all been shown to ingest folded leaves or to swallow leaves whole. These leaves are often rough and covered with stiff hairs. Their rough surfaces may directly abrade or dislodge parasites, entrap nematodes in their folds, or trigger elevated gut motility thus favoring parasite expulsion. Nematodes are sometimes expelled with such leaves, and leaf-eating behaviors are known to increase at times of the year when parasites are most common.

Self-medicating behavior should improve fitness of infected animals, should not be beneficial in the absence of parasites (assuming the plants in question are not nutritious), and should be induced by infection. One study of wooly bear caterpillars (*Grammia incorrupta*) parasitized by the larvae of tachinid flies and given the opportunity to consume curative pyrrolizidine alkaloids, met all three of these conditions (see **Figure 6.24** for one aspect). This study also indicates that self-medication need not be associated only with animals with higher cognitive abilities, and highlights the value of model systems that can be experimentally manipulated in evaluating the validity of provocative topics like self-medication as an anti-parasite strategy.

In light of the anti-parasite behaviors noted above, it is inevitable that we should contemplate our own species in this regard. Perhaps it is no accident that the odor of human feces is extremely repellent to us, that we have intense aversions to individuals who are sick, or that the notion of eating certain types of (often raw) foods is unappealing? And of course we have not proven to be reluctant when it comes to self-medication. Readers are referred to Moore (2002, 2013) for much more discussion of this interesting topic.

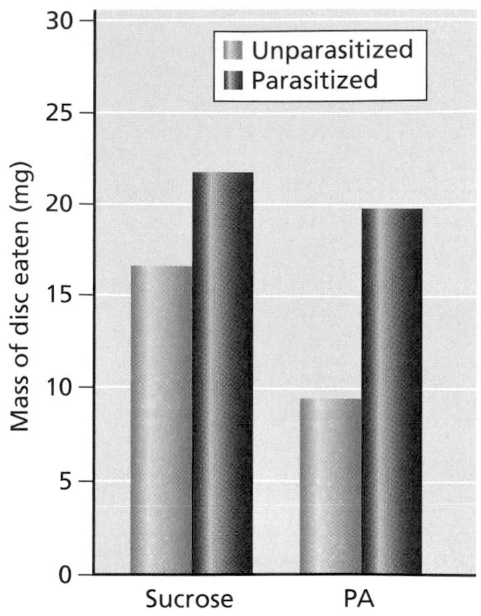

Figure 6.24 Some animals self-medicate to relieve the burden of parasites. Mass of glass fiber discs containing sucrose or self-medicinal pyrrolizidine alkaloids (PA) consumed by wooly bear caterpillars *Grammia incorrupta* parasitized or not by tachinid flies. Parasitized caterpillars ate significantly more ($P = 0.04$) of the PA-treated discs than did unparasitized caterpillars; parasitism did not significantly affect consumption of sucrose-treated discs. (From Singer MS, Mace KV, Bernays EA et al [2009] *PLoS One* 4:e4796 [doi:10.1371/journal.pone.0004796].)

Parasites influence host migratory behavior

One of the conspicuous features of the ecology of many host species—songbirds and monarch butterflies come to mind—is the long-range migration they take as part of their annual cycle. Many factors might explain such migrations, including escape from harsh conditions, availability of food, or avoidance of predators during the breeding season. Does the presence of parasites influence the migratory behavior of their hosts? One possibility is that migration is another form of parasite avoidance behavior. Migration might enable hosts to escape parasites that have accumulated in their current locality. Such migratory escape is exemplified by reindeer (*Rangifer tarandus*), which are tormented annually by warble flies, *Hypoderma tarandi* (the larvae of which are consumed by Nunamiut people; see Box 6.4). After developing over the winter in the body of the reindeer, warble fly larvae are expelled on to the ground at the time of calving in early spring. It is right after calving that reindeer migrate, with one interpretation being that in so doing they leave behind their warble flies. Adult flies, which are relatively short-lived, will emerge from pupation in the soil to find no new hosts on which to lay their eggs. Migratory escape would not necessarily benefit hosts though, if their parasites have long-lived resistant stages that could potentially survive for many months in the soil.

There are some obvious disadvantages of migration as well. A long-distance migration could bring the host into new environments and potential contact with many parasites, and it could debilitate and weaken the host's defenses against parasites. Migrating songbirds from Europe become infected with vector-transmitted blood protozoans on their tropical wintering grounds in Africa. Also, migrating birds often congregate in dense aggregations in migratory staging areas where the prospects of direct transmission of parasites would seem to be very high.

With respect to migration, consider the monarch butterfly (*Danaus plexippus*) and its interaction with the apicomplexan gregarine parasite *Ophryocystis elektroscirrha*. Some monarchs, such as those from northeastern North America regularly engage in long-distance migrations to Mexico,

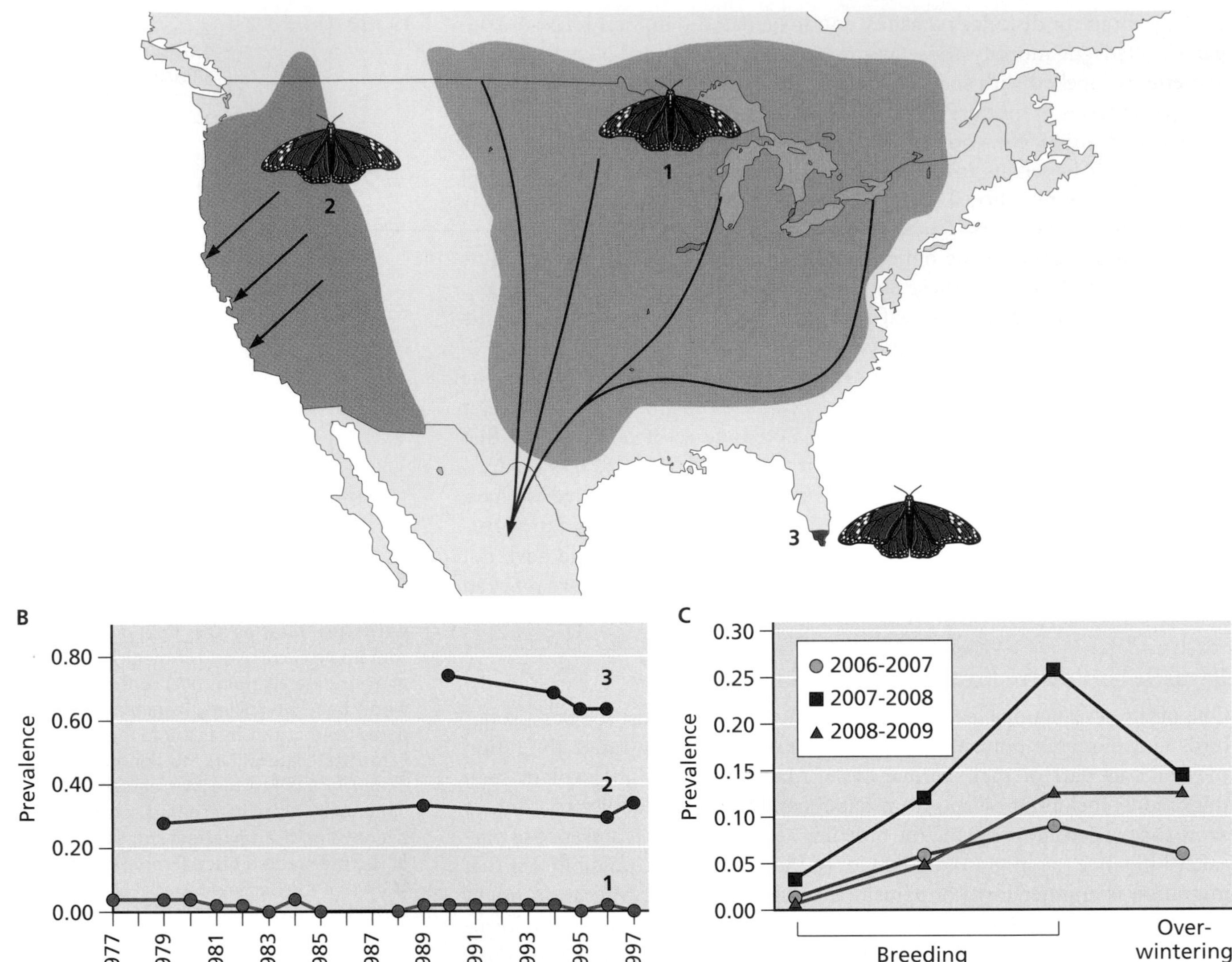

Figure 6.25 Parasites influence host migratory behavior. Migration of the monarch butterfly (*Danaus plexippus*) in relationship to infections with the gregarine *Ophryocystis elektroscirrha*. (A) Monarchs from some populations migrate long distances (1), some migrate shorter distances (2), and some are nonmigratory (3). (B) The prevalence of infection is highest among adults with longer residency, consistent with the idea that migration would be beneficial as a way to escape parasitism. (C) For migrating monarchs, parasite prevalence was also highest among butterflies sampled at the end of the breeding season than for those that reached their overwintering sites in Mexico. (From Altizer S, Bartel R, Han BA et al [2011] *Science* 331:296–302. With permission from AAAS.)

whereas other populations migrate relatively short distances (from the western U.S. to California) or not at all (Florida populations) (**Figure 6.25A**). The prevalence of the parasite increases the longer the monarchs remain on their breeding ground (Figure 6.25B), consistent with the idea that a subsequent migration to an overwintering ground would provide an escape from parasitism. Also, those butterflies that are infected fly more slowly and for shorter distances than uninfected individuals, suggesting that migratory culling occurs, helping to account for the drop in prevalence for those butterflies on their overwintering grounds (Figure 6.25C). These dynamics may help to explain why populations that migrate the furthest consistently have lower parasite prevalence than those that migrate less or not at all.

Parasites can regulate host populations, but examples are few

As discussed earlier in this chapter, regulation of a population refers to the tendency of the population to respond in a density-dependent manner, such that it increases if below a certain limit and decreases if above that limit. Regulation influences birth and death in a density-dependent manner. There are several examples of parasites acting on their hosts in a density-dependent manner, but in general, there are few bona fide examples of parasites regulating host populations. Perhaps the best-documented case comes from both descriptive and manipulative studies of the red grouse *Lagopus lagopus scoticus* in northern Britain (see Chapter 7 regarding a role of parasites in sexual selection in this species). Most red grouse populations undergo cyclic fluctuations with a periodicity of between 4 and 8 years (**Figure 6.26A**). Crashes in grouse populations are associated with high densities of the gastrointestinal nematode *Trichostrongylus tenuis* (Figure 6.26B). Additionally, and in accordance with mathematical models predicting a potential role for parasites in driving cyclical dynamics in host populations, it was shown experimentally that birds heavily infected with *T. tenuis* have reduced fecundity, and heavy worm burdens are positively associated with increased mortality of breeding birds (Figure 6.26C).

A manipulation in which grouse were treated with anthelmintic drugs was performed. Some sites served as controls and grouse were not treated

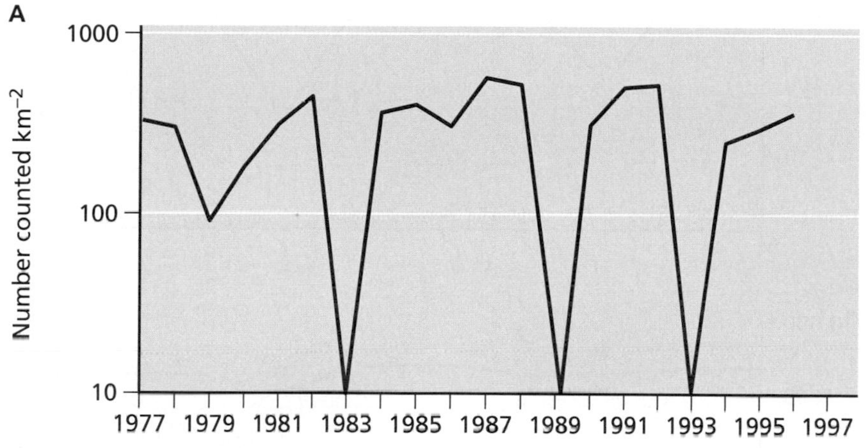

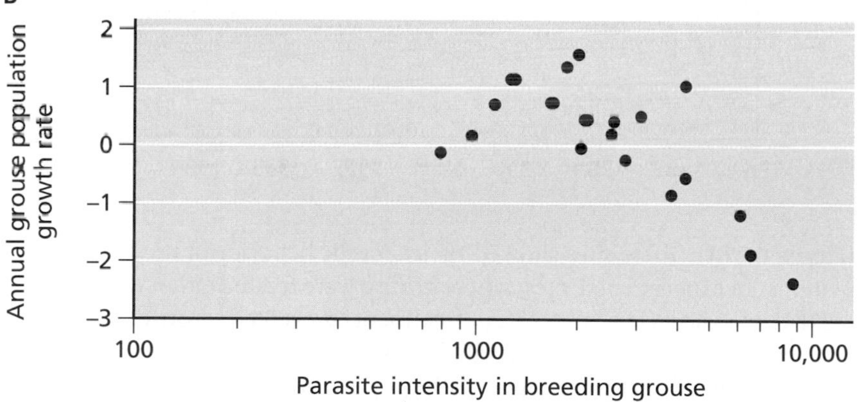

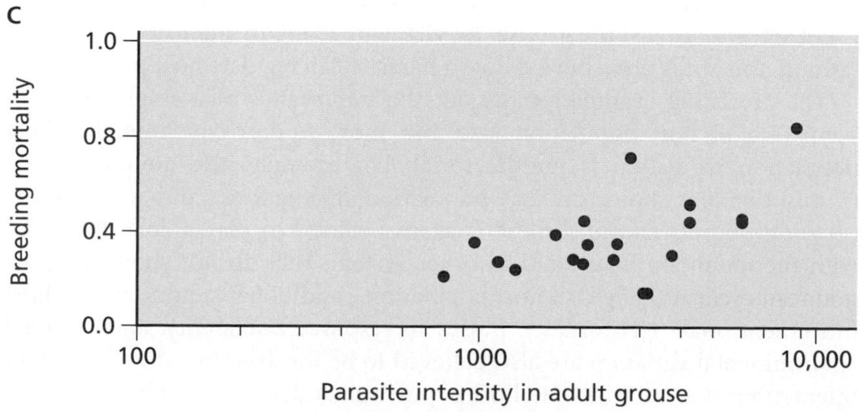

Figure 6.26 Impact of the nematode *Trichostrongylus tenuis* on populations of the red grouse *Lagopus lagopus scoticus* in northern Britain. (A) Grouse population dynamics showing the cycle in abundance, usually of 4–8 years duration. (B) Plot showing the negative relationship ($r = -0.676$) between the annual grouse population growth rate (*rt*) and parasite intensity in breeding adult grouse. (C) Plot showing positive relationship between grouse breeding mortality and mean log worm intensity ($r = 0.641$). (From Hudson PJ, Dobson AP, Newborn D et al [1998] *Science* 282:2256–2258. With permission from AAAS.)

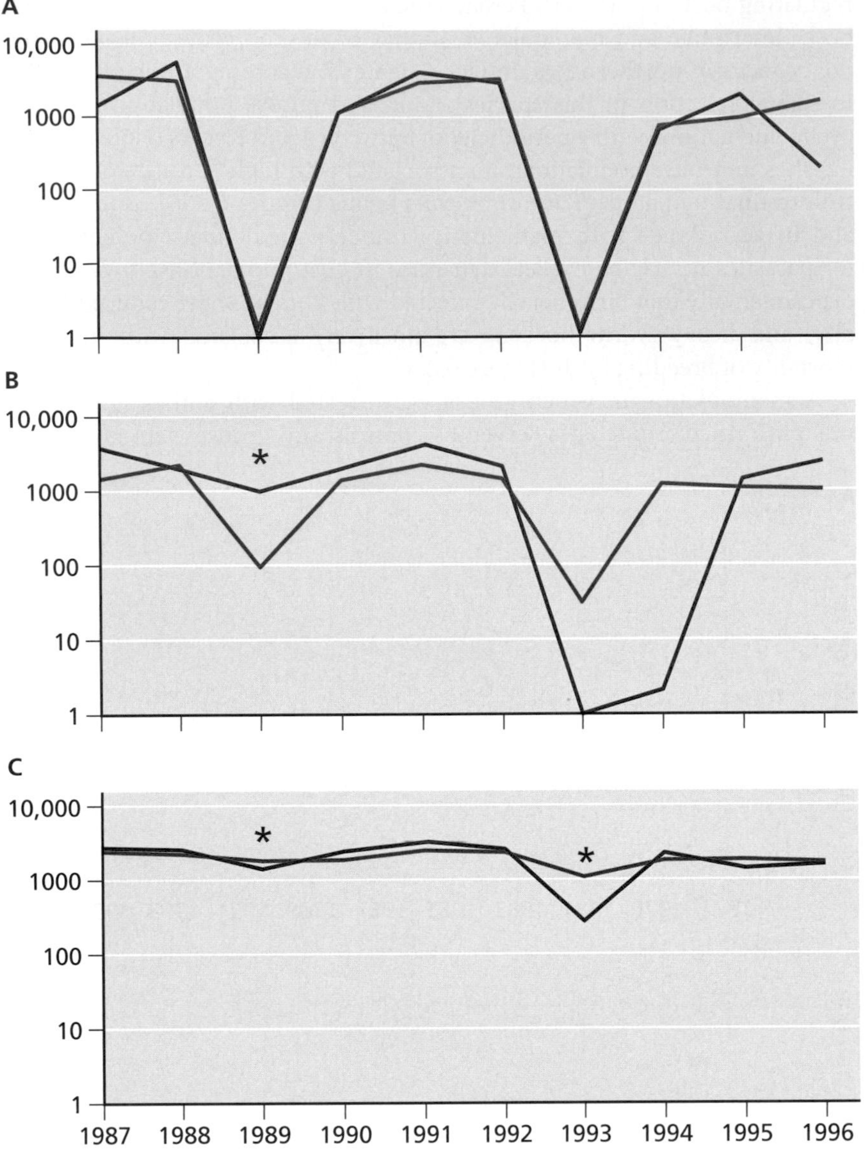

Figure 6.27 Effects of anthelmintic treatment of red grouse on their cyclical population sizes. (A) Two untreated control sites showing normal cyclical behavior. (B) Two populations with a single worm treatment each and (C) two populations with two treatments each. Asterisks represent the years of treatment when worm burdens in adult grouse were reduced. (From Hudson PJ, Dobson AP, Newborn D et al [1998] *Science* 282:2256–2258. With permission from AAAS.)

(**Figure 6.27A**). These sites showed typical cyclic behavior in the grouse populations. In experimental sites, where grouse were treated either once (Figure 6.27B) or twice (Figure 6.27C), treatment prevented grouse population crashes and thus reduced the cyclic behavior of the grouse population. The study concluded that grouse population cycles are driven by a single interaction - between grouse and nematode - and that cyclic population behavior is not intrinsic to just the grouse. By virtue of acting in a density-dependent fashion, this study also showed how a parasite can regulate host populations.

The preceding example points out the importance of a single factor in regulation of host population size, but most studies conclude that host population regulation is multifactorial. For example, the mountain hare (*Lepus timidus*) shows cyclical population fluctuations, and a nematode (*Trichostrongylus retortaeformis*) is again believed to play a role. However, with the mountain hares, and in other systems that do not show any pronounced cyclical changes in host populations, additional factors such as host movement, male territoriality, pulses in resource availability, and regional environmental variation are also believed to be involved in a complex set of interactions that contribute to host population cycles.

Parasites influence competitive interactions among hosts

We earlier discussed how parasites compete with one another. Here we focus on how parasites might influence the competitive ability of their hosts. Although most studies examine the effects of parasites on interspecific competition of host species, it should not be overlooked that parasitism can strongly influence intraspecific competition as well. For example, tadpoles of *Rana pipiens* infected with metacercariae of the trematode *Echinostoma trivolvis* had reduced growth and development compared to uninfected tadpoles, when reared with them at high densities. This suggests that when hosts were crowded and competing, parasitized hosts did less well than unparasitized hosts.

With respect to the effects of parasitism on interspecific competition between two host species, there are several ways the outcome could be affected. One is when the addition of a parasite simply augments a preexisting competitive disparity, such that a dominant competitor becomes even more so. An example is provided by the study of flour beetles. Usually *Tribolium castaneum* outcompetes *Tribolium confusum* when they are maintained together. This outcome is based on both competition for resources such as food and space and on the tendency for *T. castaneum* to preferentially consume eggs and pupae of *T. confusum*. When the beetles are exposed to infection with cysticercoids of the tapeworm ***Hymenolepis diminuta***, for which they serve as intermediate hosts, *T. castaneum* is even more likely to outcompete *T. confusum* and does so faster. *Tribolium castaneum* does so despite harboring heavier parasite burdens, probably because infection causes it to increase its consumption of *T. confusum* eggs.

Another possible outcome, parasite-mediated reversal of competitive dominance, is exemplified by the same *Tribolium* model system, but involving a different parasite, in this case the gregarine *Adelina tribolii*. Upon addition of the parasite, the tables are turned: *T. confusum* outcompeted *T. castaneum*. This outcome is perhaps not surprising simply because the parasite's effects on *T. castaneum* are far more pronounced (immature forms are often debilitated by the parasite) than on *T. confusum*.

In the preceding examples, the host species involved are competitors, but imagine a situation in which two host species do not compete for resources but simply happen to share a parasite. One example of this situation is provided by the cecal nematode *Heterakis gallinarum*, a common parasite of domestic chickens and turkeys. As touched upon in Chapters 1 and 3, this worm is interesting in a number of contexts, including its role as a vector for another parasite, the flagellated protozoan *Histomonas melagridis*. *Heterakis gallinarum* also infects ring-necked pheasants (*Phasianus colchicus*), which tolerate the parasite well, in contrast to the gray partridge (*Perdix perdix*), which is strongly and adversely affected by it. In situations that did not permit overlap or other competitive interactions between the two bird species, the intensity of infection acquired by partridges was strongly correlated with the intensity of infection in pheasants that had occupied the same areas the previous year (and thus seeded the soil with eggs of *Heterakis*). Furthermore, the intensities of infection acquired by partridges were large enough to adversely affect them. Thus a parasite originating in pheasants could be transferred to partridges at levels sufficient to harm them even though the two host species did not co-occur at the same time. The conclusion from this study was that the presence of a parasite that had a stronger impact on one host species than another may be a factor contributing to the long-term decline of partridge populations in the United Kingdom. Although it seems likely that more direct competitive interactions between pheasants and partridges may also be a factor, perhaps the most important point from this example is that parasites can play a role in the competitive displacement by a dominant (often invasive) species of a subordinate species, a theme we shall return to in Chapter 8.

Parasites can manipulate their hosts to affect the likelihood of predation

As already noted in Chapters 3 and 5, many parasites are **trophically transmitted**. That is, their transmission requires the host they are in to be eaten by the next host in their life cycle. An ever-lengthening list of studies has provided increasing documentation that trophically transmitted parasites can in fact manipulate their hosts in ways that range from subtle to spectacular (see Figures 5.15 and 5.18). In general, the parasites increase the host's conspicuousness and thus favor predation, which leads to parasite transmission to the new host (**Box 6.5**). The underlying mechanisms whereby such manipulations are achieved have been discussed in Chapter 5. Here we focus on some of the ecological implications of manipulation.

As alluded to in discussion of food webs, manipulative parasites can increase the strength of links between some prey and parasite species by increasing the likelihood of predation. Furthermore, such parasites can

BOX 6.5

Who's in Charge Here?

The ability of parasites to manipulate their hosts' behavior or appearance has, with good reason, become the stuff of legend, a favorite theme for TV shows and movies. Although most such manipulations are subtle or mundane, some are truly spectacular, with obvious ecological ramifications, and bear mention here. Recall the generalist apicomplexan *Toxoplasma gondii*, which can cause rats to lose their innate fear of cats (Figure 5.15). It also seems to influence human behavior as well, perhaps a legacy of times past when humans were regularly eaten by large felines (see Box 5.2).

Other parasites also get into the act. Adults of the digenetic trematode *Dicrocoelium dendriticum* (Figure 1A) live in the bile ducts of ruminants and other herbivorous mammals. The trematode's first intermediate hosts are numerous species of land snails that are rarely ingested by sheep. How do sheep become infected? The infected snails extrude slime balls full of cercariae that are eaten by ants (*Formica*). One of the consumed cercariae encysts in the subesophageal ganglion of the ant, precipitating a remarkable change in behavior. The infected ant will climb to the tip of a blade of grass, where it will firmly attach with its mandibles. Here it is much more vulnerable to accidental ingestion by a grazing sheep. The ant will detach, return to its colony, and repeat the behavior again the following day.

The nematode *Myrmeconema neotropicum* infects the ant *Cephalotes atratus*, causing the abdomen of the ant to swell and turn red (Figure 1B), such that it resembles a berry. Additionally, the infected ant forages outside the nest, extends its red abdomen upward, including even into a cluster of real berries. There it is likely to be eaten by a bird, which serves as a means to disperse the eggs of the parasite.

The last classic example, and one that exemplifies how the infected host becomes the extended phenotype of the parasite, is provided by *Leucochloridium paradoxum* (Figure 1C). In this case, the large brightly colored branches of the sporocysts of the trematode push their way into the tentacles of the land snail *Succinea putris*. These sporocysts contain metacercariae infective for the next host in the cycle, which is a bird. Because the tentacles are stretched so thinly, the sporocysts are clearly visible within and become all the more conspicuous both because the sporocysts regularly move back and forth within the tentacles (they are called "pumpers") and because the infected snail frequently occupies a conspicuous perch. This not only attracts the attention of parasitologists but also of birds who mistake the pulsating sporocysts for caterpillars. The bird will bite off and consume the tentacles and sporocyst within, becoming infected in the process. The snail is left unconsumed and will regrow a tentacle into which another branch of a sporocyst will push, thereby allowing a repeat performance.

Although these examples have become an important part of the lore of parasitology and provide stories that are just too good to pass up, surprisingly, they all require further documentation. Upon further quantitative study will they actually prove to function in the manner we are so keen to tell about or do they prove to be just anecdotal folklore?

References

Moore J (2002) Parasites and the Behavior of Animals. Oxford University Press.

Yanovial SP, Kaspari M, Dudley R et al (2008) Parasite-induced fruit mimicry in a tropical canopy ant. *Am Nat* 171:536–544 (doi:10.1086/528968).

Poulin R (2010) Parasite manipulation of host behavior: an update and frequently asked questions. In Advances in the Study of Behavior (Brockmann HJ ed), vol. 41, pp. 151–186. Academic Press.

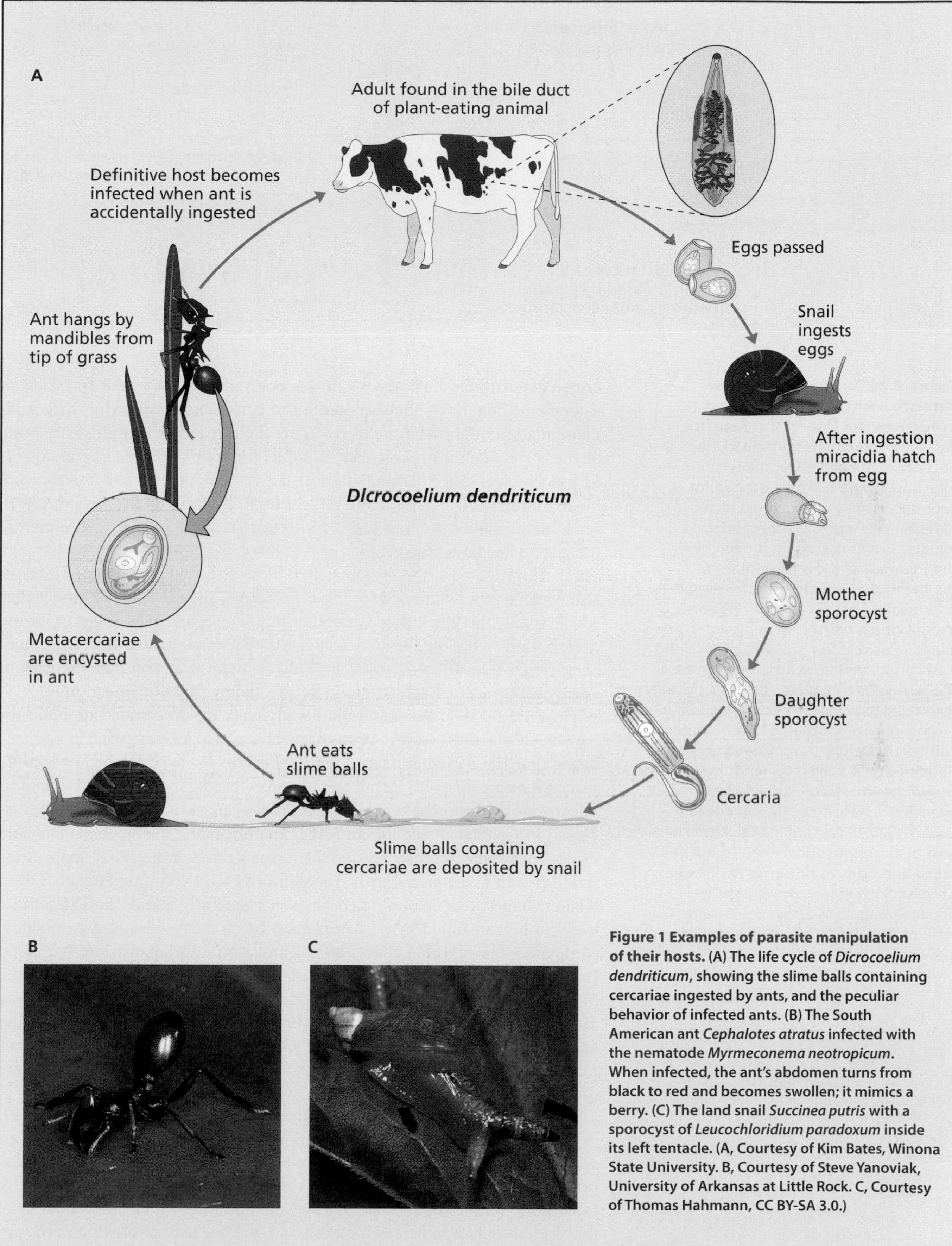

Figure 1 Examples of parasite manipulation of their hosts. (A) The life cycle of *Dicrocoelium dendriticum*, showing the slime balls containing cercariae ingested by ants, and the peculiar behavior of infected ants. (B) The South American ant *Cephalotes atratus* infected with the nematode *Myrmeconema neotropicum*. When infected, the ant's abdomen turns from black to red and becomes swollen; it mimics a berry. (C) The land snail *Succinea putris* with a sporocyst of *Leucochloridium paradoxum* inside its left tentacle. (A, Courtesy of Kim Bates, Winona State University. B, Courtesy of Steve Yanoviak, University of Arkansas at Little Rock. C, Courtesy of Thomas Hahmann, CC BY-SA 3.0.)

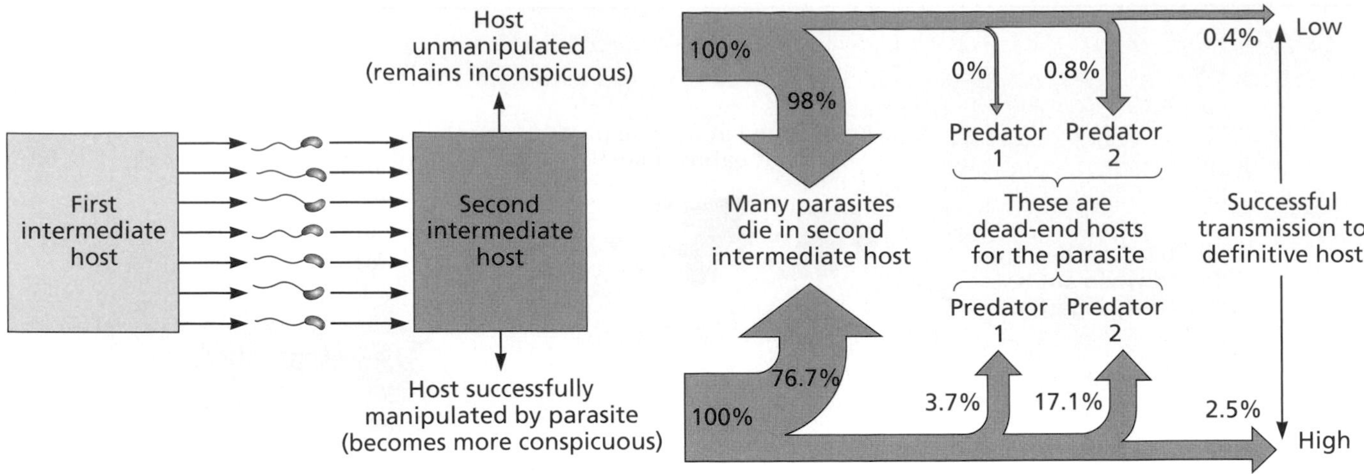

Figure 6.28 An example of how parasite manipulation of a host influences the fate of the parasite. Parasites from the first intermediate host penetrate and infect the second intermediate host. If the host has not been successfully manipulated by the parasite, it is able to burrow and so remains inconspicuous. As judged by the thickness of the arrows, many of the parasites die along with the second intermediate host. A few may be eaten by inappropriate hosts, such as predator 1 or 2, and relatively few are consumed by the appropriate definitive host. By contrast, if the second intermediate host is successfully manipulated and becomes conspicuous, a much higher proportion of the parasites will be successfully transmitted to the definitive host, even if some of these manipulated hosts are detected and eaten by inappropriate hosts. (Modified from Poulin R [2010] Parasite manipulation of host behavior: An update and frequently asked questions. In Advances in the Study of Behavior (Mitani J, Brockmann HJ, Roper T, Naguib M & Wynne-Edwards K eds), pg 151–186. With permission from Elsevier.)

create new trophic linkages within an ecosystem. **Figure 6.28** provides an example of how hosts that become more conspicuous following successful manipulation might be more likely to be eaten by either inappropriate hosts or the correct definitive host. For example, the trematode *Curtuteria australis* has as its second intermediate host the cockle *Austrovenus stuchburyi*, a bivalve mollusc. The parasite encysts in the foot of the cockle, in some cases reducing its ability to bury itself in intertidal sediments. As a consequence, the cockle becomes conspicuous and is more likely to be eaten by the parasite's definitive host, the oystercatcher. Note also that manipulated cockles also become more vulnerable to other predators, such as fish and predaceous whelk snails, thereby either strengthening or creating new linkages in the overall food web. Some evidence suggests that manipulated hosts are more energy-rich than those that are not, also providing a way that the flow of energy between trophic links could be affected by parasitic infection.

Given that so many manipulative changes in appearance or behavior have been observed to be imposed on hosts by parasites, what is the evidence that these changes are actually effective in increasing the parasite's successful transmission? Recall the example of killifish *Fundulus parvipinnis* infected with the metacercariae of *Euhaplorchis californiensis* already discussed in Chapter 5 (see Figure 5.18). The unusual, conspicuous behaviors exhibited by infected fish rendered them 30 times more likely than uninfected fish to be eaten by birds, the definitive hosts for the parasite. Other studies suggest that manipulated hosts are generally about 25%–35% more likely to be consumed by predators than hosts that were not manipulated, confirming the expected effects of manipulation on transmission. Increased consumption of manipulated hosts such as infected killifish suggests that the birds are undiscriminating dupes fated to suffer higher burden of infection. This might be selling the birds a bit short, however. If the cost of acquiring another parasite is low relative to the food value of the fish itself, then birds may come out ahead by searching for fish with abnormal behavior that are more easily caught.

Host manipulation is not without its potential downsides for the parasite. A hazard to the parasite is that its manipulated host will end up being so conspicuous that it will be eaten by the wrong host, thus causing the parasite's doom. The trematode–cockle system discussed above is again instructive (see Figure 6.28). Manipulated cockles are about 25 times more likely than uninfected cockles to be eaten by fish and whelks, neither of which can support further development of the parasite. It then seems such a manipulative strategy would be disadvantageous because more parasites are eaten by inappropriate hosts. However, this must be tempered by the realization that when the cockle is not manipulated, the probability that the parasite ends up dying with an unconsumed cockle also increases by about four-fold. Consequently, the advantages of manipulation will depend on the kinds and relative abundance of predators present. When there are proportionately more fish or whelks present, the advantage to manipulation declines. This example points out the importance of the broader ecological context in understanding the significance of manipulation.

Consider the additional cascading effects parasite manipulation can have. The bird trematode *Microphallus papillorobustus* infects as its second intermediate host the amphipod *Gammarus insensibilis*. It causes the infected gammarid to swim near the surface, where it is more liable to be ingested by a bird. The presence of the trematode in *G. insensibilis* also reduces the overall competitive advantage that this host has when interacting with the related host, *Gammarus aequicauda*, which is less affected by the parasite's manipulative ability and therefore less likely to suffer predation. In addition, when *G. insensibilis* is infected with *M. papillorobustus*, its altered position in the aquatic environment also facilitates its infection with another trematode parasite (*Maritrema subdolum*) that uses the same definitive host. Even though this latter hitchhiking species does not manipulate its host's behavior, it directly benefits from one that does.

Although we have discussed host manipulation with respect to favoring predation, it can be beneficial in other contexts as well. As just one example, recall the nematomorphs we discussed in Chapter 2. Here manipulation of the behavior of an infected arthropod host occurs, but in this case predation would mark the end of the road for the parasite. Rather, the goal is to encourage an otherwise terrestrial arthropod host to contact water, such that the adult worms can emerge, reproduce, and release their numerous progeny into the water. If a predator should by chance consume a nematomorph-infected insect, the nematomorphs within have yet another unique talent: they are good at squirming their way out of the predator's mouth. This escape behavior suggests nematomorphs have indeed been selected over time to compensate, if need be, for the alterations they impose on their host's behavior. Mermithid nematodes in mayflies can also provoke water-seeking behavior. When an infected female mayfly oviposits in a stream, the mermithids also pass their eggs into the water. If the nematode finds itself in a male mayfly, however, it is stuck because males do not return to water nor do they exhibit ovipositing behavior. The remarkable solution is that the mermithid significantly feminizes the male mayfly, provoking it both to return to water and to engage in oviposition behavior. Of course the male mayfly is not really ovipositing mayfly eggs, but rather is depositing the parasite's progeny into the aquatic environment (see also the example of *Sacculina* from Chapter 7).

Consider this final mind-boggling example of manipulation occurring in a different context, in this case to prevent predation. In this example, the braconid parasitic wasp *Glyptapanteles* sp. uses a caterpillar (*Thyrinteina leucocerae*) as a host (**Figure 6.29**). After the wasp larvae have left the caterpillar and pupated nearby, the caterpillar is inclined to wait near the pupae. The caterpillar, even though it does not feed, may pass protective silk threads over the pupae and further protect the pupae from disturbance by violent head-swings until they complete pupation. Once this is done, the caterpillar, which has both supported the larval development of the wasp and protected the wasp while it completes pupation, typically dies.

Figure 6.29 A former host protecting the brood of parasites that developed within it. A caterpillar of the geometrid moth *Thyrinteina leucocerae* guarding pupae of the braconid parasitoid wasp *Glyptapanteles* sp. (Photograph by José Lino-Neto [doi:10.1371/journal.pone.0002276].)

As a parting thought on this fascinating topic, if you take a look at the examples noted in this section, you will observe that they involve changes in host behavior that we can see. Humans are exquisitely visual creatures. When we understand something, we say, "Oh, I see." Yet from a parasite's standpoint, some manipulations might best play to other senses of their predaceous hosts. Given how notoriously unresponsive we are as compared to other animals with respect to smell or hearing, it seems there is much yet to learn by considering parasite-mediated cues (such as odor) that might attract the predator relevant to its needs. Clearly, this field will continue to attract the attention of parasitologists and the popular press alike, the latter always on the lookout for more jaw-dropping stories of real-life zombies.

6.7 ECOLOGICAL IMMUNOLOGY

Ecological immunology, or ecoimmunology, is a relatively new, fast-paced field that has at its core the idea that both biotic and abiotic features of the environment influence the evolution and function of immune responses and that the nature and extent of immune responses help us better understand an organism's ecology. **Figure 6.30** presents a conceptual framework for ecological immunology, a framework that emphasizes the value of a cross-disciplinary approach. The ideas on which ecological immunology are based permeate this book. Chapter 4 provides examples for how symbionts influence host defenses and Chapter 5 highlights that vigorous immune responses are often a double-edged sword that can result in pathology. The concept of trade-offs—for example, the notion that a strong immune response might divert energy from reproduction—is discussed in Chapters 5 and 7. Also emphasized earlier in this chapter is that hosts may deploy a variety of defenses against parasites, not always of an immunological nature. Chapter 7 also discusses the importance of parasite population genetics, concepts of parasite–host coevolution and the intersection of immunology with sexual ornamentation to affect mate choice, all part of the canon of ecological immunology. In this section, we touch on a few additional ideas that further flesh out the meaning of ecological immunology.

To provide just a few examples of the ecological inputs indicated in Figure 6.30, consider that the effectiveness of immune responses of hosts such

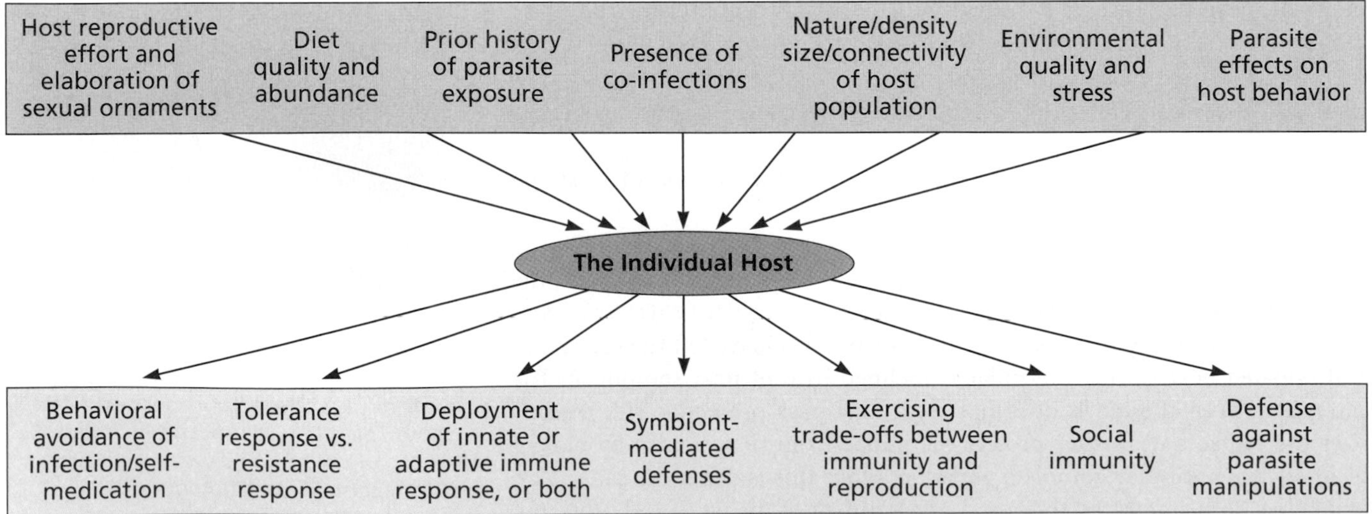

Figure 6.30 A conceptual framework for ecological immunology. An overview of ecological inputs that influence a host organism and the nature of the organism's defense or immune responses. The nature of defense and immune responses can be expected to be greatly influenced by the ecological context in which an organism occurs. This approach provides a framework to better understand the responses of organisms to their environments and provides a compelling way to think about the evolution of immune responses.

as the nematode *Caenorhabditis elegans*, water fleas (*Daphnia*), and fruit flies (*Drosophila*) to their respective parasites are all sharply influenced by environmental temperatures. Availability of nutrition also significantly influences the energy budget of organisms by determining how much energy they can afford to devote to costly defense responses. One example of the involvement of trade-offs between immune defense and both pathology and reproductive success is provided by feral Soay sheep living on islands of the St. Kilda archipelago off the coast of Scotland. Antibodies produced by sheep are effective in limiting infections by the gastrointestinal nematode *Teladorsagia circumcinta* and might be expected to reach very high levels, yet high levels of antibody production are also associated with increased production of self-reactive and damaging antibodies. This relationship suggests that the antibody response identified as maximum is not the optimum response. However, female sheep with the greatest antibody responses had the highest probability of surviving the winter (when parasites are most likely to cause mortality) (**Figure 6.31A**). Although females with high antibody titers were less likely to produce offspring, the offspring they did produce were more likely to survive the neonatal period (Figure 6.31B). The latter is suggestive of increased availability of protective maternal antibodies including those against parasites. For male sheep, those animals that had the highest antibody responses were unlikely to have successfully bred the previous year (Figure 6.31C), suggesting there was a cost to reproduction for producing high antibody levels. A reduced rate of reproduction or diminished competitive ability have often been noted in hosts that maintain high levels of resistance to parasites or that up-regulate strong anti-parasite responses (known as the cost of resistance).

With respect to immunological outcomes shown in Figure 6.30, as we noted earlier in the chapter, avoidance of exposure to infection is a common response of hosts. In contrast, the language adopted by parasitologists and infectious disease experts often emphasizes confrontation and resistance (elimination of infection). However, another trajectory is possible, one of **tolerance**. This idea has thus far received too little attention, especially from those studying the defenses of animals from infection. The essential concept is that it may be advantageous to the host to devise strategies for minimizing the harmful consequences of infection without limiting the presence of the parasite population. For example, some strains of mice are much better than others at maintaining their red blood cell (RBC) density levels and body mass in face of higher parasitemias with *Plasmodium chabaudi*. As shown in **Figure 6.32A**, the RBC density in DBA/1 mice (blue line) changes less with increasing parasite density (their line has the smallest slope value) than for other mouse strains. In Figure 6.32B, whereas mice of the NIH strain (green

Figure 6.31 The relationship between the parasite *Teladorsagia circumcinta* and immunity and fitness in Soay sheep from St. Kilda. (A) During harsh winters (blue bars), the proportion of sheep that survived was increased if they had antibody titers above a defined threshold (ANA+) as opposed to below the threshold (ANA-). The same was not true when winters were less harsh and did not provoke a crash (red bars). (B) The proportion of females that produced at least one offspring (red bars) was lower when antibody concentrations were above the threshold as compared to females below the threshold, but the proportion of females with at least one lamb that survived was considerably higher (blue bars). (C) The proportion of male sheep that sired offspring the previous year was lower when antibody concentrations were above the threshold as compared to males below the threshold (ANA-). (From Graham AL, Hayward AD, Watt KA et al [2010] *Science* 330:662–665. With permission from AAAS.)

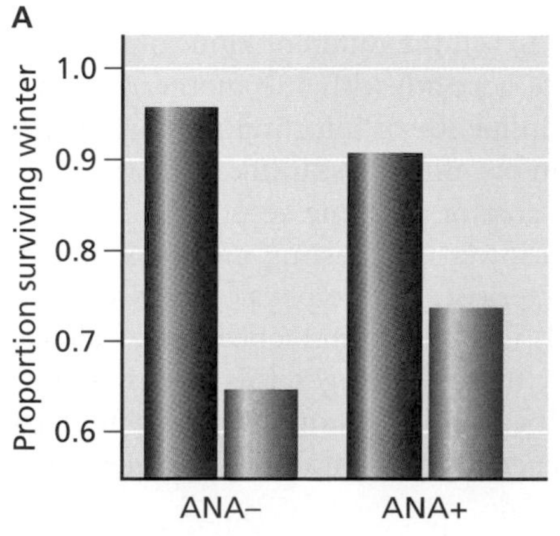

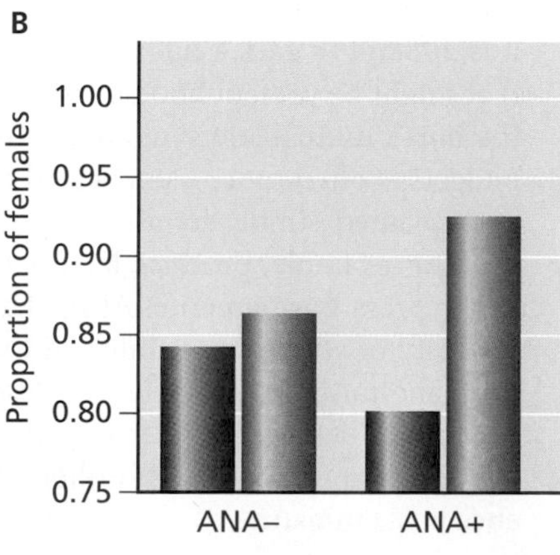

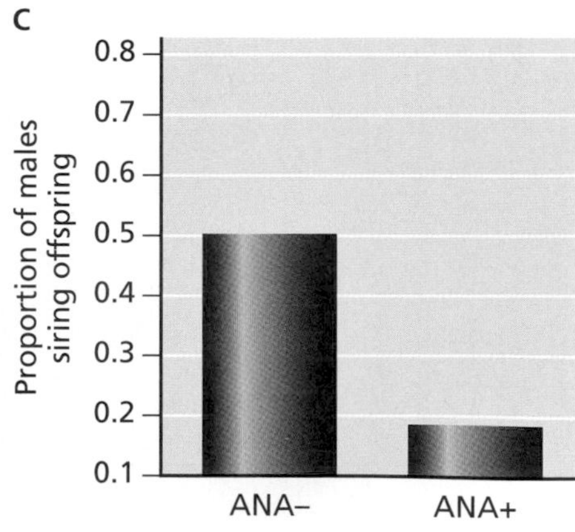

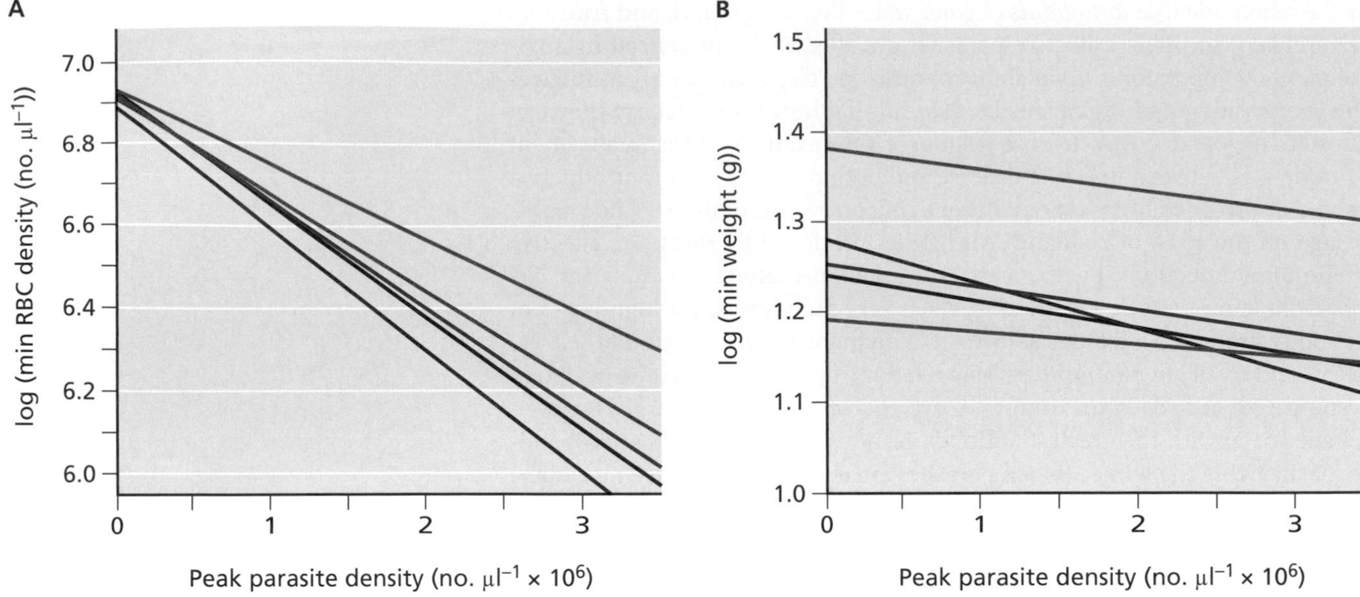

Figure 6.32 Five different inbred laboratory strains of mice differ in their tolerance to infection with *Plasmodium chaubaudi*, a rodent malaria parasite. Blue line, DBA/1 mice; green line, NIH mice; purple line, A/J mice; black lIne, CBA mice; red line, C57 mice. (A) Tolerance measured as the slope of minimum red blood cell (RBC) density (log-transformed) against peak parasite density. (B) Tolerance measured as the slope of minimum mouse weight (log-transformed) against peak parasite density. The ranking of the slopes for minimum RBC density and weight is the same for the five mouse strains, indicating that the two forms of tolerance are positively correlated. (From Råberg L, Graham AL & Read AF [2009] *Philos Trans R Soc Lond B Biol Sci* 364:37–49. With permission from Royal Society Publishing.)

line) generally lose less weight than mice of other strains and so are more vigorous by this measure, the DBA/1 mice again show the least change with increasing parasitemia. By both measures, the DBA/1 mice are more tolerant of *P. chabaudi* infections.

As a final point with respect to defense responses, the nature of these responses may be greatly influenced by the degree of sociality of the host. The social insects such as ants, bees, and termites provide the most dramatic examples. Selection by parasites has resulted in anti-parasite defenses in insect societies that are remarkably parallel to those seen in individual host organisms, further suggestive of insect societies functioning as superorganisms. A **superorganism**, well-exemplified by ant or termite colonies, consists of a social group of animals with pronounced division of labor and in which the individuals are unable to survive by themselves for extended periods of time. Insect colonies employ border defenses to prevent infection, and then if infection occurs work to minimize spread of infection among colony members. Finally, they employ defenses to prevent the infection of the reproductive individuals of the colony (see also social immunization in Chapter 4).

Ecological immunology is a flourishing field with much yet to offer. Improvements are needed with respect to defining the immune competence of hosts in a way that is more specific and relevant to the actual parasites they are encountering. Ecological immunology is also challenging because it is difficult to gain a full appreciation for all the different kinds of parasites that might be present and how they interact both with one another and with the host's mutualistic symbionts. The influence of infection on host energy budgets is often more complicated than we might first think, and it may not be explained simply by assessing the cost of immune responses. Parasites may, for example, produce factors that lower host resting metabolic rate or that repress host appetite. Also, characteristics of the physical environment (severity of winter, availability of food) will greatly influence energy allocations and immune competence. Despite these challenges, this field is exciting because it lies at the intersection of parasitology, immunity, evolutionary biology, and ecology, and it offers deeper insights into how parasites influence hosts in nature.

6.8 THE METABOLIC THEORY OF ECOLOGY AND PARASITES

Another exciting development worthy of the attention of aspiring parasite ecologists is the metabolic theory of ecology. This theory is one of the most innovative and broadly applicable recent developments in ecology. Metabolic theory postulates that the metabolic rate of organisms is of fundamental importance and ultimately governs most patterns observed in nature. The theory is based on a basic relationship between body size and metabolic rate that applies to all organisms, including parasites. The most basic relationship is embodied in Equation 6.1:

$$B = B_0 M^{3/4} \quad [6.1]$$

where B is whole organism metabolic rate (in watts or another unit of power), M is organism mass (in kg), and B_0 is a mass-independent normalization constant (given in a unit of power divided by a unit of mass, in this case, watts per kilogram). This basic relationship can be expanded to include another term to account for the increase in the rate of chemical reactions with temperature. Note that a scaling factor is a number that scales or multiplies some quantity and, in this equation, the scaling factor is raising M to the ¾ power. This ¾ value is based on the idea that metabolism is fundamentally governed by the organism's distribution networks (such as the circulatory system) for delivering nutrients and that it takes larger organisms longer to distribute nutrients, explaining their lower metabolic rates. The specific value of ¾ derives from the fact that distribution systems terminate in structures such as capillary beds and that the number of such structures is proportional to a ¾ power relationship with body size. Others argue that a ⅔ relationship is more accurate, this value based on the relatively larger surface areas that smaller organisms have relative to volume. Consequently, smaller organisms dissipate heat faster and therefore must have higher metabolic rates.

Consider that the amount of energy required for an organism parallels its metabolic rate which, as Equation 6.1 points out, is dependent on body size. The more resources required, the less abundant organisms will be, according to this relationship:

$$N = iM^{-3/4} \quad [6.2]$$

where N is population abundance and i is a normalization constant. Note how estimations of the abundance of organisms are based directly on their metabolic rate. The metabolic rate is influenced by their body size and by temperature, which for simplification we have not considered here. That is, large organisms are expected to be rarer than smaller ones.

Next, consider parasites and how to estimate their abundance. Keep in mind that parasites consume energy taken from their hosts, just as a lion consumes energy taken from a zebra. A parasite exists at a higher trophic level than its hosts. Furthermore, there is a particular efficiency with which energy is transferred from one trophic level to the next. The efficiency of energy transfer is never 100%, so the amount of energy available diminishes at each successive trophic level. To account for this, the basic relationship in Equation 6.2 has been modified as follows:

$$N = iM^{-3/4} \varepsilon^{L} \quad [6.3]$$

where ε is trophic transfer efficiency and L is trophic level. This equation accounts for the loss of energy from one trophic level to another and captures the exponential loss of abundance with trophic level. In a study of California estuarine systems (see also the discussions of trematode guilds and food webs above), it was initially shown that the relationship between abundance and body size scaled differently for free-living and parasitic

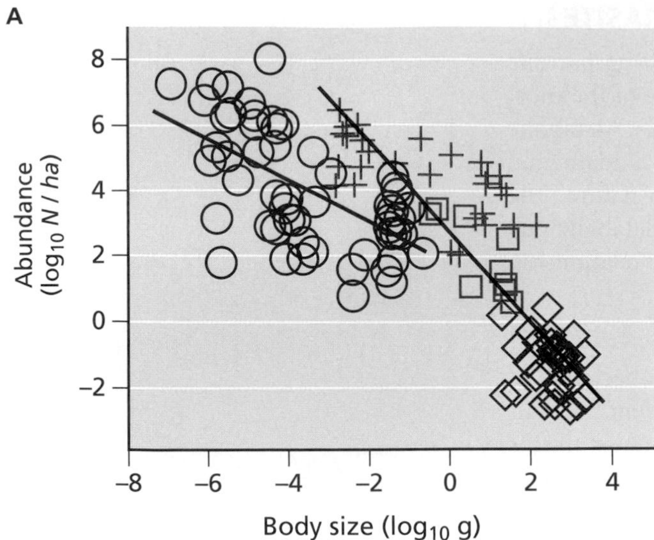

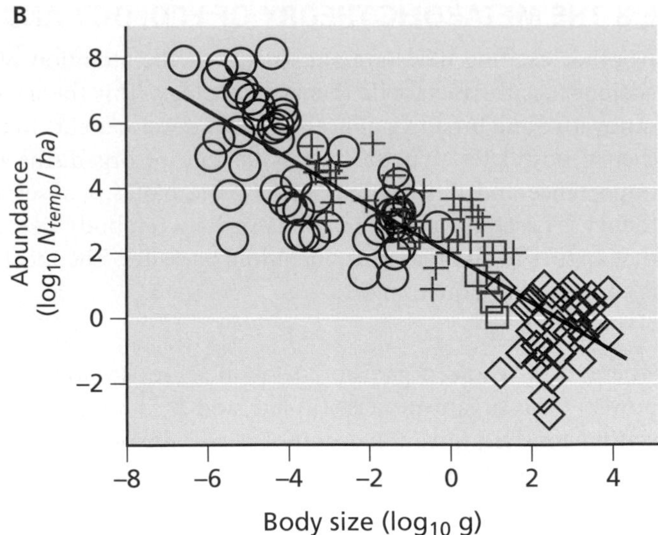

Figure 6.33 The metabolic theory of ecology applied to parasites. (A) A plot of abundance vs. body size reveals that a single regression line cannot adequately fit the data for all groups of both free-living and parasitic organisms included (circles, parasites; crosses, invertebrates; squares, fish; diamonds, birds). Notably the slopes for parasites and free-living organisms differ. (B) Similar plots but in this case controlling for trophic level. Notice that in this case all organisms scale with the same slope, the predicted −3/4 derived from metabolic scaling theory. Ha in the vertical axis refers to hectares. (From Hechinger RF, Lafferty KD, Dobson AP et al [2011] *Science* 333:445–448. With permission from AAAS.)

organisms (**Figure 6.33A**). However, when the term to account for trophic transfer was included, parasites and free-living organisms scaled similarly with the same slope, the −¾ predicted by metabolic ecology theory (Figure 6.33B). Fundamentally, abundance of both free-living and parasitic organisms scale in the same way with body mass.

Other applications of metabolic ecology theory have documented that parasite abundance scales with parasite body size within particular hosts. These studies and the example discussed above provide a glimpse into the power of metabolic ecology, building on very simple relationships between body mass, temperature, and metabolic rates to provide a predictive framework for how abundant organisms (including parasites) are in particular environments.

6.9 EPIDEMIOLOGY AND ITS RELATIONSHIPS WITH ECOLOGY

According to the World Health Organization, "**epidemiology** is the study of the distribution and determinants of health-related states or events (including disease), and the application of this study to the control of diseases and other health problems." Although epidemiology considers all sources of disease (including those related to diet or occupation), our focus will be on epidemiological investigations of parasites. Again, the word parasite is used in our preferred broad sense as introduced in Chapter 1.

The disciplines of epidemiology and ecology have much in common. Both seek to understand patterns in the distribution and abundance of organisms (such as a particular parasite species that causes a human disease), and both are interested in revealing the processes (determinants) that underlie the patterns. The goal of epidemiology is often to identify risk factors and to anticipate and prevent disease outbreaks. Epidemiological studies are often undertaken in the context of human public health, but epidemiologists also study disease outbreaks in different kinds of hosts, ranging from oak trees to honeybees to horses. For example, the term **epizootiology** refers to the study of the distribution and determinants of disease in animals. Epidemiologists are interested in a triad of factors (host, parasite, and environment) and how they intersect to create the opportunity for disease transmission.

Epidemiological investigations are often initiated following an unexpected outbreak of a disease, an epidemic. An **epidemic** is any unusual and unexpected outbreak of disease above what is expected based on recent experience. For some diseases, a single case could be considered an epidemic if it occurred in an unusual setting. For example, one case of monkeypox (generally limited to Africa) encountered in a human in North America would be considered an epidemic. In other cases, such as the common cold that occurs at high background levels, an epidemic would be defined as an outbreak significantly above that high background level. If an epidemic is occurring, two important goals are to determine what is causing it and what factors are enabling it to spread (how is it being transmitted). Also relevant is the concept of a **pandemic**, an epidemic of an infectious disease that has spread across large regions including continents or even across the entire world. The rapid and still ongoing spread of HIV throughout the world is an example of a pandemic.

Epidemiologists are also interested in diseases that are **endemic**, meaning that they enjoy a steady maintenance in a particular host population without the need for input of infected individuals from other populations. When studying such diseases, the focus is often on elucidating the factors that enable some endemic diseases to persist year in and year out.

Of central concern to epidemiologists are both incidence and prevalence of a particular infection. **Incidence** refers to the number of new cases acquired in some unit of time and is usually expressed as a proportion or rate. As stated earlier in the chapter, prevalence refers to the number of cases existing in the target population (often stated as the percentage of host individuals infected). It is a measure of how widespread a disease is.

Many epidemiological investigations proceed along the following lines. First, local health authorities notice that an unusual cluster of cases of a particular ailment has occurred and bring it to the attention of local and national epidemiologists and other public health authorities. Communication with other public health authorities is an important early step as this can help determine if the outbreak is isolated. Another early step is to determine if there are commonalities among all the cases that might point to a single source of infection. An accurate diagnosis of the ailment is critical because then appropriate treatment, vaccination, or quarantine procedures can be implemented to limit spread of the disease. Depending on the concern generated, efforts might also be initiated to model the spread of the infection. If a common source is found, measures are taken to understand why it became a source of infection in the first place and to insure it can no longer serve as a source of infection. Epidemiologists keep relevant public health authorities informed, prepare for more cases should they occur, and target individuals at high risk of infection. Educating the public about how to recognize the infection and how to prevent future infections from occurring are also important measures to take. Follow-up to ensure that improvements have been made to prevent future outbreaks is also warranted. In this way, the distribution and determinants of a disease will be revealed and what was learned can be used to implement control measures, in accordance with the basic definition of epidemiology.

Modeling is an invaluable approach to the study of infectious diseases

In addition to traditional responses to disease outbreaks described above, another important aspect of epidemiology is to develop mathematical models that enable investigators to understand more precisely the basic factors involved in the origin, spread, and probable future course of an infection. A shared tradition of model building draws the fields of ecology and epidemiology together. Such models can serve as a valuable guide to workers in the field to collect particular kinds of critical information that otherwise might not have seemed obvious. Once developed, their use can hopefully be extended to multiple locations and contexts, that is, models are generalizable.

There has been a long history of mathematical modeling of infectious diseases, starting with Daniel Bernoulli's 1766 model of smallpox outbreaks, an approach that took into account the spread of the disease and the role of immunity in preventing it. Ronald Ross, the discoverer of the role of mosquitoes in the transmission of malaria, made another notable early attempt. He modeled the transmission of malaria using a few equations to describe changes in the densities of susceptible and infected people and of susceptible and infected mosquitoes. From his model he concluded that to control malaria it was not necessary to eliminate all mosquitoes but merely to reduce them below a particular threshold value. This model is still a starting point for today's more elaborate models of malaria transmission.

In general, infectious disease modeling has continued to grow and to become ever more sophisticated, with one of the landmark developments being the publication in 1991 of Anderson and May's volume *Infectious Diseases of Humans*, which summarized much of the authors' original work on modeling transmission of both micro- and macroparasites. We discuss some of the basic concepts that guided that work. Also provided are a few examples showing how models have become more sophisticated by including factors such as indirect life cycles involving vectors and spatial and temporal considerations. Such models have become more accurate summations of reality. The models discussed are **deterministic models**, meaning they behave in a fixed way and will always return the same result if provided with the same starting parameter values. Another broad category of models consists of **stochastic models** that take into account that the real world does not behave in a deterministic way and that chance events (variation) can cause different outcomes to be recorded.

Microparasites exemplify basic modeling approaches that estimate population size and clarify transmission

Because it is difficult, if not impossible, to enumerate the actual number of individual microparasites (viruses, bacteria, or protozoans) inhabiting the body of an infected individual, a common approach for modeling microparasite infections has been to use the individual host as the basic unit of study. One basic approach (**Figure 6.34**) is to divide the host population into three categories: susceptible (S), infected (I), and recovered (R), the *SIR* model first developed in depth by Kermack and McKendrick and later expanded and popularized in treatments by Anderson and May and by Keeling and Rohani.

Inherent in this model is that recovery can occur (immunity develops). Similar models can be devised in which recovery does not occur (*SI* models). Figure 6.34 indicates the flow of hosts among susceptible, infected, and recovered categories. The number of individuals at a particular time in each of these categories is stipulated by S, I, and R, respectively. Susceptible individuals are recruited into the population by birth, at a per capita rate, b, appropriate to each of the three categories of hosts. Hosts die at a per capita natural rate of μ. Note that those infected individuals suffer an additional death risk, α, the mortality induced by exposure to the parasite, essentially a measure of the parasite's virulence. The term β is the transmission term, which refers to the probability of transmission of the parasite to a new host upon encounter. The term β is a measure of how contagious or transmissible a particular parasite happens to be. The rate at which susceptibles

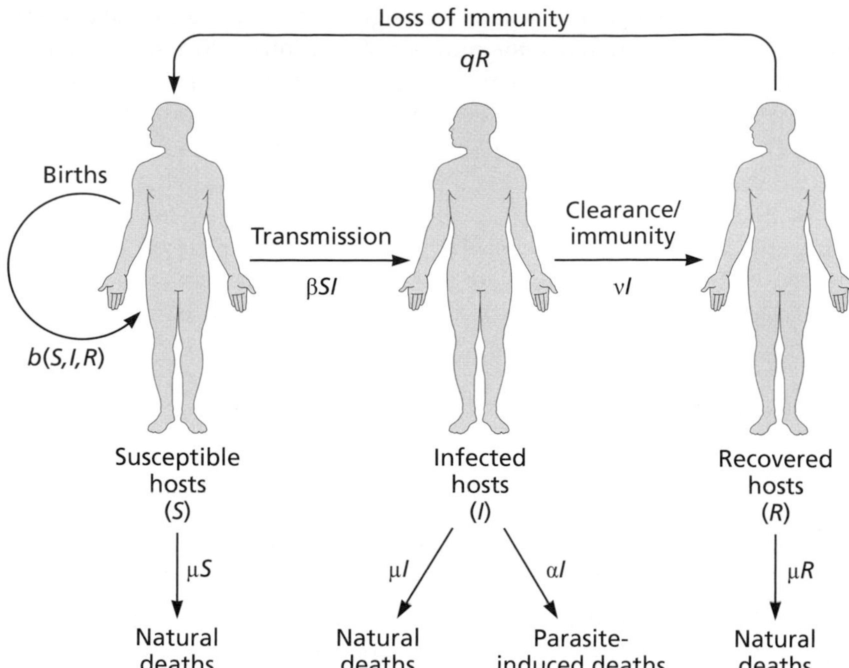

Figure 6.34 A basic *SIR* model for a microparasite. (Adapted from Schmid-Hempel P [2011] Evolutionary Parasitology, The Integrated Study of Infections, Immunology, Ecology and Genetics. Box 11.4 pp287. With permission from Oxford University Press.)

are converted into infecteds is dictated by the term βSI, which makes the assumption that transmission is dependent on the abundance of both susceptible and infected individuals. For example, the more infected individuals, the more probable a susceptible individual bumps into one of these individuals and becomes infected. The term ν refers to the host's per capita rate of recovery from infection. The likelihood of moving from the I category to the R category will depend on how many infecteds there actually are (hence the term νI). Lastly, there is a chance that immunity will be lost at a per capita rate of q and that the number of individuals moving from the R to the S category will be given by qR.

The flow diagram in Figure 6.34 identifies key rate processes that influence movement from one category to another. Notice that this model is simplified, for example, in assuming that there is a single term to describe the rate at which susceptibles become infected because we know that this process varies among individual hosts. Also, the nature of the population considered will be critical. Is it an old or a young population or is it one with lots of migration, for example? The important point is to begin by focusing on the basic form of the model; other versions of similar models take such variation into account.

The flow diagram in Figure 6.34 can be used to derive a series of three differential equations to define the rate of change in size of the three categories of individuals, S, I, and R.

$$\frac{dS}{dt} = b(S,I,R) - \mu S - \beta SI + qR \qquad [6.4]$$

$$\frac{dI}{dt} = \beta SI - \mu I - \alpha I - \nu I \qquad [6.5]$$

$$\frac{dR}{dt} = \nu I - \mu R - qR \qquad [6.6]$$

Note that the general terms and their signs in the equations make sense. For example, in Equation 6.4, the number of susceptibles will be influenced by how many births (new susceptible individuals) are occurring, thus augmenting this category [$b(S,I,R)$]. The size of the S category will be diminished by the death of some individuals (μS) and by the conversion of some into infected individuals (βSI), and it will be augmented because some individuals lose their immunity (qR).

Imagine that a parasite is introduced into a naïve population comprised entirely of individuals in the S category. In this case the entire population, designated by N, consists of susceptibles. In this case, another equation can be derived as follows:

$$R_0 = \frac{\beta N}{\mu + \alpha + \nu} \qquad [6.7]$$

R_0 refers to the basic reproductive rate of the parasite (the average number of secondary cases resulting from introduction of a primary case into a susceptible population). It makes sense that this value will be directly influenced by the size of the population, N, and by the transmission rate β. It also makes sense that it will be eroded by high values of μ, the host's per capita natural death rate, by α, the parasite-induced per capita host mortality rate (the virulence term), and by ν, the per capita rate of recovery from infection.

R_0 has become a convenient measure—perhaps too convenient—to determine the potential of spread for an infection. If $R_0 > 1$, unless purely chance events happen to wipe out early infections, the infection will spread. If $R_0 < 1$, it will die out. If $R_0 = 1$, a stable endemic state results. The higher the value for R_0, the more likely an infection will spread. Some parasites such as feline immunodeficiency virus have R_0 values slightly higher than 1, whereas others such as whooping cough are estimated in the 16–18 range. Malaria in hyperendemic areas can be as high as 80.

Models of macroparasite populations and transmission involve keeping track of individual parasites

With respect to macroparasites such as helminths or arthropods, it is actually possible to count the number of parasites per host, so a different modeling approach is taken, one based on keeping track of individual parasites as well as hosts (**Figure 6.35**). Some of the terms as described above are familiar, but there are some new ones as well, reflecting the fundamentally different circumstances of a macroparasite. In this figure, note that the emphasis is on the parasite that exists part of its time within its hosts (H) indicated by the box, and some part outside the box, in a free-living stage. The hosts have a natural mortality rate μ_H, which is augmented by parasite-induced mortality, α. Parasites within the host exist as juveniles (J) with a characteristic death rate of μ_1, or as adults (P) with their death rate of μ_2. Notice also the appearance of a term for parasite fecundity, φ, which influences the number of parasites reaching the external environment, designated by the term W. The free-living parasites might, for example, represent the egg stage of the macroparasite. The transmission rate from free-living parasites to the host is designated as β.

The flow diagram in Figure 6.35 is again the conceptual basis for deriving a series of differential equations, in this case to define rates of changes in the population sizes of hosts and in juvenile, adult, and free-living parasites. For illustrative purposes, Equations 6.8 to 6.10 are provided for hosts, adult parasites, and free-living stages of parasites.

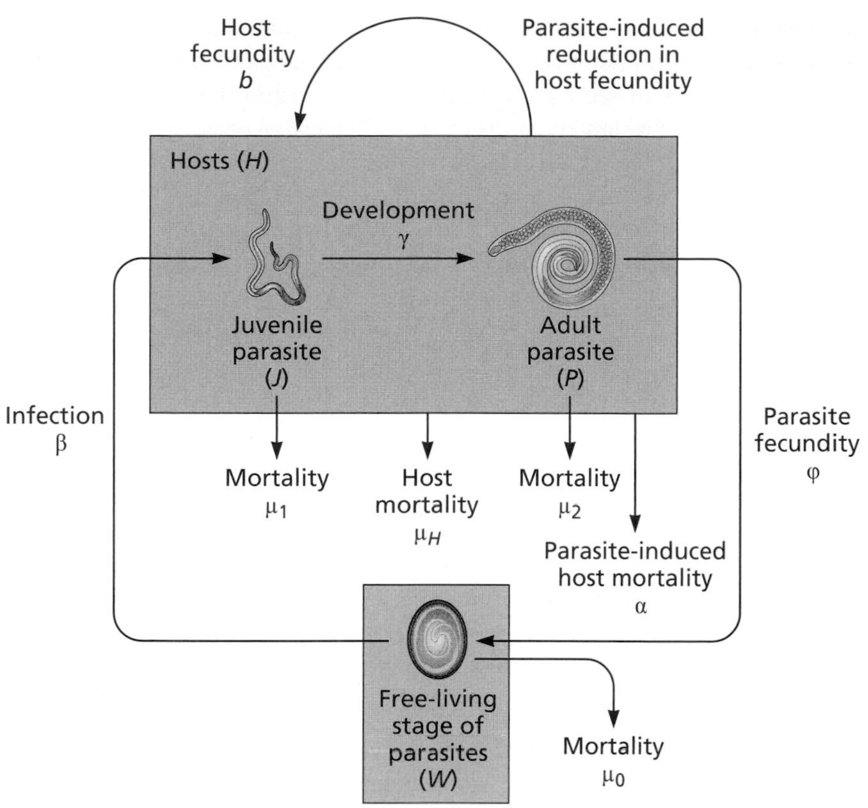

Figure 6.35 A basic macroparasite transmission model. (Adapted from Schmid-Hempel P [2011] Evolutionary Parasitology, The Integrated Study of Infections, Immunology, Ecology and Genetics. Box 11.7 pp301. With permission from Oxford University Press.)

$$\frac{dH}{dt} = (b - \mu_H)H - \alpha P \quad [6.8]$$

$$\frac{dP}{dt} = \beta WH - (\mu_2 + \mu_H + \alpha)P - \alpha\left(\frac{P^2}{H}\right)\frac{(k+1)}{k} \quad [6.9]$$

$$\frac{dW}{dt} = \varphi P - \mu_0 W - \beta WH \quad [6.10]$$

Here the terms of the equations are a bit more complicated, and another term, k, has been introduced into Equation 6.9. The variable k, refers to the degree of aggregation shown by macroparasites (see also page 204) according to the binomial distribution. Low values of k (usually less than 1) mean the parasite is strongly aggregated within the host population, whereas high values (such as 10) mean the parasites are randomly distributed among hosts. Although more complicated, the equations are still logical. To demonstrate, consider Equation 6.9. It makes sense that the size of the adult parasite population would be augmented by the transmission of the free-living parasites to the host population (the βWH term). Also, it would be expected to be diminished by the mortality of the adult parasites, by the natural mortality of the host, and by the parasite-induced mortality, and it would be expected that the strength of these effects would be enhanced the bigger the size of the parasite population [the $(\mu_2 + \mu_H + \alpha)P$ term]. Note the final term in the equation that also diminishes the size of the adult parasite population size as a result of parasite-induced host mortality. This term indicates how the rate at which hosts are killed by parasites is influenced by the average number of parasites per host, P/H, and by an index of aggregation, k. The greater the degree of aggregation of the parasite within the host, the smaller k becomes and the larger the term $(k + 1)/k$ becomes. Higher average parasite loads and higher degrees of parasite aggregation would increase the probability that a host was killed by the parasite, thereby diminishing the adult parasite population size.

Once again, a value for R_0 can be determined, as follows:

$$R_0 = \frac{\beta\varphi H}{(\mu_2 + \mu_H + \alpha)\,(\mu_0 + \beta H)} \quad [6.11]$$

Note in this expression the appearance of a term $\beta\varphi$ indicative of the parasite's effective fecundity (of getting infections to new hosts), which will enhance the value of R_0. Also included in this expression is the average life span of an infection in a host $(1/(\mu_2 + \mu_H + \alpha)$ and the average life span of a free-living parasite $(1/(\mu_0 + \beta H)$. The higher the mortality rates involved for either host or parasite stages, the lower R_0 will be.

Models for parasites with complex life cycles involving vectors become more complex

The models discussed above refer to parasites with relatively simple direct life cycles. What happens when, as with malaria parasites, a vector population also has to be considered? To give a flavor of the factors involved, consider one equation (others also exist) devised to represent R_0 for malaria transmission. Here, β refers to the rate at which the vector bites people, with the term being squared because the vector has to bite twice for transmission to occur (person to vector, vector to person). The β term clearly has an important impact on the potential for malaria to spread. The terms f_1 and f_2 refer to the fraction of people and mosquitoes, respectively, that survive long enough to be infectious. This is an important issue in malaria transmission because the parasite takes several days to develop to infectivity in the mosquito, a time period that represents a significant portion of the adult mosquito's life span. Many mosquitoes will die before the malaria parasites they have acquired complete their development and are ready to be transmitted. Therefore, f_2 may be very small and could have an overall limiting effect on transmission. N_1 and N_2 refer to the densities of human and mosquito populations, respectively. The terms $1/(b_1 + \gamma_1)$ and $1/(b_2 + \gamma_2)$ refer to the life expectancies of infectious people and mosquitoes, respectively. The term $1/b_1$ refers to the life expectancy of humans and $1/\gamma_1$ to the duration of human infectiveness. The term $1/b_2$ refers to life expectancy of mosquitoes and $1/\gamma_2$ to the duration of infectiveness in mosquitoes.

$$R_0 = \frac{\beta^2 f_1 f_2 N_2/N_1}{(b_1 + \gamma_1)(b_2 + \gamma_2)} \quad [6.12]$$

This basic expression is very instructive for highlighting the importance of certain key variables such as mosquito biting rate and the short life span of adult mosquitoes relative to parasite development time on R_0 for malaria. Of course there are several aspects of malaria transmission in the real world that complicate this basic expression. For example, recent models of malaria transmission include as many as 17 parameters and take into account considerations such as immigration rate of humans, the effect of density dependence

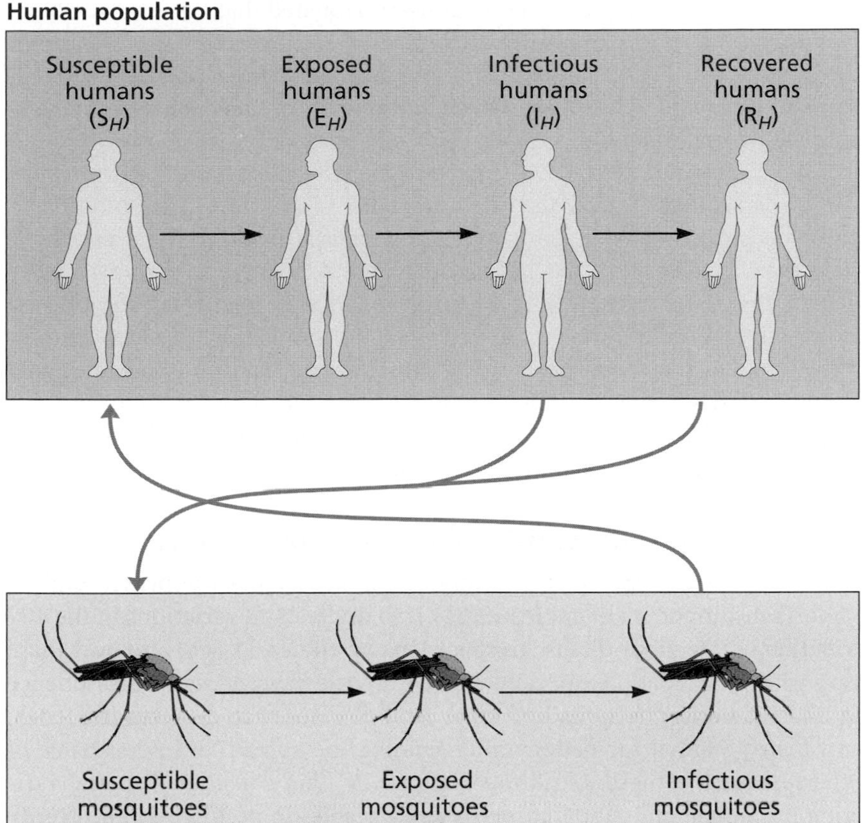

Figure 6.36 A flow diagram of a more complex model of malaria transmission. Susceptible humans, S_H are infected when they are bitten by infectious mosquitoes, I_V. Humans then progress through the exposed, E_H, infectious, I_H, and recovered, R_H, classes, before reentering the susceptible class. Susceptible mosquitoes, S_V, can become infected when they bite infectious or recovered humans. The infected mosquitoes then move through the exposed, E_V, and infectious, I_V, classes. The size of both humans and mosquito populations are also modeled, with humans having additional immigration and disease-induced death. Birth, death, and migration into and out of the population are not shown. (From Chitnis N, Hyman JM, Cushing JM et al [2008] *Bull Math Biol* 70:1272–1296. With permission from Springer Science and Business Media.)

on mosquito mortality, malaria-induced mortality in people, and loss of human immunity. Nonetheless, such models are built on similar-looking flow diagrams (**Figure 6.36**) and return some familiar conclusions. In both low and high transmission areas, one model concluded R_0 was most sensitive to mosquito biting rate, yielding the practical conclusion that interventions that diminish mosquito biting such as use of bednets treated with insecticides and indoor spraying of residual insecticides can have a large impact on reducing transmission. Recall also Figure 3.17 for another example of malaria modeling, one that emphasizes the role of rainfall and temperature in influencing the ability of mosquitoes to transmit malaria to people.

New models open the black box and estimate microparasite populations within hosts and the influences on them

As noted above, models of microparasite dynamics often use the infected host as the unit of study rather than attempting to estimate the vast populations of microparasites that occur within a host (such as the number of malaria parasites in the blood of a child). A new modeling approach adopts very simple principles developed for modeling the spread of infection between individual hosts and applies them to the spread of infection among cells within an individual host. Observations made on the known course of *Plasmodium chabaudi* infections over time in mice have been used to validate this modeling approach. Parasitemias associated with malaria are known to be influenced by host immune responses, interactions among the parasites within the blood, and availability of resources such as red blood cells to infect. A propagation number (P_e) has been estimated that takes into account how many merozoites are released when an infected red blood cell ruptures, the contact rates between merozoites and uninfected cells, and the probability that invasion of new cells will occur. Discrepancies between observed and predicted values of P_e, for example, enable an estimation of the role of immunity in eliminating infected cells. Among many results stemming from such an approach, it was shown that the availability of uninfected red blood cells could become a major limiting factor for malaria dynamics within a host. Also, the effectiveness of the immune response depended on the initial inoculation dose (**Figure 6.37**). Note for instance that at higher inoculation doses, after a strong bout of killing of infected cells at about eight days, that immunity wanes but then increases again at about days 20–25, only to wane again at about 35 days. This example points out some of the very useful insights potentially to be gained by attempting to understand more explicitly microparasite dynamics within individual hosts.

Models need to take spatial and temporal factors into account

Infectious diseases are not uniformly distributed across the planet, and an important component of epidemiology is to understand variations in the distribution of infectious disease in space (spatial epidemiology) and how these vary with time. Spatiotemporally explicit models have achieved prominence in understanding the spread of some viral diseases such as avian influenza, and their potential for better understanding the spread and persistence of other parasites is high. Recall too Figure 3.17 that considered spatial variation in rainfall and temperatures as key variables in predicting the severity of malaria.

One approach has been to use **geographical information systems (GIS)** in conjunction with remote sensing from satellites to pinpoint where particular diseases occur. GIS capture, maintain, analyze, and display all kinds of information (rainfall, school districts, satellite images of vegetation, parasite prevalence records, soil type, and so on) in a geographically (spatially) explicit way. This information can then be used to understand the underlying causality for such distributions, to target control efforts, and to ascertain how such distributions might change with, for example, changing climate.

Recent studies of schistosomiasis have provided clear evidence of the link between climatic factors and large-scale distribution patterns of the disease.

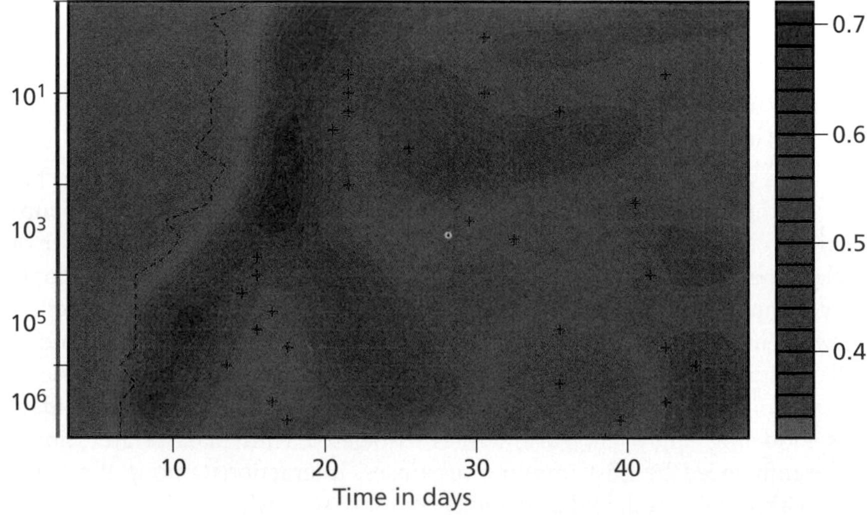

Figure 6.37 Explicit models of microparasite populations within hosts. This graph provides a two-dimensional visualization of the efficacy of immunity in *Plasmodium chabaudi*-infected mice indicated by a color representing the proportion of infected cells that are killed by immunity (see right vertical axis) over time (horizontal axis) as a function of inoculation dose (various doses shown on left side). The dashed line represents the first peak parasitemia experienced by mice. Note how the position of this line is shifted later at lower inoculation doses. The crosses indicate the time of later peaks in parasitemias. (From Metcalf CJE, Graham AL, Huijben S et al [2011] *Science* 333:984–988. With permission from AAAS.)

In Uganda, a survey of 68 communities was undertaken to establish reference points for the distribution and prevalence of ***Schistosoma mansoni***. Other layers of environmental information, including some derived from remote sensing, such as rainfall, land surface temperature, vegetation cover, and elevation were also included in the GIS. Resultant maps incorporating parasite data and other environmental information indicated that prevalence of infection was highest within 5 km of large lakes, where the frequency of water contact is high and snail populations are large. Conversely, no transmission occurred at altitudes >1400 m (probably because temperatures are too low to permit development of larval schistosomes in snails) or where total rainfall was <900 mm (inadequate amounts of water to support persistent snail populations). The results were presented at a national schistosomiasis control planning workshop and used to target control efforts more effectively.

Such studies have become more and more sophisticated through the incorporation of model-based geostatistics. An essential feature of such models is that they use measured values from a finite number of sampling localities to enable spatially explicit predictions in locations that were not sampled. This approach allows, for example, construction of maps with contours estimating risk of infection, thus providing plausible information to inform the design and implementation of control programs. Using such an approach, the distribution of co-infections with ***Schistosoma haematobium*** and hookworms was estimated and mapped across the country of Ghana (**Figure 6.38**). Both parasites were found to exist in focal locations that often overlapped, creating areas of high co-intensity. These maps provided a reliable indication of where environmental contamination was highest and where control programs could be targeted and then monitored for their effectiveness.

Some individual hosts may serve as superspreaders

As one final aspect of parasite epidemiology, consider the concept of **superspreaders**. R_0 was presented as the mean number of secondary cases resulting from introduction of a single infected individual into a susceptible population. It is important to remember though that such a value can obscure considerable individual variation in infectiousness. Some individuals are more infectious than others and are consequently much more likely to spread a new infection, that is, they are superspreaders. The outbreak of the coronavirus causing severe acute respiratory syndrome or SARS in 2005 highlighted this situation; some individuals were responsible for infecting dozens of other people. The probability that the introduction of a single infectious person will ignite an epidemic will depend considerably on the characteristics of that individual. If epidemiologists could identify superspreaders, then the prospects for averting an outbreak could be considerably enhanced.

As noted above, helminth infections are often aggregated in their hosts, and it is easy to imagine that those rare individuals with heavy worm burdens could serve as superspreaders, especially if they were placed in situations where the progeny of their helminths could readily infect others. For malaria, some people are bitten more by mosquitoes than others or are more susceptible than others. It has been estimated that 20% of the people transmit 80% of the infections (known as the 20/80 rule); these individuals play a disproportionate role in transmission. The most general point is that because of such heterogeneities in transmission efficiency, if control could be targeted at those individuals who are bitten most and who are most susceptible, its impact on reducing transmission would be much higher than if control measures were simply applied uniformly.

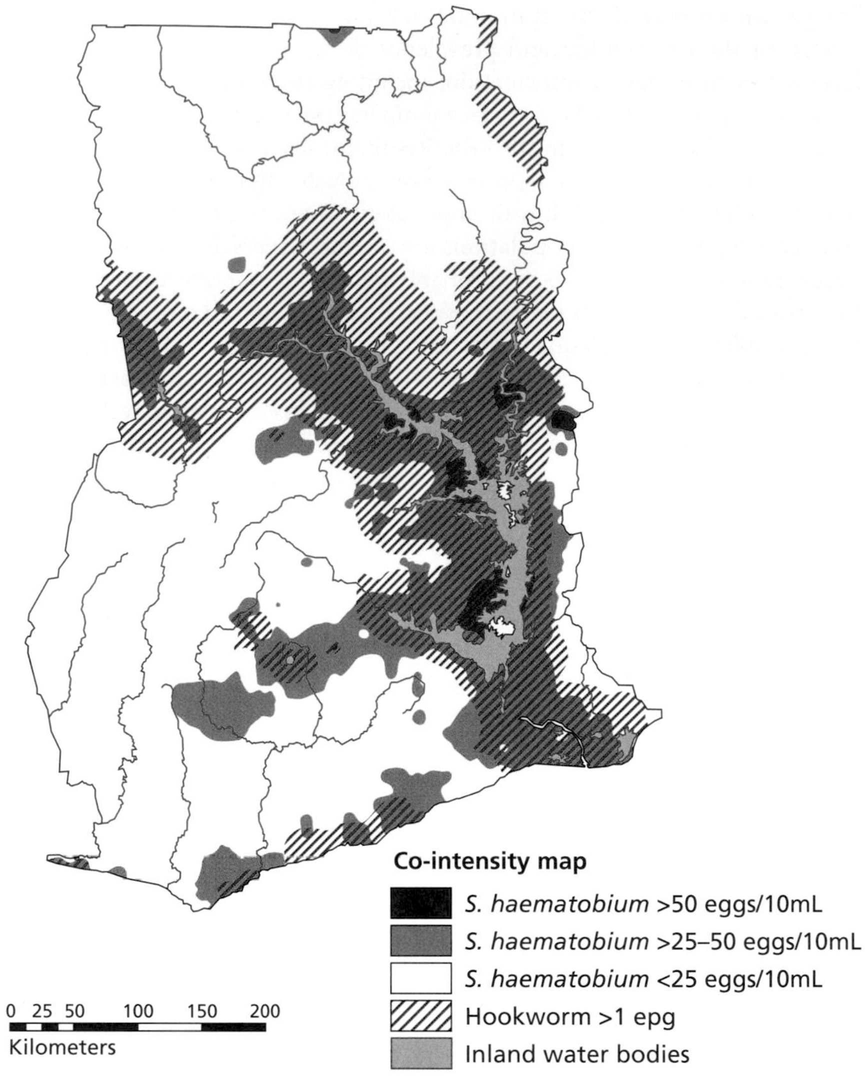

Figure 6.38 Use of model-based geostatistics to provide spatially explicit predictions in unsampled locations. Predicted areas in Ghana where both *Schistosoma haematobium* and hookworms occur in boys from 15–19 years of age, with different levels of intensity of transmission shown for *S. haematobium*. For *S. haematobium*, egg counts are for 10mL of urine, and for hookworms, epg refers to eggs per gram of feces. (Soares Magalhães RJ, Biritwum N-K, Gyapong JO et al. [2011] *PLoS Negl Trop Dis* 5:e1200 [doi:10.1371/journal.pntd.0001200].)

Summary

Several lines of investigation make the study of the ecology of parasitism a vibrant, thriving discipline and the more the impact of parasitism in the natural world is investigated—even at the scale of ecosystems—the greater its impact seems to be. An important area of study is the factors that dictate both the habitat and host specificity of parasites. In most cases, we still lack convincing underlying mechanistic explanations for these specificities, but these topics are basic to understanding how parasites interact with each other and with other symbionts such as the gut flora. The extent to which a parasite is a specialist or generalist with respect to its hosts influences its predilection to cause disease or to become problematic when introduced in new environments.

Because parasites occur both within and among host individuals, parasite population biology is particularly complex. A consistent pattern in parasite populations is that parasites are unevenly distributed among hosts. Some hosts bear particularly heavy parasite burdens, which may not only make them more vulnerable to illness or death but may also augment their role in transmitting parasites to new hosts. Parasite populations within hosts are often regulated by density-dependent factors, such as competition, that may be mediated directly as by resource competition or indirectly by the action of factors such as the host's immune system. It is important to understand how natural processes such as competition can control what parasite species are present and how common those parasites can become.

Just as the population biology of parasites is complex, so too are communities of parasites. Some host species are infected with many more parasite species than others. Also, parasite communities often appear to be stochastic in their assemblage, but we are a long way from understanding all the processes that dictate parasite community structure. In some cases though, it is clear that competitive hierarchies control the structure of parasite communities, giving us a way to predict parasite community structure. Human parasite communities are particularly intriguing, as our association with domesticated animals seems to have played a big role in influencing our parasite communities. Learning how different parasite species interact within an infected person is increasingly recognized as an important consideration in our efforts to control parasites. With respect to their role in food webs and ecosystems, recent studies suggest parasites play a far more prominent role than previously considered. For example, parasites often serve as prey for other species and in general increase the connectivity of ecosystems.

Parasites have profound impacts on the ecology of their hosts. Parasites may incite parasite-avoidance behaviors, manipulate hosts to make them more likely to be consumed by predators, and may even have major effects on regulating host population sizes. The study of parasites has played a prominent role in the development of two new, thriving subdisciplines within ecology, ecological immunology and metabolic ecology. For the former, an appreciation of how host immune response to parasites can be influenced by a number of ecological circumstances gives us a much more realistic framework with which to understand host–parasite relationships in general. The latter can reveal fundamental relationships among parasite mass, metabolic rates, and abundance that can provide enormous predictive power in understanding the global impact of parasites. There is also a strong connection between ecology and epidemiology, including an emphasis in both disciplines on modeling. Modeling offers powerful predictive insights in the study of parasite biology and transmission, and our efforts to model both micro- and macroparasite dynamics grow ever more sophisticated, including more realistic inclusion of factors that vary in space and time. Modern parasite ecology has also given us a much better platform from which to contemplate a host of global changes that pose problems that are fundamentally ecological at their core. Our understanding of the evolutionary processes affecting parasites will be greatly enhanced by an appreciation for the ecological circumstances in which parasites live.

REVIEW QUESTIONS

1. Think of the malaria parasite *Plasmodium falciparum* and all the different environments it inhabits. Starting from the moment a sporozoite enters the human bloodstream, list in order all the distinct environments malaria parasites must pass through as they complete their life cycle. You might want to consult the Rogues Gallery section to remind yourself of the malaria life cycle.

2. What is an encounter filter and a compatibility filter, and provide examples of each.

3. Why are studies to define the host specificity of parasites important?
4. Are parasites regulated in a density-dependent way in their hosts, and if so, how?
5. Are parasites important in the overall functioning of ecosystems? Identify some distinct lines of investigation that support your answer.
6. Describe circumstances in which parasites might be expected to change the behavior of their hosts. What are some of the ecological consequences of manipulation?
7. What is ecological immunology and how does it apply to parasitism?
8. How are ecology and epidemiology similar, and different?
9. Provide some ideas for how the basic mathematical models we discussed can be expanded to be made more realistic.

References

GENERAL

Bordes F, Morand S, Krasnov BR & Poulin R (2010) Parasite diversity and latitudinal gradients in terrestrial mammals. In The Biogeography of Host–Parasite Interactions (Morand S & Krasnov BR eds), pp. 89–98. Oxford University Press.

Brooks DR & McLennan DA (1993) Parascript: Parasites and the Language of Evolution. Smithsonian Institution Press.

Brown JH, Gillooly JF, Allen AP et al (2004) Toward a metabolic theory of ecology. *Ecology* **85**:1771–1789 (doi:10.1890/03-9000).

Combes C (2001) Parasitism: The Ecology and Evolution of Intimate Interactions. University of Chicago Press.

Cremer S & Sixt M (2009) Analogies in the evolution of individual and social immunity. *Philos Trans R Soc Lond B Biol Sci* **364**:129–142 (doi:10.1098/rstb.2008.0166).

Edmunds PJ & Carpenter RC (2001) Recovery of *Diadema antillarum* reduces macroalgal cover and increases abundance of juvenile corals on a Caribbean reef. *Proc Natl Acad Sci U S A* **98**: 4822–4824.

Guégan TF & Renaud F (2009) Ecology and Evolution of Parasitism. Oxford University Press.

Guernier V, Hochberg ME & Guegan JF (2004) Ecology drives the worldwide distribution of human diseases. *PLoS Biol* **2(6)**:e141. doi:10.1371/journal.pbio.0020141.

Keeling MJ & Rohani P (2008) Modeling Infectious Diseases in Humans and Animals. Princeton University Press.

Kermack WO & McKendrick AG (1927) A contribution to the mathematical theory of epidemics. *Proc R Soc Lond A Math Phys Sci* **115**:700–721.

Lafferty KD, Allesina S, Arim M et al (2008) Parasites in food webs: the ultimate missing links. *Ecol Lett* **11**:533–546 (doi:10.1111/j.1461-0248.2008.01174.x).

Mitchell CE (2003) Trophic control of grassland production and biomass by pathogens. *Ecol Lett* **6**:147–155.

Morand S & Krasnov BR (eds) (2010) The Biogeography of Host–Parasite Interactions. Oxford University Press.

Owen JP, Nelson AC & Clayton DH (2010) Ecological immunology of bird–ectoparasite systems. *Trends Parasitol* **26**:530–539 (doi:10.1016/j.pt.2010.06.005).

Poulin R (2007) Evolutionary Ecology of Parasites, 2nd ed. Princeton University Press.

Poulin R (2007) Are there general laws in parasite ecology? *Parasitology* **134**:763–776 (doi:10.1017/S0031182006002150).

Preston D & Johnson P (2010) Ecological consequences of parasitism. *Nature Education Knowledge* **1(8)**:39.

Rohde K (2010) Marine parasite diversity and environmental gradients. In The Biogeography of Host–Parasite Interactions (Morand S & Krasnov BR eds), pp. 73–88. Oxford University Press.

Scantlebury M, Waterman J M, Hillegass M et al (2007) Energetic costs of parasitism in the Cape ground squirrel *Xerus inauris*. *Proc R Soc Lond B Biol Sci* **274**:2169–2177. (doi:10.1098/rspb.2007.0690).

Schmid-Hempel P (2011) Evolutionary Parasitology: The Integrated Study of Infections, Immunology, Ecology and Genetics. Oxford University Press.

Schulenburg H, Kurtz J, Moret Y et al (2009) Ecological immunology. *Philos Trans R Soc Lond B Biol Sci* **364**:3–14 (doi:10.1098/rstb.2008.0249).

Sukhdeo MVK (2010) Food webs for parasitologists: a review. *J Parasitol* **96**:273–284.

DEFINING THE HABITATS OF PARASITES

Parasites occupy multiple habitats in succession

Wiser MF (2011) Protozoa and Human Disease. Garland Science.

Parasites have microhabitat preferences and occupy specific sites within their hosts

Clayton DH, Bush SE, Goates BM et al (2003) Host defense reinforces host–parasite cospeciation. *Proc Natl Acad Sci U S A* **100**:15,694–15,699 (doi:10.1073/pnas.2533751100).

Hayes KS, Bancroft AJ, Goldrick M et al (2010) Exploitation of the intestinal microflora by the parasitic nematode *Trichuris muris*. *Science* **328**:1391–1394 (doi:10.1126/science.1187703).

Ritter U, Frischknecht F & van Zandbergen G (2009) Are neutrophils important host cells for *Leishmania* parasites? *Trends Parasitol* **25**:505–510 (doi:10.1016/j.pt.2009.08.003).

Rohde K (1994) Niche restriction in parasites-proximate and ultimate causes. *Parasitology* **109**:S69–S84.

Whittington ID (1996) Benedeniine capsalid monogeneans from Australian fishes: Pathogenic species, site-specificity and camouflage. *J Helminthol* **70**:177–184.

Host specificity is one of parasitism's most distinctive properties

Durden LA & Musser GG (1994) The mammalian hosts of the sucking lice (*Anoplura*) of the world: a host–parasite list. *Bull Soc Vector Ecol* **19**:130–168.

Poulin R & Keeney DB (2008) Host specificity under molecular and experimental scrutiny. *Trends Parasitol* **24**:24–28 (doi:10.1016/j.pt.2007.10.002).

Encounter and compatibility filters influence the range of host species used by a parasite

Lebrun N, Renaud F & Lambert A (1988) The genus *Diplozoon* (Monogenea, Polyopisthocotylea) in southern France—speciation and specificity. *Int J Parasitol* **18**:395–400 (doi:10.1016/0020-7519(88)90150-6).

The origins and consequences of host specificity are debated

Hellgren O, Perez-Tris J & Bensch S (2009) A jack-of-all-trades and still a master of some: prevalence and host range in avian malaria and related blood parasites. *Ecology* **90**:2840–2849 (doi:10.1890/08-1059.1).

Johnson KP, Malenke JR & Clayton DH (2009) Competition promotes the evolution of host generalists in obligate parasites. *Proc R Soc Lond B Biol Sci* **276**:3921–3926 (doi:10.1098/rspb.2009.1174).

Underlying mechanisms dictating specificity are also often not known

Mandel MJ, Wollenberg MS, Stabb EV et al (2009) A single regulatory gene is sufficient to alter bacterial host range. *Nature* **458**:215-U7 (doi:10.1038/nature07660).

Martin MJ, Rayner JC, Gagneux P et al (2005) Evolution of human-chimpanzee differences in malaria susceptibility: relationship to human genetic loss of *N*-glycolylneuraminic acid. *Proc Natl Acad Sci USA* **102**:12,819–12,824 (doi:10.1073/pnas.0503819102).

Rathore D, Hrstka SCL, Sacci JB et al (2003) Molecular mechanism of host specificity in *Plasmodium falciparum* infection—role of circumsporozoite protein. *J Biol Chem* **278**:40,905–40,910 (doi:10.1074/jbc.M306250200).

PARASITE POPULATION BIOLOGY

Parasite populations are complex

Goater TM, Goater CP & Esch GW (2013) Parasitism: The Diversity and Ecology of Animal Parasites. 2nd ed. Cambridge University Press

Parasites often show aggregated (overdispersed) distributions in their hosts

Barker DE, Marcogliese DJ & Cone DK (1996) On the distribution and abundance of eel parasites in Nova Scotia: local versus regional patterns. *J Parasitol* **82**:697–701 (doi:10.2307/3283877).

Crofton HD (1971) Quantitative approach to parasitism. *Parasitology* **62**:179–193.

Gray CA, Gray PN & Pence DB (1989) Influence of social-status on the helminth community of late-winter mallards. *Can J Zool* **67**:1937–1944.

Folstad I, Nilssen AC, Halvorsen O et al (1989) Why do male reindeer (*Rangifer t. tarandus*) have higher abundance of second and third instar larvae of *Hypoderma tarandi* than females? *Oikos* **55**:87–92 (doi:10.2307/3565877).

Matthee S & Krasnov BR (2009) Searching for generality in the patterns of parasite abundance and distribution: Ectoparasites of a South African rodent, *Rhabdomys pumilio*. *Int J Parasitol* **39**:781–788 (doi:10.1016/j.ijpara.2008.12.003).

Poulin R, Blanar CA, Thieltges DW & Marcogliese DJ (2012) Scaling up from epidemiology to biogeography: local infection patterns predict geographical distribution in fish parasites. *J Biogeogr* **39**:1157–1166.

Shaw DJ & Dobson AP (1995) Patterns of macroparasite abundance and aggregation in wildlife populations: a quantitative review. *Parasitology* **111**:S111–S133.

Both density-independent and density-dependent factors influence parasite population size

Allan F, Rollinson D, Smith JE et al (2009) Host choice and penetration by *Schistosoma haematobium* miracidia. *J Helminthol* **83**:33–88 (doi:10.1017/S0022149X08073628).

Ebert D, Zschokke-Rohringer CD & Carius HJ (2000) Dose effects and density-dependent regulation of two microparasites of *Daphnia magna*. *Oecologia* **122**:200–209.

Hudson PJ, Dobson AP & Newborn D (1992) Do parasites make prey vulnerable to predation—red grouse and parasites. *J Anim Ecol* **61**:681–692 (doi:10.2307/5623).

Sirot E (1996) The pay-off from superparasitism in the solitary parasitoid *Venturia canescens*. *Ecol Entomol* **21**:305–307.

Intraspecific competition can regulate parasite populations in different ways

Fichet-Calvet E, Wang JF, Jomaa I et al (2003) Patterns of the tapeworm *Raillietina trapezoides* infection in the fat sand rat *Psammomys obesus* in Tunisia: season, climatic conditions, host age and crowding effects. *Parasitology* **126**:481–492 (doi:10.1017/S0031182003003056).

Huijben S, Nelson WA, Wargo AR et al (2010) Chemotherapy within-host ecology and the fitness of drug-resistant malaria parasites. *Evolution* **64**:2952–2968 (doi:10.1111/j.1558-5646.2010.01068.x).

Mideo N (2009) Parasite adaptations to within-host competition. *Trends Parasitol* **25**:261–268 (doi:10.1016/j.pt.2009.03.001).

Rajakumar S, Bleiss W, Hartmann S et al (2006) Concomitant immunity in a rodent model of filariasis: The infection of *Meriones unguiculatus* with *Acanthocheilonema viteae*. *J Parasitol* **92**:41-45.

Sapp KK, Meyer KA & Loker ES (1998) Intramolluscan development of the digenean *Echinostoma paraensei*: rapid production of a unique mother redia that adversely affects development of conspecific parasites. *Invertebr Biol* **117**:20-28 (doi:10.2307/3226848).

Parasite population studies often require a long-term perspective and detailed sampling

Kennedy CR (2009) The ecology of parasites of freshwater fishes: the search for patterns. *Parasitology* **136**:1653–1662 (doi:10.1017/S0031182009005794).

PARASITE COMMUNITIES

The richness of parasite communities varies among host species for reasons that are still debated

Bush AO, Aho JM & Kennedy CR (1990) Ecological versus phylogenetic determinants of helminth parasite community richness. *Evol Ecol* **4**:1–20.

Kennedy CR, Bush AO & Aho JM (1986) Patterns in helminth communities—why are birds and fish different? *Parasitology* **93**:205–215.

Kennedy CR & Guégan JF (1996) The number of niches in intestinal helminth communities of *Anguilla anguilla*: are there enough spaces for parasites? *Parasitology* **113**:293–302.

Poulin R (1995) Phylogeny, ecology, and the richness of parasite communities in vertebrates. *Ecol Monogr* **65**:283-302

Poulin R (1998) Large-scale patterns of host use by parasites of freshwater fishes. *Ecol Lett.* **1**:118–128.

Most studies suggest parasite communities are stochastic in nature

Kennedy CR (2009) The ecology of parasites of freshwater fishes: the search for patterns. *Parasitology* **136**:1653–1662 (doi:10.1017/S0031182009005794).

Parasite species within infracommunities engage in negative and positive interactions with one another

Chappell LH (1969) Competitive exclusion between two intestinal parasites of three-spined stickleback, *Gasterosteus aculeatus*. *J Parasitol* **55**:775.

Holmes JC (1961) Effects of concurrent infections on *Hymenolepis diminuta* (Cestoda) and *Moniliformis dubius* (Acanthocephala) .I. General effects and comparison with crowding. *J Parasitol* **47**:209–216 (doi:10.2307/3275291).

Holmes JC (1973) Site selection by parasitic helminths—interspecific interactions, site segregation, and their importance to development of helminth communities. *Can J Zool* **51**:333–347 (doi:10.1139/z73-047).

Kuris A (1990) Guild structure of larval trematodes in molluscan hosts: prevalence, dominance and significance of competition. In Parasite communities: patterns and processes (Esch G, Bush A & Aho J eds), pp. 69–100. Chapman and Hall.

Kuris AM & Lafferty KD (1994) Community structure—larval trematodes in snail hosts. *Annu Rev Ecol Syst* **25**:189–217 (doi:10.1146/annurev.es.25.110194.001201).

Sousa WP (1994) Patterns and processes in communities of helminth parasites. *Trends Ecol Evol* **9**:52–57 (doi:10.1016/0169-5347(94)90268-2).

Su Z, Segura M & Stevenson MM (2006) Reduced protective efficacy of a blood-stage malaria vaccine by concurrent nematode infection. *Infect Immun* **74**:2138–2144.

Telfer S, Lamblin X, Birtles R et al (2010) Species interactions in a parasite community drive infection risk in a wildlife population. *Science* **330**:243–246.

Generalizable patterns are also elusive in component communities of parasites

Kennedy CR (2009) The ecology of parasites of freshwater fishes: the search for patterns. *Parasitology* **136**:1653–1662 (doi:10.1017/S0031182009005794).

Poulin R (1997) Species richness of parasite assemblages: evolution and patterns. *Annu Rev Ecol Syst* **28**:341–358 (doi:10.1146/annurev.ecolsys.28.1.341).

Human parasites have a distinctive community ecology

Adegnika AA & Kremsner PG (2012) Epidemiology of malaria and helminth interaction: a review from 2001 to 2011. *Curr Opin HIV AIDS* **7**:221–224 (doi:10.1097/COH.0b013e3283524d90).

Degarege A, Legesse M, Medhin G et al (2012) Malaria and related outcomes in patients with intestinal helminths: a cross-sectional study. *BMC Infect Dis* **12**:291 (doi:10.1186/1471-2334-12-291).

Dessi D, Rappelli P, Diaz N et al (2006) *Mycoplasma hominis* and *Trichomonas vaginalis*: a unique case of symbiotic relationship between two obligate human parasites. *Front Biosci* **11**:2028–2034 (doi:10.2741/1944).

Horwitz LK & Smith P (2000) The contribution of animal domestication to the spread of zoonoses: a case study from the Southern Levant. *Anthropozoologica* **31**:77–84.

Kojic EM (2012) Human Papillomavirus (HPV) and trichomonas: common, concerning, and challenging sexually transmitted infections. *Med Health R I* **95**:255–257.

Perrin P, Herbreteau V, Hugot JP & Morand S (2010) Biogeography, humans and their parasites. In The Biogeography of Host–Parasite Interactions (Morand S & Krasnov BR eds), pp. 41–57. Oxford University Press.

Taylor LH, Latham SM & Woolhouse MEJ (2001) Risk factors for human disease emergence. *Philos Trans R Soc Lond B Biol Sci* **356**:983–989.

Woolhouse MEJ, Howey R, Gaunt E et al (2008) Temporal trends in the discovery of human viruses. *Proc R Soc Lond B Biol Sci* **275**:2111–2115 (doi:10.1098/rspb.2008.0294).

THE ROLE OF PARASITES IN FOOD WEBS AND ECOSYSTEMS

Parasites can be a food source for other organisms

Grutter A (1996) Parasite removal rates by the cleaner wrasse *Labroides dimidiatus*. *Mar Ecol Prog Ser* **130**:61–70 (doi:10.3354/meps130061).

Johnson PTJ, Dobson A, Lafferty KD et al (2010) When parasites become prey: ecological and epidemiological significance of eating parasites. *Trends Ecol Evol* **25**:362–371.

Kaplan AT, Rebhal S, Lafferty KD et al (2009) Small estuarine fishes feed on large trematode cercariae: lab and field investigations. *J Parasitol* **95**:477–480 (doi:10.1645/GE-1737.1).

GLOBAL PATTERNS IN PARASITE DIVERSITY

Poulin R & Morand S (2004) Parasite Biodiversity. Smithsonian Books.

Poulin R (2014) Parasite biodiversity revisited: frontiers and constraints. *Int J Parasitol.* **44**:581–589 (doi:10.1016/j.ijpara.2014.02.003)

PARASITE EFFECTS ON HOST ECOLOGY

Hosts try both to avoid infection and to actively remove parasites if they do become infected

Cotgreave P & Clayton DH (1994) Comparative analysis of time spent grooming by birds in relation to parasite load. *Behaviour* **131**:171–187 (doi:10.1163/156853994X00424).

Ezenwa VO (2004) Selective defecation and selective foraging: antiparasite behavior in wild ungulates? *Ethology* **110**:851–862.

Fritzsche A & Allan BF (2012) The ecology of fear: host foraging behavior varies with the spatio-temporal abundance of a dominant ectoparasite. *Ecohealth* **9**:70–74 (doi:10.1007/s10393-012-0744-z).

Giorgi MS, Arlettaz R, Christe P et al (2001) The energetic grooming costs imposed by a parasitic mite (*Spinturnix myoti*) upon its bat host (*Myotis myotis*). *Proc R Soc Lond B Biol Sci* **268**:2071–2075 (doi:10.1098/rspb.2001.1686).

Moore J (2002) Parasites and the behavior of animals. Oxford University Press, Oxford.

Moore J (2013) An overview of parasite-induced behavioral alterations—and some lessons from bats. *J Exp Biol* **216**:11–17(doi:10.1242/jeb.074088).

Mooring MS, Blumstein DT & Stoner CJ (2004) The evolution of parasite-defence grooming in ungulates. *Biol J Linn Soc Lond* **81**:17-37(doi:10.1111/j.1095-8312.2004.00273.x).

Stumbo AD, James CT, Goater CP & Wisenden BD (2012) Shoaling as an antiparasite defence in minnows (*Pimephales promelas*) exposed to trematode cercariae. *J Anim Ecol* **81**:1319–1326.

Suárez-Rodríguez M, López-Rull I & Macías Garcia C (2012) Incorporation of cigarette butts into nests reduces nest ectoparasite load in urban birds: new ingredients for an old recipe? *Biol Lett* **9**:20120931.

Hosts also change their diets and engage in self-medication when infected

Huffman MA, Nakagawa N, Go Y et al (2013) Primate self-medication and the treatment of parasite infection. In Monkeys, Apes, and Humans, pp. 13–23. SpringerBriefs in Biology.

Fowler A, Koutsioni Y & Sommer V (2007) Leaf-swallowing in Nigerian chimpanzees: evidence for assumed self-medication. *Primates* **48**:73–76 (doi:10.1007/s10329-006-0001-6).

Krief S, Jamart A, Mahe S et al (2008) Clinical and pathologic manifestation of oesophagostomosis in African great apes: does self-medication in wild apes influence disease progression? *J Med Primatol* **37**:188–195 (doi:10.1111/j.1600-0684.2008.00285.x).

Parasites influence host migratory behavior

Folstad I, Nilssen AC, Halvorsen O et al (1991) Parasite avoidance—the cause of post-calving migrations in Rangifer. *Can J Zool* **69**:2423–2429 (doi:10.1139/z91-340).

Waldenstrom J, Bensch S, Kiboi S et al (2002) Cross-species infection of blood parasites between resident and migratory songbirds in Africa. *Mol Ecol* **11**:1545–1554 (doi:10.1046/j.1365-294X.2002.01523.x).

Parasites can regulate host populations, but examples are few

Deter J, Charbonnel N, Cosson JF et al (2008) Regulation of vole populations by the nematode *Trichuris arvicolae*: insights from modeling. *Eur J Wildl Res* **54**:60–70 (doi:10.1007/s10344-007-0110-6).

New LF, Matthiopoulos J, Redpath S et al (2009) Fitting models of multiple hypotheses to partial population data: Investigating the causes of cycles in red grouse. *Am Nat* **174**:399–412 (doi:10.1086/603625).

Pedersen AB & Greives TJ (2008) The interaction of parasites and resources cause crashes in a wild mouse population. *J Anim Ecol* **77**:370–377 (doi:10.1111/j.1365-2656.2007.01321.x).

Scott ME & Dobson A (1989) The role of parasites in regulating host abundance. *Parasitol Today* **5**:176–183 (doi:10.1016/0169-4758(89)90140-3).

Townsend SE, Newey S, Thirgood SJ et al (2011) Dissecting the drivers of population cycles: interactions between parasites and mountain hare demography. *Ecol Model* **222**:48–56 (doi:10.1016/j.ecolmodel.2010.08.033).

Parasites influence competitive interactions amon g hosts

Craig DM (1986) Stimuli governing intraspecific egg predation in the flour beetles, *Tribolium confusum* and *T. castaneum*. *Res Popul Ecol* **28**:173–183.

Hatcher MJ, Dick JTA & Dunn AM (2006) How parasites affect interactions between competitors and predators. *Ecol Lett* **9**:1253–1271 (doi:10.1111/j.1461-0248.2006.00964.x).

Koprinikar J, Forbes MR & Baker RL (2008) Larval amphibian growth and development under varying density: are parasitized individuals poor competitors? *Oecologia* **155**:641–649 (doi:10.1007/s00442-007-0937-2).

Park T (1962) Beetles, competition and populations. *Science* **138**:1369-1375.

Tompkins DM, Greenman JV, Robertson PA et al (2000) The role of shared parasites in the exclusion of wildlife hosts: *Heterakis gallinarum* in the ring-necked pheasant and the grey partridge. *J Anim Ecol* **69**:829–840 (doi:10.1046/j.1365-2656.2000.00439.x).

Parasites can manipulate their hosts to affect the likelihood of predation

Grosman AH, Janssen A, de Brito EF et al (2008) Parasitoid increases survival of its pupae by inducing hosts to fight predators. *PLoS One* **3**:e2276.

Lafferty KD (1992) Foraging on prey that are modified by parasites. *Am Nat* **140**:854–867 (doi:10.1086/285444).

Lafferty KD & Morris AK (1996) Altered behavior of parasitized killifish increases susceptibility to predation by bird final hosts. *Ecology* **77**:1390–1397(doi:10.2307/2265536).

Lefevre T, Lebarbenchon C, Gauthier-Clerc M et al (2009) The ecological significance of manipulative parasites. *Trends Ecol Evol* **24**:41–48 (doi:10.1016/j.tree.2008.08.007).

Moore J (2013) An overview of parasite-induced behavioral alterations—and some lessons from bats. *J Exp Biol* **216**:11–17 (doi:10.1242/jeb.074088).

Mouritsen KN & Poulin R (2003) Parasite-induced trophic facilitation exploited by a non-host predator: a manipulator's nightmare. *Int J Parasitol* **33**:1043–1050 (doi:10.1016/S0020-7519(03)00178-4).

Ponton F, Biron DG, Joly C et al (2005) Ecology of parasitically modified populations: a case study from a gammarid–trematode system. *Mar Ecol Prog Ser* **299**:205–215 (doi:10.3354/meps299205).

Seppälä O, Valtonen ET & Benesh DP (2008) Host manipulation by parasites in the world of dead-end predators: adaptation to enhance transmission? *Proc R Soc Lond B Biol Sci* **275**:1611–1615.

Tompkins DM, Mouritsen KN & Poulin R (2004) Parasite-induced surfacing in the cockle *Austrovenus stutchburyi*: adaptation or not? *J Evol Biol* **17**:247–256.

Thomas F, Renaud F & Poulin R (1998) Exploitation of manipulators: "hitch-hiking" as a parasite transmission strategy. *Anim Behav* **56**:199–206.

Vance SA (1996) Morphological and behavioural sex reversal in mermithid-infected mayflies. *Proc R Soc Lond B Biol Sci* **263**:907–912.

EPIDEMIOLOGY AND ITS RELATIONSHIP WITH ECOLOGY

Modeling is an invaluable approach to the study of infectious diseases

Anderson RM & May RM (1991) Infectious Diseases of Humans: Dynamics and Control. Oxford University Press.

Chitnis N, Hyman J M & Cushing JM (2008) Determining important parameters in the spread of malaria through the sensitivity analysis of a mathematical model. *Bull Math Biol* **70**:1272–1296 (doi:10.1007/s11538-008-9299-0).

Keeling MJ & Rohani P (2008) Modeling infectious diseases in humans and animals. Princeton University Press.

Ross R (1911) The Prevention of Malaria, 2nd ed. Murray.

Microparasites exemplify basic modeling approaches that estimate population size and clarify transmission

Anderson RM & May RM (1981) The population-dynamics of micro-parasites and their invertebrate hosts. *Philos Trans R Soc Lond B Biol Sci* **291**:451–524 (doi:10.1098/rstb.1981.0005).

Models of macroparasite populations and transmission involve keeping track of individual parasites

Dobson AP & Hudson PJ (1989) Population biology of *Trichostrongylus* *tenuis*, a parasite of economic importance for red grouse management. *Parasitol Today* **5**:283–291 (doi:10.1016/0169-4758(89)90019-7).

Dobson AP & Hudson PJ (1992) Regulation and stability of a free-living host-parasite system – *Trichostrongylus tenuis* in red grouse .II. Population-models. *J Anim Ecol* **61**:487–498 (doi:10.2307/5339).

Models for parasites with complex life cycles involving vectors become more complex

Anderson RM (1993) Epidemiology. In Modern Parasitology: A Textbook of Parasitology, 2nd ed (Cox FEG, ed.), pp. 75–116. Blackwell.

New models open the black box and estimate microparasite populations within hosts and the influences on them

Day KP & Fowkes FJI (2011) Quantifying malaria dynamics within the host. *Science* **333**:943–944.

Metcalf CJE, Graham AL, Huijben S et al (2011) Partitioning regulatory mechanisms of within-host malaria dynamics using the effective propagation number. *Science* **333**:984–988.

Models need to take spatial and temporal factors into account

Brooker S & Clements AC (2009) Spatial heterogeneity of parasite coinfection: determinants and geostatistical prediction at regional scales. *Int J Parasitol* **39**:591–597.

Kabatereine NB, Brooker S, Tukahebwa EM et al (2004) Epidemiology and geography of *Schistosoma mansoni* in Uganda: implications for planning control. *Trop Med Int Health* **9**:372–380 (doi:10.1046/j.1365-3156.2003.01176.x).

Soares Magalhães RJ, Biritwum N-K, Gyapong JO et al (2011) Mapping helminth co-infection and co-intensity: geostatistical prediction in Ghana. *PLoS Negl Trop Dis* **5(6)**:e1200 (doi:10.1371/journal.pntd.0001200).

Some individual hosts may serve as superspreaders

Lloyd-Smith JO, Schreiber SJ, Kopp PE et al (2005) Superspreading and the effect of individual variation on disease emergence. *Nature* **438**:355–359 (doi:10.1038/nature04153).

Smith DL, Dushoff J, Snow RW et al (2005) The entomological inoculation rate and *Plasmodium falciparum* infection in African children. *Nature* **438**:492–495 (doi:10.1038/nature04024).

Evolutionary Biology of Parasitism

7

But if they're so successful, why haven't parasites taken over the world? The answer is simple: they have. We just haven't noticed.

Daniel Suarez, In *Daemon*

In This Chapter

Chapter 6, with its discussion of the ecology of parasitism, set the stage to launch a discussion of the evolutionary biology of parasites. Ecology has been likened to the theater in which the evolutionary play unfolds. The fundamental forces of evolution—mutation, genetic drift, gene flow, and natural selection—apply just as inexorably and in fundamentally the same ways to parasites as they do to other organisms. However, as we shall see, the dependence of parasites on their hosts, the selection imposed by the host, and the constant need to be transmitted to new hosts, all create distinctive features in parasite evolutionary biology. As often noted elsewhere in this book, we must remember the term parasite embraces an enormous variety of organisms with different phylogenetic backgrounds, bodies of different size or complexity, life cycles that range from simple to complex, generalized or specific patterns of host use, and different modes of reproduction. Furthermore, the hosts occupied differ enormously in properties such as mobility, longevity, immune capability, and spatial and temporal distributions, all features that directly influence the evolutionary biology of the parasites they harbor. Consequently, it is hardly surprising that collectively parasites exhibit a huge range of evolutionary trajectories, making grand generalizations about parasite evolutionary biology elusive.

We will first examine parasites from the perspective of **microevolution**. This refers to the evolutionary process as it occurs below the species level, particularly at the level of populations within a species, with an implied ecological time scale. This means that microevolutionary changes can often be seen within the span of a few generations. We consider the topics of parasite–host coevolution and the evolution of virulence as examples of dynamic microevolutionary processes occurring on ecological time scales. Next, we discuss **macroevolution** of parasites, which refers to large-scale patterns and processes at or above the species level that occur on an evolutionary time scale, typically measured over thousands of generations and millions of years. A macroevolutionary perspective allows us to see how parasite species have come into being and gone extinct and to reflect on some of the global patterns in parasite species diversity and distribution. We will then consider how parasitism has arisen and how the complex life cycles so typical of parasites have evolved. We conclude with a discussion of how parasites have influenced the evolutionary biology of their hosts.

Species names highlighted in red are included in the Rogues' Gallery, starting on page 429.

7.1 MICROEVOLUTION IN PARASITES

Microevolutionary studies are largely involved with monitoring changes in the distribution and frequency of genes in populations (**Box 7.1**) and with investigating the causes of those changes. You may recall that we have already discussed parasite populations both from the standpoint of being reservoirs of parasite diversity (Chapter 2) and from their complex structure (Chapter 6). Changes in the distribution and frequency of genes in populations are the essence of evolutionary change. Often changes in the abundance of variant forms (called alleles) of particular genes are followed over time by evolutionary biologists. Of particular interest is the extent to which basic evolutionary processes such as mutation, gene flow, genetic drift, and natural selection can

BOX 7.1
A Brief Primer of Concepts Important to Understanding Microevolution

Four major evolutionary forces affect allele frequencies in populations: mutation, genetic drift, gene flow, and natural selection. **Mutation** is the process whereby DNA is damaged or altered, sometimes resulting in genes with different coding sequences than the initial starting sequence. A mutation may affect a single nucleotide. Alternatively, events such as deletions or inversions may affect large parts of a chromosome. Mutation creates new genetic diversity. Mutations may be neutral, favorable, or unfavorable, and the net effect of the mutation may change depending on the particular circumstances. **Genetic drift** is the change in frequency of an allele in a population as a result of random chance. Particularly in small populations, genetic drift may cause some alleles to disappear by chance. In larger populations, there is less chance that a particular allele will be lost. Drift acts to decrease the amount of genetic variability in a population. Small populations experiencing genetic drift are likely to diverge because the alleles lost in each population will differ by chance. **Gene flow**, directly abetted by the processes of migration and dispersal, is the flow of alleles from one population to another. It generally acts to homogenize different populations. If it is restricted, then local populations may accumulate differences. The process of horizontal gene transfer (Chapter 2) can also be considered a form of gene flow. **Natural selection** occurs when particular alleles increase in a population if they have advantageous effects on an organism's fitness (survival and reproduction). Conversely, some alleles have disadvantageous effects and decrease in abundance as a result of their harmful impact on fitness. Selection often reduces genetic variation in a population by eliminating disadvantageous alleles. When the advantage of a particular allele depends on its frequency (such as having an advantage when rare, called **negative frequency-dependent selection**), selection can promote persistence of multiple alleles (see Figure 7.8). **Balancing selection** in which a heterozygote has an advantage over homozygotes can also favor persistence of multiple alleles (see Table 7.2).

When considering populations, if mating is random among all individuals, a state of **panmixia** occurs. Panmixia is indicative of unlimited movement of individuals, and thus high rates of gene flow. In reality, rates of gene flow are diminished in natural populations such that panmixia does not occur. Diminished gene flow results in subpopulations that are to some degree isolated and differentiated from one another. The extent of differentiation, or what is known as population structure, is estimated by $\boldsymbol{F_{ST}}$, an index that measures deviations in gene frequencies from an idealized, panmictic population. F_{ST} ranges in value from 0 to 1, with values close to 0 indicative of high gene flow and little differentiation among subpopulations. When subpopulations are highly differentiated, indicative of limited gene flow and the operation of genetic drift in further differentiating populations, then F_{ST} values are closer to 1. Such populations are said to have more structure than populations for which F_{ST} values are low. Differentiated populations are of interest in part because they are believed to be candidates for even further differentiation, potentially leading to speciation.

Another important general parameter with respect to microevolutionary processes is the **effective population size** ($\boldsymbol{N_e}$) as compared to the actual census size of the population (N) of a particular species. The effective population size is defined as the size of an ideal population (ideal from the standpoint of being stable and having random mating and equal sex ratios among other features) that would lose genetic variation via drift at the same rate as is observed in an actual population. N_e is almost always substantially smaller than the census size. It provides a measure of the potential impact of genetic drift. The smaller the effective population size, the greater the degree of drift. N_e has a large impact on the level of genetic diversity in a population and thus directly influences the adaptive potential of a population.

An additional concept worthy of mention is **bottlenecking**, whereby a particular species might experience a severe reduction in population size, often resulting in the loss of considerable genetic diversity. Another is the **founder effect**, referring to the initiation of a new population by very few individuals that may collectively contain only a small proportion of the total genetic variation in the larger population from which they were derived.

combine to influence the degree of genetic variability within and among populations. The more populations become differentiated from one another, the more structure they are said to possess. Understanding the microevolutionary process and how it affects parasite populations is important because it helps us to gauge the evolutionary or adaptive potential of these populations. For example, how readily might a particular parasite population evolve drug resistance or withstand a control program? As we will see, what happens at the microevolutionary scale has considerable potential to impact macroevolutionary events such as speciation as well.

The subdivided nature of their populations influences the evolution of parasites

As we have seen in Chapter 6, the population structure for a given parasite species is complicated (see Box 6.2, Figure 1). At a specific point in time, a host individual will harbor a number of parasites of a particular species, termed an **infrapopulation**. Because several host individuals are present, multiple parasite infrapopulations occur. A component population describes all the parasites of a given species in a particular ecosystem that are found within all the hosts of a particular species. An even more inclusive term, the **suprapopulation**, refers to all the parasites of a given species, in all stages of development, both free-living or in all supportive host species, that are found in a given ecosystem. Given that a particular parasite species can also occur in multiple ecosystems in different geographic localities, parasite populations are complex indeed.

How do microevolutionary processes shape the structure of these populations? Here we must bear in mind that the outcome very much depends on the details regarding the specific biological attributes of the parasite and its hosts. **Table 7.1** lists of some of the expected effects of both parasite and host properties that influence microevolutionary change and the degree of population structure of parasites. We now consider some examples of these processes.

The effective population size, N_e, influences parasite evolution

Among the important factors influencing a parasite species is the effective population size, N_e, of its infrapopulations. If N_e is small, then genetic drift and inbreeding would be expected to have more of an influence, leading to

Table 7.1 Factors that influence population genetics and microevolution of parasites

Factors That Increase Genetic Structure	Factors That Decrease Genetic Structure
Hosts immobile or asocial	Hosts, including vectors, highly mobile
Unstable, external environment	Stable homogeneous external environment
Complex life cycle with specific obligate hosts	Persistent life-cycle stages in environment or host
Patchy distribution of suitable parasite niches	Uniform availability of parasite niches
Small effective population size	Large effective population size
Highly aggregated distribution among hosts	Uniform distribution among hosts
Parasite mostly self-fertilizing	Parasite mostly outcrossing
Frequent extinctions and reestablishments	Stable populations with rare extinctions
High host specificity	Low host specificity
Low parasite mobility (including vertical transfer)	High parasite mobility (horizontal transfer)

Huyse T, Poulin R & Theron A (2005) Speciation in parasites: a population genetics approach. *Trends Parasitol* 21:469–475, Table 1.

greater differentiation among infrapopulations. N_e is lowered if the parasites have been produced by **selfing** (which refers to a hermaphroditic individual producing sperm to fertilize eggs it too has produced) or if the sex ratios in the parasite population are highly skewed. The sex ratio of ***Schistosoma*** parasites causing schistosomiasis is often significantly biased in favor of males, for example. N_e has a strong influence on the likelihood a parasite species evolves drug resistance, adapts to local hosts, forms races, or even undergoes speciation. For example, for two nematodes that infect livestock, *Ascaris suum* in pigs and *Ostertagia ostertagi* in cattle, the infrapopulation size of *A. suum* in a pig (<100) is much smaller than for *O. ostertagi* (thousands) in cattle. Consequently, *A. suum* infrapopulations are more susceptible to drift and are potentially more differentiated from one another (values for F_{ST} are much higher for this species). In contrast, *O. ostertagi* has high levels of diversity among the worms within a particular host. Both because the larger infrapopulations are less subject to random variation from genetic drift and because gene flow is high, there is relatively little differentiation among infrapopulations. It is noteworthy that the high rates of gene flow observed for this and other parasites of cattle and sheep provide considerable opportunity for rare alleles that might confer resistance to anthelmintics to spread.

The mode of parasite reproduction affects microevolutionary change

The type of reproduction characteristic of a parasite species also has major influences on its potential for microevolutionary change. In general, those parasites with sexual reproduction and high levels of outcrossing have higher variability than those parasites that engage in asexual reproduction. One example of the influence of outcrossing is provided by ***Plasmodium falciparum***. Recall that malaria parasites undergo sexual reproduction and therefore recombination in their mosquito hosts and that reproduction in their vertebrate hosts is essentially asexual. In areas of Africa and Papua New Guinea, ***P. falciparum*** enjoys a high prevalence, and in these locations it exhibits high levels of genetic variability and little differentiation among locations (low F_{ST} values) suggestive of high gene flow. In contrast, in areas in South America, prevalence and genetic diversity are lower and populations are more subdivided (high F_{ST} values). This difference arises because in high-prevalence areas in Africa and Papua New Guinea, people are much more likely to harbor multiple ***P. falciparum*** clones and to have higher mean numbers of clones than people living in low-prevalence South American locations. Mosquitoes are consequently more likely to acquire multiple parasite genotypes in high-prevalence areas (**Figure 7.1A**), and recombination among these many genotypes frequently occurs in the mosquito (Figure 7.1B). In contrast, where prevalence is low, mosquitoes become infected by fewer genotypes such that sexual reproduction might involve male and female gametes derived from the same asexual lineage. In this case, sexual reproduction essentially amounts to selfing and results in lower rates of recombination. Studies such as this point out the value of awareness of population composition and structure; ***P. falciparum*** in areas of low genetic diversity is likely to also have reduced diversity in variant surface antigens that are potentially relevant to vaccine efficacy and that might respond much differently to vaccination than hyperdiverse African populations.

Levels of genetic variability are very high for outcrossing trichostrongyle parasites such as *O. ostertagi* mentioned previously. In contrast, consider the nematode *Heterorhabditis marelatus*, an obligate parasite of soil-dwelling

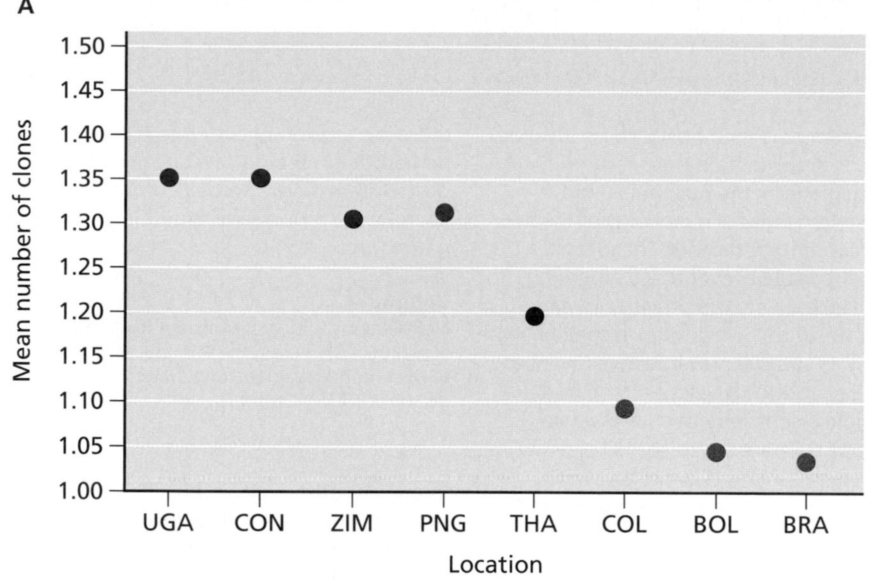

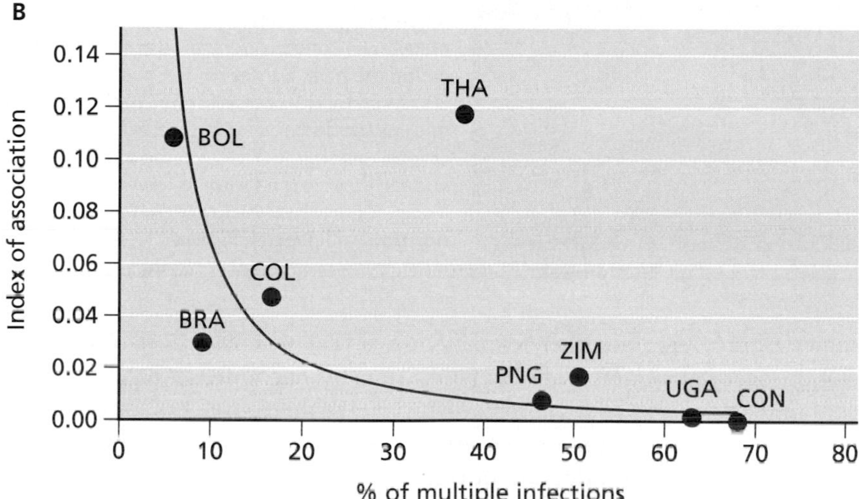

Figure 7.1 Opportunities for outcrossing influence *Plasmodium falciparum* population structure. (A) In high-prevalence areas in Africa (UGA, CON, and ZIM in red) and Papua New Guinea (PNG in blue), people are much more likely to have higher mean number of *P. falciparum* clones than from low-prevalence South American locations (COL, BOL, and BRA green). Isolates from Thailand (THA in black) show intermediate mean numbers of clones. (B) This graph indicates on the vertical axis an index of association (high levels are indicative of low levels of genetic recombination). On the horizontal axis is shown the percentage of multiple genotype infections. Note that parasites from low-prevalence areas have low effective rates of recombination. This is probably because the male and female parasites acquired by a particular mosquito when it feeds are derived from the same clone. (From Anderson TJC, Haubold B, Williams JT et al [2000] *Mol Biol Evol* 17:1467–1482. With permission from Oxford University Press.)

insects. **Box 7.2.** describes the remarkable means employed by such nematodes to infect insects. One or a small number of juvenile nematodes colonizes any particular insect host. Furthermore, once mature, the nematodes reproduce hermaphroditically—reproduction is by selfing. Consequently, the amount of genetic variability among the progeny within any infrapopulation is low. Also, the amount of gene flow is limited as the infected insects do not move much and the nematodes themselves have limited dispersal ability. As a consequence, as compared to trichostrongyles from domestic ruminants, *H. marelatus* has much lower overall genetic diversity both within populations and across the species, and it has a much more subdivided population structure.

The large liver fluke, ***Fascioloides magna*** (see also Box 6.4, Figure 1), of deer in the southeastern United States provides an example for how reproductive activities in different life-cycle stages can influence population structure. Adult flukes live in the liver of deer and pass eggs in the deer's feces that infect amphibious snails. The snails produce cercariae that encyst on vegetation and are ingested by the deer. See ***Fasciola hepatica*** in the Rogues' Gallery for more details about a closely related parasite. It was surprising to find that different flukes from a particular deer were often genetically the

BOX 7.2

Nematodes that Conspire with Specialized Bacteria to Infect Their Insect Hosts

When an infective-stage juvenile of the entomopathogenic nematode *Heterorhabditis marelatus* (family Heterorhabditidae) colonizes an insect host through a natural body opening or across the host's cuticle, it carries along with it in its gut specialized bacteria (*Photorhabdus* sp.) that are soon regurgitated into the host insect's hemocoel. The bacteria inhibit phagocytosis by host hemocytes (see also Chapter 4), thus facilitating colonization of the insect's body. The host usually succumbs from septicemia in 2 to 3 days. The nematode feeds on the bacteria and on disintegrating host tissues, producing multiple generations of nematodes within the cadaver of its host. Without high densities of *Photorhabdus* to ingest, the nematode is unable to complete its development. When the food is depleted, the nematode produces more infective juveniles, which pass out of the host and onto the soil in search of another host. They too carry within their gut the bacteria they will depend on to infect and overcome their hosts. These juveniles do not simply obtain their *Photorhabdus* inoculum from the host milieu but rather from special rectal cells of their hermaphroditic mother. The juveniles develop *in utero*, eventually emerging and killing her in the process. Remarkably, over the 2- to 3-week span in their host as many as 100,000 new infective juveniles can be produced, potentially all from a single colonizing infective juvenile (Figure 1). There are at least 12 species of *Heterorhabditis* that are known to infect a broad spectrum of insects, and there is considerable interest in using the nematodes in biological control of insect pests. Interestingly, some *Photorhabdus* bacteria are reported as emerging human pathogens.

Reference

Eleftherianos I, Ffrench-Constant RH, Clarke DJ et al (2010) Dissecting the immune response to the entomopathogen *Photorhabdus*. *Trends Microbiol* 18:552–560.

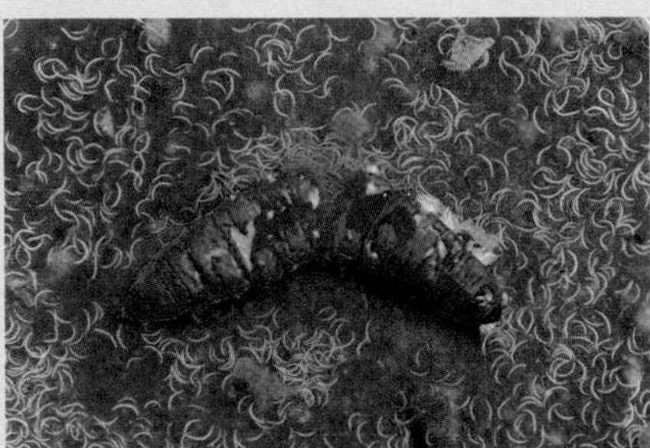

Figure 1 A collaborative effort to achieve infection. The cadaver of an insect, the greater wax moth *Galleria melonella*, that had initially been colonized by one or a small number of nematodes of the species *Heterorhabditis bacteriophora*. Note the many nematode progeny produced within the host. Success of the nematode is assured by the presence of a specialized mutualistic bacterium *Photorhabdus* it carries, one that facilitates colonization and nutritive exploitation of the host insect. (Courtesy of Peggy Greb, USDA Agricultural Research Service.)

same (**Figure 7.2**). The likely explanation is that whereas most of the vegetation deer consume is safe, they occasionally encounter a patch of grass contaminated with *F. magna* metacercariae. Furthermore, it is likely that all the metacercariae within a patch are derived from a single nearby infected snail. Insofar as the production of cercariae occurs by asexual reproduction

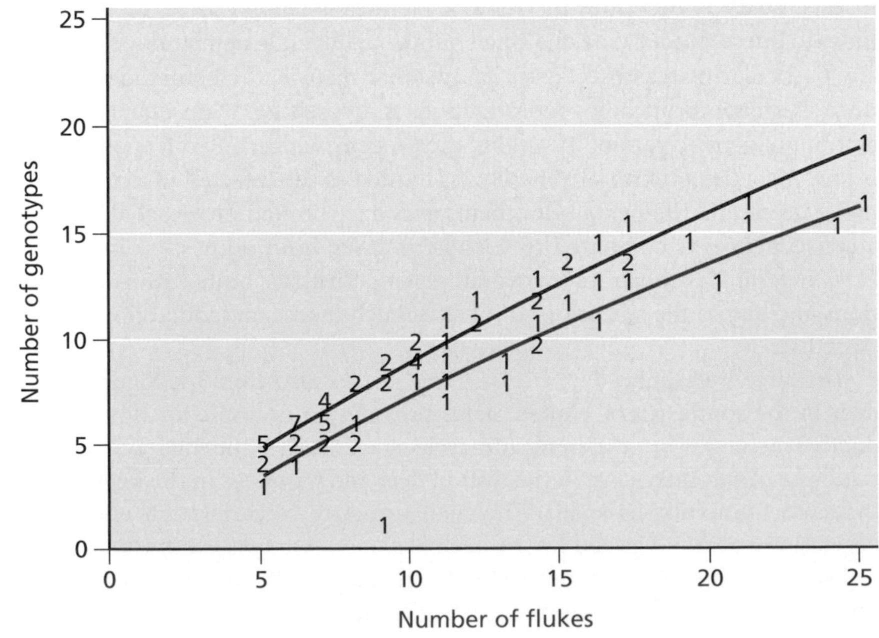

Figure 7.2 Asexual reproduction occuring in the snail host influences the population structure of *Fascioloides magna*. For 79 deer, each infected with from five to 25 flukes, the number of fluke genotypes expected per deer from simulations is shown by the red line. The actual number of fluke genotypes recovered per deer was often less than expected. For example, for the eight deer that harbored five flukes, five had flukes that all had a different genotype, but two had flukes representing only four genotypes, and one had flukes representing only three genotypes. Observed values falling on or below the lower blue line (representing 17 of 79 deer sampled) had significantly fewer different genotypes than expected by chance. (From Mulvey M, Aho JM, Lydeard C et al [1991] *Evolution* 45:1628–1640. With permission from John Wiley and Sons.)

(see Platyhelminths in the Rogues' Gallery) and all cercariae derived from the same miracidium are clone mates, then deer frequently acquire a dose of genetically identical metacercariae that will go on to develop into adult worms. In this case, the asexual reproductive activities of the parasite in the intermediate host have significant implications for the genetic composition of worms in the definitive host. The acquisition of clonally-produced progeny lowers the N_e in deer and contributes to strong differentiation of ***F. magna*** among individual deer.

Note the ***F. magna*** life cycle features metacercariae that encyst on an immobile patch of vegetation. The most common digenetic trematode cycles involve a mobile second intermediate host such as an aquatic insect or fish that becomes infected with metacercariae. The mobility and longevity of such hosts favor their infection with metacercariae originating from different snails, having the ultimate effect of insuring that when the second intermediate host is ingested, the definitive host will acquire a more diverse pool of parasites.

Stability of the host environment influences parasite microevolution

Another factor listed in Table 7.1 that directly influences parasite population structure is the stability and availability of the habitat (the host) supporting the parasite infrapopulation. If the host is short lived or only rarely available, the parasite is more likely to experience population crashes. Crashes promote loss of diversity and generate differences among infrapopulations. If the host is long lived, then the parasite populations tend to be more stable. Parasites such as ***Schistosoma mansoni*** for the most part inhabit long-lived hosts (people) and are themselves relatively long lived, creating a considerable degree of stability in their populations, which show high levels of infrapopulation diversity. Studies of this parasite in large endemic areas such as the Lake Victoria basin indicate high levels of gene flow, no doubt assisted by the mobility of infected people, which creates low differentiation among ***S. mansoni*** populations within local human populations.

The mobility of parasites impacts their evolution, as exemplified by bird lice

Dispersal of parasites from host to host with its attendant flow of genetic material potentially has a large effect on the parasite's evolution. The chewing lice (Phthiraptera) found on pigeons and doves are instructive in this regard. Lice of the genus *Physconelloides* are found on the body, often nestled in the downy feathers on the belly of birds, whereas *Columbicola* lice are found on wing feathers. Wing lice have greater powers of dispersal than body lice; they are more prone to attach to hippoboscid flies (an example of phoresy; see Chapter 1), which can transport them to new hosts. They are also more likely to be transmitted to new hosts on detached feathers or at dust baths, and they are quicker to leave a dying host than body lice. These differences in dispersal abilities have a strong effect on population structure. Note that for the wing louse *Columbicola gracilicapitis*, with respect to diversity within the *cytochrome oxidase I* (*COI*) gene, individual lice with the same haplotype (*COI* sequence) could be recovered from two different species of avian hosts (**Figure 7.3A**) and the total number of different haplotypes is relatively small, both indicative of relatively little structure within the louse species. In general, wing lice are less host specific than body lice and less likely to show patterns of **cospeciation** with their hosts (Figure 7.3B). Cospeciation means that when their host forms new species, the lice tend to do so as well, a topic we will return to in Section 7.4.

In contrast, populations of the body louse *Physconelloides eurysema* show much greater structure among populations (less gene flow, potentially more inbreeding within populations) than those of wing lice. For example, each haplotype was recovered from only one host species, and the number of haplotypes and the step differences between them were greater (Figure 7.3C).

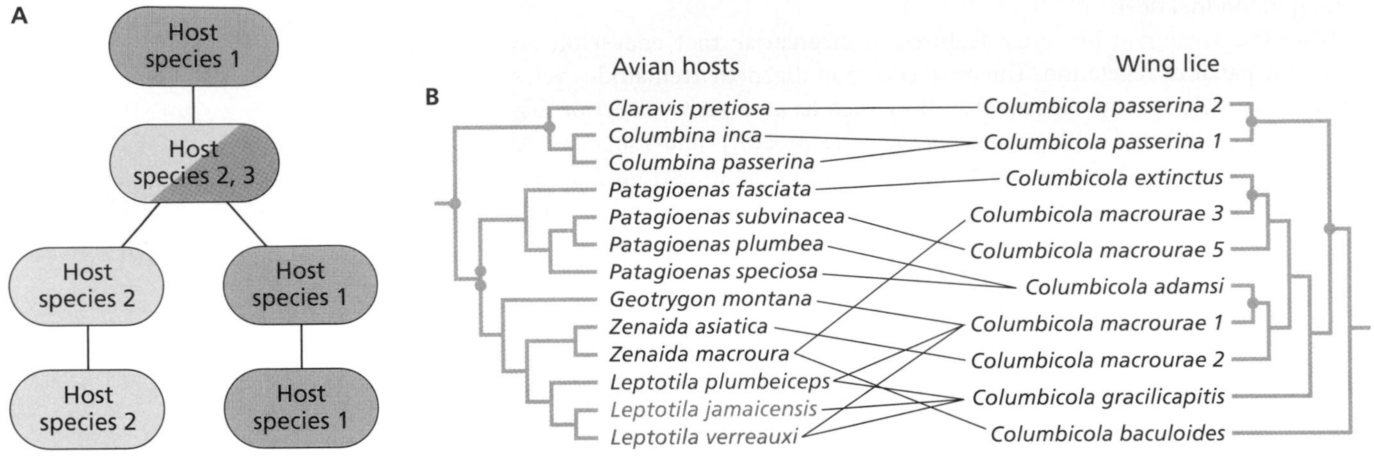

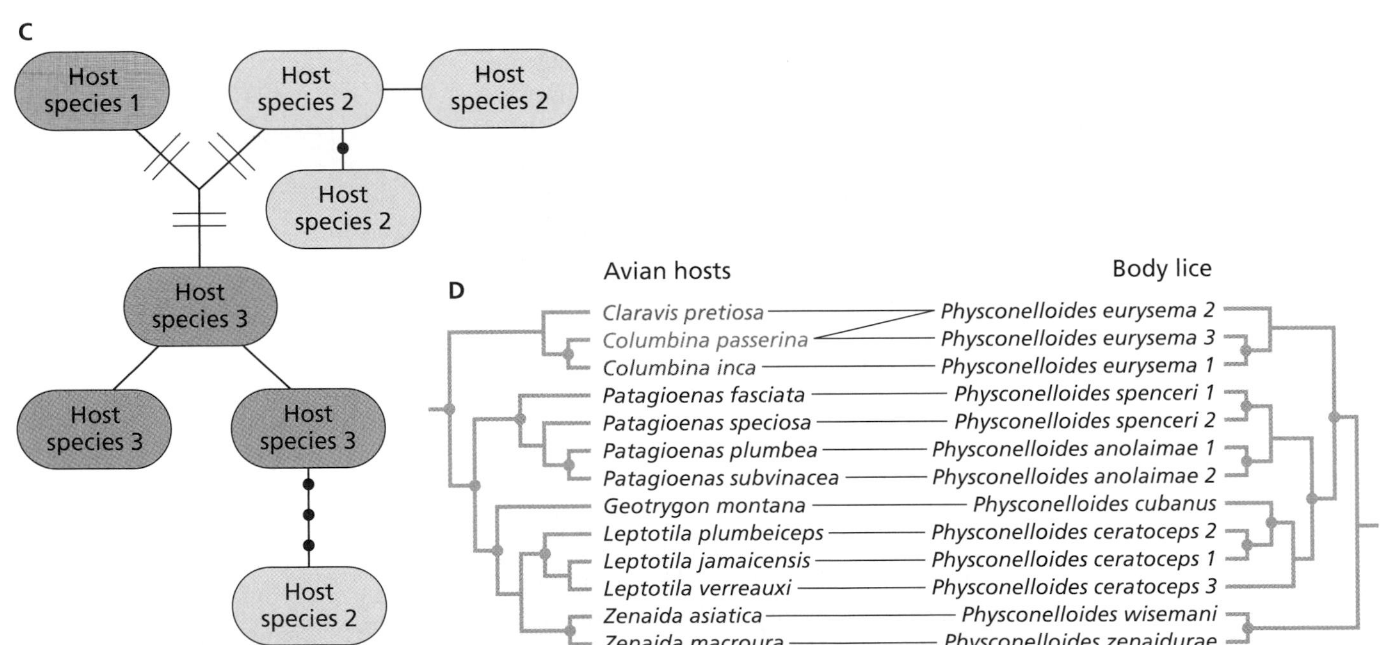

Figure 7.3 The mobility of bird lice influences their population structure and diversification. (A) A representation of haplotype (sequence) diversity for the *cytochrome oxidase I* (*COI*) gene of the wing louse, *Columbicola gracilicapitis*. Note that the color for each oval specifies a particular host species written in that same color in part B of the figure. Each oval represents one haplotype, and solid lines connect haplotypes that differ from one another by a single step. (B) Cophylogenetic study showing trees for wing lice (right) and their avian hosts (left) showing four inferred cospeciation events (dots), a number not different from what is expected by chance. (C) Haplotype diversity for three diverging lineages of the body louse, *Physconelloides eurysema*. Each tick mark along a line is indicative of a step difference. The color for each oval indicates a particular host species written in that same color in part D of the figure. The black dots refer to inferred step differences that were not recovered. (D) Cophylogenetic analysis of body lice (right) and their hosts (left). Eight inferred cospeciation events are indicated (dots), a number greater than expected by chance. (A,C, From KP Johnson, Williams BL, Drown DL et al [2002] *Mol Ecol* 11:25–38.With permission from John Wiley and Sons. B,D, From Clayton DH & Johnson KP [2003] *Evolution* 57:2335–2341. With permission from John Wiley and Sons.)

Body lice are relatively host specific and also show a pattern of **cospeciation** with their hosts (Figure 7.3D). This example thus provides a mechanism (limited dispersal ability of body lice) that helps explain a microevolutionary pattern (high population structure) that in turn bears directly on a macro-evolutionary pattern (levels of host specificity and tendency to speciate along with their host species).

Parasite microevolutionary change is strongly impacted by host mobility

Among the most important factors influencing parasite evolution, shown in several studies and already alluded to above, is the mobility of the host species. Some hosts are highly mobile and serve to transport their parasites from area to area, thus increasing gene flow; whereas other hosts are limited in their mobility, or vagility, thus limiting gene flow in their parasites. For example, trichostrongylid nematodes in domestic ruminants (like *Ostertagia* discussed above) have high within host diversity, and relatively low levels of among host diversity because of widespread transport of hosts by agriculture. Interestingly, for the related nematode *Mazamastrongylus odocoilei* from deer, large infrapopulations are also present, but more differentiation between populations is noted, likely because movement of deer is limited relative to cattle which are frequently moved from place to place by people. Widespread movement of domestic ruminants and dogs also has the effect of reducing differentiation among populations of the tapeworm ***Echinococcus granulosus***. Likewise, tick populations tend to show weak local differentiation because they are moved far and wide upon the hosts on which they reside for extended periods.

In a study of four different digenetic trematodes transmitted through snails in streams in the Pacific Northwest, three of the species that have salmon as their definitive host exhibited limited dispersal among watersheds (**Figure 7.4**). By contrast, a fourth species readily moved between watersheds because its definitive hosts are more vagile fish-eating mammals and birds. This fourth digenean species shows less population structure than the other three.

The impact of limited host dispersal capability is perhaps best shown for lice developing on pocket gophers (Geomyidae). The gophers live largely solitary lives in burrow systems and rarely contact conspecifics. Their populations are small and fragmented. Pocket gophers support a number of chewing lice (Amblycera and Ischnocera) as ectoparasites that have little dispersal ability of their own and rely on opportunities provided by host-to-host contact for their transmission. Because of the solitary habits and limited dispersal of their hosts, these opportunities are infrequent. Even among louse populations living on hosts a few meters apart underground, there are significant levels of genetic differentiation. Inbreeding, bottlenecks, and founder effects also contribute to low N_e and therefore even more differentiation among infrapopulations. The features of this system leading to strong differentiation among louse populations certainly are relevant for helping to explain the congruent phylogenies of pocket gophers and their lice (see cospeciation in Section 7.4).

A parasite's life cycle also affects the potential for evolutionary change

The nature of a parasite's life cycle can also have profound effects on the parasite's population structure. Availability of life-cycle stages or hosts in which

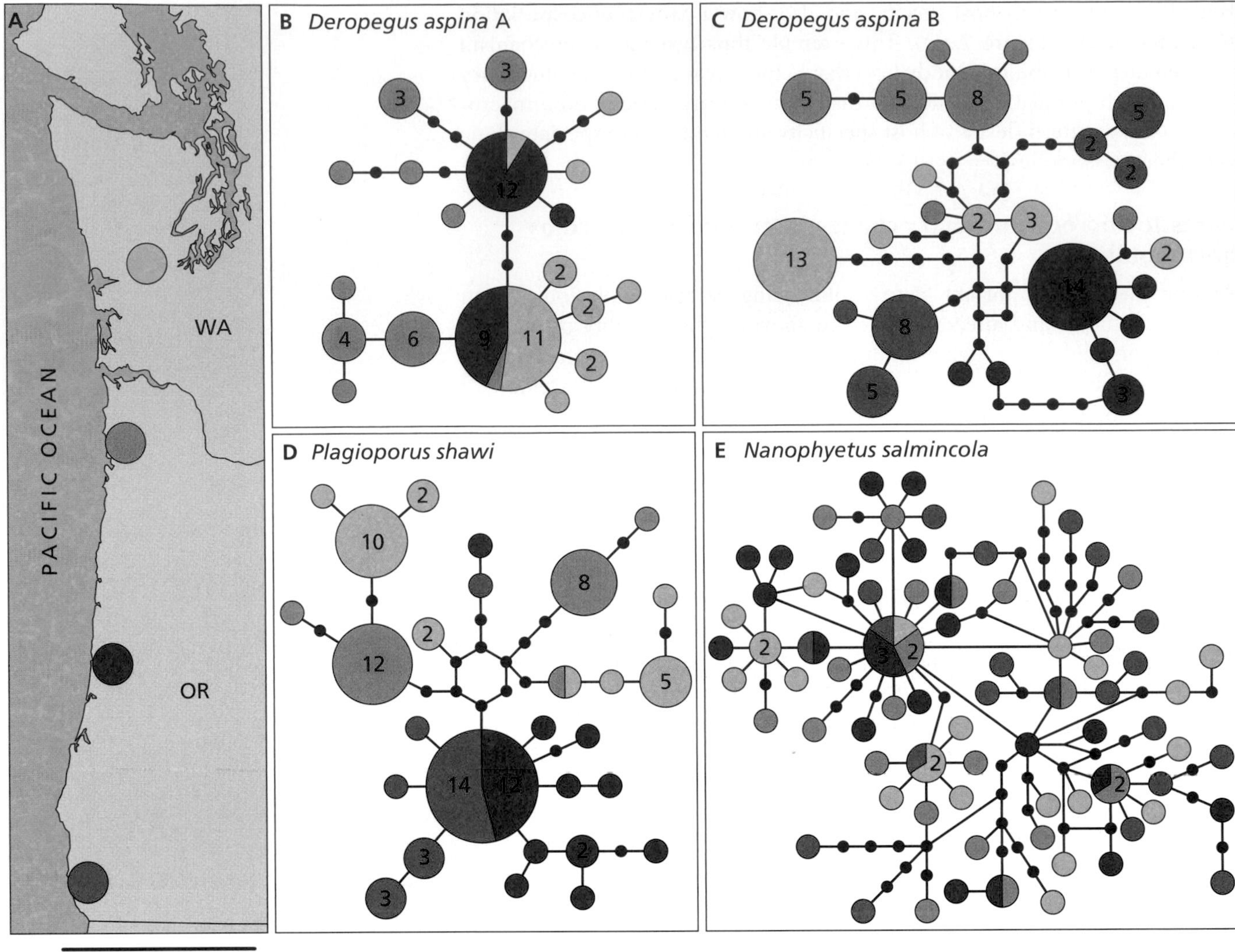

Figure 7.4 Host mobility influences parasite microevolution. (A) Digenetic trematodes of four species were sampled from each of the four localities shown (each locality indicated by a different color). For each trematode species (B to D), a haplotype map is shown, with each haplotype indicated by a circle. The size of the circle is proportional to the number of individual worms found with that haplotype (see numbers within circles). Also shown by color coding are the locations from which each haplotype was recovered. Each line connecting two haplotypes is a single mutational difference, with black circles indicating inferred haplotypes. For trematode species B through D, all with salmonid definitive hosts, note that the haplotypes tend to separate into color groups, indicative of strong geographic structure and low gene flow because fish are unable to cross between watersheds. (E) By contrast, note for *Nanophyetus salmincola*, with avian and mammalian definitive hosts able to move between watersheds, how the colors are interspersed among haplotypes indicative of high rates of gene flow among the four localities. (From Criscione CD & Blouin MS [2004] *Evolution* 58:198–202. With permission from John Wiley and Sons.)

parasites can persist for long periods in a relatively quiescent state increase the stability of parasite populations and thereby decrease genetic drift and local extinctions. The remarkable survivorship of ***Ascaris*** eggs in the external environment has much to do with the steady persistence of this parasite in human populations. The presence of long-lived resting stages, such as metacercariae in digenetic trematode cycles, favor stability and help prevent local extinctions. As noted above for ***Fascioloides magna***, the asexual reproductive process in an intermediate host may have a large impact on limiting the effective population of adult worms in a definitive host.

For parasites with complex life cycles, the genes that optimize fitness in one host may not optimize fitness in another obligate host in the life cycle. In the tick-borne apicomplexan parasite ***Theileria parva*** that causes East Coast fever in cattle in Africa, passage of the parasite through ticks results in a dramatic alteration in genetic composition of the parasite, as does passage through cattle. The parasite is continually being shaped by selection imposed by two very different host environments and immune systems. As another example, in ***Schistosoma mansoni***, deliberate selection to optimize fitness in the snail intermediate host resulted in an inverse relationship with the parasite's fitness in the definitive host (**Figure 7.5**). Such genetically based constraints have important implications for interpreting the epidemiology of parasites with complex life cycles. Once again, unique aspects of the biology of the particular parasite–host system in question can strongly influence the microevolutionary process, often in distinctive ways that defy generalization.

7.2 COEVOLUTION OF PARASITE–HOST INTERACTIONS

Parasites and hosts reciprocally affect each other's evolution

Suppose a parasite imposes a fitness cost on its host species such that an evolutionary change (increased resistance) occurs in the host. This change might then select for parasites able to infect the newly-resistant host, which again selects for a change in the host, and so on. This situation of ongoing reciprocal change is the essence of the idea of **coevolution**, not to be confused with cospeciation, which we describe in Section 7.4. There has been a tendency to use the word coevolution in a liberal sense to refer to what might simply be called parasite–host interactions, but the term is most properly used in the sense assuming reciprocity.

Coevolution is a dynamic microevolutionary process; it has the potential to change gene frequencies in both parasite and host populations and to maintain polymorphisms in key genes. It might involve traits such as parasite infectivity, host resistance, parasite host-finding ability, and parasite avoidance behavior by the host. Key to the concept is that the traits being

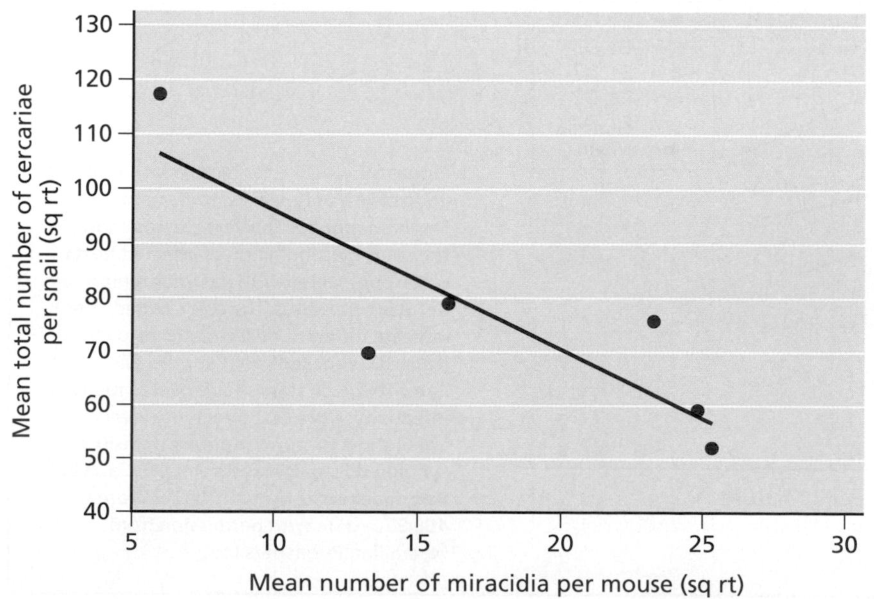

Figure 7.5 A genotype fit for one life-cycle stage may not be fit for another. For the six artificially selected lineages of *Schistosoma mansoni* examined, reproductive success in the snail (measured as mean lifetime number of cercariae produced) and in the mouse definitive hosts (measured as the mean number of miracidia produced) were inversely correlated. (From Gower CM & Webster JP [2004] *Evolution* 58:1178–1184. With permission from John Wiley and Sons.)

affected are genetically based and that changes in the gene frequencies can be documented. It is a process that has the potential to influence traits such as parasite virulence and transmissibility, and so it is important from a medical and veterinary standpoint to appreciate its potential impacts. Experimental studies exploiting the short generation times of bacteria (as hosts) and their interactions with phages (as parasites) have shown antagonistic coevolution to result in high levels of molecular evolutionary change in both.

Another similar concept to distinguish at this point is the **Red Queen hypothesis**. This is named after the character in Lewis Carroll's *Through the Looking Glass*, who notes that "it takes all the running you can do to keep in the same place." The basic idea is that biotic interactions are a fundamental driver for evolution and that for interacting species, particularly for antagonistic interactions such as a parasite and its host, a change in one is likely to select for a change in another (**Box 7.3**). This interplay gives rise to a continual

BOX 7.3

Looking Through the Mud to Clarify Red Queen Dynamics

Direct evidence for the operation of antagonistic coevolutionary interactions between host and parasite is rare, particularly when it comes to documenting the longer-term temporal dynamics. An ingenious solution to this problem was provided by the study of *Daphnia magna* (the water flea) and its bacterial parasite, *Pasteuria ramosa*, both of which produce dormant stages that survive long periods in the sediments at the bottom of lakes and ponds. These laminated sediments preserve a temporal record of the host and parasites present over long time spans (about 40 years in this study). Samples of *Daphnia* propagated from one sediment layer (a thickness representing a 2- to 4-year interval) were infected with parasites from the same, next older (lower), or next newer (higher) strata of sediments. Infectivity was consistently higher when a contemporaneous parasite was used as opposed to a past or future parasite (lower or higher strata, respectively). This pattern indicated that the parasite is able to rapidly adjust to its host over a few years (Figure 1) and also that hosts changed over time in response to the parasites. Parasite virulence (as measured by the extent it reduced *Daphnia* fecundity) actually increased over time, reflecting continued adaptation of the parasite to the host. It was concluded that the results were consistent with a model of negative frequency-dependent selection (see Box 7.1). That is, there is an advantage to being rare.

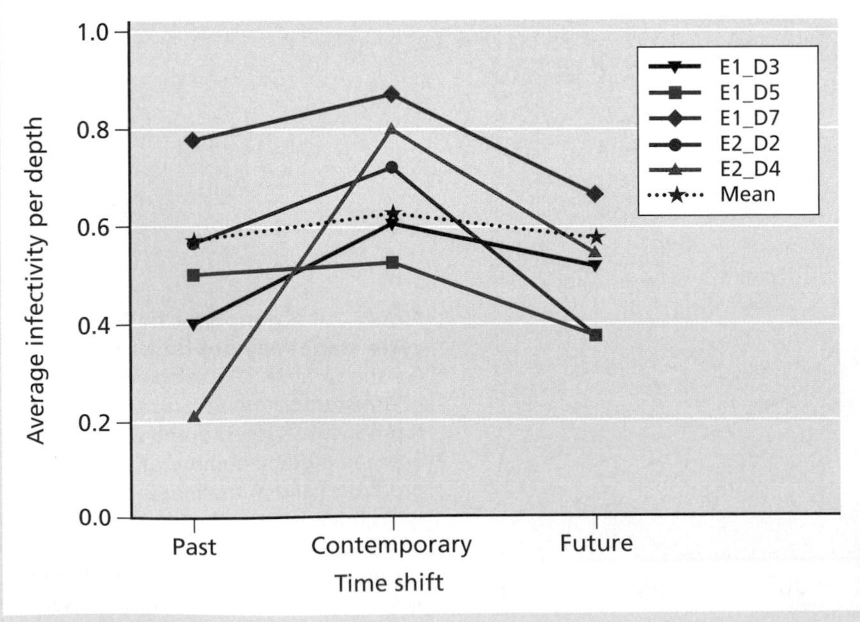

Figure 1 Evidence for long-term temporal adaptation of *Pasteuria ramosa* to *Daphnia magna*. The vertical axis shows the average proportion of infected hosts when confronted with past, contemporary, or future parasites. The color-coded lines indicate different host isolates recovered from different sediment depths. The line with black stars shows overall mean infectivity. Note that infectivity levels are highest for contemporary parasites. (Modified from Decaestecker DE, Gaba S, Raeymaekers JAM et al [2007] *Nature* 450:870–U16. With permission from Macmillan Publishers Ltd.)

natural selection for adaptation and counteradaptation (running to keep in the same place). In the parasitological literature, the Red Queen hypothesis has often been invoked more specifically, regarding the role of parasites in favoring sexual reproduction among host organisms. We will return to this topic when we discuss parasite effects on host evolution (Section 7.6).

Parasites and hosts engage in arms races

One of the possible outcomes of coevolutionary interactions between parasites and their hosts is an arms race. Although interpretation of the term varies, in our usage an **arms race** (**Figure 7.6**) occurs when a successful innovation in the parasite population results in a **selective sweep**, meaning the old trait is replaced by the new trait that increases to fixation. In response, the host population develops a new innovation to counter the parasite, and it undergoes a similar selective sweep to replace the old trait with the new one. Such a directional process continues, with both partners being continually improved by selection that favors innovations as they appear.

Although clear enough conceptually, examples of parasite–host arms races in which a selective sweep in one partner is followed by a selective sweep in the other as shown in Figure 7.6 are hard to come by, especially for eukaryotic parasites. Consider a couple of examples that seem to represent arms races. The first involves human association with African trypanosomes (see also Chapter 2). Humans are resistant to trypanosomes of cattle such as ***T. brucei brucei*** because they possess a serum factor called apolipoprotein L-1 (APOL1) that lyses the membranes of such trypanosomes (**Figure 7.7**). It seems reasonable to imagine a strong evolutionary response of humans to the threat posed by trypanosomes in Africa to account for the presence of APOL1. However, as we well know, some trypanosomes such as ***Trypanosoma brucei gambiense*** and ***T. b. rhodesiense*** have evolved means to overcome the lytic activity of APOL1, enabling them to establish long-term infections in people. For *T. b. gambiense*, which causes 97% of the cases of human African trypanosomiasis, there seem to be at least three different mechanisms for overcoming the adverse effects of APOL1. These involve the expression of a protein that stiffens trypanosome membranes and makes then less vulnerable to APOL1 lysis, an alteration in the membrane receptor involved in taking up APOL1 in the first place, and changes in biochemistry of trypanosome lysosomes to make it less likely for APOL1 to intercalate into parasite membranes. ***Trypanosoma brucei rhodesiense*** by contrast has evolved a serum resistance associated protein (SRA), which alone can prevent APOL1 lysis by binding to it and preventing its insertion into parasite membranes. All these mechanisms can be viewed as countermeasures that the human trypanosomes have taken in response to the APOL1 defense put

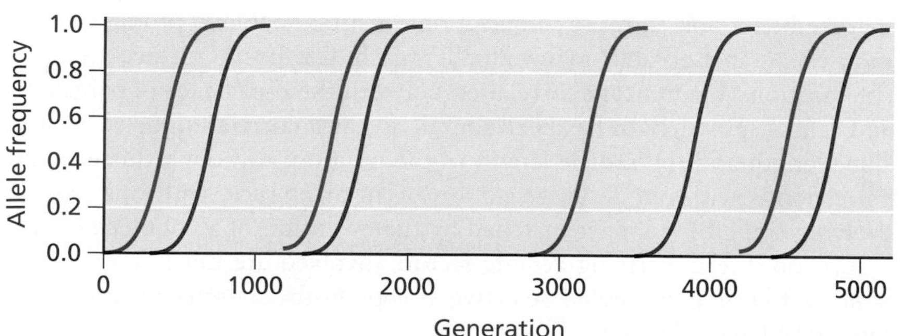

Figure 7.6 An arms race consisting of a series of selective sweeps. Genetic change is directional in this case, and change accumulates in the populations. The origin of a new host allele is followed by a selective sweep to fixation (blue line), which is followed by the origin of a new parasite counter allele by mutation that also sweeps to fixation (red line), which is followed by the rise of another host allele, and so on. At any time, there may be polymorphisms for the relevant traits in both, either, or neither host and parasite populations. Note that the time required for either parasite or host to respond is variable. (From Woolhouse MEJ, Webster JP, Domingo E et al [2002] *Nat Genet* 32:569–577. With permission from Macmillan Publishers Ltd.)

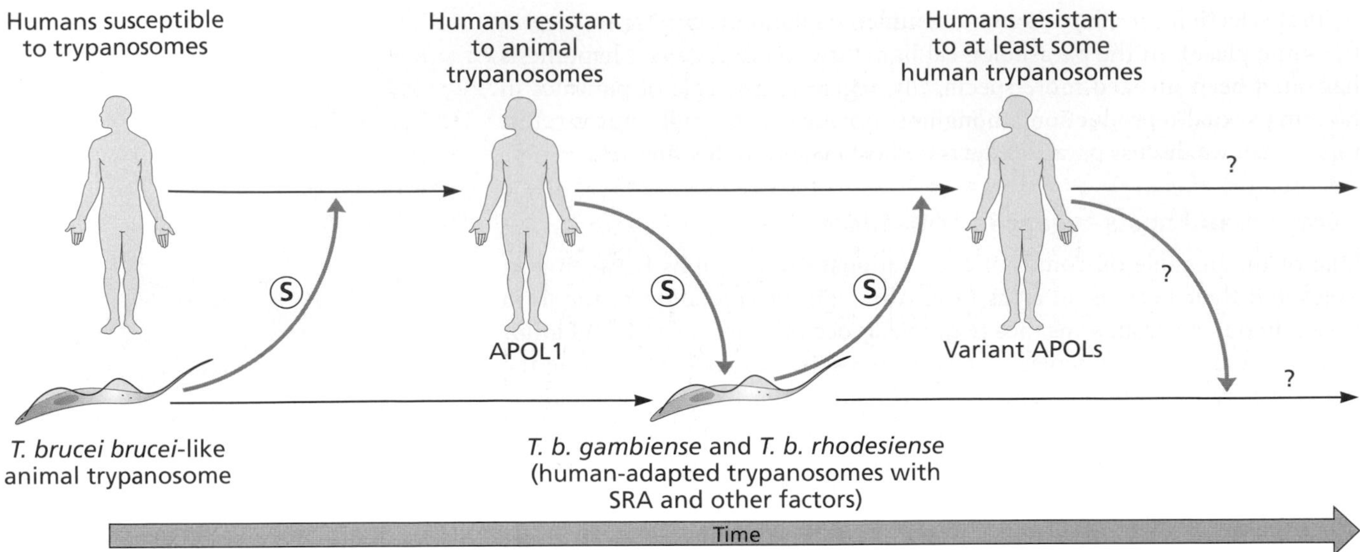

Figure 7.7 A possible arms race scenario involving humans and trypanosomes. Human populations were initially susceptible to animal trypanosomes like *Trypanosoma brucei brucei* but selection (circled S) imposed by trypanosomes favored expression of a gene encoding lytic apolipoprotein L-1 (APOL1) that conferred resistance to trypanosomes. The presence of APOL1 then selected for a change in trypanosomes to overcome its lytic effects. Consequently, a gene expressing serum resistance associated protein (SRA) was favored by selection in human-adapted trypanosomes. Selection then favored variant forms of APOL1 able to resist the effects of SRA.

up by humans. Some people in West Africa produce variants of APOL1 that prevent SRA binding to APOL1, and serum from individuals with these traits is able to kill ***T. b. rhodesiense*** parasites *in vitro*, suggesting another countermeasure taken by humans to counter trypanosome infection. Interestingly, the presence of these variant forms of APOL1 is associated with kidney disease, suggesting that intense selection as caused by an arms race with parasites can result in collateral damage.

To dispel the notion that arms races are confined to animals and their parasites, consider too the interesting example of the association between columnar cacti (*Echinopsis* and *Eulychinia*) in Chile and their parasitization by the mistletoe, *Tristerix aphyllus*, a plant holoparasite (see parasitic plants in Chapter 2). This mistletoe, unusual for its ability to infect cacti, produces a seed-containing fruit that the Chilean mockingbird (*Mimus thinca*) ingests. The bird eventually perches at the top of a cactus where it defecates the sticky mistletoe seed onto a cactus spine and essentially serves as a vector for a plant parasite. Once deposited, the mistletoe seed germinates and produces a long slender process called a radicle. Much longer than known to be produced by other mistletoes, this radicle winds along the spine toward the epidermis of the cactus where its penetrative tip (called a **haustorium**) gains entrance to the interior of the plant via the cactus' stomata. The parasite goes on to develop within the plant, eventually re-emerging to produce a flower and eventually seeds.

The consequences of infection for the cactus are considerable: its own reproduction is severely impaired or completely curtailed by mistletoe infection. The arms race here is manifested by the cacti with longer spines that are better able to resist colonization. Mistletoes able to produce longer radicles are better able to eventually reach the host's surface to effect colonization. A significant correlation between the prevalence of parasitism and cactus spine length has been found, as has a correlation between host spine length and the radicle length of the parasite. Although this example corresponds with our intuitive notion of an arms race, with one partner evolving something longer matched by the evolution of a countermeasure, and so on, the underlying genetic factors involved are not known, and it is not yet clear if successive selective sweeps in the manner envisioned in Figure 7.6 have occurred.

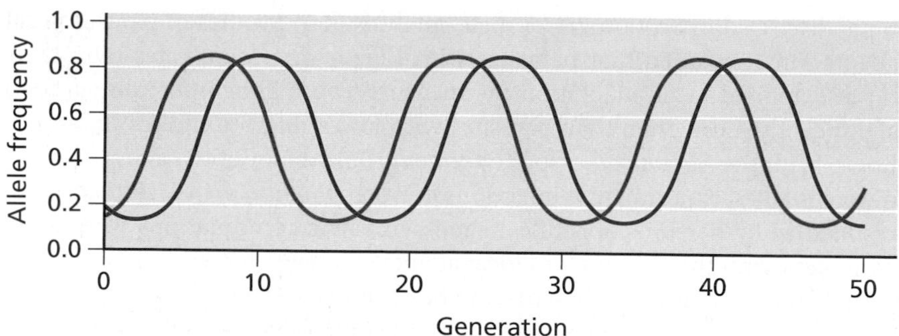

Figure 7.8 Time-lagged negative frequency-dependent selection. In this case, coevolutionary change is cyclical and nondirectional. The increasing frequency of a host allele (blue line) is mirrored (with a time lag) by an increase in frequency of a parasite allele favored by that host allele. Because the parasite becomes common, it has the effect of favoring a different, rare host allele such that the frequency of the original host allele declines, followed by a decline in the original parasite allele, and so on. Rare alleles are favored, and the same host and parasite alleles may keep recycling. At any one time, polymorphisms in both host and parasite alleles are expected. (From Woolhouse MEJ , Webster JP, Domingo E et al et al [2002] *Nat Genet* 32:569–577. With permission from Macmillan Publishers Ltd.)

In parasite–host relationships, there can be an advantage to being rare

A second form of coevolutionary trajectory is negative frequency-dependent selection (**Figure 7.8**). The *Pasteuria–Daphnia* system described in Box 7.3 behaves in this fashion. Another example of negative frequency dependence in a parasite–host system is provided by studies of the digenetic trematode *Microphallus* sp. and its interactions with the snail host *Potamopyrgus antipodarum* in New Zealand lakes. Populations of this snail are often made up of clones of individuals produced by parthenogenesis. In a long-term field study, initially rare clones almost completely replaced the most common host asexual clones. Also, the most common clone, which was initially most resistant to infection, became more susceptible to the prevalent parasite as the study progressed. In experimental populations, parasites were able to target common clones, becoming more successful in infecting them and driving down their frequency over time, a change that did not occur if parasites were not present. These results are consistent with the idea that parasite-mediated selection leads to an advantage of being rare and that the parasite present has coevolved as well to be infective to the most common host genotypes. Males can also occur in *P. antipodarum*, and higher levels of sexual reproduction occur when males are more common. Interestingly, in shallow lake waters where trematode infection rates are particularly high, sexual snails are more common. Such habitats, where parasite transmission is high can be viewed as coevolutionary hot spots, and it is in such locations that host sexual reproduction would seem to be most favored, a topic we return to in Section 7.6.

Parasites and hosts can be locally adapted, or maladapted, to one another

Another important potential outcome of coevolutionary interactions is **local adaptation** of parasites, referring to the ability of parasites to perform better in or on hosts from the same geographic area (parasite and host are sympatric) than on hosts from different geographic areas (parasite and host are allopatric). It is theorized that local adaptation is favored by the short generation times and high fecundity of parasites relative to those of their hosts and that local adaptation is more likely to occur when the migration rate (gene flow) is higher for the parasite than the host. These forces enable novel alleles to be imported into the local parasite population at a high rate, such that the parasite is able to successfully counteract host defense response. Another expectation of local adaptation is that the further removed geographically a new host is from the place of origin of a parasite, the less likely it would be successfully infected by the parasite.

One of the interesting manifestations of local adaptation is that it potentially leads to divergence in parasite populations and hence speciation, a

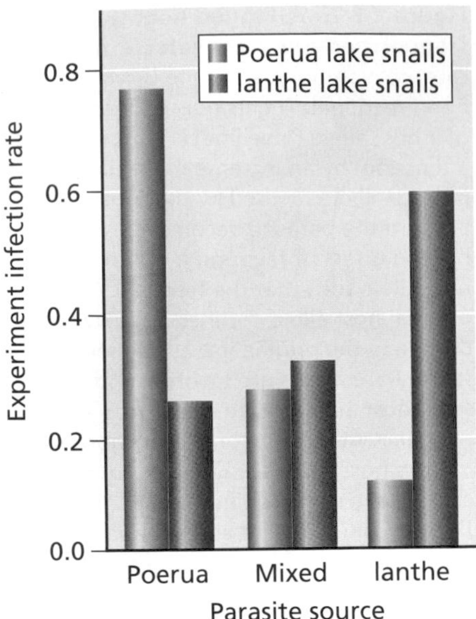

Figure 7.9 Evidence for local adaptation of a parasite to its host. Local parasite adaptation is indicated by significantly higher infection rates for sympatric parasite–host combinations compared with nonsympatric combinations. Note that when parasites are mixed according to their geographic source, they produce intermediate levels of infection in snails of both source populations. (From Lively CM & Dybdahl MF [2000]. *Nature* 405:679–681. With permission from Macmillan Publishers Ltd.)

topic discussed in Section 7.4. A meta-analysis of 57 studies of parasite local adaptation concluded that parasites with higher migration rates than their hosts exhibited local adaptation (as measured by higher infection success) significantly more often than parasites with lower migration rates than their hosts. With the *Microphallus* sp.–*P. antipodarum* system noted above, for two different lakes, experimental infections of *Microphallus* sp. in *P. antipodarum* resulted in higher infection rates when sympatric combinations were used, consistent with the operation of local adaptation (**Figure 7.9**).

In some cases, **local maladaptation** has been observed in which allopatric combinations of host and parasite are more compatible. It might occur when a parasite that has recently migrated into a new area is tested for its compatibility or in a system in which the host has higher migration rates. One example is provided by studies of the susceptibility of the lizard *Gallotia galloti* to hemogregarine (Apicomplexa) blood parasites transmitted by mites. Juvenile lizards actively migrate from their natal hatching site and consequently have high migration rates. Furthermore, they avoid colonization by the mite vector that stays on adult lizards, thus slowing down the rate of hemogregarine migration. Another possible explanation for local maladaptation is that there is a time lag in the parasite's response to its host such that, when sampled, the degree of compatibility is less. Documenting local maladaptation is more challenging than it first seems because what might appear to be local maladaptation of a parasite to its host might actually reflect a locally well-adapted host.

Some factors conspire to limit strong coevolutionary dynamics between parasites and hosts

Coevolutionary concepts are compelling and have attracted a great deal of attention, but to what extent do they actually occur in real host–parasite interactions? Although it's clear that selection imposed by a parasite can cause an evolutionary change in a host, or by a host on a parasite, it's difficult to document the reciprocity of such changes (but see Box 7.3). One of the factors potentially diminishing the likelihood of strictly dedicated and reciprocal interactions is that any host species typically has to contend with many different parasites that are bound to vary in abundance and virulence over space and time. Furthermore, a particular parasite species might infect several different host species. Accommodation of a host to one parasite might compromise its resistance to another. Also, for hosts such as humans that are long lived as compared to their parasites, how can the host population possibly adjust in sufficient time to counter the fast-evolving parasite? With hosts such as vertebrates that possess an adaptive immune system (Chapter 4), it would seem that parasite pressure has selected for a generic solution able to respond to many parasites without as much emphasis on special evolutionary accommodation to each. The existence of time lags in responses of the coevolutionary partners makes it very difficult to assess the situation during short-term study periods. Because of such considerations, strong coevolutionary associations might be most likely to occur in hosts with shorter life spans and in instances where they have particularly common and virulent parasites with which they must contend.

7.3 THE EVOLUTION OF VIRULENCE

As we noted in Chapter 1, at the core of the definition of parasitism is the notion that parasites harm their hosts. That is, they are to some extent virulent. The purpose of this section is to appreciate that this trait is influenced by natural selection and can change depending on circumstances. (It is thus another example of a trait influenced by microevolutionary processes.) We also seek to understand some of the basic forces expected to alter parasite virulence. One goal is that by understanding the basic principles whereby virulence evolves, we might someday actually be able to manipulate parasites to be less virulent. Because of the obvious importance of parasite virulence to crop production and to the health of animals and people, the study of virulence and its underlying drivers is a topic very actively pursued by evolutionary biologists and parasitologists.

In trying to understand virulence, it is important to keep two perspectives in mind. The first is that it is helpful to have a general theoretical framework from which to contemplate the evolution of virulence. The second is that the peculiarities of any particular parasite–host system being considered may resist generalization and require more specific consideration.

Virulence and transmission biology of parasites are linked

The conventional wisdom regarding virulence used to be that when a parasite first associated with a host, it tended to be virulent. Then over time, it accommodated to the host and evolved to a state of avirulence, enabling prolonged benign coexistence with its host. It is now realized that some parasites long associated with their hosts such as tuberculosis in humans are still virulent or, as has been exemplified by nematode parasites of fig wasps, are becoming more so. The old conventional wisdom has given way to the idea that natural selection should favor virulent forms if they are more immediately successful. Today's conventional wisdom regarding the evolution of virulence is largely embraced by the **trade-off hypothesis**, which recognizes virulence and transmission to a new host as linked traits (**Figures 7.10** and **7.11**). The trade-off in this case is that an increase in virulence may lead to a reduction in transmission (perhaps as a result of earlier mortality of the host) and that selection should ultimately act to favor the combinations of the two traits that maximize the overall fitness of the parasite.

The moth–mite system in Figure 7.10 exemplifies a trade-off. From the mite's perspective, the host would be best able to detect a predator if both ears were intact, but this would severely limit the opportunity for this ear-specific parasite to produce its progeny. On the other hand, colonization of both ears would increase mite progeny production but increases the chance of mortality by capture of its host by bats. That is, parasite virulence is linked to the opportunities for transmission (dependent on stability of its habitat and avoidance of bat predation), resulting in an intermediate level of virulence (only one ear infected). In places where bats have been extirpated, moths with both ears infected have been found.

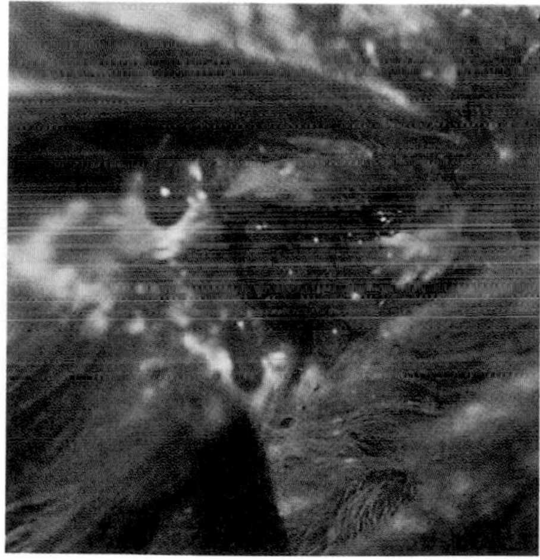

Figure 7.10 Trade-offs for mites in the ears of moths. The mite *Dicrocheles phalaenodectes* infects the ears of noctuid moths, in the process destroying the tympanic membrane and converting the ear into a crowded nursery for its progeny. Remarkably, in almost all instances, only one ear is so infested and the other ear remains functional. The ability of the moth to hear is critical because it enables it to detect the echolocation sounds emitted from predatory bats and to engage in avoidance behavior. (Treat, AE [1975] Mites of Moths and Butterflies. Comstock Publishing. Courtesy of Scot Waring.)

The relationship shown in Equation 7.1 represents the basic features of the trade-off hypothesis (recall the use of some of these terms in Chapter 6):

$$R_0 = \frac{\beta S}{\mu + \alpha + \gamma} \qquad [7.1]$$

where R_0 is the basic reproductive rate of the parasite (the number of secondary cases resulting from a primary case), S is the number of susceptible hosts in the population, β is the parasite's transmission rate (often influenced by host density), μ is the host's natural death rate, α is the parasite-induced host mortality rate (this is the virulence term), and γ is the host's rate of recovery from infection.

The term βS refers to the number of infections resulting from the initial infection per unit time. The terms μ, α, and γ all influence the duration of the

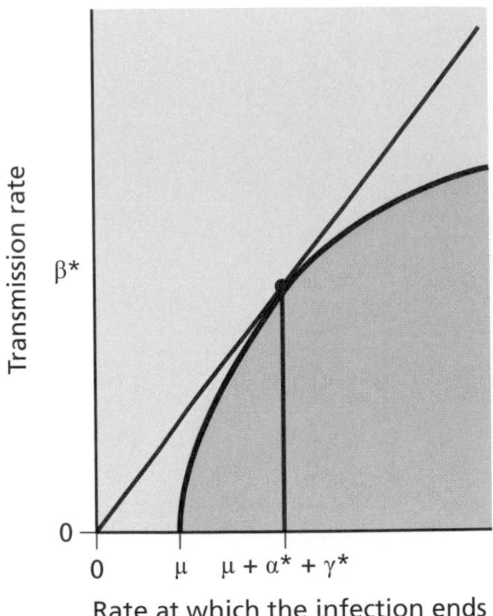

Figure 7.11 A curve showing the trade-off between transmission and virulence. The vertical axis shows the transmission rate β and the horizontal axis the rate at which infections end (infections of short duration appear further to the right on this axis). The strategy that will evolve (where R_0 is maximized) is given by the red diagonal line passing through the origin that is the tangent of the trade-off curve. The tangent point and the red vertical line dropped from it define the optimal levels of transmission (β*), virulence (α*), and recovery rates (γ*). (From Alizon S et al [2009] *J Evol Biol* 22:245–259. With permission from John Wiley and Sons.)

infection: when μ, α, and γ are large (infections are of short duration), the value of R_0 would be diminished. Assuming all the terms were independent of one another, then R_0 would be high when α was low. However, it is reasonable to expect the terms of the equation are not independent of one another. For example, if α were low because the parasite was reproducing poorly, then the likelihood of transmission, β, might be adversely affected. Or conversely, if parasite reproduction was high such that α became high, then the host might die before transmission could occur.

These relationships can also be visualized as a trade-off curve, with one graph axis representing the transmission rate and the other measuring the rate at which infections are lost (Figure 7.11). A typical trade-off curve (the thick green curve) is actually the boundary of the set of possible combinations of transmission and the rate at which infection ends (the set is shown by the green shaded area). If the parasite population starts to evolve at the interior of this set, evolution will favor strains with higher transmission (β) and longer infections (that is, low virulence α and low recovery γ) until the population hits the boundary (the thick green curve). Increasing rates of transmission β can only occur by accelerating mortality α or the recovery rate γ (infections shorten and the curve flattens out to the right).

The simplified version of the trade-off model in Equation 7.1 and figure 7.11 indicates an optimum virulence is found by maximizing R_0, the number of secondary infections produced by a primary infection. R_0 is basically a measure of the parasite's lifetime reproductive success. The essential trade-off referred to in most cases is between how fast and how long the parasite can be transmitted. High rates of progeny production are beneficial, but they would be expected to increase the virulence to the host and possibly cause host mortality, induce strong immunity and limit infection duration, or limit the host's mobility, all potentially limiting the interval that transmission is possible. Conversely, a lower rate of progeny reproduction might extend host longevity, diminish the immune response and enable transmission to persist longer, but fewer offspring are produced per unit time. Depending on the peculiarities of the system in question, any level of virulence could potentially evolve. Note also that the trade-off model refers to a host–parasite system in equilibrium (one that has been established for a long time such that selection on parasite virulence can approach an optimum value).

The trade-off hypothesis requires a nuanced approach

Several studies have provided empirical support for the trade-off hypothesis, but many additional considerations need be taken into account. For example, it is difficult to measure transmission and virulence (particularly indirect effects, such as on host competitive ability). Another consideration is that multiple parasite clones might infect a particular host, and these are likely to differ in their rates of reproduction. In such a case, it is hard to imagine that within host competition among parasites would not occur such that the faster growing forms would come to predominate and thus cause more harm to the host, potentially resulting in its death. This idea has been called the **short-sighted hypothesis**. One example is provided by the infection of mice with a mixture of clones of the rodent malaria parasite *Plasmodium chabaudi* that are known to differ in virulence. During the acute phase of the infection, the more rapidly reproducing (virulent) strains predominate. This ability is expected to translate into greater transmission success to mosquitoes for fast reproducing isolates because most transmission is believed to occur in this early, acute stage of the infection.

One practical implication from such studies relates to human malaria, where the diversity of parasites infecting a single person is often high (see Figure 7.1). If this diversity could be reduced and intrahost competition diminished, then perhaps the overall virulence of the infection could be reduced. The expectation of higher virulence levels in such a case might be altered if there was a high degree of relatedness among the clones, such that there was a degree of cooperativity among the parasites favoring lower levels of reproduction and longer-term infections. A study of ***Schistosoma mansoni*** cercariae production in snails did not find a relationship between the level of relatedness of the parasite clones infecting snails and the extent to which host exploitation occurred, as measured by the number of cercariae produced. This study suggests that in some systems where parasite populations are diverse and well mixed, opportunities for sharing hosts with kin will be limited such that kin recognition systems may not have evolved.

Several additional nuances enter into consideration regarding the evolution of virulence. Equilibrium conditions are assumed for the basic form of the trade-off hypothesis, but often such conditions are far from being the case. Parasites are emerging into new circumstances all the time. In some cases, such as the ongoing HIV pandemic, the parasite is still expanding on a global scale. In a newly established host–parasite system, the partners may not be well adapted to one another and virulence is not settled at an optimum level and is likely to evolve rapidly. Also, the manner in which selection would favor parasite virulence should vary dramatically depending on temporal and spatial availability of host populations. For example, if hosts are distributed in small isolated populations, then a virulent parasite might burn through the local population and go extinct if it does not encounter other host populations. When hosts are plentiful though, there should be fewer constraints on maintaining a high level of virulence.

As noted above, virulence is influenced by the genetic composition of the host as well, and hosts vary considerably in immune competence, which can also be further influenced by environmental factors such as nutritional status. The role of host immunity and the parasite's avoidance of it are understudied, yet their effects are likely to be considerable. For example, virulence may be more a consequence of an inappropriate host immune response than a parasite-produced toxin. We may also be discounting the possibility of tolerance (see Chapter 6), whereby the host evolves to minimize the damage associated with parasite infection as opposed to limiting parasite numbers. Tolerance may favor parasites with lower virulence because the host is not contesting the parasites' numbers or their persistence.

Finally, virulence might involve a third party such as a virus infecting a parasite, a factor not taken into consideration in typical modeling efforts. For instance, consider *Leishmania guyanensis*, responsible for causing a pathogenic condition known as mucocutaneous leishmaniasis. This illness occurs in some instances when the parasite metastasizes to nasopharyngeal tissues and causes their disfiguring destruction. Why does this happen? We now know the most invasive forms of ***Leishmania*** are infected with a virus (**Figure 7.12**). The virus produces a double-stranded RNA that is bound by Toll-like receptor 3 (TLR-3, see Chapter 4) on the phagosome membrane of infected macrophages. Binding triggers a cytokine reaction that results in an extensive inflammatory response. ***Leishmania*** variants lacking the virus are confined to the skin and induce much smaller inflammatory responses. How is the presence of the virus helpful to the parasite? Apparently, presence of the virus is associated with an improved ability of the parasites to withstand toxic oxygen radicals produced within infected host cells, enabling the

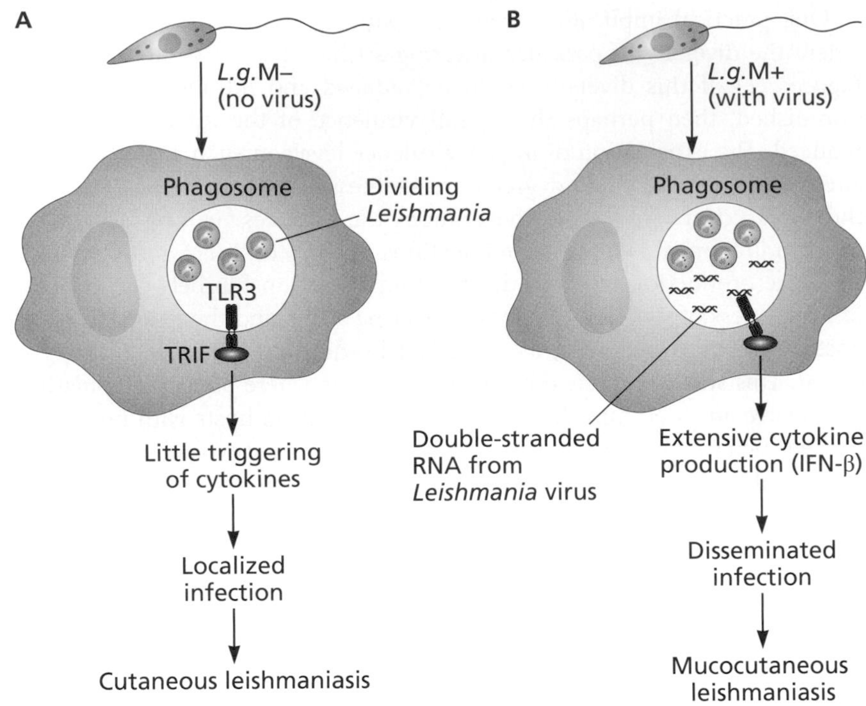

Figure 7.12 A virus influences the level of virulence of *Leishmania* in people. (A) *Leishmania guyanensis* parasites not infected with the *Leishmania* RNA virus 1 (*L.g.*M–) trigger little production of cytokines, remain localized and cause cutaneous leishmaniasis. (B) In infections initiated by virus-infected parasites (*L.g.*M+), double-stranded viral RNA triggers an extensive inflammatory response, including production of IFN-β, a cytokine capable of causing organ damage such as the tissue destruction and the disfigurement characteristic of mucocutaneous leishmaniasis. Note that TLR-3 on the phagosome membrane interacts with double-stranded RNA within the phagosome. (Modified from Olivier M [2011] *Nature* 471:173–174. With permission from Macmillan Publishers Ltd.)

parasite to reproduce more effectively even in the face of the intense host inflammatory response. This example is instructive for highlighting the role of a third party—the ***Leishmania*** virus—in influencing the virulence of a parasite and in reminding us that the damage done by a parasite is often the direct consequence of an immune response it has engendered (Chapter 5).

The mode of transmission influences virulence

Parasites that are transmitted by vectors (such as ***Plasmodium*** transmitted by mosquitoes) can rely on the vector to transport them from one host (for example, a vertebrate such as a human) to another. In contrast, parasites with direct life cycles rely on an ambulatory host to introduce parasite progeny into a variety of environments to increase the likelihood of transmission. For this reason, vector-borne parasites might be expected to be more virulent to their vertebrate host and suffer no penalty for immobilizing their host. A host sickened by a virulent parasite may also be less able to defend itself from the biting activity of its vector (virulence facilitates transmission to a new host). Immobilization is potentially a sign of being sick, and it would also seem to correlate with poor ability to defend from vector biting attacks. Anecdotally, people suffering malaria attacks are not particularly mobile or adept at deterring mosquitoes from biting. What does a quantitative test tell us about the idea?

In general agreement with these ideas, willow ptarmigan *Lagopus lagopus* infected with high intensities of vector-borne parasites, particularly *Leucocytozoon lovati*, were more likely to remain immobile than flee upon disturbance by hunters than birds infected with a directly transmitted parasite such as *Eimeria* (**Figure 7.13**).

Consider also those parasites that are transmitted vertically (from mother to offspring, for example), in contrast to those that are horizontally transmitted. In the case of vertical transmission, given that the parasite's own

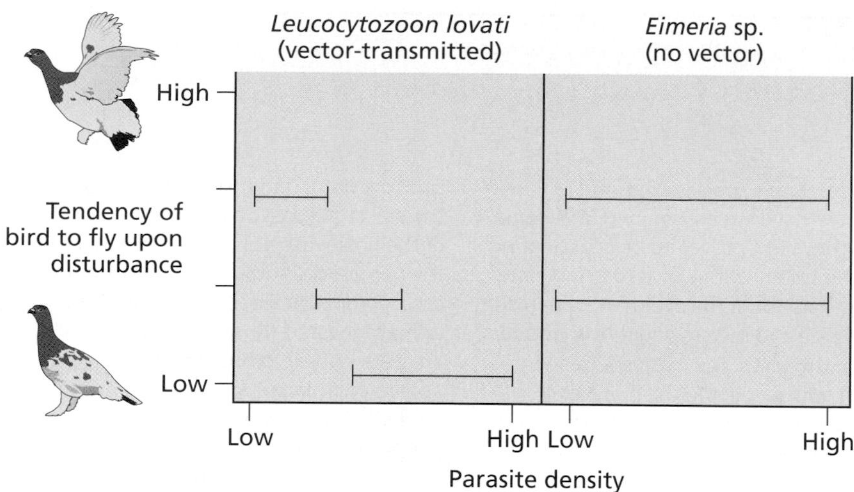

Figure 7.13 Birds with vector-borne parasites are more likely to be immobile than birds with directly transmitted parasites. The vertical axis provides four categories of the tendency of a ptarmigan to fly upon disturbance, ranging from low (meaning the bird will remain in its spot) to progressively higher (meaning the birds will flush and fly at progressively greater distances from the hunter). On the horizontal axis is the parasite density for two common ptarmigan parasites, the vector-borne *Leucocytozoon lovati* (left), and the directly transmitted *Eimeria* sp. (right). The horizontal bars show the range of parasite densities for several birds. Note how birds infected with *L. lovati* are relatively immobile at high parasite densities, but high densities of *Eimeria* infections have no such effect. (From Holmstad PR , Jensen KH & Skorping A et al [2006] *Int J Parasitol* 36:735–740. With permission from Elsevier.)

transmission is directly related to the host's fecundity and dependent on the host's ability to be healthy enough to reproduce, the expectation is that vertically transmitted parasites evolve to be more benign. The impact of opportunities for horizontal vs. vertical transmission upon the virulence has been documented in a model involving fig wasps infected by nematodes (**Box 7.4**). In accordance with theory, it was shown that nematodes with a greater opportunity for horizontal transmission were more virulent to their wasp hosts than nematodes restrained by their host's life history to vertical transmission.

7.4 MACROEVOLUTIONARY PARASITOLOGY

Macroevolution refers to evolutionary questions relating to the species level or above. Even though the formation of new species (speciation) has its origins in microevolutionary processes, because it culminates in the formation of new species, it is considered in this section. A macroevolutionary perspective considers not only the birth of new species but also their extinction, and it contemplates why some groups of parasites may have expanded considerably in species diversity, whereas others have not. Macroevolution often involves the study of global scale patterns above the species level that have developed over a time scale measured in millions of years. As parasites typically leave no fossil record (but see **Box 7.5**), such investigations are often challenging and rely extensively on what we can learn from present-day parasite–host associations.

New parasite species are potentially formed in at least three different ways

In Chapter 2 we defined a species as a group of individuals with similar properties that are able to interbreed with one another and produce fertile offspring and that don't regularly interbreed with other species. As noted in that chapter, multiple species definitions exist, and no one definition applies to the multitudes of different kinds of parasites. Essential to the process of speciation is that a group of organisms with a common gene pool that is recognized as a particular species is divided into two populations, such that the populations no longer have gene pools in common. Assuming that the gene pools remain distinct and are not later conjoined by the processes

BOX 7.4

Nematodes of Fig Wasps—Models to Study Parasite Transmission and the Evolution of Virulence

The association of figs (*Ficus*) and their pollinating wasps is a classic example of a mutualism. Specific fig species are pollinated by specific fig wasps. A gravid foundress wasp oviposits in the developing flower of the host tree and in the process pollinates the flowers. The female wasp dies shortly thereafter. As the fruit ripens and seeds develop, her progeny develop, each offspring feeding on a single seed. Wasps of the new generation emerge, mate, gather pollen, and leave the flower to repeat the cycle. Also involved in this classic mutualism, however, are nematodes of the genus *Parasitodiplogaster* that parasitize the fig wasps. Female wasps harbor these nematodes in their body cavities. Shortly after the death of the female wasp, the nematodes emerge from her body and reproduce in the fig flower, producing juveniles that will colonize the wasp's emerging female progeny. The nematodes enter the body of the newly emerged wasps and both feed on its contents and use the wasp for transmission to the next fig flower.

About 11 different fig species, each with a different fig wasp, and each wasp species with a specific *Parasitodiplogaster* species, have been examined in Panama, and they have proven to be a fascinating model system with which to investigate questions about the evolution of parasite virulence and its association with mode of transmission. The wasp species differ with respect to the number of foundresses that colonize a host flower. In some species, only a single wasp colonizes a flower, meaning that the progeny derived from the nematodes in her body will only have that particular wasp's progeny to eventually infect. In this case, nematode transmission is essentially vertical. In such a case, heavy exploitation of the female wasp by the nematode in its parasitic phase would eventually limit the number of wasp progeny present in the flower and hence the opportunities for transmission. These nematodes are predicted to be less virulent, and that is generally what has been found (Figure 1).

In other cases, multiple foundresses can colonize a single flower and any nematodes colonizing that flower have the opportunity for transmission to the progeny of multiple wasps (horizontal transmission). Under such circumstances, the consequences of limiting the fertility of the host wasp are less detrimental to the nematode. Additionally, in the latter case, competition occurs among the nematodes, and those nematodes that are better at exploiting host nutrients are able to grow larger and become more virulent, as measured by the number of progeny their host produces. For wasp species with multiple foundresses, their associated nematodes have proven to be more virulent, again in accordance with theoretical expectations. Thus a careful long-term study of an obscure group of parasites of fig wasps has provided one of the best empirical confirmations of the important theoretical expectation that transmission biology influences the level of virulence.

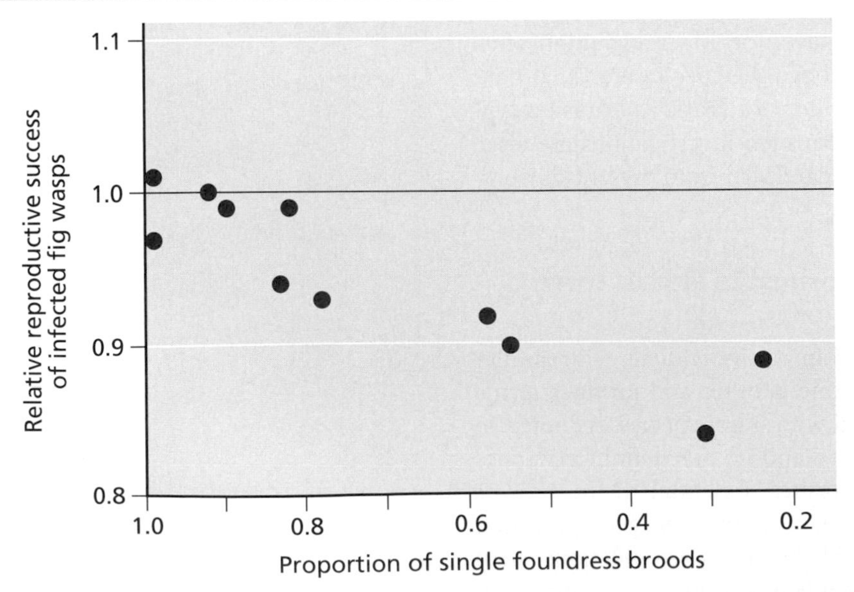

Figure 1 Virulence is influenced by mode of transmission. Virulence is measured on the vertical axis as the lifetime reproductive success of nematode-infected female fig wasps relative to uninfected female fig wasps (the higher the value, the lower the virulence). On the horizontal axis is shown the proportion of single foundress broods encountered in 11 fig wasp species. As the proportion of single foundress broods decreases (multifoundress broods increase), there are increased opportunities for horizontal transmission of nematodes infecting the wasps. Those species of wasp characterized by increased opportunities for horizontal transmission of their species-specific parasitic nematodes harbor more virulent species of nematodes. (Modified from Herre EA [1993] *Science* 259:1442–1445. With permission from AAAS.)

BOX 7.5
Parasites in the Fossil Record

Because parasites are small and soft-bodied, they have not left an abundant fossil record, but the quantity and quality of the record is steadily improving thanks to the ingenious approaches and sharp eyes of paleontologists. Several examples have been found of fossil hosts that bear anomalies similar to those that parasites produce in modern-day hosts. Fossils of male crabs from the Cretaceous Period (145 to 66 million years ago) exhibit broadened, feminized abdomens, similar to those found on extant crabs infected with *Sacculina* (see also Figure 7.26). Fossil crinoids (sea lilies) from the Carboniferous Period (359 to 299 million years ago) bear marks similar to those produced by living myzostomids (see also Figure 2.20). Both fungi and algae in 400-million-year-old rocks show signs of defense responses (enlarged cells and callosities) known to be mounted by living representatives when under attack by parasitic fungi. Ammonites bear on the inside surfaces of their shells blisterlike structures similar to those produced by digenetic trematodes in modern-day bivalves. Characteristic death-grip scars on the veins of 48-million-year-old fossil leaves resemble those produced by ants infected with a fungus (*Ophiocordyceps unilateralis*). This fungus manipulates the behavior of ants, which after infection have been likened to zombies because they leave their colony and clamp onto leaf veins where they eventually die. With the ants in this position, the fungus is able to use the wind to broadcast its spores to infect more ants. The oldest presumed fossil metazoan parasite is of a wormlike tube found in the mantle cavity of a 520-million-year-old brachiopod from the Cambrian Period.

The fossil record also reveals cases of organisms that strikingly resemble living parasites but that can't necessarily be considered parasitic because they haven't been "caught in the act." For example, late Cambrian (570 to 505 million years ago) limestone contains phosphatized fossils that resemble pentastomid larvae (Figure 1). As noted in Figure 2.19, pentastomids are today an exclusively parasitic group and are very unusual crustaceans. If the fossils are in fact pentastomes, then their presence in the Cambrian suggests that the divergence of pentastomes from branchiuran crustaceans took place much earlier than expected.

Coprolites (fossil dung samples) are also a potential source of fossil parasites. Entities that look like the cysts of parasitic protists or the eggs of helminths have been reported from dinosaur (iguanodons) coprolites from the Early Cretaceous Period. In the case of dinosaurs, smooth-edged pits and holes found in the jaws of *Tyranosaurus rex* specimens have been likened to the ulcers produced by the flagellate *Trichomonas gallinae* in the jaws and beaks of extant birds. In addition, flea-like organisms may have parasitized dinosaurs (Figure 10.2).

Amber has provided some dramatic examples of parasitism frozen in time (Figure 2). One of the oldest known ectoparasite infestations of one terrestrial animal by another, found in amber that is 20 to 135 million years old, is of a mite attached to the abdomen of a biting midge. The mite was probably sucking blood from the abdomen of the midge that had recently taken a blood meal from a vertebrate host. It is actually a case of hyperparasitism if the blood-feeding midge itself is considered to be a parasite. An example of what has been interpreted as a fossil kinetoplastid (*Paleoleishmania neotropicum*) has been described from the ruptured body of a sand fly (*Lutzomyia adiketis*) from mid-Tertiary Period amber 20 to 30 million years old. Other examples of host–parasite interactions immortalized in amber are provided by Poinar and Poinar (1999).

Given that methods for finding, preparing, and analyzing fossil specimens are constantly being improved, no doubt the future holds exciting discoveries regarding parasites in the fossil record.

continued

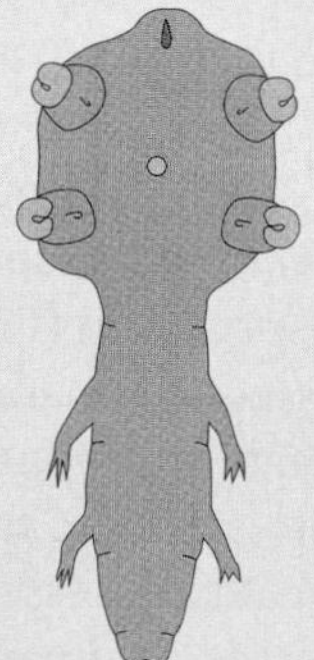
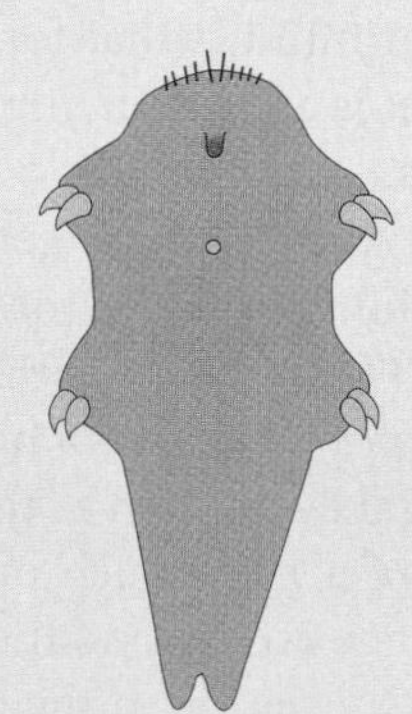
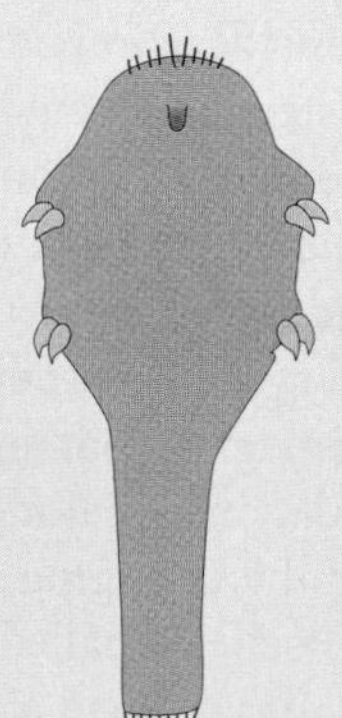
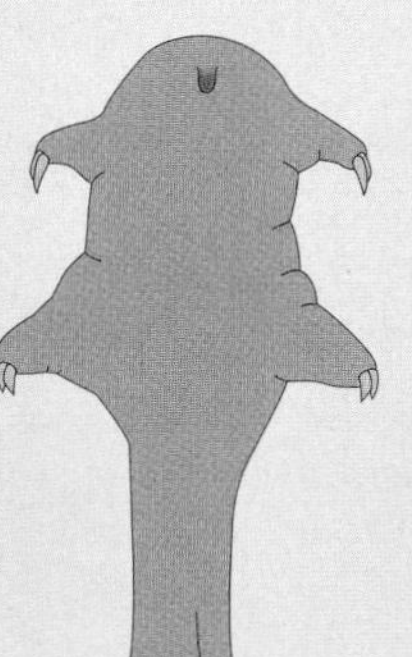

Figure 1 Ancient fossils look like modern-day pentastome larvae. Illustration of a putative ancestral pentastomid-like organism *Bockelericambria peltrue* (leftmost figure), from the lower Paleozoic (redrawn from Walossek and Müller [1994]) along with representative larvae of other extant pentastomid lineages. (Modified from Almeida WO and Christoffersen ML [1999] *J Parasitol* 85:695–704. With permission from Allen Press Publishing.)

BOX 7.5

Parasites in the Fossil Record (continued)

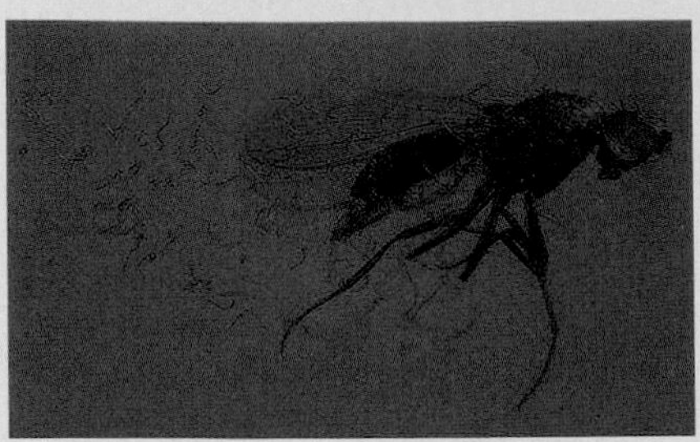

Figure 2 Insights about parasitism from amber. A drosophilid fly parasitized by allantonematid nematodes captured in amber from the Dominican Republic, 15–45 million years old. The nematodes' exit from the fly was likely triggered by the fly's fall into amber. (Courtesy of George Poinar, Oregon State University.)

Skepticism is required in interpreting this record as often the parasites themselves are lacking or are being perceived at the limits of meaningful resolution. It may not always be a safe assumption that the damage evident in fossil hosts has been caused by parasites. With the recent stunning successes in sequencing genomes of now extinct humans or mastodons, it is a matter of time before sufficiently well preserved, extinct parasites will also yield near complete genome sequences. Such sequencing may be particularly likely to occur for lice, which are often found on feathers or hair samples from extinct hosts. Whether fossil parasite specimens will ever come to light to provide clarification for pivotal events early in the evolution of major parasite groups remains a matter for speculation, but it might be premature to bet against this possibility.

References

De Baets K, Klug C & Korn D (2011) Devonian pearls and ammonoid–endoparasite co-evolution. *Acta Palaeontol Pol* 56:159–180.

Hughes DP, Wappler T & Labandeira CC (2011) Ancient death-grip leaf scars reveal ant-fungal parasitism. *Biol Lett* 7:67–70.

Poinar, G (2008) *Lutzomyia adiketis* sp n. (Diptera: Phlebotomidae), a vector of *Paleoleishmania neotropicum* sp n. (Kinetoplastida: Trypanosomatidae) in Dominican amber. *Parasit Vectors* 1:1–8.

Poinar Jr G & Poinar R (1999) The Amber Forest. Princeton University Press.

Poinar Jr G & Poinar R (2004) *Paleoleishmania proterus* n. gen., n. sp., (Trypanosomatidae: Kinetoptastida) from Cretaceous Burmese amber. *Protist* 155:305–310.

Taylor TN, Remy W & Hass H (1992) Parasitism in a 400-million-year-old green alga. *Nature* 357:493–494.

Wolff EDS, Salisbury SW, Horner JR et al (2009) Common avian infection plagued the tyrant dinosaurs. *PLoS One* 4:e7288.

of migration, crossbreeding, and gene flow, then it is likely the populations will diverge. Divergence might result from accumulated changes arising by chance (genetic drift), by the accumulation of different mutations, or by different selective regimes experienced by the two populations. If this divergence is accompanied by the development of isolating mechanisms that preclude cross-fertilization between the two diverging populations such that their gene pools remain distinct, then speciation has occurred.

Speciation is a complicated topic that is still actively debated among evolutionary biologists. What about speciation as it occurs in parasites? Although the same general principles that govern speciation in other organisms will operate in parasites, as frequently noted in this book, one of the major considerations to keep in mind is that for a parasite, for much or all of its existence, the environment is another living organism that can strongly influence the speciation process.

Three basic mechanisms for speciation are generally postulated to occur: **allopatric** (or **vicariant**) **speciation**, **peripatric** (or **peripheral isolates**) **speciation**, and **sympatric speciation** (**Figure 7.14**). For allopatric speciation (Figure 7.14A), imagine that an initial population of a host and associated parasites was divided by a major geological event such as the emergence of a mountain range or the changing course of a river. Such events, quite extrinsic to the parasite itself, are called vicariance events, and they have created the essential prerequisites defined above for speciation. If the host and associated parasite populations remain separate for long enough, they will accumulate differences which may become sufficient such that if they ever again intermingle the parasites associated with each host population may remain distinct. The same is possibly true for the host populations as well. Speciation in the host accompanied by speciation in the parasite is called cospeciation, or

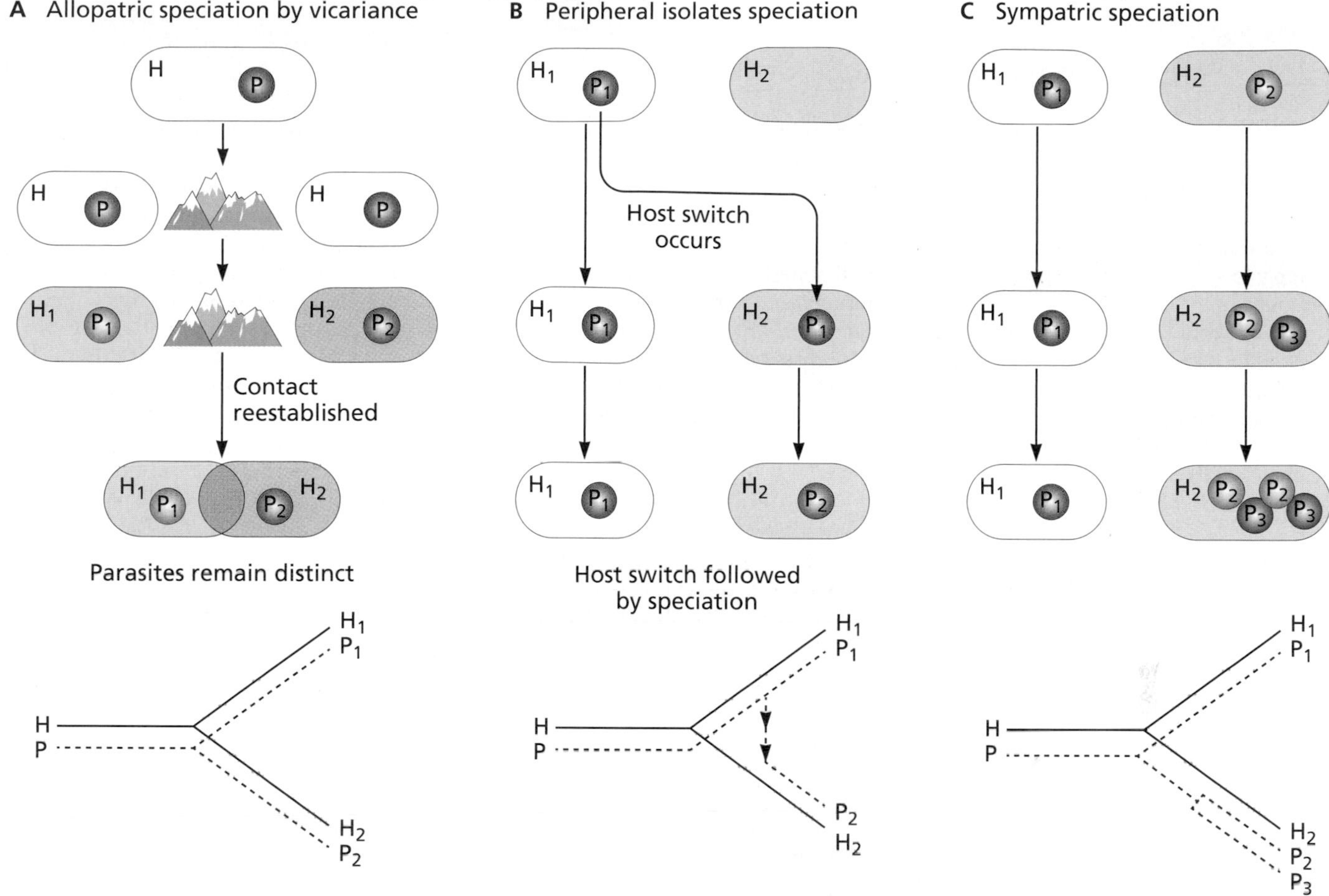

Figure 7.14 Three possible modes of speciation in parasites. (A) Allopatric speciation by vicariance, such as the formation of a mountain range. In the case shown, what was initially a single host and associated parasite population has been divided into two parts, followed by divergence. When contact was reestablished, both separated host and parasite populations remain distinct as separate species. Also shown are phylogenetic trees for both host (solid lines) and parasites (dashed lines). The trees are congruent and represent a case of cospeciation. (B) For peripatric (peripheral isolates) speciation, the parasite switches into a new host (the individual parasites found in the new host are the peripheral isolate) and this results in a parasite speciation event. (C) Sympatric speciation occurs when species arise in the absence of a geographic or physical barrier. In host species 1, no sympatric speciation occurs, but in host species 2, parasite species 2 gives rise to parasite species 3. (Modified from Huyse T, Poulin R & Theron A [2005] *Trends Parasitol* 21:469–475. With permission from Elsevier.)

association by descent. If this pattern is repeated, then a situation is created whereby the phylogenetic trees of parasite and host have congruent branching patterns (also referred to as parallel cladogenesis, or cophylogeny). This congruent pattern is also the essence of **Fahrenholz's rule** (Box 7.6), namely that parasite and host phylogenies will be congruent as a consequence of repeated cospeciation events. The parasite species so formed are likely to be very host specific. An example of parasite and host cospeciation is also shown in Figure 7.3D, with respect to the body lice found on doves.

In peripatric speciation (Figure 7.14B), a small population at the periphery of a large geographic range may become isolated sufficiently such that genetic drift or inbreeding causes it to diverge from the other individuals in the broader range. This divergence potentially could lead to formation of a distinct species, if this new peripheral population remained isolated and were denied gene flow from the larger part of its original range. For parasites, this process might be exemplified by **host switching**, or association by colonization, where one or a few parasites colonize a new host species (essentially forming a peripheral population). If followed by divergence and isolation, then a new parasite species–host combination has formed.

Sympatric speciation (Figure 7.14C) refers to the formation of a new species without any physical or geographical barriers being present to isolate the populations present. For parasites, it is frequently considered that this could happen within a particular host species. The consequence is that two closely related parasite species would end up inhabiting the same host species. This is also called **synxenic speciation** or **duplication**, and is discussed further in a following section.

BOX 7.6

Some Famous Parasitological Rules or Laws

Each is associated with a particular parasitologist, and each has become entrenched in the parasitological literature. Today their value is both historical and as useful ideas that *may* apply to some parasites in particular groups.

Dollo's law. As most generally stated, evolution is not reversible. As it is applied to parasitism, organisms can evolve parasitism from a free-living state but because of the specializations involved with parasitism, they do not evolve from parasitic to a free-living state (see also Chapter 2).

Eichler's rule. The more genera of parasites a host harbors, the larger the systematic group to which the host belongs.

Emery's rule. Social parasites invade or lay their eggs in the nest of a host organism and develop on food provided in that nest. Social parasites among insects tend to be parasites of species or genera to which they are closely related.

Fahrenholz's rule (which also owes development to **Nitzsch**, **Kellogg** and **Harrison**). Parasite phylogeny mirrors host phylogeny (see also Szidat's Rule below).

Harrison's rule. Parasites on large-bodied species of hosts are often bigger than those on small-bodied hosts.

Manter's rules. (1) Parasites evolve more slowly than their hosts (but see the microevolution section above). (2) The longer the association with a host-group, the more specificity exhibited by the parasite group. (3) A host species harbors the largest number of parasite species in the area where it has resided the longest. Consequently, if the same or two closely related species of hosts exhibit a disjunct distribution and possess similar parasite faunas, the areas in which the hosts occur must have been contiguous at a past time.

Szidat's rule. The more primitive the host, the more primitive the parasites that it harbors.

Different outcomes can be expected when parasites or their hosts diversify

With these basic speciation mechanisms in mind, let us consider some additional possibilities that can be identified regarding speciation of a parasite relative to and influenced by what is happening with their hosts. All of these events have the effect of reducing the amount of congruence between parasite and host phylogenies that is noted in the cospeciation scenario in Figure 7.14A. As shown in **Figure 7.15A**, the host may undergo a speciation event that is not followed by a parasite speciation event (called **failure to speciate**). Ongoing gene flow between the parasites in the two host species would contribute to the failure of the parasites to speciate. Another possibility is that a cospeciation event involving both host and parasite might occur but is later followed by extinction of the parasite in one of the new host lineages, a situation referred to as **drowning on arrival** (Figure 7.15B). In yet another outcome, the parasite may fail to colonize one of the newly emerging host lineages, a situation called **missing the boat** (Figure 7.15C). This outcome might occur if a peripheral host population in the process of diverging into a separate species happened by chance to be free of a parasite that occupied other host populations in other parts of its range.

The remaining possibilities occurring in Figure 7.15 refer to situations in which a parasite undergoes a host switch, shifting from its original host species into a new one. In contrast to what is shown in Figure 7.14B, where a host switch is accompanied by speciation of the parasite, in some cases the parasite retains the ability to infect the original host along with the new host. That is, the parasite has become less host specific and broadened its host range (Figure 7.15D). In some cases following a host switch, the parasite might go extinct in its original host but persist in its new host either as the same species (Figure 7.15E) or as a distinct species from the one in the host from which it arose (Figure 7.15F). A legitimate host switch must be distinguished from **straggling**, where a particular parasite manages to colonize a new host but the association is accidental and temporary.

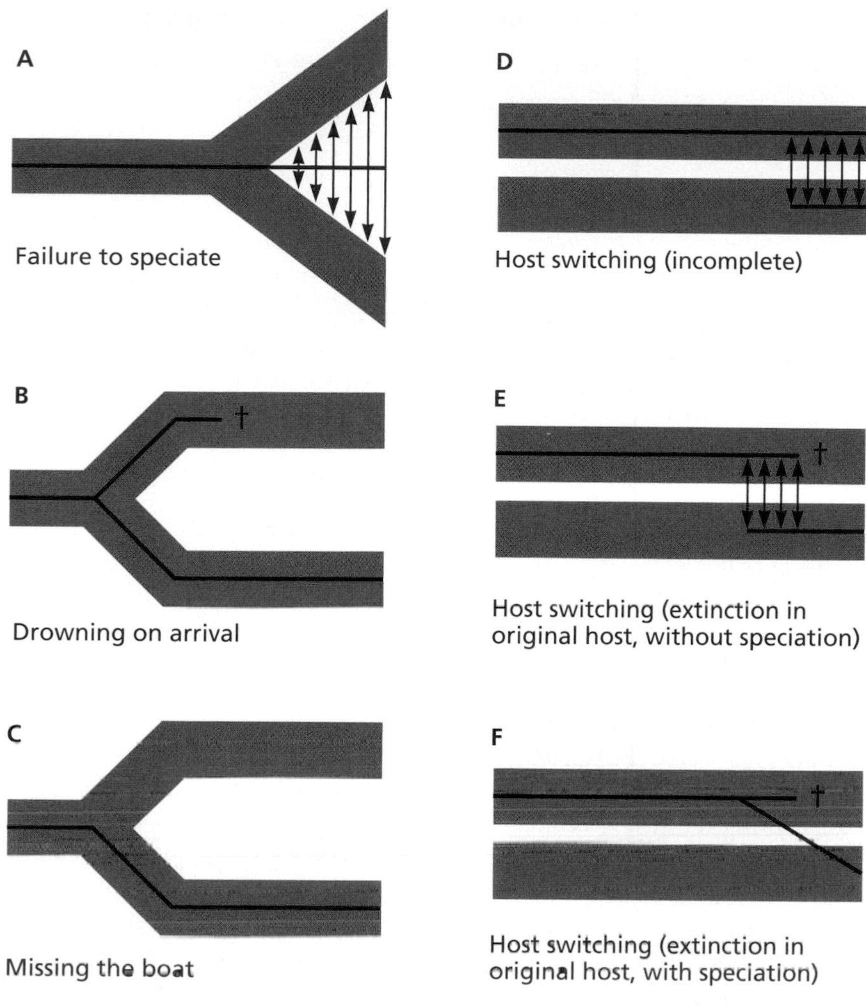

Figure 7.15 Some different evolutionary outcomes involving diversification of parasite and host lineages. Yellow lines represent hosts and black lines parasites. Black arrows indicate gene flow occurring between parasite populations. The crosses indicate extinctions.

What does the evidence suggest about how parasites have speciated?

Bearing in mind that because we lack a comprehensive fossil record for parasites, the patterns noted in present-day parasite host associations are our window to the past. Crucial to understanding speciation is development of phylogenetic hypotheses, with their attendant historical insights for both parasites and their hosts (see Chapter 2). An understanding of both morphological features and molecular sequence data are instrumental to this effort. A consideration of the ecology of both parasite and host and the associated microevolutionary processes that result are also of great value in reconstructing how parasites speciate.

There is no better example to gain an understanding of parasite speciation than that provided by chewing lice (Amblycera and Ischnocera) that infest pocket gophers (Geomyidae) (**Figure 7.16**). This model system has already been mentioned in the context of how limited gopher vagility favors development of strong population substructure in their lice. Suppose a particular population of pocket gophers was to become subdivided geographically such that isolation, divergence, and speciation occurred. What would be the fate of the lice on the gophers in each population? Given that the lice spend their entire lives on their hosts, that they have very little dispersal ability of their own, and that they must rely on the limited opportunities of host-to-host

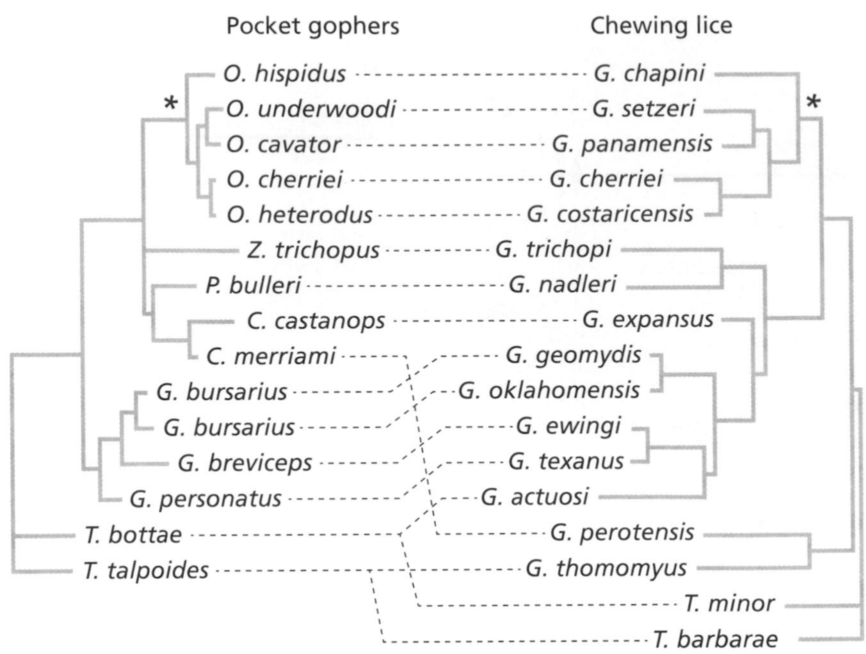

Figure 7.16 Cospeciation of gophers and their lice. Phylogenies of pocket gophers (left side) and their chewing lice (right side). Shown are composite trees based on multiple phylogenetic methods. Significant similarity in branching structure between these trees was documented. As one example, notice how the branching pattern for the group of five gopher species identified by the asterisk (upper left) matches the branching pattern for the five species of lice identified by the asterisk (upper right). (Adapted from Balbuena JA et al [2013] *PLoS One 8*:e61048. doi:10.1371/journal.pone.0061048.)

contact for their transmission, then it follows that the lice in the two host populations would be isolated as well. This separation might result in an accumulation of genetic differences, reproductive isolation, and speciation. Considering the strong ecological constraints inherent in this system, it is not surprising that gopher chewing lice, and many other lice that infest other mammals and birds (but not all lice—see dove wing lice shown in Figure 7.3B), provide the best examples for the process of cospeciation among eukaryotic parasites (Figure 7.16). Other examples of cospeciation include the pinworms that infect primates and the associations between European bats and their ectoparasitic mites. This pattern is most likely to occur for parasites such as lice that are essentially transmitted vertically in or on their hosts. Once again, overgeneralization can be hazardous; further examination of the specifics of the biology of parasites and hosts will help to clarify such situations.

The preponderance of a growing number of molecular phylogenetics studies suggests that a strict pattern of congruence between parasite and host phylogenies is the exception rather than the rule. Many phylogenies show a combination of processes, such as cospeciation, failure of the parasite to speciate with the host, and host switching. For example, in a study of European bats and their ectoparasitic mites (**Figure 7.17**), an overall correspondence between parasite and host phylogenies can be observed, indicative of a signal for cophylogeny. However, this figure also points out several instances where there is a diminished degree of congruence between the two phylogenies that results from a failure of the parasite to speciate along with the host and the occurrence of apparent host shifts.

In a study of avian malaria parasites (***Plasmodium*** and *Haemoproteus*), it was found that when comparing parasite taxa that were less genetically divergent, a pattern of cospeciation with their hosts could be detected, but as the parasite lineages attained greater divergence (by looking at nodes deeper in the tree), a cospeciation signal was swamped by evidence of host switching and long-distance dispersal events. Studies of diverse groups of eukaryotic parasites such as tapeworms of sharks and rays (Trypanorhyncha) or mammals (Taeneidae); nematodes of the genus ***Trichinella***; schistosomes;

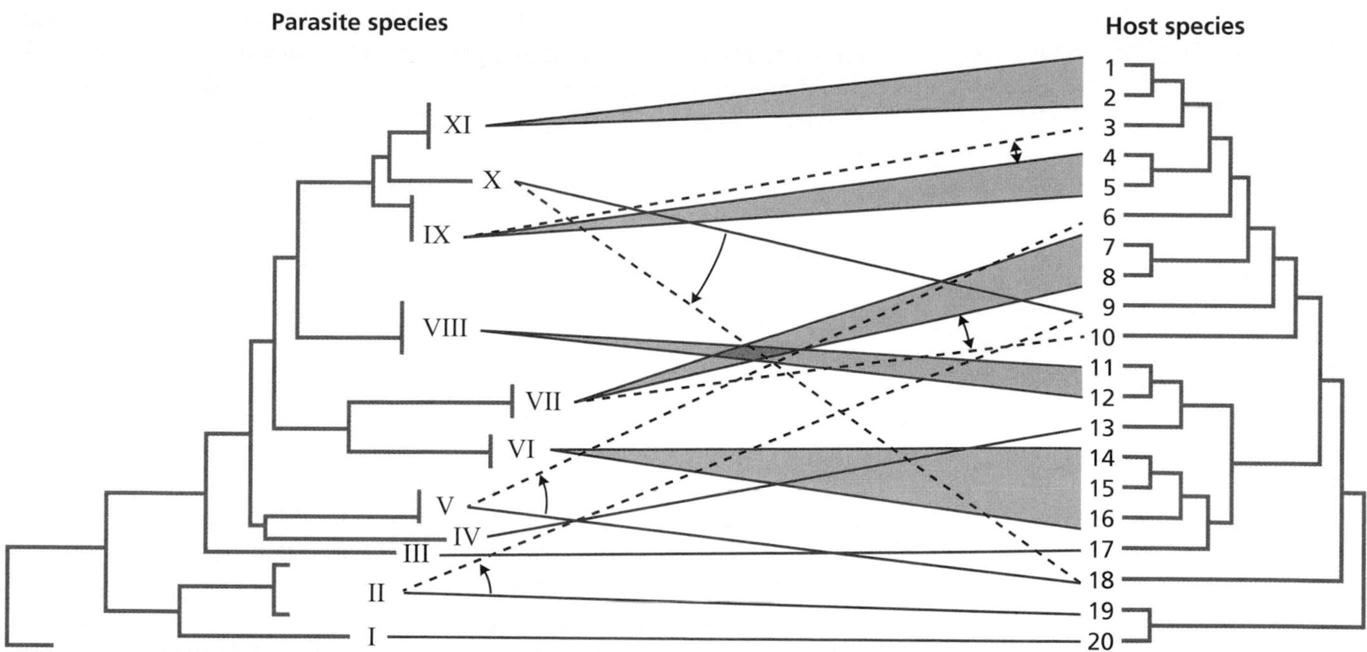

Figure 7.17 Complex evolutionary relationships between bats and their mites. A molecular phylogeny for eleven groups of ectoparasitic bat mites (left, see roman numerals) and European bat cladogram (20 species, right) are shown. Blue lines connecting parasites and hosts indicate associations observed in the field. Red dashed lines represent inferred host switch events and arrows indicate the most likely direction of the switch. Green triangles represent possible failure to speciate events between or among closely related host species. (From Bruyndonckx N, Dubey S, Ruedi M et al [2009] *Mol Phylogenet Evol* 51:227–237. With permission from Elsevier.)

monogenean parasites on fish; and malaria parasites have all revealed extensive evidence for host switching as a prominent event leading to the formation of new parasite lineages. Although it is easy to see how host switching can lead to parasite speciation, it is harder to envision how parasite species would arise via sympatric speciation. This is our next topic to consider.

Does sympatric speciation occur in parasites?

The idea that speciation can occur without any physical or geographical barriers being present to isolate the populations present—that it occurs in an area of sympatry—is a beguiling one, but one that continues to generate controversy among evolutionary biologists. Opinions vary regarding the extent to which it occurs, from never to sometimes to frequently. The essential idea regarding sympatric speciation for parasites is that an intrinsic change occurs in some members of a population, and this change could potentially cause these parasites to become genetically isolated, even if there is no barrier or geographic separation with other members of the parasite species lacking this intrinsic change. Some evolutionary biologists have considered parasites to be good subjects for studying sympatric speciation because parasites commonly switch from one host species into another sympatric host species, possibly leading to speciation.

However, what comprises an area of sympatry for free-living organisms might be quite different from a parasite's point of view. For a parasite, the area it occupies is the host. Imagine, then, that a parasite that has undergone no intrinsic change and is unable to choose its own host happens by accident to colonize a new host species. Even if that host species occupies the same geographic area as the original host species, the parasite is unlikely to encounter other members of the parental population from which it came. This colonizing parasite is essentially allopatric from its originating population. Consequently, it could be argued that such a host switch, assuming it is repeated and followed by divergence and speciation, does not really qualify as an example of sympatric speciation. Instead, as noted above, a switch to a new species might more appropriately be considered a form of peripatric speciation: a peripheral isolate of the parasite (in a new host) has arisen.

For sympatric speciation to be claimed for parasites, it is generally considered that it must occur within sympatric members of a single host species. As noted previously, within host speciation has also been called duplication or synxenic speciation. For example, related monogenean parasites, which often have very specific site preferences on the gills or other body parts of fish, have been offered as exemplars of sympatric or synxenic speciation.

Another interesting example of what appears to be repeated within host speciation is provided by *Haemoproteus* parasites within a particular bird species from Europe, the blackcap, *Sylvia atricapilla* (**Figure 7.18**). In this case, the blackcap has been shown to harbor 17 distinct *Haemoproteus* lineages (called a species flock) that are not found in other sympatric bird species. Furthermore, these parasite lineages show evidence of remaining genetically distinct from one another. This is true even though multiple lineages occur within the same bird, and they would thus be mixed and potentially subject to recombination in the biting midge (*Culicoides*) vector. The nature of the isolating mechanisms allowing these lineages to remain distinct is not clear, though it does not appear that different lineages originated in different parts of the host's range and blackcaps themselves do not exhibit a great deal of genetic structure across their range.

In another possible form of sympatric speciation, **allochronic speciation**, sympatric populations of the parasite remain separate by virtue of having different periods of transmission. The example typically cited is of strains of ***Schistosoma mansoni*** specialized to infect either humans or rats that are found on the island of Guadeloupe. In the case of the strain adapted to humans, infected snails release cercariae at midday when people are likely to be in the water and in contact with cercariae. The strain adapted to rodents, by contrast, releases its cercariae from snails in the evening such they are present when nocturnally active rodents are in the water. By virtue of

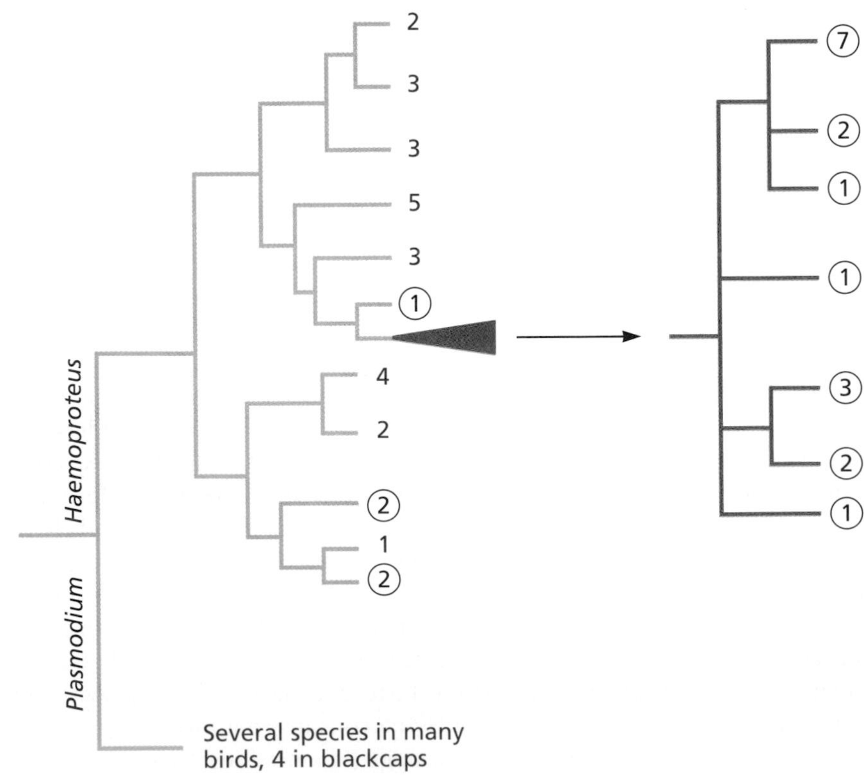

Figure 7.18 A flock of parasite species within a single host species. The blackcap (*Sylvia atricapilla*) parasite flock placed in a phylogenetic context. The tree on the left provides a simplified overview of the major lineages of *Haemoproteus* found among several bird species. The uncircled red numbers represent the number of *Haemoproteus* species found uniquely in other bird species. The circled numbers indicates those found in blackcaps. The green triangle indicates a part of the tree that is shown in an expanded form by the tree on the right. Although other bird species not shown may also be infected by *Haemoproteus* from this expanded part of the tree, note there are an additional 17 *Haemoproteus* species (the sum of the circled numbers) that infect blackcaps. These constitute the blackcap parasite species flock. (Simplified from Perez-Tris J, Hellgren O, Krizanauskiense A et al [2007] *PLoS One* 2: [doi:10.1371/journal.pone.0000235.g002].)

producing cercariae at different times, isolation of the two sympatric strains is favored. Note that the strains involved have by no means proceeded to reproductive isolation and the barriers to gene flow are likely not absolute: some cercariae shed at midday may persist until the evening hours and infect rodents, or vice versa. Some of the *Haemoproteus* lineages recovered from blackcaps may also circulate in birds at different times of the year and thus exemplify allochronic speciation. Other intriguing flocks of parasite species occurring within a single host species have been noted, such as pinworm nematodes from the cloaca of tortoises or strongyloid nematodes in horses, elephants, and kangaroos. In some cases these flocks may arise by sympatric speciation, but in others the species within the flocks are not each other's closest relatives, suggesting other modes of origin.

A final means whereby sympatric speciation could occur is when certain members of a parasite population undergo a pronounced genetic change, such as a chromosomal rearrangement, that directly affects the ability of the altered parasite to copulate with its conspecifics. A remarkable example that might be explained by such a situation is of the fish *Brycinus nurse* from Africa, which is colonized by at least eight closely related species of gill-inhabiting monogeneans (*Annulotrema* sp.). These monogenean species, which are hermaphroditic outcrossers, differ from one another in their copulatory structures such that one species is unable to inseminate another. Because both parasites and host are known only from this area, it seems that the parasite species would have essentially evolved in sympatry (same host and even the same gill habitat). As a general caution here for all examples of sympatric speciation, it is sometimes difficult to exclude alternative possibilities. For example, perhaps the fish formerly had a broader range and the different monogeneans actually speciated peripatrically, but as the fish's habitat shrank, different fish populations commingled enabling the diverged monogenean species to commingle also and persist in sympatry on the same host individual.

The study of sympatric speciation is bedeviled by how to define sympatry, and parasites impose unique challenges in this regard. Are two populations of the same parasite species sympatric if they are found in the gut of a different host species, but the two host species share an overlapping geographic range? Or are they sympatric if one parasite population is found in the upper small intestine and another in the lower small intestine of the same host species? Are these examples of sympatry or what might be called microallopatry? Some biologists even argue the term sympatric speciation should be abandoned. In our view, a switch into a different host species does not necessarily provide the prerequisites for sympatric speciation because from the parasites' point of view they are allopatric (in a different host species such that barriers exist). The examples of apparent within host speciation described above are intriguing and deserving of further study. It is noteworthy that it is common to find pairs of closely related parasite species (congeners) within the same host species. The study of sympatric speciation in parasites may prove to shed distinctive new light on the general topic of speciation for all evolutionary biologists.

Host switches can enable radiations in parasites

There is a temptation to conclude that parasites, because they tend to become specialized to particular hosts, essentially become evolutionary dead ends. By becoming totally committed to their hosts, they are not likely to give rise to dramatic new radiations of species. Yet how do we square this thinking with the observation that enormous numbers of species are found in certain parasite lineages? How have parasites become so numerous in species? One key reason is that parasites are quite capable of switching to new hosts and thus are able to enter new adaptive space where host defenses and natural enemies may be lacking. Several factors can favor switches from one host to another. An ecological change allowing two host species normally held separate to commingle could occur, thus providing a new opportunity for parasite exchange (the encounter filter has opened; see Chapter 6). Also, environmental change acting in concert with other factors such as host density or condition might conspire to make a particular host species more susceptible to colonization, or the presence of one parasite may sufficiently damage or suppress the host's immune system so as to favor colonization by other parasites to which it is normally resistant (the compatibility filter opens).

Once a new host lineage has been colonized, descendants of the original parasite colonist might later colonize other related host species or potentially cospeciate with them. These events can result in a **radiation**, a relatively sudden expansion in the number of parasite species, typically resulting from elaboration of an innovation (switch to new host lineage or new form of immune evasion, for example). One example of the importance of host switching and the possibilities for subsequent radiation is provided by malaria parasites (***Plasmodium*** and closely related genera), for which each major clade is associated with a shift in the type of insect vector used (**Figure 7.19**). Because the biology, abundance, and blood-feeding habits of different groups of vectors vary, each such shift to a new vector would create opportunities to establish new cycles of infection. For example, the adoption of relatively common and mobile biting midges (Ceratopogonidae) as vectors by the ancestors of *Parahaemoproteus* likely led to extensive diversification of this parasite genus in birds all over the world. Similarly, the adoption of mosquitoes (Culicidae) by the ancestors of ***Plasmodium*** parasites has favored the spread of this parasite genus into a huge variety of lizards, birds, and mammals.

Similar examples of switches of both schistosomes and liver flukes into different lineages of snail intermediate hosts have also been associated with expansion into new geographic areas and subsequent diversification of species. Host switching accompanied by prolific speciation has also been a prominent trend for monogeneans of the genus *Gyrodactylus*, for which as many as 20,000 species have been estimated to exist. In this latter case, rampant host switching has occurred, with switching to fish of a new family often followed by repetitive speciation events on the new fish hosts. As a consequence, the phylogenies of *Gyrodactylus* and their fish hosts are totally discordant and independent.

Parasites go extinct, sometimes along with their hosts

Just as the formation of new species is a natural part of the evolutionary process, so too is extinction of species. Over long spans of evolutionary time, species are continually being formed and going extinct. Most of the species that have ever existed are now extinct, and this is likely true of parasites as well, though this would be hard to confirm given the meager nature of the parasite fossil record (see Box 7.5).

The topic of extinction with respect to parasites would at first glance appear to be dominated exclusively by thoughts of how we might promote or enhance their extinction. However, from another perspective, parasites collectively represent a vast amount of the world's biodiversity, worth saving in its own right. We shall return to a discussion of the topic of extinction with respect to parasite conservation biology in Chapter 8 and to parasite control

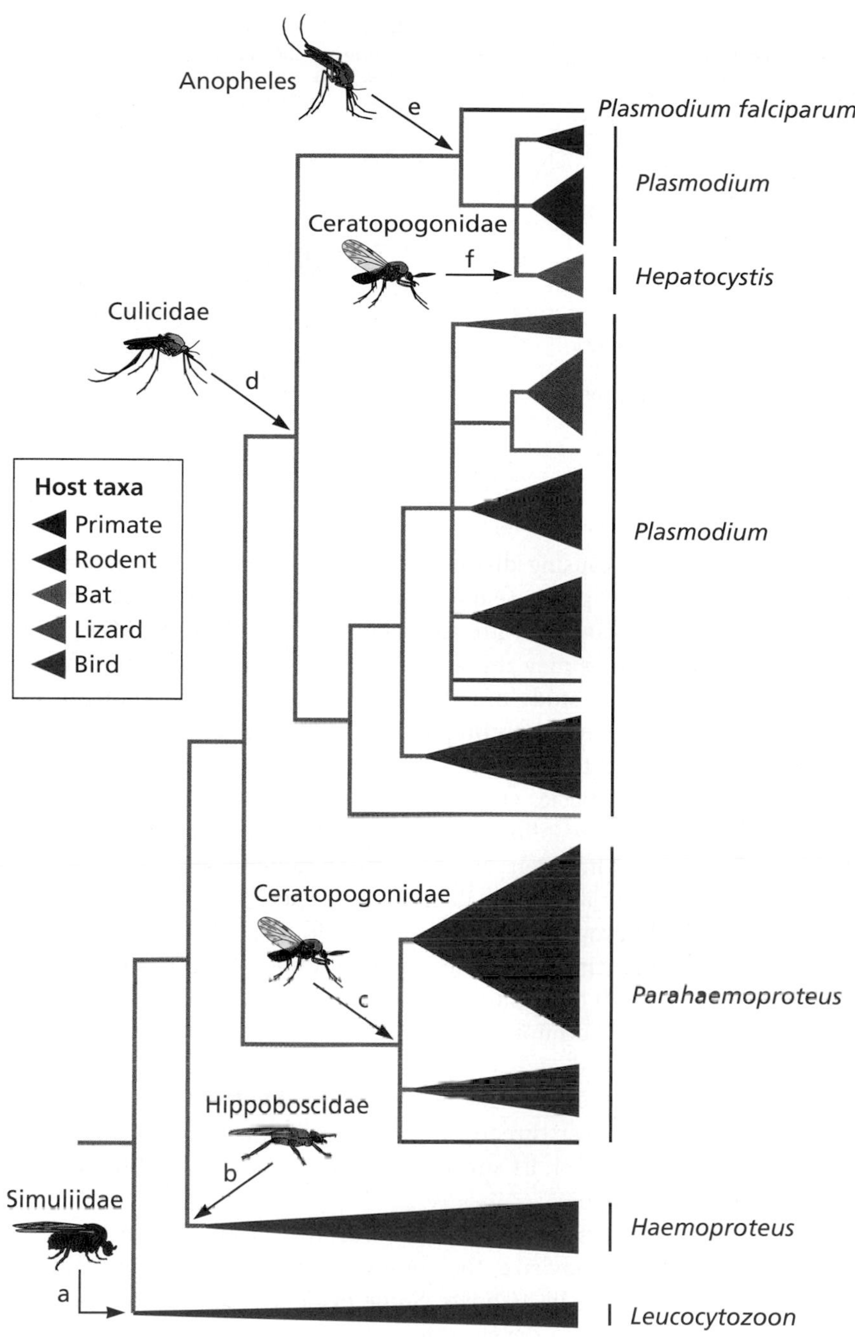

Figure 7.19 Switches into new groups of vectors have diversified haemosporidian parasites. This summary shows a phylogenetic analysis of five genera of haemosporidian parasites (*Plasmodium* and relatives), indicating the relative number of parasite taxa included in the study for each clade (size of colored triangles). The colors of the triangles correspond to the groups of vertebrates parasitized. The letters a to f and arrows show six major switches into new vectors and are accompanied by an image of the vector group involved. The lineage representing the virulent human malaria parasite *Plasmodium falciparum* is labeled at the top of the phylogeny. (From Martinsen ES, Perkins S & Schall JJ [2008] *Mol Phylogenet Evol* 47:261–273. With permission from Elsevier.)

in Chapter 9. For now though, our goal is to consider extinction of parasite species as part of the evolutionary process and to examine some of the basic factors that might influence the probabilities of extinction.

The likelihood of extinction in parasites is inevitably strongly influenced by what is going on with their hosts. One factor influencing the likelihood of parasite extinction is the degree of specificity a parasite shows for its hosts. For a specialized parasite with one host species, if its host goes extinct, the parasite will surely follow. Generalist parasites able to infect several host species have a hedge against extinction because there is less likelihood of all their hosts going extinct. Although many parasite species are confined to a single host species, this is by no means always the case. For example,

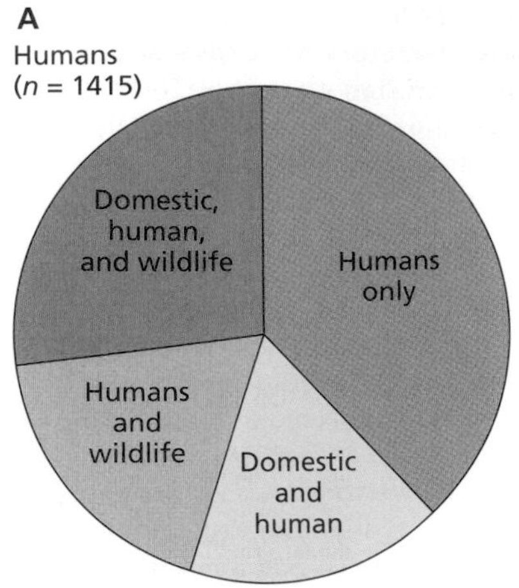

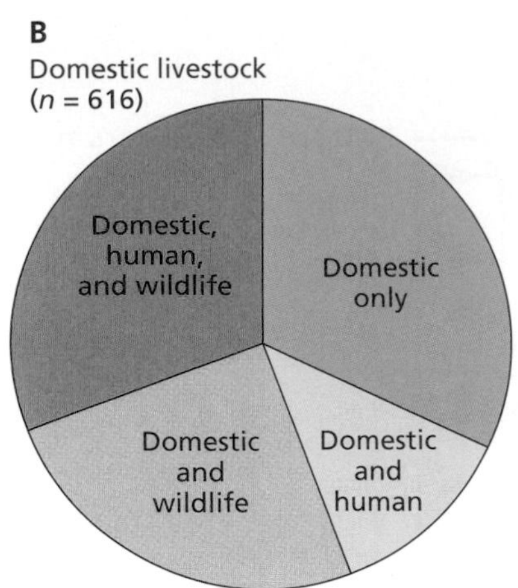

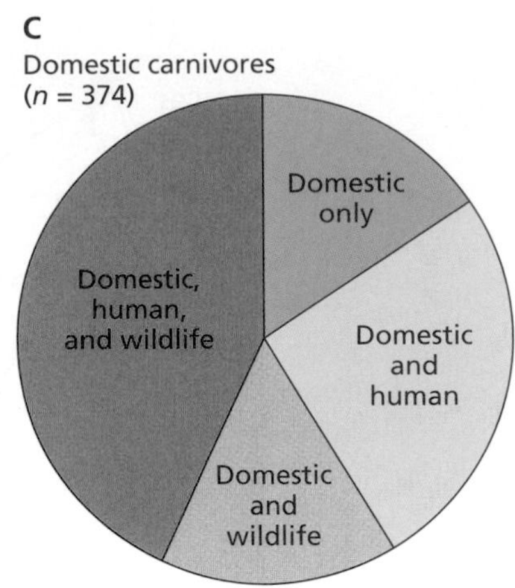

Figure 7.20 It is not uncommon for parasites to infect multiple host species. Patterns of host usage for (A) human, (B) domestic livestock, and (C) domestic carnivore parasites. Noneukaryotic parasites such as viruses and bacteria are included in this analysis. (Modified from Cleaveland S, Laurenson MK & Taylor LH [2001] *Philos Trans R Soc Lond B Biol Sci* 356:991–999. With permission from the Royal Society.)

among 1415 parasites causing disease in humans, over 60% are capable of infecting multiple host species (**Figure 7.20**). For domestic livestock, of 616 parasites considered, about 70% are able to infect multiple hosts. More thorough surveys of parasites may reveal some to be less host specific than we first thought, but further study might also reveal that what we thought was a generalist species is actually a complex of cryptic species (see Chapter 2). Consequently, it is hard to know the impact that host specificity per se will have on future parasite species extinctions.

Another factor that may influence the probability of extinction is the complexity of the parasite's life cycle. If a parasite has an obligatory complex life cycle involving sequential use of different host species, then if any one of those species should drop out, the parasite would be unable to complete its life cycle and also go extinct. This effect may to some extent be counteracted by adjustments in which the parasite simplifies its life cycle. For example, the nematode *Camallanus cotti* normally cycles between fish as definitive hosts, where larvae are produced, and copepods, where the larvae become infective to fish. If maintained over time in the absence of copepods, the parasite is able to adjust and infect fish directly, thus truncating the life cycle and favoring its persistence. However, its survival to maturity is increased if parasites are acquired from copepods.

Other general factors likely to influence the probability of parasite extinction are parasite body size, the abundance of its host species, and the prevalence of the parasite in its hosts. Some evidence suggests that parasite species of intermediate size develop more intense infections in their hosts and consequently may be less prone to extinction. Small host populations are also more prone to extinction, carrying with them their specific parasites. Finally, if a parasite species was only ever able to achieve a low prevalence rate in its hosts (perhaps limited by host immune responses or other environmental limitations), then purely by chance, it might fail to reach a threshold of success and go extinct. This latter possibility serves to exemplify the point that a parasite species might go extinct even if its host does not. This cause of extinction might be especially likely to occur if the free-living stages of parasites were vulnerable to environmental contaminants that prevented them from establishing new infections.

Interactions may also occur among the factors predisposing to extinction. For example, for metazoan parasites of freshwater fish, generalist parasites (advantage against extinction) often show patterns of low prevalence (favors extinction) in their hosts. Small host populations in conjunction with low

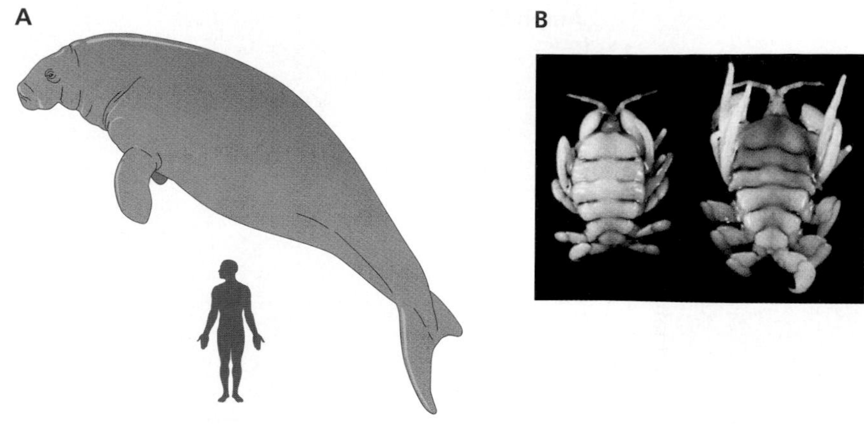

Figure 7.21 Host extinction inevitably followed by parasite extinction. (A) The Steller's sea cow, *Hydrodamalis gigas*, was described by Georg Steller in 1741, and 27 years following its discovery by Europeans, it had been hunted to extinction. (B) Steller found parasites in and on the sea cow including a distinctive species of ectoparasite that was later described as a new genus based on his early descriptions. It was similar to the specimen of *Cyamus ovalis*, the whale louse, shown. The whale louse is actually an amphipod crustacean and is not a true louse, which is an insect. The original parasite specimens have been lost and, because the host is now extinct, we will never know the identities or intricacies of the parasites of this large, distinctive mammal. (B, courtesy of Vicky Rowntree, University of Utah.)

parasite success (prevalence) in such populations would be a potentially deadly combination for associated parasites. Given that many host species have already gone extinct (**Figure 7.21**) or will soon do so, that many others exist in small, vulnerable fragmented populations, and that we are experiencing rapid rates of global change, then high rates of parasite extinction seem likely. We will return to the theme of parasite extinction in Chapters 8 and 10.

Macroevolutionary patterns among parasites are not yet very clear

We now have an understanding of some of the basic processes involved in the evolution of parasites, including the speciation process. Bearing in mind that the number of species present is a function of the rate at which they are formed vs. the rate at which they go extinct, what kinds of patterns can we detect in the present-day abundances of parasite species? These patterns reflect the operation of evolutionary forces that have been at play for millions of years. Have general features emerged that provide profound insights into parasitism? The search for large-scale patterns and trends lies at the heart of macroevolutionary studies. Keep in mind also the discussions of the species richness in communities of parasites we discussed in Chapter 6.

Early efforts to detect patterns often assumed the status of rules that have become part of the parasitology canon but, no doubt in accordance with the wishes of the original formulators, should not be considered as universal or infallible but simply as possibilities among a series of possible explanations (see Box 7.6). Because these studies are inherently comparative, as a general precaution, it is important to consider the role of phylogenetic history for the species being compared. For example, two related host species of sharply different body size may harbor many of the same parasites simply because of their shared ancestry. This might mask or overwhelm the ability to discern an effect of differing body size on acquisition of parasites by the two species.

One unmistakable pattern that has emerged is that parasite species are not distributed randomly among host species. Some host species have more parasites and some less than we might expect by chance. Furthermore, it is often the case that if a host species of a certain host group tends to accumulate species of one parasite group, it is also likely to harbor parasites of other groups as well (**Figure 7.22**). It is also the case that some groups of parasites have many more species than others. For example, among monogeneans, many more species are found in two families, the Dactylogyridae and Gyrocotylidae, than in other families. What explains these patterns?

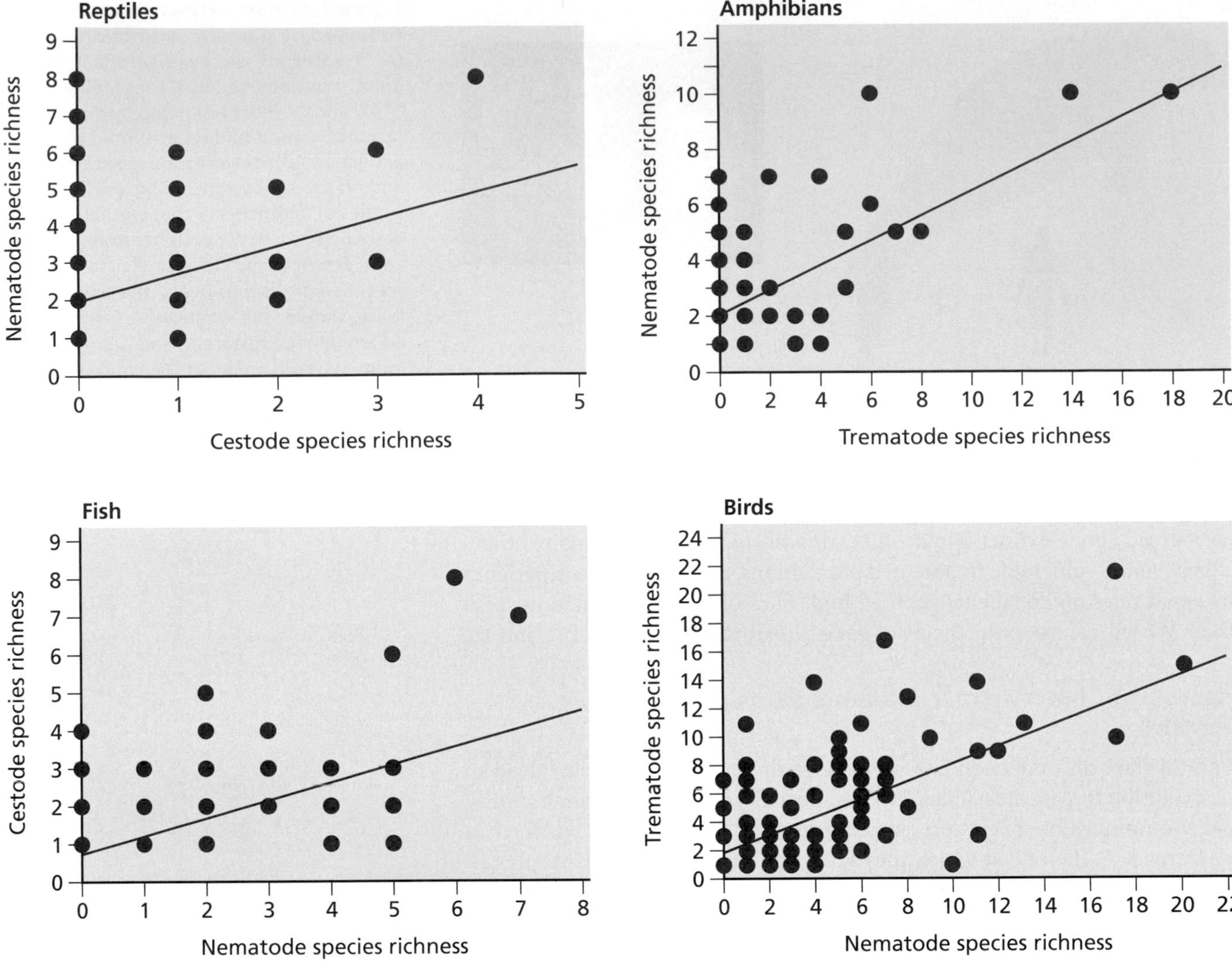

Figure 7.22 Some host species accumulate more helminth parasites of multiple groups than others. Graphs are shown for reptiles, amphibians, fish, and birds. For each graph, each dot represents a different host species, and the plots show the species richness (the number of parasite species harbored) for two different parasite groups (for example, nematodes and cestodes) for that species. The lines indicate positive relationships in species richness between the two parasite groups (Data from Bush, Aho JM, Kennedy CR [1990] and Poulin R [1995]. From Poulin R & Morand S [2004]. Parasite Biodiversity. Smithsonian Books. With permission from Smithsonian Institution Scholarly Press.)

There are at least three general categories of factors that might influence the number of parasite species recovered from a host species: host-related factors, factors intrinsic to the parasites themselves, and geographic and historical factors as noted in the following list, which is modified and summarized from Poulin and Morand. For many of the factors identified, the status of our knowledge remains uncertain, suggesting we have a long way to go to understand underlying causes of macroevolutionary patterns with respect to parasites.

Host Factors

- A more speciose host taxon should harbor more parasites than a less speciose sister group taxon (more opportunities for cospeciation to occur)—results mixed
- Parasite species richness and host geographic range should be positively correlated—data becoming more convincing
- Longer-lived hosts (or more persistent host populations) may harbor more species of parasites than shorter-lived hosts (or less persistent host populations)—difficult to assess because of the long time spans involved

- Larger-bodied hosts should have more parasite species than smaller-bodied hosts (larger-bodied hosts eat more food, provide more opportunities for infection, tend to live longer, and harbor more distinct niches)—results becoming more convincing
- Terrestrial vs. aquatic hosts—usually aquatic vertebrate hosts have more parasite species than terrestrial counterparts, but causality is uncertain
- Hosts with more varied diet should harbor more parasite species—results mixed
- The higher the host's metabolic rate (and the more it consequently eats), the more parasite species it will have—some preliminary support
- Host genetic variability and parasite species richness are related—some positive support from monogeneans in African fish, some evidence against (helminth species richness and North American fish heterozygosity)
- Polyploid host species harbor more parasite species than diploid host species—some evidence
- Hosts with high population density have more parasites than hosts with low population density—becoming more convincing

Parasite Factors

- Parasitism per se may be a lifestyle more prone to diversification—some evidence, as from platyhelminths
- Parasites with small body size may be more prone to speciate than large species (small species perceive more intrahost nitches and can subdivide the environment more finely)—some evidence, especially for endoparasites as compared to ectoparasites. Size covaries with a number of traits that effectively serve to maximize R_0—some evidence consistent with this
- Life-cycle complexity may increase the opportunities for speciation—some evidence consistent with this (digeneans more speciose than aspidogastreans)
- Highly specialized parasites that are specific on a narrow range of hosts should be more likely to speciate than generalists—some evidence

Geographical and Historical Factors

- Vicariance events, periodic range contractions, and dispersal all favor speciation—some evidence
- On an area basis, the more host species present, the more parasite species present (more species of unionid mussels present where there are more species of fish hosts)—evidence becoming strong
- Parasite species per host increasing in the tropics—results mixed
- Other gradients such as distance from coral reefs or from hydrothermal vents or distance from shallow to deep water—some evidence (fewer parasites in deep than shallow water)
- Abiotic characteristic of habitats influence species diversity of parasites, such as low pH, or great depth may limit intermediate host populations or limit contact—some evidence

7.5 SOME DISTINCTIVE ASPECTS OF PARASITE EVOLUTION

Organisms have repeatedly adopted parasitism by more than one route

One simple yet fundamental question we have not considered is, how do parasites come into being in the first place? Although we have not been present to witness such events and our account is necessarily speculative, we can be sure of a few things. First, parasites can exist only once hosts are in place to infect. Second, our growing understanding of the phylogenetic relationships among parasites (see Chapter 2) suggests that parasitism has originated repeatedly throughout the history of bacteria, protists, red algae, plants, fungi, and animals. Third, adopting parasitism offers considerable advantages with respect to providing a place to live, largely devoid of predators, with untapped energy sources and mobility provided courtesy of the host.

We can imagine parasitism originating in a few different ways. One common notion is that conditions experienced during free-living existence select for traits that may eventually serve the organism well if it adopted parasitism. The process whereby a trait evolves for one purpose but later is used for another is called **exaptation**. Note that the concept of exaptation does not imply that an organism is somehow foreseeing its future but rather that the current circumstances of its existence select for attributes that would also happen to be valuable if it adopted another very different type of lifestyle (such as parasitism).

Nematodes are a popular subject for consideration of the origins of parasitism because at least seven separate origins of parasitism have occurred in this phylum. For example, some nematodes (Order Rhabditida) live in decaying plant material or dung, environments that are low in pH and oxygen, prone to high temperature, and awash with lytic enzymes from bacteria. If accidentally ingested, these tolerances may also enable the nematode to survive in the gut of an animal. Any progeny produced by such a nematode may be more likely to survive in a similar environment, and parasitism might quickly be adopted. Another possible example of exaptation is provided by the nematode *Pristionchus pacificus*, which is specifically associated with scarab beetles. It colonizes a beetle and after the beetle dies, the nematode devours the bacteria associated with the decomposing host. This nematode has a large increase in detoxification enzymes as compared to *Caenorhabditis elegans*, and these enzymes have been considered as a possible exaptation to parasitism. The possession by some free-living nematodes of a dauer-stage larva similar in appearance to the infective larval stage of parasitic nematodes may also be an exaptation to parasitism, an idea strengthened by the discovery that the endocrine signaling pathway used by free-living nematodes when entering the dauer stage is the same as the one used by parasitic nematodes as they become infective-stage larvae.

Another possibility is that the journey to parasitism is gradual, involving a transitional series of lifestyles proceeding from free-living, to some degree of dependence on a host, to facultative parasitism, finally on to obligate parasitism. Rhabditid nematodes have also been offered as an example of this possible transition as some species are free-living, some have either superficial or invasive phoretic associations with invertebrate hosts, some regularly undertake both parasitic and free-living life cycles (*Rhabdias* and *Strongyloides*), and some are obligate parasites. In a study to examine the ability of rhabditid nematodes to survive either in simulated vertebrate gut conditions or in the gut of frogs, it was shown that species with phoretic associations with invertebrates were not better than free-living species in surviving vertebrate gut conditions and that free-living species were able to survive well for at least 72 hours in the frog gut. These findings suggest that a prior history of some degree of parasitism may not be needed to make this transition, and that—probably not surprisingly—an organism tending to parasitism in one host does not necessarily fare better when confronted with a very different type of host.

Among the myriads of organisms living in close association with other organisms, there are several organisms that straddle the boundaries with parasitism. Many organisms are harmless commensals unless and until the host becomes debilitated in some way, after which they assume the status of facultative parasites. This phenomenon is particularly true of ciliate protists living on the surface of fishes. For example, *Brooklynella hostilis* normally glides over the surface of the gills of fish feeding harmlessly on sloughing gill epithelium. If the fish becomes stressed, as in captivity, the ciliate can attack the gills and denude them of cells, killing the fish in the process. A transition for such an organism to obligatory parasitism is easy to envision. Mites in particular provide many examples of species that are trending toward parasitic lifestyles, and multiple origins of parasitism in mites from predatory ancestors are likely. The mite *Kennethiella trisetosa* is phoretic on a wasp parasitoid (*Ancistrocerus antilope*) that lays its eggs on caterpillars (**Figure 7.23**). At the time of the wasp's oviposition, the mite disembarks and not only feeds from the caterpillar but also extracts hemolymph from the larvae of the wasp developing in the caterpillar. Examples like this demonstrate that the paths to parasitism are many and often potentially devious in their course.

Some parasites are derived from their hosts

Another way in which parasitism can arise in some circumstances is by colonizing a close relative, perhaps even another individual of the same species. In social insects, social parasites generally originate from closely related forms that serve them as hosts, which has come to be known as Emery's rule (see Box 7.6). Such parasites are known as **agastoparasites**. A similar consideration applies to red algae, where they have traditionally been called **adelphoparasites**. See Chapter 2 for a description of the peculiar nature of parasitism in red algae, one that includes injection of organelles from the parasitic red alga into its host's cells. The parasite essentially transforms the host, cell by cell, and the host will eventually produce spores and gametes that contain the parasite's nuclei. A molecularly based phylogenetic study (**Figure 7.24**) of the parasitic red algal genus *Asterocolax* and associated host species revealed that often the closest relative to the host is the

A

B

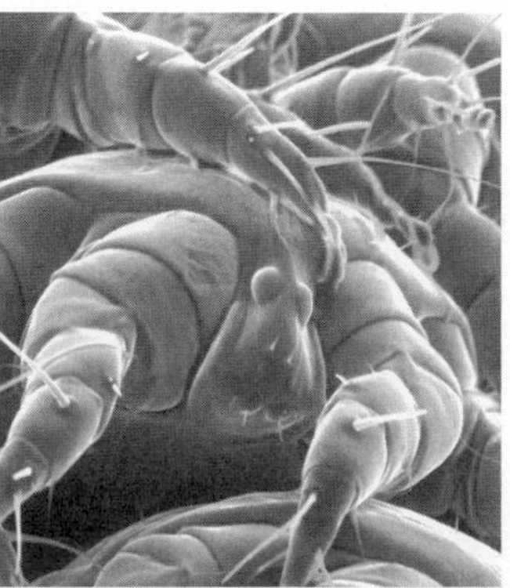

Figure 7.23 Phoretic mites trending towards parasitism (A) An *Ancistrocerus antilope* wasp. The arrow and part (B) show a large cluster of immature nonfeeding *Kennethiella trisetosa* mites (deutonymphs) clustered in protected locations (the acarinarium, or mite chamber) on the right side of the wasp's thorax. Once the wasp oviposits on a caterpillar, some of the mites disembark and transform into feeding stages that feed both on the caterpillar and the developing wasp larvae. (A, courtesy of David Cowan. B, courtesy of Rob Eversole, Western Michigan University.)

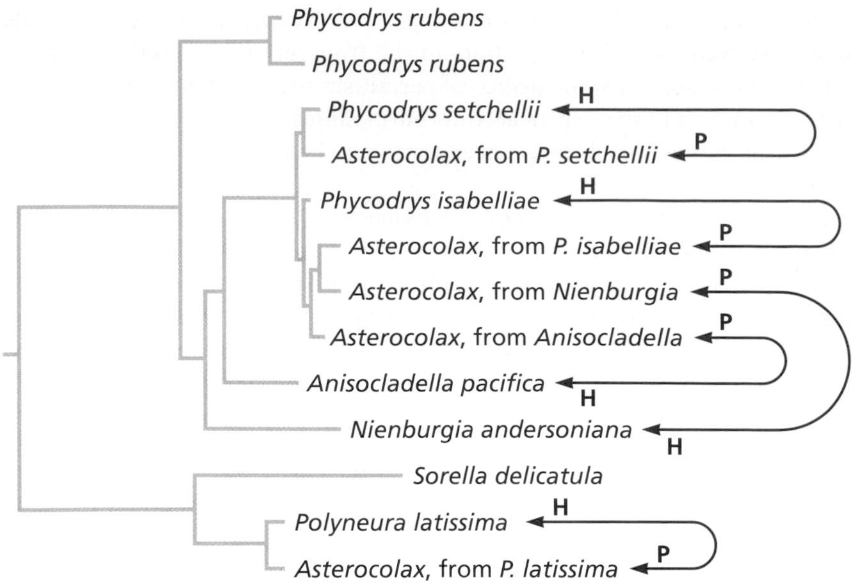

Figure 7.24 Red algal parasites are related to and arise from their hosts. A phylogenetic tree based on internal transcribed spacer regions (ITS 1, ITS 2, and intervening 5.8S rDNA) of *Asterocolax* species, their hosts, and closely related nonhosts of the *Phycodrys* group. Note the red arrows showing a parasite (P) derived from a closely related host (H). [(From Goff LJ et al [1997] *Evolution* 51:1068–1078. With permission from John Wiley and Sons.)

parasite (see also Figure 2.10). In this case, it seems that the close relationships between parasite and host facilitates the hijacking of the host's cellular machinery, resulting in a host that produces its parasite's progeny.

In earlier sections, we have discussed the route to parasitism as if it is a one-way trip, which is surely the predominant trend, in support of Dollo's law on the irreversibility of parasitism (see Chapter 2 and Box 7.6). As noted in Chapter 2, however, there are examples, and again mites figure prominently, to suggest that parasites in some cases can revert to free-living existence. The general point is that the associations between intimately interacting species are often dynamic and may favor different outcomes depending on the circumstances. Clearly, one outcome, commonly adopted over evolutionary time, has been parasitism.

Selection can favor the evolution of complex parasite life cycles

One of the hallmarks of parasitology—to the vexation of the student attempting to master the discipline—is the diversity and complexity of parasite life cycles. Here we ask a typical evolutionary question: why have they evolved to be so diverse and complex (see also Chapter 3)? To some extent, life cycles are constrained by the core attributes of the parasite lineages involved. For example, all nematodes undergo a stereotypical pattern of development starting with eggs, passing through four larval stages, and then becoming adults. Consequently, no nematode has a life cycle involving both sexual reproduction by adults and asexual reproduction by larval stages, as occurs in some flatworm parasites. Nematodes, however, have taken their basic developmental plan and modified it in countless successful ways, ranging from simple direct cycles like that in ***Trichuris*** to complex cycles like that in ***Wuchereria***, involving both vertebrate definitive host and invertebrate vectors.

Life cycles have also been profoundly influenced by predictable behaviors of the hosts involved and by predictable ecological associations among those hosts. Parasites have exploited reliable trophic links (frogs predictably eat dragonflies, or canids eat bovids), in some cases by altering host behavior to increase the likelihood of predation and hence transmission (see Chapter 6). In what has been called upward incorporation, a one-host life cycle might be expanded into a two-host cycle if a new host higher up in the food chain eats the original host. Upward incorporation confers an advantage to the parasite by avoiding mortality when its original host is eaten and by ending up in what is likely to be a bigger and more prosperous host. Similarly, downward incorporation can occur, in which a new intermediate host is added at a lower trophic level. This change may increase the likelihood of transmission to the definitive host by exploiting a predictable trophic link. It is also possible that at some point the further acquisition of more new hosts following trophic links might become too tenuous and not be favored. For example, all the different hosts might not reliably overlap in space and time. Another common strategy has been the production of life-cycle stages able to persist either in the external environment or within the tissues of a host for a long period, awaiting transmission to a new host. The resistant eggs of ***Ascaris*** or the metabolically quiescent metacercariae stages of digenetic trematodes are examples.

Over evolutionary time, many natural experiments on parasite transmission have been undertaken, and no doubt most have failed. Today we see the results of the successful experiments. We also see evidence for the evolution of innovations that have enabled some groups to become very successful. For example, there are an estimated 18,000 living digenetic trematode species as compared to only about 80 species in their sister group, the aspidogastreans. There are two essential innovations that separate these two parasite groups. In the digeneans, larval development in the snail host is accompanied by asexual reproduction and the production of thousands of progeny, each capable of going on to produce eventually an adult worm in another host (**Figure 7.25**). Furthermore, these asexually produced progeny, the cercariae, typically leave the molluscan host and engage in behaviors that directly favor transmission. By contrast, an aspidogastrean developing in the molluscan host does not undergo asexual reproduction and can only await predation by a vertebrate host, if the life cycle even involves one. Do not conclude, however, that the aspidogastreans are in some way inferior; they are often found in ancient lineages of hosts (see Szidat's rule in Box 7.6) and have probably deployed their relatively simple but obviously effective life cycles for over 400 million years.

The life cycles we see today reflect phylogenetic constraints, the exploitation of predictable trophic links, a tendency to buy time in the form of resistant resting stages, and the use of innovations such as extra rounds of reproduction or production of stages that actively seek hosts. They reflect the action of natural selection operating on all the life-cycle stages (potentially with contradictory effects on different stages as shown in Figure 7.5), such that those particular parasites able to maximize their production of progeny will be favored in their competition with conspecifics. The natural experimentation with parasite life cycles is ongoing. Some parasite propagules will always find themselves in an unexpected host or a new locality. In most cases, they may simply die, but if the circumstances are right, transmission will occur and new variations on an old theme might arise.

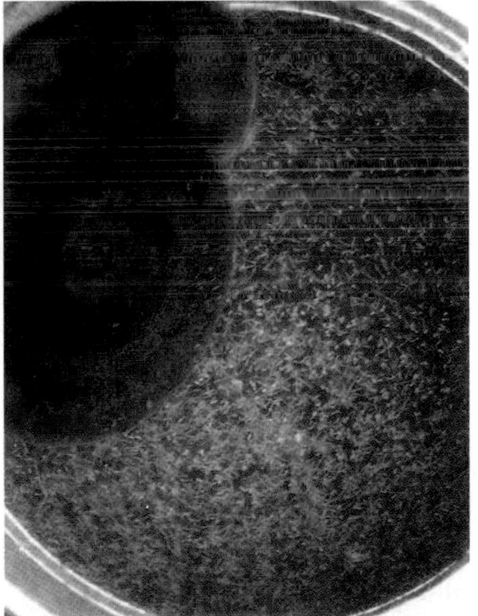

Figure 7.25 Production of many cercariae by asexual reproduction—a successful life cycle innovation by digenetic trematodes. The freshwater snail *Biomphalaria sudanica* infected with larvae of a digenetic trematode (a strigeid fluke), releasing thousands of actively swimming cercariae into the surrounding water.

Sometimes complex life cycles are simplified secondarily

The new life-cycle variations do not inevitably add additional hosts or further complexity (recall the example of the truncated life cycle of *Camallanus* nematodes described earlier). There are many examples among the digenetic trematodes where the basic three-host life cycle has been simplified secondarily to a two-host cycle or, in a few cases, to a one-host cycle. In one instance, the digenetic trematode *Plagioporos sinitsini* can engage in a three-host life cycle (snail, arthropod, fish), a two-host version (snail, fish), and even a one-host version (snail only) in which adult flukes can be found. In this latter case, at some point in the past, circumstances favored variant flukes that precociously developed to maturity while still in the molluscan host. Why some parasites are capable of such evident plasticity in their life cycles whereas most others are not, remains to be determined. As a consolation to the difficulty of mastering their complexity, the life cycles of parasites do serve as an amazing reminder of how natural selection can favor parasites that exploit predictable trophic linkages or adopt novel strategies for enhancing their transmission such that they can attain the habitats in which to maximize their reproductive success.

Parasites often have simplified bodies or genomes but also have other talents not seen in free-living organisms

Many parasites have evolved bodies that are greatly simplified. In some cases, they are virtually unrecognizable as compared to their free-living relatives. This simplification is referred to as **sacculinization**, a term inspired by the remarkable parasitic barnacle *Sacculina* that exhibits this very phenomenon (**Figure 7.26**; see also Chapter 2 for parasites with cryptic phylogenetic affinities). In a most un-barnacle fashion, *Sacculina* lacks appendages and any traces of segmentation (Figure 7.26A), and it possesses only reproductive organs and a nervous system. The host is inevitably sterilized as a result of infection. The externa of the female parasite (Figure 7.26B) is colonized by *Sacculina* males, which are tiny by comparison and hence they are called dwarf males. Fertilization occurs and the externa can release the parasite's nauplius larvae each year for as long as a decade. One of the most bizarre features of rhizocephalan biology is the behavior of both female and male crabs that have been parasitized (the latter become feminized in both behavior and appearance;). The crabs actively protect the externa as if it were its own brood. This behavior is particularly unusual for male crabs, which normally don't undertake brooding behavior.

Adopting extreme morphological alterations or simplifications is one response to adoption of a parasitic lifestyle. Another is to hijack nutrients and biochemical precursors from their hosts or from symbionts, such that their metabolic pathways, indeed the entire genomes of parasites, are often much reduced and simplified (see Chapter 3). For example, one of the smallest known eukaryotic genomes (2.9 Mbp) belongs to an intracellular parasite, the microsporidian ***Encephalitozoon cuniculi***. No doubt even smaller genomes exist.

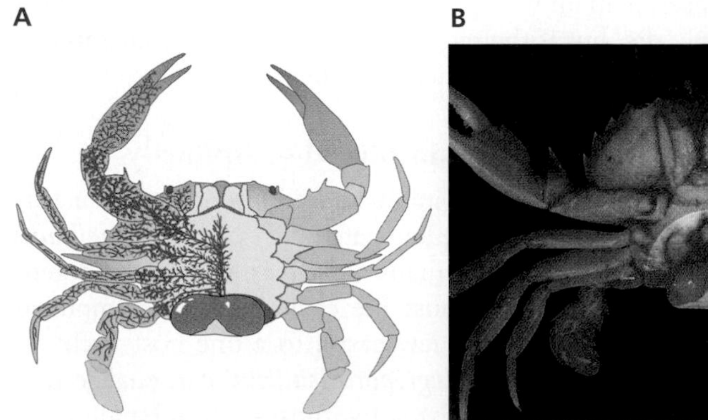

Figure 7.26 Parasites, though often morphologically simplified, are still complex creatures. *Sacculina* and its relatives are a group of highly specialized crustaceans (Rhizocephala) related to barnacles, and that infect other crustaceans, especially decapods (crabs). The female produces a branching rootlike structure (A, left-hand side of figure) that extends throughout the host's body (shown in right-hand side of figure), even to the tips of the legs. A portion of the female parasite called the externa (B) comes to lie outside the host's body typically in the position of the host's own brood pouch, where it is colonized by *Sacculina* males. (B, © Hans Hillewaert, CC BY-SA 4.0.)

However, parasites often have extraordinary capabilities lacking in free-living organisms. They must orchestrate complex life cycles involving as many as four or five different hosts (recall *Alaria* from Chapter 3), managing to colonize each host in succession. They may possess complicated sensory structures, the functions of which we largely remain ignorant. They must also be able to find their preferred locations within the host, locate a mate, and in many cases produce prolific numbers of progeny. They must also be up to the task of evading or otherwise thwarting their host's immune responses and, as in the case of *Sacculina*, be able to manipulate the endocrinology, energy budget, and behavior of the host to favor production of their own progeny. Not surprisingly, some parasites such as ***Trichomonas vaginalis*** have genomes and numbers of protein-encoding genes much larger than expected (the ***T. vaginalis*** genome size is 160 Mbp with nearly 60,000 predicted protein-encoding genes) given their relatively simple body plans and life cycles. Rather than conclude parasites have evolved to be simplified or degenerate, it is clear that parasites have been shaped in distinctive and often astonishing ways by natural selection to succeed in the context of living in or on another living organism.

7.6 PARASITE EFFECTS ON HOST EVOLUTION

Parasites make a profound evolutionary commitment to their hosts. Although hosts are surely not dependent on their parasites, because they are constantly assaulted by a variety of common and often virulent parasites, it is hardly surprising that the evolutionary biology of host organisms is also shaped by selection imposed by parasites. The process of coevolution that we discussed in Section 7.2 certainly exemplifies this. In this section, we provide some additional examples of how parasites have affected host evolution and note how their impact has been far-reaching, ranging from immediate effects on survival, formation of social systems, influence on host speciation and diversification, and even the choice of mates and reproduction. Of course, the elaboration of an immune system to defend the body from infection is one manifestation of the evolutionary pressure imposed by parasites. This topic has also been discussed in Chapter 4.

Parasites select for genetic changes and genetic diversity in their hosts

We need look no further than our own species to see how selection imposed by parasites can favor changes in genes and their frequencies (see Figure 7.7). ***Plasmodium falciparum***, the most virulent of the malaria species infecting humans, still kills over a half million children per year in Africa. Before the availability of modern drugs and control measures, an estimated 20% of African children may have perished in their first decade of life from ***P. falciparum***. Additionally, the parasite has a major impact on human reproductive success, often causing inflammation and restricted blood flow in the placenta. As we have noted in Chapter 5, this parasite, unlike its close relatives, has a unique ability to initiate the formation of knobs on the surface of infected erythrocytes that favor the sequestration of infected cells on the inner wall of endothelial cells, including in vital organs like the brain and kidneys.

It is now well appreciated that the pathology imposed by ***P. falciparum*** has resulted in selection for alternative forms of human hemoglobin that confer some protection against malaria (**Table 7.2**). For example, the sickle-cell form of hemoglobin (HbS) differs in sequence from the normal hemoglobin

Table 7.2 Polymorphisms in hemoglobin and erythrocytes that are protective against *Plasmodium falciparum*

Name of Polymorphism	Gene Affected	Proposed Mechanism of Protection
HbS	*HBB*	Increased sickling of parasitized erythrocytes, enhanced clearance in spleen
		Does not support parasite growth well under conditions of low oxygen tension
		Reduced cytoadherence of parasitized erythrocytes
		Enhanced innate and acquired immunity
HbC	*HBB*	Reduced cytoadherence of parasitized erythrocytes
HbE	*HBB*	Erythrocyte membranes more resistant to invasion
α-thalassemia	*HBA1/HBA2*	Reduced binding of uninfected to infected erythrocytes
G6PD deficiency	*G6PD*	Enhanced phagocytosis of infected erythrocytes

HBA1 = hemoglobin A1 gene, encodes α-chain of hemoglobin; *HBA2* = hemoglobin α-2 gene, also encodes α-chain of hemoglobin; *HBB* = hemoglobin β gene, encodes β-chain of hemoglobin.
Modified from Wellems TE, Hayton K & Fairhurst RM (2009) The impact of malaria parasitism: from corpuscles to communities. *J Clin Invest* 119: 2496-2505, Table 1. With permission from American Society for Clinical Investigation.

β-chain by just one amino acid. Although individuals homozygous for the sickle-cell trait are at high risk for developing sickle-cell anemia (also called sickle-cell disease) and often suffer early mortality, individuals who are heterozygous have enhanced resistance to ***P. falciparum*** as compared to individuals homozygous for the normal allele. The erythrocytes of heterozygous individuals, once infected, have a greater probability of assuming a jagged, misshapen form (sickling), such that they are more readily cleared and eliminated in the spleen (**Figure 7.27**). Furthermore, the aberrant molecules of hemoglobin interfere with trafficking of parasite proteins to the infected erythrocyte surface and diminish the ability of infected cells to stick to endothelial surfaces, thus alleviating one of the major sources of pathogenicity of this parasite. Remarkably, the human sickle-cell trait has been selected for increased frequency independently at least five separate times in different malarious localities across Africa and Asia, and its distribution corresponds closely with the geographic regions with the most intense malaria transmission (**Figure 7.28**), particularly as a result of ***P. falciparum***.

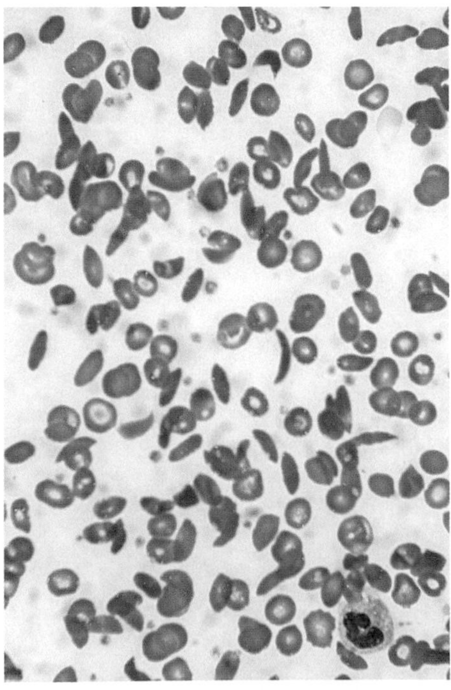

Figure 7.27 Peripheral blood smear of a patient with sickle-cell anemia. This blood film shows irreversibly sickled cells (note slender sickle-shaped erythrocytes). (Courtesy of Ed Uthman, CC BY 2.0.)

Several other human polymorphisms attributed to malaria are also known, including some likely driven by another common human malaria species, ***Plasmodium vivax***. This species lacks ***P. falciparum***'s propensity to make infected erythrocytes sticky. ***Plasmodium vivax*** is only able to infect young erythrocytes (reticulocytes) that (usually) express the Duffy blood group antigen on their surface, which is bound by a parasite surface molecule (PvDBP) as an essential part of the entry process into erythrocytes. The high prevalence of individuals in West Africa who have lost the Duffy antigen is believed to account for the lack of ***P. vivax*** in that area. In Papua New Guinea, where ***P. vivax*** is common, Duffy negativity has also appeared, and individuals heterozygous for this trait are less susceptible to ***P. vivax***. To accentuate the dynamic nature of host–parasite associations, in Madagascar, ***P. vivax*** is capable of infecting the erythrocytes of Duffy-negative individuals, suggesting the parasite is evolving to overcome the lack of this particular receptor. This geographic area includes both Duffy-negative and Duffy-positive individuals. One possibility is the presence of the latter has permitted ***P. vivax*** to flourish, providing an opportunity for new genetic variants to arise, some of which have a mutation favoring their entry into Duffy-negative cells. The potential of ***P. vivax*** to evolve alternative invasion strategies carries with it an important lesson: vaccination based on any single parasite target, such as PvDBP, may encounter significant problems owing to the potential of parasites to evolve alternative means to achieve infection.

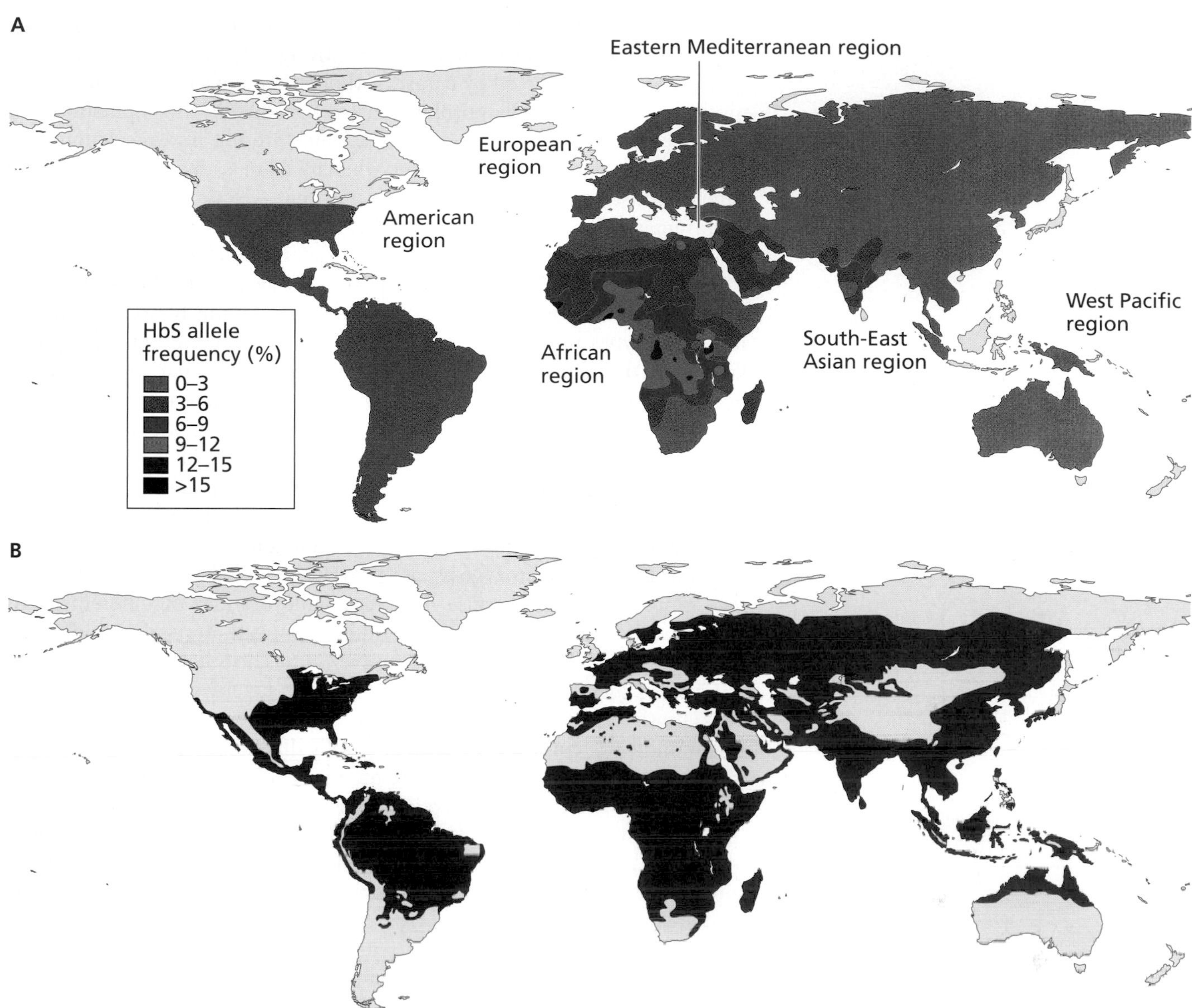

Figure 7.28 The distribution of the sickle-cell trait mirrors that of heavy *Plasmodium falciparum* transmission. (A) The distribution and frequency of the sickle-cell hemoglobin (HbS) allele is shown. (B) The historical global distribution of malaria (red) caused by all species before interventions to control malaria, with *P. falciparum* most intensively transmitted in sub-Saharan Africa and southern Asia. (A, From Rees DC, Williams TN, Gladwin MT et al [2010] *Lancet* 376:2018–2031. With permission from Elsevier. B, Adapted from Hay SI, Guerra CA, Tatem AJ et al [2004] *Lancet Infect Dis* 4:327–36. With permission from Elsevier.)

The malaria-driven polymorphisms are vivid examples of what seems to be a common occurrence: evidence of the impact of parasitism in altering the genetic composition of their host populations. Genes of the immune system are often under intense selection. This is determined both by finding striking polymorphisms in host immune genes and by analyses that indicate particular loci are subject to selection. For example, one signature of selection is the discovery of extraordinarily high rates of nucleotide changes that lead to actual alterations in amino acids and consequently immune protein structure (nonsynonymous mutations). Such nonsynonymous changes at key immune loci predominate over nucleotide changes that don't lead to amino acid changes (synonymous changes).

Parasites affect the evolution of host MHC genes

In vertebrates, gene products of the major histocompatibility complex (MHC; see Chapter 4) are involved in antigen presentation to T lymphocytes and thus play an important anti-parasite role. MHC genes show a strong predominance of nonsynonymous over synonymous mutations in their peptide-binding regions and have extensive allelic diversity such that populations are highly polymorphic (over 1000 MHC alleles have been found in human populations). This polymorphism indicates the strong role of selection, generally considered to be mediated by parasites, though opinions differ regarding the mechanism by which polymorphisms are maintained in the population. The three mechanisms generally considered, which are not necessarily mutually exclusive, are: (1) balancing selection, whereby there is an advantage of being heterozygous; (2) negative frequency-dependent selection, whereby there is an advantage of being rare, although this advantage will likely fluctuate with time; and (3) selection for different MHC alleles best able to counter particular parasites, with the realization that the abundance of different parasites changes both temporally and spatially. Many associations between particular MHC genotypes and resistance to parasites have been found in natural populations of a diversity of vertebrate hosts. For example, comparisons of rodents made across species have associated high helminth species richness with increased MHC class II polymorphisms. In humans, regional differences in MHC class I diversity have been associated with intracellular pathogen richness. Experimental studies have also shown that the frequency of MHC alleles associated with protection against parasites can quickly increase following imposition of a parasite burden on some populations.

Some degree of heterozygosity at the MHC loci is believed to be an advantage to an individual because it increases the number of different MHC alleles available and thereby the breadth of parasite-derived peptides that can be recognized (see Box 4.2). However, it has also been considered that if the MHC loci collectively represent too much heterozygosity, there would be a negative effect on immune function. This negative effect arises because the immune system deletes clones of T cells that are potentially reactive to its own MHC molecules. If there are too many MHC alleles, then too many T cells clones are deleted, potentially leaving the T cell repertoire bereft of diversity. This consideration would suggest that in some cases, intermediate levels of MHC heterozygosity are optimal and would be favored. Empirical support for the notion that the lowest levels of parasite infection are found in individuals of intermediate MHC diversity has come from sticklebacks (**Figure 7.29**), pythons, and bank voles. A general consensus has yet to emerge though regarding what level of MHC diversity is best suited to limit parasite infections; it may prove to depend on the particular system under study. As we shall soon see, the MHC may also play a role in mate choice and host speciation.

Parasites play a role in host selection of mates

Sexual selection can be thought of as two forms of intraspecific competition. The first is intrasex competition, such as between males fighting for access to females, with the most robust and vigorous male prevailing. In this case, the males are said to elaborate weapons used in combat. With respect to male–male competition, males impaired by high parasite loads may be smaller in size, produce smaller weapons such as antlers, have less stamina and strength, and even have their genitalia effectively blocked by parasites, all of which would make them less able to compete for access to mates. Healthy individuals consequently would tend to win the right to reproduce.

The second form of intraspecific competition is intersex competition, in which one sex, usually the female, exercises choice in picking a mate. In this case, a male often elaborates an ornament by which the female judges the male's quality. Parasites might affect mate choice directly, as when a female avoids a potential mate that is visibly infected (as with lice or other ectoparasites) and could transfer parasites to her or her offspring. Alternatively,

A

B

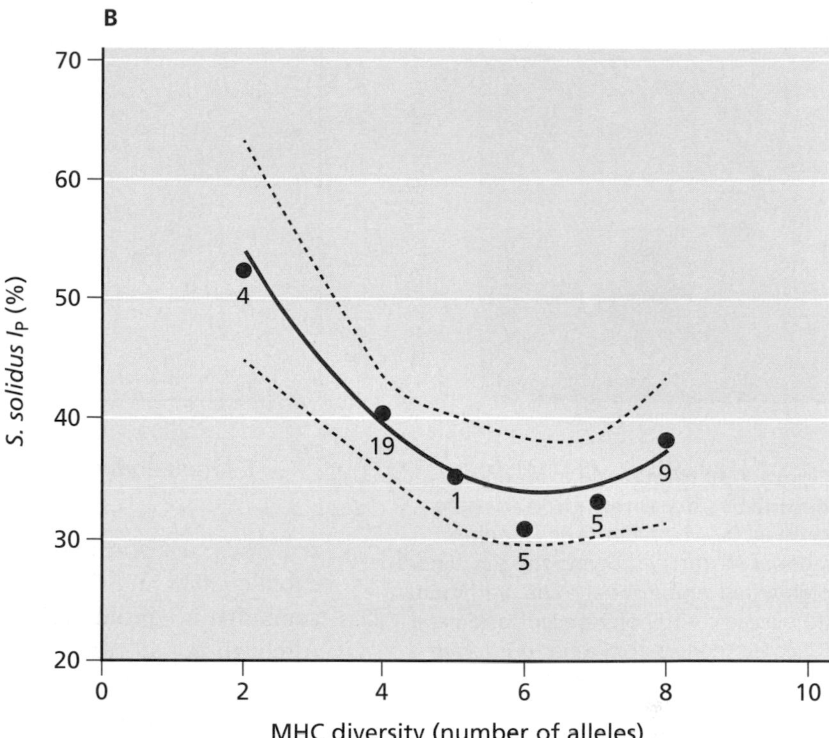

Figure 7.29 Intermediate levels of MHC diversity promote optimal parasite defense in some model systems. (A) Plerocercoids of the cestode *Schistocephalus* from the abdominal cavity of a stickleback fish. (B) Three-spined sticklebacks (*Gasterosteus aculeatus*) were exposed to *Schistocephalus solidus* tapeworms. Exposed sticklebacks were later dissected to screen for parasite infection and to take measurements of immune function. Shown on the vertical axis is a parasite index (*Ip*, a measure of parasite biomass relative to host biomass) for the fish (n = 43). On the horizontal axis is the diversity of MHC class IIB alleles. Numbers next to data points indicate sample size and broken lines show 95% confidence limits. Note that the minimum number of parasites was found at intermediate levels of MHC diversity. (From Kurtz J, Kalbe M, Aeschlimann PB et al [2004] *Proc R Soc Lond B Biol Sci* 271:197–204. With permission from the Royal Society.)

a female might choose a male that is visibly healthy and vigorous and thus able to provide for her young. However, it is the hypothesis that parasites indirectly influence male ornamentation and consequently female choice that has become one of the most popular topics of study involving parasites among evolutionary biologists. This is called the **parasite-mediated sexual selection hypothesis**, first proposed by Hamilton and Zuk (1982). According to this hypothesis, males produce an elaborate ornament, such as a long tail or a courtship dance, which is expensive to produce and thus provides some honest indication of a male's fitness. A burden of parasitism borne by the male would be expected to diminish the energy available to him to grow and maintain the ornament.

A further consideration is likely applicable as well. Elaboration of the ornament often requires androgens such as testosterone, which are typically immunosuppressive in their physiological effect, so an ornamented male then might become more vulnerable than normal to infection. This is known as the **immunocompetence handicap hypothesis**. From the perspective of the choosy female, if a male is evidently healthy and still able to make a superior ornament, then this trait serves as a reliable indicator that the male has good genes that will be conferred to her offspring if she chooses to mate with him. Additionally, colorful male traits may owe their presence to carotenoid pigments that must be acquired from the diet. These pigments also play a role in immune function, so a male faces an additional trade-off with respect to how to deploy these valuable compounds: maintain defense or elaborate an ornament. Again, the basic idea is that males with superior resistance genes will not have to siphon carotenoids off to aid their defense and can use them instead to make the ornament.

A number of correlative studies support the predicted inverse relationships between the quality of ornamental traits and parasite loads. One example is provided by the red grouse (*Lagopus lagopus scotticus*) and its interactions with the intestinal nematode *Trichostrongylus tenuis* (**Figure 7.30**). In

A

B

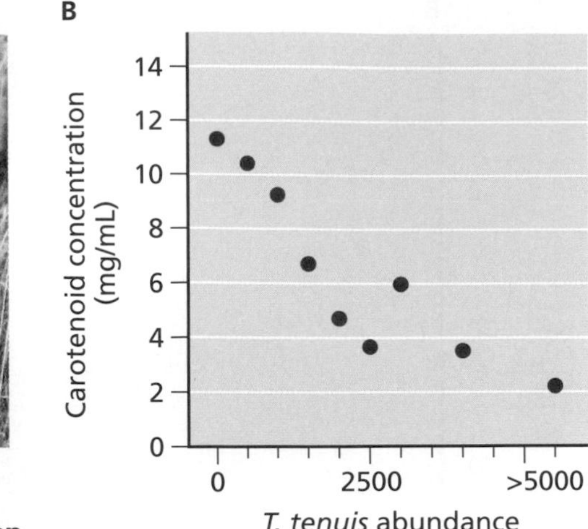

C

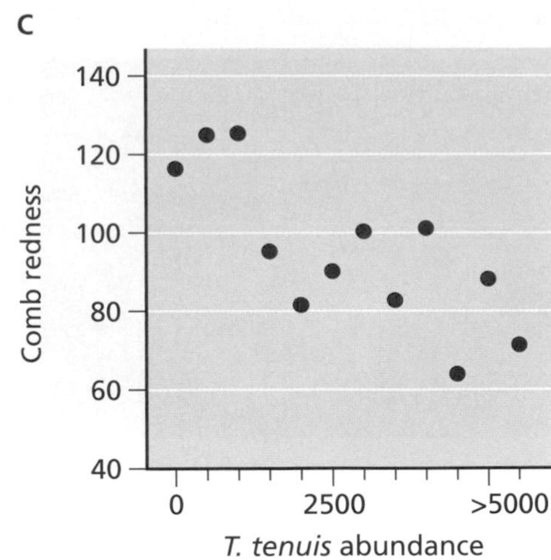

Figure 7.30 Nematode infection diminishes ornamentation in male grouse. (A) The red grouse (*Lagopus lagopus scotticus*) showing the carotenoid-pigmented supraorbital comb. Variation in (B) plasma carotenoid concentration and (C) comb redness according to *T. tenuis* abundance (number of worms per grouse). (A, courtesy of Dick Daniels. B, C, Modified from Mougeot F, Perez-Rodriquez L, Martinez-Padilla J et al [2007] *Funct Ecol* 21:886–898 With permission from John Wiley and Sons.)

red grouse, as worm burdens increase, there is a decline in both the carotenoid content and the coloration of the carotenoid-pigmented supraorbital comb that is a prominent ornament mediating female choice in these birds. Although not all results support the Hamilton–Zuk hypothesis, and it may be that ornaments provide good signals of male conditions only in some years (such as in harsh years when parasites are especially common), a growing body of evidence supports the idea that parasites affect elaboration of ornaments and, indirectly, female choice of males.

Mate choice can also be based on factors other than ornaments, and one factor considered prominently in this regard is the MHC. As noted above, some degree of heterozygosity at MHC loci is considered advantageous because it increases the number of different MHC alleles available and thereby the breadth of parasite-derived peptides that can be recognized. Consequently, might it be advantageous to pick a mate with MHC alleles that would favor offspring who are either maximally or optimally heterozygous and therefore better protected from parasites? Perhaps, though, it might be advantageous to select a mate with a particular MHC allele effective against a very common local parasite?

Here too evolutionary parasitologists have yet to reach consensus and studies of MHC diversity and mate choice are ongoing. Although some studies have failed to find any effects of MHC on mate preference, several studies in fish, reptiles, birds, and mammals including primates suggest that MHC influences mate choice. For example, females of Savannah sparrows (*Passerculus sandwichensis*) avoid males sharing similar MHC alleles, and house sparrow (*Passer domesticus*) females avoid males with low MHC diversity. Experiments in both house sparrows and sticklebacks (*Gasterosteus aculeatus*) suggest that female choice is dictated not so much by maximizing MHC heterozygosity among progeny as by selecting a mate that can provide an intermediate level of heterozygosity (see also Figure 7.29). Human studies have returned equivocal results, some suggesting a role for MHC in mate preference and some not. One conclusion from many studies is that the outcome may depend on the genetic structure of the populations involved. For example, individuals from isolated populations with low levels of variation may be more in need of MHC-mediated mate choice as a way to attain heterozygosity.

You may have wondered how it is possible to know what MHC alleles a prospective mate possessed. At least in some cases, the answer seems to be on the basis of odors associated with particular MHC alleles, although other odor-producing genes may also be involved that augment MHC gene effects, even in humans. The role of parasites in host mate choice remains a fascinating topic that will no doubt continue to provoke many more studies.

Host speciation may be facilitated by parasites

The idea that parasites can be a factor favoring diversification and speciation of host lineages is not new, but there are few empirical studies to document that it actually occurs or with meaningful frequency. The lack of conclusive studies is not surprising given the length of time needed for the speciation process to occur and the difficulties in documenting the involvement and relevance of parasites in promoting reproductive isolation and speciation. In this section, we first briefly make the point that pressure from parasites can serve as a force to cause variation within host species and thus create a situation in which differentiation into separate species becomes more plausible. Then three possibilities are discussed which suggest how parasite-mediated speciation might occur.

If parasites are to cause speciation in the host, we might expect some prerequisites to be met. One is that the pressure imposed by parasites varies among host populations, such that it would then be expected that host gene frequencies in response to parasitism would also vary. Second, the differences in infection among populations with respect to parasite pressures experienced should be fairly constant, such that differences can accumulate and divergence among host populations increases. Also, the parasites should have fitness consequences for the host that are strong enough to override potentially conflicting consequences of other factors.

For example, malaria might be encountered only in parts of the host's range where the appropriate species of mosquitoes were present or where different suites of parasites might be encountered in different foraging habitats, such as limnetic or benthic parts of a lake. Strong local adaptation in the form of immunological accommodations of particular host populations to the distinctive parasites they encounter would occur, such that across a broader host range host populations differed significantly with respect to the nature of their responses. Differences among populations of the same host species with respect to their immune defenses have been noted for *Drosophila* against parasitoids, Darwin's finches, sticklebacks coping with eye flukes, and marine amphipods infected with trematodes. These localized differences might be further reinforced by **philopatry** (limited host dispersal) resulting from both a strong preference to interact only with individuals that have similar immune accommodations and from avoidance of out-group individuals that harbor exotic parasites to which they are not accommodated (**Figure 7.31**). Differentiation resulting from spatially variable antagonistic interactions with parasites would in this case provide the substrate for further diversification of host lineages.

If parasites mediate divergence among populations of the host species, then we might envision host speciation eventually occurring by the mechanisms shown in **Figures 7.32** to **7.34**. The first (Figure 7.32A) is that hosts immigrating into areas new to them but occupied by divergent hosts may be maladapted to the parasites present and may be selected against, thereby reinforcing the differences between host populations. Similarly, hybrids resulting from crosses between host individuals from different areas may prove to be less resistant to parasites found in either area (Figure 7.32B).

A second possibility whereby parasites might promote host speciation is that they may influence the distribution of a host gene that has multiple

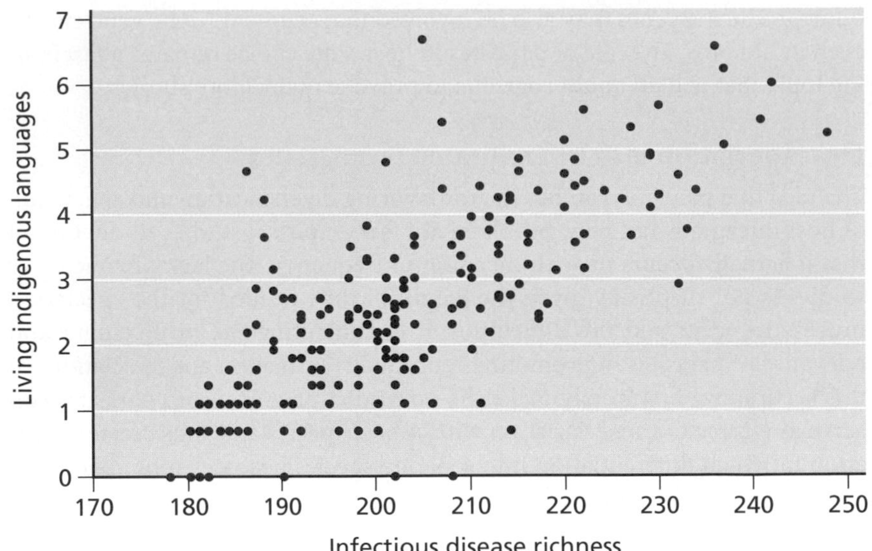

Figure 7.31 Parasites as a wedge that drive differentiation of host populations. A positive correlation is shown between human parasite richness and the natural log of indigenous living languages per country found across the world. The presence of a distinct language is taken as an indicator of isolation of the population involved, leading to differentiation among populations. The researchers undertaking this study hypothesized that people respond to the greater pressure imposed by parasites by limiting their dispersal, both because of a strong preference to interact only with individuals who have similar immune accommodations and to avoid contagion with out-group individuals that harbor exotic parasites. Parasites thus serve as a wedge to cause populations to become isolated and diverge, and hence their different languages. (From Fincher CL & Thornhill R [2008] *Oikos* 117:1289–1297. With permission from John Wiley and Sons.)

effects, say one that influences immune function and that also affects mate choice (Figure 7.33). This situation of multiple effects caused by the same gene is called **pleiotropy**.

A third possibility is ecologically based sexual selection (Figure 7.34). In this case, populations with different parasite pressures first diverge in their use of mating cues (these different cues better signal resistance to the different infections). Later this divergence is reinforced by sexual selection in which individuals that better resist the local parasites are chosen.

Although an exhaustive discussion of all these possibilities is beyond our scope, consider once again the MHC loci, which we have noted play an important role in parasite defense and are implicated in mate choice (see Figure 7.33). If parasite pressures vary from one region to another, then it is reasonable to expect that the complement of MHC alleles present in any area would also vary. Females might be expected to pick males with MHC alleles best able to deal with the local parasites. That is, females in this case

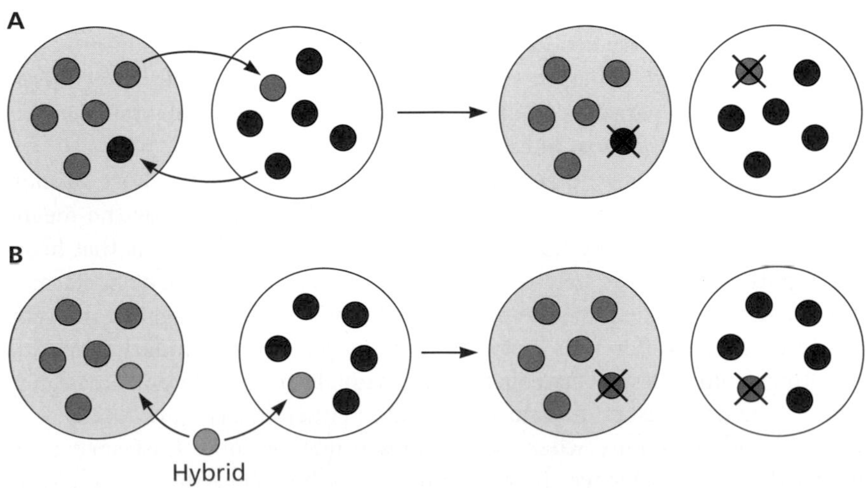

Figure 7.32 Influence of parasite-mediated reduction of viability or fecundity on host speciation. Parasites may promote host speciation when reproductive isolation of hosts is reinforced by the reduced viability or fecundity of immigrants or hybrids. The two large circles (gray and white) represent host populations experiencing different parasite pressures such that the individual hosts within each population (small blue or red circles) have differentiated. In (A), immigrants from one host population into the other suffer high infection levels resulting in reduced survival or fecundity. This further reinforces the differentiation between the two host populations. In (B), this scenario is similar except hybrids (purple) between divergently adapted parent populations have higher infection levels and reduced survival or fecundity in either of the parental habitats. Hybrids suffer more because their intermediate defense profiles do not match the parasite pressure of the parental habitats. (Modified from Karvonen A & Seehausen O [2012] *Int J Ecol* Article ID 280169 [doi:10.1155/2012/280169].)

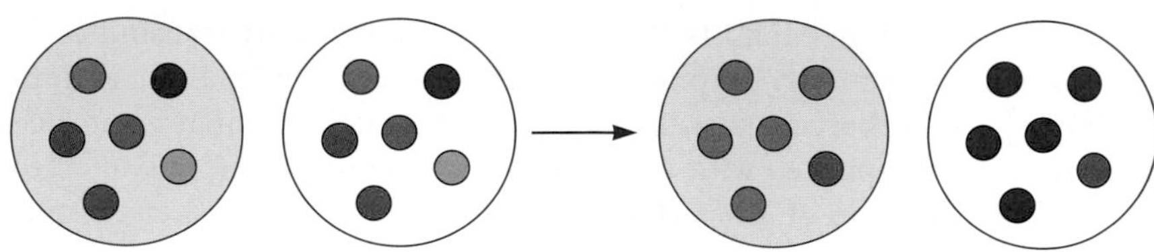

Figure 7.33 Multiple parasite effects on MHC and host speciation. Parasites may favor host speciation by promoting reproductive isolation of hosts as a result of pleiotropic effects of MHC on mate choice. Imagine two host populations (large gray and white circles) initially contain individuals with diverse but similar MHC profiles (small circles of different colors). If the two populations experience different parasite pressures, then different MHC alleles will be favored in the two populations. Reproductive isolation between the two populations increases because of the pleiotropic effects of the MHC on mate choice. (Modified from Karvonen A & Seehausen O [2012] *Int J Ecol* Article ID 280169 [doi:10.1155/2012/280169].)

assortatively choose males that are best adapted for local conditions. MHC genes could thus serve as examples of pleiotropic traits, or what have been called **magic traits**, meaning they are both under divergent selection (from parasites) and affect mate choice and thus could promote host reproductive isolation and speciation.

This possibility has been most studied with three-spined sticklebacks (*Gasterosteus aculeatus*) that experience different suites of parasites depending on the habitat in which they live: either benthic or limnetic. Fish from these two habitat types have different class II MHC alleles, with females from benthic habitats having more. Females from benthic habitats also have strong odor-based assortative mate preferences, all consistent with the possibility of parasite-driven MHC traits playing a role in further diverging host populations. Similarly, MHC divergence in a closely related and sympatric pair of cichlid fish species from Lake Malawi has been proposed to be a result of local host–parasite coevolutionary associations and to have influenced odor-mediated mate choice, and ultimately found to have favored speciation.

Another possible parasite-facilitated mechanism of host speciation also deserves mention, one that also involves hybridization but in a different way from that suggested in Figure 7.32. Hybridization can contribute to the formation of new host species by **allopolyploidy**, when chromosomes are contributed by different parental species to form a new species. Allopolyploidy has been postulated to account for the origin of most species in the anuran genus *Xenopus*. Hybrids in this case often have increased resistance for parasites as compared to either parental species, potentially providing a selective advantage to favor the persistence of new species of recent hybrid origin.

Can infection directly cause speciation?

Another factor potentially driving host speciation, one that has not yet received the full attention it deserves, is the involvement of symbionts. As we have noted throughout this book, symbionts are a ubiquitous feature of organismal life and are bound to influence all aspects of the biology of their hosts. In some cases, these symbionts dramatically alter their host's reproductive biology, so much so they are considered reproductive parasites. Consider for a moment α-proteobacteria of the genus *Wolbachia* that infect many arthropods and nematodes, with species of fruit flies (*Drosophila*) being

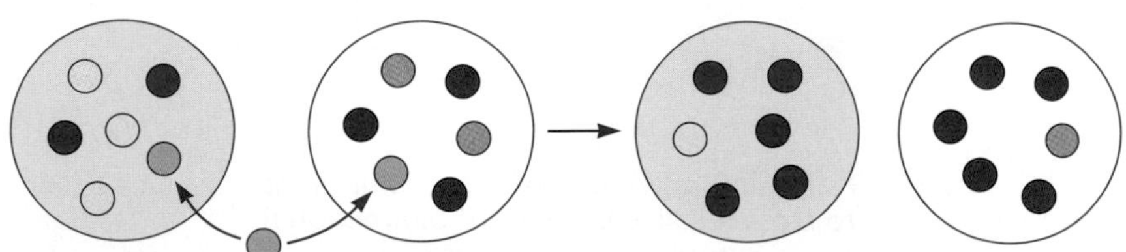

Figure 7.34 Ecologically based sexual selection driven by parasites. Imagine two host populations (large gray and white circles) experience different environments and different parasite pressures. Individual hosts within the two populations (small circles) diverge in their use of mating cues because different cues better signal heritable resistance to the different infections. Sexual selection will favor individuals that better resist parasites in their respective environments (dark blue and red circles) over more heavily infected individuals (light blue and red circles). Hybrids (purple) not optimally adapted to either environment, do not persist. This selection facilitates divergent adaptation and favors reproductive isolation between the host populations. (Modified from Karvonen A & Seehausen O [2012] *Int J Ecol* Article ID 280169 [doi:10.1155/2012/280169].)

commonly studied model hosts. These bacteria are vertically transmitted in the host female's gametes. In some instances they cause cytoplasmic incompatibility (CI). In this case, sterility can result if a male host that is infected with *Wolbachia* copulates with a female who is uninfected (or is infected with a different strain of *Wolbachia*). The offspring suffer 100% embryonic mortality. By killing the offspring of females lacking the *Wolbachia* causing CI, females who do carry CI will be favored, and thus CI-causing *Wolbachia* will spread relentlessly in the population.

Because CI causes reproductive isolation between hosts with different infection states, or with different strains of *Wolbachia*, it is not surprising *Wolbachia* is suspected of influencing speciation of its hosts. **Figure 7.35** shows one way this might occur. In this scenario, if two host populations each acquire different cytoplasmic *Wolbachia* infections, these infections could spread to fixation within each population by unidirectional CI, which would confer a fitness advantage on infected females: crosses between infected males and uninfected females would not yield progeny, whereas infected females can cross with either infected or uninfected males. If the two populations come into secondary contact, bidirectional CI would preclude successful crosses in either direction. For this scenario to work, a couple of conditions have to be met. First, the infection polymorphism (for instance, some host individuals with one and some with another *Wolbachia* strain, or some with and some without *Wolbachia*) has to persist stably, a condition for which experimental evidence has been provided. The second condition is that during this period of isolation a sufficient barrier to gene flow must exist between the two populations to allow them to diverge such that they achieve reproductive isolation even if the *Wolbachia* infections should be lost. Evidence consistent with this second condition remains equivocal.

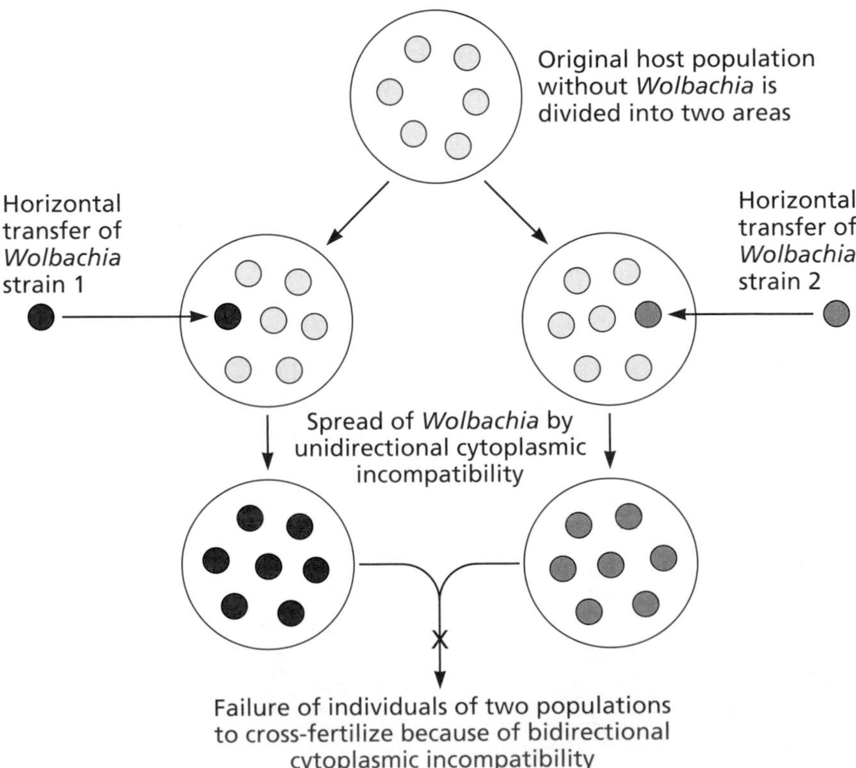

Figure 7.35 One view demonstrating how host speciation might be favored by reproductive parasites. A population (large circle) of uninfected individuals (small open circles) is divided into two populations, and each population subsequently acquires a different cytoplasmic *Wolbachia* infections (red or blue colored circles). These would be acquired from an infected individual host outside the population by horizontal transfer. These infections spread to fixation within each population by unidirectional cytoplasmic incompatibility. Upon secondary contact of individual hosts in these populations, bidirectional cytoplasmic incompatibility prevents successful crosses in either direction, giving rise to species that have not diverged genetically. (Modified from Brucker RM & Bordenstein SR [2012] *Trends Ecol Evol* 27:443–451. With permission from Elsevier.)

Interestingly, once again mate choice may come into play here as there is evidence from *Drosophila* hosts that, not surprisingly, uninfected females are strongly selected to avoid mating with *Wolbachia*-infected males. Furthermore, there is even a possibility of involvement of another magic trait, as the symbionts in some cases are known to produce odors that act as sex attractants for their hosts. Although it may remain arguable if *Wolbachia* infections actually lead to host speciation, it is less arguable that in a broader sense they clearly contribute to reproductive isolation and thus favor the possibility of host speciation. It is noteworthy too that other organisms beside *Wolbachia* can cause CI and potentially favor isolation of host populations. In general, we need remain mindful of the possibility of symbiont-mediated effects on host speciation.

The example of *Wolbachia* reminds us once again of the pervasively important role of symbionts in host evolution, a role that has become more obvious during the genome age. Acquisition of new symbionts can dramatically change a host species—they have been likened to megamutations—and can spread rapidly in their new hosts (see Box 7.2 and Chapter 4 for discussions of how symbionts can influence parasite infectivity or protect their hosts).

Parasites are believed to favor the evolution of sexual reproduction in their hosts

One of the most enduring mysteries in evolutionary biology is why sexual reproduction should be so common. Whereas all individuals in an asexually reproducing species can produce progeny, in a sexually reproducing species, only half the individuals (namely the females) can do so. Why should a female make males that can not themselves directly produce offspring? The advantages of sexual reproduction must be very strong to counterbalance this so called twofold disadvantage of sex.

One of several hypotheses proposed to solve this mystery relates to the Red Queen hypothesis discussed earlier; the pressure imposed by fast evolving parasites results in selection on hosts to adopt sexual reproduction. Recombination of genetic material resulting from sexual reproduction is expected to produce more diverse progeny, some of which can withstand the onslaughts of parasites. If asexual reproduction were the only possibility, then hosts would be relatively defenseless to ensure their progeny would not be attacked by parasites of the same constitution as attacked them. This supposition might be called the parasite hypothesis or Red Queen hypothesis as it relates specifically to the evolution of sex.

Experimental support is provided by studies of the model organism, the nematode *Caenorhabditis elegans* and an often lethal bacterial parasite *Serratia marcescens*. Normally in wild-type *C. elegans*, about 80% of the reproduction is by selfing that is achieved by hermaphrodites that make up most of the population. The remaining 20% is sexual, involving hermaphrodites being inseminated by relatively rare males. If wild-type *C. elegans* is forced to grow in the presence of *S. marcescens*, which is itself free to evolve in response to changes in the worms, after only 30 generations the proportion of sexual reproduction in *C. elegans* increased to 80%–90%. Furthermore, nematode lines incapable of outcrossing with males went extinct in less than 20 generations when forced to coexist with *S. marcescens* (**Figure 7.36**). Because both worms and bacteria can be frozen and later revived, it was possible to document that ancestral worms are 2–3 times more vulnerable to varieties of the bacteria that had interacted with the worms for 30 generations than they were to the ancestral bacteria. Worms that had interacted with bacteria for 30

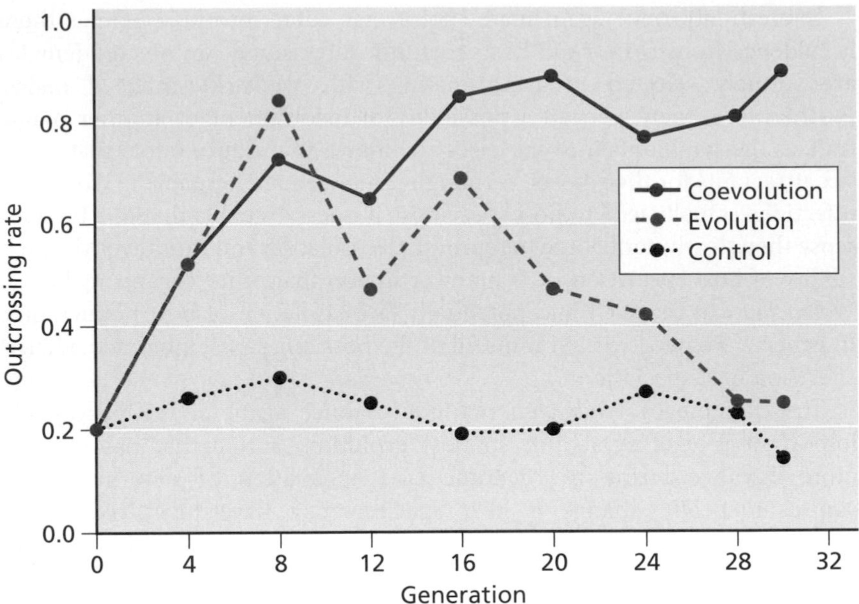

Figure 7.36 Biparental sex is favored in a nematode host when confronted with a parasite. The nematode *Caenorhabditis elegans* usually reproduces hermaphroditically (selfing), as opposed to sexually (outcrossing with another individual). But what happens when it has to contend with a coevolving parasite, the bacterium *Serratia marcescens*? In the experiments shown, outcrossing rates were measured over 30 worm generations. Control wild-type worms were not exposed to *S. marcescens* (dotted line) and maintained the background rate of outcrossing of about 20%. The dashed line shows the response of *C. elegans* to a fixed strain of *S. marcescens* not able to respond to the nematode. The red solid line shows the *C. elegans* population exposed to actively evolving *S. marcescens*. Note in this case the outcrossing rate increased dramatically. (From Morran LT, Schmidt OG, Gelarden IA et al [2011] *Science* 333:216–218, Fig. 1. With permission from AAAS.)

generations were much better at withstanding their contemporaneous bacteria than their ancestors. This study provides experimental confirmation from one model system that coevolving parasites select for biparental sex in their hosts, in accordance with predictions of the Red Queen hypothesis.

Parasites can cause extinction of host species

One possible impact of parasitism on host evolution is to cause extinction of a host species. In fact, this outcome has already been observed. *Partula turgida*, a tree snail endemic to the Society Island of Raiatea in the South Pacific, was last observed in the wild in 1991, when the remaining snails were collected for captive propagation. The captive population crashed in 1996, and the last known specimen apparently succumbed to infection with a microsporidian (*Steinhausia* sp.). Other species of the same snail genus occurring only in captivity also seem to be at risk. This example reminds us that host species with very small geographic ranges or with very limited genetic diversity are at high risk to perturbations, including parasitism. This may occur in spite of—or even because of—the deliberate attempts to save them by captive propagation. Because widespread extinction is a great concern in our modern world, we will return to this topic in Chapter 8.

Summary

The intimate, persistent, and often strongly antagonistic interactions between parasite and host provide a rich source of material for exploring evolutionary principles, a source that is still understudied with much left to teach us. The huge diversity of present-day parasites suggests that parasitism has always been a major player in Earth's evolutionary history. Furthermore, it is fascinating that the evolutionary trajectories taken in various parasite–host systems are so different. These differences suggest that the peculiar attributes of the biology of both interacting partners can drive these outcomes in divergent and often unexpected ways that, on the one hand, resist generalization but, on the other, enable us to gain a deeper appreciation of how evolution works.

Parasites, like all organisms, are shaped by basic microevolutionary processes—mutation, genetic drift, gene flow, and natural selection. The extent of gene flow among parasite populations can be greatly influenced by characteristics of the parasite's life cycle, their discontinuous presence among their hosts, or the extent of the mobility of both parasite and host. Coevolution, the process whereby parasites and hosts reciprocally affect each other's evolution, is a particularly dynamic example of why the study of parasite evolution is fascinating; this process can lead to arms races or negative frequency dependent selection. Better knowledge of parasite evolutionary biology can also help us understand how virulence is determined; revealing the connection between parasite reproductive rate and transmission is a central concept.

An important part of the study of parasite evolution is to learn how populations diverge, because this potentially leads to speciation and helps explain why there are so many species of parasites in the world. In some cases, parasite and host species seem to have diverged congruently, an outcome called cospeciation. In other cases, parasites have switched into novel hosts leading to dramatic expansions of parasite species. Parasites have played a strong role in influencing the evolutionary biology of their hosts by directly influencing fitness, mate choice, and the tendency for host populations to speciate. Sexual reproduction seen so pervasively among organisms seems to be one response whereby hosts can keep up with the rapid pace of evolutionary change demanded by their parasites.

REVIEW QUESTIONS

1. With respect to parasites, identify some factors expected to increase the genetic differentiation (structure) among populations.

2. Recall the example of *Plasmodium falciparum* and its varying abundance in different parts of the world. Explain how this abundance affects the amount of genetic recombination and gene flow that occurs in this species, and why understanding gene flow and recombination frequency matters.

3. What is coevolution, and what are some of its outcomes? How does coevolution differ from "evolution"? Hint: don't confuse this term with cospeciation.

4. What is virulence and what is meant by the trade-off hypothesis with respect to parasite virulence?

5. How are new species of parasites formed?

6. Why do parasites often have such complex life cycles?

7. What concrete evidence can be mustered that parasites have an impact on host evolution?

8. How do parasites influence mate selection?

9. Do hosts have sexual systems of reproduction because of parasites?

References

GENERAL REFERENCES

Brooks DR & McLennan DA (1993) Parascript. Parasites and the Language of Evolution. Smithsonian Institution Press.

Goater TM, Goater CP & Esch GW (2013) Parasitism: The Diversity and Ecology of Animal Parasites. Second edition. Cambridge University Press

Combes C (2001) Parasitism. The Ecology and Evolution of Intimate Interactions. University of Chicago Press.

Price PW (1980) Evolutionary Biology of Parasites. Princeton University Press.

Schmid-Hempel P (2011) Evolutionary Parasitology: The Integrated Study of Infections, Immunology, Ecology and Genetics. Oxford University Press.

Thompson JN (2005) The Geographic Mosaic of Coevolution. University of Chicago Press.

Thompson JN (2009) The coevolving web of life. *Am Nat* **173**:125–140.

MICROEVOLUTION IN PARASITES

The subdivided nature of their populations influences the evolution of parasites

Criscione CD, Poulin R & Blouin MS (2005) Molecular ecology of parasites: elucidating ecological and microevolutionary processes. *Mol Ecol* **14**:2247–2257.

The effective population size, N_E, influences parasite evolution

Beltran S & Boissier J (2010) Male-biased sex ratio: why and what consequences for the genus *Schistosoma*? *Trends Parasitol* **26**:63–69 (doi:10.1016/j.pt.2009.11.003).

Blouin MS, Dame JB, Tarrant CA et al (1992) Unusual population genetics of a parasitic nematode—mtDNA variation within and among populations. *Evolution* **46**:470–476.

Blouin MS, Yowell CA, Courtney CH et al (1995) Host movement and the genetic-structure of populations of parasitic nematodes. *Genetics* **141**:1007–1014.

Nadler SA, Lindquist RL & Near TJ (1995) Genetic-structure of midwestern *Ascaris suum* populations—a comparison of isoenzyme and RAPD markers. *J Parasitol* **81**:385–394.

The mode of parasite reproduction affects microevolutionary change

Anderson TJC, Haubold B, Williams JT et al (2000) Microsatellite markers reveal a spectrum of population structures in the malaria parasite *Plasmodium falciparum*. *Mol Biol Evol* **17**:1467–1482.

Blouin MS, Liu J & Berry RE (1999) Life cycle variation and the genetic structure of nematode populations. *Heredity* **83**:253–259.

Mulvey M, Aho JM, Lydeard C et al (1991) Comparative population genetic-structure of a parasite (*Fascioloides magna*) and its definitive host. *Evolution* **45**:1628–1640.

Rauch G, Kalbe M & Reusch TBH (2005) How a complex life cycle can improve a parasite's sex life. *J Evol Biol* **18**:1069–1075.

Stability of the host environment influences parasite microevolution

Steinauer ML, Hanelt B, Agola LE et al (2009) Genetic structure of *Schistosoma mansoni* in western Kenya: the effects of geography and host sharing. *Int J Parasitol* **39**:1353–1362.

Parasite microevolutionary change is strongly impacted by host mobility

Lymbery AJ, Constantine CC & Thompson RCA (1997) Self-fertilization without genomic or population structuring in a parasitic tapeworm. *Evolution* **51**:289–294.

Nadler SA & Hafner MS (1989) Genetic differentiation in sympatric species of chewing lice (Mallophaga, Trichodectidae). *Ann Entomol Soc Am* **82**:109–113.

A parasite's life cycle also affects the potential for evolutionary change

Barrett LG, Thrall PH, Burdon JJ et al (2008) Life history determines genetic structure and evolutionary potential of host–parasite interactions. *Trends Ecol Evol* **23**:678–685.

Katzer F, Ngugi D, Schnier C et al (2007) Influence of host immunity on parasite diversity in *Theileria parva*. *Infect Immun* **75**:4909–4916.

Prugnolle F & De Meeûs T (2008) The impact of clonality on parasite population genetic structure. *Parasite* **15**:455–457.

COEVOLUTION OF PARASITE–HOST INTERACTIONS

Parasites and hosts reciprocally affect each other's evolution

King KC, Delph LF, Jokela J et al (2009) The geographic mosaic of sex and the red queen. *Curr Biol* **19**:1438–1441.

Parasites and hosts engage in arms races

Genovese G, Friedman, DJ, Ross MD et al (2010) Association of trypanolytic ApoL1 variants with kidney disease in African Americans. *Science* **329**:841–845 (doi:10.1126/science.1193032).

Medel R, Mendez MA, Ossa CG et al (2010) Arms race coevolution: the local and geographic structure of a host–parasite interaction. *Evo Edu Outreach* **3**:26–31.

Raper J & Friedman DJ (2013) Parasitology: molecular one-upmanship. *Nature* **501**:322–323.

Uzureau P, Uzureau S, Lecordier L et al (2013) Mechanism of *Trypanosoma brucei gambiense* resistance to human serum. *Nature* **501**:430–434. (doi: 10.1038/nature12516).

Parasites and hosts can be locally adapted, or maladapted, to one another

Gandon S & Michalakis Y (2002) Local adaptation, evolutionary potential and host–parasite coevolution: interactions between migration, mutation, population size and generation time. *J Evol Biol* **15**:451–462.

Greischar MA & Koskella B (2007) A synthesis of experimental work on parasite local adaptation. *Ecol Lett* **10**:418–434.

Lemoine M, Doligez B & Richner H (2012) On the equivalence of host local adaptation and parasite maladaptation: an experimental test. *Am Nat* **179**:270–281 (doi:10.1086/663699).

Oppliger A, Vernet R & Baez M (1999) Parasite local maladaptation in the Canarian lizard *Gallotia galloti* (Reptilia: Lacertidae) parasitized by haemogregarian blood parasite. *J Evol Biol* **12**:951–955 (doi:10.1046/j.1420-9101.1999.00101.x).

THE EVOLUTION OF VIRULENCE

Virulence and transmission biology of parasites are linked

Davies TH (1969) Ear mites of the genus *Dicrocheles* (Acarina: Mesostigmata) found on noctuids in Hawkes Bay. *N Z Entomol* **4**:26–32.

Donoghue HD, Spigelman M, Greenblatt CL et al (2004) Tuberculosis: from prehistory to Robert Koch, as revealed by ancient DNA. *Lancet Infect Dis* **4**:584–592.

Ebert D & Bull JJ (2008) The evolution and expression of virulence. In Evolution in Health and Disease (Stearns SC & Koella JC eds), pp 153–170. Oxford University Press.

Ebert D & Herre EA (1996) The evolution of parasitic diseases. *Parasitol Today* **12**:96–101.

The trade-off hypothesis requires a nuanced approach

Bell AS, De Roode JC, Sim D et al (2006) Within-host competition in genetically diverse malaria infections: Parasite virulence and competitive success. *Evolution* **60**:1358–1371.

Frank SA (1996) Models of parasite virulence. *Q Rev Biol* **71**:37–78.

Steinauer ML (2009) The sex lives of parasites: investigating the mating system and mechanisms of sexual selection of the human pathogen *Schistosoma mansoni*. *Int J Parasitol* **39**:1157–1163 (doi:10.1016/j.ijpara.2009.02.019).

The mode of transmission influences virulence

Ewald PW (1994) Evolution of Infectious Disease. Oxford University Press.

MACROEVOLUTIONARY PARASITOLOGY

New parasite species are potentially formed in at least three different ways

Page RDM (2003) Tangled Trees. Phylogeny, Cospeciation and Coevolution. University of Chicago Press.

What does the evidence suggest about how parasites have speciated?

Brant SV & Loker ES (2013) Discovery-based studies of schistosome diversity stimulate new hypotheses about parasite biology. *Trends Parasitol* **29**:449–459 (doi:10.1016/j.pt.2013.06.004).

Brooks DR & Glen DR (1982) Pinworms and primates—a case-study in coevolution. *Proc Helminthol Soc Wash* **49**:76–85.

Hafner MS, Sudman PD, Villablanca FX et al (1994) Disparate rates of molecular evolution in cospeciating hosts and parasites. *Science* **265**:1087–1090.

Hoberg EP, Alkire NL, de Queiroz A et al (2001) Out of Africa: origins of the *Taenia* tapeworms in humans. *Proc R Soc Lond B Biol Sci* **268**:781–787.

Nadler SA & Hafner MS (1989) Genetic differentiation in sympatric species of chewing lice (Mallophaga, Trichodectidae). *Ann Entomol Soc Am* **82**:109–113.

Olson PD, Caira JN, Jensen K et al (2010) Evolution of the trypanorhynch tapeworms: parasite phylogeny supports independent lineages of sharks and rays. *Int J Parasitol* **40**:223–242 (doi:10.1016/j.ijpara.2009.07.012).

Paterson AM & Banks J (2001) Analytical approaches to measuring cospeciation of host and parasites: through a glass, darkly. *Int J Parasitol* **31**:1012–1022 (doi:10.1016/S0020-7519(01)00199-0).

Paterson S, Vogwill T, Buckling A et al (2010) Antagonistic coevolution accelerates molecular evolution. *Nature* **464**:275–278.

Ricklefs RE & Fallon SM (2002) Diversification and host switching in avian malaria parasites. *Proc R Soc Lond B Biol Sci* **269**:885–892.

Zarlenga DS, Rosenthal BM, La Rosa G et al (2006) Post-Miocene expansion, colonization, and host switching drove speciation among extant nematodes of the archaic genus *Trichinella*. *Proc Nat Acad Sci U S A* **103**:7354–7359 (doi:10.1073/pnas.0602466103).

Does sympatric speciation occur in parasites?

Euzet L & Combes C (1998) The selection of habitats among the monogenea. *Int J Parasitol* **28**:1645–1652.

McCoy KD (2003) Sympatric speciation in parasites—what is sympatry? *Trends Parasitol* **19**:400–404.

Theron A & Combes C (1995) Asynchrony of infection timing, habitat preference, and sympatric speciation of schistosome parasites. *Evolution* **49**:372–375.

Host switches can enable radiations in parasites

Bakke TA, Harris PD & Cable J (2002) Host specificity dynamics: observations on gyrodactylid monogeneans. *Int J Parasitol* **32**:281–308.

Beveridge I, Chilton NB & Spratt DM (2002) The occurrence of species flocks in the nematode genus *Cloacina* (Strongyloidea: Cloacininae), parasitic in the stomachs of kangaroos and wallabies. *Aust J Zool* **50**:597–620.

Zietra MS & Lumme J (2003) The crossroads of molecular, typological and biological species concepts: two new species of *Gyrodactylus* Nordmann, 1832 (Monogenea: Gyrodactylidae). *Syst Parasitol* **55**:39–52.

Parasites go extinct, sometimes along with their hosts

Levsen A & Jakobsen PJ (2002) Selection pressure towards monoxeny in *Camallanus cotti* (Nematoda, Camallanidae) facing an intermediate host bottleneck situation. *Parasitology* **124**:625–629.

Morand S & Poulin R (2002) Body size–density relationships and species diversity in parasitic nematodes: patterns and likely processes. *Evol Ecol Res* **4**:951–961.

Poulin R (1998) Large-scale patterns of host use by parasites of freshwater fishes. *Ecol Lett* **1**:118–128.

Macroevolutionary patterns among parasites are not yet very clear

Bush AO, Aho JM & Kennedy CR (1990) Ecological versus phylogenetic determinants of helminth parasite community richness. *Evol Ecol* **4**:1-20.

Kamiya T, O'Dwyer K, Nakagawa S et al (2014) What determines species richness of parasitic organisms? A meta-analysis across animal, plant and fungal hosts. *Biol Rev* **89**: 123–134.

Poulin R (1995) Phylogeny, ecology, and the richness of parasite communities in vertebrates. *Ecol Monogr* **65**:283–302.

Poulin R & Morand S (2004) Parasite Biodiversity. Smithsonian Books.

SOME DISTINCTIVE ASPECTS OF PARASITE EVOLUTION

Organisms have repeatedly adopted parasitism by more than one route

Blaxter ML, De Ley P, Garey JR et al (1998) A molecular evolutionary framework for the phylum Nematoda. *Nature* **392**:71–75.

Dieterich C & Sommer RJ (2009) How to become a parasite—lessons from the genomes of nematodes. *Trends Genet* **25**:203–209.

Dowling APG & O'Connor BM (2010) Phylogeny of Dermanyssoidea (Acari: Parasitiformes) suggests multiple origins of parasitism. *Acarologia* **50**:113–129.

Lom J (2005) Ciliophora (ciliates). In Marine Parasitology (Rohde K ed), pp. 37-41. CSIRO Publishing.

Ogawa A, Streit A, Antebi A et al (2009) A conserved endocrine mechanism controls the formation of dauer and infective larvae in nematodes. *Curr Biol* **19**:67–71.

Warburton EM & Zelmer DA (2010) Prerequisites for parasitism in rhabditid nematodes. *J Parasitol* **96**:89–94.

Selection can favor the evolution of complex parasite life cycles

Parker GA, Chubb JC, Ball MA et al (2003) Evolution of complex life cycles in helminth parasites. *Nature* **425**:480–484 (doi:10.1038/nature02012).

Sometimes complex life cycles are simplified secondarily

Barger MA & Esch GW (2000) *Plagioporos sinitsini* (Digenea: Opecoelidae): A one-host life cycle. *J Parasitol* **86**:150–153.

Parasites often have simplified bodies or genomes but also have other talents not seen in free-living organisms

Corradi N, Pombert JF, Farinelli L et al (2010) The complete sequence of the smallest known nuclear genome from the microsporidian *Encephalitozoon intestinalis*. *Nat Commun* **1**:77 10.1038/ncomms1082).

PARASITE EFFECTS ON HOST EVOLUTION

Parasites select for genetic changes and genetic diversity in their hosts

Cyrklaff M, Sanchez CP, Kilian N et al (2011) Hemoglobins S and C interfere with actin remodeling in *Plasmodium falciparum*-infected erythrocytes. *Science* **334**:1283–1286 (doi:10.1126/science.1213775).

Menard D, Barnadas C, Bouchier C et al (2010) *Plasmodium vivax* clinical malaria is commonly observed in Duffy-negative Malagasy people. *Proc Nat Acad Sci U S A* **107**:5967–5971.

Parasites affect the evolution of host MHC genes

De Bellocq JG, Charbonnel N & Morand S (2008) Coevolutionary relationship between helminth diversity and MHC class II polymorphism in rodents. *J Evol Biol* **21**:1144–1150.

Eizaguirre C & Lenz TL (2010) Major histocompatibility complex polymorphism: dynamics and consequences of parasite-mediated local adaptation in fishes. *J Fish Dis* **77**:2023–2047.

Kloch A, Babik W, Bajer A et al (2010) Effects of an MHC-DRB genotype and allele number on the load of gut parasites in the bank vole *Myodes glareolus*. *Mol Ecol* **19**:255–265.

Madsen T & Ujvari B (2006) MHC class I variation associates with parasite resistance and longevity in tropical pythons. *J Evol Biol* **19**:1973–1978.

Prugnolle F, Manica A, Charpentier M et al (2005) Pathogen-driven selection and worldwide HLA class I diversity. *Curr Biol* **15**:1022–1027.

Parasites play a role in host selection of mates

Hamilton WD & Zuk M (1982) Heritable true fitness and bright birds—a role for parasites. *Science* **218**:384–387.

Horak P, Saks L, Karu U et al (2004) How coccidian parasites affect health and appearance of greenfinches. *J Anim Ecol* **73**:935–947.

Huchard E, Knapp LA, Wang JL et al (2010) MHC, mate choice and heterozygote advantage in a wild social primate. *Mol Ecol* **19**:2545–2561.

Jackson JA & Tinsley RC (2003) Parasite infectivity to hybridising host species: a link between hybrid resistance and allopolyploid speciation? *Int J Parasitol* **33**:137–144.

Milinski M (2006) The major histocompatibility complex, sexual selection, and mate choice. *Annu Rev Ecol Evol Syst* **37**:159–186.

Piertney SB & Oliver MK (2006) The evolutionary ecology of the major histocompatibility complex. *Heredity (Edinb)* **96**:7–21.

Polak M, Luong LT & Starmer WT (2007) Parasites physically block host copulation: a potent mechanism of parasite-mediated sexual selection. *Behav Ecol* **18**:952–957.

Roberts T & Roiser JP (2010) In the nose of the beholder: are olfactory influences on human mate choice driven by variation in immune system genes or sex hormone levels? *Exp Biol Med* **235**:1277–1281.

Vergara P, Mougeot F, Martinez-Padilla J et al (2012) The condition dependence of a secondary sexual trait is stronger under high parasite infection level. *Behav Ecol* **23**:502–511 (doi:10.1093/beheco/arr216).

Yamazaki K & Beauchamp GK (2007) Genetic basis for MHC-dependent mate choice. Genetics of sexual differentiation and sexually dimorphic behaviors. *Adv Genet* **59**:129–145.

Host speciation may be facilitated by parasites

Blais J, Rico C, van Oosterhout C et al (2007) MHC adaptive divergence between closely related and sympatric African cichlids. *PLoS One* **2**:e734.

Bryan-Walker K, Leung TLF & Poulin R (2007) Local adaptation of immunity against a trematode parasite in marine amphipod populations. *Mar Biol* **152**:687–695.

Eizaguirre C, Lenz TL, Kalbe M et al (2012) Rapid and adaptive evolution of MHC genes under parasite selection in experimental vertebrate populations. *Nat Commun* **3**:621 (doi:10.1038/ncomms1632).

Kalbe M & Kurtz J (2006) Local differences in immunocompetence reflect resistance of sticklebacks against the eye fluke *Diplostomum pseudospathaceum*. *Parasitology* **132**:105–116.

Kraaijeveld AR & Godfray HCJ (1999) Geographic patterns in the evolution of resistance and virulence in *Drosophila* and its parasitoids. *Am Nat* **153**:S61–S74.

Lindstrom KM, Foufopoulos J, Parn H et al (2004) Immunological investments reflect parasite abundance in island populations of Darwin's finches. *Proc R Soc Lond B Biol Sci* **271**:1513–1519.

Matthews B, Harmon LJ, M'Gonigle L et al (2010) Sympatric and allopatric divergence of MHC genes in threespine stickleback. *PLoS One* **5**:e10948.

Rafferty NE & Boughman JW (2006) Olfactory mate recognition in a sympatric species pair of three-spined sticklebacks. *Behav Ecol* **17**:965–970.

Scharsack JP, Kalbe M, Harrod C et al (2007) Habitat-specific adaptation of immune responses of stickleback (*Gasterosteus aculeatus*) lake and river ecotypes. *Proc R Soc Lond B Biol Sci* **274**:1523–1532.

Thomas F, Guldner E & Renaud F (2000) Differential parasite (Trematoda) encapsulation in *Gammarus aequicauda* (Amphipoda). *J Parasitol* **86**:650–654.

Can infection directly cause speciation?

Bordenstein SR (2003) Symbiosis and the origin of species. In Insect Symbiosis (Bourtzis K & Miller T eds), pp 283–304. CRC Press.

Engelstadter J & Hurst GDD (2009) The ecology and evolution of microbes that manipulate host reproduction. *Annu Rev Ecol Evol Syst* **40**:127–149.

Himler AG, Adachi-Hagimori T, Bergen JE et al (2011) Rapid spread of a bacterial symbiont in an invasive whitefly is driven by fitness benefits and female bias. *Science* **332**:254–256.

Parasites are believed to favor the evolution of sexual reproduction in their hosts

Gomulkiewicz R, Thompson JN, Holt RD et al (2000) Hot spots, cold spots, and the geographic mosaic theory of coevolution. *Am Nat* **156**:156–174.

Hamilton WD (1975). Innate social aptitudes of man: an approach from evolutionary genetics. In Biosocial Anthropology (Fox R ed), pp. 133-135. Malaby Press.

Koskella B & Lively CM (2009) Evidence for negative frequency-dependent selection during experimental coevolution of a freshwater snail and a sterilizing trematode. *Evolution* **63**:2213–2221.

Lively CM (1987) Evidence from a New Zealand snail for the maintenance of sex by parasitism. *Nature* **328**:519–521.

Parasites can cause extinction of host species

Cunningham AA & Daszak P (1998) Extinction of a species of land snail due to infection with a microsporidian parasite. *Conserv Biol* **12**:1139–1141.

Parasites and Conservation Biology

8

The conservation biologist knows that each imperiled species is a masterpiece of evolution, potentially immortal except for rare chance or human choice, and its loss a disaster.

E. O. Wilson (2000)

In This Chapter

Our world is increasingly imperiled by threats posed by overpopulation, habitat destruction, pollution, climate change, and a globalized economy that favors introductions of exotic species. Consequently, the topic of conservation biology becomes ever more relevant (**Figure 8.1**). **Conservation biology** refers to the study of the status of the Earth's biodiversity, with a primary goal to preserve from extinction as many species in as natural a setting as possible. Preservation of the integrity and extent of natural habitats and ecosystems is often key to the process of saving the species that inhabit these environments. Conservation biology is a discipline that engages not only a wide variety of biologists and other scientists, but also economists and resource management specialists. It requires an interdisciplinary approach, one that often interfaces science with public policy and politics. Parasitologists have an important role to play in conservation biology, for parasites figure into strategies to save the world's biota in several ways.

You might wonder why conservation biology is important. So what if we lose a few species? There are many pragmatic reasons to care about conservation (see Figure 8.1). Endangered species might have as yet unfathomed practical benefits, such as being sources of new antibiotics, pharmaceuticals, or materials with unique structural properties. Also, intact ecosystems perform many largely unappreciated but beneficial services. **Ecosystem services** can be thought of as the ability of natural environments to provide food and clean water, to regulate climate and infectious diseases, to support nutrient cycles and necessary functions such as pollination, and to provide recreational and aesthetic benefits. If we begin to poke holes in the species composition, we risk compromising those services. For example, the complexity and balance in natural ecosystems may make it harder for introduced species to survive or for new diseases to emerge. Many endangered species also have unique historical or cultural significance for indigenous peoples or are tourist attractions in national parks. Finally, there is a powerful moral reason—as embodied in the quote at the beginning of this chapter—as to why the diversity of life forms on our planet is both a cause for celebration and concern. It is up to us to preserve this astounding diversity for future generations to better understand and appreciate.

Building on the platform of parasite ecology and evolutionary biology we have presented in previous chapters, we first examine the ways in which

Species names highlighted in red are included in the Rogues' Gallery, starting on page 429.

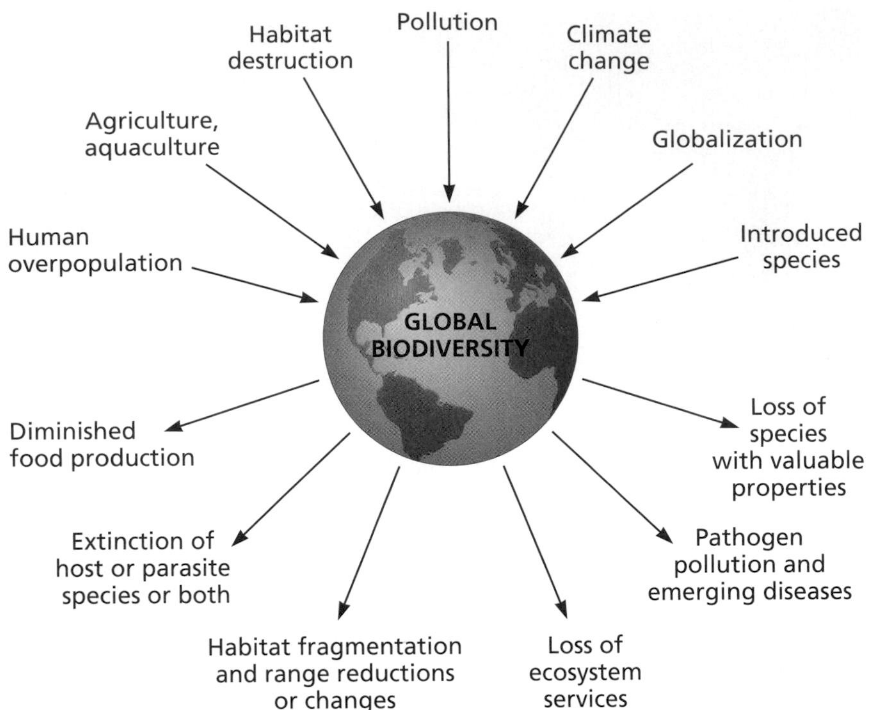

Figure 8.1 An overview of current factors influencing the world's biodiversity, with some of the potential consequences of the changes that are occurring.

parasites can negatively affect and alter the diversity of host species, particularly host species that are threatened. For sure, parasites can pose major problems for conservation biologists. We then discuss how the topic of parasitism relates to the pervasive and ongoing process of introductions of exotic species. A term to be aware of, and one that we will return to later in the chapter, is **pathogen pollution**. It is the introduction of pathogenic parasites into a new or naïve host species or population. Pathogen pollution is one aspect of a broader topic of concern not only for conservation biologists but for everyone, namely emerging infectious diseases. **Emerging infectious diseases** are caused by parasites previously unknown to science that suddenly make their presence known or by previously known parasites that become more common (increasing incidence and prevalence), that have expanded their geographic range (perhaps by human-assisted introductions), that have acquired new host species, or that have undergone evolutionary changes that have increased their virulence or transmissibility.

To avoid drawing too simple a conclusion that parasites are always bad, though, we also provide some examples of how parasites can be useful tools to monitor and assess the overall condition of natural habitats. We end the chapter by addressing the need to preserve parasite diversity. Although we have often gone to extraordinary lengths to extirpate parasites, and many might conclude that the only good parasite is a dead parasite, we must also remember that parasites themselves are a conspicuous part of the world's biological diversity. Parasites have unique, fascinating, and useful attributes, and many are today exceptionally vulnerable to extinction. Consequently, the case is made that parasites deserve just as much consideration for preservation as the charismatic megafauna that currently dominates conservation efforts.

Because we are often left trying to reconstruct catastrophic events that happened without our even knowing they were underway, it is helpful to have good baseline information from which to draw inferences. For parasites, such baseline information, at a minimum, would consist of knowledge about which parasite species normally infect which host species in particular locations. Even this rudimentary information is often lacking. Indeed, a significant challenge to conservation biology in general is to catalog the full diversity of species on Earth (see discussions in Chapter 2) before many slip into extinction without our ever knowing they existed. To paraphrase Wilson (2000), in order to care deeply about something important, it is first necessary to know it even exists.

8.1 SOME THEORY ABOUT PARASITES AND CONSERVATION BIOLOGY

As noted elsewhere in this book, parasites by definition have a negative impact on the fitness of their hosts. If you were charged with the preservation of an endangered host species, then parasites could very well become another major concern in your efforts to preserve the species. Although infectious diseases have been listed as among the top five causes of species extinctions, of 833 known extinction events since the year 1500, infectious diseases are listed as a contributing factor for less than 4%. Similarly, infectious disease is listed as a contributing factor in less than 8% of 2852 critically endangered listings of plants and animals. We first take a look from a theoretical point of view as to why parasites are often considered to pose less of a threat to host species extinctions, but then consider some situations where their impact may be greater than expected.

Theory often predicts parasites will not extirpate their hosts, but by no means always

Theory often predicts that in a parasite–host system, the parasite will coevolve with and thus coexist with the host rather than wipe it out. If the parasite diminishes the host population to a low level, then the prediction is often that the parasite will disappear before the host. That is, there is a threshold host population size below which the parasite will drop out. The parasite may for instance have trouble infecting new hosts if host density is very low. This general idea explains why parasites are often considered to be less likely than other causes, such as habitat destruction or fragmentation, to be responsible for species extinctions.

However, in some cases, even if there is a threshold population size of the host for the parasite, host extinction might still occur. For example, if a large host population is infected with a lethal parasite, transmission may be sufficiently fast owing to the commonness of the host such that all hosts become infected and the host population is extirpated as a result. **Figure 8.2** shows this possibility, using a basic Susceptibles (S) and Infecteds (I) model (see the flow chart in this figure and recall our discussion of modeling in Chapter 6). This graph is for a short-term epidemic in which births and deaths not resulting from disease in the host population can be ignored. The graph incorporates the following parameters into a couple of equations. The expressions $\mathrm{d}S/\mathrm{d}t$ and $\mathrm{d}I/\mathrm{d}t$ refer to changes in the number of susceptible hosts (S), or infected hosts (I) with time, respectively. We have encountered previously the terms β, the transmission rate, and α, parasite-induced mortality. The two equations are:

$$\frac{\mathrm{d}S}{\mathrm{d}t} = -\beta SI \qquad [8.1]$$

$$\frac{\mathrm{d}I}{\mathrm{d}t} = \beta SI - \alpha I \qquad [8.2]$$

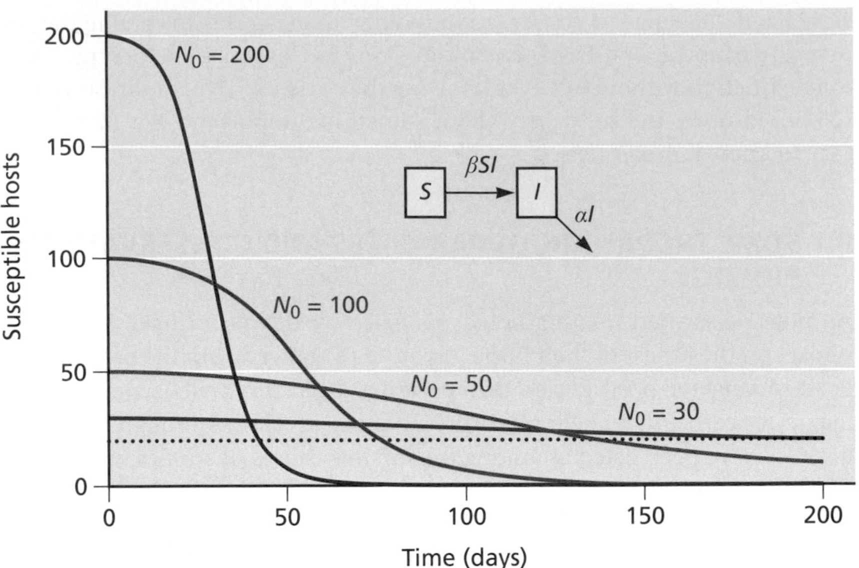

Figure 8.2 A virulent, rapidly transmitted parasite might cause extinction of a large host population. Even though a threshold host population size of 20 (dotted line) exists for parasite persistence, it does not prevent the host from going extinct when a virulent parasite colonizes a sufficiently large population in which transmission can be rapid. All hosts become infected before the population is reduced to below the threshold level that would otherwise cause the parasite to drop out. (Fisher MC et al [2012] *Nature* 484:186–194. With permission from Macmillan Publishers Ltd.)

Basically, the number of susceptibles will be diminished by conversion to infecteds, and the number of infecteds will be diminished by their mortality from infection. In Figure 8.2, the host threshold size is set at 20, $\beta = 0.001$ per individual per day, and parasite-induced death rate $\alpha = 0.02$ per day. Note that at smaller starting host population sizes (30 or 50), susceptible hosts persist, but at higher populations (100 or 200), susceptible individuals are extirpated. Note too that the presence of a threshold size does not prevent the host populations from going extinct under certain circumstances (large host population, lethal parasite, fast transmission among hosts).

Persistent parasite infectious stages may also favor demise of host populations

A second factor that can contribute to parasite-mediated host extinction is if the parasite in question has a free-living stage that can remain infectious to the host for a long time. In **Figure 8.3**, the fraction of the host population

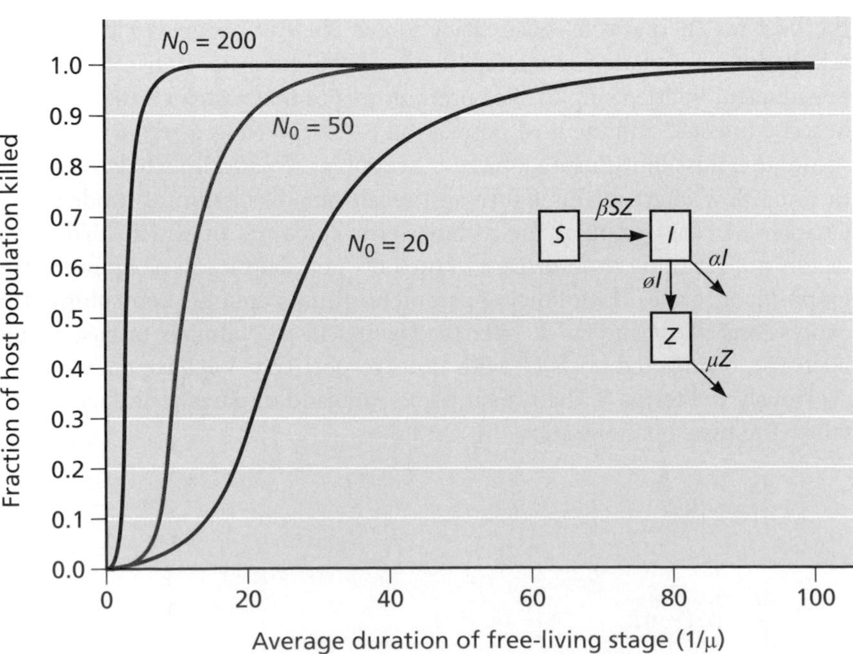

Figure 8.3 Persistent parasite infectious stages influence the likelihood of host species persistence. The longer the free-living parasite stages persist, the more certain host extirpation becomes, an effect increased in larger starting host populations N_0. (Fisher MC et al [2012] *Nature* 484:186–194. With permission from Macmillan Publishers Ltd.)

killed as a function of the duration the free-living stage's survival ($1/\mu$) is shown. In this example, $\beta = 5 \times 10^{-6}$, and the parasite-induced death rate, $\alpha = 0.02$. In addition, ϕ the rate of release of parasite propagules from the host = 10; Z = the number of parasite propagules; and μ = mortality rate of the free-living parasite stage. The equations relating these variables are more complicated in this case and are not shown, but the underlying flow chart is shown in the figure. Note that the longer the parasite can survive in the environment, the higher the proportion of the host population killed.

The presence of a parasite-tolerant host species may endanger a susceptible one

A final factor to consider that can be critical to determining if a susceptible host species (species B in this case) can survive a parasite is if another host species (species A) is present that is tolerant of the parasite. Species A serves as a continuous source of infection for the susceptible host species. In **Figure 8.4**, species A is tolerant, and species B suffers mortality at a rate α_B. As shown in the graph, species B will go extinct at higher densities of A. Also the higher the transmission rate, β, the more likely species B is to go extinct.

Imagine from Figure 8.4 that the susceptible species is one that has a relatively small population size, so much so that it is considered endangered. The presence of the abundant, less-affected reservoir host species can maintain the parasite even as the endangered host species becomes rarer or even goes extinct. For this reason, parasite-mediated extinctions are considered more likely when a generalist parasite (one able to infect multiple host species) colonizes a naïve susceptible host species, especially when that host population is small and lacks diversity. It is then not surprising, as we shall see, that host species on islands are particularly vulnerable to extinctions mediated by generalist parasites maintained by other hosts, often ones that have been introduced. In addition, do not forget the scenarios presented in Figures 8.2 through 8.4 from the parasite's point of view. If the parasite is host specific, it too risks extinction if the numbers of its only host species drop too low.

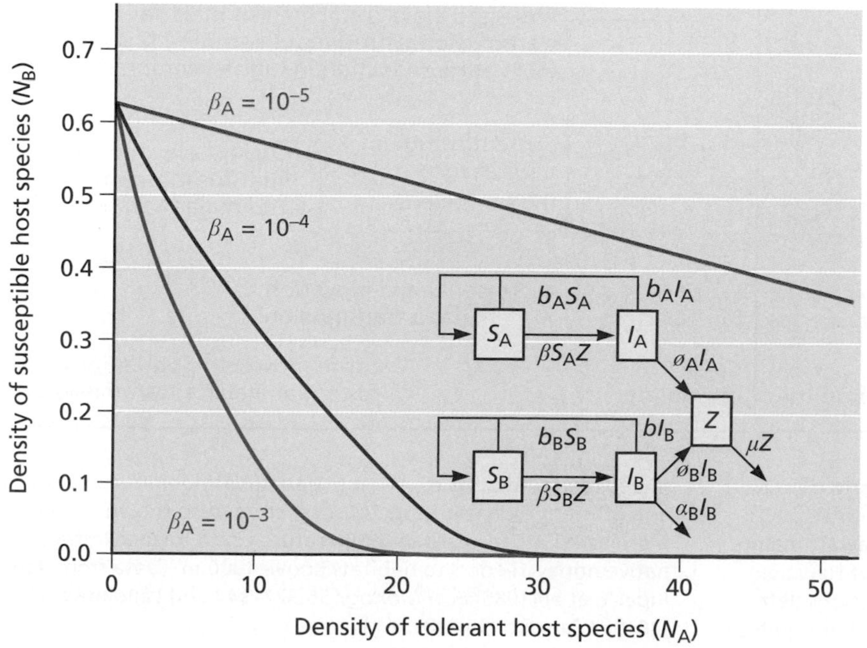

Figure 8.4 Presence of a host species that reliably produces a parasite can cause the demise of a second species vulnerable to that parasite. In this case, the presence of a tolerant host species, A, can serve as a continuing source of infectious stages of the parasite for a susceptible species, B. The more individuals of species A present and the higher the transmission rate, the more likely the susceptible species is wiped out. (Fisher MC et al [2012] *Nature* 484:186–194. With permission from Macmillan Publishers Ltd.)

8.2 PARASITES INFLUENCE EFFORTS TO PRESERVE HOSTS

With this theoretical foundation, we will now look at several actual situations in which the effects of parasites in causing extinctions have been or are considerable, or may prove to be.

Parasites can cause extinction of host species

One of the most frequently cited examples of the role of parasites in species extinctions is provided by the loss of as many as 17 honeycreeper (subfamily Drepanidinae) bird species endemic to the Hawaiian Islands (**Figure 8.5A**), with 14 more species listed as endangered or critically endangered. Although a suite of interacting factors such as over-collecting, habitat loss, and introduced predators or competitors may have contributed to their decline, a significant reduction in their abundance began after the arrival of mosquitoes, avian malaria (*Plasmodium relictum*) and avian poxvirus (*Avipoxvirus*) to the islands. The islands were free of mosquitoes until 1826 when the southern house mosquito, ***Culex quinquefasciatus*** was introduced in water casks as visiting ships stopped to collect fresh water supplies. The dates of arrival of both avian pox and malaria are unknown, but the former probably arrived by the late 1800s, as assessed by amplification of viral sequences from museum specimens dating from this time. Avian pox is transmitted mechanically both through direct contact and by the bite of mosquitoes. The mosquito-transmitted *Plasmodium relictum* was probably present in Hawaii by the early twentieth century, believed to have been imported when at least 50 species of birds originating from around the world were introduced to the islands. The role of these birds in also maintaining malaria reminds us once again of the potential devastating role of tolerant hosts, as pointed out in Figure 8.4. Malaria parasites were first detected in endemic Hawaiian birds in the 1930s. An indication of the potentially devastating impact of introduced *P. relictum* and avian pox on these naïve and long-isolated native bird populations came

A

B

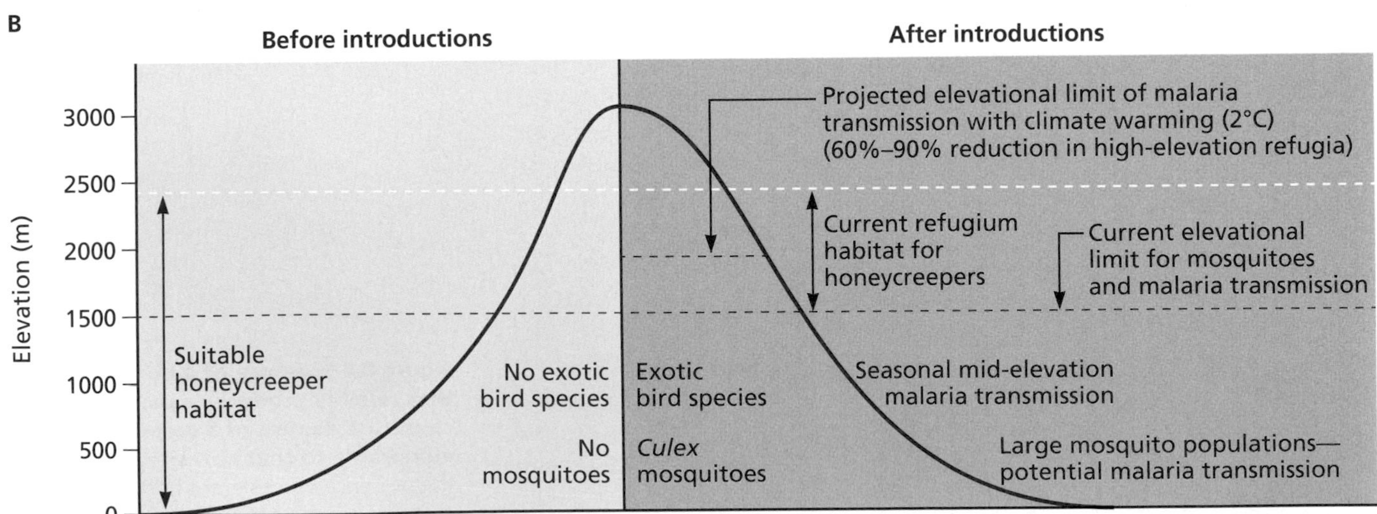

Figure 8.5 Avian malaria and elevational range of Hawaiian honeycreepers. (A) Example of an extinct Hawaiian honeycreeper, *Drepanis pacifica*, the Hawai'i mamo. (B) A graph showing how the elevational range of Hawaiian honeycreepers has changed from before introductions (left side of figure) to after introductions (right side) of mosquitoes (*Culex quinquefasciatus*), avian malaria (*Plasmodium relictum*), and efficient reservoir hosts for *P. relictum*. Also shown is how a projected 2°C increase in temperature could further restrict native honeycreepers to habitats above 1900 m. (Data from Van Riper C et al [1986] *Ecol Monogr* 56:327–344 and LaPointe DA [2010] *J Parasitol* 96:318–324.)

in the 1950s when experimental exposures to local mosquitoes showed the birds readily succumbed to malaria and avian pox. Similar problems have been noted for endemic birds of the Galapagos Islands.

Although evidence for a direct role of these agents in native bird extinctions is scant, native honeycreepers have been shown to be susceptible to both avian pox and malaria infections, and the birds became rare in the Hawaiian lowlands (Figure 8.5B) where mosquito populations and biting rates are high and the potential for *P. relictum* transmission is high (assuming susceptible bird species are present). At higher elevations (>1500 m) where mosquito populations are absent or rare, at least some of the endemic bird species have persisted. Interestingly, another factor contributing to higher elevations acting as refugia is that malaria sporozoite development in mosquitoes slows below 15°C and ceases below 13°C. The 15°C isotherm in Hawaii closely follows the 1500-m contour line. The negative correlation between infected mosquito and native bird abundances indirectly implicates *P. relictum* in extirpating birds from lower elevations.

Several factors continue to work against the survival of endemic Hawaiian birds. One is that some native bird species are able to support persistent malaria infections without succumbing and thus serve as an ongoing source of infection for mosquitoes. A second factor is that malaria and avian pox infections often co-occur in the same individual bird. The two pathogens may interact to further weaken the host and, at the very least, make it harder to disentangle ultimate causes of mortality. A third factor involving yet another introduced species, feral pigs, also works against the endangered birds because the pigs knock over tree ferns and eat the starchy pith at the center of the trunks, leaving trough-shaped cavities where water collects and mosquito larvae can thrive. A fourth factor is that global warming could favor an upward elevational shift of the mosquito belt, either totally eliminating or further reducing the size of high-elevation refugia (Figure 8.5B). To counter these depressing developments, some endangered endemics such as the Hawaii ʻamakihi (the honeycreeper, *Hemignathus virens virens*) have become more common, even in the mosquito-rich lowlands and may have sufficient genetic variation to favor the emergence of individuals resistant to the pathogenic effects of malaria or avian pox. Whether rarer species with less genetic diversity at their disposal can do the same is unclear.

A less spectacular but still instructive example of parasite-mediated extinction also discussed in Chapter 7 is worth a mention here too. The tree snail *Partula turgida* was endemic to the Society Island of Raiatea. In well-intended and pragmatic efforts to save this species, the last remaining specimens known from the wild were collected for laboratory propagation in 1991. This population declined and by 1996 the last known specimen died in the London Zoo. As the final nail in the coffin, infection of the laboratory-propagated individuals by the microsporidian *Steinhausia* sp. resulted in the demise of the snail. Once again, a small population of hosts endemic to an island was involved. One lesson here is that long-term propagation of such a species in laboratory circumstances may unwittingly create peculiar conditions of exposure to a generalist parasite that can have unfortunate consequences.

Parasites work in concert with other stressors to affect hosts

One factor leading us to underestimate the potential impact of parasites is that they may work in conjunction with a number of environmental stressors to affect host species. These include general organic pollution, heavy metals, habitat degradation, elevated temperature, and crowding—factors that have been called anthropogenic stressors. Recall that the Hawaiian honeycreepers have been affected by habitat loss as well as parasites. Together parasites and these stressors may exert a combined effect on a host species greater than that achieved by any single factor. Suppressed immune competence resulting either from the environmental stressors or the parasites themselves may contribute to this effect. Although caution is required in overgeneralizing, a common outcome of field and lab studies is to document that parasitism and other stressors have additive effects on the potential to cause mortality of a host species or to affect certain biomarkers such as blood cell count or level of oxidative stress differently than if acting alone.

Consider this simple experimental example. If guppies (*Poecilia reticulata*) are exposed to increasing levels of zinc up to 240 μg/l, they suffer negligible mortality (**Figure 8.6**). However, if zinc-exposed guppies are infected with a small standard dose of the **monogenean** ectoparasite *Gyrodactylus turnbulli*, guppy mortality increases linearly with increasing zinc concentration, with up to 80% mortality at high zinc concentrations.

A more poignant example of how parasites acting in conjunction with other stressors can affect hosts is demonstrated by the plight of the world's amphibian species. Thirty percent of all amphibian species are listed as threatened, with some species such as the golden toad (*Bufo periglenes*) of Central America (**Figure 8.7A**) already extinct. Climate change, land-use changes, and infection with the chytrid fungus *Batrachochytrium dendrobatidis* (see Box 8.1 for more about this killer fungus) all are contributing and, in many places, are overlapping in their impact (Figure 8.7B). Amphibian declines are likely to accelerate in the future because multiple drivers of extinction may jeopardize amphibians more than previous monocausal assessments suggested. For this reason, parasites may have more of an impact on tipping the balance against a host than might normally be considered.

The impact of parasitism is influenced when hosts occur in small or fragmented populations.

Endangered host species are often greatly reduced in numbers in the areas they occupy and may be forced to exist in smaller and fragmented populations. How do characteristics of hosts' populations affect the impact that parasites may have upon them? As we have noted above (see Figure 8.2), such populations might experience fewer parasites than larger populations, in part because the rareness of the host may diminish transmission past a threshold point. A comparison of 36 threatened and 81 nonthreatened primate species,

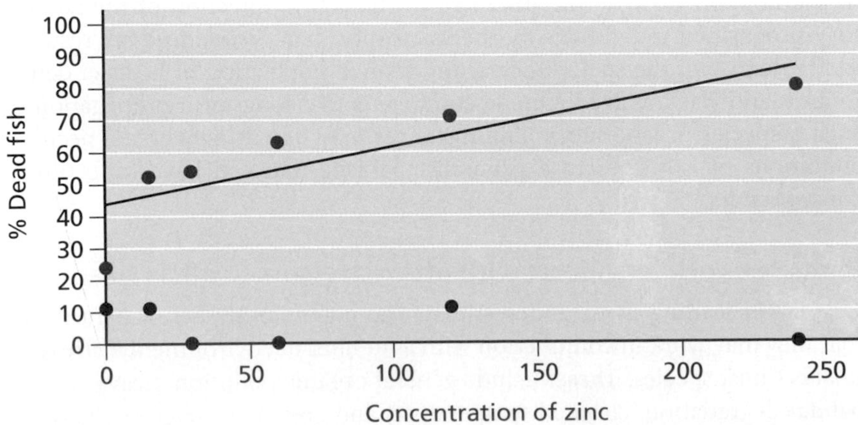

Figure 8.6 Parasites can act in concert with other stressors to kill hosts. Shown is the mortality of uninfected fish (red circles) or fish infected with three *Gyrodactylus turnbulli* (blue circles) exposed to increasing concentrations of zinc (Zn). (Gheorgiu C et al [2006] *Parasitology* 132:225–232. With permission from Cambridge University Press.)

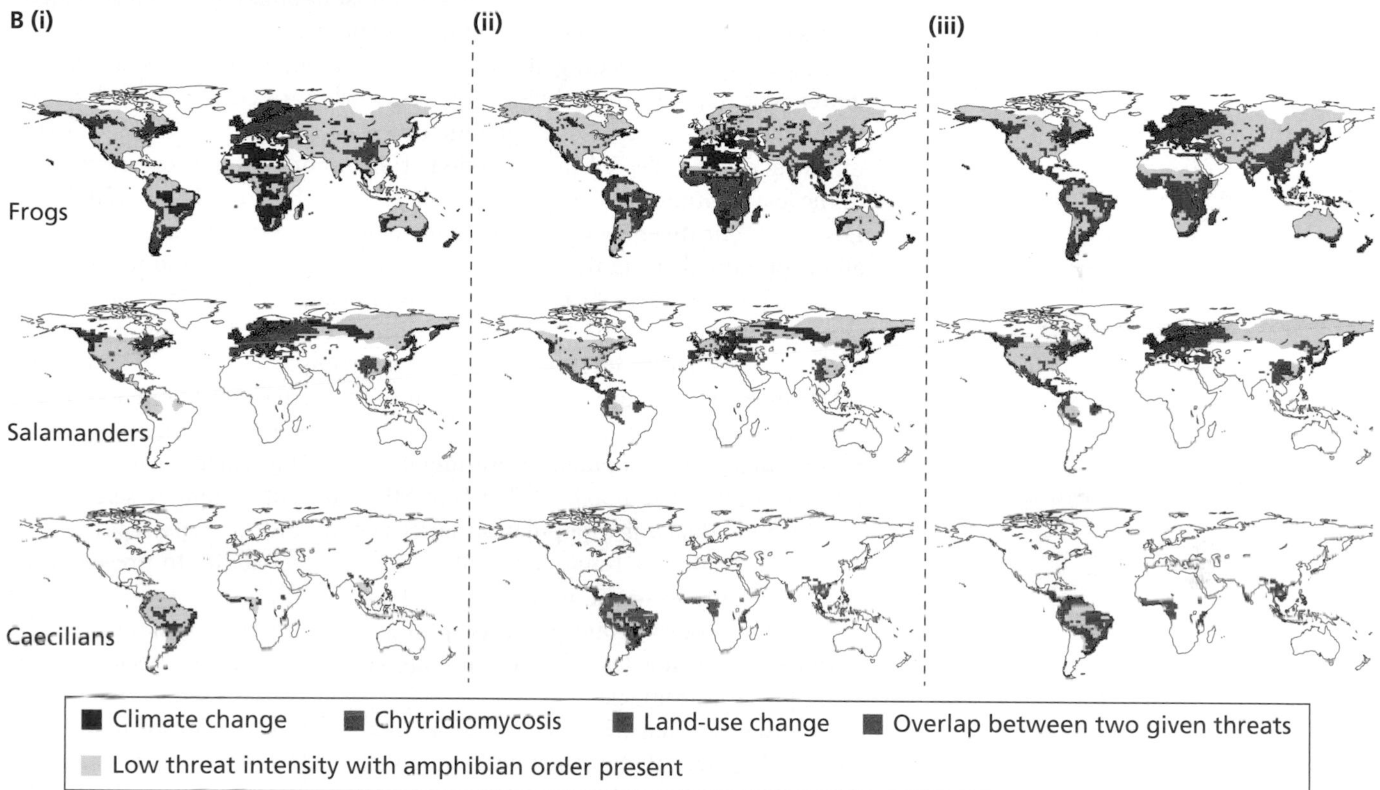

Figure 8.7 Extinction and decline of amphibians mediated by a parasite acting in conjunction with other stressors. (A) The extinct golden toad (*Bufo periglenes*) of Central America, believed to have been extirpated by chytriodiomycosis caused by *Batrachochytrium dendrobatidis*. (B) Spatial distribution and pairwise overlap of the three main factors threatening global amphibian biodiversity, projected for the year 2080. (i) Climate change and chytridiomycosis. (ii) Climate change and land-use change. (iii) Chytridiomycosis and land-use change. Colors indicate areas of particularly high threat intensity. Note areas of overlapping threats shown in yellow. (A, Courtesy of Charles H Smith, US Fish and Wildlife Service. B, From Hof C et al [2011] *Nature* 480:516–519. With permission from Macmillan Publishers Ltd.)

after controlling for host phylogeny and uneven sampling, found that threatened primates had a lower richness of viruses, protists, and helminths.

This finding supports the idea that smaller, isolated populations support fewer parasite species. It was concluded that human activities and host characteristics that increase the risk of extinction also may lead to the simultaneous loss of parasites. The loss of parasites as a consequence of extinction of hosts is an example of coextinction, a topic we return to later.

Not all studies reach the same conclusion, though. For example, the endangered red colobus monkey (*Procolobus rufomitratus*) exists in both fragmented and unfragmented forest habitats in Kibale National Park, Uganda. Studies of seven nematodes, one cestode, and three protists revealed that monkeys from fragmented populations had higher parasite prevalence and greater numbers of multispecies infections than monkeys from unfragmented habitats. Infectious stages of the parasites were also found to be more numerous in the fragmented habitats. Hosts forced to use small habitat patches may intensify contamination of such habitats with parasite infective stages (see Figure 8.3). Also, hosts in populations that are small may be stressed and consequently may be immunocompromised, or because their numbers are small, as we discuss next, they may lack the genetic diversity to evolve tolerance to parasites over time.

Parasites can strongly affect hosts with reduced genetic variation

Some host populations by virtue of suffering losses in numbers consequently experience loss of genetic diversity. This diversity is needed to withstand both short-term environmental perturbations and, on longer time scales, to evolve in response to environmental changes (recall the example of the Hawaii ʻamakihi mentioned earlier). Small host populations can lose genetic diversity as a result of random events (genetic drift; see Chapter 7) or as a consequence of inbreeding if the diversity of potential mates is limited. **Inbreeding** refers to reproduction by mates who are closely related, which leads to homozygosity, loss of genetic diversity, and increased chances that offspring can be affected by deleterious traits. **Inbreeding depression** refers to the loss of fitness in a population from the breeding of related individuals. Loss of genetic diversity might have a general effect that ultimately limits the ability of individuals in the population to mount effective immune responses or it might more specifically affect immune genes directly responsible for recognizing and resisting parasites. For instance, recall from Chapter 4 that proteins encoded by loci of the major histocompatibility complex (MHC) play a major role in antigen presentation in the vertebrate adaptive immune response. MHC genes are highly polymorphic and many variant alleles are present in a typical population. Although we noted in Chapters 4 and 7 that optimal (rather than maximal) levels of MHC diversity might be selected for within individuals, if a population lacks genetic diversity, its repertoire of MHC alleles may be limited, potentially limiting its ability to present antigens derived from some parasites. The relative roles of general inbreeding depression, and of reduced diversity in specific immune loci such as the MHC are often not teased apart in conservation studies of species endangered by parasite attack.

Is it the case that when overall genetic diversity is low, a host population may lack genetically distinct individuals able to resist a particular parasite? Several examples suggest this is a very real concern for endangered host species. With respect to vertebrates, the Gila topminnow (*Poeciliopsis occidentalis occidentalis*) has suffered marked reduction in numbers in its native habitat, the Gila River of Arizona and New Mexico, and is listed as an endangered species. Competition from introduced fish species such as the mosquito fish (*Gambusia affinis*) is at least in part responsible for its decline. Also introduced into this habitat is the monogenean fluke, *Gyrodactylus turnbulli*, a parasite of guppies from Trinidad. The most homogenous of the topminnow populations show the greatest prevalence of infection with *G. turnbulli* and are also most likely to succumb to infection. Inbred fish are also more likely to succumb to infection than outbred fish (**Table 8.1**). In topminnow populations known to be homozygous at MHC loci, survivorship following infection was lower than for populations with MHC heterozygosity. The low overall variability of the Gila topminnow in general makes this species more vulnerable to the exotic parasite than related host species that are more diverse.

Consider too an example provided by the isolated island population of Soay sheep *Ovis aries* that we also discussed in Chapter 6. These sheep suffer

Table 8.1. Inbred topminnows *Poeciliopsis occidentalis occidentalis* are more vulnerable than outbred individuals to the monogenean fluke *Gyrodactylus turnbulli*. The mean number of flukes on fish that had flukes, for both inbred and outbred fish, from Cienega Creek, Arizona, at five-day intervals is listed. *N* is the sample number of fish. The dash indicates there were no fish with flukes. Mortality was 69% for inbred fish and 31% for outbred fish.

Population	Measure	Number of Flukes				
		Day 5	Day 10	Day 15	Day 20	Day 25
Inbred	Mean	5.5	12.3	29.8	35.3	50
	N	8	7	5	3	1
Outbred	Mean	4.9	5.9	3.3	–	–
	N	11	9	3	–	–

Simplified from Hedrick PW et al (2001) Parasite resistance and genetic variation in the endangered Gila topminnow. *Anim Conserv* **4**:103–109, Table 4. With permission from John Wiley and Sons.

from the stomach nematode *Teladorsagia circumcincta* (previously known as *Ostertagia circumcincta*). Relatively inbred individuals, as assessed by microsatellite profiles, are more susceptible to infection, and when heavily infected are more likely to die in harsh winter months (**Figure 8.8**). This study also showed that by selectively killing more homozygous individuals in a population, parasites can promote and maintain genetic diversity of their hosts. The role of parasites as drivers of diversity is discussed further in a later section.

Similar dynamics may also be at work for invertebrate hosts, with an unfortunate example being provided by the widespread population reductions of bumblebees in both North America and Europe (see also Box 1.1 regarding honeybees). Because of their role as pollinators, the decline of bumblebees has several potential cascading effects on ecosystems. One study showed that populations of the bumblebee species *Bombus muscorum* were more likely to harbor large populations of the kinetoplastid flagellate *Crithidia bombi* if their overall level of genetic diversity was low. It was concluded that as host populations lose heterozygosity, the impact of parasitism would increase, pushing the bee populations closer to extinction. The decline of some species of bumblebees in North America has also been associated with low levels of genetic variation and the success of the introduced microsporidian parasite, *Nosema apis*, which may have escaped from commercial to wild bees.

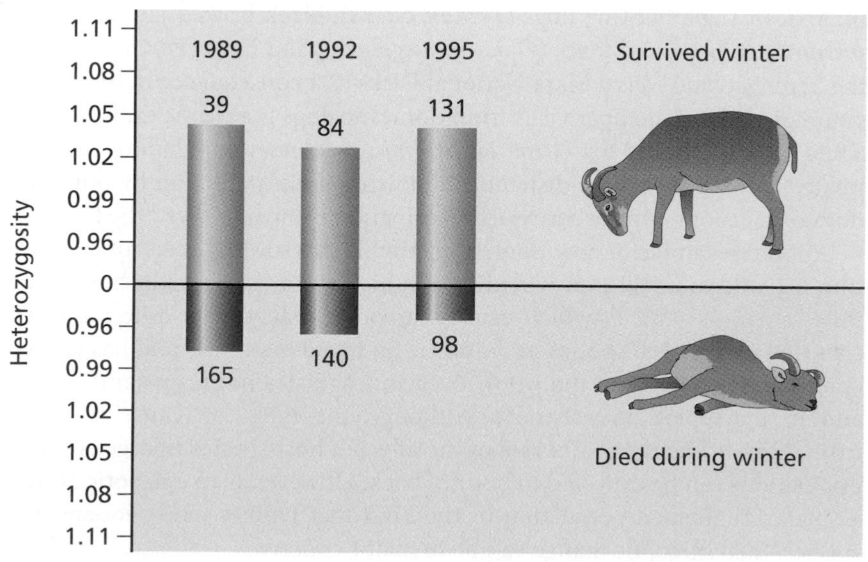

Figure 8.8 Genetic heterozygosity and winter survivorship of sheep infected with the nematode *Teladorsagia circumcincta*. Mean individual heterozygosity of Soay sheep that either died (red columns) or survived (blue columns) years of high overwinter mortality. Numbers above or below columns indicate sample size. (Adapted from data found in Coltman DW et al [1999] *Evolution* 53:1259–1267. With permission from John Wiley and Sons.)

Whether *N. apis* has merely been favored by low genetic diversity in some bee populations or if it is actually responsible for the decline in some bee species remain subjects for future work.

Captive host populations are often very vulnerable to parasites

The loss of habitat may require that some rare host species be propagated in captivity, and this carries with it its own risks and complications. Recall the extinction of the tree snail *Partula turgida* discussed above. Similar concerns apply for the remaining 20 or so species of *Partula* occurring solely in captivity. Two other examples highlight particular difficulties with maintaining endangered hosts in zoos. The first pertains to the masked bobwhite quail (*Colinus virginianus ridgwayi*) of the American southwest. Quail rearing efforts have been frustrated by the acquisition of life-threatening mosquito-borne infections of ***Plasmodium*** and *Haemoproteus*. The presence of other captive bird species or of wild bird species using the zoo area, both serving as reservoir hosts and sources of infections for mosquitoes, increase the risk of acquiring such infections.

Ring-tailed lemurs (*Lemur catta*) point out another difficulty. The lemurs are endemic to the relatively isolated island of Madagascar. They are listed as vulnerable to extinction because of widespread habitat loss on their native island. Zoo-reared lemurs have proven to be extraordinarily susceptible to infections with ***Toxoplasma gondii***, a generalist parasite species with which they have had no previous contact. Transmission of ***T. gondii*** to lemurs in zoos may be facilitated by contamination of food with oocysts originating from either zoo-maintained wild cat species or by feral house cats on zoo grounds.

Parasites are frequently transferred from abundant host species to rare relatives, including from humans to our great ape cousins

An additional scenario that can cause problems for endangered and restricted wild host species occurs when they live in proximity to related domesticated species that are members of large, thriving populations that have acquired and experienced parasites from many areas (recall Figure 8.4). As noted in Figure 7.20, about 80% of domesticated livestock and carnivore parasites can infect multiple host species, not surprisingly often close relatives. The demise of African Cape hunting dogs (*Lycaon pictus*), black-backed jackals (*Canis mesomelas*), bat-eared foxes (*Otocyon megalotis*), and lions (*Panthera leo*) in the Serengeti and Masai Mara National Parks as a consequence of the acquisition of canine distemper virus from domestic dogs is a classic example. The Santa Catalina Island fox (*Urocyon littoralis catalinae*) population was decimated by 90% by canine distemper following its introduction by domestic dogs or raccoons. Only a concerted vaccination campaign saved the foxes.

Another example of how domestic animals can sow the seeds of destruction for wild animals is provided by the highly contagious skin-burrowing mite ***Sarcoptes scabiei***, which causes sarcoptic mange. The mite has been reported in over 100 species of domestic and wild mammals and has caused epizootics in carnivores the world over, in ungulates in Europe and Africa, and in marsupials in Australia. Although the mite can cause localized extinctions, if the range and size of the affected host species is large enough, populations can be expected to bounce back. However, if an epizootic afflicts a small, fragmented population of the kind that typifies many endangered species, then sarcoptic mange can be of major concern.

Among the mammals afflicted and potentially killed by the scabies mite are mountain gorillas, *Gorilla gorilla beringei*, and chimpanzees, *Pan troglodytes*. The mites are widely believed to be acquired by wild apes when they have repeated close encounters with people. Along with mites, human respiratory viruses such as measles, respiratory syncytial virus, and human metapneumonia virus can also infect and kill great apes. Human feces also provide a source of infection for apes, spreading human gut bacteria and helminths and even a possible case of polio.

Orangutans (*Pongo* sp.) have suffered population reductions and habitat loss in their native range in Borneo and Sumatra and are listed as endangered. They are often infected with intestinal nematodes of the genus *Strongyloides*, which can develop proliferative infections and may represent a major cause of death in captive individuals. In Borneo, both wild and captive orangutans and a human caretaker were shown to share *S. fuelleborni fuelleborni* worms, suggestive of the possibility they traded worms back and forth. *S. stercoralis*, primarily a human parasite, was also found in one orangutan there.

Our ape cousins can also be a source of human infections, as in the case of the human immunodeficiency virus (HIV-1), which was believed to have been derived from a simian immunodeficiency virus (SIV) transmitted when humans contacted blood from chimpanzees killed for bushmeat in the early twentieth century. As discussed in Chapter 2, our most recent evidence suggests malaria (***Plasmodium falciparum***) too has been acquired from gorillas, although chimpanzees and bonobos (*Pan piniscus*) both harbor ***Plasmodium*** species similar to ***P. falciparum***. An ancestor of our pubic louse, ***Pthirus pubis***, was also likely acquired from gorillas It is likely we have been trading infections with related ape species for a long time, though the concern in the modern world is that humans have become so abundant relative to our fellow apes. Consequently, the flow of pathogens is very likely to be in a direction that could further contribute to their demise in the wild. The emphasis in this section has been on the mostly one-way flow of parasites from common to rare species. We will return to the subject of spillover of parasites from introduced hosts in a later section.

Farming can pose parasite problems for wild host species

Intensive human agriculture and aquaculture efforts often have parasite-related effects that ripple on to affect nearby wildlife species. One prominent example is the intensive rearing of salmon in net pens in marine waters around Canada and Europe, often situated in or near wild salmon migratory routes. In British Columbia, the salmon louse *Lepeophtheirus salmonis* (**Figure 8.9**) is an indigenous parasitic copepod (a type of crustacean) that lives on mucus, skin, and blood of both wild and farmed salmon. Following two free-swimming nonfeeding stages, a copepodid stage attaches to a fish to

A

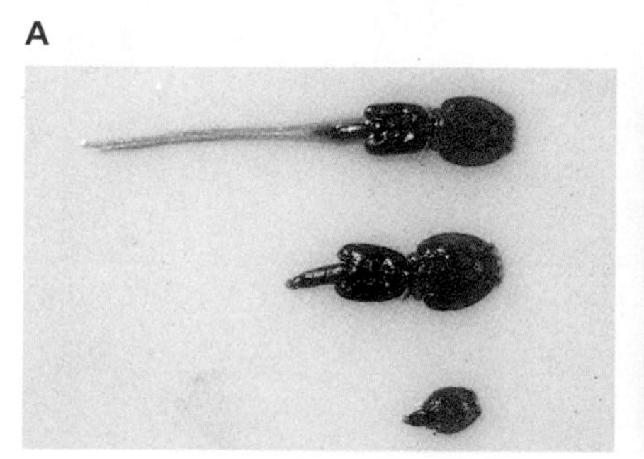

B

Figure 8.9 The salmon louse *Lepeophtheirus salmonis*. The salmon louse is a parasitic copepod, which is a type of crustacean. (A) Shown from top are a mature female with egg strings, a female without egg strings, and an immature stage. Each egg string can contain from 500–1000 eggs and a single female may produce several strings in her life. (B) Note the presence of *L. salmonis* on this young fish. (A, Courtesy of Thomas Bjørkan, CC By-SA 3.0. B, courtesy of Alexandra Morton.)

initiate an infection and begins to feed. After passing through four chalimus stages that feed on the fish's skin, the louse then passes through two preadult stages and then becomes either a male or female copepod. Females in particular are likely to feed on the host's blood. Infestations can diminish growth or kill salmon. Transmission of salmon lice is favored by the close contact among the farmed fish because copepodids can readily swim to and infect fish, and preadult and adult lice can also move between fish. Predation can also transfer the lice from prey to predator. These large populations of lice from pens are suspected of colonizing wild salmon stocks that swim near the penned salmon and salmon louse abundance on farms has been negatively associated with productivity and survival of wild salmon stocks. Thus the abnormal concentration of salmon in pens has created a situation that favors parasite transmission and threatens native salmonids. These artificially high densities of penned fish in turn require implementation of a number of control methods such as leaving pens fallow, applying chemicals that kill the louse, and re-situating pens such that native stocks are not exposed to parasites and thus even further endangered.

Parasites of an iconic symbol—the giant panda—point out our need to know more

One of the very symbols of the conservation movement, the giant panda (*Ailuropoda melanoleuca*), serves to conclude this section because it points out many of our concerns and uncertainties regarding parasites and endangered species. Giant pandas have in recent years been threatened increasingly by infection with an ascarid nematode, *Baylisascaris schroederi* (**Figure 8.10**), the larvae of which undergo extensive migrations (called visceral larval migrans) through sensitive panda tissue and can cause death. The adult worms living in the panda's intestine can also cause obstructions. Even in a high profile case like this, we are not sure why *B. schroederi* is now more of a problem for pandas than in the past. Are new ecological circumstances, perhaps reduced or fragmented habitats, involved? Are there other hosts—perhaps introduced hosts—that facilitate maintenance of the parasite? Also, as there are cases of other hosts, such as the North American Allegheny wood rat *Neotoma magister*, being imperiled by related *Baylisascaris* parasites, it serves to remind us that the parasites of relevance to conservation biology are diverse. We have highlighted roles of viruses, fungi, or ***Plasmodium***

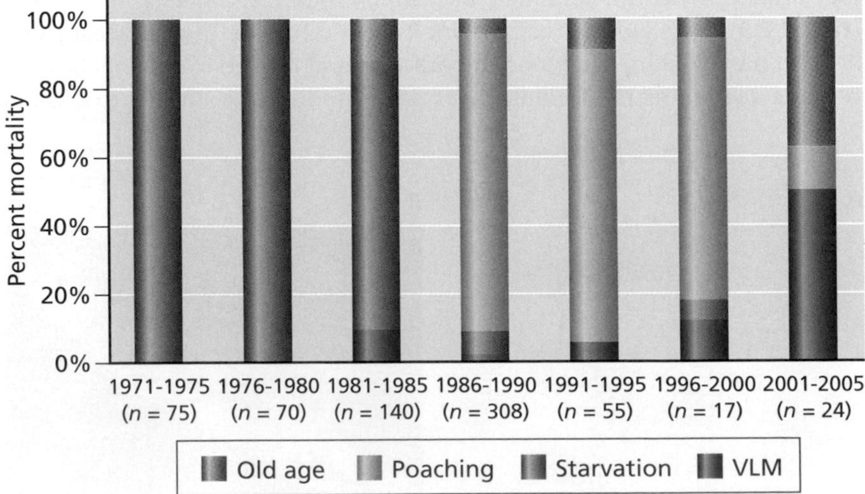

Figure 8.10 The increasing involvement of parasites in panda mortality. Note that by 2001–2005, half the pandas investigated were believed to have succumbed to visceral larval migrans (VLM) caused by *Baylisascaris schroederi*. (From Zhang JS et al [2008] *Ecohealth* 5:6–9. With permission from Springer Science + Business Media.)

parasites in threatening host species, but even macroscopic helminths can get into the act as well.

8.3 DANGERS RESULTING FROM SPECIES INTRODUCTIONS

The rate at which species of all kinds are transported beyond their native ranges is alarmingly high, aided by our globalized economy. Intercontinental transportation of exotic organisms occurs by air, land, and sea, for purposes as diverse as recycling of tires, aquaculture, distribution of food, the ornamental animal and plant trade, recreational boating, biological control operations, or laboratory use. Introductions are both deliberate, as is often the case for aquaculture, or accidental, as in the ballast water of ships or as airplane stowaways. Introductions often cause catastrophic and unforeseen effects and represent a major threat to maintaining the integrity and diversity of natural ecosystems, so much so that we must resign ourselves to the reality that the world's ecosystems have been irrevocably changed. After habitat loss, introduction of exotic species has been ranked as the second most immediate cause of the loss of biodiversity worldwide. One study notes that invasive species are often generalists; they frequently displace specialist species and consequently have an additional effect of diminishing the rate at which new, specialized species are formed. Among the many biological entities frequently introduced into new settings are parasites, and as noted in the introduction, this has been called pathogen pollution. One potential ramification of introductions, whether stemming from pathogen pollution directly or from more indirect changes imposed on native parasites or their hosts, is the emergence of new infectious diseases (**Box 8.1**).

BOX 8.1

Challenges posed by emerging fungal infectious diseases for conservation biologists

Among the most prominent of today's emerging parasites are several kinds of fungi. Emerging fungal diseases affect groups of animals as diverse as soft corals, bees, amphibians, bats, and people. From 1995 to 2010, fungi have increased from 1% to 7% of emerging disease alerts. Emerging rust fungi affecting both wheat and coffee may well adversely affect what appears on your dinner table in coming years.

Of the many emerging fungal diseases of animals, two stand out as particularly worthy of concern. Ironically, in a world that is warming, both thrive in cooler locations. The first is *Batrachochytrium dendrobatidis* (Figure 1A), a chytrid fungus (see Figure 8.7), first shown in 1997 to cause skin infections (chytridiomycosis) in amphibians. This organism has been called "possibly the most deadly invasive species on the planet (excluding humans)" (Rohr et al, 2008). It is the first chytrid shown to infect vertebrates. The chytrid kills amphibians by inhibiting electrolyte transport across the skin, disrupting sodium and potassium levels, and causing cardiac arrest.

Nearly half of the world's amphibian species are now in decline as a result of chytridiomycosis, although other factors such as iridoviruses, climate change, and habitat fragmentation certainly play roles too. How did we arrive at this very sad state of affairs? The spread of *B. dendrobatidis* was aided by the global trade in North American bullfrogs (*Rana catesbeiana*) and African clawed frogs (*Xenopus laevis*). Both species are infected with *B. dendrobatidis* yet are tolerant of its effects, and they serve to broadcast the chytrid to new habitats. Furthermore, the worldwide spread has been dominated by a single lineage, one believed to be more virulent than other known lineages.

Can we stop the insidious spread of the chytrid and protect our remaining amphibians? Natural selection will favor the emergence of resistant variants in some amphibian species, but other species with smaller and genetically homogeneous populations are much more vulnerable. One encouraging finding is that bacteria sometimes found on the skin of amphibians produce antifungal peptides that can limit chytrid growth. The chytrid cannot grow in 25–37°C water, prompting suggestions that reintroductions of amphibians might occur in habitats that allow amphibians access to warmer water. Anti-chytrid drugs are also being tested but, as yet, there are no consistently reliable cures.

The second fungus of concern burst on the scene in 2006 when hibernating bats in New York caves were found with a fungus growing on their muzzles, ears, and wing membranes. This affliction was dubbed white-nose syndrome

BOX 8.1 (continued)

Challenges posed by emerging fungal infectious diseases for conservation biologists

A

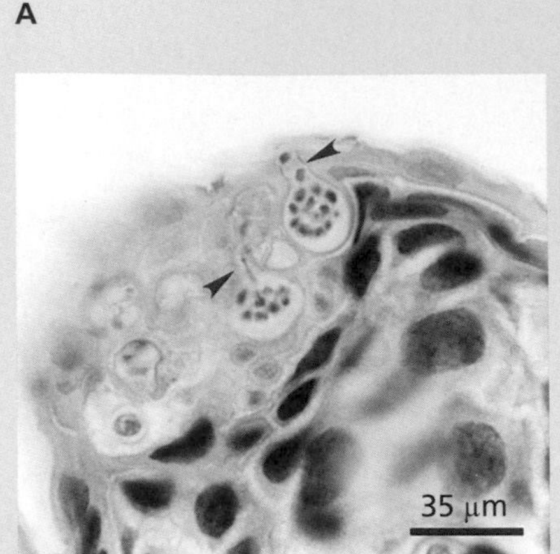

B

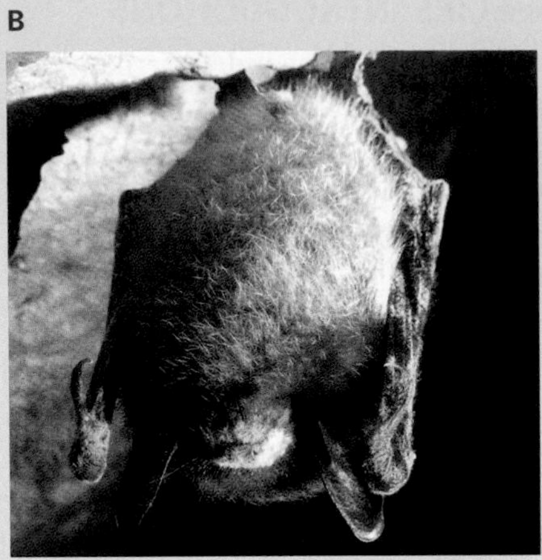

Figure 1 Two prominent emerging fungi. **(A) Two flask-shaped, spore-containing bodies (sporangia) of *Batrachochytrium dendrobatidis*, each containing numerous zoospores, are seen in this section of the skin of *Atelopus varius*, the Costa Rican variable harlequin toad. Exiting zoospores can be seen in the discharge tubes (arrows) of each sporangium. The zoospores swim in water and can initiate infection in the skin of another animal. (B) White-nose syndrome caused by *Pseudogymnoascus destructans* in a little brown bat, *Myotis lucifugus*. (A, Courtesy of the CDC. B, Courtesy of Marvin Moriarty, US Fish and Wildlife Service.)**

(WNS), now known to be caused by the ascomycete fungus, *Pseudogymnoascus destructans* (formerly called *Geomyces destructans*). Related fungi are common in soils of temperate and high-latitude ecosystems. By 2013, WNS had spread to 22 states in the USA and five Canadian provinces. As many as 5.5 million bats of several species are now believed to have died, and one species, the little brown bat *Myotis lucifugus* (Figure 1B), faces local extinction. The fungus grows on the cool (7°C) skin of bats in underground hibernacula. Fungal infections are believed to disturb the bat's state of torpor, raising their body temperatures, and burning up fat reserves needed to complete hibernation. As a consequence, they become emaciated and die.

Why should North American bats suddenly be succumbing to fungal infections? In one study, bats were inoculated with *P. destructans* isolates from Europe and, in contrast to control bats, developed WNS. This finding suggests the pathogen may have invaded from Europe, possibly on the boots or equipment of bat researchers or cavers. With their ability to undertake long-distance movements, bats can rapidly spread this infection to new locations in North America. Once again we are left wondering how to cope. Answers are scarce—we are still in the midst of an expanding epizootic.

What general lessons do these two examples teach us? First, human-facilitated range expansions have played a role in both cases. Also, an evolutionary change resulting in increased virulence seems to have occurred in *B. dendrobatidis*. Fungi often have broad host ranges (*B. dendrobatidis* can infect over 500 species), and some host species can serve as persistent sources of infection (see Figure 8.4). Finally, fungi can survive as resistant spores or possibly as free-living mycelia in habitats without hosts (see Figure 8.3). When hosts reappear, the fungi are then poised to initiate new infections.

References

Blehert DS (2012) Fungal disease and the developing story of bat white-nose syndrome. *PLoS Pathog* 8(7):e1002779 (doi:10.1371/journal.ppat.1002779).

Fisher MC, Henk DA, Briggs CJ et al (2012) Emerging fungal threats to animal, plant and ecosystem health. *Nature* 484:186–194.

Rohr JR, Raffel TR, Romansic JM, McCallum H & Hudson PJ (2008) Evaluating the links between climate, disease spread, and amphibian declines. *Proc Natl Acad Sci USA* 105:17436–17441.

Warnecke L, Turner JM, Bollinger TK et al (2012) Inoculation of bats with European *Geomyces destructans* supports the novel pathogen hypothesis for the origin of white-nose syndrome. *Proc Natl Acad Sci USA* 109:6999–7003.

Parasites can be introduced with their hosts and have spillover effects

It is not uncommon for parasites to be introduced along with their hosts into new environments, which is an example of pathogen pollution. In their new habitat, the introduced parasites may then begin to exert effects, known as spillover effects, on indigenous host species (**Figure 8.11**). The term spillover is also used to refer to the transfer of a parasite from a domestic or commercial population of hosts into a wild host population, such as described above for the transfer of *Nosema apis* from commercial to wild bumblebee populations or for transfer of *Lepeophtheirus salmonis* from penned to wild salmonids.

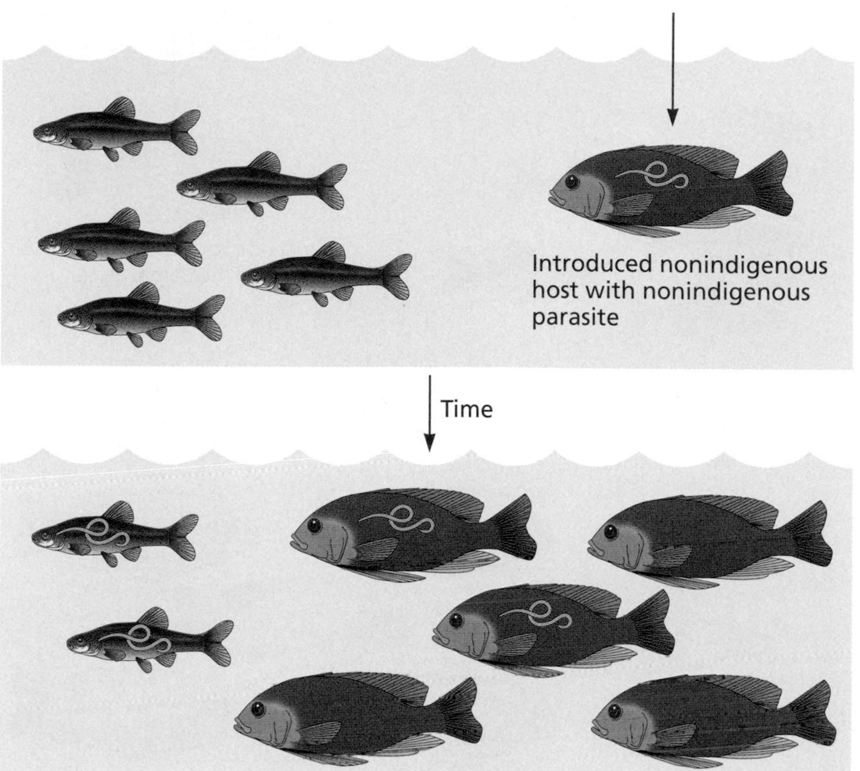

Figure 8.11 Parasite spillover. In the top image, a nonindigenous host containing a nonindigenous parasite is introduced into a new habitat containing an indigenous host species lacking the parasite. As shown in the bottom image, the nonindigenous parasite spills over into the indigenous host, potentially even assuming greater abundance in the indigenous host species and causing that host to become less abundant.

Spillovers are one of the more direct manifestations of introductions and have frequently been implicated in the decline of native host species. For example, European eel (*Anguilla anguilla*) populations have been adversely affected by the introduction into Europe in the early 1980s of East Asian eels (*Anguilla japonica*) carrying the nematode *Anguillicola crassus*. The adult nematode colonizes the eel swimbladder and achieves much higher prevalence and intensity of infections in European eels than it typically does in Asian eels. Infected fish may stop feeding, suffer swimbladder collapse, lose buoyancy, be unable to complete migrations necessary for spawning, and die. Interestingly, although the parasite *A. crassus* survived the introduction into Europe, its native eel host *A. japonica* did not.

As with the eel swimbladder nematode, the native host species most affected by spillover are often those closely related to the introduced host species. The study of emerging diseases indicates that this is by no means always the case. For emerging human diseases, ungulates and carnivores have been shown to be more likely originating host species than primates. Also, an already broad host range as opposed to phylogenetic relatedness of introduced and native hosts has been shown to be more important in dictating success in interspecific pathogen jumps.

Along with introductions of exotic parasites that can spill over into native host species, in some cases introductions of new host species can enable native or endemic parasites to assert themselves in new and often surprising ways, as discussed next.

Introduced hosts can favor indigenous parasites and cause spillback effects

Spillback refers to situations in which a newly introduced host species may act as a good host for an indigenous parasite, so much so that the parasite thrives and then comes to spillback (**Figure 8.12**) and has an even greater

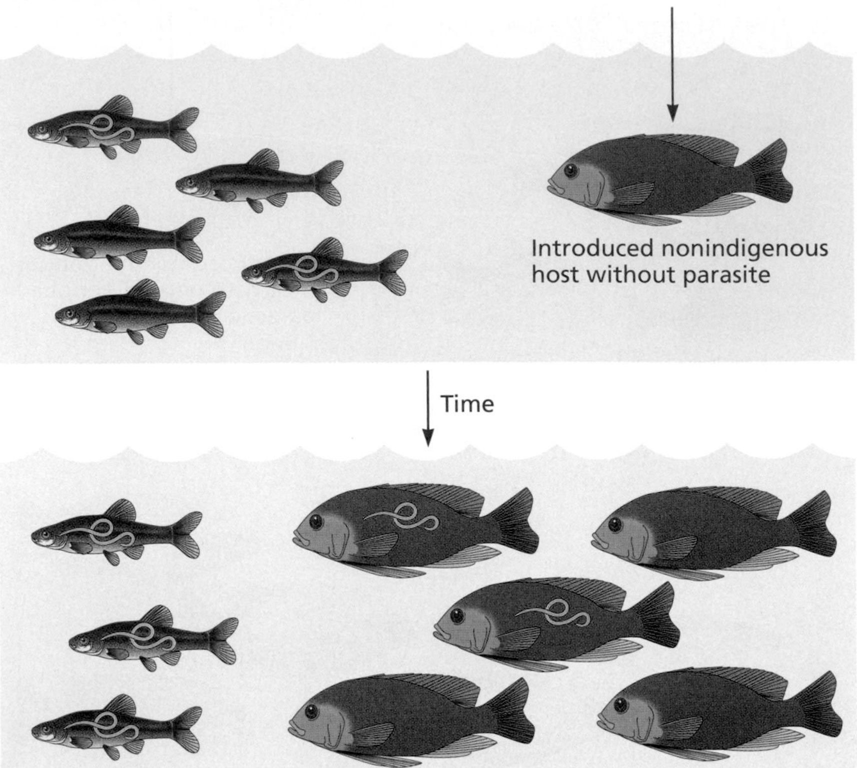

Figure 8.12 Parasite spillback. The introduction of a nonindigenous host (in this case, without its parasites) has the effect of favoring the transmission of an indigenous parasite that then becomes more abundant and has a greater impact on its indigenous host species than it did before the nonindigenous host was introduced. Note that the nonindigenous host may prosper as a consequence.

detrimental effect on its original indigenous host species. Acquisition of native parasites by introduced hosts is common. One analysis of 40 animal introductions showed a mean of 6.3 native parasite species acquired per host, with many of the parasites involved being generalists as would be expected. Furthermore, once acquired, the newly introduced host species often serve as competent hosts, supporting both the reproduction of parasites and their transmission. Introduced hosts often enjoy rapid population growth and dissemination, and they can be very effective in favoring the indigenous parasite.

One of the most notorious and well publicized of all introductions was of the cane toad, *Rhinella marina*, a South American native, into Australia in 1935, ostensibly to eliminate beetles and other pests of sugar cane fields. It has since become an extreme nuisance species in Australia, in part because it is highly toxic and has led to the decline of a number of native predatory reptiles. In addition, it seems to be involved in parasite spillback. Two species of myxosporean parasites of the genus *Cystodiscus* that are apparently indigenous to native Australian amphibians can be supported and propagated in cane toads. Because the cane toads are common, their increased numbers led to greater prevalence and negative effects of the *Cystodiscus* species on their native frog hosts. These frogs, like so many amphibians worldwide, are declining, and parasite spillback involving cane toads seems to be at least one contributor in Australia.

A story involving spillover and possibly spillback as well is unfolding today among aquatic birds of the Mississippi River following the introduction there in about 2002 of the invasive snail *Bithynia tentaculata*. This snail serves as both first and second intermediate host for two trematodes, *Sphaeridiotrema globulus* and *Cyathocotyle bushiensis*. The former fluke species is likely native to North America because it is also transmitted by native gastropod hosts in areas where *B. tentaculata* does not occur, but *C. bushiensis* was introduced along with *B. tentaculata* (and thus represents an example of spillover). *C. bushiensis* is acquired in large numbers by American coots (*Fulica americana*) and *S. globulus* by lesser scaup (*Aythya affinis*), as these birds forage on *B. tentaculata*, many of which are infected. These birds can ingest lethal doses of these two flukes in even a 24-hour foraging period. Annually recurring die-offs of lesser scaup and coots involving thousands of birds have occurred. It seems that the introduced snail host has greatly favored the transmission of the native fluke species *S. globulus* that is now spilling back into its native definitive hosts. At the same time, the introduced fluke *C. bushiensis* is also spilling over into what are new avian definitive hosts for it. The two phenomena are unfortunately having a devastating impact on the avian definitive host species for these flukes. Caution is required in interpreting this situation however, as a second species of *Sphaeridiotrema*, *S. pseudoglobulus*, is now known to be present. Whether this latter trematode is an exotic species and whether it plays a role in avian mortality remain to be determined.

Sometimes introduced nonhost organisms can influence indigenous parasite transmission

In some cases, introduction of an exotic species can influence indigenous parasite transmission, even if it does not serve as a competent host. For instance, if the introduced species is one that will readily acquire life-cycle stages of a native parasite (either by direct penetration of the parasite or perhaps by trophic transmission) yet does not enable the full development or transmission of the parasite, it effectively acts as a **sink** for the parasite. Although examples are scarce, this kind of thinking has been the basis of some efforts to reduce the transmission of human schistosomiasis. For example, in Puerto Rico, it was shown in laboratory trials that infection rates of *Biomphalaria glabrata* with ***Schistosoma mansoni*** could be reduced from 90% to about 1% if the exotic snail *Marisa cornuarietis* was present. The latter snail acted as a sponge or decoy for schistosome miracidia and interfered with infection of *B. glabrata*. This effect was also duplicated in natural ponds, but both lab or field trials required *M. cornuarietis* to be 6–10 times more abundant than *B. glabrata* to be effective. However, this is a large, omnivorous snail that could have many effects in aquatic habitats where introduced, not the least of which could be consumption of native species and aquatic food plants such as rice seedlings.

Another scenario is that an introduced species may compete with or prey upon an indigenous parasite's normal hosts and cause the parasite to decline. An example is provided by the North American Louisiana red swamp crayfish *Procambarus clarkii*, which was introduced into parts of Africa for aquaculture. This species also has a generalist omnivore diet and will readily consume schistosome-transmitting snails and their egg masses (*Marisa* snails mentioned above can also have these effects). One study conducted in Kenya showed that the presence of *P. clarkii* in aquatic habitats significantly diminished populations of the snail *Bulinus africanus*, and the likelihood that schoolchildren living near these habitats acquired infections of urinary schistosomiasis (***Schistosoma haematobium***) (**Figure 8.13**).

This last scenario describes how introduction of a nonhost species might influence transmission of an indigenous parasite. Introduction of trout into New Zealand streams has displaced native fish such as *Galaxias depressiceps* into more restricted and shallow habitats. This displacement may increase the exposure of *G. depressiceps* to trematode infection by concentrating them in locations where cercariae shedding from host snails is favored by warmer

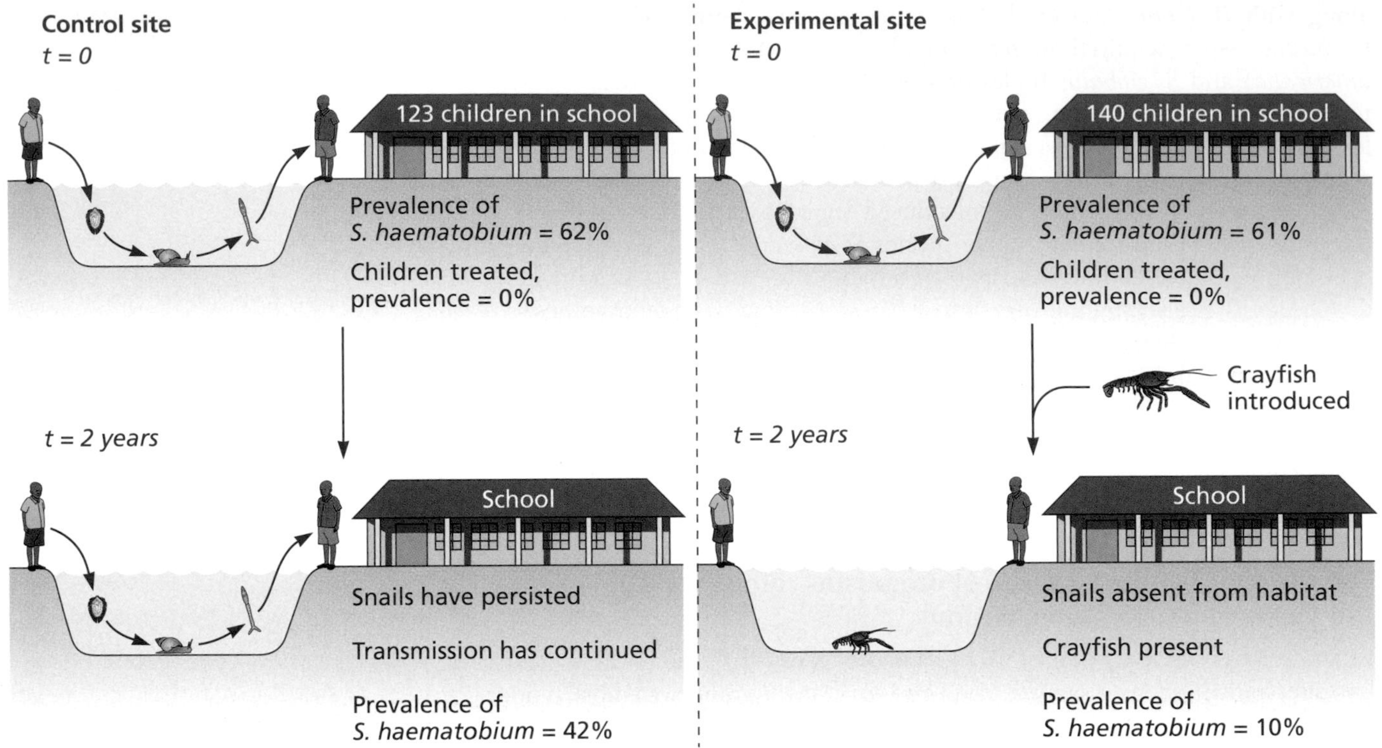

Figure 8.13 The impact of an introduced nonhost species on schistosomiasis transmission. Shown on the left is a control site (Matuu, in rural Kenya), where *Schistosoma haematobium* is transmitted in a small water body by *Bulinus* snails. At the time the study was started ($t = 0$), the prevalence of infection was 62% in children in the nearby school. Children were then treated with praziquantel and the prevalence was reduced to 0%. Two years later, transmission of *S. haematobium* was assessed again. *Bulinus* snails were still present, and the prevalence of *S. haematobium* had returned to 42%. In the experimental site (Kataluni) on the right, the prevalence of *S. haematobium* was also high (61%) in children at $t = 0$, and schistosome-transmitting *Bulinus* snails were present. Following treatment of the children, crayfish (*Procambarus clarkii*) were introduced into the snail habitat. Two years later, crayfish were still present, *Bulinus* snails were absent, and the prevalence of *S. haematobium* in school children was significantly lower (10%) than at the control site. (Data from Mkoji GM et al [1999] *Am J Trop Med Hyg* 61:751–759.)

temperatures. These cercariae in turn penetrate the native fish host and can cause spinal deformations and mortality. In general, several of the examples in this discussion reveal that introductions of parasites, their hosts, or even of nonhost species that affect indigenous transmission cycles may have further, more cryptic, downstream effects that could affect entire communities of organisms, what have been called **knock-on effects**. Knock-on effects are often hard to detect or document, consequently the impact of introductions may prove to be even greater than we currently know.

Invading hosts can benefit by leaving their natural enemies, such as parasites, behind

Exotic species frequently undergo explosive population increases following their introduction, often at the expense of indigenous relatives. Why might this occur? According to the **enemy release hypothesis**, if an invading host species has managed to leave its parasites behind, then it may have greater success in its new environment, which includes giving it a competitive edge over closely related native species that are still contending with their natural enemies. One study (**Figure 8.14**) documented that for 26 separate introductions, a host species harbored on average about 16 parasite species in its native range, of which only an average of three were transferred to the new habitat. On average, the introduced host acquired four additional parasite species from the new habitat. Overall, the introduced host species suffered from half as many parasite species in its new home as compared to its native home. Furthermore, the intensity of infection is less in the new habitat.

There are several reasons for this reduced level of parasitism in introduced hosts. Some parasites may happen to be absent in the founders of the introduced host population (they missed the boat—see Chapter 7). It also is possible that some parasites present in the introduced host individuals may quickly die out if the host initially is rare or scattered in distribution in its new habitat, making transmission difficult. Also, for parasites with complex

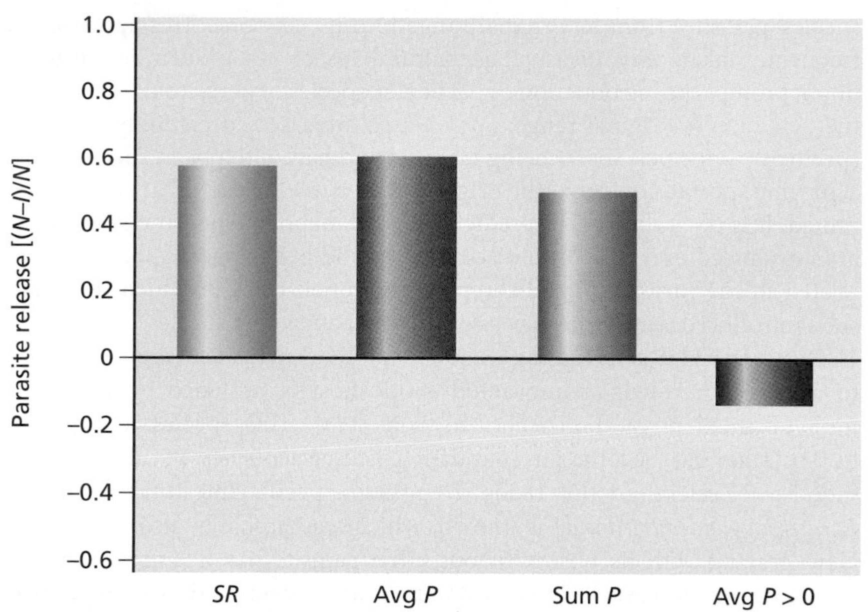

Figure 8.14 Support for the enemy release hypothesis. This graph summarizes the parasite release noted for 26 different host species. The parasite burdens were quantified in their native (*N*) and introduced (*I*) ranges. The more positive the value of the parasite release index (*N* – *I*)/*N* shown on the vertical axis, the greater the disparity in parasite burdens between *N* and *I*. Shown is the average index value calculated on the basis of standardized parasite species richness (*SR*), average prevalence (Avg *P*), summed prevalence (Sum *P*), and average prevalence on a per parasite species basis (parasites with zero prevalence excluded). (Torchin ME et al [2003] ***Nature*** 421:628–630. With permission from Macmillan Publishers Ltd.)

life cycles, needed hosts may be missing in the new habitat, such that the new parasite can never become established. Last, the introduced parasites may be sufficiently host specific such that they are unable to use alternative hosts in their new habitat; this final point helps us to understand why widely introduced parasites are often generalists.

Given all these factors, it can be imagined that the introduced host would possess an advantage of parasite release over its native counterparts and potentially outcompete them. Recall too from Chapter 6 the discussion of apparent competition mediated by parasites. One example of an introduced bird species that may have benefitted from enemy release is the common starling, *Sturnus vulgaris*. A founding population of 60–120 birds derived from England and released in New York City in 1890–1891 has spread across the North American continent and now numbers 200 million birds. North American starlings bear only nine of the 44 (20%) parasite species they carry in their ancestral Eurasian home.

The advantage experienced by an invading, relatively parasite-free host might diminish with time if subsequent invasions of the host succeeded in transporting more of its indigenous parasites, or if these indigenous parasites were deliberately introduced as a control measure. Repeated invasions of the black rat *Rattus rattus* into new habitats may explain why up to 38% of its native parasites have been transported.

Another factor potentially diminishing the success of an invasive species is if parasites of a related native host species are able to switch into the invasive host species. As we have noted in the discussion of spillback, this switch does indeed happen. Such a switch might be rendered even more likely if the introduced host becomes extremely common. For example, of seven examined species of exotic molluscs in North America (many of these species now extremely common on this continent), six have acquired trematode species native to North America.

Invasive hosts can potentially be controlled by parasites from their original range

One obvious way to diminish the advantage that an invasive species might enjoy from leaving its parasites behind is to deliberately introduce that species' parasites from its original range into its new one. This approach is frequently taken with invasive agricultural insect pests. Such an invasion might prompt the introduction of particular species of parasitoid wasps from the pest insect's original range. Such wasps are assessed carefully for their specificity to the pest insect being targeted and have often been successful in achieving specific control of the original pest. Nonetheless, great care is warranted. It is often difficult to predict what will happen when the parasitoids are introduced into new habitats containing indigenous host species closely related to the invasive pest. Examples have often been provided in which such introduced parasitoids have switched to native host species.

One particularly intriguing example of how parasites might be used to achieve control of an unwanted exotic host is provided by the common brushtail possum (*Trichosurus vulpecula*). This species, indigenous to Australia, has become an abundant pest species since its introduction into New Zealand in the 1850s. A possum-specific intestinal nematode *Parastrongyloides trichosuri* is known, which was probably also brought to New Zealand with its host. Obviously the parasite has not prevented the spread of the possums in New Zealand, so what good is it? One thought is to genetically engineer the parasite to carry and express genes encoding proteins essential to possum reproduction. The parasite could thus serve as a very specific self-perpetuating vector for delivering these proteins into possums. These proteins would then either incite immune responses or serve as anti-hormones in the infected host that could eventually disrupt its reproduction. This is surely an ingenious use of a parasite to control an invasive pest, but it has also been noted that the genetically modified nematode could potentially be introduced back into Australia where the exotic genes it carries could potentially disrupt the reproductive activities of brushtail possums, as well as other possum species, in their native habitat. We will have to wait to see if the nematode can be successfully modified with an appropriately disruptive gene and if the approach of using parasites for control of harmful exotics will gain acceptance in this model system or potentially in others.

Introductions of parasites or hosts often fail

Introductions of either parasites or their hosts frequently fail, though we are less likely to be aware that such failures have occurred because they often leave no evidence behind. Parasite introductions might fail, for instance, because the colonist is unable to breach defenses of indigenous host species. Likewise, host introductions may fail because they are ill prepared for indigenous parasites. Introduction of the American rainbow trout *Oncorhynchus mykiss* into Europe is one such example. This deliberate introduction was thwarted by the presence of the myxozoan parasite ***Myxobolus cerebralis***, which was likely normally transmitted by the indigenous European brown trout, *Salmo trutta*. Rainbow trout readily succumbed to whirling disease in Europe. Unfortunately, when this parasite was eventually introduced into North America with brown trout, the parasite readily afflicted naïve indigenous rainbows and other salmonids, which have suffered outbreaks of whirling disease as a consequence.

Translocations of endangered host species can have unforeseen consequences

In some cases, deliberate introductions or transportations of organisms are undertaken with the best of intentions, such as to repopulate areas where native species have been previously eliminated or to move endangered species into areas where they may be more protected. Recent efforts to move both white rhinos (*Ceratotherium simum simum*) and black rhinos (*Diceros bicornis michaeli*) into areas in Kenya where they once existed have resulted in high mortality rates, and parasites were at least partially to blame. The rhinos were found to be emaciated and heavily infested with ticks. Sick rhinos were beset by tsetse flies, harbored multiple species of trypanosomes and filarial worms, and had theileriosis. In such cases, it is hard to know if the effects of arthropod or blood parasites were directly responsible for death or if they were secondary influences because the hosts had already been stressed by transfer or were in marginal habitat. Although successful introductions have surely occurred, examples such as this one also point out that despite the best intentions, consequences of translocations may arise that are unforeseen and that can set conservation efforts back.

Can invasional meltdown occur?

The term **invasional meltdown** refers to the possibility that as the number of introductions of exotic species increases, the ease with which they and other exotics become established also increases. Invasional meltdown may be because one introduced species can facilitate rather than antagonize the establishment of another. This facilitation may be particularly relevant for parasites where the introduction of a host species—or perhaps the introduction of multiple host species for parasites with complex life cycles—could in turn favor the introduction of a parasite dependent on those hosts. One vivid example of the role of multiple introductions in favoring the establishment of a parasite in a new location is provided by the liver fluke ***Fasciola hepatica*** in the Altiplano region of Bolivia (**Figure 8.15**). It has now been established that ***F. hepatica*** from the Altiplano, which was probably introduced by Spanish colonizers, is very similar to isolates obtained from the Iberian Peninsula in Europe. Somewhat more surprising was the discovery that the snail intermediate host efficiently transmitting *F.* ***hepatica*** on the Altiplano proved to be *Galba truncatula*, also a European species again likely introduced by European colonialists. Given that the sheep and cattle are the major vertebrate reservoirs of infection and that these two species were also introduced by Europeans into South America, then it seems colonization of the area was accompanied by everything needed for *F.* ***hepatica*** to thrive. One of the very unfortunate consequences of these multiple introductions is the emergence of *F.* ***hepatica*** as a human disease (mostly affecting children) on the Altiplano, with human infection rates and intensities of infection there higher than for any other known locality.

8.4 PARASITES AS INDICATORS OF ENVIRONMENTAL HEALTH

Parasites can help us monitor ecosystem integrity

As we have noted above, one of the keys to preserving species diversity is to maintain the integrity or health of the ecosystems in which the species live. A healthy ecosystem is one that persists through time, retaining productivity and diversity, and one that is resilient to change. How can such an elusive parameter as ecosystem health be measured? One approach is to gauge the extent to which indigenous parasites are maintained by the ecosystem in question. As we noted in Chapter 6, many of the trophic links in natural food webs are composed of parasites and their hosts. Linkages involving parasites are common because each host species will likely harbor one or more parasite species, and in addition, many parasites have complex life cycles that

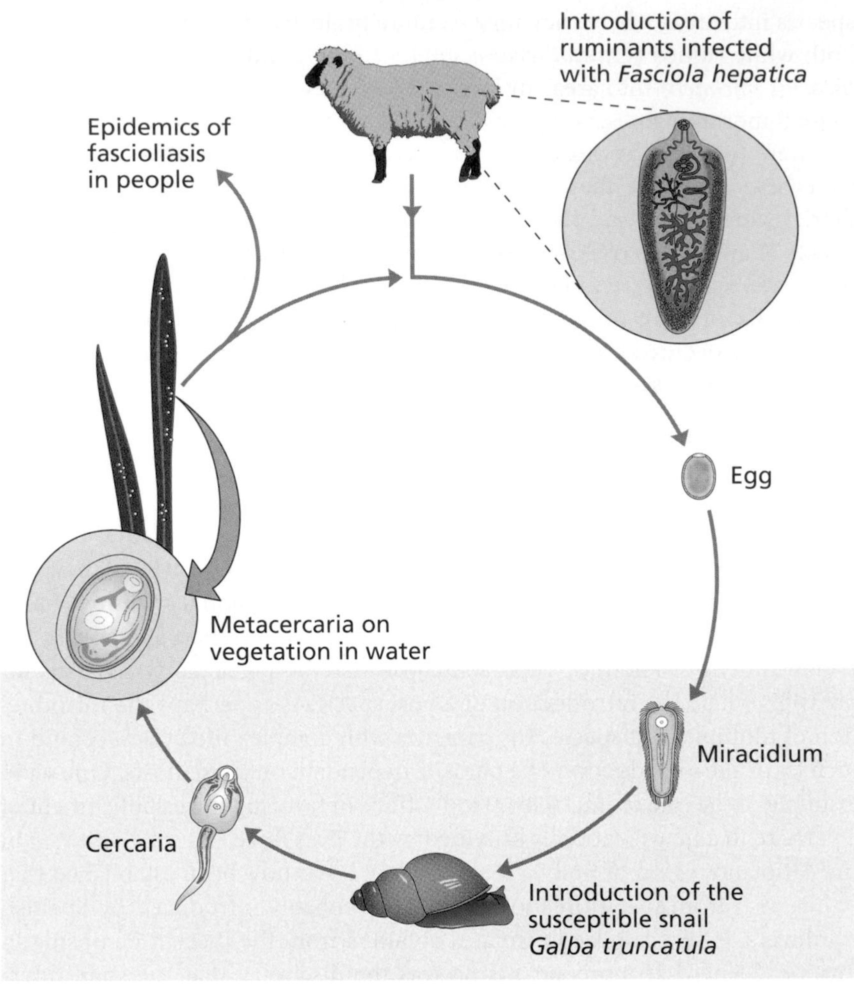

Figure 8.15 Invasional meltdown involving a parasite. All the hosts needed for the sheep liver fluke *Fasciola hepatica* to thrive in the Altiplano region of Bolivia were introduced by Spanish colonialists; sheep infected with *F. hepatica* were introduced along with an excellent first intermediate host, *Galba truncatula*. The remaining stages of the life cycle are either free living or, with respect to the metacercariae, they encyst on vegetation or in water. The parasite has caused epidemics among people living on the Altiplano and elsewhere in South America. (Mas-Coma S et al [2001] *Parasitology* 123:S115–S127. With permission from Cambridge University Press.)

require as many as three or even four different host species to complete. The continued presence of such kinds of parasites in the environment serves as a proxy to indicate all the component hosts are present and the trophic linkages among them are intact.

Furthermore, suppose, for example, that a parasite species cycled between snails and a specific shorebird species. It might actually be easier to assess the continued presence of the shorebird in the ecosystem by examining the snails to see if they are infected with the larval stages of the parasite. This alternative means of monitoring the shorebird's presence might be more reliable, particularly so if the bird is elusive and hard to spot or only present at certain times of the year. A survey to account for all the vertebrate hosts present in an ecosystem might miss the birds at one time point, whereas parasites are acquired by the snails and continue to be produced by them over time, thus providing a more integrative indication of the presence of the bird in the environment in the recent past.

In addition to providing information about the host species present and the extent to which trophic links are intact, parasites can also be used to monitor changes occurring in ecosystems. Both field and laboratory studies indicate that environmental impacts resulting from hurricanes, eutrophication, oil spills, or increased levels of heavy metals or pesticides can have significant effects on levels of parasite populations. We can then think of parasites as being potential biological indicators of environmental impact. **Biological indicators** are species whose presence, abundance, and biological roles can be used to reveal and monitor the extent of ecosystem integrity. One study of the parasites of fish (perch and roach) in Finnish lakes revealed that lakes with high levels of pulp and paper mill effluents had fewer digeneans and myxozoan parasites as compared to lakes with lower effluents. Furthermore, during an eight-year recovery period, some but not all effects on parasite abundance could be reversed.

Use of parasites to evaluate restoration processes has also been documented for Californian salt marsh habitats that support large numbers of digenetic trematodes cycling through the California horn snail *Cerithidea californica* (**Figure 8.16**). Salt marshes that had been degraded had lower prevalence (12%) and species diversity (4.5 species) of digenean infections in snails as compared to unaffected marshes (7 parasite species with overall prevalence of 28%). Over six years of restoration, digenean prevalence rose dramatically in restored areas to 48% and species richness jumped to 9 species, whereas the control habitats remained steady at 26% prevalence with 7.8 species on average. The changes were attributed to the increased use of birds (definitive hosts for the parasites involved) of the restored habitats.

Examples in which parasites serve as indicators or sentinels of environmental problems or changes also unfortunately occur. Marine mammals ranging from sea otters to harbor seals to killer whales along the Pacific Coast of North America have been found in recent years to be infected with a variety of parasites customarily associated with humans, pets, and farm animals from terrestrial environments. These parasites exemplify what have been called pollutogens. **Pollutogens** are infectious agents that originate outside a particular ecosystem and are able to develop within a host found in that ecosystem, yet they do not require that host for reproduction. For instance, co-infections of two related apicomplexans, *Toxoplasma gondii* and *Sarcocystis neurona*, have emerged as a major cause of death in southern sea otters, *Enhydra lutris nereis*. Others kinds of pathogens associated with terrestrial environments such as *Neospora caninum*, another apicomplexan normally problematic in both cattle and its canine definitive hosts, have also

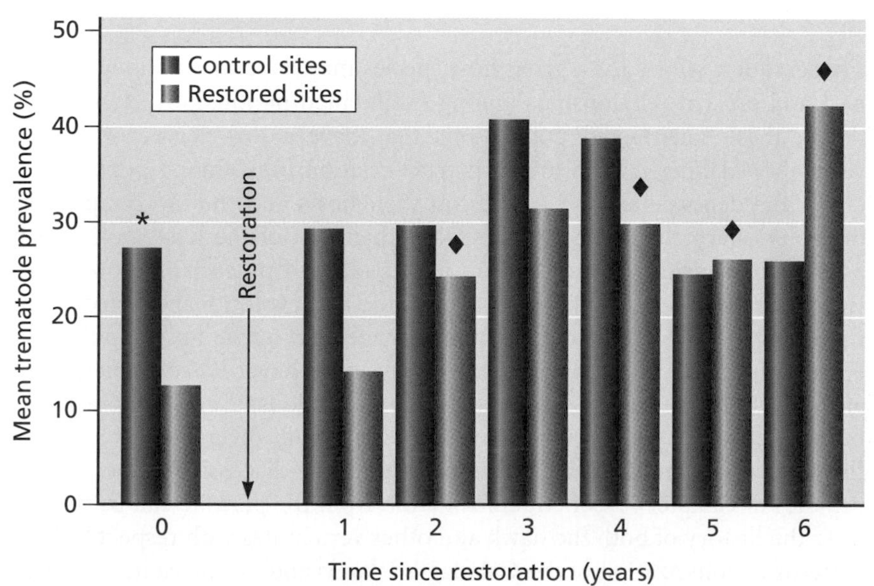

Figure 8.16 Parasites can be used to monitor habitat restoration. The presence of parasites can serve to indicate that certain trophic linkages are intact. Shown are changes in the mean prevalence of larval trematodes that parasitize snails in salt marsh habitats over time from control sites and restored sites. An asterisk (*) above a bar indicates a year when control means were significantly higher than means for restored sites. Diamonds (♦) represent years when restored sites had significant relative increases in mean prevalence as compared to restored sites before restoration began. (From Huspeni TC & Lafferty KD [2004] *Ecol Appl* 14:795–804. With permission from the Ecological Society of America.)

appeared in marine mammals. Several underlying factors combine to create this situation. First, these marine species have had little past exposure to such parasites and so are ill prepared to deal with them. Also, marshlands, which are believed to filter out many parasites, have diminished in extent, allowing parasite-laden freshwater runoff more direct access to the oceans. The ongoing range expansion along the Pacific Coast of the introduced Virginia opossum *Didelphis virginianus*, the definitive host of *S. neurona*, has provided a new and ready source of this parasite as well. The number of feral and domestic cats continues to increase, and treatment of cat litter entering sewage systems does not destroy ***T. gondii*** oocysts. Wild felids also contribute distinct strains of ***T. gondii*** oocysts that find their way into marine waters. Sea otters and other marine mammals have been likened to canaries in a mine, with the parasites they are newly encountering serving as effective indicators of a changing marine ecosystem. In general, changing marine environments (some would say degrading marine environments) pose enormous challenges for the future (see Chapter 10).

One last topic regarding parasites as indicators of environmental health is the discovery that parasites, particularly adult acanthocephalans and to a lesser extent cestodes, accumulate remarkably high levels of toxic metals while residing in their hosts. For example, mean concentrations of lead and cadmium in adults of the acanthocephalan *Pomphorhynchus laevis* were 2700 and 400 times higher, respectively, than in organs of its fish host, the chub *Leuciscus cephalus*, and 11,000 and 27,000 times higher than in the surrounding water. Consequently, the acanthocephalans could potentially be used as very sensitive biological indicators of heavy metal pollution. Experimental exposure of infected and uninfected fish to lead shows that acanthocephalans absorb bile–metal salts and thus reduce the amount of these compounds that are reabsorbed by the fish and accumulate in fish organs. The concentration of metals by helminths is probably a side product of their need for uptake of bile with its content of cholesterol and fatty acids, which they are unable to synthesize *de novo*, or of the avidity of helminth eggshells for metals.

8.5 PARASITES AS INFERENTIAL TOOLS TO PRESERVE HOST BIODIVERSITY

Parasites can provide information useful to preserving their hosts

Conservation efforts for a given host species may depend on knowing both past and present relationships among fragmented populations. As we have noted, these fragmented populations may have a low degree of genetic variability making it hard to reconstruct relationships among populations. Given the dependence of parasites on their hosts and the more rapid rate of evolutionary change in parasite DNA than that of the host species, it is possible to consider parasites as an additional and potentially more sensitive means to infer their hosts' demographic and evolutionary history. This information may contribute to future management of the host populations. For example, studies of the threatened Galapagos hawk *Buteo galapagoensis* and its ectoparasites have shown the ischnoceran louse *Degeeriella regalis* to have much more genetic variation (1.5% maximum divergence) relative to the hawk (0.2% maximum divergence). Recall the discussion of population genetics in Chapter 7. Genetic information from the parasite can be used to infer the history of both the hawk and other vertebrates with respect to their patterns of colonization of the Galapagos Islands and can be cautiously used in conjunction with information about the host to guide management decisions, as indicated in the following example.

An analysis of mitochondrial DNA sequence data, which portrays different sequences in a haplotype network (**Figure 8.17**), shows that the host DNA is relatively invariant with respect to island source. In contrast, *D. regalis* found on birds from the island of Española are relatively divergent, suggesting that hawks from that island may deserve special management status. The figure also shows that the value of particular parasites in guiding such decisions varies. *D. regalis* is vertically transmitted from parents to offspring during brooding and less prone to switching among individual birds. Consequently, its population has more genetic structure. The amblyceran louse *Colpocephalum turbinatum* (which is more prone to switching between individual hawks) and especially the hippoboscid fly *Icosta nigra* (which can fly and thus is even more mobile) provide less information of

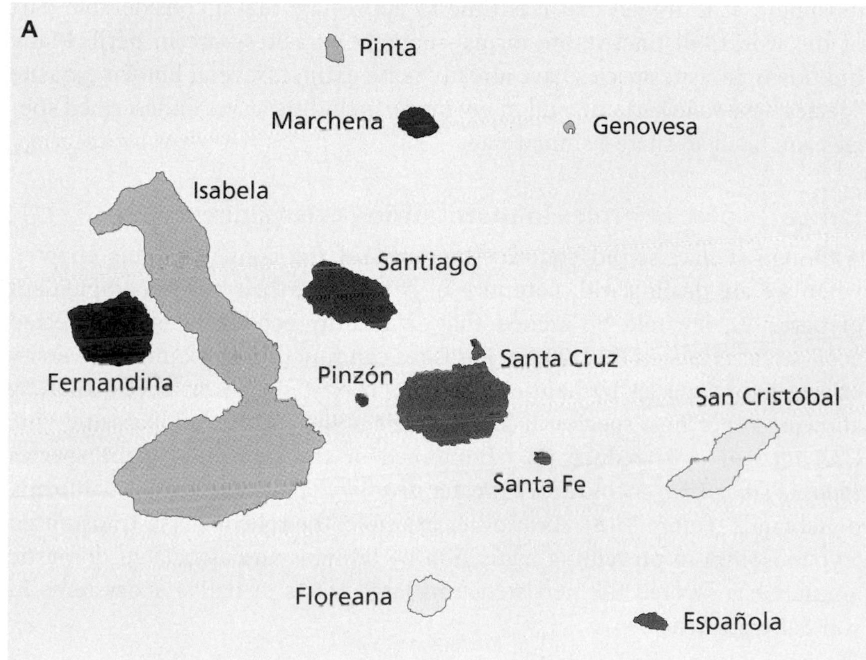

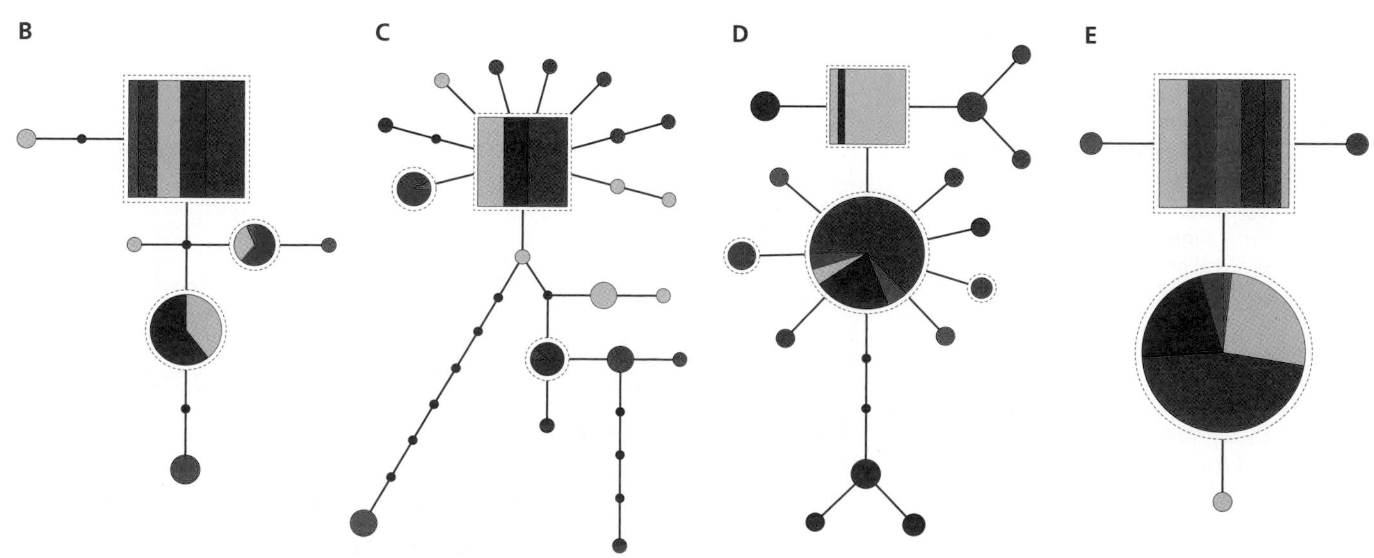

Figure 8.17 Parasites can be sensitive indicators of the presence of distinctive host populations. (A) Map of the Galapagos Islands, with each island given a different color. Haplotype networks of combined mitochondrial DNA (mtDNA) sequence data for the Galapagos hawk *Buteo galapagoensis* (B) and mtDNA sequence data from each of three ectoparasite species of the Galapagos hawk: (C) *Degeeriella regalis*; (D) *Colpocephalum turbinatum*; and (E) *Icosta nigra*. Geographical locations are color coded as in the accompanying map. Each connection (line) between haplotypes represents one mutational step and small black circles are inferred (extinct or unsampled) haplotypes. Sampled haplotypes are represented by circles or squares. Squares represent the putative oldest haplotypes. If more than one island harbored a haplotype, its frequency in each is indicated by the pie chart or the proportionately divided rectangles. In general note how *D. regalis* provides more differentiation for revealing island source as compared to the host and how distinctive the Española Island population of this parasite is. (Whiteman NK et al [2007] *Mol Ecol* 16:4759–4773. With permission from John Wiley and Sons.)

use to conservation efforts for the Galapagos hawk. As another conservation implication, given that parasites such as *D. regalis* are much more variable among islands than the hawk, introduction of the relatively invariant hawks onto new islands must be done cautiously lest they encounter versions of the parasite with distinctive infective attributes to which the transferred hawks may be poorly prepared to respond.

8.6 THE NEED TO PRESERVE PARASITE DIVERSITY

Chapter 2 discussed the remarkable extent of the world's parasite biodiversity and Chapter 7 considered extinction as part of the evolutionary process, including factors contributing to extinction of parasites in particular. In this section we revisit these intersecting themes by addressing the need to preserve parasite diversity. In general, people have been slow to embrace the thought of deliberately preserving parasite diversity. After describing some pragmatic reasons for maintaining parasites as an integral part of the intact biosphere, it is argued that it is time to appreciate that a considerable part of the world's distinctive life forms—namely parasites—are in peril. Many unknown parasite species have already gone extinct, several known parasite species have followed suit, and many more, including as yet undescribed species, are likely to share a similar fate.

Parasites play key roles in maintaining ecosystem health

Although it may sound contrary to much of the content of this chapter, when we are dealing with natural ecosystems and their native complement of parasites, it could be argued that "a healthy ecosystem is an infected ecosystem" (Hudson et al, 2006). Parasites can function to maintain diversity of free-living species by holding a common host species in check, thereby allowing other host species to survive. The ability of the holoparasitic vine *Cuscuta salina* to reduce the abundance of the dominant plant species *Plantago maritima* encourages a greater diversity of plant species in California marshlands (**Figure 8.18**). As another example, the role of tsetse-transmitted trypanosomes in preventing habitation by humans and associated domestic animals has favored the persistence of large tracts of native ecosystems in sub-Saharan Africa.

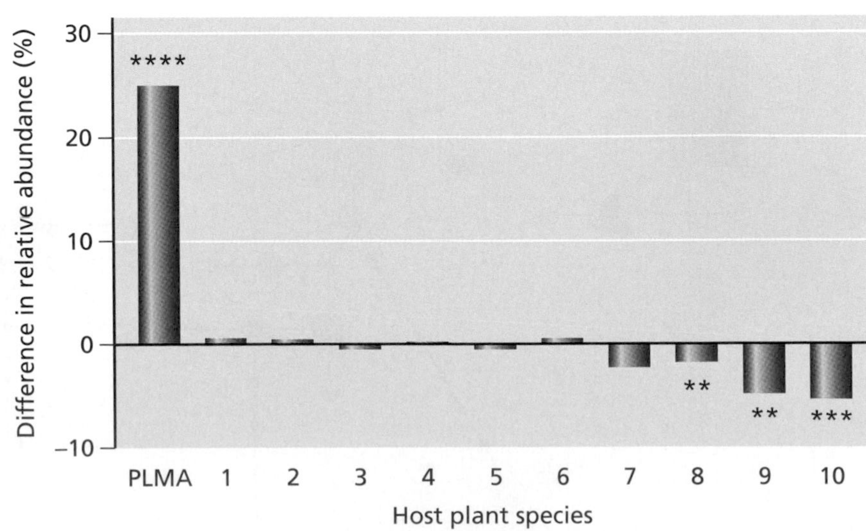

Figure 8.18 Removal of a parasite infecting a dominant plant species has a negative effect on plant community diversity. Each bar represents the mean difference (±SE) in relative abundance of 11 plant species observed following two years of removal of the plant holoparasite *Cuscuta salina*. Significant responses to *C. salina* removal are shown: ** ($P < 0.01$), *** ($P < 0.001$), and **** $P < 0.0001$). Because *C. salina* mostly infects the plant *Plantago maritima* (PLMA), this plant responds very positively to removal. Note though that three plant species became significantly less abundant after *P. aritima* was released from parasitism by *C. salina*. (From Grewell BJ [2008] *Ecology* 89:1481–1488. With permission from the Ecological Society of America.)

Also, as mentioned earlier and in Chapter 6, parasites play key roles in the trophic **connectance** found in natural ecosystems. Connectance refers to the ratio of the number of observed links to the number of potential links in a food web. Parasites in some cases are involved in up to 78% of the trophic links (either as prey items for predators or trophically acquired by hosts from the host's prey). It is generally considered that connectance plays a role in cohesiveness, stability, and resilience of ecosystems.

Parasites are drivers of biodiversity

We have already noted how parasites, by selectively killing more homozygous individuals in a Soay sheep population, promote and maintain the genetic diversity of their hosts. This suggests a larger role for parasites—as drivers of biodiversity. It has been argued that the adoption of particular host lineages by some groups of parasites and the consequent effects of these lineage-specific parasites on the hosts has been a factor favoring the emergence of immunological novelties across animal phyla. Furthermore, it has long been recognized that parasites drive polymorphisms in genes of both innate and adaptive immune systems and that immune genes (such as those of the MHC) are often among the most polymorphic and rapidly evolving. For example, the richness of the intracellular parasite fauna has been associated with regional differences in MHC class I diversity. The many mechanisms that parasites have undertaken to thwart host immune systems (Chapter 4) provide a spectacular example of diversity in its own right.

As discussed in detail in Chapter 7, host species often develop strong local accommodations to the particular parasites they encounter. For example, with the three-spined stickleback *Gasterosteus aculeatus*, fish derived from lakes where the eye fluke *Diplostomum pseudospathaceum* is common were more resistant to the parasite than fish from rivers where this parasite is absent. Furthermore, fish from lake populations could generate greater oxidative respiratory bursts and had bigger spleens (both indicative of greater immunocompetence) than river fish. Thus differences among fish population seem to be driven by differing parasite pressures. As noted previously, parasites can also affect mate choice; females may prefer males with more high quality ornaments, and ornament quality might be influenced by the success of local males in dealing with their local parasites. Ornament-based female choice could lead to further divergence among hosts from different localities. Infectious speciation provides yet another possible mechanism whereby parasites can increase host diversity. Processes like these that increase host diversity are quite in contrast to other processes underway in today's world, such as eutrophication and introduction of exotic species, which are implicated in diminishing the diversity in the natural world.

Parasites are a source of pharmacological and therapeutic novelties

Parasites have been likened to an as yet largely unexplored and untapped pharmacopeia of molecules that may have benefits in treating human diseases. For instance, a surprisingly long list of compounds able to inhibit various steps in the blood coagulation pathway have now been isolated and characterized from a variety of blood-feeding parasites. These parasites, all of which benefit from an ability to keep their host's blood freely flowing, include the canine hookworm ***Ancylostoma caninum***, ticks, black flies, and mosquitoes.

Many molecules produced by parasitic helminths also have immunomodulatory properties. As discussed in Chapter 5 relative to the hygiene hypothesis, intestinal helminths may play an important natural role in educating and modulating the immune system to be less responsive to "self." Similarly, parasite-derived immunomodulatory compounds could be developed and exploited as new therapies for inflammatory diseases such as Crohn's disease, asthma, or autoimmune (type 1) diabetes. That parasites possess such molecules makes sense because parasites have evolved as master regulators of immune responses. Furthermore, it can be argued that all of these compounds have already been extensively tested for efficacy in the human host, so are likely to prove active and selective when tested in a therapeutic context. To provide a few examples, ***Schistosoma mansoni*** produces a chemokine-binding protein, smCKBP, that has the ability to suppress inflammation. ***Toxoplasma gondii*** produces a molecule called C-18 that mimics the ligand that binds to the CCR5 receptor on the surface of dendritic, antigen-presenting cells, thereby preventing proper activation of these cells in response to parasite antigen. As CCR5 also serves as a co-receptor for HIV-1, might C-18 block HIV-1 infection of its target cells, $CD4^+$ lymphocytes? This effect has in fact been shown (**Figure 8.19**), raising the possibility that recombinant and modified forms of C-18 could serve as an HIV therapy.

The discovery process for such compounds has just begun, and a potentially huge range of specialized and therapeutically beneficial compounds are encoded within the genomes of parasites and await discovery. It is also possible that parasites that inhabit very particular microenvironments (such as ***Schistosoma haematobium*** living in veins in the wall of the urinary bladder) could be modified to secrete therapeutically useful recombinant proteins and thus serve to deliver therapy in a target-specific manner over a long period of time. A similar role for viruses has already been envisioned.

Slip sliding away—parasite diversity is being lost

We've made the case that there are utilitarian reasons to preserve parasite diversity. Here we argue from a different point of view that parasites are intrinsically fascinating and unique, and they too have the right to survive (**Box 8.2**). We are often very keen to preserve predators, animals that kill numerous prey animals in succession. Parasites typically do not kill their hosts, but the idea of going out of our way to preserve them is slow to catch on. Remember, however, that parasites represent a significant amount of the diversity of life on earth, and we owe it to posterity to preserve as much of

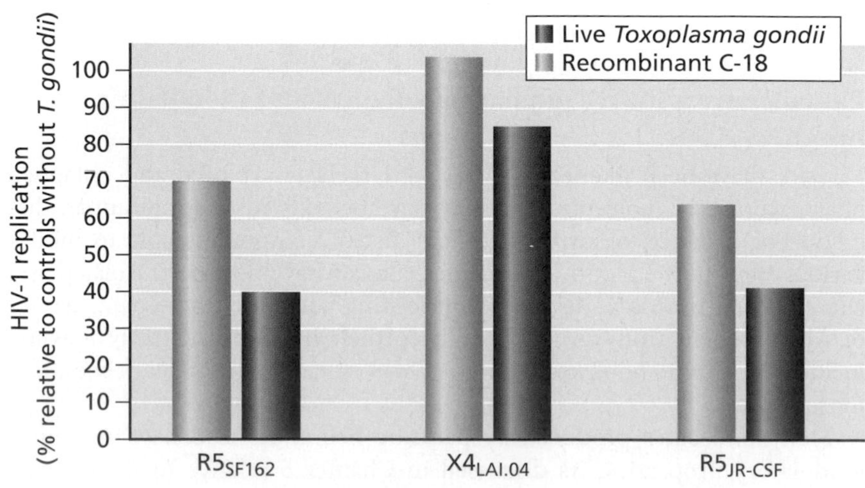

Figure 8.19 A parasite-produced molecule C-18 can inhibit HIV-1 replication. This graph shows that live *Toxoplasma gondii* (purple bars), or to a lesser extent recombinant C-18 (blue bars) derived from *T. gondii*, can inhibit replication of different isolates of HIV-1 ($R5_{SF162}$, $X4_{LAI.04}$, or $R5_{JR\text{-}CSF}$). The columns represent the mean of HIV-1 total replication in treated tissues expressed as percent of HIV-1 replication in matched untreated control tissues (set at 100%). (Sassi A et al [2009] *Microbes Infect* 11:1106–1113. With permission from Elsevier.)

this diversity as possible. Parasites provide their own unique perspective on what makes life fascinating to behold, which is poignantly brought to mind when we lose parasite species to extinction.

In some cases parasites have been quite deliberately targeted for extinction. It is entirely possible that one of the most spectacular of human parasites, the Guinea worm, ***Dracunculus medinensis*** (**Figure 8.20A**) will be extinct before 2020, marking the first known eradication of a human-infecting eukaryotic parasite by the deliberate actions of the public health community.

BOX 8.2

An unusual endangered group of parasites

There are several prominent groups of parasites we all recognize such as lice, tapeworms, and flukes, and many species within these groups are rare and endangered. There are also many other kinds of organisms that represent spectacular forms of biodiversity that are also endangered and that we may fail to appreciate are parasitic in their life style. Among them are freshwater bivalve molluscs of the family Unionidae, called unionids or pearly naiads (Figure 1A). Collectively they represent one of the most endangered groups of organisms in the world. Although these organisms spend their potentially very long adult lives as free-living, bottom-dwelling filter feeders, nearly all species undergo an obligatory larval stage of development as parasites living on the skin, fins, or gills of fish, or rarely on other aquatic vertebrate species such as salamanders. To enable colonization of their hosts, female unionids produce large numbers of **glochidia** larvae (page166), each of which has an adhesive thread and teeth on its shell valves that enable it to attach to its host. The female unionids then employ remarkable means, some spectacularly devious (Figure 1B), to ensure their glochidia end up successfully colonizing fish. Once on a fish, the glochidia undergo a period of development lasting from weeks to months. The juvenile unionids then fall off their hosts and grow to become free-living adults; without their parasitic phase, they are unable to mature. There are about 1000 species of unionids in the world, and approximately 300 species in North American waters alone. Sadly, about 70% of all unionid species are listed as threatened, endangered, or of special concern.

Although adult unionids are threatened by a host of environmental factors that includes pollution, dredging of stream bottoms, introduction of exotic bivalves such as zebra mussels, and construction of dams, their success is also tied to the diversity, abundance, and behavior of their specific fish hosts. A particular unionid species often has primary fish host species, on which its prospects for glochidial attachment and eventual metamorphosis are relatively high, and marginal fish host species on which its rate of development is lower but can still occur. If primary or even marginal fish host species disappear from a stream, then eventually so will the dependent unionid species. In addition, if the preferred fish host has a limited dispersal ability, then the opportunities made available to that unionid species to colonize new habitats, to maintain connectivity with other populations, and to survive localized extinction of the fish host are lessened, and survival of the unionid is much more problematic. Unionids serve to remind us of certain fundamental realities for all parasites: (1) if their host species are extirpated, then they are soon to follow suit, and (2) many of other factors such as introductions of exotic species or deteriorating environmental quality can contribute to the demise of parasites.

A

B

Figure 1 Unionid bivalves and infection of their hosts. The endangered unionid bivalves use remarkable ploys to infect their hosts. (A) An adult unionid, the Ouachita kidneyshell *Ptychobranchus occidentalis* that lives in rivers of the Ozark Plateau. (B) Ovisacs produced by females of *P. occidentalis* look like larval fish and attract the attention of a potential fish host. Each ovisac contains hundreds of glochidia and is attached to a substrate by a transparent tail. When the potential host fish bites the ovisac, the glochidia are released, usually through the ovisac's false eyespots which tend to rupture. The glochidia then colonize the gills of the fish host. (A, courtesy of Ed Miller, Great Plains Nature Center. B, with permission from Barnhart MC, Missouri State University

References

McNichols KA, Mackie GL & Ackerman JD (2011) Host fish quality may explain the status of endangered *Epioblasma torulosa rangiana* and *Lampsilis fasciola* (Bivalvia:Unionidae) in Canada. *J North Am Benthol Soc* 30:60–70 (doi:10.1899/10-063.1).

Schwalb AN, Cottenie K, Poos MS et al (2011) Dispersal limitation of unionid mussels and implications for their conservation. *Freshw Biol* 56:1509–1518 (doi:10.1111/j.1365-2427.2011.02587.x).

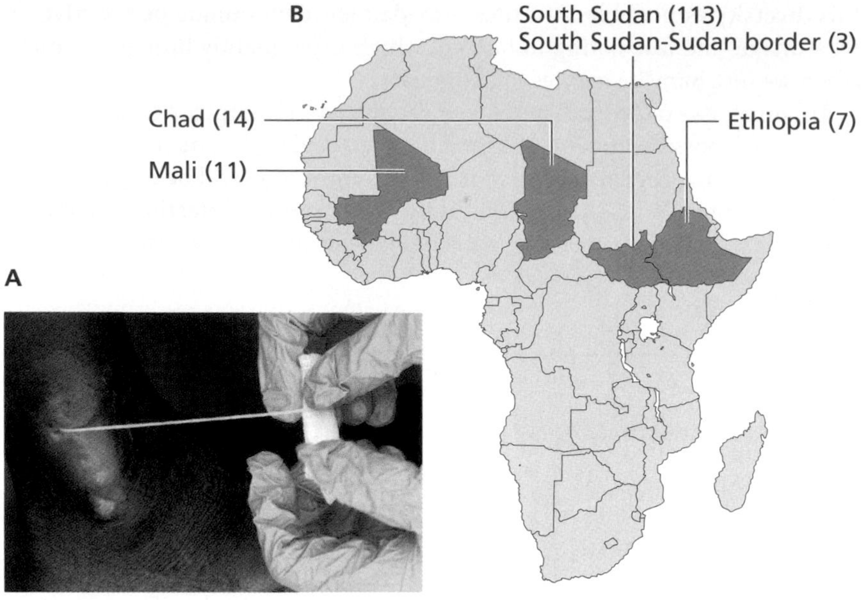

Figure 8.20 Good riddance, or farewell, to one of humanity's most distinctive parasites. (A) A female Guinea worm being pulled from the ankle. (B) A map of Africa showing the four remaining countries where dracunculiasis still occurs (Mali, Chad, South Sudan, and Ethiopia). Shown for each country is the number of cases known to exist, as of 2013. Prior to initiation of international control efforts in 1986, there were more than 3.5 million cases in 21 Asian and African countries. (A, © The Carter Center/L Gubb.)

Thanks to the concerted efforts of the Carter Center and other international health organizations only about 148 cases of this infection occurred in 2013 in four African countries (South Sudan, Mali, Ethiopia, and Chad). (See Figure 8.20B and Chapter 9 for a discussion of control measures.) In 2014, only 80 provisional cases have been reported, the majority occurring in South Sudan. The implementation of simple methods such as filtering drinking water, using water from boreholes, treating water sources with a copepod-killing compound, and preventing infected people from contacting drinking water sources have all contributed to eradication. An additional factor favoring the demise of this species is that it is primarily a human parasite, although dogs can be infected.

Furthermore, the Guinea worm lives on an annual cycle. New infections are acquired each year, and then adult worms mate within a human host and adult females emerge through the skin to release their larvae into aquatic environments later in the same 12-month period. Adult females do not persist for many years in their human hosts. Given that infections in copepods are short-lived (maybe a few months), if the annual cycle of transmission could be broken everywhere, then the infection would be gone for good. However, before we get too confident, it should be noted that dracunculiasis reappeared in Chad in 2010 after an apparent absence of ten years. It appeared in both a small number of people and a large number of dogs. There is also some evidence to suspect that paratenic hosts like fish or frogs may be involved in trophically connecting infections in copepods with dogs or people. Whether this represents an emerging pathway for human infection, or whether it is a transmission pattern distinctive to Chad requires further study, but reminds us that caution is required and its implications for the global eradication effort will have to be carefully considered.

It would be hard to argue too persuasively for letting the Guinea worm persist in nature because it causes true misery, pain, and suffering for those unfortunate enough to be infected. Consider though that this worm has been part of the human experience for thousands of years, and many of our ancestors have been afflicted with these "fiery serpents." When this worm is exterminated, we will have lost a palpable connection with our past. A spectacular example of biodiversity—an unusually large nematode with a remarkable ability to exit through the skin of the human body—will have been lost. Although other species of ***Dracunculus*** are known, this is the only species with a preference for infecting humans. Given its imminent demise, we must do what we can at the least to insure that samples of the worm, including its genome, are preserved for posterity to study.

Other deliberate, human-mediated parasite extinctions may have also been perpetrated. For example, when the last surviving 22 California condors *Gymnogyps californianus* were taken into captivity in 1987, they were treated for parasites, including fumigation with carbaryl insecticide. Although it is possible the condor-specific louse *Colpocephalum californici* (**Figure 8.21**) may already have been extinct when these condors were collected, it is likely fumigation had the effect of eliminating the louse from these few remaining birds. Although the intentions were certainly noble—to favor survival of one species, a magnificent and critically endangered bird—the result was to extirpate another, a louse that is known in nature only from the condors. A similar scenario may have followed rediscovery of one extant natural population of the endangered black-footed ferret *Mustela nigripes* in Wyoming in 1981. Prior to moving these few remaining individuals to the lab and after establishment of a lab colony, surviving ferrets and their prairie dog prey were fumigated, with one consequence being the elimination from the ferrets of lice of the genus *Neotrichodectes*. This louse may have been a distinct species specific to black-footed ferrets, although this is not known with certainty. In all likelihood, it will now never be known: the louse has not been seen on the ferrets since 1997.

Several other parasites known to be extinct include about a dozen species of lice from now extinct avian species and, as mentioned in Figure 7.21, a cyamid whale louse (a crustacean) from the extinct North Pacific Steller's sea cow (*Hydrodamalis gigas*). These examples remind us of the reality of the coextinction process discussed in Chapter 7: host specific parasites will go extinct if their host species is extirpated. Even if the host species lives on, if its populations become too fragmented or small, the parasite might still go extinct (see the discussion of Figure 8.2). Coextinctions involving pairs of mutualists or host–parasite units may be the most common forms of loss of biodiversity. For example, out of 700 species of hard ticks (**Ixodidae**), one estimate places the number of endangered species at 63 species, with one

Figure 8.21 The endangered California condor (*Gymnogyps californianus*) and its now extinct louse (*Colpocephalum californici*). This louse was only known from the California condor and may have gone extinct as a result of efforts to rid the bird of its parasites. (A, courtesy of Stacy Spensley, CC BY 2.0. B, with permission from Vincent Smith.)

species already extinct. The endangered tick species fall into this category because the hosts on which they depend are also endangered. The infestation of endangered hosts by equally endangered parasites poses a dilemma for conservation biologists, especially given that ticks often transmit diseases to their endangered host species. Should efforts to save the hosts require us to eliminate the ticks? One factor to consider is that even if its indigenous ticks are eliminated, the host might later acquire exotic species of ticks to which it is not adapted and which could cause even greater harm.

Other factors favoring parasite extinctions are the complex life cycles involving multiple host species exhibited by some parasite species; lack of any intermediate or definitive host species would spell doom. Also, poor dispersal ability of some parasite species may predispose them to extinction.

Finally, as emphasized in Chapter 7 and in other contexts in this chapter, interactions frequently occur among the factors predisposing to extinction. For example, a host-specific parasite achieving a low prevalence in a host species with small or fragmented population sizes has a much poorer chance of survival than a generalist parasite attaining high prevalence values in large host populations.

Summary

It is increasingly appreciated that parasites are of major relevance to the study of conservation biology. The latter discipline is devoted to the study of the world's biodiversity and how to preserve it, this in face of major challenges imposed by habitat destruction, pollution, climate change, and many other factors. Parasites can directly cause the extinction of rare host species, but in cases when the parasite is host specific, it is generally considered that the parasite will itself go extinct before its host. In cases where a parasite is a generalist and can be maintained in common reservoir host species, then the parasite can potentially cause the extinction of rare hosts in which it also occurs. Emerging fungi, such as those responsible for causing chytridiomycosis in amphibians and white-nose syndrome in bats are particularly of concern in causing host extinctions. A combination of factors—parasites, habitat loss, species introductions, and pollution—often work in concert to affect host populations. Small, genetically homogeneous and fragmented host populations, such as those found on islands or zoos are particularly vulnerable to extinction mediated by parasites.

Introductions of species, often aided and abetted by our globalized economy, play havoc with conservation biology. Introduced parasites can spillover into indigenous hosts in their new habitats. Alternatively, introduced hosts may prove efficient in transmitting indigenous parasites, causing spillback effects that are detrimental to the indigenous host species. Introduced hosts may benefit from release from the enemies that naturally infect them in their area of origin, and cause considerable effects on related host species in their new homes. Some introductions can have unforeseen impacts on established host–parasite interactions if, for example, the new species is a predator or competitor of an established host. Parasites too provide a possible example of the phenomenon of invasional meltdown, when numerous introductions (as of host species), then make it possible for other species (such as a new parasite) to invade.

Although parasites are often viewed as villains, they can be used to monitor the integrity of ecosystems, track the progress of restoration programs, or provide unique insights into the management of host species. Parasites often act to help maintain balance and resilience in ecosystems and are a source of unique pharmaceutically relevant compounds. Purely from an aesthetic point of view, parasites are spectacular forms of biodiversity worthy of preservation for their own sake. Because so many of their hosts are today vulnerable to extinction, parasites too are vulnerable, in what is called coextinction, particularly if they are found only in rare hosts. The future holds many exciting opportunities for young scientists wishing to explore the interface between parasitology and conservation biology. In the face of the inexorable pressure of human population growth and all that comes with it, inventive approaches will be needed to accomplish the goals of conservation biology. At the same time, we also need to be able to control the adverse effects of parasites on the health of food plants, domestic animals and human beings, using precise methods that are efficient, selective and not harmful to the environment.

REVIEW QUESTIONS

1. Why is it often supposed that parasites will not be a factor in causing extinction of host species?

2. Why do we see present day examples, such as with chytrid *Batrachochytrium dendrobatidis*, that seem to be responsible for causing extinctions? Identify some particular properties of this species that might lead to this result

3. One of the most famous examples of how parasites can cause extinctions of host species, and thereby pose considerable concerns for conservation biologists, is provided by the indigenous birds of Hawaii and their exposure to avian malaria.
   a) What is particularly noteworthy about the honeycreepers found in Hawaii?
   b) Why are they so susceptible to avian malaria?
   c) What additional factors have favored the continued persistence of avian malaria in Hawaii?
   d) What factors have enabled some indigenous bird species to survive?
   e) How does climate change figure into this example?
   f) Is there any hope for the native birds? If so, why?

4. What is spillover? Distinguish it from spillback.

5. What advantages do invading hosts have with respect to parasitism, and how might we diminish these advantages to help control the invasive host?

6. How can parasites help us to understand if an ecosystem is intact, or is losing species?

7. Consider *Dracunculus medinensis*, a human-infecting nematode now on the verge of extinction. Can you think of any factors that might prevent its extinction? Can you make an argument explaining why it would be desirable to figure out how to preserve this species?

8. What factors favor parasite extinction?

References

GENERAL REFERENCES

Clavero M & Garcia-Berthou E (2005) Invasive species are a leading cause of animal extinctions. *Trends Ecol Evol* **20**:110–110 (doi:10.1016/j.tree.2005.01.003).

Cunningham AA, Daszak P & Rodriguez JP (2003) Pathogen pollution: defining a parasitological threat to biodiversity conservation. *J Parasitol* **89** (suppl): S78–83.

Daszak P, Cunningham AA & Hyatt AD (2000) Wildlife ecology—emerging infectious diseases of wildlife—threats to biodiversity and human health. *Science* **287**:443–449 (doi:10.1126/science.287.5452.443).

Molnar JL, Gamboa RL, Revenga C et al (2008) Assessing the global threat of invasive species to marine biodiversity. *Front Ecol Environ* **6**:485–492 (doi:10.1890/070064).

Smith KF, Sax DF & Lafferty KD (2006) Evidence for the role of infectious disease in species extinction and endangerment. *Conserv Biol* **20**:1349–1357 (doi:10.1111/j.1523-1739.2006.00524.x).

Wilson EO (2000) On the future of conservation biology. *Conserv Biol* **14**:1–3.

SOME THEORY ABOUT PARASITES AND CONSERVATION BIOLOGY

Fisher MC, Henk DA, Briggs CJ et al (2012) Emerging fungal threats to animal, plant and ecosystem health. *Nature* **484**:186–194.

PARASITES INFLUENCE EFFORTS TO PRESERVE HOSTS

Parasites can cause extinction of host species

Atkinson CT & Lapointe DA (2009) Ecology and pathogenicity of avian malaria and pox. In Conservation Biology of Hawaiian Forest Birds: Implications for Island Avifauna. (TK Pratt, CT Atkinson, PC Banko et al eds), pp. 234–252. Yale University Press.

Cunningham AA & Daszak P (1998) Extinction of a species of land snail due to infection with a microsporidian parasite. *Conserv Biol* **12**:1139–1141 (doi:10.1046/j.1523-1739.1998.97485.x).

De Castro F & Bolker B (2005) Mechanisms of disease-induced extinction. *Ecol Lett* **8**:117–126 (doi:10.1111/j.1461-0248.2004.00693.x).

Woodworth BL, Atkinson CT, LaPointe DA et al (2005) Host population persistence in the face of introduced vector-borne diseases: Hawaii amakihi and avian malaria. *Proc Natl Acad Sci USA* **102**:1531–1536 (doi:10.1073/pnas.0409454102).

Parasites work in concert with other stressors to affect hosts

Marcogliese DJ & Pietrock M (2011) Combined effects of parasites and contaminants on animal health: parasites do matter. *Trends Parasitol* **27**:123–130 (doi:10.1016/j.pt.2010.11.002).

The impact of parasitism is influenced when hosts occur in small or fragmented populations

Gillespie TR & Chapman CA (2008) Forest fragmentation, the decline of an endangered primate, and changes in host–parasite interactions relative to an unfragmented forest. *Am J Primatol* **70**:222–230 (doi:10.1002/ajp.20475).

Parasites can strongly affect hosts with reduced genetic variation

Brown MJF (2011) Conservation: the trouble with bumblebees. *Nature* **469**:169–170 (doi:10.1038/469169a).

Radwan J, Biedrzycka A & Babik W (2010) Does reduced MHC diversity decrease viability of vertebrate populations? *Biol Conserv* **143**:537–544 (doi:10.1016/j.biocon.2009.07.026).

Whitehorn PR, Tinsley MC, Brown MJF et al (2011) Genetic diversity, parasite prevalence and immunity in wild bumblebees. *Proc R Soc Lond B Biol Sci* **278**:1195–1202 (doi:10.1098/rspb.2010.1550).

Captive host populations are often very vulnerable to parasites

Pacheco MA, Escalante AA, Garner MM et al (2011) Haemosporidian infection in captive masked bobwhite quail (*Colinus virginianus ridgwayi*), an endangered subspecies of the northern bobwhite quail. *Vet Parasitol* **182**:113–120 (doi:10.1016/j.vetpar.2011.06.006).

Spencer JA, Joiner KS, Hilton CD et al (2004) Disseminated toxoplasmosis in a captive ring-tailed lemur (*Lemur catta*). *J Parasitol* **90**:904–906 (doi:10.1645/GE-249R).

Parasites are frequently transferred from abundant host species to rare relatives, including from humans to our great ape cousins

Cleaveland S, Laurenson MK & Taylor LH (2001) Diseases of humans and their domestic mammals: pathogen characteristics, host range and the risk of emergence. *Philos Trans R Soc Lond B Biol Sci* **356**:991–999).

Graczyk TK, Mudakikwa AB, Cranfield MR et al (2001) Hyperkeratotic mange caused by *Sarcoptes scabiei* (Acariformes : Sarcoptidae) in juvenile human-habituated mountain gorillas (*Gorilla beringei*). *Parasitol Res* **87**:1024–1028.

Labes EM, Wijayanti N, Deplazes P et al (2011) Genetic characterization of *Strongyloides* spp. from captive, semi-captive and wild Bornean orangutans (*Pongo pygmaeus*) in Central and East Kalimantan, Borneo, Indonesia. *Parasitology* **138**:1417–1422 (doi:10.1017/S0031182011001284).

Pence DB & Ueckermann E (2002) Sarcoptic mange in wildlife. *Rev Sci Tech* **21**:385–398.

Roelke-Parker ME, Munson L, Packer C et al (1996) A canine distemper virus epidemic in Serengeti lions (*Panthera leo*). *Nature* **379**:441–445 (doi:10.1038/379441a0).

Williams JM., Lonsdorf EV, Wilson ML et al (2008) Causes of death in the Kasekela chimpanzees of Gombe National Park, Tanzania. *Am J Primatol* **70**:766–777 (doi:10.1002/ajp.20573).

Farming can pose parasite problems for wild host species

Krkosek M, Connors BM, Morton A et al (2011) Effects of parasites from salmon farms on productivity of wild salmon. *Proc Natl Acad Sci U S A* **108**:14,700–14,704 (doi:10.1073/pnas.1101845108).

Parasites of an iconic symbol—the giant panda—point out our need to know more

Bauer C (2013) Baylisascariosis—infections of animals and humans with 'unusual' roundworms. *Vet Parasitol* **193**:404–412 (doi:10.1016/j.vetpar.2012.12.036). (Epub 2012 Dec 27.)

DANGERS RESULTING FROM SPECIES INTRODUCTIONS

Parasites can be introduced with their hosts and have spillover effects

Taraschewski H (2006) Hosts and parasites as aliens. *J Helminthol* **80**:99–128 (doi:10.1079/JOH2006364).

Wielgoss S, Taraschewski H, Meyer A & Wirth T (2008). Population structure of the parasitic nematode *Anguillicola crassus*, an invader of declining North Atlantic eel stocks. *Mol Ecol* **17**:3478–3495.

Woolhouse MEJ, Haydon DT & Antia R (2005) Emerging pathogens: the epidemiology and evolution of species jumps. *Trends Ecol Evol* **20**:238–244.

Introduced hosts can favor indigenous parasites and cause spillback effects

Bergmame L, Huffman J, Cole R et al (2011) *Sphaeridiotrema globulus* and *Sphaeridiotrema pseudoglobulus* (Digenea): species differentiation based on mtDNA (barcode) and partial LSU-rDNA sequences. *J Parasitol* **97**:1132–1136 (doi:10.1645/GE-2370.1).

Hartigan A, Dhand NK, Rose K et al (2012) Comparative pathology and ecological implications of two Myxosporea parasites in native Australian frogs and the invasive cane toad. *PLoS One* **7**:e43780 (doi:10.1371/journal.pone.0043780).

Hartigan A, Fiala I, Dykova I et al (2011) A suspected parasite spill-back of two novel *Myxidium* spp. (Myxosporea) causing disease in Australian endemic frogs found in the invasive cane toad. *PLoS One* **6**:e18871 (doi:10.1371/journal.pone.0018871).

Herrmann KK & Sorensen RE (2011) Differences in natural infections of two mortality-related trematodes in lesser scaup and American coot. *J Parasitol* **97**:555–558 (doi:10.1645/GE-2693.1).

Kelly DW, Thomas H, Thieltges DW et al (2010) Trematode infection causes malformations and population effects in a declining New Zealand fish. *J Anim Ecol* **79**:445–452 (doi:10.1111/j.1365-2656.2009.01636.x).

Paterson RA, Rauque CA, Fernandez MV et al (2013) Native fish avoid parasite spillback from multiple exotic hosts: consequences of host density and parasite competency. *Biol Invasions* **15**:2205–2218 (doi:10.1007/s10530-013-0445-8).

Sometimes introduced nonhost organisms can influence indigenous parasite transmission

Laracuente A, Brown RA & Jobin W (1979) Comparison of 4 species of snails as potential decoys to intercept *Schistosome* miracidia. *Am J Trop Med Hyg* **28**:99–105.

Leprieur F, Hickey MA, Arbuckle CJ et al (2006) Hydrological disturbance benefits a native fish at the expense of an exotic fish. *J Appl Ecol* **43**:930–939 (doi:10.1111/j.1365-2664.2006.01201.x).

Invading hosts can benefit by leaving their natural enemies, such as parasites, behind

Prenter J, MacNeil C, Dick JTA et al (2004) Roles of parasites in animal invasions. *Trends Ecol Evol* **19**:385–390 (doi:10.1016/j.tree.2004.05.002).

Torchin ME, Lafferty KD, Dobson AP et al (2003) Introduced species and their missing parasites. *Nature* **421**:628–630 (doi:10.1038/nature01346).

Invasive hosts can potentially be controlled by parasites from their original range

Cowan PE, Ralston MJ, Heath DD et al (2006) Infection of naive, free-living brushtail possums (*Trichosurus vulpecula*) with the nematode parasite *Parastrongyloides trichosuri* and its subsequent spread. *Int J Parasitol* **36**:287–293 (doi:10.1016/j.ijpara.2005.11.004).

Gilna B, Lindenmayer DB & Viggers KL (2005) Dangers of New Zealand possum biocontrol research to endogenous Australian fauna. *Conserv Biol* **19**:2030–2032 (doi:10.1111/j.1523-1739.2005.00286.x).

Gruenberg A & Bisset S (1998) *Parastrongyloides trichosuri* as a potential vector for genes expressing proteins which could interfere with reproductive success, growth, or longevity of brushtailed possums. *Misc Roy Soc New Zeal* **45**:10–12.

Introductions of parasites or hosts often fail

Hedrick RP, McDowell TS, Marty GD et al (2003) Susceptibility of two strains of rainbow trout (one with suspected resistance to whirling disease) to *Myxobolus cerebralis* infection. *Dis Aquat Org* **55**:37–44 (doi:10.3354/dao055037).

Translocations of endangered host species can have unforeseen consequences

Obanda V, Kagira JM, Chege S et al (2011) Trypanosomosis and other co-infections in translocated black (*Diceros bicornis michaeli*) and white (*Ceratotherium simum simum*) rhinoceroses in Kenya. *Scientific Parasitol* **12**:103–107.

Can invasional meltdown occur?

Simberloff D & Von Holle B (1999) Positive interactions of nonindigenous species: invasional meltdown? *Biol Invasions* **1**:21–32 (doi:10.1023/A:1010086329619).

PARASITES AS INDICATORS OF ENVIRONMENTAL HEALTH

Parasites can help us monitor ecosystem integrity

Colegrove KM, Grigg ME, Carlson-Bremer D et al (2011) Discovery of three novel coccidian parasites infecting California sea lions (*Zalophus californianus*), with evidence of sexual replication and interspecies pathogenicity. *J Parasitol* **97**:868–877 (doi:10.1645/GE-2756.1).

Costanza R & Mageau M (1999) What is a healthy ecosystem? *Aquat Ecol* **33**:105–115 (doi:10.1023/A:1009930313242).

Khalil M, Furness D, Polwart A et al (2009) X-ray microanalysis (EDXMA) of cadmium-exposed eggs of *Bothriocephalus acheilognathi* (Cestoda: Bothriocephalidea) and the influence of this heavy metal on coracidial hatching and activity. *Int J Parasitol* **39**:1093–1098 (doi:10.1016/j.ijpara.2009.02.023).

Sures B (2008) Host-parasite interactions in polluted environments. *J Fish Biol* **73**:2133–2142 (doi:10.1111/j.1095-8649.2008.02057.x).

Valtonen ET, Holmes JC & Koskivaara M (1997) Eutrophication, pollution and fragmentation: Effects on the parasite communities in roach and perch in four lakes in central Finland. *Can J Fish Aquatic Sci* **54**:572–585.

Vidal-Martinez VM, Pech D, Sures B et al (2010) Can parasites really reveal environmental impact? *Trends Parasitol* **26**:44–51 (doi:10.1016/j.pt.2009.11.001).

PARASITES AS INFERENTIAL TOOLS TO PRESERVE HOST BIODIVERSITY

Parasites can provide information useful to preserving their hosts

Bollmer JL, Kimball RT, Whiteman NK et al (2006) Phylogeography of the Galapagos hawk (*Buteo galapagoensis*): a recent arrival to the Galapagos Islands. *Mol Phylogenet Evol* **39**:237–247 (doi:10.1016/j.ympev.2005.11.014).

Whiteman NK & Parker PG (2005) Using parasites to infer host population history: a new rationale for parasite conservation. *Anim Conserv* **8**:175–181 (doi:10.1017/S1367943005001915).

THE NEED TO PRESERVE PARASITE DIVERSITY

Parasites play key roles in maintaining ecosystem health

Hudson PJ, Dobson AP & Lafferty KD (2006) Is a healthy ecosystem one that is rich in parasites? *Trends Ecol Evol* **21**:381–385.

Lafferty KD, Allesina S, Arim M et al (2008) Parasites in food webs: the ultimate missing links. *Ecol Lett* **11**:533–546 (doi:10.1111/j.1461-0248.2008.01174.x).

Nichols E & Gomez A (2011) Conservation education needs more parasites. *Biol Conserv* **114**: 937–941.

Parasites are drivers of biodiversity

Blais J, Rico C, van Oosterhout C et al (2007) MHC adaptive divergence between closely related and sympatric African cichlids. *PLoS One* **2**:e734.

Kalbe M & Kurtz J (2006) Local differences in immunocompetence reflect resistance of sticklebacks against the eye fluke *Diplostomum pseudospathaceum*. *Parasitology* **132**:105–116.

Lazzaro BP & Little TJ (2009) Immunity in a variable world. *Philos Trans R Soc Lond B Biol Sci* **364**:15–26.

Loker ES (2012) Macroevolutionary immunology: a role for immunology in the diversification of animal life. *Front Immunol* (doi:10.3389/fimmu.2012.00025).

Prugnolle F, Manica A, Charpentier M et al (2005) Pathogen-driven selection and worldwide HLA class I diversity. *Curr Biol* **15**:1022–1027.

Van Doorn GS, Edelaar P & Weissing FJ (2009) On the origin of species by natural and sexual selection. *Science* **326**:1704–1707.

Parasites are a source of pharmacological and therapeutic novelties

Fallon PG & Alcami A (2006) Pathogen-derived immunomodulatory molecules: future immunotherapeutics? *Trends Immunol* **27**:470–476 (doi:10.1016/j.it.2006.08.002).

Golding H, Aliberti J, King LR et al (2003) Inhibition of HIV-1 infection by a CCR5-binding cyclophilin from *Toxoplasma gondii*. *Blood* **102**:3280–3286 (doi:10.1182/blood-2003-04-1096).

Golding H, Khurana S, Yarovinsky F et al (2005) CCR5 N-terminal region plays a critical role in HIV-1 inhibition by *Toxoplasma gondii*-derived cyclophilin-18. *J Biol Chem* **280**:29,570–29,577 (doi:10.1074/jbc.M500236200).

Johnston MJG, MacDonald JA & McKay DM (2009) Parasitic helminths: a pharmacopela of anti-inflammatory molecules. *Parasitology* **136**:125–147 (doi:10.1017/S0031182008005210).

Koh CY & Kini RM (2009) Molecular diversity of anticoagulants from haematophagous animals. *Thromb Haemost* **102**:437–453 (doi:10.1160/TH09-04-0221).

Maizels RM, Balic A, Gomez-Escobar N et al (2004) Helminth parasites—masters of regulation. *Immunol Rev* **201**:89–116 (doi:10.1111/j.0105-2896.2004.00191.x).

Smith P, Fallon RE, Mangan NE et al (2005) *Schistosoma mansoni* secretes a chemokine binding protein with antiinflammatory activity. *J Exp Med* **202**:1319–1325 (doi:10.1084/jem.20050955).

Slip-sliding away—parasite diversity is being lost

Bimi L, Freeman AR, Eberhard ML et al (2005) Differentiating *Dracunculus medinensis* from *D. insignis*, by the sequence analysis of the 18S rRNA gene. *Ann Trop Med Parasitol* **99**:511–517 (doi:10.1179/136485905X51355).

Dunn RR, Harris NC, Colwell RK et al (2009) The sixth mass coextinction: are most endangered species parasites and mutualists? *Proc R Soc Lond B Biol Sci* **276**:3037–3045 (doi:10.1098/rspb.2009.0413).

Eberhard ML, Ruiz-Tiben E, Hopkins DR et al (2014) The peculiar epidemiology of dracunculiasis in Chad. *Am J Trop Med Hyg* 90: 61-70.

Gompper ME & Williams ES (1998) Parasite conservation and the black-footed ferret recovery program. *Conserv Biol* **12**:730-732 (doi:10.1046/j.1523–1739.1998.97196.x).

Martin JW & Heyning JE (1999) First record of *Isocyamus kogiae* Sedlak-Weinstein, 1992 (Crustacea, Amphipoda, Cyamidae) from the eastern Pacific, with comments on morphological characters, a key to the genera of the Cyamidae, and a checklist of Cyamids and their hosts. *Bull South Calif Acad Sci*: **98**:26–38.

Mey E (2005) *Psittacobrosus bechsteini*: a new extinct chewing louse (Insecta, Phthiraptera, Amblycera) off the Cuban macaw *Ara tricolor* (Psittaciiformes), with an annotated review of fossil and recently extinct animal lice. *Anz Ver Thüring Ornithol* **5**:201–217.

Mihalca AD, Gherman CM & Cozma V (2011) Coendangered hard-ticks: threatened or threatening? *Parasit Vectors* **4**:71 (doi:10.1186/1756-3305-4-71).

The Challenge of Parasite Control

9

Good plumbing has done more for good health than good medicine.
William Trager

In This Chapter

The numbers tell the story. Parasitic disease still imposes an unacceptable burden on human welfare. Preeminent among those diseases is no doubt malaria. According to the World Health Organization (WHO) there were an estimated 207 million new cases of the disease in 2013, resulting in approximately 627,000 deaths. Ninety percent of these deaths occurred in sub-Saharan Africa. Of those, over 90% occurred in children under the age of five. Protozoa such as ***Giardia***, ***Cryptosporidium***, and ***Entamoeba*** cause an estimated 60 million cases of childhood diarrhea annually. ***Ascaris***, ***Trichuris***, and hookworms, the "big three" soil-transmitted helminths, are each responsible for well over 500 million infections each year. Schistosomiasis is estimated to infect at least 200 million people. As with malaria, tropical Africa bears a disproportionate share of this burden. As noted elsewhere, in addition to simply killing their victims, many of these diseases negatively affect cognitive development and severely impair the ability of infected individuals to achieve their full socioeconomic potential.

In light of the continuing importance of parasitic disease, the need for appropriate control measures remains an ongoing concern. In this chapter, we will investigate the many strategies that have been used in parasite control efforts as well as novel approaches, some of which are still on the drawing board. All control measures have associated advantages and problems, and we will consider when a particular line of attack is likely or unlikely to achieve satisfactory results. Indeed, because many parasitic diseases are complex and involve biological, human behavioral, and socioeconomic factors, more than one strategy often is required to lift the burden of parasitic disease. As we inspect such strategies, we will consider both the successes and failures in an effort to better elucidate what works and what does not.

The techniques we use to prevent and treat parasitic infections fall into three broad categories:

- manipulations or interventions designed to reduce parasite transmission
- use of anti-parasitic drugs
- vaccines

Species names highlighted in red are included in the Rogues' Gallery, starting on page 429.

Strategies in the first category are used to reduce contact between parasites and potential hosts. Anti-parasitic drugs can play two principal roles in control efforts: they may be used prophylactically to prevent infection or disease in high-risk individuals, or they can be used therapeutically to treat those showing signs and symptoms of disease. Vaccines, when successful, prevent disease in exposed individuals. In this chapter, we will discuss each of these three general categories and identify some of the relevant concepts and problems associated with each. Although we will consider appropriate examples as we proceed, this chapter does not provide an exhaustive compendium of control strategies. Rather, the goal is to better understand what those concerned with public health are up against when they decide to take on a parasitic disease, what weapons they have at their disposal, and what the most judicious use of those weapons might be.

9.1 STRATEGIES TO REDUCE PARASITE TRANSMISSION

The measure of success in any attempt to reduce parasite transmission is the sustained reduction in parasite incidence in a given area. Of course, when incidence is reduced, there should be associated reductions in prevalence and intensity, which should lead to lower rates of morbidity and mortality.

In more developed parts of the world, strategies used to reduce contact between parasites and their human hosts have essentially eradicated many parasitic diseases. Mosquito control, for instance, eliminated the scourge of malaria in Europe and North America. Adequate sanitation in tandem with clean food and water supplies has made food and waterborne disease uncommon. Yet in less developed parts of the world, poor infrastructure, a lack of will on the part of governments, and extreme poverty often conspire to make even basic sanitation nothing more than a distant dream (**Figure 9.1**). In addition, although certain diseases such as malaria attract considerable attention and research funding, many parasitic diseases tend to receive relatively low priority in the public health community. Many of these diseases, often collectively referred to as **neglected tropical diseases** (**Table 9.1**), are at least in theory preventable, if only the underlying socioeconomic factors could be appropriately addressed.

Figure 9.1 Nightsoil collection. Human feces, known as nightsoil, is still used as fertilizer in many regions of the world. Here, a nightsoil collector in Bangalore, India is at work in a public latrine. The use of nightsoil and other unsanitary practices allows many parasitic infections to flourish in less developed parts of the world. (Courtesy of Claude Renault, CC BY 2.0)

Parasite transmission may be reduced in various ways

Strategies to reduce transmission run the gamut from very simple (**Figure 9.2**) to cutting-edge high tech (for example, the release of transgenic mosquitoes). No single method has proven effective in all cases. Certainly, first and foremost, a careful consideration of a parasite's life cycle and mode of transmission is required if the disruption of transmission has any chance of success. Other biological parameters such as the intensity and periodicity of transmission will influence the selection of appropriate control measures. Yet unfortunately, even the most meticulously researched control program can be scuttled for reasons that are generally considered to be nonbiological and consequently outside the typical parasitologist's area of expertise. These include degree of government or nongovernment organization (NGO) support, political unrest, and even human nature (for example, social and cultural factors that might lead to non-compliance or a lack of interest in fostering control). Designing a program that successfully addresses all the biological as well as socioeconomic issues becomes no small feat.

Because the disruption of transmission requires a clear understanding of how a particular parasite is transmitted in the first place, we will organize

Table 9.1 Neglected tropical diseases caused by eukaryotic parasites. The World Health Organization recognizes 17 different neglected diseases of the tropics caused by various parasites. Only those diseases caused by eukaryotic parasites are listed. Other neglected diseases include those caused by bacteria (for example, yaws and trachoma) and viruses (dengue fever). Only the major involved parasite species are listed. In several instances (for example, echinococcosis caused by *E. multilocularis*) other related species are responsible for a relatively small percentage of cases.

Disease	Major Involved Parasite	Parasite Type
Chagas' disease	*Trypanosoma cruzi*	Protozoan-kinetoplastida
Dracunculiasis	*Dracunculus medinensis*	Nematode
Echinococcosis	*Echinococcus granulosus*	Cestode
Food-borne trematodiases	*Clonorchis sinensis* *Fasciola hepatica* *Opisthorchis viverrini* *Paragonimus westermani*	Trematode
African trypanosomiasis	*Trypanosoma brucei* complex	Protozoan-kinetoplastida
Leishmaniasis	*Leishmania species*	Protozoan-kinetoplastida
Lymphatic filariasis	*Wuchereria bancrofti* *Brugia malayi*	Nematode
Onchocerciasis	*Onchocerca volvulus*	Nematode
Schistosomiasis	*Schistosoma species*	Trematode
Soil-transmitted helminthiases	*Trichuris trichiura* *Ascaris lumbricoides* *Necator americanus* *Ancylostoma duodenale*	Nematode
Cystercercosis	*Taenia solium*	Cestode

Source: www.who.int/neglected_diseases/diseases/en/. With permission from WHO.

our discussion by modes of transmission, all of which were introduced in Chapter 3 (pages 77–85).

Parasites using trophic transmission can be controlled by insuring food safety

Humans become exposed to a wide range of parasites through the consumption of food items containing infective life-cycle stages. An example is the beef tapeworm (***Taenia saginata***), contracted by eating raw or undercooked beef infected with tapeworm cysticerci. Likewise, the lung fluke (***Paragonimus westermani***) is transmitted to fish-eating mammals, including humans, when they eat undercooked or raw freshwater crabs containing metacercariae. Freshwater fish that serve as second intermediate hosts may contain ***Clonorchis sinensis*** (the Chinese liver fluke) metacercariae. An estimated 35 million people in East Asia are infected with this trematode from the consumption of undercooked or raw fish.

Obviously the key to successfully disrupting transmission for these parasites is through the application of modern farming practices and vigilant food safety inspection. Such practices, as well as the increased use of frozen meats, are believed to be the principal reason for the decline in ***Toxoplasma gondii*** seroprevalence in many developed countries over the past several decades. Meats containing cysts, especially mutton and pork, are recognized as the principal source of infection for this apicomplexan parasite. Contaminated meat is now suspected to be an even more important transmission route than

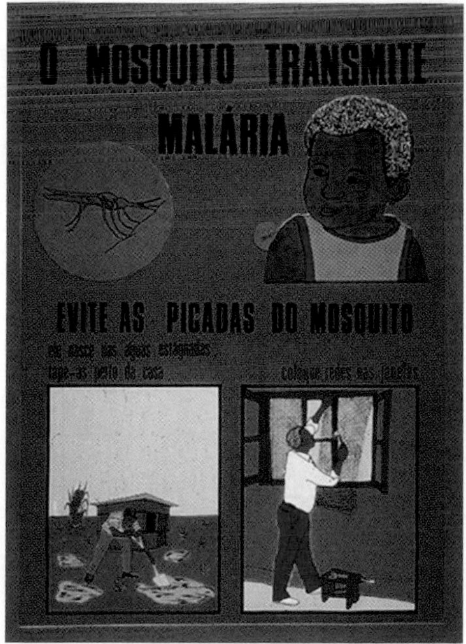

Figure 9.2 Malaria control in Brazil. This poster illustrates the role of public education in disease control and highlights some simple measures, such as installing window screens and eliminating mosquito breeding sites, that can contribute to a reduction in parasite transmission.

is exposure to infective oocysts in cat feces. Yet even in the most developed of countries, cultural factors, including the popularity of cats as house pets, and simple dietary preferences can conspire to undermine otherwise rigorous control efforts. In the United States, for instance, over 10% of the population may be infected. Seroprevalence in France is over 80%, probably because of the popularity of rare or raw meat dishes.

A further example of how simple alterations in food handling practices can decrease the incidence of a trophically transmitted parasite involves anisakiasis, caused by consumption of nematodes in the genus *Anisakis* (family Anisakidae). Marine mammals serve as the usual definitive host, becoming infected as they consume infected fish or squid (**Figure 9.3**). Humans may also become infected if they consume infected raw seafood, such as sushi or ceviche. Although adequate cooking or freezing of seafood kills the infective larvae, marinating or salting does not necessarily do so.

Following consumption, L3 larvae attempt to penetrate the lining of the stomach or intestine. Symptoms such as violent abdominal pain, nausea, and vomiting can begin within several hours, although some infected individuals remain asymptomatic. Diagnosis can be made via gastroscopic examination, followed by removal of larvae.

Anisakiasis was first described in 1955 in the Netherlands. It would be five years, however, during which a cluster of cases occurred, before the causative agent was identified as *Anisakis*. Epidemiological investigation into the apparently new disease indicated that infection resulted from the consumption of raw herring.

There were two reasons for the Dutch emergence of anisakiasis infection. First, raw salted herring, locally known as green herring, became popular in the Netherlands in the years following World War II. Second, the manner in which commercial fishermen processed fish was changing. Traditionally, Dutch fishermen cleaned their fish at sea. This process eliminated almost all larvae, which were imbedded in the lining of the gastrointestinal tract. However, beginning in the 1950s, freshly caught fish were placed on ice to await cleaning on shore. Several days might pass before the ships returned to port. During this time, at least some of the larvae migrated into muscle tissue. Later cleaning no longer eliminated the risk of human infection. Once this became apparent, legislation was enacted that required all fish to be frozen prior to human consumption. This practice has virtually eliminated anisakiasis in the Netherlands.

Proper sanitation is the key to controlling parasites transmitted via the fecal-oral route

Famed parasitologist William Trager is probably best remembered for developing the methodology to culture ***Plasmodium falciparum*** in the laboratory. He is also credited with the quotation that opened this chapter, which in a nutshell sums up how parasites using fecal–oral transmission to infect humans are controlled. To put it another way, proper sanitation, which includes the hygienic disposal of human and animal waste and the provisioning of safe, uncontaminated food and water, disrupts the transmission cycle of those parasites that rely on fecal–oral transmission. If assiduously applied, simple sanitation practices can virtually eliminate such parasites in human populations. The same principles apply to parasites such as schistosomes and hookworms, which release eggs into the environment along with the definitive host's feces and later re-enter their human host through direct penetration of the skin.

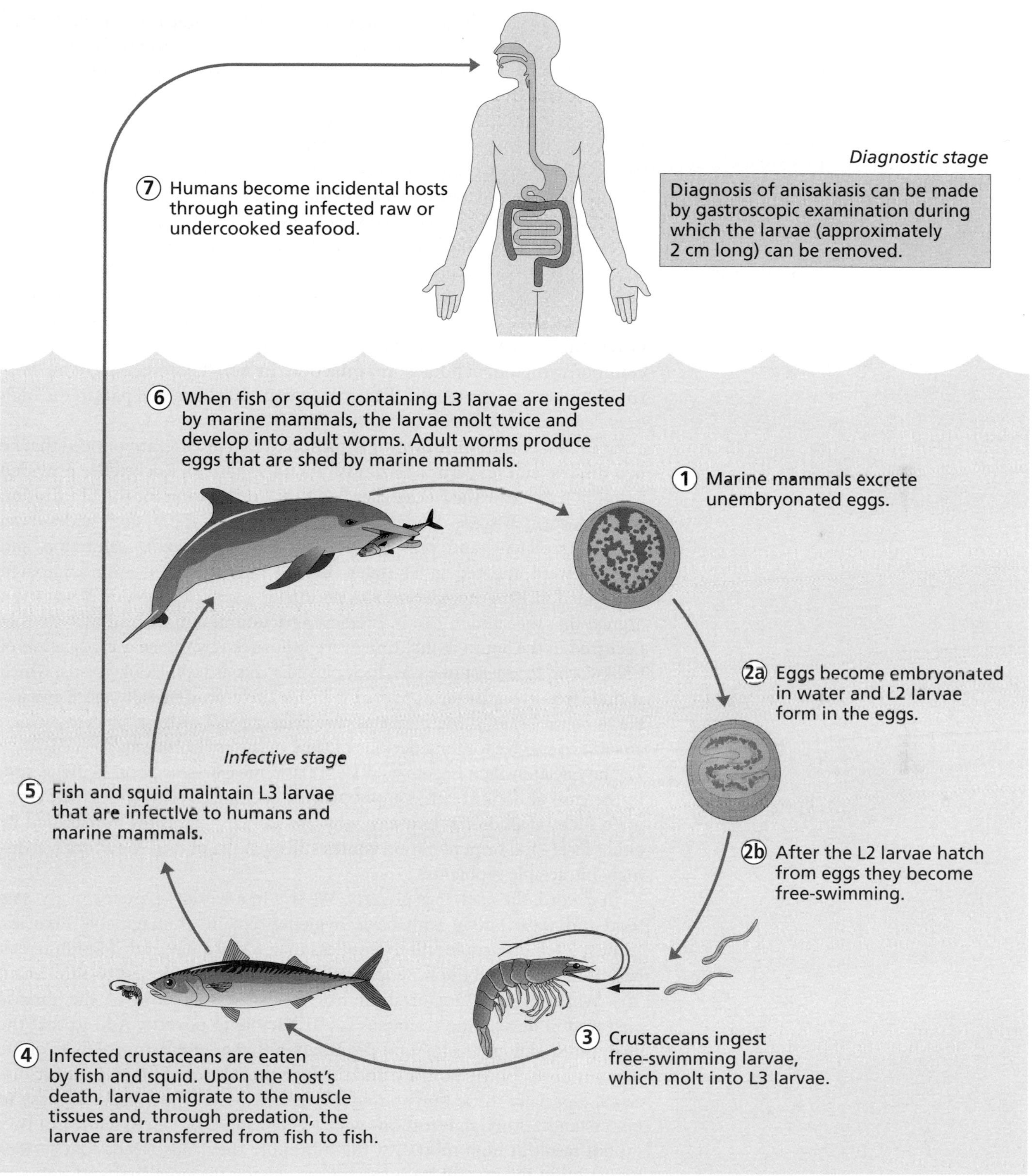

Figure 9.3 Life cycle of *Anisakis* spp. The *Anisakis* life cycle involves at least three hosts. Marine mammals that serve as definitive hosts become infected when they consume the second intermediate host, saltwater fish or squid containing infective L3 larvae. Eggs passed in the feces of the definitive host ultimately hatch, releasing L2 larvae. The larvae are consumed by crustaceans, which serve as the first intermediate host, and the second intermediate host, a fish or squid, becomes infected as it consumes the crustacean. Predatory fish or squid that consume the second intermediate host act as paratenic hosts. Humans who consume raw fish or squid containing infective L3 larvae can likewise become infected. (Courtesy of CDC.)

Disease control through improved sanitation is not a particularly new idea. An especially illustrative example from the early twentieth century is provided by the efforts to eliminate hookworm disease, caused by ***Necator americanus***, in the American South. Hookworm eggs, released in the feces of infected humans, hatch and release larvae that develop in the soil. Infective L3 larvae penetrate the skin of a human, ultimately returning to the intestine, where they develop into adult worms. Because of the prevailing high humidity and temperatures, hookworm larvae were able to thrive in the soil of the southern United States. Furthermore, in the days before indoor plumbing was common, especially in rural areas, waste disposal often meant an outhouse with a moist, shaded dirt floor—ideal for the survival of hookworm larvae. Complicating the situation further was the lack of shoes for many poor, rural individuals. Because of the anemia and impaired cognitive development that results in many infected individuals, severe lethargy is a common symptom of hookworm infection. In fact, the stereotype of the lazy, rural southerner, common at the time, owes itself in no small part to the high prevalence of hookworm infection.

In 1902, Charles Stiles, investigating hookworm disease, announced that he had discovered the "germ of laziness". In 1909, John D. Rockefeller provided $1 million to create the Rockefeller Sanitary Commission for the Eradication of Hookworm Disease. Under the auspices of the commission, widespread testing, treatment, and prevention programs emphasizing sanitation and hygiene were initiated in 11 states. By the time the eradication campaign was ended in 1914, hookworm was no longer a serious problem. It was even argued that the sudden rise in literacy, agricultural output, and income that occurred in the South at that time were influenced by the near eradication of hookworm. In at least one way, luck played a role as well. Hookworms, which include free-living larvae as part of their life cycle, are generally more amenable to control than those parasites that release resistant eggs or cysts, which are able to survive for long periods of time in the environment.

The question then becomes, if such basic measures as good hygiene and, in the case of hookworms, simply providing children with shoes, can eliminate such infections in humans, why are so many parasites transmitted by either fecal–oral or penetration routes still such major and sometimes seemingly intractable problems?

In a word, the answer is poverty. We live in a world where for many, safe food and water, along with basic hygiene, remain unimaginable luxuries. Almost 3 billion people still live on less than $2.00 a day and 2.4 billion lack basic sanitation. A billion people still have no regular access to safe water. The World Health Organization has estimated that 45% of the disease observed in developing countries is attributable to poverty. Add to that the civil unrest and environmental degradation that are unfortunate facts of life in many developing countries, and the high prevalence of many parasitic diseases, especially those transmitted via the fecal–oral route becomes easy to understand. Although infections with ***Ascaris*** and ***Giardia***, to name just two, do not result in high mortality, the morbidity they cause leads to a serious reduction in income for people who are already marginalized and living in severe poverty. The vicious cycle that exists between poverty and poor health exacerbates an already grim situation.

Various other factors influence the success of control efforts

Irrespective of poverty, certain parasites are more amenable to control efforts than others. Guinea worm (***Dracunculus medinensis***) provides an excellent example as well as a genuine success story. The detailed life cycle of this remarkable nematode is provided in the Rogues' Gallery. In this section, we briefly recount some of the details relevant to control.

Humans become infected after inadvertently swallowing infected, freshwater copepods that serve as intermediate hosts. These small crustaceans often inhabit sites that humans use for drinking water. Once swallowed, larvae penetrate the small intestine and enter connective tissues. Development to adulthood, followed by mating, ensues. Males die after mating, while females migrate to the skin, most often on the legs. By this time (about a year after initial infection), the uterus of the ovoviviparous female is full of embryos. The body wall of the female, which is under high pressure from the large number of embryos, will eventually rupture and release L1 larvae. A hypersensitivity response to these larvae causes the formation of a painful blister on the skin. The blister may eventually rupture, providing an exit for the larvae. If an infected person enters water (perhaps to relieve the severe burning sensation caused by the blister), the uterus of the female is stimulated to contract, expelling larvae into the water. The larvae must next be consumed by a copepod, where they develop into the infective stage for humans. Because of its unusual mode of egress from its human host, we cannot say that the Guinea worm relies on fecal–oral transmission. We discuss it here because its transmission via contaminated water means that basic principles relevant for other parasites using fecal–oral transmission apply equally well.

In the mid-1980s the Carter Center, established by former President Jimmy Carter, announced its goal of Guinea worm eradication (**Figure 9.4**). That goal is now close to realization (see Figure 8.20). To date, only two pathogens, smallpox virus and the virus causing the cattle disease rinderpest, have been completely eradicated in the wild. If the Guinea worm joins this list, it would be the first animal parasite to be intentionally driven to extinction. Eradication in South Asia, the Middle East, and much of Africa is already complete. In 1986 there were 3.5 million cases in 21 African and Asian countries. In 2013 there were only 148 cases and these were confined to four countries in Africa: South Sudan, Ghana, Chad and Mali. Almost all of these cases were in South Sudan, highlighting the manner in which civil strife can undermine disease control efforts. In fact, the original target date for eradication—1995—came and went largely because of Sudan's lengthy civil war.

Why is it even possible to contemplate the complete eradication of ***D. medinensis***, while other helminths still resist all attempts to control them? Two principal factors related to the biology of the parasite are involved.

First, ***D. medinensis*** has an Achilles' heel in its life cycle; Guinea worms are the only helminth parasite that is almost entirely dependent on the consumption of contaminated drinking water to achieve transmission to its definitive host. Furthermore, its transmission is highly focal. In an ironic twist, with Guinea worm we observe a parasite with obligatory aquatic stages that does best under drought conditions. In arid North Africa, where the parasite now makes its last stand, few or no new cases are seen during relatively wet years. But during a drought, as rivers and many water bodies dry up, both humans and copepods increasingly depend on the few isolated water sources that remain. All the elements necessary for increased transmission—humans (including those previously infected), intermediate hosts, and water—converge. Copepods thrive in the warm water, and humans may have no other options when it comes to finding water for drinking, cooking, and bathing. Consequently, control and eradication of Guinea worm can focus on preventing consumption of contaminated water. The strategies to do this in the case of ***D. medinensis*** are numerous, simple, and often very cheap. A simple piece of muslin can be used to filter copepods out of drinking water

Figure 9.4 Guinea worm eradication. As part of a public education campaign in South Sudan, a poster depicts a woman warning a Guinea worm patient not to step into fresh water. [© The Carter Center /L. Gubb.]

and small handheld filters can be used to insure that any water consumed is copepod-free. Treating water with Temephos, a chemical that kills copepods that is relatively safe for vertebrates, also can interrupt transmission.

Second, although other members of the genus may infect other vertebrates, ***D. medinensis*** relies almost exclusively on humans as definitive hosts; except for some unusual cases involving dogs, there are no animal reservoirs. This fact alone greatly simplifies control efforts. Controlling a **zoonosis**, an animal pathogen that can also be transmitted to and cause disease in humans, is a daunting task. Reservoir animals must be eliminated, vaccinated, or treated if transmission is to be interrupted, and all of these options are generally undesirable or unfeasible at best. But the Guinea worm, like the already eradicated smallpox virus and poliovirus, which is targeted for eradication, has nowhere to hide outside of humans. Thus, barring additional civil or political strife in its last African redoubts, or unexpected wrinkles in the life cycle involving alternative definitive or paratenic hosts, we may be witnessing the fiery serpent's swan song.

The control of vector-borne diseases focuses on reducing human–vector contact

For those parasitic diseases transmitted by vectors, the goal of control efforts can be summed up simply: reduce the likelihood that infected vectors and susceptible humans will come into contact. Such efforts were first attempted late in the nineteenth century when it was realized that filariasis, yellow fever, malaria, and other infectious diseases were spread through the bites of infected arthropods (see Figures 3.3 and 3.4). Window screens and bed nets were used as barriers to prevent contact between vectors and humans. Environmental modifications such as swamp drainage to reduce mosquito oviposition sites or the application of kerosene or other toxic compounds to water bodies to kill larvae were employed to reduce vector density.

The discovery of DDT radically altered vector control efforts

Although the use of chemical compounds to kill insect pests dates back into antiquity, the entire notion of vector control changed dramatically in 1939, when the Swiss chemist Paul Muller discovered that dichlorodiphenyltrichloroethane (commonly called DDT) was highly effective against a variety of insect species (**Figure 9.5**). As World War II concluded, DDT was already in use for the control of malaria and louse-borne typhus. Following the war, its use skyrocketed as it became readily available for the control of agricultural pests. In 1948, Muller won the Nobel Prize in medicine for his discovery and by 1955, the World Health Organization announced its plans for the worldwide eradication of malaria, relying largely on the use of DDT, as well as the anti-malarial drug chloroquine.

A

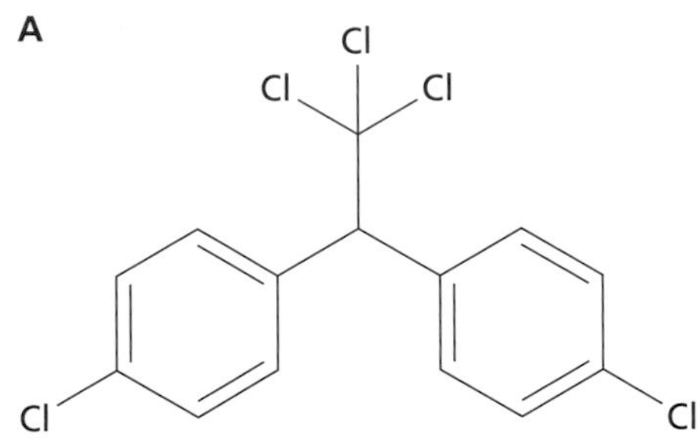

B

Figure 9.5 DDT. (A) The structure of dichlorodiphenyltrichloroethane (DDT). (B). A US soldier is demonstrating DDT-hand spraying equipment in 1944. DDT was used to control the spread of lice carrying typhus, and it was considered harmless to humans unless ingested. (B, Courtesy of CDC.)

DDT is an example of an organochlorine insecticide, a highly hydrophobic compound that kills insects by opening sodium ion channels on neurons. The result is unregulated neuron firing, spasms, and death. However, some insects carry mutations in the genes for the sodium ion channel proteins. DDT is less effective at altering the function of some of these mutated proteins, setting the stage for resistance. Other insects, alternatively, are able to resist the effects of DDT by up-regulating certain P450 genes. These genes encode proteins that act enzymatically to oxidize certain molecules, including exogenous toxins such as organochlorines.

The World Health Organization program was initially a stunning success. Malaria was eradicated from large parts of the Pacific, North Africa,

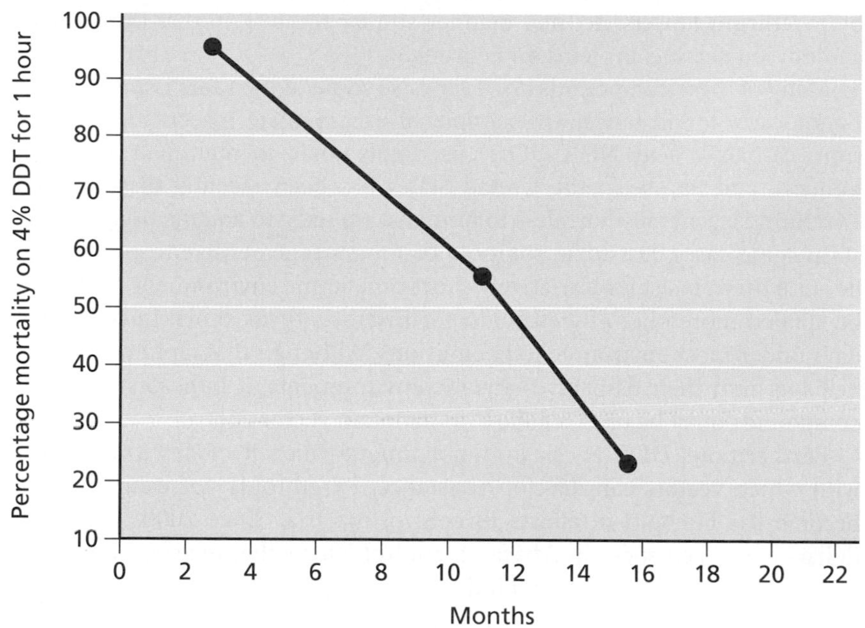

Figure 9.6 Mosquito resistance to DDT. Increase in the frequency of pesticide resistance in mosquitoes (*Anopheles culicifacies*) after regular spraying with DDT. A sample of mosquitoes was captured at each time indicated and the number that were killed by a standard dose of DDT (4% DDT for one hour) in the laboratory was measured. Whereas almost all mosquitoes were still sensitive to the pesticide three months after the start of DDT exposure, by 16 months only about 20% of mosquitoes remained sensitive. (Adapted from Curtis CF, Cook LM & Wood RJ [1978] *Ecol Entomol* 3[4]:273–287. With permission from John Wiley and Sons.)

and the Caribbean. It was completely eradicated from Taiwan, and dramatic decreases in mortality were reported across the Indian subcontinent. The intense use of the insecticide (40,000 tons used annually worldwide) imposed a strong selection pressure on exposed mosquitoes, however, and before long, widespread mosquito resistance undermined eradication efforts (**Figure 9.6**). Consequently, malaria resurged and in some areas transmission even increased above precontrol levels. Transmission remained low only in those relatively well-developed areas with an established health care infrastructure or where transmission was low or seasonal to begin with. In 1969, the goal of eradication was quietly abandoned, with a new focus on control and patient treatment.

Compounding the problem were the environmental and health issues associated with DDT. Trepidation about such hazards arose as early as the 1940s, but they were largely ignored in light of the astounding benefits the insecticide was providing. In 1962, however, Rachel Carson published *Silent Spring*, in which the environmental impact of DDT was described in lyrical detail. Concern over DDT's use escalated and over the next several years evidence that supported Carson's contention accumulated. DDT was finally banned in the United States for most purposes in 1972. Similar bans were instituted in most other countries. Although it is still in limited use today, DDT is no longer viewed as a viable option for vector control in the vast majority of situations.

Because of its highly lipophilic nature, DDT is stored in many body tissues and it readily bioaccumulates as it moves up food chains. The effect is most pronounced in predatory birds, in which DDT interferes with reproductive success (**Figure 9.7**). In humans and other mammals, DDT is known to interfere with endocrine signaling. It may also be both a mutagen and a carcinogen.

Figure 9.7 The effect of DDT. Because it is lipophilic, DDT accumulates in the tissues of animals that consume it. Birds that feed on fish or other animals are particularly vulnerable because of the position they occupy at the apex of food chains. Because of the increasing concentration with increased trophic level, such birds accumulate high levels of the pesticide in their tissues. Calcium carbonate deposition in eggshells is inhibited, resulting in eggs with thin shells that cannot bear the weight of the incubating parent bird. Such eggs are broken prematurely, resulting in greatly impaired reproductive success. This affected egg was photographed in 1970, shortly before bans on widespread DDT usage were instituted. (Photo by Dave Siddon, wildlifeimages.org.)

Newer insecticides provide alternatives to DDT

Many alternatives to DDT exist and insecticides are still widely used to reduce vector density. Examples include the organophosphates such as malathion and the pyrethroids such as deltamethrin. Organophosphates work by inhibiting acetylcholinesterases (enzymes that hydrolyze the neurotransmitter acetylcholine) in insects. Pyrethroids, similar to compounds produced by pyrethrum flowers, act in a manner similar to DDT in that they disrupt sodium ion channel proteins on neurons.

Many of these compounds however raise some of the same issues as DDT. Bendiocarb, for instance, an example of a carbamate insecticide (derived from carbamic acid–NH_2COOH), is highly toxic to mammals including humans, and its use in the United States has been recently discontinued. Pyrethroids, generally harmless to humans, are toxic to aquatic organisms.

In some cases, less toxic compounds are more expensive to produce or, because they persist for a relatively short time in the environment, they must be applied more often. Pyrethroids, for instance, break down in only a few days under most environmental conditions. Although this rapid breakdown helps to limit their damage to aquatic environments, it indicates that these compounds must be used regularly to maintain vector control.

Furthermore, DDT is certainly not unique when it comes to the speed with which vectors can develop resistance. Pyrethroids, for example, have become the linchpin in efforts to control malaria. Since 2000, billions of dollars have been spent in Africa on control efforts that focus on the use of pyrethroid-treated bed nets and indoor spraying. In fact, pyrethroids are currently the only insecticides approved by the World Health Organization for impregnating bed nets because of their low cost, efficacy, and relative safety. These recent control efforts have proven successful, greatly reducing the death rate from malaria in Africa over the last decade. Although the possibility of resistance was recognized from the start, the public health community thought that the scale up in bed net and spraying campaigns was worth the risk, considering the number of lives that might be saved. That possibility, however, has unfortunately become reality. The intense selection pressure on mosquitoes has translated into increasing resistance to pyrethroids, as well as the emergence of resistance in many new areas (**Figure 9.8**).

The World Health Organization is now formulating a plan to combat the growing problem of resistance to this mainstay insecticide. Central to the plan will be the rotation of pyrethroids with other insecticides to reduce the selection pressure on mosquitoes to any one specific insecticide. In areas where pyrethroids are being used on bed nets, a different insecticide will be recommended for indoor spraying. Ultimately, however, entirely new insecticides are needed, especially those that can be used on bed nets. The development of new insecticides, however, may take time. Compared to more high profile drug and vaccine research, vector control is notoriously underfunded, which helps to explain why those interested in control currently have limited options.

Transmission of vector-borne parasites can be reduced through environmental manipulation

In the past, vector-borne diseases were often controlled by draining water bodies, altering irrigation practices, managing vegetation, and employing other environmental manipulations designed to negatively affect vectors and thereby reduce their populations (**Box 9.1**). In recent years, however, environmental manipulation for vector control has been largely relegated to the back burner. In schistosomiasis control programs, for example, transmission control of any kind has been largely replaced by morbidity control through chemotherapy. Regarding malaria, as discussed above, vector control efforts focus almost entirely on the use of insecticides. Manipulations that make habitats less suitable for ***Anopheles*** mosquitoes are largely the fortuitous consequence of economic development or improved housing. The story is similar for other vector-borne parasitic diseases.

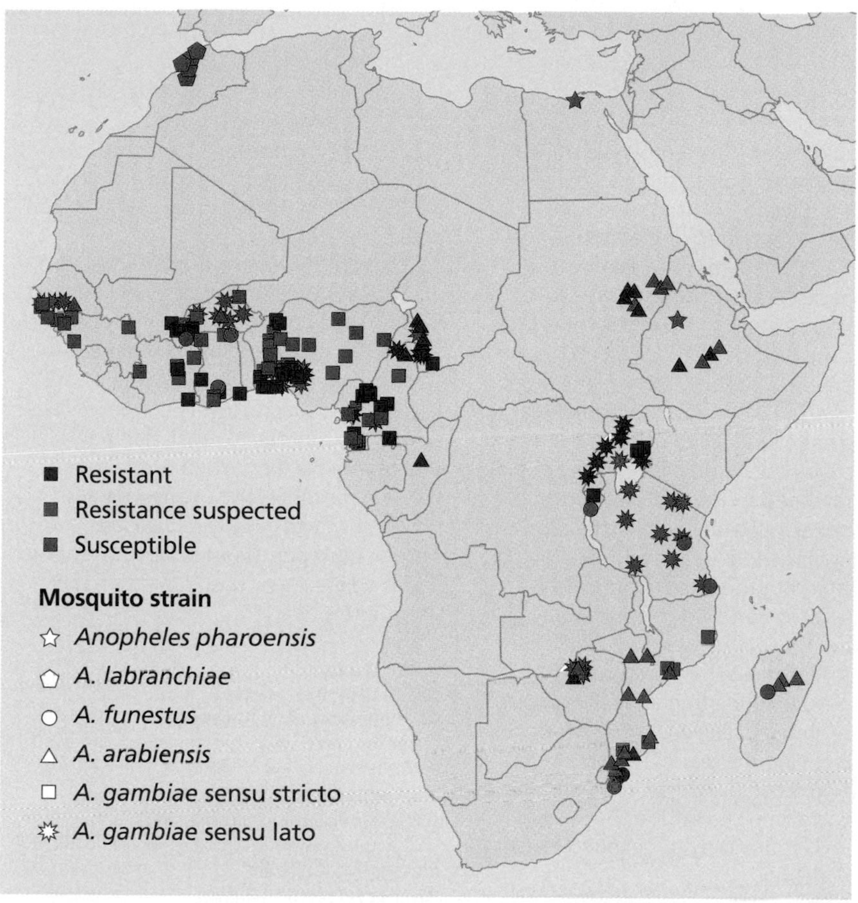

Figure 9.8 Mosquito resistance to pyrethroids. These data were compiled by the WHO in 2010, indicating where resistance to pyrethroid insecticides has emerged. The situation is believed to have worsened since then. (Adapted from Butler D [2011] *Nature* 475:19. With permission from Macmillan Publishers Ltd.)

This lack of emphasis on environmental manipulation may be a missed opportunity. Although there is little information on the manner in which parasitic disease burden might be alleviated through environmental control, a World Health Organization study in 2006, suggested that the effect could be considerable. A panel of experts considered specific diseases and the environmental variables that might reasonably be manipulated to reduce disease prevalence. The panel concluded that for selected parasitic diseases the reduction in prevalence could range from 10% for onchocerciasis to 100% for schistosomiasis (**Table 9.2**).

Biological control offers the possibility of low-cost, sustainable control

In light of the unresolved problems of vector resistance to pesticides and the sustained and often logistically complex efforts required to maintain effective environmental manipulation, **biological control** may offer a relatively low-cost and possibly self-sustaining means of disrupting transmission of parasitic diseases. Biological control is the purposeful introduction of a biological agent to control or eliminate an organism that is believed to be a pest. Most control agents either prey upon or compete with the pest. At least in principle, once an appropriate biological control agent is successfully introduced into a particular area, there is little or no further cost or effort required. The controlling agent reduces the target pest to low, acceptable levels. Ideally, the controlling agent survives so that reapplication of the agent is only necessary occasionally. In addition, unlike pesticides to which target organisms can evolve resistance, each evolutionary change on the part of the target is matched by a counter evolutionary shift by the controlling agent.

BOX 9.1

Traps for Tsetse (*Glossina*)

Parasitologists are rarely consulted for fashion tips. But should you be considering a trip to the African "fly belt," here is a little advice as you pack. Leave the blue jeans and your favorite little black dress at home. Instead, we suggest lighter hues. Maybe a nice pair of khakis or a yellow blouse. Not only will you look tasteful and elegant, but you'll be less likely to come down with trypanosomiasis.

Tsetse, it turns out, see color, and when seeking a host on which to feed, dark colors, especially black and blue, are particularly attractive to a hungry fly. Red is somewhat less appetizing. Green and especially yellow or white are least appealing.

The tsetse's color preferences have led to the development of effective traps. These traps reduce fly numbers and therefore disrupt trypanosome transmission. Such trapping has become an important component in trypanosomiasis control programs. Tsetse traps may consist of a single sheet of material with alternating blue and black sections (Figure 1) or they may be more elaborate conical or pyramidal shaped traps. The traps may be made of cotton, polyester, or plastic. When the more simple sheets are used, the material is treated with insecticide, killing the flies when they land. In more complex traps, the flies are generally lured inside, where they are either killed by insecticide or where they are channeled into a collection chamber from which they cannot escape. The original attempts to catch tsetse often employed traps in the shape of cattle, but subsequent studies have shown that such traps do not increase trapping efficiency. Although shape may not matter to a hungry tsetse, the shade of the color and manner in which the black and blue areas are distributed does. Deep black is preferable to off-black, whereas electric blue (technically known as phthalogen blue) has a reflectivity that is most effective at attracting flies. Traps designed to draw the flies inside should be blue on the outside and black on the inside. Both black and blue are equally attractive to tsetse, but the flies are more likely to actually land on black. If traps were black on the outside, or if they were entirely black, flies would frequently land on the outside of the trap and fail to enter. Consequently, blue is used to attract the flies, and the black interior tempts them to enter and alight. Traps should be placed in woodlands or bush areas where flies are likely to be found. Tsetse feed most commonly at dawn or dusk in relatively open areas. Thus, the ideal trap location is not shaded during the times of peak fly activity.

Trapping programs require sustained effort. Traps must be properly maintained, as they tend to get blown over, ripped, or stolen. Over time, as the color fades, the traps become less effective. Ideally traps should be inspected every couple of days to adjust supports, to repair damage, and to remove flies. When used properly, tsetse traps represent an effective and low-cost way to greatly reduce the burden of trypanosomiasis in a part of the world where such control measures are desperately needed.

Figure 1 A tsetse trap. A sheet trap designed to catch tsetses. Sheet traps are treated with insecticide to kill alighting flies. The efficiency of tsetse traps can be increased by placing an open container filled with cattle urine near the trap, as the odor helps to draw tsetses into the area. (From Hirayama K [2008] Schistosomiasis and Immunity. With permission from WHO/TDR.)

References

Green CH (1986) Effects of colours and synthetic odours on the attraction of *Glossina pallidipes* and *G. morsitans morsitans* to traps and screens. *Physiol Entomol* 11:411–421.

Green CH (1989) The use of two-coloured screens for catching *Glosslna palpalis* (Robineau-Desvoidy) (Diptera: Glossinidae). *Bull Entomol Res* 79:81–93.

Green CH (1990) The effect of colour on the numbers, age and nutritional status of *Glossina tachinoides* (Diptera: Glossinidae) attracted to targets. *Physiol Entomol* 15:317–329.

Biological control has long been used to control agricultural pests. A variety of potential control agents, including viruses, bacteria, and predatory or parasitoid insects, have been evaluated and used, and an extensive body of literature exists detailing both the successes and failures of such efforts. When used to control either vectors or intermediate hosts of parasitic diseases, the goal, of course, is to disrupt the parasite life cycle and thereby decrease transmission to tolerable levels.

An example is the control of ***Aedes aegypti***, the most important vector of dengue in Vietnam, using predatory freshwater copepods (genus *Mesocyclops*) that consume mosquito larvae. Likewise, the larvae of *Toxorhynchites* mosquitoes are voracious predators of mosquito larvae, including those of vectors such as ***A. aegypti***. Because adult *Toxorhynchites* do not take blood meals and therefore do not transmit diseases, their use for vector control has been advocated, as has the use of various fish species

Table 9.2 Attributable fractions and disease burden resulting from the environment. The estimates listed represent the fraction of total disease burden that could in theory be eliminated through alterations in the environment. The data are based on a systematic literature review on environmental risks conducted by over 100 experts in the environmental components of disease. These experts provided quantitative estimates for 85 diseases. Only those diseases caused by eukaryotic parasites are listed.

Disease	% Estimate of Attributable Fraction
Malaria	42
Soil-transmitted helminthiases	100
Schistosomiasis	100
Chagas' disease	56
Lymphatic filariasis	66
Onchocerciasis	10
Leishmaniasis	26

(Adapted from Pruss-Ustun A & Corvalon C [2007] *Epidemiology* 18:167–178. With permission from Wolters Kluwer Health.)

known to prey on mosquito larvae. Under certain conditions such as small, artificial water bodies in which mosquitoes frequently breed and in rice paddies, a measure of successful control has been demonstrated. The fungus *Lagenidum giganteum* and the bacterium *Bacillus thuringiensis*, both of which are pathogenic for arthropods, have been used to control mosquito and black fly larvae. The molluscan intermediate hosts of schistosomiasis are thought to have been controlled through the introduction of competitor gastropod species such as *Melanoides tuberculata* and *Marisa cornuarietis* on several Caribbean islands. *Biomphalaria glabrata* was historically an important intermediate host for ***Schistosoma mansoni*** in Puerto Rico, the Dominican Republic, and elsewhere. The successful eradication of this species on Guadeloupe, for instance, was attributed in part to the introduction of *M. cornuarietis* from Venezuela (**Figure 9.9**). It has even been suggested that under certain conditions, parasitic trematodes within the genus *Echinostoma* can be used to control schistosomiasis. Echinostome eggs can be produced

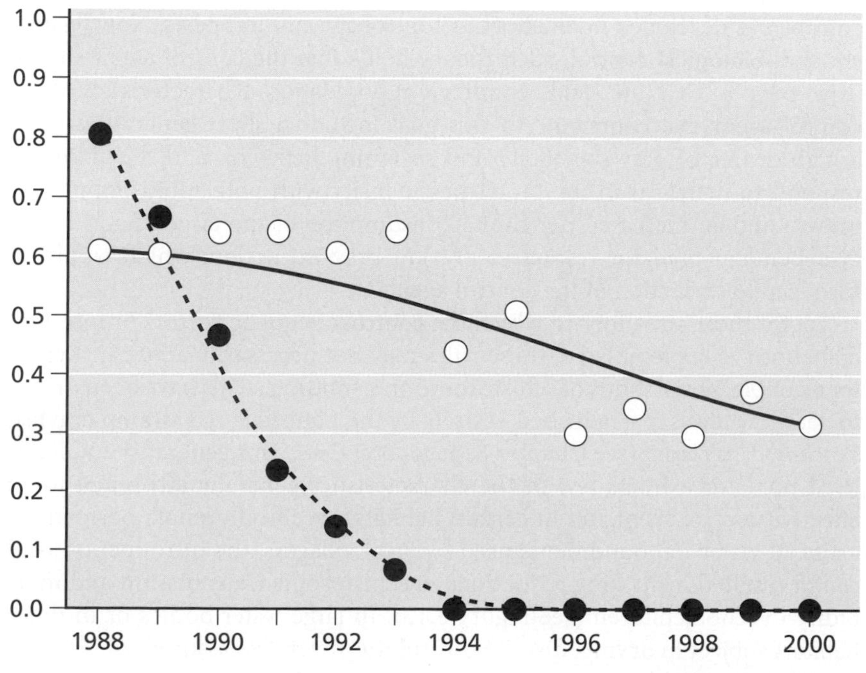

Figure 9.9 Biological control of a schistosome-transmitting snail. *Biomphalaria glabrata* previously served as an intermediate host for *Schistosoma mansoni* on certain islands in the Caribbean area. In late 1987, the neotropical prosobranch snail *Marisa cornuarietis* was introduced on the island of Guadeloupe into a series of small ponds containing *B. glabrata*. Numbers of habitats in which *B. glabrata* survived were determined annually from 1988 until 2000. The *Y* axis indicates the proportion of ponds in which *B. glabrata* was present during this time. White circles represent control ponds without *M. cornuarietis*. Black circles represent experimental ponds into which *M. cornuarietis* was introduced. The ability of *M. cornuarietis* to eliminate *B. glabrata* in experimental ponds was attributed to its consumption of both *B. glabrata* egg masses and the aquatic vegetation on which these masses were typically laid. (Adapted from Pointier JP & David P [2004] *Biol Control* 29:81–89. With permission from Elsevier.)

in large numbers in the lab and subsequently added to water bodies known to be foci of schistosome transmission. The echinostome miracidia infect the same snails harboring schistosome larval stages. The echinostome larvae, however, produce generations of rediae within their molluscan intermediate hosts. Schistosomes produce only sporocysts. Because the rediae actually feed on the sporocysts (see Figure 6.15), the echinostome can effectively clear the snail of its schistosome infection. However, although this may be an appealing strategy, it remains unclear how feasible sustained control using echinostomes can be under field conditions.

It is clear that biological control is no panacea, and biological control can rarely function as the sole method of control except under specific conditions. The role of *Gambusia affinis* in mosquito control is illustrative. This fish, which feeds on mosquito larvae, has been a workhorse of mosquito control efforts for over a century. Because it tolerates a wide range of environmental conditions and because of its reputation as a voracious mosquito predator, *G. affinis* is probably now the most widely distributed freshwater fish in the world. Yet its reputation may be undeserved. In many areas it has caused widespread disruption of natural ecosystems, and it has proved to be very detrimental to native fish, amphibians, and other aquatic organisms. Furthermore, its ability to control mosquitoes is at best equivocal. In many habitats, native fish have already maximally reduced mosquito populations and the addition of *G. affinis* has provided no additional control. Furthermore, in some resource-limited environments, *G. affinis* might actually increase the ability of adult mosquitoes to transmit disease. When predators such as *G. affinis* remove a portion of the mosquito larvae in such habitats, there may be reduced intraspecific competition among the surviving larvae. These survivors may develop into larger, more fertile adults, and both of these characteristics are indicative of a higher vector capacity (see Chapter 3, pages 81–82). Yet in some situations *G. affinis* has apparently delivered as advertised (**Figure 9.10**). The eradication of malaria from parts of South America and from the Ukraine and Southern Russia during the mid-twentieth century is believed to have resulted from the introduction of *G. affinis*.

Figure 9.10 *Gambusia affinis.* A monument to *G. affinis* erected at Sochi, a Russian town on the Black Sea. The fish is commemorated because of its pivotal role in the eradication of malaria from southern Russia in the mid-twentieth century.

Part of the problem regarding the use of biological control may stem from underlying assumptions about what constitutes an effective agent for the control of vector-borne diseases. Much of the theory regarding biological control was developed in an effort to control agricultural pests. Now termed **classical biological control**, such theory holds that the control agent should drive the pest to a low, stable equilibrium population. Characteristics of the control agent that contribute to this goal include a short generation time, a high degree of prey specificity and searching behavior, and the ability to respond to increases in the target pest density with both rapid population growth and an increased per capita consumption of the target. Such density-dependent predator responses prevent prey extinction, which would in turn lead to extinction of the control agent.

Yet for those situations in which the control of aquatic vectors or intermediate hosts is contemplated, these rules may not necessarily apply. In Kenya, for example, populations of schistosome-transmitting snails have been driven to very low levels or eradicated entirely by the Louisiana red swamp crayfish *Procambarus clarkii* (see Chapter 8, pages 335–336) This generalist and long-lived predator exhibits few of the characteristics generally attributed to an effective control agent. Yet in certain habitats, specifically small, permanent, artificial water impoundments used by local villagers, the introduction of *P. clarkii* resulted in striking reductions in schistosome transmission and morbidity in schoolchildren (see Figure 8.13). In large water bodies or in those habitats subject to drying, disease control was much less assured.

In a recent and potentially exciting development, it has been shown that the ***Aedes aegypti*** mosquitoes responsible for dengue transmission can be rendered resistant to the virus by infecting them with certain strains of the endosymbiotic bacteria *Wolbachia*. Infected mosquitoes efficiently passed the bacteria on to offspring and, in field trials, mosquitoes carrying the bacteria rapidly replaced those mosquitoes lacking the symbiont. Furthermore, because the *Wolbachia*-infected mosquitoes are resistant to dengue virus serotype II, one of the serotypes associated with dengue hemorrhagic fever, it is postulated that *Wolbachia* may provide a novel means of disease control. Other vector-borne diseases may be amenable to a similar control strategy. For example, a strain of *Wolbachia* that can infect the ***Plasmodium*** vector *Anopheles stephensi* has very recently been identified. The infection persisted over many generations, and when infected females were introduced into uninfected laboratory mosquito populations, the bacterial symbionts were able to establish themselves in these populations. Furthermore, the *Wolbachia*-infected mosquitoes were highly resistant to a subsequent infection with ***Plasmodium falciparum***. Yet optimism regarding the use of *Wolbachia* in this manner should be tempered by the recent finding that certain strains of the bacteria actually increase, rather than decrease, susceptibility of mosquitoes to *Plasmodium*.

The production of transgenic vectors provides hope as a means to reduce vector capacity

As we have seen thus far, success in controlling vectors has been something of a mixed bag; there are successes as well as setbacks and failures. In malaria control, for instance, both insecticide spraying and insecticide impregnated bed nets have reduced the burden of disease considerably in many endemic areas. In too many cases, however, a convergence of logistic problems and the need for sustained effort combined with the problem of insecticide resistance have conspired to allow the disease to rebound.

Clearly, new strategies for vector control are required. One promising approach is the development of transgenic vectors with reduced vector capacity (see Chapter 3, pages 81–82). **Transgenesis** refers to the deliberate introduction of exogenous genetic material into a living organism. The newly acquired genes, called *transgenes*, endow the recipient organism with new properties that will be transmitted to progeny. If vector capacity can be reduced with this technique, the hope is that transgenic vectors can be released into areas of endemicity, resulting in reduced transmission and morbidity. Such genetic methods, usually used in tandem with more conventional control, have been under investigation since the 1950s. The operational application of genetically altered vectors to reduce transmission has proven to be one of the most difficult challenges of vector control.

One of the principal hurdles when introduced into the field has been that genetically modified vectors are frequently at a competitive disadvantage as compared to their wild-type conspecifics. The genetic manipulation itself often renders transgenic vectors less fit than more vector-competent and robust wild-type vectors. Consequently, some of the same issues that plague more conventional control efforts, such as the need for regular reintroduction and sustained effort, are relevant for modified vectors as well.

Nevertheless, over the last two decades there have been tremendous advances in the fields of vector genomics and proteomics and in the ability to genetically manipulate medically relevant vectors. Consequently, encouraging results in the laboratory are beginning to show promise in the field and in some cases the use of transgenic vectors to control disease transmission is already operational.

Furthermore, disease control via the use of transgenic vectors is not necessary in all areas of disease transmission. In areas of low to moderate transmission of malaria, for example, currently available techniques may be sufficient to achieve either control or even local eradication. In such areas, control efforts may be better served by focusing research on the means to improve sustainability of more conventional methods. Yet even in some areas of low transmission, tried-and-true methods such as the use of impregnated bed nets and focal, indoor spraying may not result in effective control. Many ***Anopheles*** species, for instance, either feed or rest outside rather than in homes, rendering certain control methods far less effective. In such circumstances a role for transgenic mosquitoes may still be warranted. Certainly in high transmission regions the need for new genetic interventions is especially keen.

In general, the strategy behind the use of genetically modified vectors falls into one of two categories: reduction strategies and replacement strategies. **Reduction strategies** focus on the introduction of vectors with lowered fitness. Often these vectors continue to mate regularly but such matings do not result in the production of progeny. Consequently, if the vector population is reduced, the probability of contact between humans and vectors and therefore disease incidence declines as well. In **replacement strategies** the objective is to introduce genetically altered vectors that are resistant to infection with a particular pathogen. If such vectors replace wild-type vectors in the field, disease incidence will decline as the vector population becomes increasingly resistant. In either case, the goal is ultimately the same: the reduction of vector capacity, such that $R_0 < 1$ (see Chapter 6, page 246). If achieved, as transmission declines, morbidity and mortality decrease as well.

To date, the primary reduction strategy has been the **sterile-male technique**. The basic idea is to release large numbers of lab-reared sterile males, which then mate with wild females (**Figure 9.11**). Such females cannot produce fertile eggs, and the wild target population consequently declines. Historically, males have been irradiated to render them sterile. Although the

Figure 9.11 Screwworm eradication using sterile males. (A) An adult screwworm (*Cochliomyia hominivorax*) marked as part of a screwworm surveillance program. Females lay their eggs in wounds on livestock and wildlife. Emerging larvae feed on the flesh of these animals, causing a condition called myiasis. (B) Screwworm eradication using the sterile-male technique began in the mid-twentieth century. The map shows the year that successful eradication was achieved in various areas in North and Central America. (A, Courtesy of Agriculture Research Service, USDA. B, Food and Agriculture Organization of the United Nations, 2001.)

A

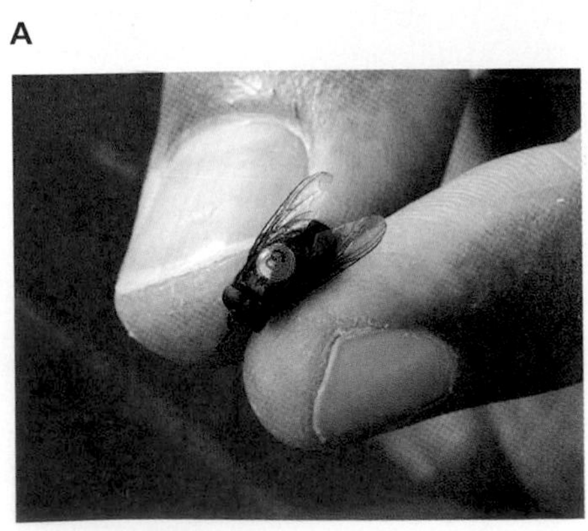

B

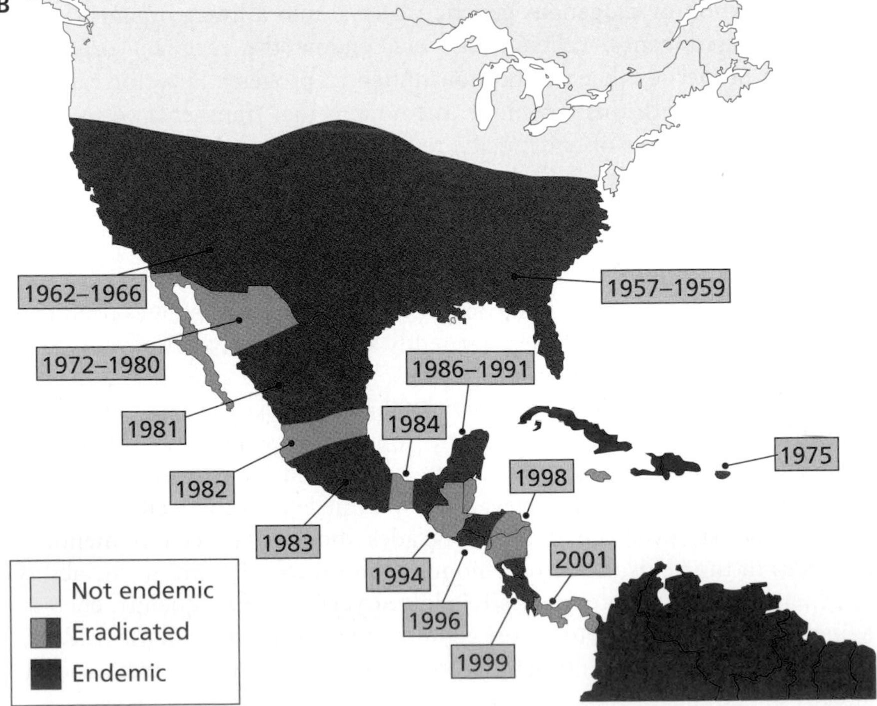

technique has been used successfully with certain agricultural pests, it has not proven to be effective with mosquitoes because irradiated male mosquitoes have not been able to compete successfully for mates in the field with wild-type males. Genetic manipulation, however, may overcome this problem. For example, male ***Anopheles gambiae*** mosquitoes that produce no sperm have recently been described. These spermless males were developed through the use of RNA interference (RNAi) to silence a gene necessary for sperm production (**Figure 9.12**). Although they produce no sperm, these males continue to mate with wild-type females, which then produce unfertilized eggs that are incapable of hatching.

Other encouraging results have been announced in which transgenic male ***Aedes aegypti***, carrying a lethal dominant mutation have been developed (**Figure 9.13**). The mutation renders the males dependent on tetracycline for their survival. Males were raised to adulthood in the presence of tetracycline in the laboratory. These males were then released into the wild on a 10-hectare plot in the Cayman Islands. In the absence of tetracycline, the males then mated with wild-type females before they died. The dominant lethal mutation expressed in progeny killed the larvae before they could pupate. The population fell as a result. This is the first successful field trial of a strain of genetically modified mosquitoes. Because ***A. aegypti*** is the most important vector of dengue, the technique could eventually be of value in controlling this viral disease.

A number of replacement strategies have been considered. Any technique in which a gene that interferes with the pathogen has been introduced might be envisioned. The transgene product might be toxic to the pathogen, or transgene expression might result in the use or degradation of resources that are crucial for pathogen development or replication in the vector. Ideally, the introduction of such a gene would have a low fitness cost for the vector, a goal

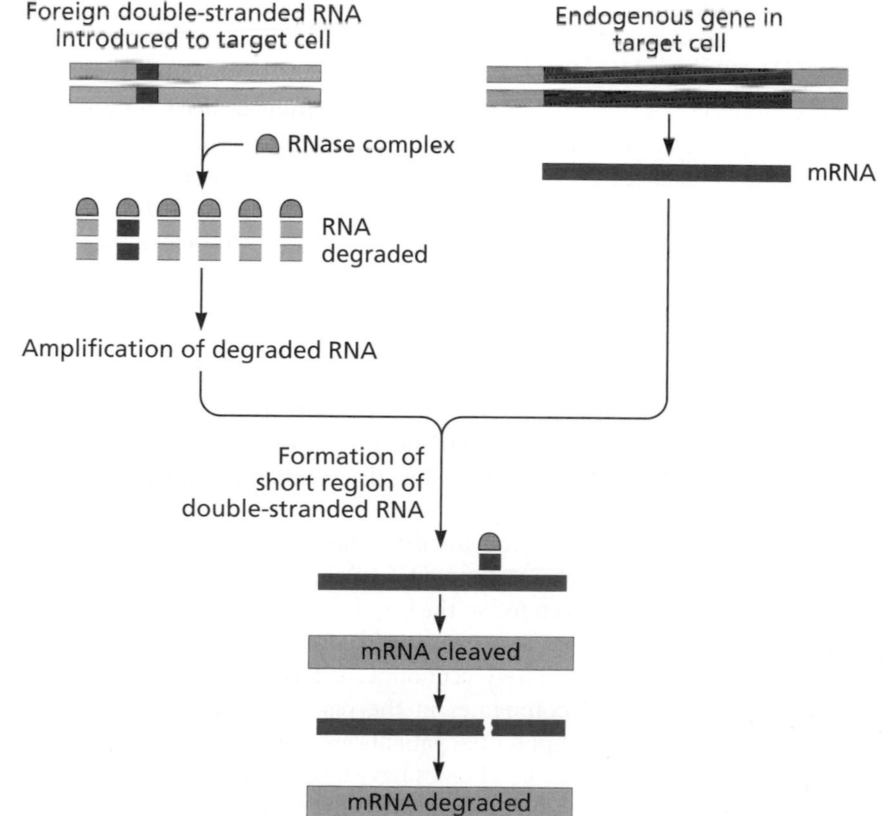

Figure 9.12 RNA interference. The fate of foreign double-stranded RNA (dsRNA) is shown on the left. Such RNA, usually observed in viral infections, is recognized by a large protein complex that has RNase activity. The dsRNA is degraded into short fragments that are approximately 23 nucleotide pairs long. These fragments are sometimes amplified by an RNA-dependent RNA polymerase and, in this case, can be efficiently transmitted to progeny cells. If foreign RNA has a nucleotide sequence similar to that of a cellular gene, mRNA produced by this gene will also be degraded by the pathway shown. In this way, the expression of a cellular gene can be shut off by introducing dsRNA into the cell that matches the nucleotide sequence of the gene.

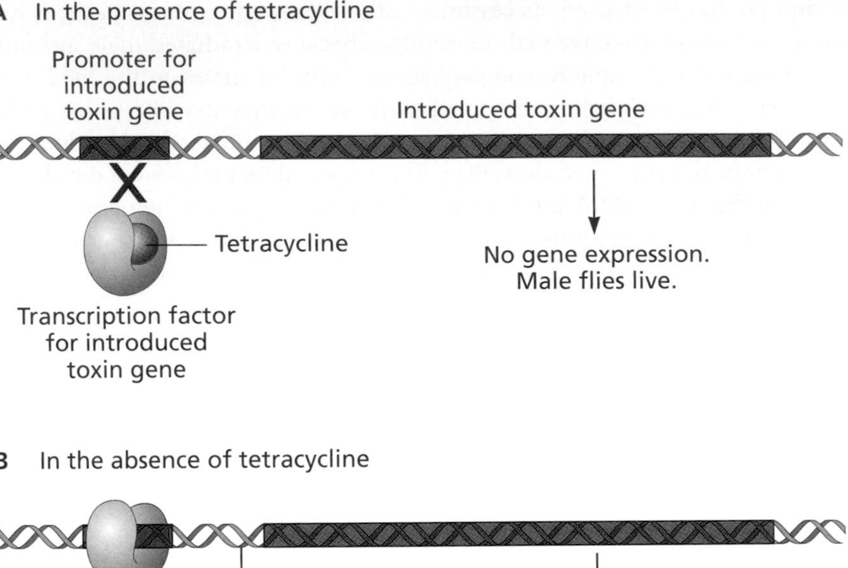

Figure 9.13 Vector control using transgenic males. In this example, a gene for a lethal dominant mutation (in this case, a toxin) is introduced into male mosquitoes. (A) As long as tetracycline is present, the males survive because the antibiotic binds to and interferes with the ability of the transcription factor for the toxic product gene to bind the promoter for this gene. (B) In the absence of tetracycline, the transcription factor binds the promoter, the introduced gene is expressed, and the vector dies.

that might be achieved by insuring that transgene expression only occurs in specific tissues or for specific lengths of time. Other creative ideas, such as the introduction of genes expressing exogenous odor receptors in the vector's antennae, which might redirect an anthropophilic vector to alternative hosts, illustrate the variety of ways in which vector capacity might be reduced using transgenetic vectors.

9.2 ANTI-PARASITIC DRUGS

There is nothing new about anti-parasitic drugs. Indeed, the Chinese have known about the antimalarial properties of *Artemisia annua* (sweet wormwood) for over 2000 years. Likewise, Spanish Jesuits in Peru first learned that bark from the chinchona tree could be used to treat fevers caused by malaria as far back as the early 1600s. It was some time, however, before there was any understanding of how these drugs functioned. In an odd silver lining to an otherwise disastrous war, Chinese scientists first discovered the active ingredient of *A. annua* (artemisinin) in 1971. The impetus for the research came from a request to the Chinese government from Ho Chi Minh at the height of the Vietnam War in 1968. It was not until 1820 that quinine was isolated from chinchona bark. A synthetic variant of quinine, chloroquine, was first produced in 1934. The use of both artemisinin and chloroquine continues today.

Yet even though a few anti-parasitic drugs have been available for a considerable time, more modern progress in terms of new drug development has been exasperatingly slow. Even today, the list of anti-parasitic drugs is short—far shorter than the list of available antibiotics to treat bacterial infections.

Some of the reasons are purely economic. It has been difficult to interest major pharmaceutical companies in the research and development of anti-parasitic drugs, as most potential patients are poor and live in less developed parts of the world. Such companies have largely focused on drugs with potentially large profit margins. The potential profits to be earned from a new heartburn medication, for instance, far exceed those that might be expected from a new drug to treat hookworm infection.

Yet there has been some recent progress; the last few decades have seen the development of several new drugs such as albendazole, ivermectin, and Malarone®. Yet even for many of these more contemporary drugs, the bottom line has contributed significantly to their development. The research and development of ivermectin, for example, was largely driven by a lucrative market for canine heartworm prophylaxis in developed countries. It is a happy accident that the drug is now also widely used to treat a variety of human filarial worm infections in less developed parts of the world. Similarly, it is debatable whether or not Malarone® would have been developed as an antimalarial drug were it not for the impetus provided by the US military, or the large numbers of westerners who were available to purchase it as a prophylactic when they ventured into malarious areas.

Other reasons for the paucity of effective anti-parasitic drugs are more biological in nature. Good drugs against any parasite, whether viral, bacterial, or eukaryote, must kill or inhibit the growth of the parasite without severely compromising the host. This objective may be achieved in one of two ways. First, the drug may target a structure or biochemical pathway in the parasite that is not present in the host. Such targeting, known as **selective toxicity**, at least in principle allows the parasite to be negatively affected without any adverse side effects on the host. Unique prokaryotic-associated targets such as cell wall components or the prokaryotic 70S ribosome, which differs significantly in size and structure from eukaryotic 80S ribosomes, are common.

The second manner in which drugs may act reflects the fact that eukaryotic protozoans or helminths offer far fewer unique targets, simply because they are eukaryotic. In most ways, the cells of these parasites are not much different from our own, meaning that compounds toxic to the cells of eukaryotic parasites are likely to be toxic to our cells as well. Such toxicity is a primary reason why the development of safe and effective molecules to combat these infections has been so difficult. Many anti-parasitic drugs, therefore, rely on the identification of a safe **therapeutic index**, a comparison of the amount of the drug that is toxic to host to the amount that will kill the parasite. The precise calculation of therapeutic index is determined by the equation:

$$\text{Therapeutic index} = \frac{\text{Toxic dose}}{\text{Effective dose}}$$

Consequently, a relatively high therapeutic index is consistent with a drug that can still provide the desired anti-parasitic effect safely. However, finding the safe middle ground where the dose is high enough to kill the parasite but low enough to be used to without harming the host is not always possible.

Yet anti-parasitic drugs still offer hope of treatment or cure in many cases. With increasingly sophisticated technology, there is reason for optimism for the future. In this section, we will explore some of the pertinent issues associated with anti-parasitic drugs and assess the prospects for new drug development.

Various factors influence the selection of the best anti-parasitic drug in different situations

Selection of a particular anti-parasitic drug is not as straightforward as simply picking the compound most likely to kill the parasite in question. Rather, a physician or public health professional must consider a number of important variables that will influence drug choice. What are the likely drug side effects? Does the patient have any known drug allergies? Has the parasite been definitively identified? These and other factors must be considered before selecting the best drug in a specific situation. Often, the deciding factor is not a medical one. Some drugs might be inappropriate in certain situations because of their high cost. In undeveloped, rural areas where electrical supply is uncertain, it may be necessary to opt for drugs that do not require refrigeration. We will next consider a few of the most relevant issues when selecting an appropriate anti-parasitic drug.

Different drugs may be appropriate for treatment and for prophylaxis

Drugs may either be used to prevent infection in the first place (**prophylaxis**) or to treat infected individuals. When the goal is prophylaxis rather than treatment, drugs must meet a higher safety standard than they do when used for treatment. That is, they must show higher selective toxicity or the therapeutic index must be higher than it is for drugs used to treat sick individuals. Certain less serious side effects such as headache or nausea may be acceptable for a short time in a sick patient. They are likely to be unacceptable in otherwise healthy people who simply wish to avoid becoming infected. This point is especially true if the healthy individual will be in an at-risk area for a prolonged period of time.

Prophylaxis against malaria has a long history (**Figure 9.14**). Today, one of the most frequently used antimalarial drugs for this purpose is a combination of atovaquone and proguanil, known commercially as Malarone®. Atovaquone is a ubiquinone analog, which preferentially inhibits the electron transport chain in ***Plasmodium*** and other apicomplexans. Laboratory results demonstrate that it is 200 times more active against the ***Plasmodium*** electron transport chain than it is against that of mammals, presumably because of differences in the electron transport chain components of apicomplexans and vertebrates. Thus, with a high therapeutic index, atovaquone is reasonably safe to use. Proguanil is metabolized within the host to cycloguanil, which inhibits dihydrofolate reductase, an enzyme used in the conversion of folate to the deoxynucleoside thymidine. As such, proguanil inhibits nucleic acid synthesis. Its relative safety may result from its preferential inhibition of the parasite enzyme.

Malarone® is not without side effects. Sleep disturbances are commonly reported, as are canker sores and headaches. Yet these are usually infrequent or mild enough that Malarone® is now widely used for malaria prophylaxis.

When an anti-parasitic drug is used for treatment, the goal may be elimination of the parasite within the host or at least the prevention of serious complications associated with chronic infections. Albendazole, for instance, is a mainstay for the treatment of **geohelminth** (nematodes that are primarily transmitted through contaminated soil) infection (**Figure 9.15**). Its principal mode of action involves its ability to inhibit the polymerization of tubulin into microtubules. Side effects associated with the use of this drug include dizziness, headache, vomiting, and temporary hair loss. More rarely, it can cause seizures, as well as kidney and liver disorders.

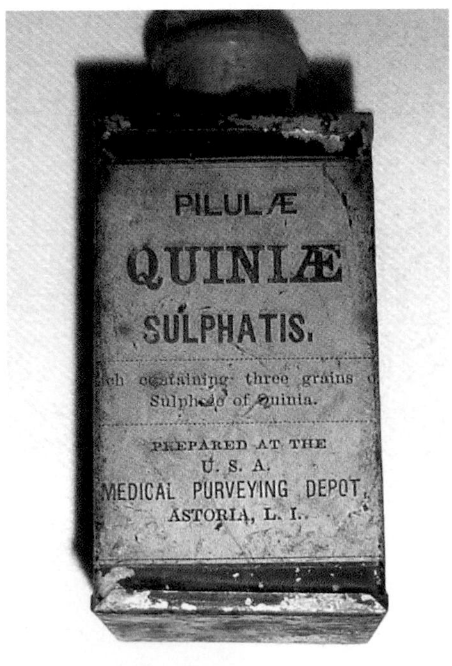

Figure 9.14 Protection against malaria. A quinine tin from the 1860s. Extracted from the chinchona tree, quinine was widely used during the second half of the nineteenth century.

Drugs may be used to either treat or protect individuals or to protect a population

For the physician confronted with a patient who is either likely to become exposed to a specific parasite or is already infected with one, drug use is primarily focused on either protecting or treating the individual. A public heath official or an epidemiologist may have a different perspective. The goal in this

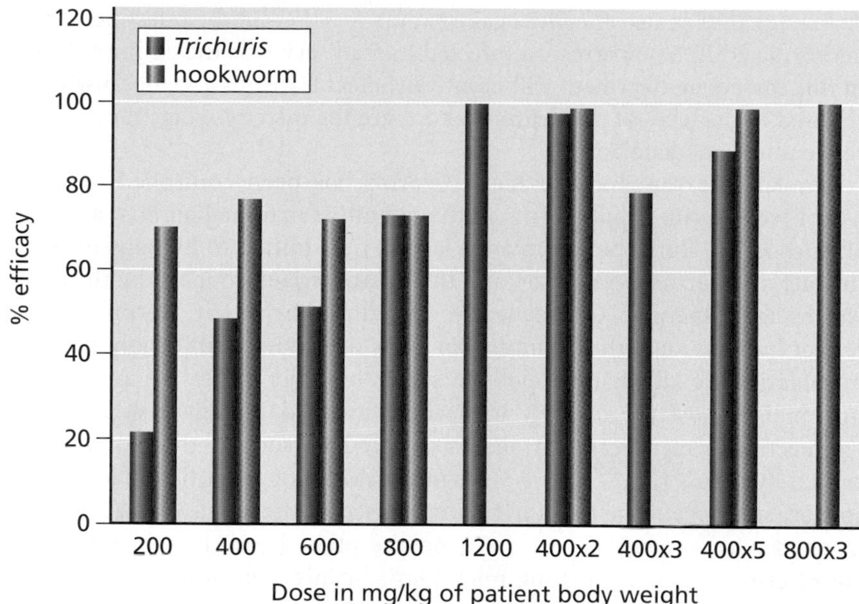

Figure 9.15 Anti-helminthic activity of albendazole. Relative efficacy of albendazole in humans for the treatment of two nematode parasites. The *Y* axis shows % efficacy; the *X* axis is dose in mg/kg of patient body weight. Doses labeled as x2, x3, or x5 were administered multiple times daily. (Adapted from Geary TG et al [2010] *Int J Parasitol* 40:1–13. With permission from Elsevier.)

case may be to reduce overall transmission of a parasite and thereby protect a particular population or community from infection. The topic of epidemiology (disease in human populations) as it applies to parasites was discussed in Chapter 6. Here we consider a few additional aspects of parasite epidemiology related to the use of anti-parasitic drugs.

When the objective is population protection, either a sizable percentage or even the entire population may be treated as a means to reduce parasite transmission and thereby reduce morbidity and mortality. Such a strategy may be effective if the parasite uses only human reservoirs. It is for this reason that such great strides have been made in the control of West African onchocerciasis. If, however, a particular parasitic disease is zoonotic or if humans are dead-end hosts, mass chemotherapy may do little or nothing to reduce parasite transmission.

If mass application of anti-parasitic drugs is deemed appropriate as a means of reducing transmission, who exactly should receive such drugs? In most cases, the treatment of symptomatic individuals, although certainly of value from the standpoint of those individuals, will have minor or no impact on transmission. In too many cases, by the time symptoms develop, transmission has already occurred during the asymptomatic phase of the disease. Consequently, even those individuals with asymptomatic infections require treatment. Because diagnosis of asymptomatic patients is not always easy, it may be more realistic and cost effective to simply treat everyone in a population, whether or not they are infected. This of course raises the same issues of toxicity that were discussed in the previous section.

Certain drugs are active only against specific parasite life-cycle stages

Not all parasite life-cycle stages are equally vulnerable to a specific drug. Praziquantel, for instance, kills adult schistosomes but is less effective against migrating shistosomulae. Praziquantel can also be used to treat cysticercosis, caused when humans become infected with cysticerci, the cyst-like larval stage of the pork tapeworm ***Taenia solium***. Although treatment may certainly be indicated for the infected individual, the destruction of cysticerci in a human will not affect parasite transmission because, in this case, the infected human is a dead-end host.

Praziquantel is also effective against adult ***T. solium*** in the intestine. When harboring adult tapeworms, an infected human serves as the definitive host. In this case, drug treatment will negatively affect transmission, because eggs, released in the feces of the definitive host, are the infective stage for the subsequent intermediate host.

Similarly, transmission of filarial worms has been controlled by mass use of ivermectin. Adult worms in the definitive mammalian host are usually not killed. Both the infective L3 larvae, transmitted to humans through the bite of an arthropod vector, and the microfilariae produced by the adult worms are vulnerable. Consequently, if used prophylacticly, ivermectin can render humans and other mammalian hosts refractory to infection. Because mirofilariae are killed, arthropods in search of a blood meal on an infected mammalian host are unlikely to become infected if their host is taking ivermectin. In many cases of filariasis, of course, such as the elephantiasis caused by ***Wuchereria bancrofti***, ivermectin will not assist the infected person, as pathology is the result of trauma caused by the adult worm. A patient infected with ***Onchocerca volvulus***, on the other hand, may benefit from ivermectin because circulating microfilariae cause the most serious symptoms of infection, including blindness. **Box 9.2** describes a new combination therapy involving ivermectin that may improve treatment of canine heartworm caused by the filarial parasite *Dirofilaria immitis*.

BOX 9.2

New Treatments for Man's Best Friend

It's a sad day for dog owners if they learn that their pet has canine heartworm disease. The disease is caused by the filarial nematode *Dirofilaria immitis*. Dogs become infected when mosquitoes bearing infective L3 larvae feed upon them. Following an extensive migration through the circulatory system, developing worms ultimately reach the pulmonary arteries (Figure 1). Mature female worms are usually about 30 cm in length, whereas males average 23 cm. About six months postinfection, females begin to give birth to microfilariae, the infective stage for mosquitoes.

Infected dogs show few symptoms early in an infection. Light infections may remain asymptomatic. In heavy infections, lethargy, loss of appetite, weight loss, and a chronic cough are typical. Ultimately, if untreated, the dog may die of congestive heart failure.

Fortunately, infection is easily prevented with prophylactic oral ivermectin. The drug is 99% effective in killing larval worms before they mature into adults. However, should an unprotected dog become infected, treatment is not so simple. Ivermectin has limited effect on adult worms. High doses may eventually kill adults, but treatment can take two years or more. The only currently available alternative for killing adult worms is melarsomine dihydrochloride, an arsenic-based compound. This drug is fairly toxic and must be administered in a complex treatment regimen. Exercise restriction is required for a least a month, as abrupt worm death in the pulmonary arteries can cause severe cardiac stress. During exercise, dead worms may dislodge and end up in the lungs, where they can cause respiratory failure.

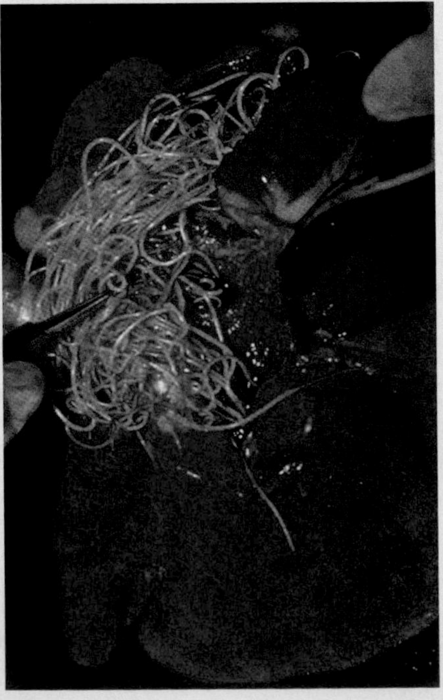

Figure 1 Canine heartworms. *Dirofilaria immitis* in the pulmonary artery of a dog that died of heartworm disease. The worms provoke a strong inflammatory response that can result in embolisms and arterial lesions. Dying worms often are carried to the lungs, where they can provoke similar inflammation. When adult worms are killed with melarsomine dihydrochloride, the release of *Wolbachia* and their metabolic products exacerbates the inflammatory response. (Courtesy of Ray Dillon, Auburn University College of Veterinary Medicine.)

Fortunately, there may be a relatively benign alternative. Like most filarial worms and insects, *D. immitis* harbors bacteria in the genus *Wolbachia*. As we

Drug use may affect the immune status of the population

There is no doubt that the mass use of antimalarials to reduce malaria transmission or the similar use of ivermectin to break the cycle of transmission of ***O. volvulus***, just to cite two examples, has contributed substantially to the overall public health in certain parts of the world where the burden of parasitic disease looms large. But a note of caution is called for—one that sometimes gets drowned out in the excitement generated by a successful drug campaign. As an example, consider the Garki Project.

The Garki Project was a large and long-term study on the epidemiology and control of malaria conducted by the World Health Organization and the Nigerian Government from 1969 to 1976. Specifically, regular mass prophylaxis with antimalarial drugs was combined with mosquito control in an effort to break the malaria transmission cycle. The results were dramatic. Within two years there was a substantial reduction in prevalence and a corresponding decrease in morbidity and mortality. However, it was also noted that because exposure to ***Plasmodium*** became so infrequent, the natural immunity that builds up in those individuals who are regularly exposed to malaria declined. Immunity to malaria, although imperfect, does offer considerable protection, and those who are exposed to regular infection generally suffer less severe symptoms compared to immunologically naïve

have noted repeatedly throughout this text, these Gram-negative intracellular organisms are often involved in complex endosymbiotic relationships with their hosts. In the case of filarial worms, antibiotics used to kill the bacteria quickly curtail larval development and growth within adult females, as well as adult survival, suggesting that the bacteria are vital to the survival and reproduction of the worms. This finding has led to the idea that heartworm might be treated with antibiotics. Numerous studies now suggest that doxycycline combined with other more conventional therapies may provide a way to minimize toxicity while maximizing therapeutic effectiveness.

The value of such combination therapy has been demonstrated in both experimentally and naturally infected dogs. Treatment with both doxycycline and ivermectin, for instance, eliminates microfilariae within nine weeks (Figure 2). Presumably this is a result of the destruction of adult worms.

Thus it has been suggested that the use of the antibiotic along with ivermectin prior to the administration of the arsenic compound or the use of the antibiotic and ivermectin alone might provide a cure with far fewer dangerous and unpleasant side effects.

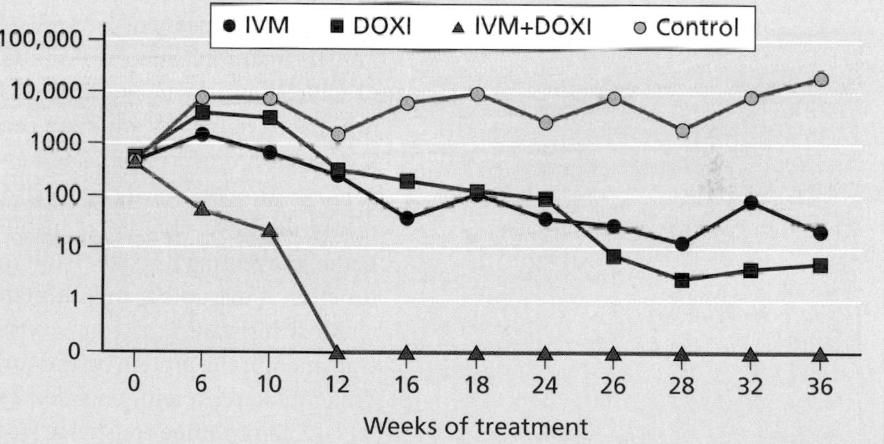

Figure 2 Antimicrofilarial activity of ivermectin and doxycycline. Mean microfilarial counts obtained when experimentally infected dogs were treated with ivermectin (IVM), doxycycline (DOXI), or ivermectin plus doxycycline. (Adapted from McCall JW et al [2008] ***Vet Parasitol*** **158:204-214. With permission from Elsevier.)**

As knowledge of *Wolbachia* and its relationship with *D. immitis* and other filarial worms becomes better understood, improved protocols can be expected that both increase efficiency and decrease toxicity. It is likely that the same strategy may be useful for the treatment of human diseases caused by filarial worms. In the meantime, don't forget the monthly dose of ivermectin during mosquito season for your canine friends.

References

McHaffie J (2012) *Dirofilaria immitis* and *Wolbachia pipientis*: a thorough investigation of the symbiosis responsible for canine heartworm disease. *Parasitol Res* 110:499–502.

McCall JW, Genchi C, Kramer L et al (2008) Heartworm and *Wolbachia*: therapeutic implications. *Vet Parasitol* 158:204–214.

Grandi G, Quintavalla C, Mavropoulou A et al (2010) A combination of doxycycline and ivermectin is adulticidal in dogs with naturally acquired heartworm disease (*Dirofilaria immitis*). *Vet Parasitol* 169:347–351.

individuals. Thus, when a person in the study area did on occasion become infected, the course of disease was especially serious.

The example of the Garki Project highlights the need for sustainability in large-scale efforts to reduce disease transmission. Without such sustainability, shorter-term gains can eventually be swamped by longer-term negative consequences. The fact that that the burden of most parasitic diseases disproportionately falls on the poorest and least developed parts of the world, where logistics, economics, and political unrest often conspire to undermine even the most carefully thought out control program, underscores this point.

Of course, when it comes to such longer-term negative consequences, arguably the most important of all is drug resistance. In the next section we will consider this constantly lurking threat to effective anti-parasite chemotherapy.

The use of anti-parasitic drugs can lead to resistance

Hovering like a dark cloud over all efforts to combat infectious disease with chemotherapy is the problem of drug resistance. We now hear regularly about bacterial infections such as methicillin-resistant *Staphylococcus aureus* (MRSA), for instance, that no longer respond to antibiotics that were once effective. Eukaryotic parasites may also develop resistance to specific drugs. When they do, control efforts and the treatment of infected individuals can become especially problematic.

There is nothing particularly mysterious about how drug resistance develops. Like other genetically determined characteristics, resistance evolves through the process of natural selection. As an example, consider the well-known and problematic resistance of ***Plasmodium falciparum*** to chloroquine.

Chloroquine acts by diffusing into the parasite's digestive vacuole, where it binds to heme. As you may recall from Chapter 5 (pages 173–175) heme is a waste product of parasite hemoglobin digestion. Because heme is toxic, ***P. falciparum*** ordinarily converts heme to nontoxic hemozoin, which collects in the digestive vacuole as insoluble crystals. But chloroquine can bind to heme, preventing its conversion to hemozoin. As heme builds up, membrane function is disrupted, and ultimately the parasite undergoes lysis.

Most resistant ***P. falciparum*** strains have, however, a modified form of a transmembrane protein on the surface of their digestive vacuole (**Figure 9.16**). This mutated protein, encoded by a gene called *PfCRT* (***Plasmodium falciparum*** chloroquine resistance transporter), rapidly pumps chloroquine back out of the digestive vacuole before it can bind to heme.

Throughout its long evolutionary history, occasional ***Plasmodium*** parasites no doubt experienced a mutation in their *PfCRT* genes, equipping them with an altered transporter protein. Because they were not exposed to chloroquine, the mutation offered them no selective advantage and such parasite strains eventually died out. If, however, we fast-forward a few million years to a time when chloroquine use became widespread, some parasites, those that by chance happen to excrete the drug, are at a tremendous advantage. Other parasites, lacking the favorable mutation, are not so lucky. In the face of chloroquine exposure they fail to replicate and, over time, most members of the population will have descended from those cells that could originally excrete chloroquine. As long as chloroquine exposure continues, any parasites undergoing random mutations that enhance their ability to excrete the drug continue to proliferate at the expense of other less-resistant parasites. Finally, a completely chloroquine-resistant strain emerges. The key phrase here is *as long as chloroquine exposure continues*. Once chloroquine

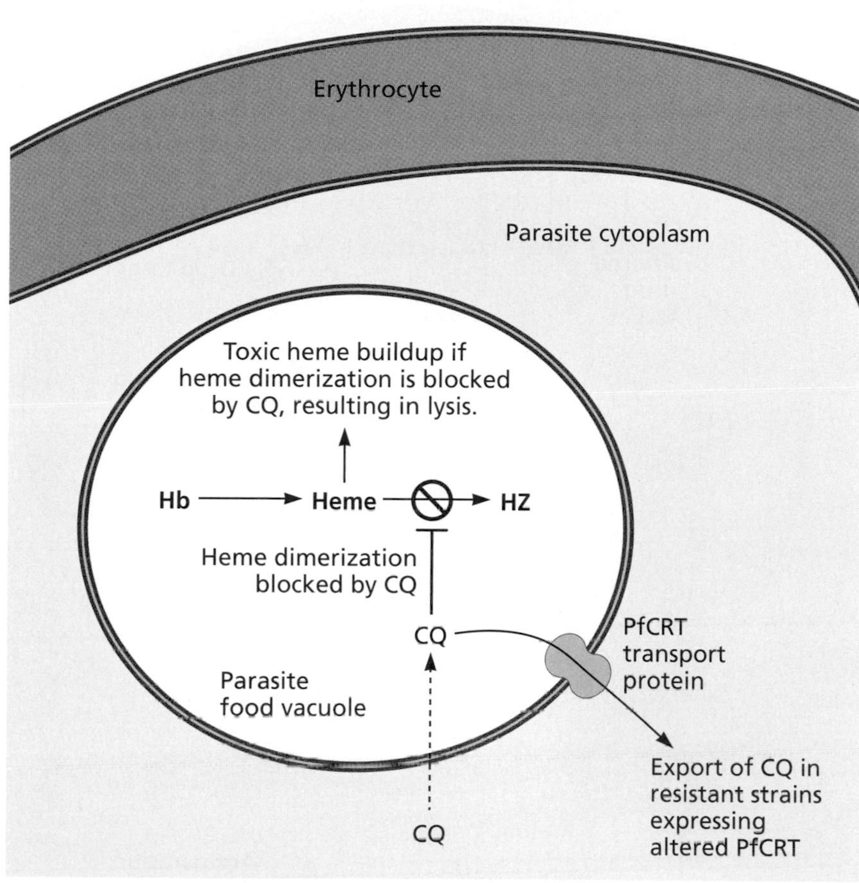

Figure 9.16 Action of and resistance to chloroquine. Erythrocytic stages of *Plasmodium* digest hemoglobin (Hb) in their food vacuole, using digested peptides to meet their own need for amino acids. Heme is released as a byproduct of this digestion. Heme is toxic to the parasite, and it is rapidly dimerized by a *Plasmodium*-encoded enzyme into nontoxic hemozoin (HZ). Chloroquine (CQ) accumulates in the food vacuole, where it inhibits heme dimerization. The resulting buildup of heme subsequently causes parasite lysis. In some resistant strains, an altered form of the PfCRT transport protein allows the rapid excretion of chloroquine from the vacuole before it can interfere with hemozoin formation.

is removed from the environment, resistant parasites lose their competitive edge and may cease to predominate.

Genetic alterations can cause resistance in diverse ways

Mutant, up-regulated, or duplicated transporter genes are thought to play a role in the resistance of other parasites as well. Examples include the resistance of some ***Leishmania*** strains to pentavalent antimonial drugs. Likewise, when *Caenorhabditis elegans* was selected for resistance to ivermectin *in vitro*, the resistance was traced to rapid drug excretion. This may be the underlying mechanism behind the observed resistance to ivermectin in certain scabies-causing mites and in strongyloid nematodes that plague livestock. In other cases, if the gene for a particular membrane receptor experiences a mutation, the receptor may become inactive. If this receptor is ordinarily involved in transporting a specific drug into a parasite, the drug may lose its effectiveness, rendering the parasite resistant. This process is believed to be how certain strains of African trypanosomes have developed resistance to diminazene. To be effective, this drug must enter the parasite and move to the nucleus, where it binds to and interferes with nucleic acid synthesis.

Other mechanisms also may result in resistance (**Figure 9.17**). In some cases, for instance, the drug itself is inactive but is metabolically converted to an active toxic form. Resistant parasites may interfere with such conversion. In other cases, the structure of the ultimate drug target is altered. Parasites may achieve resistance if the drug is unable to bind and interfere with the now altered target. Resistance to metronidazole and albendazole, two of the most widely used anti-parasitic drugs, illustrates these alternative strategies.

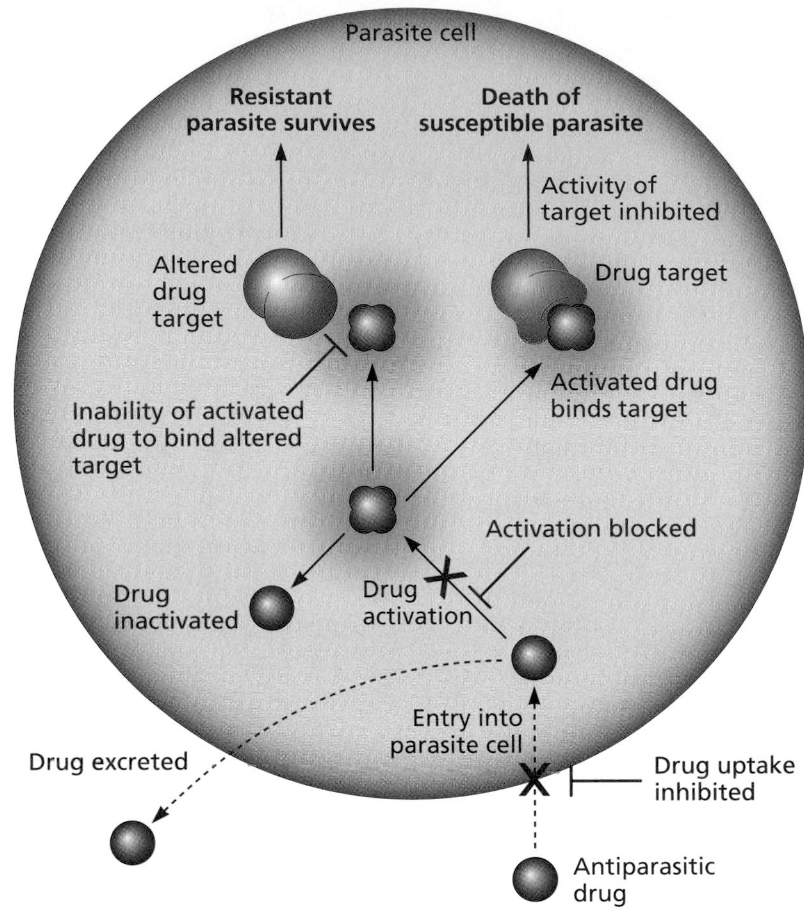

Figure 9.17 Mechanisms of drug resistance. A simplified scheme illustrating some of the principal biochemical mechanisms that parasites may use to develop drug resistance. Some of the more common resistance mechanisms are highlighted in red.

Metronidazole, frequently marketed under the name Flagyl®, is used to treat infections with anaerobic bacteria and several medically important protozoa, including ***Entamoeba histolytica***, ***Giardia lamblia***, and ***Trichomonas vaginalis***. Metronidazole is absorbed into the cells of these parasites by diffusion. It is subsequently reduced by interacting with ferredoxin (**Figure 9.18**).

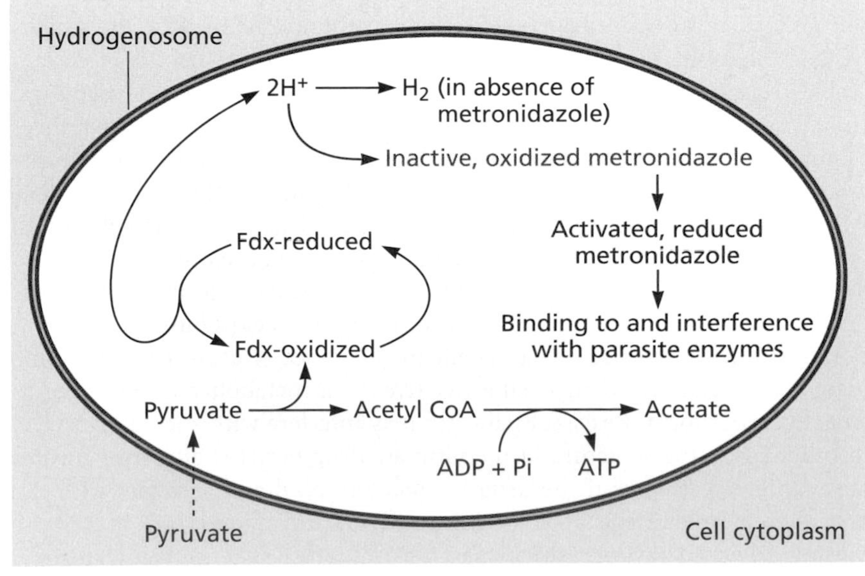

Figure 9.18 Activation of metronidazole in *Trichomonas*. Parasites in this genus convert pyruvate into acetate in an organelle called the hydrogenosome. Normally in this process, as pyruvate is oxidized to acetyl CoA, ferredoxin (Fdx) is reduced. The reduced ferredoxin is subsequently oxidized, as $2H^+$ is converted to molecular hydrogen (H_2; hence the name hydrogenosome). As acetyl CoA is coverted into acetate, ATP is generated by substrate-level phosphorylation. Inactive oxidized metronidazole can accumulate in the hydrogenosome, where it can be reduced by ferredoxin. In its now active reduced form, the drug can form covalent linkages with parasite enzymes, rendering them inactive.

It is the reduced form of the drug that exhibits anti-parasitic activity. In its reduced form, metronidazole can form covalent linkages with many enzymes, rendering them inactive.

Resistant strains of both ***G. lamblia*** and ***T. vaginalis*** have been reported. In the case of resistant ***T. vaginalis***, and perhaps ***G. lamblia*** as well, the proximal electron donor (ferredoxin) is lacking. In these cells, metronidazole remains in its inactive oxidized state.

Albendazole is a widely used broad-spectrum anthelmintic, used for both nematodes and cestodes as well as for certain trematodes such as ***Fasciola hepatica***. Once absorbed by the parasite, albendazole binds to β-tubulin subunits, preventing their polymerization into microtubules. The inability to form cytoplasmic microtubules leads to impaired glucose uptake and depletion of glycogen stores. In consequence, ATP production falls, resulting in parasite immobilization and eventual death.

Resistance to albendazole has been observed in strains of ***Wuchereria bancrofti***. In resistant worms, there is point mutation in the gene encoding β-tubulin. Because of this single-nucleotide substitution, phenylalanine is replaced by tyrosine at position number 200 of the protein. Albendazole is unable to bind this altered protein, rendering it ineffective. Thus resistance, in this case, is the result of target alteration into a form with which the drug cannot interact.

Resistance poses a considerable problem for disease control programs

Regardless of the mechanism, drug resistance regularly threatens to undermine disease control efforts. The emergence of chloroquine-resistant ***Plasmodium*** is perhaps the best example of this phenomenon. In the early 1950s, the World Health Organization announced a goal of malaria eradication, which would rely heavily on the use of **mass drug administration**, that is, the administration of drugs to whole populations in high-transmission areas, whether or not individuals were infected at the time. In retrospect, the consequences were predictable. Chloroquine-resistant ***P. falciparum*** was first detected in South America and Southeast Asia in the late 1950s. Through the 1960s and 1970s, resistant strains spread across Asia and Latin America and were first reported in East Africa in 1978. During the 1980s, resistance spread across Africa and, in 1989, chloroquine-resistant ***P. vivax*** emerged in Papua New Guinea, from where it spread into Southeast Asia (**Figure 9.19**). As resistance spread, the World Health Organization was forced to abandon its goal of malaria eradication in favor of one to reduce morbidity and mortality by treating clinical cases only. As previously discussed, the simultaneous development of DDT resistance in mosquitoes only exacerbated the problem.

In other cases in which control efforts rely almost entirely on a single drug, the possibility of resistance is always a concern. In large parts of Africa, for example, the morbidity caused by onchocerciasis declined dramatically in the late twentieth century, thanks initially to the efforts of the Onchocerciasis Control Program, and most recently of the African Programme for Oncocerciasis Control. Through the efforts of this international consortium, the risk of blindness was eliminated for millions of people. Central to its success was the widespread use of ivermectin, a single dose of which eliminates microfilariae in the skin and effectively sterilizes adult females for a year or more. Throughout this period, little or no resistance to ivermectin was detected. Yet a 2007 study, identifying a doubling of the prevalence rate in Ghana between 2000 and 2005, raises concerns in this

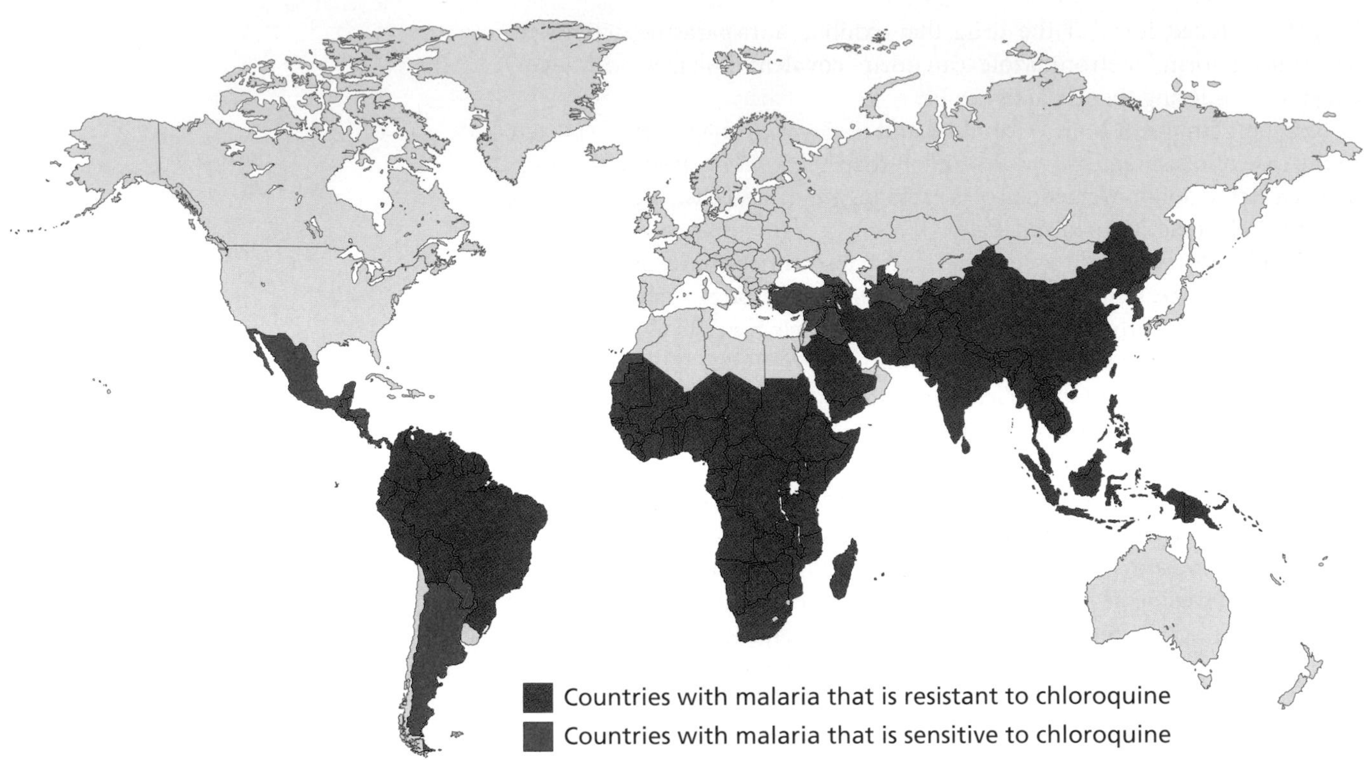

Figure 9.19 Chloroquine resistance. Resistance to this antimalarial was first observed in South America and Southeast Asia in the 1950s. Since then, resistance has spread widely across Africa and South Asia. (Courtesy of CDC.)

regard. Currently, although alternative drugs are under investigation, ivermectin remains the only drug used to treat onchocerciasis, which remains a significant health issue in parts of Africa.

Similar worries overshadow schistosomiasis control, which relies almost exclusively on the use of praziquantel. Although there remains no firm clinical evidence of schistosome resistance to this potent anthelmintic, low cure rates in some African studies at least hint at the possibility of emerging resistance. Now that the WHO has called for the elimination of human schistosomiasis, the projected use of praziquantel will increase dramatically. This raises an even greater concern about the emergence of resistance.

Drug resistance can be prevented or reversed

Drug resistance emerges when, by using drugs, we create an environment for the pathogen that favors the selection of resistance. Similarly, drug resistance can be reversed when we create environments in which such resistance loses its selective advantage. The example of chloroquine use in the Central African nation of Malawi is illustrative. Owing to heavy chloroquine use, the prevalence of the resistant *PfCRT* genotype in ***Plasmodium falciparum*** reached 85% in 1992. In 1993, chloroquine was removed from the market in Malawi because of its limited efficacy. After chloroquine was withdrawn, chloroquine-resistant parasites then lost their selective advantage over sensitive strains and a decline in resistance ensued. By 2000, the prevalence of the resistant genotype was only 13%.

Of course, when there are currently no drug alternatives, as is the case with praziquantel and schistosomiasis, simply abandoning the drug and waiting for resistance to subside is not really a viable solution. Options are better when there are multiple drugs to choose from. **Combination therapy**, the use of more than one drug simultaneously to prevent the development of resistance, is already a common treatment method for certain bacterial and viral infections. It is now commonly used in malaria control as well.

To understand how combination therapy can prevent the emergence of resistance, consider a hypothetical situation in which a parasite can be treated with three drugs (A, B, and C). If an infection is treated with all three drugs simultaneously, it is extremely unlikely that resistance to all three drugs will arise (**Figure 9.20**). However, as shown in the figure, if only one of the three drugs is used, the eventual likelihood of a parasite strain that is multi-drug resistant increases considerably. Consider that before 1996, an infection with HIV was basically a death sentence, in large measure because of the almost inevitable resistance that developed to the available antiviral drugs. With the introduction of new drug classes in the mid-1990s and the implementation of combination therapy, HIV has become a largely treatable and chronic infection.

Not any drug combination will do. To decrease the likelihood of resistance, combined drugs should typically act on different targets. This is the case with the anti-HIV drugs mentioned above. As previously described, it is also true of the antimalarial Malarone®, which consists of both atovaquone and proguanil.

Artemisinin, originally isolated from a Chinese herb as previously described, has been used to treat malaria for over a thousand years. Until recently, all strains of ***P. falciparum*** were susceptible to artemisinin. Its use is carefully monitored and it is only used in combination with other drugs. Such

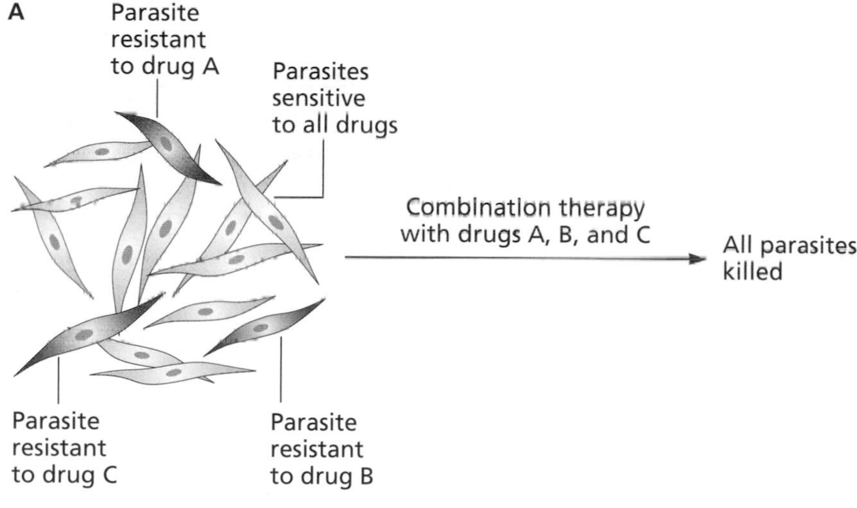

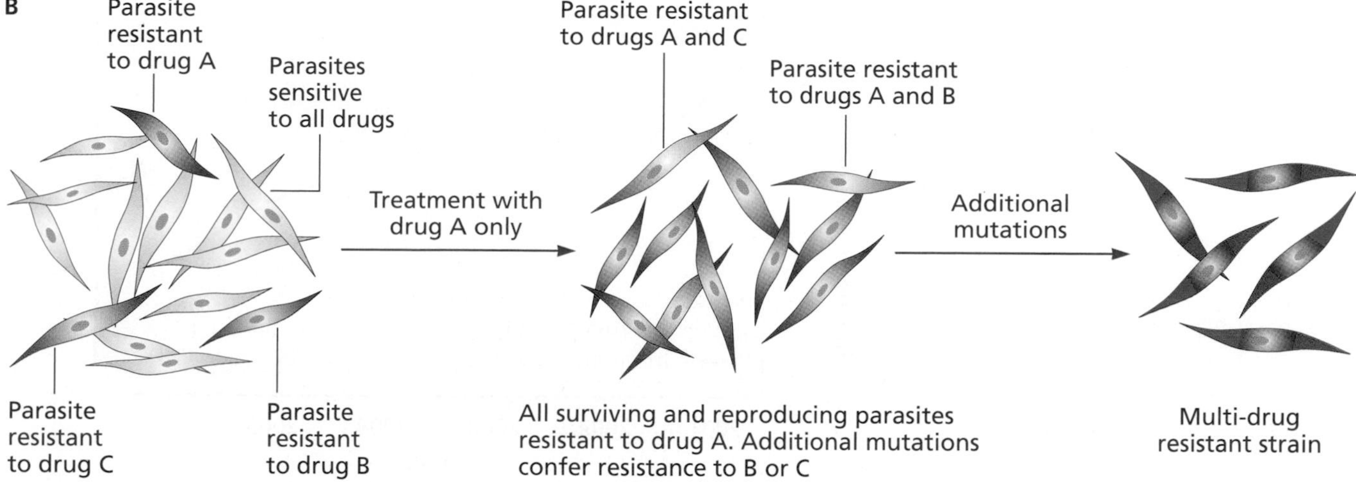

Figure 9.20 Combination therapy and drug resistance. (A) In any population of parasites there is a high probability that small numbers of organisms experience mutations rendering them resistant to one of three available drugs: drugs A, B, or C. Treatment with all three drugs will eliminate all parasites, as it is extremely unlikely that any parasites are resistant to all three drugs. (B) If only one of the three drugs is used (drug A in this example), the portion of the parasites that are resistant to this drug will survive and multiply. Further mutations may occur that eventually result in resistance to either two or even all three of the drugs. Treatment of this strain that is now multi-drug resistant becomes problematic, requiring other less efficacious or more problematic second-line drugs.

combinations are now considered to be the mainstay of malaria treatment and along with the use of insecticide-treated bed nets and focal insecticide use, **artemisinin-based combination therapy** (ACT) has substantially reduced the burden of malaria in sub-Saharan Africa (**Figure 9.21**). Its efficacy has

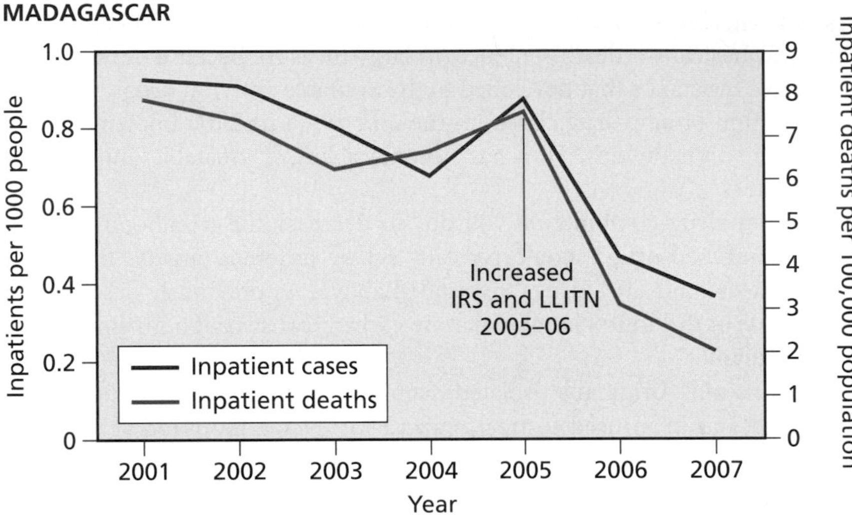

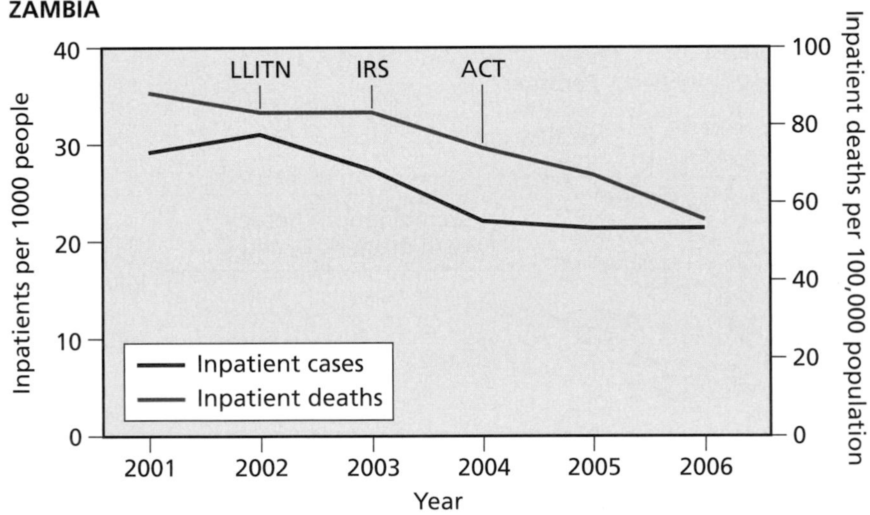

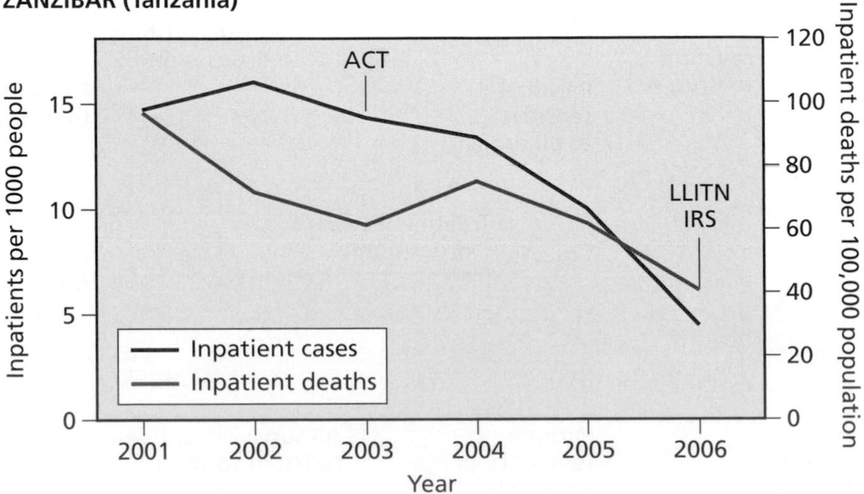

Figure 9.21 Malaria on the decline. Trends between 2001 and 2006 in three representative African countries in terms of malaria cases and deaths that resulted from malaria control programs based on indoor residual spraying (IRS); long-lasting, insecticide-treated bed nets (LLITN); and artemisinin-based combination therapy (ACT). The dates when these various measures were implemented are indicated. (Adapted from Eastman RT & Fidock DA [2009] *Nat Rev Microbiol* 7:864–874 [doi:10.1038/nrmicro2239]. With permission from Macmillan Publishers Ltd.)

been tested with a variety of antimalarial drugs. Its use with piperaquine is currently recommended by the World Health Organization for the treatment of uncomplicated malaria caused by ***P. falciparum***. Related to chloroquine, piperaquine presumably works via a similar mechanism. It was originally developed in the mid-1960s in China as an alternative to chloroquine in the face of spreading chloroquine resistance. Usage declined in the 1980s as piperaquine-resistant strains of ***P. falciparum*** arose. It remains highly effective, however, when paired with artemisinin. Yet even the above described precautions may not be enough to prevent resistance, and the recent report of artemisinin-resistant ***P. falciparum*** in Southeast Asia is cause for concern. Other innovative strategies for reducing the problem of drug resistance will be considered in Chapter 10.

Concerns about resistance highlight the need for new anti-parasitic drugs

With resistance to many anti-parasitic drugs already a reality, there is an urgent need for new drugs. Such drugs alone will not solve the problem of resistance; parasites might ultimately develop resistance to them as well. New anti-parasitic compounds could dramatically improve public health in the short term, however, and may at least buy time until longer-lasting solutions are developed and implemented. Ideally, if used judiciously, new drugs might stave off resistance indefinitely, especially if they rely on novel mechanisms of drug action. Their usefulness in preventing resistance might be further enhanced if used in combination, as described above.

Because of the emergence of multi-drug resistant strains, the need for novel antimalarials is especially acute. One fruitful avenue in this regard may be to target the parasite's liver stage. Currently, almost all antimalarial drugs and almost all current research focus on blood stages of the parasite. The feasibility of new drugs that might zero in on liver stages is supported by recent findings on differentially expressed genes in ***Plasmodium*** blood and liver stages. Such differential expression suggests that the two stages provide different targets on which new drugs might act. Drugs against liver stages would be especially welcome for the treatment of ***P. vivax*** and ***P. ovale***, both of which can remain dormant in the liver as hypnozoites for months or years after an initial infection, only to re-emerge, causing a relapse in the patient. **Box 9.3** describes another potential approach for new antimalaria drugs.

As previously mentioned, for some diseases such as onchocerciasis and schistosomiasis current treatment relies on a single drug. Even though resistance against ivermectin and praziquantel is not yet a major problem, the hints of emerging resistance described above underscore the need for new alternatives. For onchocerciasis, clinical trials are currently under way for moxidectin, an anthelmintic currently used in veterinary medicine. A derivative of a metabolic waste produced by soil bacteria in the genus *Streptomyces*, moxidectin works by selectively binding chloride-ion channels on neurons. Consequent disruption of neurotransmission results in paralysis and ultimately in death of the parasite. For schistosomiasis, it has been suggested that a unique schistosome enzyme, thioredoxin glutathione reductase, might serve as a target for a new drug that could provide an alternative to praziquantel. The enzyme is crucial to the schistosome's survival, as it allows the worm to evade reactive oxygen products generated by the host's immune response. A new anti-schistosome drug that interfered with the enzyme would thus render schistosomes more vulnerable to immune destruction.

BOX 9.3

The Apicoplast: A Potential New Drug Target

Apicomplexans, of course, take a backseat to nobody, when it comes to doling out misery in the form of parasitic diseases. *Toxoplasma*, *Cryptosporidium*, *Eimeria*, and most notably *Plasmodium*, all belong to this unsavory group. Yet the evolutionary origins of apicomplexa suggest a less malevolent past. The apicomplexans belong to the large eukaryotic super kingdom Chromalveolata, which also includes kelps, diatoms, dinoflagellates, and other photosynthetic algae. In fact, as described in Chapter 2 (pages 35–36), most apicomplexans still retain a remanant chloroplast, highlighting their photosynthetic ancestry. The structure, known as the apicoplast, was acquired by the primitive protozoan ancestor of the apicomplexans via secondary endosymbiosis (Figure 1) (see also Figure 2.8). It no longer functions as a photosynthetic organelle in apicoplasts, which raises the question, why would parasites hang on to such an apparently useless relic structure?

Cryptosporidium, for example, has lost the apioplast over evolutionary time. In *Plasmodium* and *Toxoplasma*, the apicoplast isn't useless at all. Chloroplasts are involved in more than photosynthesis; many molecules of biological importance are synthesized there. At least some of these biosynthetic pathways have been retained by the apicoplast.

Furthermore, it is known that the apicoplast is essential for parasite survival. Exactly why it is so crucial is less clear. Most evidence suggests that at least in *Plasmodium*, the organelle's vital role is related to either fatty acid synthesis or the synthesis of isoprenoid precursors. Isoprenoids are a large group of biological molecules including cholesterol, dolichol, and ubiquinone. They are all synthesized from a 5-carbon precursor molecule called isopentenyl pyrophosphate (IPP). Although *Plasmodium* can meet its requirements for fatty acids by acquiring them from the host, IPP can only be supplied by the apicoplast.

This reliance on the apicoplast was recently demonstrated in a series of experiments that showed that when the apicoplast is lost as a result of exposure to clindamycin, *Plasmodium* cannot survive. Yet such parasites can be rescued with exogenous IPP. Other exogenous compounds such as fatty acids result in no such parasite survival, demonstrating that IPP is the lone essential molecule produced by the apicoplast.

IPP synthesis in the apicoplast might be an attractive target for new antimalaria drugs, especially because IPP in apicomplexans is produced in a different manner than it is in animals. Whereas apicomplexans synthesize IPP via the DOXP pathway (named after a key intermediate compound: D-xylose 5-phosphate) (see Figure 1), animals use a different series of reactions known as the mevalonate pathway. Thus, targeting the DOXP pathway may permit any future drugs to meet the criterion of selective toxicity, which is often such a problem when it comes to the development of anti-parasitic drugs.

References

Nair SC & Striepen B (2011) What do human parasites do with a chloroplast anyway. *PLoS Biol* 9(8):e1001137 (doi:10.1371/journal.pbio.1001137).

Yeh E & DeRisi JL (2011) Chemical rescue of malaria parasites lacking an apicoplast defines organelle function in blood-stage *Plasmodium falciparum*. *PLoS Biol* 9(8):e1001138 (doi:10.1371/journal.pbio.1001138).

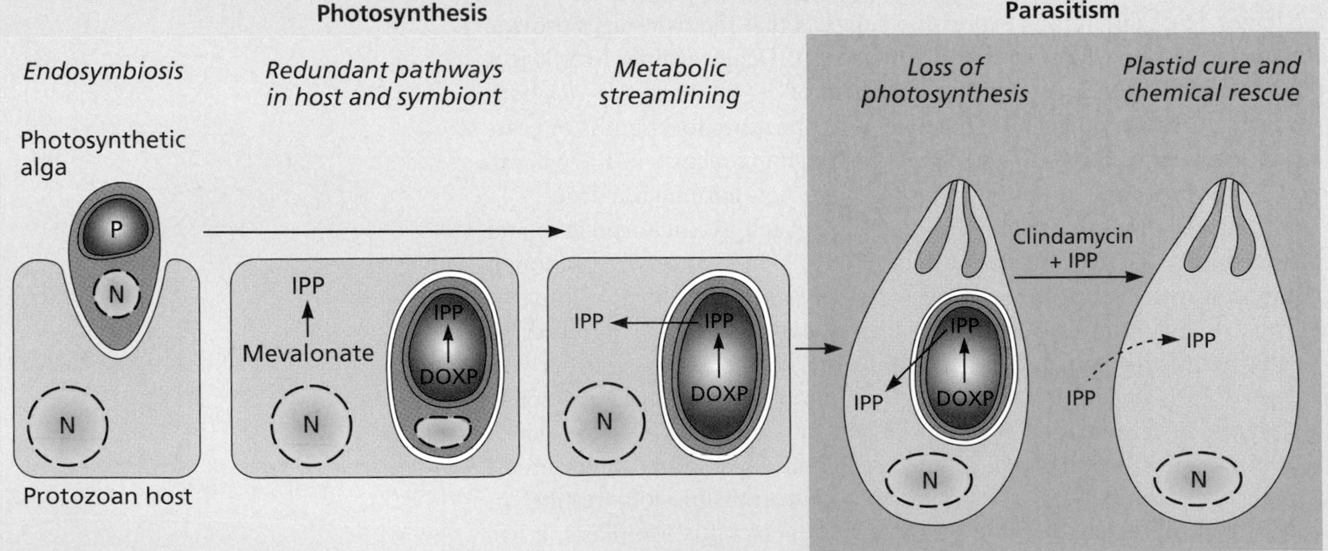

Figure 1 The Apicoplast. The ancestral protozoan first acquired a chloroplast through secondary endosymbiosis when it took up a photosynthetic algal cell. Algae had previously acquired chloroplasts via primary endosymbiosis with photosynthetic *Cyanobacteria*. Originally, there would have been two redundant pathways for the synthesis of IPP: the mevalonate pathway, observed in animals and fungi, and the DOXP pathway, still observed in the *Plasmodium* apicoplast. Over evolutionary time, the mevalonate pathway was lost, leaving *Plasmodium* dependent on the DOXP pathway. When the apicoplast is experimentally removed by exposing *Plasmodium* to clindamycin, the parasite cannot survive unless exogenous IPP is provided, demonstrating the apicoplast's crucial role in IPP synthesis. (Adapted from Nair SC & Striepen B [2011] *PLoS Biol* 9[8]:e1001137.)

New drugs are also needed to replace older more toxic medications

The threat of drug resistance is not the only reason to welcome the arrival of new anti-parasitic drugs. As previously discussed, the difficulty of achieving selective toxicity against eukaryotic parasites is an important reason that we have so few safe and effective anti-parasitic drugs to begin with. In certain cases, the drugs that are available carry substantial baggage in the form of side effects or complex treatment regimens. All the drugs currently used to treat the later stages of African trypanosomiasis, for instance, require either injection or intravenous administration, sometimes for as long as two weeks. Many of them cause serious side effects. The problem is almost as grim when considering treatment options for American trypanosomiasis (Chagas' disease). Currently, there are only two available drugs, both of which can cause adverse reactions. Nifurtimox can cause symptoms ranging from anorexia to nerve damage. Benzidazole may sometimes cause severe skin inflammation. New treatment options against either African trypanosomiasis or Chagas' disease have been excruciatingly slow in coming, but new and safer drugs that can be taken orally are now at least in trials, offering hope that new treatment options may be on the horizon.

The manner in which new drugs are discovered has changed considerably

Where will new anti-parasitic agents come from? In the past, **empirical methodology** was used to identify new and potentially useful drugs. Information that is obtained empirically indicates that it was derived through experimentation or observation rather than from a theoretical premise. Put more simply, empirical methodology relies to a large degree on trial and error. In terms of empirical drug discovery, whole, living parasites were exposed to a systematic screen of particular chemicals. If the parasite lived, the chemical was deemed unpromising. If the parasite died, optimization of the chemical became the focus, as researchers sought new synthetic analogs to the chemical, with characteristics such as improved persistence and safety. Many of the most promising empirical compounds have come originally from some of the same types of soil microbes that have proven to be a rich source of antibacterial agents as well. When drugs are discovered empirically, the mechanism by which the drug actually works is secondary; the principal goal is to find molecules that exhibit anti-parasitic activity. The drug mechanism, if it is ever understood, is elucidated later.

In recent years there has been a shift from traditional empirical screening to approaches in which potential drug targets are first characterized, and then compounds that specifically interfere with these potential targets are sought.

The increasingly available genomic and proteomic information available for parasites now allows researchers to identify specific potential targets in advance. Often these targets are unique parasite molecules that may be disrupted in the parasite, without undue risk to the host (see the previous discussion of thioredoxin glutathione reductase as an example of a target for a future anti-schistosome drug). In other cases, the potential target may not be unique to the parasite, but it may be sufficiently different in structure from the host molecule so that targeting it is at least a possibility. The process of developing such new medications based on a priori knowledge of the biological target is called **rational drug design**.

Potential targets can first be identified through the use of one or more modern molecular techniques. One such technique is to mine the genomes of model organisms such as *Caenorhabditis elegans* to identify novel nematode-specific targets. Of course, because *C. elegans* is a free-living nematode, it may lack specific genes or gene products related to parasitism and is thus of somewhat limited utility. But the number of available parasite genomes is increasing rapidly. We can therefore expect the search for target genes to become more specific and more likely to yield promising results.

RNA interference is increasingly used as a strategy to identify potential targets for new drugs. Using this methodology (see Figure 9.12), specific genes can be knocked out in a particular parasite of interest. If parasites experiencing such knockouts are less viable, this finding suggests that the product of the gene in question may be a satisfactory target. Alternatively, once a promising gene is identified, chemical mutagenesis can be used to generate mutants in which the gene has been altered. Impaired survival in such a mutant strain is suggestive that the gene might code for a promising target.

Once a potential target is identified, small molecules can be screened as potential inhibitors of the target. This task has been considerably simplified with the advent of high-throughput screening technology. Using such technology, thousands of compounds can be screened in a relatively short time. If the potential target is well characterized, the process may be simpler, as the number of compounds to be screened may be considerably reduced. For example, if the potential target is a type of kinase, it can be screened against libraries of known kinase inhibitors, which currently number close to one thousand.

Potentially effective drugs usually require chemical modification prior to their use

If a promising anti-parasite molecule is identified, it is unlikely that this molecule can be used as an effective drug as is. Rather, this initial compound, known as the **lead molecule**, must undergo substantial modification and optimization before its use can be considered, in even initial phase I clinical trials (trials designed to assess drug safety and tolerability, along with other pharmacological parameters). This optimization phase may attempt to improve the lead molecule's potency, penetration, stability, solubility, and its spectrum of activity.

To assist the researcher in this regard, certain attributes of lead molecules, when compared to their eventual commercial counterparts, have been identified in known and currently available drugs. On average, lead molecules are lower in molecular weight and less lipophilic than corresponding commercial drugs. They tend to have fewer functional groups able to form hydrogen bonds and a lower **Andrews binding energy** (an estimated binding energy that a molecule would establish with a hypothetical molecular partner, if all functional groups contributed to binding). Such information provides those interested in drug optimization with a starting point as to how the lead molecule might be modified.

To provide just one example, consider the recent modification of amphotericin B into a potentially safe and oral drug for the treatment of leishmaniasis. In the past, this disease has been treated with pentavalent antimonial drugs (antimony-containing compounds). Such drugs carry all the disadvantages too often attributed to many anti-parasitic drugs—they can only be given by injection and they carry a substantial risk of serious side effects. Furthermore, in many parts of the world, the parasite has become resistant to pentavalent antimonials, although the level of resistance varies according to ***Leishmania*** species. Consequently, in many cases amphotericin B has become the treatment of choice.

Amphotericin B is not exactly a benign alternative, however. Usually used to combat serious fungal infections, it is administered intravenously for up to forty days, and it causes many side effects that include red blood cell lysis, chills, fever, and pain. However, by making the molecule more lipophililc, some of these problems have been apparently overcome, at least in ***Leishmania***-infected mice. The newer, novel, and lipid-based amphotericin B is taken orally, it has increased solubility and improved stability, and it seems to cause fewer side effects.

Economic issues often affect the rate at which new drugs are developed

As we have discussed, developing a new and effective anti-parasitic drug for which other issues such as safety and toxicity have been fully addressed is no easy task. To further complicate things, certain nonmedical concerns must also be considered. Indeed, the probability that any potential drug will see the commercial light of day often hinges more on economic concerns and market conditions than anything else.

In short, pharmacological companies seek to make a profit with the drugs that they market. The likelihood of making a profit depends in part on the cost of bringing a potential medication to market. This cost is based on many factors, including the cost of the starting material, the number of synthetic steps needed to produce the drug, and the difficulty of producing it on a commercial scale. The decision to proceed with the development of a promising lead chemical might be based on any of these factors.

Furthermore, there must be a market for the medication in question. Yet most potential customers live in less-developed parts of the world and often subsist on a few dollars a day. Likewise, the governments in such countries generally are economically strapped as well, and they are in no position to buy large quantities of new anti-parasitic medications to help insure the health of their populace. With little return on their investment, pharmaceutical companies have slight commercial incentive to discover and develop drugs for humans. The potential for a favorable bottom line is generally greater for drugs required to maintain the health of livestock. The vast numbers of cattle, sheep, poultry, and other animals in developed countries and the need for ranchers and farmers to keep them healthy insures a constant market for new veterinary drugs. The demand for veterinary medications is augmented by the multibillion-dollar pet industry in the developed world. Even for effective veterinary drugs, however, a market is not necessarily assured. Often, to insure profitability, the focus of pharmaceutical companies will be on broad-spectrum medications that might be used, for example, against a variety of intestinal nematodes or ectoparasites. There is generally less enthusiasm in such companies for developing drugs that will eliminate or control only a single parasite.

Sometimes an anti-parasitic drug developed with animals in mind is found to be effective for the treatment of parasitic infections in humans. Ivermectin is the classic example. Yet in other instances, drugs developed to treat animals are simply impractical for the treatment of human infections, even if they are safe and effective. If, for instance, cattle, chickens, or other animals to be eventually consumed by humans are to be treated, medications should have short persistence in the body. This requirement is intended to alleviate concerns over the consumption of animal products that still contain quantities of medication. For human drugs, however, where many of the people to be treated live in remote rural areas with limited or no health care facilities, longer drug persistence may be required. In such situations, the repeated treatment that is necessary when drugs have a short half-life is unfeasible.

If all the problems and limitations of drug development that have been discussed thus far are considered, it is not too surprising that new anti-parasitic drugs seem to arrive at a snail's pace. Yet reasons for some optimism remain. In recent years, pharmaceutical companies have ramped up their investment in philanthropic drug development, providing required medications at little or no cost to those in need. Increasingly, these companies are partnering with academic institutions and other organizations interested in global health such as the World Health Organization to provide new low-cost, safe, and effective drugs to those who need them most. Such **integrated drug discovery**, in which the best minds from industry, academia, and the public health community collaborate, may offer the best hope for confronting the burden of parasitic disease in the poorest parts of the world. The need for such cooperation in the future will be further explored in Chapter 10.

9.3 VACCINES

If low-cost, safe, and effective vaccines were available to prevent infection with specific parasites in the first place, the issues of control and drugs for these parasites would become moot. Vaccines are largely harmless preparations (at least in theory) to which individuals are exposed, with the goal of establishing immunological memory against a particular parasite. Vaccines, consequently, represent a form of immune manipulation, which if successful, prevent future infection. Vaccines are not the only option when it comes to immune manipulation. Researchers have investigated, for instance, whether cutaneous leishmaniasis can treated by injecting IL-12 directly into the lesions. Such a strategy helps shift the immune response from a less protective humoral to a more protective cell-mediated response (see Chapter 4, pages 131–132). Yet in general, when discussing immune manipulation, we are talking about vaccines. It is consequently on vaccines that we will focus our attention in this section.

The development of effective vaccines against a variety of pathogens has been a key reason, along with improved sanitation and the development of anti-parasite chemotherapy, for the remarkable decrease in the burden of infectious disease since the nineteenth century. Vaccination was crucial to the successful effort to eradicate smallpox in the 1970s (**Figure 9.22**), and

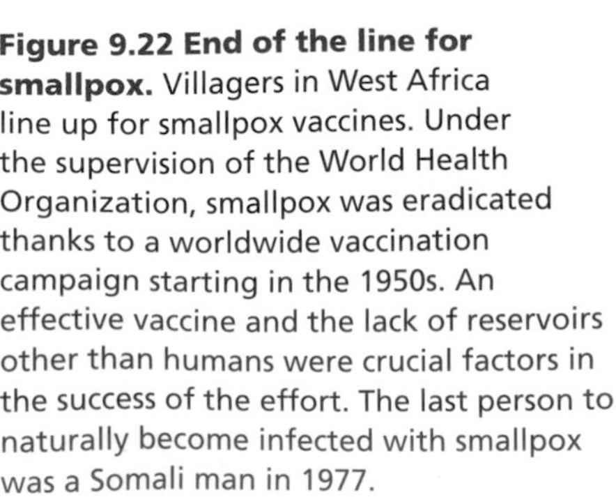

Figure 9.22 End of the line for smallpox. Villagers in West Africa line up for smallpox vaccines. Under the supervision of the World Health Organization, smallpox was eradicated thanks to a worldwide vaccination campaign starting in the 1950s. An effective vaccine and the lack of reservoirs other than humans were crucial factors in the success of the effort. The last person to naturally become infected with smallpox was a Somali man in 1977.

Table 9.3 Infectious diseases preventable through vaccination. Effective vaccines are available for the human diseases listed. Note that none of these diseases is caused by eukaryotic pathogens. Not all of these vaccines are available in all countries and some have drawbacks that limit their usefulness.

Disease	Pathogen
Anthrax	Bacterium
Cervical cancer	Virus
Diphtheria	Bacterium
Hepatitis A	Virus
Hepatitis B	Virus
Haemophilus influenzae type b (Hib)	Bacterium
Human papillomavirus	Virus
H1N1 flu (Swine flu)	Virus
Influenza (Seasonal flu)	VIrus
Japanese encephalitis (JE)	Virus
Lyme disease	Bacterium
Measles	Virus
Meningococcal	Bacterium
Monkeypox	Virus
Mumps	Virus
Pertussis (whooping cough)	Bacterium
Pneumoccocal pneumonia	Bacterium
Poliomyelitis (polio)	VIrus
Rabies	Virus
Rotavirus	Virus
Rubella (German measles)	Virus
Shingles	Virus
Smallpox	Virus
Tetanus	Virus
Tuberculosis	Bacterium
Typhoid fever	Bacterium
Varicella (chickenpox)	Virus
Yellow fever	Virus

(Courtesy of CDC.)

vaccines remain front and center in the ongoing efforts to eradicate polio and measles. **Table 9.3** lists other diseases brought to heel by vaccines.

Nevertheless, many parasites have resisted our best efforts to develop effective vaccines. HIV and gonorrhea are just two examples. Eukaryotic parasites, whether protozoan or helminth, have proven to be particularly obdurate. A few vaccines against eukaryotic parasites are licensed for use in animals (**Table 9.4**). Despite concentrated and innovative research efforts, there are no such vaccines currently available for clinical use in humans. In this section, we will describe the basic strategy of vaccination in general and discuss some of the problems and prospects for vaccines against eukaryotic parasites. To start, consider what any successful vaccine must be able to accomplish.

Table 9.4 Anti-parasitic vaccines available for veterinary use. A partial list of some of the vaccines that are commercially produced or distributed by governmental organizations. These vaccines vary in their effectiveness and not all of them are available in all countries.

Parasite	Host	Vaccine type
Eimeria	Poultry	Killed, attenuated, and subunit
Toxoplasma gondii	Sheep	Attenuated
Neospora caninum	Cattle	Killed
Babesia canis	Dogs	Subunit
Babesia spp.	Cattle	Attenuated
Theileria annulata	Cattle	Attenuated
Giardia lamblia	Dogs	Killed
Tritrichomonas foetus	Cattle	Killed
Leishmania donovani	Dogs	Subunit
Taenia ovis	Sheep	Subunit
Dictyocaulus viviparus	Cattle	Attenuated
Boophilus microplus	Cattle	Subunit

Vaccines must be safe and inexpensive, while inducing long-term immunity

Even the most effective vaccine is unacceptable if it induces serious side effects. This is particularly true if the chance of natural infection is rare. Cost is another consideration, especially if large numbers of individuals are to be vaccinated or if the vaccine is to be used in poor, underdeveloped areas. Even when issues of safety and cost are satisfactorily resolved, a significant number of immunological obstacles must be overcome if the vaccine is to have real protective value.

It is no coincidence that most effective vaccines (see Table 9.3) are those that provoke a strong humoral response, resulting in high titers of antibodies that neutralize either the parasite itself or a toxin produced by the parasite. Such an immune response (see the online review of vertebrate immunology) is frequently successful against extracellular bacteria and even many viruses. For some intracellular pathogens such as HIV and *Mycobacterium tuberculosis* even a vigorous humoral response is not necessarily protective. The same is certainly true for intracellular eukaryotic parasites such as ***Leishmania***. For these pathogens, a robust cell mediated response mediated by Th-1 cells is required. Generating the necessary cell-mediated response has proven to be a major obstacle in vaccine development. Other characteristics of effective vaccines are listed in **Table 9.5**.

Other issues can hamstring the development of effective vaccines. One concern is the route through which a vaccine is administered. To effectively stimulate a protective immune response, the key antigenic peptides in the vaccine must be presented to $CD4^+$ T cells by antigen-presenting cells. Because of the need to prevent proteolysis of the antigens before they reach the antigen-presenting cells, many vaccines are given by injection rather than orally to avoid the low pH of the stomach and proteases found there. Vaccine administration via injection may not be of concern if the vaccine is designed to protect against a pathogen normally found in the blood or the lymphatic fluid. For other parasites, for example, those of intestinal lumen or those inhabiting the lungs, this route of administration may reduce vaccine

Table 9.5 Characteristics of effective vaccines. There are a number of criteria that must be met before using a new vaccine.

Characteristics of Effective Vaccines	
Safety	Vaccine itself must not cause illness or death
Protective	Vaccine must protect against illness resulting from exposure to live pathogen
Sustained protection	Protection against disease must last for several years
Induction of humoral and cell-mediated responses	Ideally, vaccines has antigens that are activating for both B and T cells
Practical considerations	Vaccines must have low cost per dose, be biologically stable, have ease of administration, and have minimal side effects

efficacy because it does not produce the necessary mucosal immunity. Ideally, a vaccine will mimic a normal infection as closely as possible, and inducing immunity at the normal portal of entry for the parasite is an important consideration.

Another question with particular relevance for many eukaryotic parasites is whether or not vaccines are really feasible for parasites that tend to establish long-term chronic infections. If the parasite establishes a chronic infection in the first place, it is likely that effective clearance by the immune system does not occur. Schistosomes, for instance, live for years within the circulatory system, where they should be easily accessible to the immune system. Yet, for reasons discussed in Chapter 4 (pages 151–152), they persist. Could any traditional vaccine achieve what a normally occurring immune response cannot? This is not to suggest that an anti-schistosome vaccine is impossible. It simply suggests that it may be necessary to approach such a vaccine in novel ways.

Vaccines against eukaryotic parasites are particularly problematic

The chronic nature of infections caused by many protozoa and helminths is only one of the problems posed by parasitic eukaryotes. Such parasites, simply because they are eukaryotes, are more complex than bacteria or viruses, and this complexity presents a couple of thorny problems. One such problem is the myriad of ways that different parasites have evolved to subvert, hide from, or otherwise interfere with an effective immune response. Certainly, viruses or bacteria also have adaptations to thwart effective immunity and many employ various strategies to prevent immune clearance. When it comes to the variety and complexity of such immune interference, however, eukaryotes win top prize. We have discussed many examples of this phenomenon (for example, see Chapter 4, pages 143–152). Recall, for instance, the manner in which African trypanosomes alter their surface antigens. As the host develops an effective immune response to one such antigen, antigenic switching produces new clones with different antigenic properties, rendering the earlier immune response ineffective. It may never be possible to develop an effective vaccine against all possible surface antigens, and any potential vaccine might need to be coupled with a technique to stop antigenic variation in the first place.

Another problem is that many eukaryotic parasites undergo considerable developmental change within their host, and often antigens present in one stage are different from those expressed in other stages. ***Plasmodium***, which develops from sporozoite to merozoite and finally to gametocyte within the mammalian host, is a prime example. Of course, this seeming difficulty can potentially be used against the parasite because it increases the number of antigens against which immunity might be generated.

Vaccines can be categorized into several types

Regardless of parasite type—viral, bacterial, or eukaryotic—vaccines tend to fall into one of several categories. Until recently, almost all vaccines could be accurately described as either killed or live, attenuated vaccines (**Table 9.6**). **Killed vaccines** (also called inactivated vaccines), as the name suggests, consist of inactivated virus or dead organisms. The Salk polio vaccine and the injected, inactivated influenza vaccines are examples of killed vaccines. **Live, attenuated vaccines** contain a weakened strain of the parasite, which although able to infect and replicate in the host, is unable to cause disease. The oral Sabin polio vaccine and the more recent attenuated influenza vaccine inhaled as a nasal mist both consist of an attenuated virus. Both types of vaccines have advantages and disadvantages. An attenuated vaccine more closely mimics natural disease and is more likely to stimulate both humoral and cell-mediated immunity. Furthermore, such vaccines are administered through the normal portal of entry and are therefore more likely to generate immunity at the exact site where the parasite is found in the body. There is, however, a low but often significant probability that the attenuated organisms in a live vaccine will revert to virulence if they undergo random mutation. Attenuated polio virus in the Sabin vaccine, for instance, reverts to virulence in about one in each 2.6 million vaccinated individuals, causing symptoms of polio.

Modern molecular biology techniques can be used to render attenuated vaccines even safer. The ***Plasmodium*** sporozoite, for instance has frequently been considered as a target for vaccine development. Genes in the ***P. falciparum*** genome have been identified that encode proteins crucial for the invasion of liver cells by sporozoites. By deleting these genes, sporozoites

Table 9.6. Killed vs. live vaccines. Most currently licensed vaccines for use in humans are either killed (inactivated) or live (attenuated) vaccines. Some of the advantages and disadvantages of each type are listed.

	Killed (Inactivated) Vaccines	Live (Attenuated) Vaccines
Advantages	Does not cause disease	May be administered via normal portal of entry (injection not required)
	Safe for immunodeficient individuals	Mimics a normal infection, resulting in stronger immunity May be transmitted to others increasing the overall immunity of the population
Disadvantages	Results in less robust immunity	May revert to virulent form, causing disease in vaccinated individuals
	Must be injected	Unsafe for immunodeficient individuals

have been produced that are still immunogenic but unable to complete the ***Plasmodium*** life cycle.

Besides being safer, inactivated vaccines are easier and less expensive to produce. Their biggest drawback, as mentioned earlier, is that they do not closely mimic a natural infection and are consequently less effective than live vaccines. Furthermore, because killed preparations are pH labile and susceptible to proteases in the stomach, they usually must be injected, even if the parasite in question is not bloodborne. The efficacy of killed vaccines can often be increased by including **adjuvants** in the vaccine preparation. An adjuvant is a compound that increases the immune response to the antigens with which it is mixed. Often these adjuvants are ligands for Toll-like receptors (TLRs) that readily bind TLRs on antigen-presenting cells (see the online review of vertebrate immunology).

Several newer vaccine types have been developed, some of which are acellular vaccines in that they do not include whole organisms. An example is the **subunit vaccine**, in which the vaccine consists of particular immunostimulatory antigens only. The hepatitis B vaccine, for instance, is composed only of viral surface proteins. Our understanding of the stimulatory role of T cells in a humoral response has resulted in the development of **conjugate vaccines** (Figure 9.23A). These vaccines rely on a combination of antigens

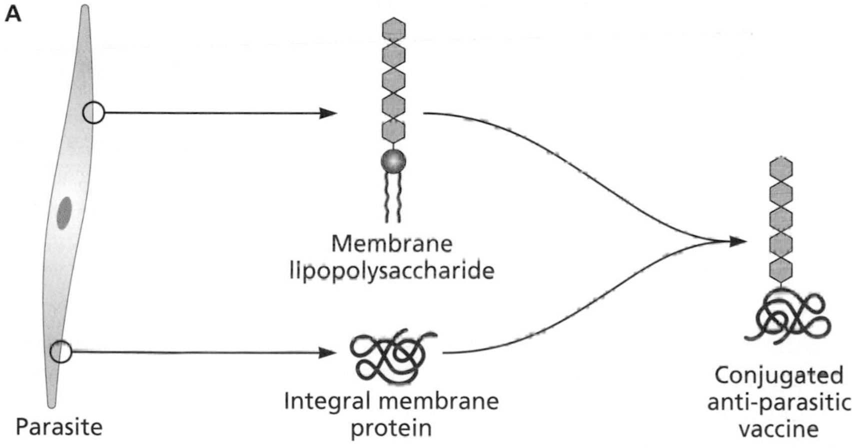

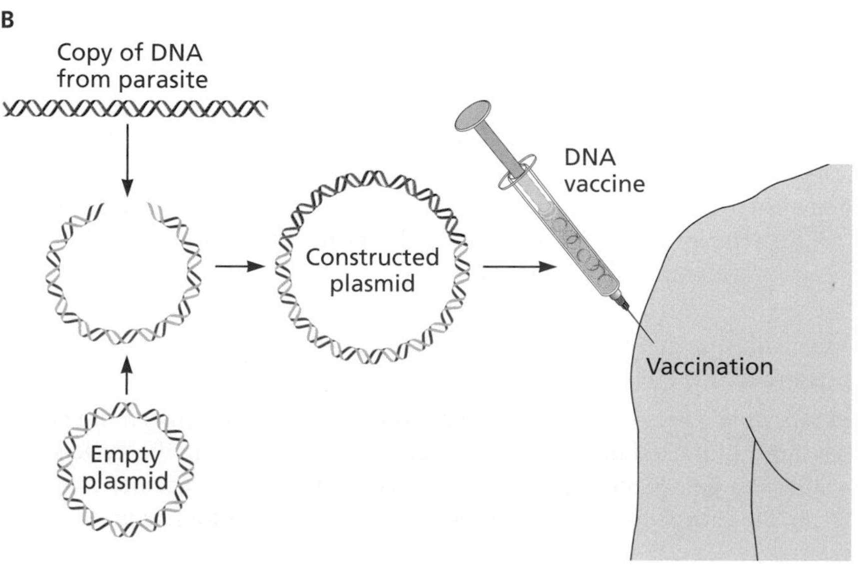

Figure 9.23 Alternative vaccine types. (A) A conjugate vaccine. This acellular vaccine consists of antigens that stimulate both humoral and cell-mediated immunity, in this case, both a membrane lipopolysaccharide and a membrane protein from the parasite. (B) DNA vaccines contain purified pathogen DNA, which encodes highly antigenic pathogen proteins. These parasite genes are inserted into a bacterial plasmid, which is then used as a vaccine preparation. Host cells may absorb this foreign DNA and, in at least some cases, incorporate it into their own DNA. The foreign proteins, when expressed, stimulate an immune response against these proteins that will result in protective, immunological memory against the actual parasite, when and if it infects the vaccinated individual. (B, © Vical Incorporated, used with permission.)

that stimulate both B and T cells. The *Haemophilus influenzae* vaccine, for instance, combines polysaccharides found in the bacterial capsule with peptides recognized by antigen-specific T cells. The result is a much stronger antibody response then could be elicited with the polysaccharides alone.

DNA vaccines represent yet another approach. Largely experimental, such vaccines are based on the premise that cells will, at least on occasion, take up exogenous DNA (see Figure 9.23B). If parasite DNA for particular antigens is used in vaccine preparation, host cells may not only absorb this DNA, but they may express the foreign antigens that it encodes. These antigens then prime cells of the immune system, establishing memory that may be effective in preventing disease if the actual parasite is encountered in the future. Parasite genes cloned into bacterial plasmids could then be used in such vaccine preparations.

Having described some of the basic principles of vaccination therapy, let us now take a look at where vaccine development stands in preventing infections from a few of the most important medical or veterinary eukaryotic parasites.

An effective malaria vaccine has been the object of intensive investigation

The development of an antimalaria vaccine has been something of a holy grail in vaccine research. Malaria is notoriously difficult to control through more conventional means, and the emergence of multi-drug resistant ***Plasmodium*** strains only highlights the need for a safe, economical, and effective vaccine. Modern efforts to develop such a vaccine accelerated in the 1970s when William Trager developed *in vitro* methods for the maintenance of the ***P. falciparum*** life cycle. This development suddenly provided researchers with an essentially limitless amount of blood-stage parasite material for experimental purposes. Yet in spite of decades of rigorous and concentrated research, there is still no such vaccine that is used clinically. We will consider some of the obstacles to be overcome, as well as some of the reasons that many remain optimistic about the prospects for a malaria vaccine.

There is reason for optimism because individuals in endemic countries who are naturally exposed to ***Plasmodium*** develop a measure of acquired immunity. Furthermore, this immunity can be transferred to previously unexposed individuals with purified anti-***Plasmodium*** antibodies. However, we must temper such optimism because naturally acquired immunity is not sterilizing; parasites may be reduced in number to the point where symptoms resolve, but the parasites are not eradicated completely. Thus, even individuals with low parasitemia continue to serve as reservoirs of infection. Additionally, should they leave an endemic area and no longer continue to be regularly exposed to natural boosters, their acquired immunity diminishes to the point where they are once again vulnerable to disease. Vaccine development is further complicated by the complex ***Plasmodium*** life cycle and the different antigens displayed during different stages of this cycle. Many of these antigens are highly polymorphic.

Vaccines against different life-cycle stages offer different potential benefits

Plasmodium's life cycle does, however, come with a silver lining: more potential targets for candidate vaccines. Vaccines against specific life-cycle stages would also be beneficial for different reasons. In theory, vaccines might be formulated against all stages, in order to achieve more complete protection.

An anti-sporozoite vaccine might prevent infection in the first place (**Figure 9.24**). Such a vaccine would therefore not only prevent disease in a vaccinated individual, but it would eliminate this person as a potential reservoir. This is a tall order, however, because sporozoites only remain in the blood for 30 to 60 minutes before they invade liver cells. Furthermore, an anti-sporozoite vaccine would have to be effective enough to prevent most or even all sporozoites from reaching the liver, as even a few escapees might result in an eventual blood parasitemia high enough to cause disease.

An anti-blood stage vaccine has the advantage in that it does not need to be as effective as the anti-sporozoite vaccine (see Figure 9.24). This anti-disease vaccine simply needs to hold parasite levels below the disease threshold (see Chapter 5, page 160) to prevent the onset of clinical malaria. Thus, it would protect the vaccinated individual from the worst consequences of malaria infection. Unless the vaccine resulted in the elimination of all parasites however, the protected individual might still act as a reservoir, meaning that overall disease transmission may remain unaffected.

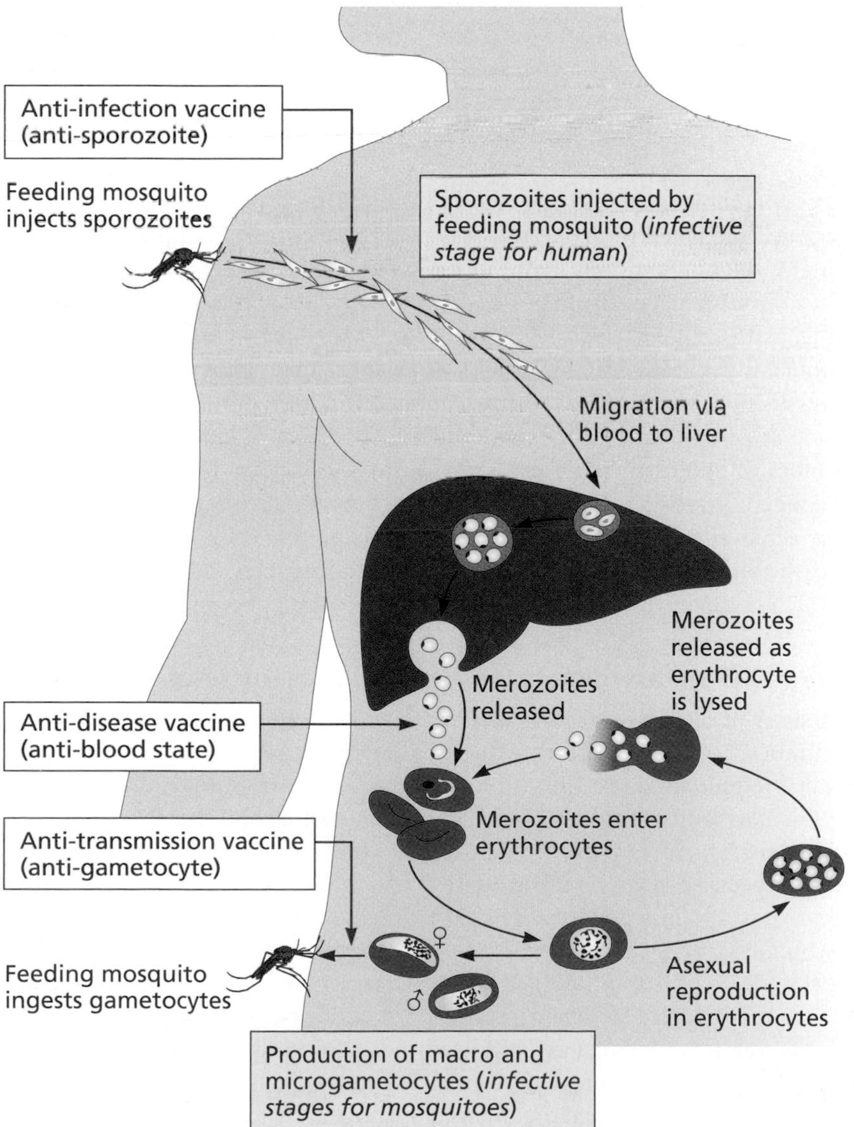

Figure 9.24 Antimalaria vaccines. Only those stages in in the *Plasmodium* life cycle that occur in the mammalian host are shown. This complex life cycle offers several targets for vaccine development. Anti-sporozoite vaccines target the infective stage for humans and can be considered anti-infection vaccines. The goal of a vaccine that targets merozoites would be to hold the number of infected erythrocytes at a low number that does not result in symptoms of disease. The individual is still infected, although ideally the infection is subclinical. Thus, such a vaccine would represent an anti-disease vaccine. Should a vaccine target gametocytes, the infected individual might still develop clinical malaria, but there would ideally be no transmission of gametocytes to mosquito vectors. As such it would be an anti-transmission vaccine, which might reduce overall transmission in a population.

This objection might be addressed with an anti-gametocyte vaccine. Yet such a vaccine would not help the vaccinated individual; the individual would still be susceptible to malaria. It would, however, reduce transmission and help to protect the community as a whole.

Recent candidate vaccines may be used clinically in the near future

Many ***Plasmodium*** antigens have been evaluated as potential vaccine targets, and almost all have come up short. Currently, over 30 are still under consideration, and two in particular appear to be especially promising.

One such candidate targets a protein found on the surface of ***P. falciparum*** called *Plasmodium falciparum* reticulocyte binding protein homolog 5 (PfRH5). ***P. falciparum*** is notorious for using a variety of receptor–ligand interactions to gain access to erythrocytes, and until now, none of these interactions were known to be required by all parasite strains. However, PfRH5, which binds a protein called basigin on erythrocytes, apparently is essential for merozoites of all strains investigated to date. Early trials have shown that merozoite invasion of erythrocytes can be completely blocked by neutralizing antibodies and that unlike other potential targets on blood-stage parasites, PfRH5 appears to have little diversity. Researchers have indicated that barring unforeseen problems, clinical trials with an anti-PfRH5 candidate vaccine could begin within a few years.

A different approach and the one probably closest to clinical introduction is offered by the recombinant vaccine called RTS,S. This anti-infection vaccine combines an anti-sporozoite humoral response, with an anti-liver-stage cell-mediated response, as a way to help protect those most at risk from malaria—young children living in endemic areas. The RTS,S vaccine combines the circumsporozoite protein (CSP) found on sporozoites with protein antigens from the capsid of the hepatitis B virus and an adjuvant. The CSP stimulates humoral immunity against sporozoites, while the viral antigen allows for cell-mediated immunity against liver-stage parasites. In recent trials on 15,000 infants and young children, there was a 46% reduction in the number of clinical malaria cases for up to 18 months. This may not seem like much. After all, most potential vaccines are only released clinically when they achieve over 90% protection. It is estimated that even if only 46% effective, the burden of falciparum malaria in young African children is such that the RTS,S vaccine might save millions of lives over a decade.

A few anti-eukaryote vaccines are available for veterinary use

Although antimalaria vaccines receive most of the attention, vaccine research continues for a wide variety of diseases caused by eukaryotes. Although we await the first such vaccine to be licensed for use in humans, vaccines are available to protect animals against certain parasitic conditions.

Eimeria tenella is a serious pathogen of chickens and can cause high mortality, especially in young birds. Only chickens are affected by this coccidian, which has developed resistance against most available drugs. Yet a vaccine that offers at least partial protection has been available since 1952, when Coccivac®, the first commercial vaccine against a eukaryotic pathogen, was released. This attenuated vaccine consisted of oocysts from several ***E. tenella*** strains, variants of which are still used today. Recently, the first subunit vaccine to target a eukaryote, CoxAbic®, has entered the market. This vaccine is composed of purified macrogametocyte antigens. Vaccinated birds respond by generating high titers of IgY (the avian equivalent to mammalian IgG).

Because large amounts of antibodies in vaccinated female chickens are transferred into eggs and ultimately chicks, newly hatched birds are protected precisely at the time they are most vulnerable to disease.

Leishmaniasis has been the object of vaccine research for decades, and a vaccine is feasible because a primary infection renders the host resistant to subsequent infections. A number of attenuated, killed, and subunit vaccines have proven effective in the laboratory using animal models, but results have been less than satisfactory in human field trials. LeishMune®, a subunit vaccine consisting of purified glycoproteins from ***L. donovani*** promastigotes, is available to protect dogs from canine visceral leishmaniasis. TrichGard®, a killed vaccine consisting of ***Tritrichomonas foetus*** trophozoites is another commercially available vaccine that can be used to protect cattle from spontaneous abortions caused by infection with this protozoan parasite. Dogs can be protected from ***Giardia lamblia*** with a killed vaccine known as GiardiaVax®.

Vaccines against helminth parasites are being investigated

In addition to the protozoan parasites discussed above, many researchers consider certain helminthic diseases to be amenable to control through vaccination, at least in theory. The rationale is straightforward; for at least some helminths, natural exposure results in at least some immunity. For many helminthic infections, such partial immunity may be enough; even incomplete protection may keep worm burden below the threshold intensity associated with clinical signs and symptoms.

Schistosomiasis provides a good example. The most severe manifestations are observed only in those patients who are heavily infected with many worms. Light infections tend to be asymptomatic. Thus, an anti-schistosome vaccine may be useful even if it does not result in sterilizing immunity. Furthermore, when infected individuals are chemotherapeutically cured, resistance to reinfection is eventually observed in at least some individuals, suggesting that vaccination may be a feasible option.

Attenuated cercariae and schistosomules have been used to establish such resistance in experimental animal models (**Figure 9.25**). Specific potential antigens that have been investigated as vaccine candidates include paramyosin (a contractile protein found in many invertebrate phyla), glutathione-S-transferase (a large group of enzymes involved in the detoxification of various compounds), Sm14 (a fatty acid binding protein), and

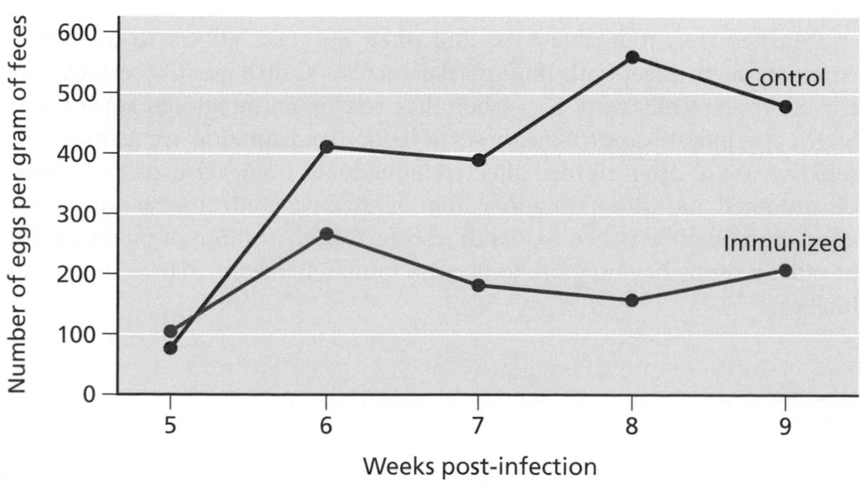

Figure 9.25 An anti-schistosomiasis vaccine in an animal model. Because they are easily infected with *Schistosoma japonicum*, miniature pigs serve as a useful animal model for vaccine research. Immunized pigs were vaccinated with cercariae that were attenuated with radiation. The pigs were subsequently challenged with 200 wild-type *S. japonicum* cercariae. Control animals received no vaccination prior to challenge. Although vaccination did not prevent infection, it did result in significantly lower egg output. Such a vaccine may thus be a way to reduce morbidity from schistosomiasis in light of the important role that eggs play in the pathological response. (From Hirayama K [2008] Schistosomiasis and Immunity. With permission from WHO/TDR.)

in particular, Sm-p80, which plays an important role in immune evasion by the parasite. In both mouse and baboon models, DNA vaccines consisting of Sm-p80 DNA resulted in a significant reduction in adult worm numbers and in egg production. Similar results were obtained with an Sm-p80 subunit vaccine. The reduction in eggs is particularly noteworthy because so much of the pathology seen in schistosomiasis results from the host response to eggs lodged in various tissues. Human trials are now in preparation. Furthermore, veterinary vaccines, such as current work on a vaccine against *S. japonicum* infection in water buffalo, are being investigated. By reducing infection in important reservoirs such as water buffalo in China, transmission to humans could be consequently reduced.

Summary

Because of the continued importance of many parasitic diseases, the need to seek out more effective means of control is ongoing. In general, strategies to control these diseases fall into three basic categories. First are those control measures designed to reduce transmission including the simple provisioning of safe food and drinking water to prevent infection via trophic or fecal–oral transmission and sophisticated genetic manipulation to reduce the vector competence of arthropod vectors. Second is the use of antiparasitic drugs either to treat infected individuals or to provide prophylactic protection. Third are the ongoing efforts to develop new vaccines. There are advantages and disadvantages associated with all strategies in any of these three categories.

Transmission control is often successful, but it can be limited by logistical difficulties in rural settings, problems in getting people to change their habits, or, in the case of vector-borne parasites, the development of pesticide resistance in the involved vectors. Chemotherapy has in many cases resulted in dramatic improvements in public health, but there are no effective drugs for many parasitic infections and many currently used drugs are difficult to administer and cause significant side effects. There is therefore a desperate need for new therapeutic agents. The usefulness of those drugs presently available is also all too often undermined by the development of resistance and even the best drugs are unable to prevent reinfection. Vaccines offer hope in this respect, but to date, that hope is largely unrealized. There are a number of thorny problems to overcome in the development of vaccines against eukaryotic parasites, including the difficulty of generating the proper cell-mediated immunity, and the complex evasive strategies used by many parasites.

Parasites are intractable foes, and often our best efforts to control or eliminate them meet with only partial success. Consequently, serious control efforts typically work best when they rely on an integrated approach in which appropriate control measures to reduce transmission are teamed with selective use of drugs to treat affected individuals. Long-term success hinges on sustained and thorough application of effective control measures. Failed efforts can often be traced to loss of resolve, war, disruption of public health, or other reasons having more to do with human behavior than with parasite biology.

REVIEW QUESTIONS

1. Compare the relative difficulty of control via a reduction in transmission for the three most important human schistosomes, *Schistosoma mansoni*, *S. japonicum*, and *S. hematobium*. Which factors render each species relatively easy or difficult to control?

2. Yellow fever, caused by the yellow fever virus, utilizes a variety of primates as reservoirs. In the early twentieth century, when it was first realized that mosquitoes transmitted yellow fever, mosquito control was implemented in both Panama and Cuba. Yellow fever was effectively eradicated from Cuba. In Panama, while the number of cases declined dramatically, eradication efforts were never completely successful. Explain this difference, considering that Panama had large populations of wild primates, while Cuba did not.

3. In considering the control of parasite-transmitting mosquitoes through the use of insecticides, what is the logic of rotating different insecticides?

4. Why is a higher therapeutic index considered to be a favorable attribute when considering an anti-parasitic drug? Why in general is a higher therapeutic index more important when considering a drug for prophylaxis as opposed to one for treatment of a sick individual?

5. Imagine a newly developed drug that is highly lethal against all stages in the life cycle of *Toxoplasma gondii*. The drug has been found to be safe and effective both for human and for veterinary use. Although the drug can eliminate infection in both humans and cats, its use in humans will have very little impact on overall prevalence of human toxoplasmosis. Its use in cats, alternatively, could, at least in theory, reduce prevalence in both humans and cats. Explain this difference.

6. Suppose that resistance of schistosomes to praziquantel were to become a reality. Currently, praziquantel is the only drug widely used to treat humans infected with schistosomes. Imagine that a new anti-schistosome drug has been developed that is just as safe and effective as praziquantel. How could these drugs be used together to reduce drug resistance to either drug?

7. Explain the following statement; "Part of the difficulty in developing a vaccine against many eukaryotic parasites, is that the vaccine would have to do something that does not occur in the course of a natural infection".

References

GENERAL REFERENCES

Albonico M, Crompton DWT & Savioli L (1998) Control strategies for human intestinal nematode infections. *Adv Parasitol* **42**:277–341.

Garba A, Toure S, Dembele R et al (2006) Implementation of schistosomiasis control programmes in West Africa. *Trends Parasitol* **22**:322–326.

Molyneux DH (2006) Control of human parasitic diseases: control and overview. *Adv Parasitol* **61**:1–45 (doi:10.1016/S0065-308x(05)61001-9).

Savioli L, Smith H, Thompson A et al (2006) *Giardia* and *Cryptosporidium* join the neglected disease initiative. *Trends Parasitol* **22**:203–208.

STRATEGIES TO REDUCE PARASITE TRANSMISSION

Parasites using trophic transmission can be controlled by insuring food safety

Audicana MT & Kennedy MW (2008) *Anisakis simplex*: from obscure infectious worm to inducer of immune hypersensitivity. *Clin Microbiol Rev* **21(2)**:360–379 (doi:10.1128/CMR.00012-07).

Jones JL, Kruzon-Morgan D, Sanders-Lewis K & Wilson M (2007) *Toxoplasma gondii* infection in the United States, 1999–2004, decline from the prior decade. *Am J Trop Med Hyg* **77**:405–410.

Kijlstra A & Jongert E (2008) Control and risk of human toxoplasmosis transmission by meat. *Int J Parasitol* **38**:1359–1370.

Proper sanitation is the key to controlling parasites transmitted via the fecal–oral route

Bleakley H (2007) Disease and development: evidence from hookworm eradication in the American South. *Q J Econ* **122**:73–117.

Holveck JC, Ehrenberg JP, Ault SK et al (2007) Prevention, control, and elimination of neglected diseases in the Americas: pathways to integrated, inter-programmatic, inter-sectoral action for health and development. *BMC Public Health* **7**:6 (doi:10.1186/1471-2458-7-6).

Eberhard M, Ruiz-Tiben E, Hopkins DR et al (2014) The peculiar epidemiology of *dracunculiasis* in Chad. *Am J Trop Med Hyg* **90**:61–70.

Hopkins DR, Ruiz-Tiben E, Weiss A et al (2013) *Dracunculiasis* eradication: and now, South Sudan. *Am J Trop Med Hyg* **89**:5–10

McCarthy JS (2009) Control of intestinal helminths in indigenous communities. *Microbiol Aust* **30**:200–201. World Health Organization. World Health Report; 2010.

Various other factors influence the success of control efforts

Molyneux D (2010) Eradicating Guinea worm disease: a prelude to neglected tropical disease elimination. *Lancet* **376**:947–948 (doi:10.1016/S0140-6736(10)61437-7).

The discovery of organic pesticides radically altered the manner in which vectors were controlled

Wasserstrom CG (1981) Agricultural production and malaria resurgence in Central America and India. *Nature* **293**:181–185.

World Health Organization (1979) DDT and its derivatives. pp 194. World Health Organization, Geneva.

Newer insecticides provide alternatives to DDT

Butler D (2011) Mosquitoes score in chemical war. *Nature* **475**:19.

Transmission of vector-borne parasites can be reduced through environmental manipulation

The malERA consultative group on vector control (2011) A research agenda for malaria eradication: vector control. *PLoS Med* **8**:e1000401 (doi:10.1371/journal.pmed.1000401).

Muturi E, Burguess P & Novak RJ (2008) Malaria vector management: where have we come from and where are we going? *Am J Trop Med Hyg* **78**:536–537.

Prüss-Üstün A & Corvalan C (2007) How much can disease burden be prevented by environmental interventions? *Epidemiology* **18**:167–178.

Biological control offers the possibility of low-cost, sustainable control

Bian G, Joshi D, Dong Y et al (2013) *Wolbachia* invades *Anopheles stephensi* populations and induces refractoriness to *Plasmodium* infections. *Science* **340**:748–751.

Combes C (1982) Trematodes: antagonism between species and the sterilizing effects on snails in biological control. *Parasitology* **84**:151–175.

Hughes GL, Vega-Rodriguez J, Xue P et al (2012). *Wolbachia* strain WAlbB enhances infection by the rodent malaria parasite *Plasmodium berghei* in *Anopheles gambiae* mosquitoes. *Appl Environ Microbiol* **78**:1491–1495.

Lacey LA & Orr BK (1997) The role of biological control of mosquitoes in integrated vector control. *Am J Trop Med Hyg* **50**:97–115.

Murdoch WW, Chesson J & Chesson PL (1985). Biological control in theory and practice. *Am Nat* **125**:344–366.

Service MW (1981) Ecological considerations in biological control strategies against mosquitoes. In Biocontrol of Medical and Veterinary Pests (M Laird ed), pp. 173–195. Praeger.

Walker T, Johnson PH, Moreira LA et al (2011) The WMel *Wolbachia* strain blocks dengue and invades caged *Aedes aegypti* populations. *Nature* **476**:450–455.

The production of transgenic vectors provides hope as a means to reduce vector capacity

Alphey L (2014) Genetic control of mosquitoes. *Annu Rev Entomol* **59**:205–214.

Harris AF, Nimmo D, McKemey AR et al (2011) Field performance of engineered male mosquitoes. *Nat Biotechnol* **29**:1034–1037.

Terenius O, Marinotti O, Sieglaff D & James AA (2008) Molecular genetic manipulation of vector mosquitoes. *Cell Host Microbe* **4**:417–423.

Thailayil J, Magnusson K, Godfray HCJ et al (2011) Spermless males elicit large-scale female responses to mating in the malaria mosquito *Anopheles gambiae*. *Proc Natl Acad Sci U S A* **108**:13,677–13,681.

ANTI-PARASITIC DRUGS

Woods DJ & Williams TM (2007) The challenges of developing novel anti-parasitic drugs. *Invert Neurosci* **7**:245–250.

Drugs may be used to either treat or protect individuals or to protect a population

Foy BD, Kobylinski KC, Marques da Silva I et al (2011) Endectocides for malaria control. *Trends Parasitol* **27**:423–428.

Certain drugs are active only against specific parasite life-cycle stages

Derbyshire ER, Mota MM & Clardy J (2011) The next stage in antimalarial drug discovery: The liver stage. *PLoS Pathog* **7**:e1002178 (doi.10.1371/journal.path.1002178).

Drug use may affect the immune status of the population

Molineaux L & Gramiccia G (1980) The Garki project: research on the epidemiology and control of malaria in the Sudan savannah of West Africa. pp 311. World Health Organization, Geneva.

The use of anti-parasitic drugs can lead to resistance

Martin RE, Marchetti RV, Cowan AI et al (2009) Chloroquine transport via the malaria parasite's chloroquine resistance transporter. *Science* **325(5948)**:1680–1682.

Genetic alterations can cause resistance in diverse ways

Borst P & Ouellette M (1995) New mechanisms of drug resistance in parasitic protozoa. *Annu Rev Microbiol* **49**:427–460.

Doenhoff MJ, Cioli D, Utzinger J et al (2008) Praziquantel: mechanisms of action, resistance and new derivatives for schistosomiasis. *Curr Opin Infect Dis* **21**:659–667.

Greenberg RM (2013) New approaches for understanding mechanisms of drug resistance in schistosomes. *Parasitology* **14**:1534–1546.

Stelma FF, Sall S, Daff B et al (2007) Oxamniquine cures *Schistosoma mansoni* infection in a focus in which cure rates with praziquantel are unusually low. *J Infect Dis* **176**:304–307.

Resistance poses a considerable problem for disease control programs

Greenwood B (2007) The use of antimalarial drugs to prevent malaria in the population of malaria endemic areas. *Am J Trop Med Hyg* **70**:1–7.

Gryseels B, Mbaye A, De Vlas SJ et al (2001) Are poor responses to praziquantel for the treatment of *Schistosoma mansoni* infections in Senegal due to resistance? An overview of the evidence. *Trop Med Int Health* **6**:864–873.

McCarthy J (2005) Is antihelminthic resistance a threat to the program to eliminate lymphatic filariasis? *Am J Trop Med Hyg* **73**:232–233.

Osei-Atweneboana MY, Eng JKL, Boakye DA et al (2007) Prevalence and intensity of *Onchocerca volvulus* infection and efficacy of ivermectin in endemic communities in Ghana: a two-phase epidemiological study. *Lancet* **369**:2021–2029.

Drug resistance can be prevented or reversed

Aung PP, Nkhoma S, Stepniewska K et al (2012) Emergence of Artemisinin-resistant malaria on the Western border of Thailand: a longitudinal study. *Lancet* **379:** 1960–1966. (doi:10.1016/S0140-6736(12)60484-X).

Cammack N (2011) Exploiting malaria drug resistance to our advantage. *Science* **333**:705–706.

D'Alessandro U (2009) Progress in the development of piperaquine combinations for the treatment of malaria. *Curr Opin Infect Dis* **22**:588–592.

The four artemisinin-based combinations (ABC) study group (2011) A head-to-head comparison of four artemisinin-based combinations for treating uncomplicated malaria in African children: a randomized trial. *PLoS Med* **8(11)**: e1001119 (doi:10.1371/journal.pmed.1001119).

Kublin et al (2003) Reemergence of chloroquine-sensitive *Plasmodium falciparum* malaria after cessation of chloroquine use in Malawi. *J Infect Dis* **187**:1870–1875.

Concerns about resistance highlight the need for new anti-parasitic drugs

Derbyshire ER, Mota MM & Clardy J (2011) The next opportunity in antimalarial drug discovery: the liver stage. *PLoS Pathog* **7**:e1002178 (doi:10.1371/journal.ppt.1002178)

Kuntz AN, Davioud-Charvet E, Sayed AS et al (2007) Thioredoxin glutathione reductase from *Schistosoma mansoni*: an essential parasite enzyme and a key drug target. *PLoS Med* **4(6)**:e206 (doi:10.1371/Journal.pmed.0040206).

New drugs are also needed to replace older more toxic medications

Leslie M (2011) Drug developers finally take aim at a neglected disease. *Science* **333**:933-935.

Torreele E, Trunz BB, Tweats D et al (2010) Fexinidazole—A new oral Nitroimidazole drug candidate entering clinical development for the treatment of sleeping sickness. *PloS Negl Trop Dis* **4**:e923 (doi:10.1371/journal.pntd.0000923).

The manner in which new drugs are discovered has changed considerably

Woods DJ & Williams TM (2007) The challenges of developing novel anti-parasitic drugs. *Invert Neurosci* **7**:245–250.

Potentially effective drugs usually require chemical modification prior to their use

Hann MM, Leach AR & Harper G (2001) Molecular complexity and its impact on the probability of finding leads for drug discovery. *J Chem Inf Comput Sci* **41**:856–864.

Wasa KM, Wasan EK, Gershkovich P et al (2009) Highly effective oral amphotericin B formulation against murine visceral leishmaniasis. *J Infect Dis* **200**:357–370.

VACCINES

Vaccines must be safe and inexpensive, while inducing long-term immunity

Levine MM & Levine OS (1997) Influence of disease burden, public perception and other factors on new vaccine development, implementation and continued use. *Lancet* **350**:1386–1392.

Ulmer JB & Liu MA (2002) Ethical issues for vaccines and immunization. *Nat Rev Immunol* **2**:291–296.

Vaccines against eukaryotic parasites are particularly problematic

Grange JM & Stanford JL (1996) Therapeutic vaccines. *J Med Microbiol* **45**:81–83.

Jenikova G, Hruz P, Anderson MK et al (2011) Alpha 1 Giardin-based live heterologous vaccine protects against *Giardia lamblia* in a murine model. *Vaccine* **29**:9529–9537.

Kaye PM & Aebischer T (2011) Visceral Leishmaniasis: immunology and prospects for a vaccine. *Clin Microbiol Infect* **17**:1462–1470.

Kedzierski L (2010) Leishmaniasis vaccine: where are we today? *J Glob Infect Dis* **2**:177–185.

Kumar R, Goto Y, Gidwani K et al (2010) Evaluation of the *ex vivo* human immune response against candidate antigens for a visceral leishmania vaccine *Am J Trop Med Hyg* **82**:808–813.

Plebanski M, Proudfoot O, Pouniotis D et al (2002) Immunogenetics and the design of *Plasmodium falciparum* vaccines for use in malaria-endemic populations. *J Clin Invest* **110**:295–301.

Rivero FD, Saura A, Prucca CG et al (2010) Disruption of antigenic variation is crucial for effective parasite vaccine. *Nat Med* **16**:551–558.

Vaccines can be categorized into one of several types

Anderson RM, Donnelly CA, Gupta S et al (1997) Vaccine design, evaluation and community-based use for antigenically variable infectious agents. *Lancet* **350**:1466–1470.

Van Duin D, Medzhitov R, Shaw AC et al (2005) Triggering TLR signaling in vaccination. *Trends Immunol* **27**:49–55.

Wolff JA & Budker V (2005) The mechanism of naked DNA uptake and expression. *Adv Genet* **54**:3–20.

An effective malaria vaccine has been the object of intensive investigation

Corradin C & Engers H (2014) Malaria vaccine development: over 40 years of trials and tribulations. Future Medicine Ltd. London.

Greenwood BM, Fidock DA, Kyle DE et al (2008) Malaria: progress, perils and prospects for eradication. *J Clin Invest* **118**:1266–1276.

Langhorne J, Ndungu FM, Sponass AM et al (2008) Immunity to malaria: more questions than answers. *Nat Rev Immunol* **9**:725–732.

The MalERA Consultative Group on Vaccines (2011) A research agenda for malaria eradication: vaccines. *PLoS Med* **8(1)**:e1000398 (doi:10.1371/Journal. Pmed. 1000398).

Vaccines against different life-cycle stages offer different potential benefits

Douglas AD, Williams AR, Illingworth JJ et al (2011) The blood stage malaria antigen PfRH5 is susceptible to vaccine-inducible cross-strain neutralizing antibody. *Nat Commun* **2**:601.

Hill AV (2006) Pre-erythrocytic malaria vaccines: Towards greater efficacy. *Nat Rev Immunol* **6**:21–32.

Plassmeyer ML, Reiter K, Shimp RL et al (2009) Structure of the *Plasmodium falciparum* CSP: A leading malaria vaccine candidate. *J Biol Chem* **284**:26,951–26,963.

Recent candidate vaccines may be used clinically in the near future

Agnandji ST, Lell B, Soulanoudjingar SS et al (2011) First results of phase 2 trial of RTS,S/AS01 malaria vaccine in African children. *N Engl J Med* **365**:1863–1875.

Bejon P, Cook J, Bergmann-Leitner E et al (2011) Effect of the pre-erythrocytic candidate malaria vaccine RTS,S/AS01E on blood stage immunity in young children. *J Infect Dis* **204**:9–18.

Crosnier C, Bustamante LY, Bartholdson SJ et al (2011) Basigin is a receptor essential for erythrocyte invasion by *Plasmodium falciparum*. *Nature* **480**:534–537.

White MT, Bejon P, Olutu A et al (2014) A combined analysis of immunogenicity, antibody kinetics, and vaccine efficacy from phase 2 trials of the RTS,S malaria vaccine. *BMC Medicine* **12:**117 doi:10.1186/s12916-014-0117-2.

A few anti-parasitic vaccines are available for veterinary use

Chapman HD (2000) Practical use of vaccines for the control of coccidiosis in the chicken. *Worlds Poult Sci J* **56**:7–20.

Olson ME, Ceri H, Morck DW et al (2000) *Giardia* vaccination. *Parasitol Today* **16**:213–217.

Sharman PA, Smith NC, Wallach MG et al (2010) Chasing the golden egg: vaccination against poultry coccidiosis. *Parasite Immunol* **32**:590–598.

Trigo J, Abbehusen M, Netto EM et al (2010) Treatment of canine visceral leishmaniasis by the vaccine Leish-111f+MPL-SE. *Vaccine* **28**:3333–3340.

Uehlinger FD, O'Handley RM, Greenwood SJ et al (2007) Efficacy of vaccination in preventing giardiasis in calves. *Vet Parasitol* **146**:182–187.

Zheng J, Ren W, Pan Q et al (2011) A recombinant DNA vaccine encoding *Cryptosporidium andersoni* oocyst wall protein induces immunity against experimental *Cryptosporidium parvum* infection. *Vet Parasitol* **179**:7–13.

Vaccines against helminth parasites are being investigated

Babayan SA, Allen JE, Taylor DW et al (2011) Future prospects and challenges for vaccines against filariasis. *Parasite Immunol* **34:** *243-253.* (doi:10.1111/J.1365–3024.2011.01350X).

Bungiro R & Cappello M (2011) Twenty-first century progress toward the global control of human hookworm infection. *Curr Infect Dis Rep* **13**:210–217.

Harris NL (2011) Advances in helminth immunology: optimism for future vaccine design? *Trends Parasitol* **27**:288–293.

Hewitson JP & Maizels RM (2014) Vaccination against helminth parasite infections. *Expert Review of Vaccinations* **13**:473–487.

McManus DP & Loukas A (2008) Current status of vaccines for schistosomiasis. *Clin Microbiol Rev* **21(1)**:225–242.

McWilliam HE, Piedrafita D, Li Y et al (2013) Local immune response of the Chinese water buffalo *Bubalus bubalus* against *Schistosome japonicum* larvae: crucial insights for vaccine design. *Plos/Neg Trop Dis* doi:10,1371/journal.pntd.0002460.

Siddiqui AA, Siddiqui BA & Ganley-Leal L (2011) Schistosomiasis vaccines. *Hum Vaccin* **7(11)**:1192–1197.

The Future of Parasitology

10

...the future's uncertain and the end is always near.
Jim Morrison, Roadhouse Blues

In This Chapter

The previous chapters have provided many examples of how parasites have influenced our world and continue to do so. Although attempts to predict the future often merely provide a good laugh for those who will actually live there, we would be remiss if we did not conclude by discussing some of the many opportunities worthy of the attention of those contemplating a future studying parasitology. The world we inhabit is changing dramatically, so much so that we may experience a state shift, a degree of change possibly quite unpredicted and unprecedented, comparable in magnitude to that experienced during the last glacial–interglacial transition period or to previous mass extinction events. In this chapter, we describe several areas in need of further investigation, after first briefly providing some widely accepted points of view regarding what our world will look like in the next few decades.

10.1 OUR FUTURE WORLD

Our future world will be one even more dominated by humans than it is at present, so much so that the age we live in is now informally called the Anthropocene (the prefix anthropo- refers to humans). The human population, as of this writing, about 7.27 billion strong, is projected to increase to 8 billion by 2030 and to 9–9.5 billion by 2050 (**Figure 10.1**). This is a most sobering prospect, as the demands placed on all the Earth's resources—fresh water, land for growing food and providing living space, energy and minerals—will be stretched to what can only be concluded are unsustainable levels. For example, a 100% to 110% increase in global crop demand is predicted from 2005 to 2050, this as a result of the increased numbers of mouths to feed and the increasing per capita consumption of calories (including calories from animals fed on grain products) in developing countries.

Food security will become harder to insure because the carrying capacity of the world will be pushed to the limit, with but small tolerances for unexpected outcomes such as droughts or floods. Furthermore, the health of both food plants and animals will be compromised by parasites, many of which have become resistant to our chemical defenses. It seems inevitable that inadequate nutrition if not outright starvation will continue to be a frequent outcome for those living in impoverished areas with marginal rainfall,

Species names highlighted in red are included in the Rogues' Gallery, starting on page 429.

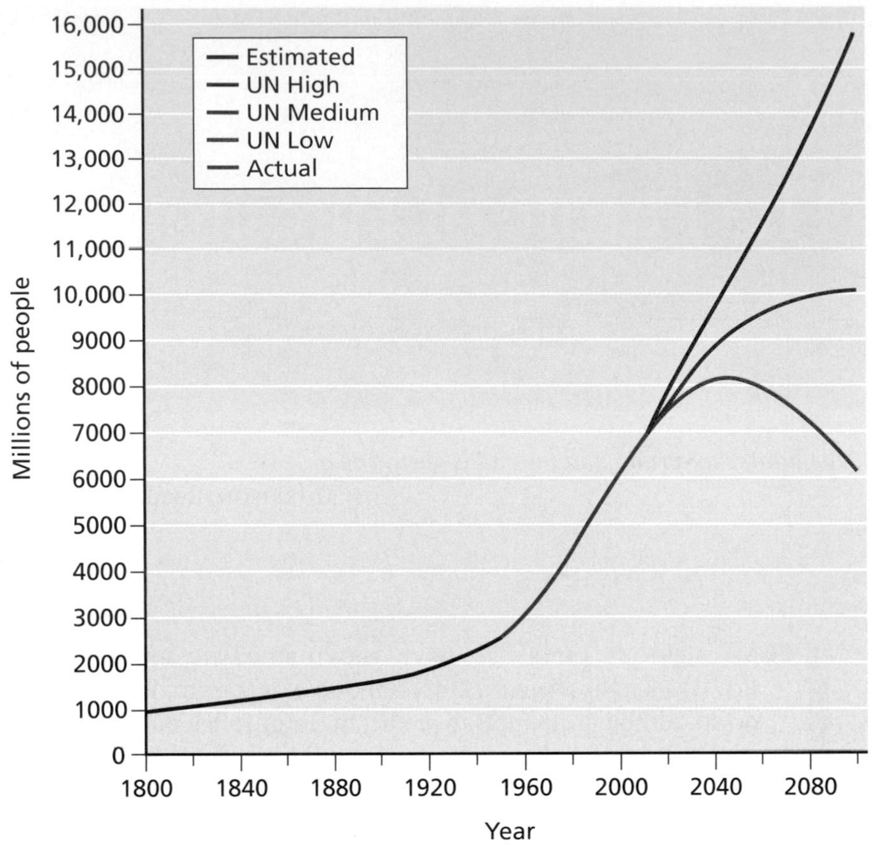

Figure 10.1 The growing human population. Shown are three different UN projections for human population growth in the next century, one estimating as many as 16 billion people by 2100, a middle projection of about 10 billion, and a low projection of about 6 billion that represents a decline from the present number of about 7.27 billion. Note that all three projections indicate an increasing human population for at least another 30 years.

for example. Although there has been a recent and unprecedented decrease in the number of people living in extreme poverty from about 1.3 billion in 2005 to 900 million in 2010, with further projected decreases to 600 million in 2015, much of the decrease has been achieved in developing Asian countries. Other regions such as central Africa remain obdurate to change. Also, the disparity in incomes between rich and poor nations continues to grow, such that in the future a privileged few will live in affluence and many will live in poverty or near poverty with fundamental concerns about inadequate food, disease, provision of basic human rights, and quality of life.

Tensions among human populations—if today's current situation is any indication—will continue, encouraged by shortages of resources. The ensuing disruption of access by health care workers will provide distinct opportunities for infectious diseases to persist in recalcitrant enclaves. Climate change caused by human activities will likely be exacerbated by continued consumption of fossil fuels and will have major impacts on the distribution of animal and plant species, including parasites, as will be discussed further in Section 10.2. As noted in Chapter 8, the predictions for loss of biodiversity are dire; it is estimated that a combination of factors such as habitat loss, introductions of exotic species including disease agents, and increasing levels of pollution will collude to eliminate species at a rate 10 to 100 times present rates that are already 100 to 1000 times higher than background extinction rates. The world's ecosystems will at the very least be significantly altered and their integrity likely to decline. These changes have prompted prominent ecologists to warn that we may be approaching a tipping point on a planetary scale, with an attendant need to do more to detect early warning signs and to address the root causes.

Against this rather bleak set of prospects, there are some reasons to be optimistic. Along with a decline in extreme global poverty, startling improvements in health care, global communications, and information transfer, an increased emphasis on sustainable use of resources, and development of new technology for energy production offer hope for diminishing suffering and providing a future that is more realistically sustainable. With this broad background in mind, how will the discipline of parasitology be affected and what opportunities does it present?

10.2 SOME FUTURE CHALLENGES FOR PARASITOLOGISTS

There is always something new to be found under the parasitological sun

As noted throughout this book, parasites are an enormously diverse group of organisms, and we have yet to see the full scope of their diversity. Many parasitologists are engaged in research to identify new species of parasites. By fully characterizing parasite diversity, parasitologists are then better able to unravel the intricacies of their evolutionary histories. As discussed in Chapter 8, we live in the early to middle stages of what has been called the sixth mass extinction, with this episode of mass extinction different from previous ones because it is caused principally by *Homo sapiens*. The ongoing loss of biodiversity, including the threat of coextinction, provides a powerful incentive to characterize the parasites of the world before they are lost forever. New parasite species are continually being found and described (**Figure 10.2**), and this effort must continue.

In addition, new species of parasites will continue to find us. One example is the apicomplexan ***Cyclospora cayetanensis***. Human infections were virtually unknown before 1990, but in recent years they have become more commonly diagnosed among travelers or even in inhabitants of developed countries who have eaten imported fruits or vegetables contaminated with human fecal material containing the infective oocysts of this organism. A good reason to deliberately search for parasite diversity is that by doing so we encounter many new species that help us identify and put into context those previously unknown species that suddenly burst on the scene and are implicated in causing unexpected disease outbreaks.

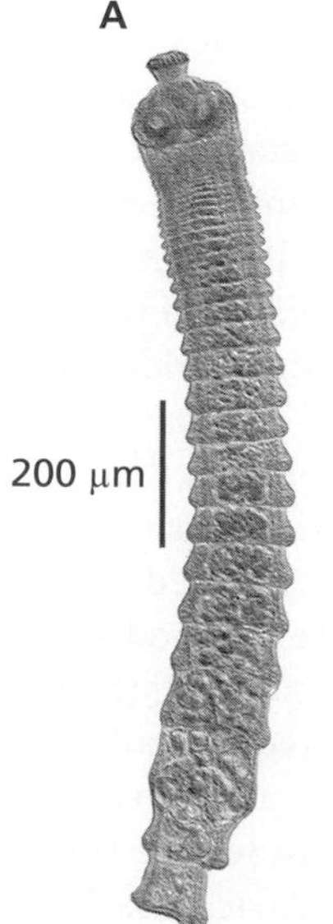

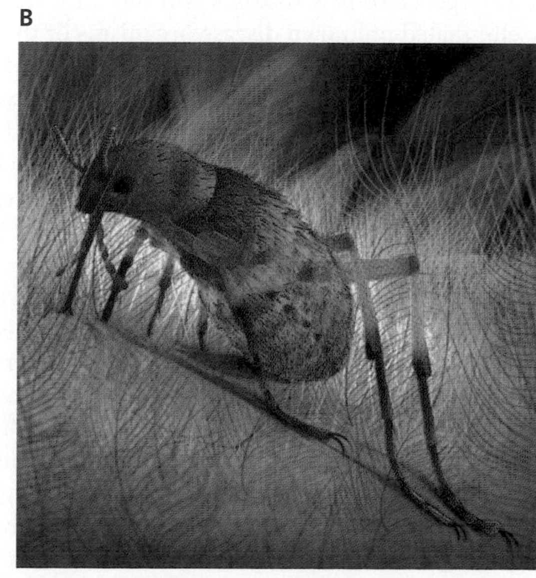

Figure 10.2 Two examples of newly described parasite species, one still extant and one from the fossil record. (A) Shown is just one example among many of a newly-described parasite, a tapeworm *Rodentolepis gnoskei* from the shrew *Suncus varilla minor* from Malawi. (B) Illustration of the pseudoflea *Pseudopulex jurassicus* recently recovered from Mesozoic deposits. A reconstruction showing it parasitizing a feathered dinosaur, *Pedopenna daohugouensis*, from the Middle Jurassic (≈165 mya) in China. Note the long proboscis and large size of these flealike fossils (17–22 mm long) as compared to most modern fleas, which are under 6 mm long. (A, From Greiman SE & Tkach VV [2012] *Parasitol Int* 61:343–350. With permission from Elsevier. B, Drawing by Wang Cheng. From Poinar G [2012] *Curr Biol* 22:732–735. With permission from Elsevier.)

Another major objective for parasitologists will be to obtain more complete genome sequences for a broader variety of parasites, with the goal of amassing complete sequence information for several species representing all major lineages of parasites. As genome-sequencing technology becomes faster, cheaper, and easier to analyze, it is not hard to imagine that descriptions of a new parasite species will routinely be accompanied by a completed genome sequence. Genomic information will be put to many uses, with one obvious example being to provide definitive molecular phylogenetic interpretations of the evolutionary history of parasites. This will continue a trend of remarkably improved understanding of the origins and diversification of parasite lineages (Chapter 2). For example, thanks to the completion of several ***Plasmodium*** genomes, new molecular phylogenies based on as many as 45 genes have provided powerful new interpretations of the origins of ***Plasmodium*** parasites in primates and rodents. As we have noted before, the genomes of parasites also collectively represent an enormous store of potentially useful genetic information that will eventually be exploited for such practical purposes as modulation of immune responses, suppression of growth of other parasites, or provision of innovative new structural materials. By preserving a broad diversity of parasites in museums (**Box 10.1**), most of them presently poorly known, we can eventually unlock this vast treasure trove of genomic information.

We need to better understand the ecological and evolutionary roles of parasites

As discussed in Chapters 6 and 7, a growing number of ecologists and evolutionary biologists have turned their attention to understanding the role of parasites in ecosystems, and to illuminating their wide-reaching evolutionary effects. As we are a long way from consensus on what these roles and effects are, the study of parasite ecology and evolutionary biology will continue to provide fertile ground for aspiring parasitologists.

With respect to ecology, parasitologists have an important role to play in trying to understand and predict the outcomes of the myriad of environmental changes underway in our fast-paced world. Between 1940 and 2004, over 300 emerging diseases afflicting humans were recognized. Is this trend likely to continue? On the one hand, it could be argued that as biodiversity diminishes and humans push into every corner of the earth, the likelihood of the emergence of new diseases will diminish; we will have either unwittingly eliminated unknown disease organisms or their reservoir hosts, or we would have already contacted obscure parasites and provided opportunities for their transfer to new hosts or expansion into new circumstances. On the other hand, it can be argued that drug pressure and other control measures or that ongoing introductions of parasites into new hosts or locations serve as evolutionary accelerants that greatly favor the emergence of new drug resistant strains of parasites or that promote recombination or hybridization events among parasites to create new lineages.

Similarly, conservation biologists have presented a popular view that biodiversity generally reduces wildlife and human disease transmission (the dilution effect, discussed in Chapter 8). Ecosystems with reduced biodiversity, what has also been called trophic downgrading, are believed to be more vulnerable to disease outbreaks. Loss of biodiversity could affect disease transmission in an unpredictable way. For example, a reduction in the abundance of competitor or predatory species might favor an increase in the populations of host or vector species of a particular parasite. The transmission of ***Schistosoma haematobium*** in Lake Malawi in Africa has been

BOX 10.1
Preserving Parasite Diversity for the Future in Museums

One of the major challenges for the future will be to preserve and curate parasite diversity in a way that is permanent, accessible, and enables diverse avenues of study. It is already the case that some parasites are known only because specimens of them were once deposited in museums; many have not been seen in the wild since that time. Also, by preserving historically invaluable collections of specimens, museums provide much needed baseline information against which the effects of changing climate or other factors can be measured. Museums also provide specimens for systematics work, including classical anatomical studies, but most also provide specimens appropriate for collection of sequence data and other modern analyses. Included among the specimens to be curated should be well-known parasites of medical and veterinary significance that have been or will be subjected to eradication programs (the Guinea worm comes to mind) and that potentially have been significantly altered as a consequence of control efforts. Having such specimens available for future study will enable parasitologists to learn how control efforts have changed parasites and to improve control efforts in the future.

We can only begin to guess what new techniques might be applied to museum specimens by scientists of the future, and the value of having a strong, sustainably supported system of museums preserving parasite diversity for the future cannot be overstated in this era of diminishing natural diversity. The availability of representative and thorough collections of parasites in museums will enable us to better interpret the composition of the world's ecosystems prior to their widespread disruption and will better enable us to foresee the nature of future changes.

References

Hoberg EP (2010) Invasive processes, mosaics and the structure of helminth parasite faunas. *Rev Sci Tech* 29(2):255–272.

Fernandez-Triana J, Smith MA, Boudreault C et al (2011) A poorly known high-latitude parasitoid wasp community: unexpected diversity and dramatic changes through time. *PLoS One* 6(8)23719 (doi: 10.1371/journal.pone.0023719).

Figure 1 Museums provide needed historical perspective regarding parasite diversity. Museums hold vast collections of parasites, many of them preserved in formalin or ethanol in jars or small tubes such as the ones shown here. Many parasite specimens, especially helminths such as the stained and cleared pieces of a tapeworm from a bear in the foreground, are permanently mounted on microscope slides. On such slides, they can be examined in detail and compared to descriptions of similar species previously recorded in the scientific literature. Some specimens may also be frozen in liquid nitrogen to permit future studies of either the parasite's nucleic acids or proteins. With continually improving techniques for extracting intact nucleic acids from preserved specimens, museum specimens represent an enormous source of novel genetic information. To go along with the parasite specimens, there is frequently information provided about the host species and the date and place from which it was recovered. Such integrated sources of data are invaluable in helping parasitologists discern significant long-term biotic trends that might otherwise go unnoticed. (Courtesy Division of Parasitology, Museum of Southwestern Biology, University of New Mexico.)

increased by the reduction of snail-eating fish as a consequence of overfishing, thereby favoring expansion of the schistosome-transmitting snail *Bulinus nyassanus*.

As shown in **Figure 10.3**, there are several additional ways that the transmission of parasites with complex life cycles could be adversely affected by the diversity of other species living in the same environment. These other species might serve as unproductive decoy hosts, as discussed in Chapter 8, or might even be hyperparasites of the particular parasite in question. If some of these ancillary species were to drop out, then the transmission of the parasite in question might suddenly be favored.

Much work remains to fully understand how parasite transmission in a particular environment is influenced by biodiversity and it may prove difficult to generalize the relationship between lower biodiversity and increased disease transmission. For example, in some cases, dwindling biodiversity may impair the ability of parasites infecting the hosts involved to complete

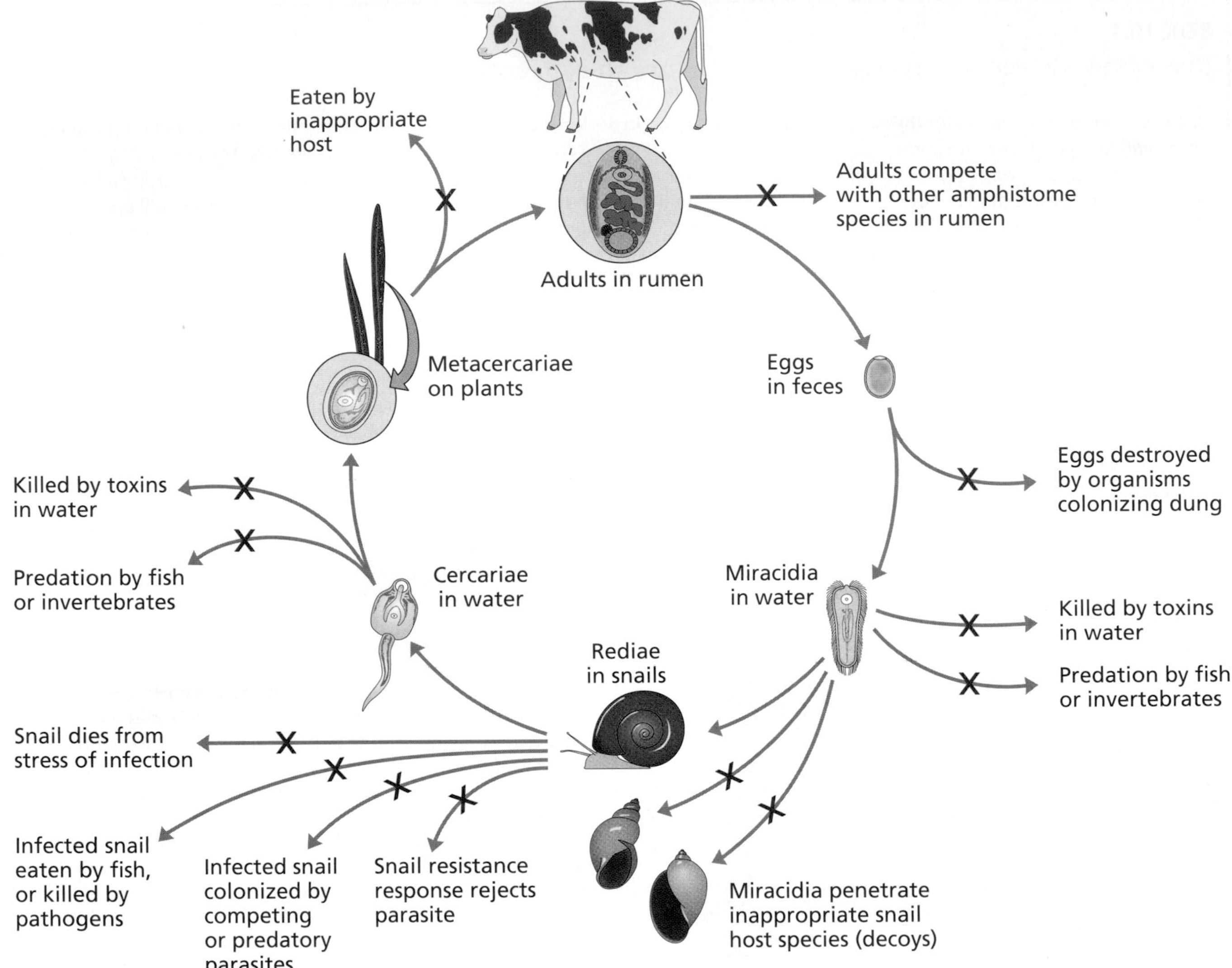

Figure 10.3 Interference by other organisms with transmission of a particular parasite. Shown is the life cycle of an amphistome trematode, the adults of which are found in the rumen of cows and goats. The familiar trematode life cycle involves several stages as shown in the central, circular diagram. Note the many arrows marked by a red "X" that indicate how other organisms or properties in the amphistome's environment could interfere with its transmission. If competitors or predators of the amphistome were removed, then its transmission might increase.

their life cycles, thereby lowering parasite abundance and decreasing disease transmission. Clearly, more empirical and theoretical work beyond a few model systems is needed to understand the future consequences of biodiversity loss for overall levels of disease transmission in the field. It is likely that significant new insights will emerge. For example, are the host species most likely to persist in the face of declining biodiversity also those most likely to be susceptible to parasite infection, such that they carry disproportionately high parasite burdens?

With respect to evolutionary studies of parasite biology, several topics are in need of further study. One of direct relevance is achieving a more complete understanding of the factors influencing parasite virulence, particularly including human manipulations that might unwittingly increase or decrease virulence levels. Intensive farming with attendant needs for parasite control are likely to have major effects on parasite biology (see also the later discussion of drug control), including virulence. For example, commercial rearing of salmon, which involves keeping fish at high densities and using anti-parasite treatment, may be selecting for an increase in virulence of the salmon louse *Lepeophtheirus salmonis*. Salmon farming favors lice that mature faster and produce more progeny, the latter occurring at the expense of weight gain of the salmon.

Studies of vaccination of rodents against malaria (*Plasmodium chabaudi*) using a particular antigen target (AMA-1) surprisingly did not result in selection on the parasite to change that antigen, but unexpectedly did favor more virulent clones of the parasite that were not controlled by AMA-1 induced immunity. Surprising outcomes like this remind us that caution is warranted in anticipating the eventual effects of malaria vaccination in people and that parasites may respond in unexpected ways.

Studies of the extent to which parasites have cospeciated with their hosts or that have undergone host switches leading to exploitation of new host lineages will continue to be relevant. These studies are important both for understanding factors that determine large-scale patterns of parasite species diversity and for better appreciating the underlying factors driving disease emergence, which often involves unpredictable shifts into new host species. Although many factors influence emergence, in some cases it has been favored by evolutionary changes in the parasites resulting from mutation (many occurrences of mutated genes encoding drug targets in ***Plasmodium***, for example), recombination (exemplified by influenza or SARS), horizontal gene transfer (as by many bacteria in inheritance of virulence or drug resistance genes on plasmids), or even hybridization (as with the plant oomycete *Phytophthora alni* that causes root rot in alders and is believed to have originated by hybridization of two related *Phytophthora* lineages). Better understanding of the nature of evolutionary change in parasite lineages will improve our understanding of disease emergence. Another major reason to study the evolution of parasites and their hosts is to more fully appreciate the extent to which host attributes such as immune responses have been drivers of parasite diversity, and conversely, the extent to which parasite pressure may have favored diversification, including speciation, of host lineages (see Chapter 7).

One more reason to explore the long-term evolutionary effects of parasitism is to better understand the biology of our own species. In general, there is an increasing appeal to better integrate the study of evolution into the biomedical enterprise so that our modern health problems can be viewed in a broader evolutionary context and then become more amenable to sustainable solutions. Our long association with parasites is believed to have influenced the evolution of several human attributes including sexual dimorphism in size and lifespan, birth weight, nausea responses, body odors, IQ, sleep patterns, and even diversity of languages and religions (see Chapter 7). Many of these associations—although certainly intriguing and quotable—are tenuous and require much further study and clarification before they can be taken seriously.

Revealing how parasite and host molecules interact is needed to clarify many fundamental aspects of parasitism

What are the molecular mechanisms actually involved in mediating some of the most iconic features of parasitism: host infectivity, host specificity, virulence, and transmission strategies? The study of viruses and bacteria has led the way in revealing specific molecular mechanisms that help explain these features, and the increased availability of genomes of eukaryotic parasites, followed by studies to assess the functional relevance of parasite genes, are beginning to reveal the determinants of infectivity, specificity, virulence, and transmission mode for these organisms as well.

For example, one recent study compared the apicomplexans ***Toxoplasma gondii*** and ***Neospora caninum***. The former is recognized as a classic generalist capable of infecting an extraordinary diversity of warm-blooded vertebrate species, mostly by horizontal transmission, with cats as definitive hosts. ***N. caninum*** is another example of a relatively recently discovered parasite long mistaken for ***T. gondii***. It develops in dogs but has a predilection for vertical transmission in bovines. A comparison of genome sequences and transcriptional profiles for the two species revealed that although there were many similarities, there were some conspicuous differences. Of particular note was the discovery that a certain family of genes encoding surface antigens (the SRS gene family) was expanded in ***N. caninum*** relative to ***T. gondii***. This observation was surprising because it had been expected that the broad host range of ***T. gondii*** was related to its possession of many SRS genes. However, ***N. caninum*** actually expresses fewer of these genes than does ***T. gondii***, implying that differences in numbers of SRS genes actually expressed may be critical to host range determination. Also, another gene, *ROP18*, was found to be inactive in ***N. caninum*** and, as a consequence, this species (unlike ***T. gondii***) is less able to inactivate host defenses as the parasite enters a host's cells. Inactive *ROP18* is believed to limit the virulence of ***N. caninum***, in keeping with its largely vertical mode of transmission in cattle, in contrast to the more prominent horizontal transmission utilized by the more virulent ***T. gondii*** (see also the discussion in Chapter 7). Loss of *ROP18* may also impose limits on the intermediate host range of ***N. caninum***.

Studies like this point the way for future investigations, highlighting the need to more explicitly identify the factors influencing infectivity, host range, and virulence, and to relate them ultimately to transmission modes. A close look at this study also points out the need for bioinformatics skills in comparing and analyzing large data sets. Such skills will be called for increasingly in future studies of host–parasite systems (**Box 10.2**).

Climate change will affect parasites, but we know little about how

Average global surface temperatures have risen at a rate of 0.13°F (0.072°C) per decade since 1901 (**Figure 10.4**), and, worldwide, the decade 2000–2009 was the warmest on record. Anthropogenic climate warming is increasingly acknowledged as a reality.

How will it affect the distribution and abundance of parasites and the health problems they cause? Each parasite species (and the vectors for that species, if they exist) has its own optimum range of temperatures. Global warming will definitely change parasite distributions, but the nature of the changes (contractions or expansions of existing ranges, or shifts into new locations) are harder to determine and will have to be rigorously considered on a case-by-case basis.

The interactions between climate change and infectious diseases are complex and pose a great challenge for future parasitologists to decipher. Keep in mind that the rises in global temperature will not be uniform (temperate and arctic regions may be affected more than temperatures in the tropics) and that it may be more newsworthy to emphasize the potential increases as opposed to the possible decreases in infectious disease that result from climate change. It is very tempting, for example, to conclude that increasing global temperatures are bound to increase the range of tropical diseases, particularly those that are vector-borne, into temperate latitudes and thus increase global areas in which malaria, dengue, and other diseases might occur. Vector-borne parasites are believed to be particularly responsive to warming climates because invertebrate vectors are directly exposed to the effects of increasing temperatures that can increase rates of vector development and of development of parasites within vectors. Furthermore, global warming may remove temperature barriers that have prevented spread of

BOX 10.2

Bioinformatics, Big Data, and Parasitology

The parasitologist of the future will need to be conversant with and adroit in handling at least some components of the ever-expanding field of **bioinformatics**. Bioinformatics is the application to biology of computer science and information technology. It commonly deals with application of algorithms, often to large biological data sets, to discern patterns or to draw inferences from the data. Large data sets, such as complete genome sequences, are referred to as **big data** in the parlance of modern biology. Big data includes any large data set too large or complex to be analyzed by traditional means. Bioinformatics is involved with three-dimensional reconstruction of molecules, image analysis, modeling and simulations, and identification of biological signals or significant trends among complex backgrounds. It is also used to reveal and decipher networks, whether they occur among molecules within a parasite cell or involve parasites in a food web within an ecosystem. Bioinformatics engages many other disciplines including web technologies, artificial intelligence, mathematics, and statistics, and is very much oriented to the development of better tools that can be used to interpret patterns in biological data. Bioinformatics often proceeds hand in glove with **systems biology**, which addresses complex biological interactions taking a holistic rather than a reductionist approach. A systems approach might involve gaining an overview of entire networks of metabolic interactions occurring within cells and deriving insights regarding emergent properties that stem from these network overviews.

Bioinformaticians work in particular computing environments such as that provided by MATLAB (MathWorks), which enables development of algorithms, data analysis, visualization of data on graphs, analysis of images, numerical computation of data, simulations, and creation of interfaces with other widely used programs. For example, MATLAB has been used for analyzing the motion of larvae of *Oesophagostomum dentatum*, the nodular worm of pigs. This application showed that drug-sensitive and drug-resistant larvae move differently, thus providing a novel way to detect drug resistance.

Another example of the application of bioinformatics is provided by a study defining the transcriptome of a plant parasitic nematode *Pratylenchus thornei*, a major pest of cereals such as wheat and barley. The **transcriptome** is the set of all RNA molecules from an organism; in this case the emphasis is on mRNAs, those RNAs eventually to be translated into proteins. As a first step, a complementary DNA (cDNA) library was made using the mRNA from *P. thornei* as a template. Then 787,275 short cDNA sequence reads were obtained by a high throughput sequencing method (454 sequencing). This study (Figure 1) used two programs (CAP3 and MIRA) to assemble these short sequence reads into 34,312 contigs. A **contig** (from the word contiguous) is a set of overlapping cDNA segments (the sequencing reads) that in this context provide longer stretches of sequence and thus help to identify particular genes. The contigs were then subjected to **annotation**, a process to identify where the genes reside in the genome and, most particularly, to provide insights as to what they do. To do the annotation, sequences were compared to previously obtained sequences using the BLAST (Basic Local Alignment Search Tool) algorithm. A BLAST search enables comparison of a query sequence with what is in a library of previously obtained sequences that reside in a resource such as GenBank. If the query sequence is similar to other genes that are well known and have known functions, then valuable insights are gained as to the query gene's likely function.

Other kinds of analyses such as GO (Genome Ontology) annotation can be used. This type of annotation enables unknown query sequences to be related to similar molecules and categorized with respect to where they occur in a cell, their molecular function, or the biological processes in which they are involved. KEGG (Kyoto Encyclopedia of Genes and Genomes) analysis can also be undertaken. KEGG is a collection of databases dealing with genomes and metabolic pathways that also allows networks of metabolic interactions to be constructed. Studies such as these highlight the power of applying bioinformatics tools, typically in combination, to make novel insights into parasite biology, including in this case identifying new genes involved in parasitizing plants and that could be targeted in control efforts.

References

Baozhen C, Deutmeyer A, Carr J et al (2011) Microfluidic bioassay to characterize parasitic nematode phenotype and anthelmintic resistance. *Parasitology* 138(1):80–88 (doi:10.1017/S0031182010001010).

Nicol P, Gill R, Fosu-Nyarko J et al (2012) De novo analysis and functional classification of the transcriptome of the root lesion nematode, *Pratylenchus thornei*, after 454 GS FLX sequencing. *Int J Parasitol* 42(3):225–237 (doi:10.1016/j.ijpara.2011.11.010).

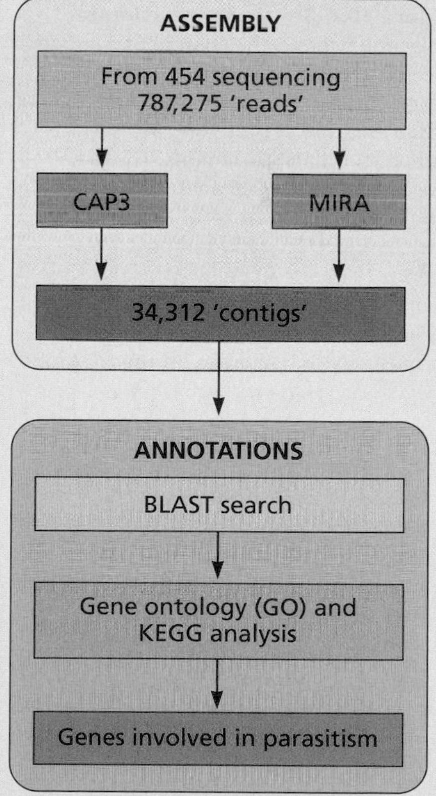

Figure 1 An example of a bioinformatics pipeline. Overview of the steps used to assemble and annotate the transcriptome of the plant parasitic nematode *Pratylenchus thornei*, as described by Nicol et al (2012).

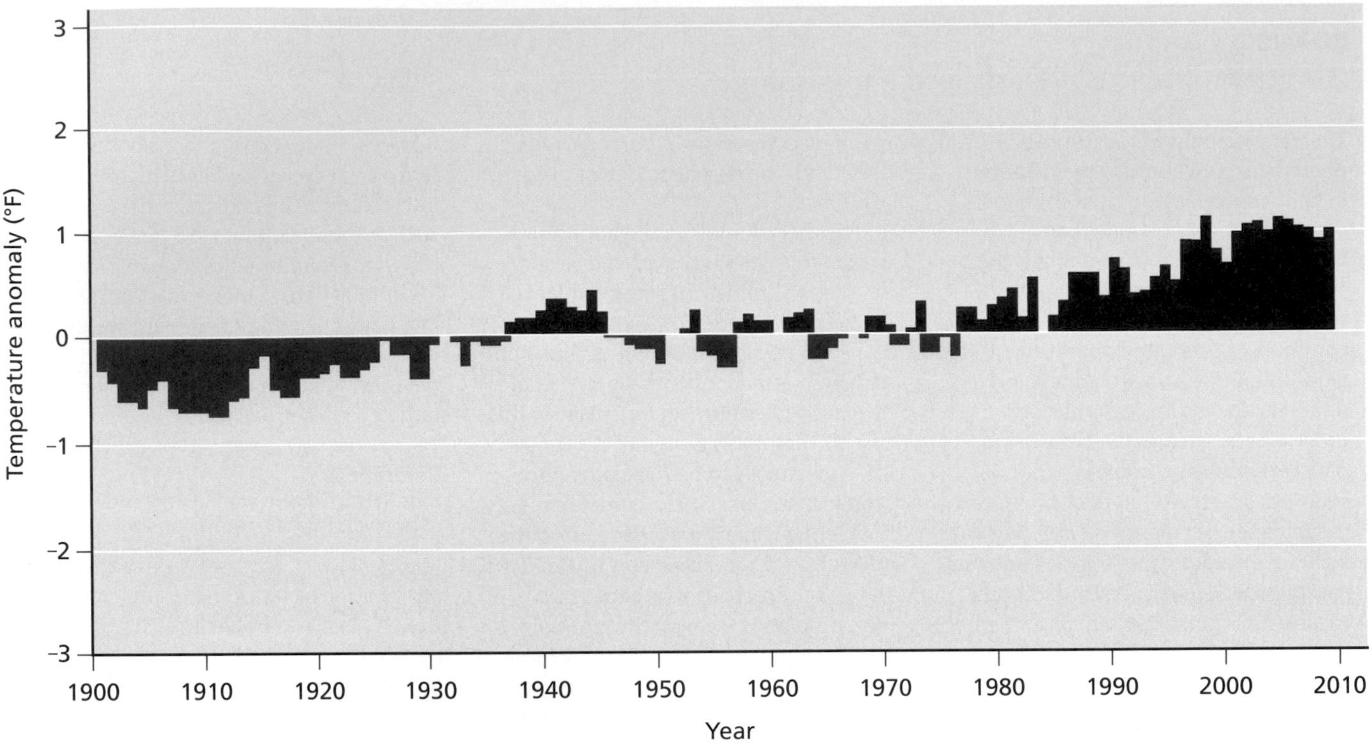

Figure 10.4 Our warming climate. This graph shows the change of average worldwide temperatures since 1901. The data come from a combination of land-based weather stations and sea surface-temperature measurements. The graph uses the 1901 to 2000 average as a baseline (the line centered at zero) for depicting the change. Note how temperatures before 1935 are all below the long-term average (blue), whereas those after about 1975 are all above (red), and become increasingly so. (Courtesy of the NOAA.)

tropical vectors into more temperate climates in the past. These are indeed alarming prospects, especially because human populations in temperate areas lack immunological experience with such diseases.

The case of malaria is instructive for pointing out many of the complications that arise in contemplating the effects of future climate change. Increases in malaria transmission in African highlands have been attributed to climate warming, sounding an alarm that other range expansions could follow. On the one hand, increasing temperatures are expected to favor increased mosquito biting rates and to speed up malaria development times in mosquitoes. These parameters, which we have noted earlier (Chapter 6), are important in malaria epidemiology. However, as temperatures increase further, these and other critical rates, such as mosquito fecundity and survivorship, begin to fall (**Figure 10.5**). A trade-off between increased development rates vs. lower survival times with rising temperature is often encountered among parasites or their vectors. These factors suggest the optimal temperature for malaria transmission may be lower than predicted by several climate models.

Additionally, if range expansions of malaria occur into temperate regions, it may have relatively little consequence on public health because countries in these regions are generally more affluent and have better infrastructure in place to handle vector-borne diseases like malaria. Furthermore, projections of the impact of climate change need to take into account the specific ecological context of the affected areas. Some important factors include the extent to which agriculture is practiced and the land has been drained to minimize mosquito habitats, the extent to which human populations are sequestered in urban environments with few vectors or few opportunities for vectors to bite people (at last, a benefit of staying indoors to watch TV), the presence of predators of the vectors or of competing parasites, the annual activity cycles of host species that may limit options for parasite transmission, and the presence of barriers to dispersal of vectors (such as lack of contiguous aquatic habitats). For parasites with complex life cycles, the need for many different

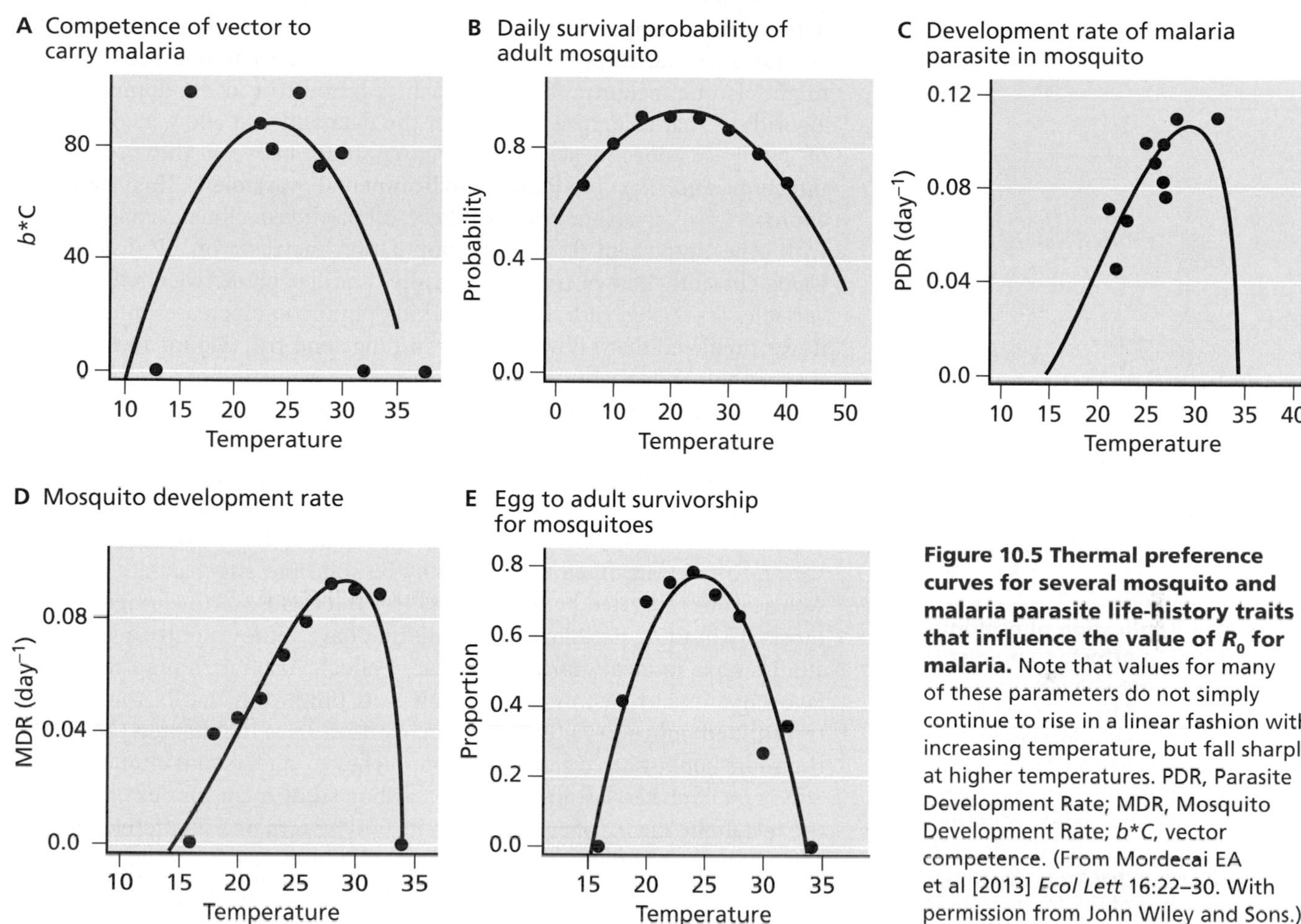

Figure 10.5 Thermal preference curves for several mosquito and malaria parasite life-history traits that influence the value of R_0 for malaria. Note that values for many of these parameters do not simply continue to rise in a linear fashion with increasing temperature, but fall sharply at higher temperatures. PDR, Parasite Development Rate; MDR, Mosquito Development Rate; *b*C*, vector competence. (From Mordecai EA et al [2013] *Ecol Lett* 16:22–30. With permission from John Wiley and Sons.)

host species to be present to enable the life cycle to establish may also diminish the likelihood that a disease can spread into new areas.

Furthermore, many ecologists feel that climate change, which could cause species to shift into new areas where they are not well adapted, will have an overall impact of reducing biodiversity. Is this a likely outcome for parasites as well? If climate change has the effect of reducing the diversity of species, then as discussed above with respect to the dilution effect, disease prevalence could rise. As this issue is studied more, it will be interesting to see if the complexities and idiosyncrasies associated with each parasite species and its transmission continue to defy meaningful generalizations about climate change or if underlying rules can be identified that make the future prediction of the impact of climate change more generalizable. It seems reasonable to expect that the conditions suitable for transmission for many parasites will shift, frequently poleward or to higher altitude, in response to global warming. Whether the parasites actually colonize these new locations remains to be seen.

Innovative approaches are being brought to bear on the study of climate effects on parasite distributions. One example applies ecological niche modeling to the distribution of lymphatic filariasis (LF) caused by ***Wuchereria bancrofti***. Ecological niche modeling in this context is the process of extracting associations with environmental parameters in areas where LF is known to be present, and using them to predict LF presence in unknown areas. This information is then used to define the environmental requirement of LF and then to predict suitable habitats in areas not yet surveyed. Climate data such as temperature and precipitation are important variables, but several other variables such as presence of appropriate intermediate hosts might also be included. This approach is innovative in developing flexible algorithms that recognize patterns in the data and that allow incorporation of nonlinear dependencies (see Figure 10.5) between the presence of infection and the predictive environmental variables. This particular approach also incorporates the effect of predicted climate change along with other important factors like population increases on LF distribution. Using currently known transmission sites and five predictive environmental variables associated with them and taking population increases into account, it was predicted that LF will expand in range and risk (**Figure 10.6**). As also stressed in the discussion of malaria above, continuation of ongoing control efforts may prevent such changes from ever actually occurring.

Studies of wildlife parasites in the Arctic, one of the areas of the globe subject to the fastest warming trends, may prove to be particularly instructive for gauging the effects of climate change. One modeling study of the caribou-infecting nematode *Ostertagia gruehneri* shows how a continuous spring-to-fall transmission season may be split into two separate transmission seasons separated by an interval. This interval is warm enough to cause mortality in those parasites on the soil that have yet to colonize a host. Field studies have in fact confirmed this very effect. Also, warming seems to be favoring annual transmission of a musk-ox lungworm that historically was transmitted only every other year. Such studies have highlighted the need to use more sophisticated modeling approaches to understand climate change effects on parasites. For instance, the caribou study incorporated elements of the metabolic theory of ecology with its well-known and predictive relationships between temperature, metabolism, and body size (see Chapter 6) into standard parasite transmission models to predict how temperature changes will influence R_0. Use of relationships from the metabolic theory help to estimate model parameters even for parasites that are not well known.

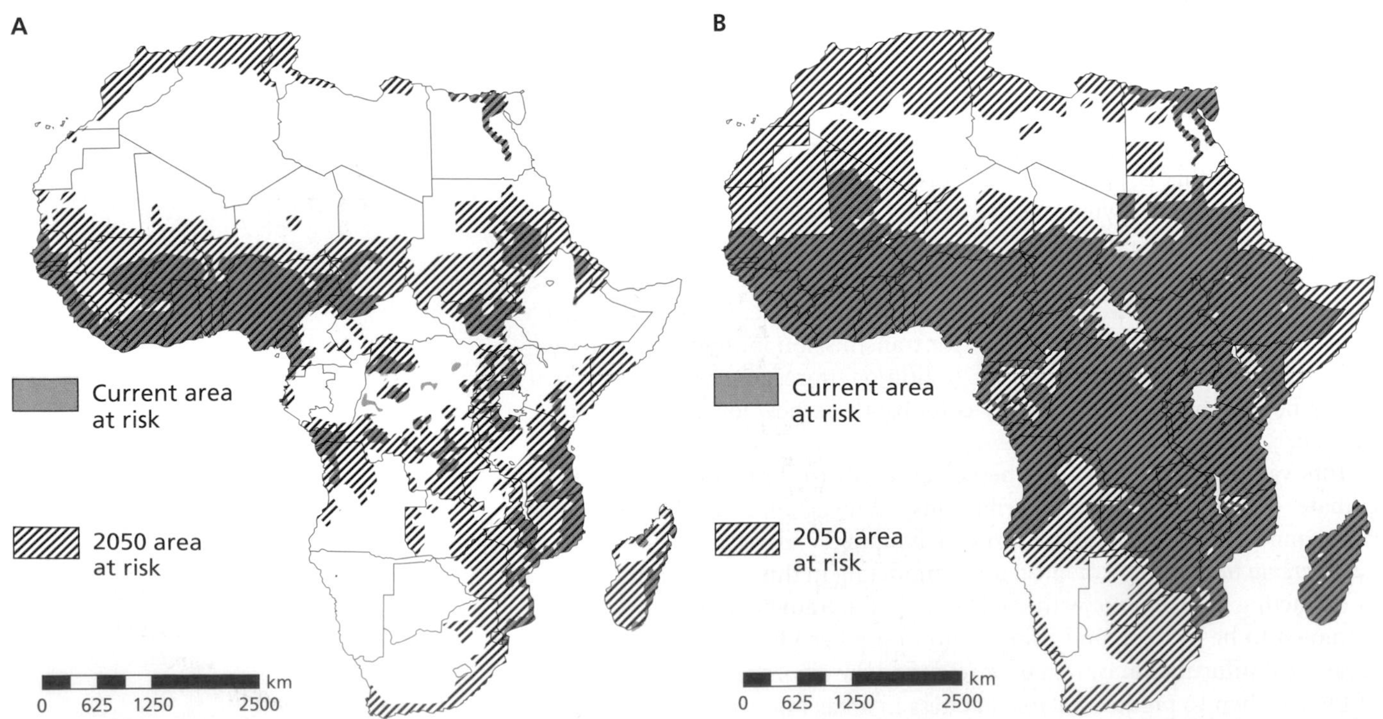

Figure 10.6 Application of ecological niche modeling efforts to predict the impact of climate change on lymphatic filariasis (LF) in Africa. Two different predictions for the distribution of LF in 2050 (striped) as compared to values in 2000 (gray). The two different versions are based on different climate predictions, and begin with different estimates of areas at risk in 2000. Note that both scenarios show an increased area of risk for transmission in 2050. (From Slater H & Edwin M [2012] *PLoS One* 7:2012 [e32202].)

10.3 CONTROLLING PARASITES IN THE FUTURE

Improved understanding of immunity should enable development of new anti-parasite vaccines, but so far the parasites are winning

The study of immunoparasitology had contributed a great deal to our understanding of basic immune system functioning. The discussions of the dichotomy between cell-mediated immunity and humoral immunity, and of the role of eosinophils in immunity (see online review of vertebrate immunity) are just two examples. However, we have yet to be able to exploit our basic knowledge of immunity to develop effective vaccines for any human eukaryotic parasite. Vaccine development remains a goal of primary importance for parasitology because the public health impact of vaccination is beyond dispute. As drug resistance of parasites looms as a constant threat (see below), it would be helpful to have vaccines as an alternative control measure.

Malaria provides a good example (see Chapter 9). A tremendous amount of research has brought a new malaria vaccine to clinical trial and, although it can protect about 50% of the recipients from more severe clinical episodes (which is a significant accomplishment), much remains to be done before there is a malaria vaccine that will have the effect of preventing and eventually eliminating transmission in endemic areas. As discussed in detail in Chapter 9, what we have learned is that malaria parasites, and indeed other eukaryotic parasites such as helminths, present considerably more difficulties to the immunologist than do many viral or bacterial pathogens with respect to development of effective vaccines. To be woven into the already complex tapestry of vaccine research should be increased appreciation of the evolutionary biology of parasites. For example, it has recently been noted that parasites may be selected to display immunodominant antigens that provoke immune pathways that lead to ineffectual or strain-specific immune responses. Instead, we need to find and target parasite antigens that are perhaps disguised or more conserved that provoke more effective and less strain-specific killing responses. As noted earlier in the chapter, vaccination efforts might also unwittingly favor more virulent forms of malaria parasites. If a vaccine is effective, it is likely that the selection on parasites to overcome it will be strong, and so it seems likely that a continuing effort will be required to maintain effective vaccines.

With respect to helminths, given that they are still collectively responsible for the annual loss of 39 million disability adjusted life years (DALYs) and control based only on use of chemotherapy is difficult, it would be a benefit to have effective helminth vaccines available. Studies of basic helminth immunology have made great strides in revealing remarkably complex immune pathways that function to expel worms (see Chapter 4). Even though the mechanistic understanding of anti-helminth immune responses has improved dramatically, some individuals are known to be naturally resistant to helminths, and it is clear that mammals can develop resistance upon repeated exposure to helminths, nonetheless progress towards developing practical anti-helminth vaccines has been slow. We still need to learn more about the factors regulating the initiation of protective responses and on the exact mechanisms and mediators involved in worm killing or expulsion.

Helminth vaccine naysayers might argue that given that it is evidently difficult for mammals to develop strong, protective immune responses to helminths, the likelihood of developing worthwhile vaccines for them is low. According to this view, helminth vaccine research is but a costly diversion that will either never come to fruition or be developed too late, after other control measures have already done the job. However, to counter these points, it can also be argued the effort to develop helminth vaccines needs to be supported, for two important reasons: (1) we know more than ever before and so are in a position to make better progress, and (2) the increasing failure of helminths such as those of livestock to respond to anthelmintic treatment increases the urgency for vaccine development. The likelihood of emergence of resistance to anthelmintic drugs for human helminths will likely increase as future drug-based control efforts are stepped up.

Chemotherapy-based control is an arms race between human ingenuity and parasite evolvability

As discussed in Chapter 9, the use of anti-parasite drugs has saved millions of lives, has spared millions more from the most severe pathological consequences of infection, and has a preeminent position in our armamentarium against parasites. We must continue to depend on anti-parasite chemotherapy in the foreseeable future, especially until other means of control can be devised or are more frequently applied. Parallel considerations apply to the use of insecticides to control vector populations.

It is also clear that continued heavy and in some cases near-exclusive reliance on chemotherapy is not a sustainable long-term control strategy. Treatment generally does not prevent reinfection, and drug resistance is a constant threat. Drug resistance has, for example, become especially rampant and worrisome for treating parasites of livestock, with attendant consequences for the ability to feed the expanding world population and sustain an expected quality of life. The detection of strains of ***Plasmodium falciparum*** less responsive to artemisinin, first in 2006 from Cambodia and then in 2008 from the border region between Thailand and Myanmar, underscores what has been a continual problem for malaria control: one drug after another has lost efficacy against malaria parasites.

The World Health Organization recently announced an exciting new program to eliminate human schistosomiasis as a public health problem by 2025. The use of a single drug, praziquantel, has thus far had enormously positive effects on decreasing the morbidity caused by schistosome parasites, but there is considerable concern that a largely unidimensional approach to schistosomiasis control will fail if resistance to praziquantel ever becomes prominent.

What to do given the evident ability of parasites to evolve resistance to drugs? Implicit in all the strategies discussed below is the need for a thorough and improved understanding of the evolutionary and population biology of parasites, including an understanding of how they respond to human-imposed selection from drug pressure or other control measures. An evolutionary perspective provides one important suggestion to guide all considerations of drug use and to manage resistance, and that is to impose as little selection as possible on parasites, because selection encourages emergence and spread of drug resistance. One way to lessen the force of selection, as the study of livestock helminths has repeatedly demonstrated, is to maintain some parasites in refugia not exposed to drug pressure, such that parasite genes associated with susceptibility to drugs are retained in the parasite population. These genes can serve as a continuing source to replenish susceptibility in worm populations and thus slow the development of resistance. With respect to sheep, leaving older ewes infected with worms untreated or treating only animals obviously anemic may provide a susceptibility refugium for helminths without significantly compromising the productivity of the sheep. This is an example of targeted selective treatment and is often included in integrated control programs, which are discussed below. Another valuable approach coming from the livestock industry and elsewhere, such as with the used of artemisinin-based combination therapy for malaria, is a strategy to preserve the useful and precious life span of anti-parasite drugs by either rotating or using simultaneously multiple drugs from different classes with different modes of action (see Figure 9.20). The underlying assumption is that this strategy will impose differing selective pressures on parasites, thus slowing the rate of strong directional selection on any one resistance gene and reducing the likelihood that any one parasite can simultaneously evolve resistance to more than one effective drug.

Approaches such as these require an improved understanding of the exact targets and mechanisms of drug resistance and more precise means to detect changes in these targets so we can respond rapidly to them. With respect to malaria, as has also come to light in studies of drug resistance in HIV-1, one possible approach is to combine two different drugs that target the same parasite enzyme but that would then select for their own mutually incompatible combinations of mutations. In other words, it is hard for the parasite to remain resistant to both compounds when both are administered at the same time. Mutants of the enzyme dihydrofolate reductase that confer high level resistance to pyrimethamine render malaria parasites more susceptible to the drug WR99210, and vice versa, because of conflicting requirements for accommodation in the active site.

Availability of parasite genomes and development of high-throughput screening and genomewide association analyses to find alterations resulting from drug pressure will prove to be an enormous asset in this quest. Improvement of mathematical models is needed to better predict (1) spatial and temporal fluctuations in parasites abundance, including those resulting from drug control, and (2) the impact of drug resistance. Recent modeling efforts indicate that rotation of drug treatments have more of an impact on controlling sheep nematodes (*Trichostrongylus colubriformis, Haemonchus contortus*, and *Teladorsagia* (*Ostertagia*) *circumcincta*) and preventing the emergence of drug resistance than does the use of refugia that leave up to 10% of adult sheep untreated. In addition, the simultaneous deployment of combinations of old and new drugs was more effective than rotation of drugs.

One factor regarding drug resistance that will be of particular interest to evolutionary biologists is the trade-off between resistance and fitness. One example of a likely indication of the cost of resistance is provided by malaria parasites in regions such as Malawi where, 12 years following the withdrawal of chloroquine in favor of use of other drugs for use in malaria treatment, chloroquine-sensitive parasites have returned (Chapter 9). Another example is provided by ***Giardia lamblia***, for which isolates resistant to metronidazole are relatively rare. Resistant isolates are rare because they frequently exhibit a deficit in their ability to adhere to the host's intestinal epithelium and consequently to establish infections (**Figure 10.7**). This defect is related to impaired glucose metabolism, a trait that also prevents the activation of metronidazole to its active form.

Cases like these raise the general question of how often trade-offs between resistance and fitness might occur and escape our attention. Also, how likely is it that the parasite will undergo further mutations that overcome the fitness costs of drug resistance? The costs imposed by resistance also remind us that drug-resistant parasites are often outcompeted by drug-sensitive parasites when they co-occur within a host. The superior competitive ability of drug-sensitive parasites has led to an alternative view of how such mixed infections should be treated to minimize emergence of resistance, a view that may come to dominate our future conceptions of treatment of diseases such

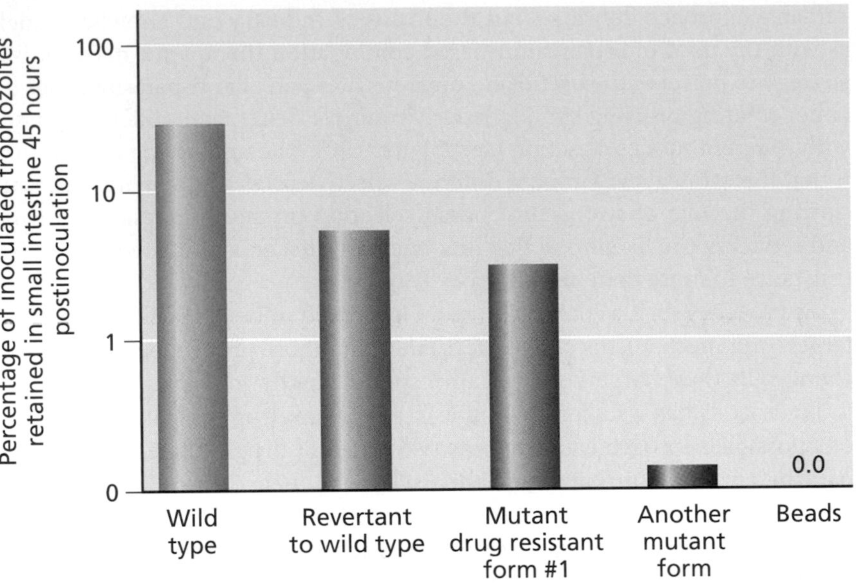

Figure 10.7 Drug resistance and parasite fitness. This graph shows the retention of trophozoites of *Giardia lamblia* in the small intestine of suckling mice that were inoculated with 10^7 trophozoites. Mice were inoculated with wild type *G. lamblia*, a mutant form that had reverted to wild type, a mutant drug resistant form, another mutant, or beads. Trophozoites and beads were enumerated at the designated times. Data are means of the percentages of the initial inoculum (3 to 6 animals per data point). Note how the wild-type and revertant trophozoites were retained longer in the intestine than drug resistant forms or beads.

as malaria. Instead of maintaining a prolonged course of treatment that will put drug-sensitive parasites at a continual disadvantage relative to resistant ones and that increases the probability of the emergence of drug resistance, if shorter courses of treatment can be found that are still clinically effective and that preserve the inherent reproductive advantage of drug-sensitive parasites (that impose no more selection on the parasites than necessary), then these should be sought.

Even if we can effectively manage resistance to the drugs currently available, it would be very reassuring to have new drugs at the ready if needed. Continued emphasis should consequently be placed on the development of new categories of anti-parasite drugs. The recent development of compounds such as monepantel, emodepside, and derquantel to treat helminths of sheep and companion animals and spiroindolones to treat malaria are just a few examples. Development of new methods of screening compounds, including compounds already approved for use in other contexts, should be encouraged. Along with this should go ways of retrieving or preserving the usefulness of drugs already available. A significant setback for malaria control was the development of chloroquine resistance in ***P. falciparum***. As noted in Chapter 9 (pages 378–379), this resulted from a mutation in a membrane-associated transporter that enabled malaria parasites to pump chloroquine out of the parasite's digestive vacuole. Some channel-blocking drugs such as verapamil, by interfering with the mutated transporter protein, interfere with the export of chloroquine, thus potentially restoring its effects. Just one example of the sophisticated and powerful approaches available in the future for retrieving the activity of drugs such as chloroquine, and possibly other drugs to which parasites have developed resistance, is to express the target transporters of parasites in easily manipulated cell types where they can be more easily studied (**Figure 10.8**).

Integrated control may provide the best prospects for sustainable parasite control and is built on a thorough knowledge of parasite biology

Throughout the long history of the attempts to control insect pests or infectious diseases, it has been emphasized repeatedly that the most sustainable approach to control is one that capitalizes on intimate knowledge of the target

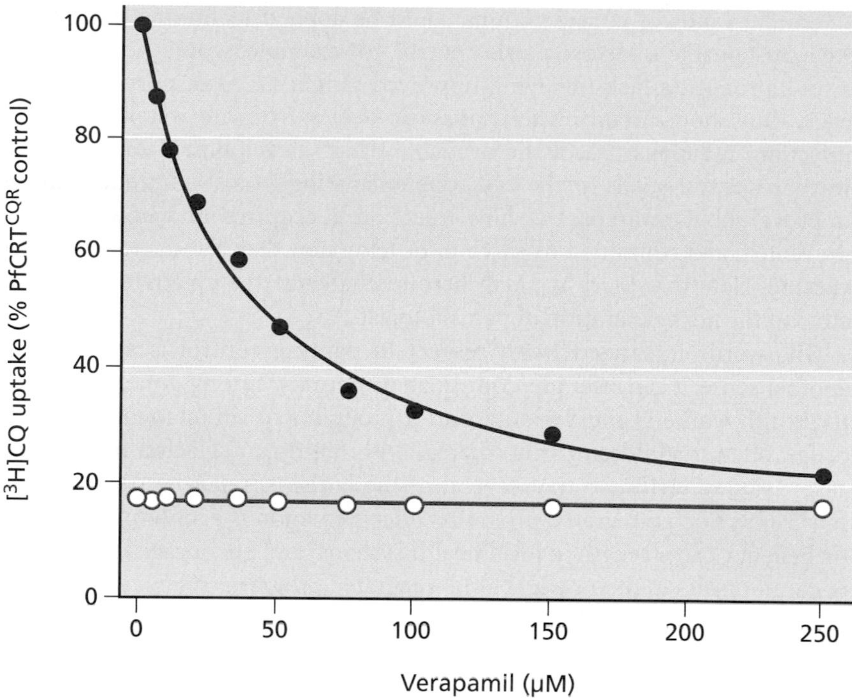

Figure 10.8 Devising new ways to study and retrieve valuable anti-malarial drugs. Transport properties of the *Plasmodium falciparum* chloroquine resistance transporter (PfCRT) were studied by inserting them into the membrane of easily manipulated cells (oocytes) from lab-reared frogs (*Xenopus*). Some oocytes received the version of the transporter from parasites resistant to chloroquine (solid red circles) and some received the version of the transporter in parasites susceptible to chloroquine (open green circles). Notice how application of increasing amounts of the channel blocker verapamil prevented transport of labeled chloroquine by oocytes expressing the chloroquine-resistant version of the transporter, essentially reversing the chloroquine-clearing action of the mutant form of the transporter. Uptake is shown as mean transport from three to five separate experiments, within which measurements were made from 10 oocytes per treatment. (From Martin RE et al [2009] *Science* 325:1680–1682. With permission from AAAS.)

organisms and that employs a variety of control strategies in an integrated manner. For example, control of livestock helminths should fully exploit the knowledge of their biology, including features such as the annual increase in passage of helminth eggs onto pastures in the spring time (the spring rise). Exposure to infection could be minimized by various approaches including (1) implementing sustainable stocking densities and rotations among different pastures to avoid unhealthy levels of pasture contamination, (2) exploiting other organisms such as fungi that can trap and kill the infective larval stages of sheep nematodes on pasture, (3) maintaining refugia for protection of drug sensitivity, in part by selective treatment of sick animals (**Figure 10.9**), (4) minimizing drug use, (5) rotating drug use as needed to preserve efficacy, (6) being prepared to accept some minor losses to preserve drug efficacy, (7) considering use of feed supplements that make animals more resistant to infection or that minimize the consequences of infection, (8) encouraging development of new control approaches such as vaccines, and (9) maintaining an effective system of educational exchange such that new control approaches can be readily disseminated to farmers in the field.

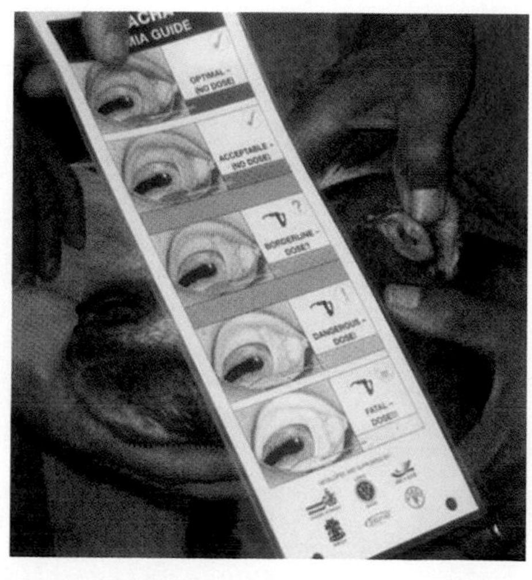

Figure 10.9 An integrated approach to sustainable parasite control. This photograph shows an example of one aspect of integrated control of parasites of sheep that involves targeted treatment of sheep or goats that have become particularly anemic as a result infection with the nematode *Haemonchus contortus* (also known as the barber pole worm because of its striped appearance). This trichostrongylid nematode attaches to the mucosal lining the rumen of its hosts and sucks blood. In this case, the FAMACHA© system a simple, inexpensive, and reproducible method using eyelid color to measure the degree of anemia resulting from *H. contortus*. Nonanemic animals, even though they might be infected, need not be treated, thus saving money and helping to maintain refugia from drug treatment and prolonging the life span of anthelmintics. This system was developed by Faffa Malan and is distributed under the auspices of the South African Veterinary Association (Gareth F. Bath, Project Coordinator). (Reprinted with permission from www.sare.org.)

Similar kinds of considerations could be applied to human parasites as well. Sustainable schistosomaisis control, for example, would be favored by a multipronged attack involving improved sanitation to dispose of parasite eggs, thus short-circuiting transmission; delivery of safe water to prevent infection; judicious use of the available drugs; development of innovative ways to control snails or the cercariae emanating from them; education to improve public awareness of how infection is acquired and how it can be prevented; and continued research to develop new drugs and a long-awaited vaccine. Note that there is much here to challenge the creativity and work ethic of the next generation of parasitologists.

The word "integrated" with respect to parasite control is also used in another sense: it can refer to a coordinated approach among donors, in-country health workers, and various control programs to simultaneously deliver drugs, often to children living in areas with multiple neglected tropical diseases (**Figure 10.10**), so that several of these diseases can be treated at the same time. Such integrated programs offer tremendous economic and logistic benefits, can strengthen local health systems, and encourage involvement of communities in their own health. Integrated programs offer a way forward in the future for control of several intransigent parasites, as discussed more fully in the next section.

Major programs are underway to eliminate many parasites as public health problems

Thanks in large part to leadership provided by the World Health Organization (World Health Organization, 2013), many of the most iconic human parasites have been targeted for elimination or eradication in the next two decades (**Table 10.1**). The major tool to be employed in most cases is repeated mass drug administration in the affected populations. **Elimination** in this context refers to a cessation of transmission in a particular area from deliberate

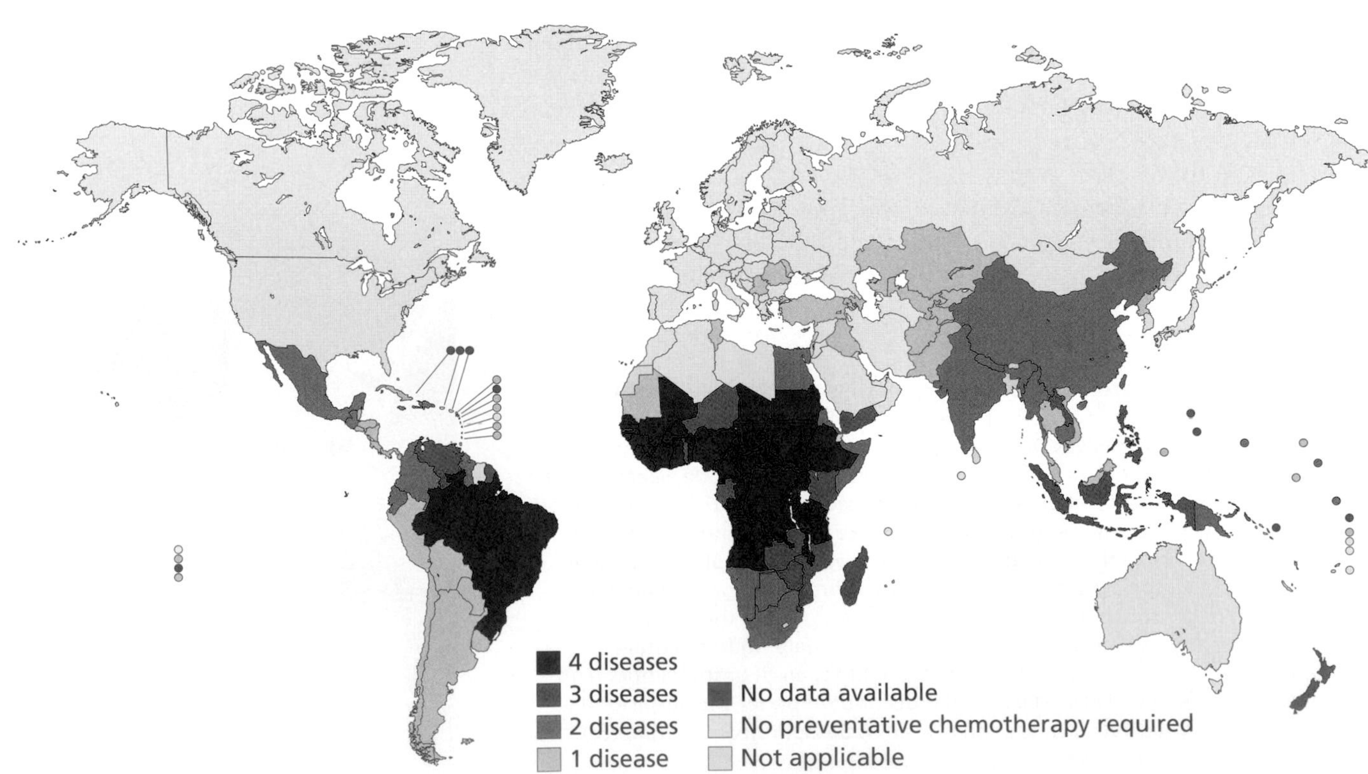

Figure 10.10 Many countries suffer from multiple species of parasites. Shown are 56 nations in which two or more neglected tropical diseases occur, often with multiple infections within a single individual. These countries are ideal targets for integrated programs in preventive chemotherapy. The diseases include lymphatic filariasis, onchocerciasis, schistosomiasis, hookworms, ascariasis, trichuriasis, and trachoma. (From Crompton DWT ed. [2013] WHO/HTM/NTD/2013.1. With permission from WHO.)

Table 10.1. Targets and milestones for eliminating and eradicating several parasitic diseases from 2015–2020, established by the World Health Organization.

Disease	2015				2020			
	Eradication	Global Elimination	Regional Elimination	Country Elimination	Eradication	Global Elimination	Regional Elimination	Country Elimination
Chagas Disease			✓ Transmission through blood transfusion interrupted				✓ Intra-domiciliary transmission interrupted in the Americas	
Human African trypanosomiasis				✓ In 80% of foci		✓		
Visceral leishmaniasis							✓ Indian subcontinent	
Dracunculiasis	✓							
Lymphatic filariasis						✓		
Onchocerciasis			✓ Latin America	✓ Yemen				✓ Selected African Countries
Schistosomiasis			✓ Eastern Mediterranean, Caribbean region, Indonesia, Mekong River				✓ Americas, Western Pacific Region	✓ Selected African Countries

Modified from World Health Organization (2013) Sustaining the drive to overcome the global impact of neglected tropical diseases (David WT Crompton ed). pp134 With permission from WHO.

control efforts, such that the incidence of new cases in that area is zero. **Eradication** refers to a permanent global reduction to zero of the incidence of new infections with no danger of reintroduction. Eradication is essentially an extinction event, unless of course the targeted organism is maintained in confined and protected settings such as government laboratories.

These control efforts, as noted in the previous section, rely on innovative partnerships between pharmaceutical companies that donate drugs, public health ministries of individual countries to be targeted, contributions by governments of wealthy countries, private donors and foundations, and of course grass roots participants in local communities. Such partnerships have increased in numbers in recent years with the Roll Back Malaria program, the Global Programme to Eliminate Lymphatic Filariasis, and the Neglected Tropical Disease Control Program being three examples. Such programs offer tremendous opportunities for parasitologists interested in lessening the burden of neglected tropical diseases. It is important to train people with the appropriate backgrounds to understand how such partnerships are forged and maintained. It is also important that these partnerships have available individuals with appropriate training and experience in operational aspects of control programs, so they can work effectively, often in difficult circumstances, to implement control programs. Problems such as the emergence of drug resistance or the outbreak of civil wars have and will beset these programs and challenge the workers involved, and they will require development of new tools and technologies to meet these challenges. One important tool that is needed is better diagnosis of parasite infections, discussed next.

We will need improved methods to detect low levels of parasite infection and transmission in the future

Thanks to continued generous contributions from donors and development of improved networks to implement integrated control, as discussed above, it is likely that the prevalence of many parasite infections will fall dramatically in the coming decades. Although this is a wonderful and desired outcome, it carries with it a few concerns: (1) an increased tendency to write-off and forget the controlled parasites because they have fallen below the threshold of public health concern; (2) neglect of the associated public health infrastructure, including failure to train health workers who know about the parasites in question and who can quickly recognize their re-emergence and employ measures to control them; and (3) the greater difficulty of detecting the controlled parasites because of their lower prevalence and intensity of infection. As a consequence, outbreaks of these parasites can be expected and, unless detected and dealt with quickly, precontrol prevalence and intensity levels could be quickly reestablished.

To prevent sliding backward, continued development of improved diagnostic techniques is needed. These would preferably be noninvasive, meaning, for instance, that samples of urine or sputum could be tested instead of having to draw a patient's blood. Simple, easy-to-read tests that could be administered under field conditions and that do not involve complicated sample preservation would also be ideal. Also desirable are tests that are both **specific** (they do not give positive results if the patient is not infected or is infected with some other species of parasite) and **sensitive** (they accurately detect the patients who are truly positive).

One example of a new generation of needed diagnostic tools is the Kalazar Detect strip test used to diagnose infections of visceral leishmaniasis caused by ***Leishmania donovani***, *L. chagasi*, or *L. infantum* (**Figure 10.11**). This test uses preformed strips coated with a particular ***Leishmania*** antigen that can detect the presence of diagnostic anti-***Leishmania*** antibodies in a patient's serum or urine. If urine is used, the test is noninvasive. The test can be run at room temperature, read within 10 minutes, and has >95% sensitivity and specificity.

Should circumstances require that parasite nucleic acids need to be detected, procedures such as loop-mediated isothermal amplification assays

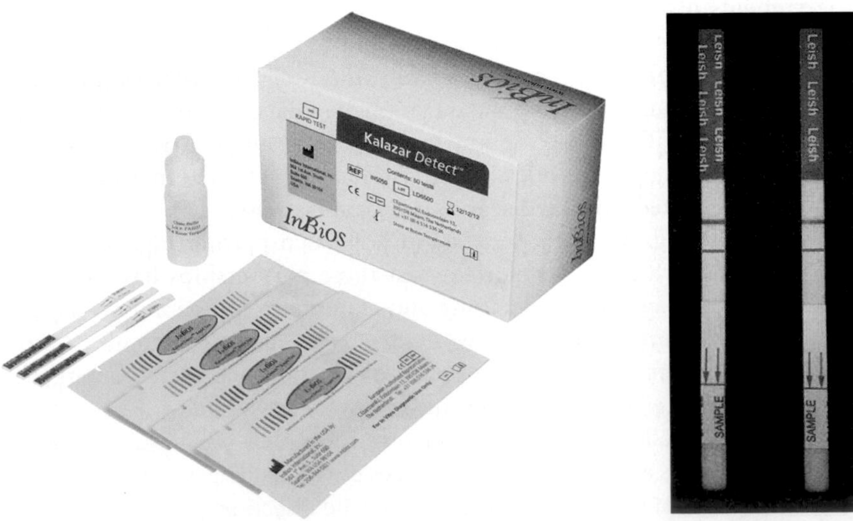

Figure 10.11 An example of a sensitive, specific, potentially noninvasive, and easy-to-read test to diagnose visceral leishmaniasis. The Kalazar Detect test uses a recombinant form (meaning it was engineered to be made in quantity by bacteria) of an antigenic 39-amino acid repeat region called rk39 from a protein from *Leishmania chagasi*. This antigen is precoated on a test line on a membrane strip. A serum or urine sample from a patient suspected of having visceral leishmaniasis is added to the membrane, which also includes protein A bound to colloidal gold. Protein A binds to antibodies in the patient's sample. As the patient's antibodies diffuse across the membrane, if anti-rk39 antibodies are present, they will accumulate at the test line containing the diagnostic parasite antigen. The line turns red because of the accumulation there of antibody–protein A-colored colloidal gold complexes. The second red line is a control to check that the sample has diffused far enough onto the strip. It is loaded with antibodies that bind to protein A. (A, B With permission from InBios International.)

(LAMP assays) may be in order. LAMP assays allow the amplification of parasite nucleic acids, thereby making them easier to detect without the need for complex thermal cyclers or other laboratory equipment. This kind of test provides one way forward for health workers in endemic areas with limited access to expensive lab equipment.

Although it may not seem as glamorous as developing a new malaria vaccine, the heroes of future efforts to control parasites will be those at the front lines who are able to detect infections and respond quickly and appropriately to them. Continued and improved diagnosis and surveillance will become even more important than they already are.

Provision of improved living conditions, including education, will further discourage parasite transmission

The annual United Nation's Millennium Goals report is a good reminder of where we stand with respect to many key issues influencing the human condition, including health. Among the key determinants of good health are income and education level, as well as provision of clean water and adequate food, shelter, and basic sanitation. Although the costs associated with providing these determinants are high, the benefits are many. As noted in Chapter 9, amebiasis, giardiasis, cryptosporidiosis, schistosomiasis, cysticercosis and soil-borne helminth diseases such as ascariasis, are just some of the parasitic diseases that would be diminished by provision of sanitation and clean water. As additional benefits, water associated with human wastes can be processed and made safe for drinking, and the wastes themselves can be processed to provide safe fertilizers for crops, and combustible gases for cooking food. These measures would help alleviate our unsustainable use of fertilizers and firewood. Better housing, including provision of window screens and bed nets, would help prevent a number of vector-borne diseases such as Chagas' disease, leishmaniasis, malaria, onchocerciasis, and filariasis. Better sanitation and housing would also help prevent many viral and bacterial diseases as well. In addition to seeking the means to develop and wield improved disease-specific tools, it is important to remember that achieving the more general goals spelled out in the Millennium Goals project represents a daunting challenge worthy of the consideration of parasitologists of the future.

Also included among the Millennium goals is improved education, which is important in at least three ways. The first is the broad goal of doing everything possible to improve the overall education of the world's population as a way of improving economic prospects and increasing the overall standard of living.

The second goal is to promote more educational programs relating to parasites in general and to particular parasitic diseases in specific. Successful control programs often include an educational component whereby people being served by the program can better understand the nature of the parasite being controlled and why particular measures are being used in control. There is a great, untapped potential for reducing behaviors that put people at risk of acquiring certain parasitic infections. For example, over 600 million cases of food-borne trematode infections occur in Asia, some of which, such as ***Opisthorchis viverrini***, are associated with serious consequences such as life-threatening cholangiocarcinoma. A vigorous and sustained educational program targeting schoolchildren and focusing on the risks of eating raw or poorly cooked fish could do a great deal to alter traditional behavior patterns that make these people vulnerable to these diseases. Not to be underestimated in such endeavors is the value of cell phones and other modern electronic devices in delivering the message to affected people, often even in remote areas. Devising effective education programs related to parasite control programs is yet another way for parasitologists to make an impact.

The third way education can make a difference is in the offering of courses in general parasitology, which continues to be a tremendous vehicle for parasitologists to get their basic messages across to a wider public audience. Our own experience has convinced us that students find parasite biology to be intrinsically fascinating, a fascination that is only further embellished by the obvious health relevance and occurrence in exotic locations of parasites. Furthermore, the study of parasite life cycles and their associated ecological and evolutionary implications provides outstanding opportunities for students to get into the field and to learn about the subtleties of the natural world.

Summary

For aspiring parasitologists, there is a great deal of work to be done and much yet to learn. Many parasites are still unknown to science. Specialized expertise is needed to identify newly emergent and previously unknown parasitic infections. Cataloging and preserving parasite diversity in museums, both to provide an invaluable historical perspective for environmental change and as a repository of the genetic novelty inherent in parasites, is also an important endeavor. A number of fundamental biological questions regarding parasites remain largely unanswered. How do parasites influence natural ecosystems? What molecules govern the success of parasites in their hosts? What is the fundamental mechanism at the molecular level that governs the phenomenon of host specificity?

Given the aggressive agenda for parasite control provided by the World Health Organization and associated international consortia, in many ways it can be said we live in the age of parasite control. Yet this control will not come easily—parasites are by their very nature resilient and hard to control. Furthermore, changing global climates make it harder to anticipate future parasite distributions. Development of new drugs or of novel ways to use old drugs is urgently needed, as is an ability to glean from the burgeoning knowledge of immunology new and successful ways to vaccinate against parasites. For control to be successful, better tools are needed to detect and monitor infections, as is better outreach and grass roots education about the biology of parasites and how to control them. Last, and perhaps most important, new insights are needed to help us appreciate the implications of widespread ecological changes that are, and will continue to be, the hallmark of the Anthropocene epoch in which we all now live. Future studies will contribute to determining if we are approaching state changes in the Earth that portend unforeseen and unfavorable changes for the future, and help avert them. All of these lines of work—whether the scale is microscopic or global—afford great opportunities for students contemplating a career involving the ever-fascinating world of parasitism to make a positive difference in our world.

REVIEW QUESTIONS

1. An important issue for the future is food security. Discuss how parasites pose major challenges to our ability to adequately feed everyone in a future world with well over 7 billion people.

2. Of what use is having complete genome information for parasites?

3. Molecular parasitology also has a great deal to offer us in the future. Provide some examples.

4. How is the study of climate change relevant to parasitology? For instance, if the climate is warming, can we expect that vector-borne parasitic diseases of the tropics will invade more developed countries?

5. Drug resistance is a huge problem in our ongoing battle to control parasites. Outline some approaches for how might we overcome it.

6. Why does integrated control represent a smart way forward with respect to coping with the burden of parasites?

7. What is the difference between elimination and eradication of a particular parasite?

8. What do parasite control programs need to succeed?

References

OUR FUTURE WORLD

Barnosky AD, Hadly EA, Bascompte J et al (2012) Approaching a state shift in Earth's biosphere. Nature **486**:52–58.

Chandy L & Gertz G (2011) Poverty in Numbers: The Changing State of Global Poverty from 2005 to 2015. The Brookings Institution.

Tilman D, Balzer C, Hill J et al (2011) Global food demand and the sustainable intensification of agriculture. *Proc Natl Acad Sci USA* **108(50)**:20,260–20,264 (doi:10.1073/pnas.1116437108).

UN Millennium Development Goals (http://www.un.org/millenniumgoals/pdf/%282011_E%29%20MDG%20Report%202011_Book%20LR.pdf).

SOME FUTURE CHALLENGES FOR PARASITOLOGISTS

There is always something new to be found under the parasitological sun

Mackinnon M J & Marsh K (2010) The selection landscape of malaria parasites. *Science* **328(5980)**:866–871 (doi:10.1126/science.1185410).

Silva JC, Egan A, Friedman R et al (2011) Genome sequences reveal divergence times of malaria parasite lineages. *Parasitology* **138(13)**:1737–1749 (doi:10.1017/S0031182010001575).

We need to better understand the ecological and evolutionary roles of parasites

Barclay VC, Sim D, Chan BHK et al (2012) The evolutionary consequences of blood-stage vaccination on the rodent malaria *Plasmodium chabaudi. PLoS Biol* **10(7)**:e1001368 (doi:10.1371/journal.pbio.1001368).

Bradley J, Cardinale J, Duffy E et al (2012) Biodiversity loss and its impact on humanity. *Nature* **486**:59–67.

Estes JA, Terborgh J, Brashares JS et al (2011) Trophic downgrading of planet Earth. *Science* **333(6040)**:301–306 (doi:10.1126/science.1205106).

Johnson PTJ & Thieltges DW (2010) Diversity, decoys and the dilution effect: how ecological communities affect disease risk. *J Exp Biol* **213**:961–970.

Jones KE, Patel NG, Levy MA et al (2008) Global trends in emerging infectious diseases. *Nature* **451**:990–993.

Keesing F, Belden LK, Daszak P et al (2010) Impacts of biodiversity on the emergence and transmission of infectious diseases. *Nature* **468(7324)**:647–652 (doi:10.1038/nature09575).

Loker ES (2012) Macroevolutionary immunology: a role for immunity in the diversification of animal life. *Front Immunol* **3**:25 (doi:10.3389/fimmu.2012.00025).

Mennerat A, Hamre L, Ebert D et al (2012) Life history and virulence are linked in the ectoparasitic salmon louse *Lepeophtheirus salmonis. J Evol Biol* **25(5)**:856–861 (doi:10.1111/j.1420-9101.2012.02474.x).

Mennerat A, Nilsen F, Ebert D et al (2010) Intensive farming: evolutionary implications for parasites and pathogens. *Evol Biol* **37(2–3)**:59–67 (doi:10.1007/s11692-010-9089-0).

Stauffer JR, Madsen H, McKaye K et al (2006) Schistosomiasis in Lake Malawi: relationship of fish and intermediate host density to prevalence of human infection. *Ecohealth* **3(1)**:22–27 (doi:10.1007/s10393-005-0007-3).

Thomas F, Daoust SP & Raymond M (2012) Can we understand modern humans without considering pathogens? *Evol Appl* **5(4)**:368–379 (doi:10.1111/j.1752-4571.2011.00231.x).

Woolhouse MEJ, Haydon DT & Antia R (2005) Emerging pathogens: the epidemiology and evolution of species jumps. *Ecol Evol* **20(5)**:238-244 (doi:10.1016/j.tree.2005.02.009).

Wood CL, Lafferty KD, DeLeo G et al (2014) Does biodiversity protect against human infectious disease? *Ecology* **96**: 817–832.

Revealing how parasite and host molecules interact is needed to clarify many fundamental aspects of parasitism

Reid AJ, Vermont SJ, Cotton JA et al (2012) Comparative genomics of the apicomplexan parasites *Toxoplasma gondii* and *Neospora caninum*:

coccidia differing in host range and transmission strategy. *PLoS Pathog* **8(3)**:e1002567 (doi:10.1371/journal.ppat.1002567).

Climate change will affect parasites, but we know little about how

Altizer S, Ostfeld RS, Johnson PT et al (2013) Climate change and infectious diseases: from evidence to a predictive framework. *Science* **341**:514–519.

Kutz SJ, Jenkins EJ, Veitch AM et al (2009) The Arctic as a model for anticipating, preventing, and mitigating climate change impacts on host–parasite interactions. *Vet Parasitol* **163(3)**:217–228 (doi:10.1016/j.vetpar.2009.06.008).

Lafferty KD (2009) The ecology of climate change and infectious diseases. *Ecology* **90(4)**:888–900 (doi:10.1890/08-0079.1).

Molnar PK, Kutz SJ, Hoar BM et al (2013) Metabolic approaches to understanding climate change impacts on seasonal host-macroparasite dynamics. *Ecol Lett* **16(1)**:9-21 (doi:10.1111/ele.12022).

Mordecai EA, Paaijmans KP, Johnson LR et al (2013) Optimal temperature for malaria transmission is dramatically lower than previously predicted. *Ecol Lett* **16(1)**:22–30 (doi:10.1111/ele.12015).

Rohr JR, Dobson AP, Johnson PTJ et al (2011) Frontiers in climate change-disease research. *Ecol Evol* **26(6)**:270–277 (doi:10.1016/j.tree.2011.03.002).

Slater H & Edwin M (2012) Predicting the current and future potential distributions of lymphatic filariasis in Africa using maximum entropy ecological niche modeling. *PLos One* **7(2)**:e32202 (doi:10.1371/journal.pone.0032202).

United States Environmental Protection Agency (2012) Climatic change indicators in the United States, 3rd ed. http://www.epa.gov/climatechange/pdfs/climateindicators-full-2012.pdf.

CONTROLLING PARASITES IN THE FUTURE

Improved understanding of immunity should enable development of new anti-parasite vaccines, but so far the parasites are winning

Harris NL (2011) Advances in helminth immunology: optimism for future vaccine design? *Trends Parasitol* **27(7)**:288–293 (doi:10.1016/j.pt.2011.03.010).

Hewiston JP, Maizels RM (2014) Vaccination against helminth parasite infections. *Expert Rev Vaccines* **13**:473-487. (doi:10.1586/14760584.2014.893195)

Little TJ, Allen JE, Babayan SA et al (2012) Harnessing evolutionary biology to combat infectious disease. *Nat Med* **18(2)**:217–220.

Nossal GJV (2011) Vaccines of the future. *Vaccine* **29**:D111–D115 (doi:10.1016/j.vaccine.2011.06.089).

Chemotherapy-based control is an arms race between human ingenuity and parasite evolvability

Dobson RJ, Barnes EH, Tyrrell KL et al (2011) A multi-species model to assess the effect of refugia on worm control and anthelmintic resistance in sheep grazing systems. *Aust Vet J* **89**:200–208 (doi:10.1111/j.1751-0813.2011.00719.x).

Epe C & Kaminsky R (2013) New advancement in anthelmintic drugs in veterinary medicine. *Trends Parasitol* **29**:129–134.

Goldberg DE, Siliciano RF & Jacobs Jr WR (2012) Outwitting evolution: fighting drug-resistant TB, malaria, and HIV. *Cell* **148**:1271–1283 (doi:10.1016/j.cell.2012.02.021).

Hastings MD & Sibley CH (2002) Pyrimethamine and WR99210 exert opposing selection on dihydrofolate reductase from *Plasmodium vivax*. *Proc Natl Acad Sci USA* **99(20)**:13,137–13,141.

Gray DJ, McManus DP, Li Y et al (2010) Schistosomiasis elimination: lessons from the past guide the future. *Lancet Infect Dis* **10**:733–736 (doi:10.1016/S1473-3099(10)70099-2).

Knox MR, Besier RB, Le Jambre LF et al (2012) Novel approaches for the control of helminth parasites of livestock VI: summary of discussions and conclusions. *Vet Parasitol* **186**:143–149 (doi:10.1016/j.vetpar.2011.11.054).

Laufer MK, Thesing PC, Eddington ND et al (2006) Return of chloroquine antimalarial efficacy in Malawi. *N Engl J Med* **355**:1959–1966 (doi:10.1056/NEJMoa062032).

Martin RE, Marchetti RV, Cowan AI et al (2009) Chloroquine transport via the malaria parasite's chloroquine resistance transporter. *Science* **325**:1680–1682 (doi:10.1126/science.1175667).

Mueller IB & Hyde JE (2010) Antimalarial drugs: modes of action and mechanisms of parasite resistance. *Future Microbiol* **5**:1857–1873 (doi:10.2217/FMB.10.136).

Read AF, Day T & Huijben S (2011) The evolution of drug resistance and the curious orthodoxy of aggressive chemotherapy. *Proc Natl Acad Sci U S A* **108**:10,871–10,877 (doi:10.1073/pnas.1100299108).

Rottmann M, McNamara C, Yeung BKS et al (2010) Spiroindolones, a potent compound class for the treatment of malaria. *Science* **329**:1175–1180 (doi:10.1126/science.1193225).

Tejman-Yarden N, Millman M, Lauwaet T et al (2011) Impaired parasite attachment as fitness cost of metronidazole resistance in *Giardia lamblia*. *Antimicrob Agents Chemother* **55**:4643–4651 (doi:10.1128/AAC.00384-11).

World Health Organization. Status report on artemisinin resistance. January 2014. Global Malaria Programme. http://www.who.int/malaria/publications/atoz/update-artemisinin-resistance-jan2014/en/

Integrated control may provide the best prospects for sustainable parasite control and is built on a thorough knowledge of parasite biology

Hotez PJ, Molyneux DH, Fenwick A et al (2007) Current concepts—control of neglected tropical diseases. *N Engl J Med* **357**:1018–1027 (doi:10.1056/NEJMra064142).

Larsen JWA (2014) Sustainable internal parasite control of sheep in Australia. *Small Ruminant Res* **118**:41–47.

Molento MB (2009) Parasite control in the age of drug resistance and changing agricultural practices. *Vet Parasitol* **163**:229–234 (doi:10.1016/j.vetpar.2009.06.007).

Major programs are underway to eliminate many parasites as public health programs

World Health Organization (2013) Sustaining the drive to overcome the global impact of neglected tropical diseases. (www.who.int/iris/bitstream/10665/77950/1/9789241564540_eng.pdf).

We will need improved methods to detect low levels of parasite infection and transmission in the future

Poon LLM, Wong BWY, Ma EHT et al (2006) Sensitive and inexpensive molecular test for falciparum malaria: detecting *Plasmodium falciparum* DNA directly from heat-treated blood by loop-mediated isothermal amplification. *Clin Chem* **52**:303–306.

Singh D, Pandey K, Das VNR et al (2013) Evaluation of RK-39 strip test using urine for diagnosis of visceral Leishmaniasis in an endemic region of India. *Am J Trop Med Hyg* **88**:222–226 (doi:0.4269/ajtmh.2012.12-0489).

Provision of improved living conditions, including education, will further discourage parasite transmission

Nichols E & Gomez A (2011) Conservation education needs more parasites. *Biol Conserv* **144**:937–941 (doi:10.1016/j.biocon.2010.10.025).

Ziegler AD, Andrews RH, Grundy-Warr C (2011) Fighting liverflukes with food safety education. *Science* **331**:282–283 (doi:10.1126/science.331.6015.282-b).

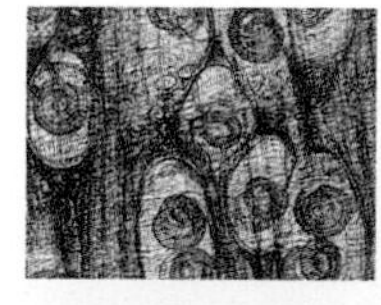

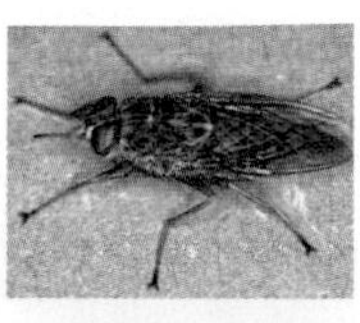
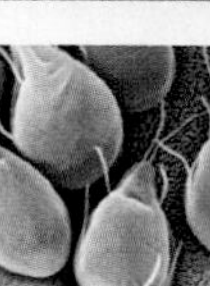
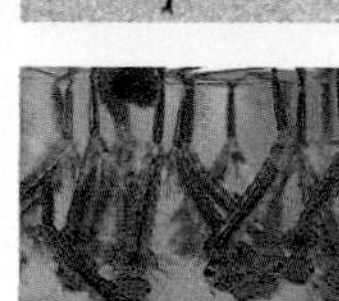

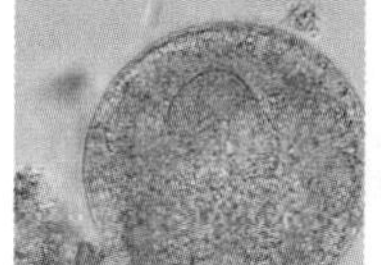
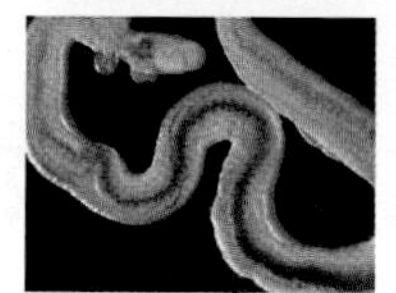
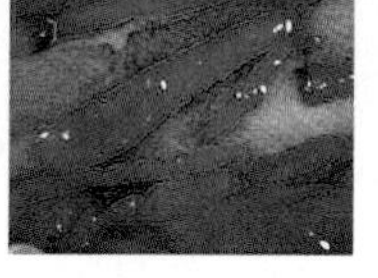

Rogues' Gallery of Parasites

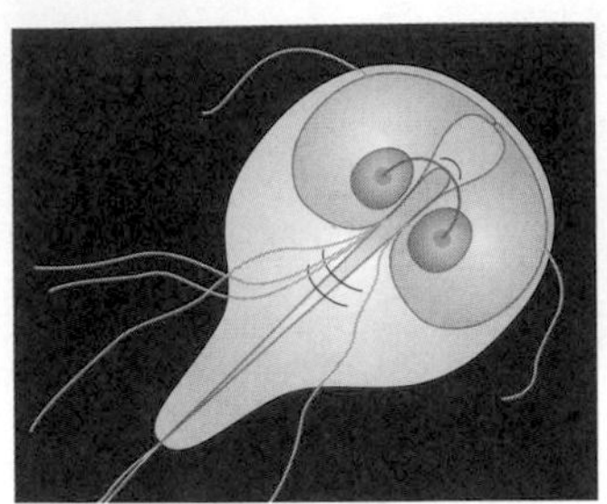

The Protozoa

The Protozoa are a diverse group of unicellular eukaryotic organisms. Their classification has yet to be fully resolved. Other than typical eukaryotic features such as the presence of membrane-bound organelles, there are few morphological features common to all protozoa. There is no precise, universally accepted definition for the term protozoa, and although once used to define a phylum of animal-like single-celled organisms, it has no modern taxonomic validity. Rather it is used commonly to describe a number of polyphyletic phyla. In some cases the term protozoa is used interchangeably with the term protista, although in some classification schemes, Protista refers to an even larger, more diverse, and polyphyletic group, which includes protozoa, along with unicellular algae, diatoms, and other unicellular eukaryotes.

Protozoa typically range from 10 to 200 micrometers in length. Most are motile and most are heterotrophic, although autotrophy is not uncommon. In various species, certain subcellular organelles are lacking or differ structurally and functionally from those found in other eukaryotes. Because they are unicellular and cannot rely on specialized cell types, all life processes must take place at the cellular level. Consequently, protozoan cells often display a level of complexity not seen in the cells of multicellular organisms. Another feature common to many protozoa is a life cycle containing different morphological stages. Like other aspects of their biology, protozoan life cycles are extremely diverse. Asexual reproduction is a common feature in most life cycles. Sexual reproduction is likewise common, and many organisms alternate between asexual and sexual life-cycle stages. Various protozoa may be free-living or parasitic. In many protozoa there is an active feeding and reproducing stage known as the trophozoite and a quiescent metabolically inactive cyst stage. The cyst stage is typically observed in those protozoa that spend at least a portion of their life cycle outside the body of a host, generally as a way to survive during unfavorable environmental conditions.

Because so many aspects of protozoan taxonomy remain unresolved, there is no single taxonomic scheme upon which all researchers agree. Molecular analysis suggests that some protozoa are actually more closely related to multicellular plants, animals, or fungi than they are to other protozoa. Historically, protozoa have been divided into four major groups based on their mode of locomotion. Hence, reference is frequently made to the flagellates, the amebas, the ciliates, and the sporozoa (which lack specific locomotory structures). Yet it is now evident that features such as flagella or ameboid movement have evolved more than once and that these traditional groups are polyphyletic. Recent molecular data have suggested that all eukaryotes might be best subdivided into 5 subdomains. These consist of the **Opisthokonta** (animals, fungi, and choanoflagellates), the **Archaeplastida** (examples include red and green algae and plants), the **Amoebozoa** (examples include lobose amebas such as *Entamoeba* and slime molds), **Excavata** (diplomonads such as *Giardia*; kinetoplasts such as *Trypanosoma* and *Leishmania*; and parabasalids such as *Trichomonas*), and the **SAR** (**Stramenopila-Alveolata-Rhizaria** lineage. Included in the SAR lineage are the apicomplexans, the ciliates and giant kelps (see Figure 2.6). Yet because of ongoing ambiguity and controversy surrounding the phylogeny of protozoa, many currently used classifications represent a compromise between available molecular data, which may better reveal evolutionary relationships, and traditional groupings, which often better meet the practical necessity for a common nomenclature. No doubt in the future, as whole genomes and expressed protein data continue to become available for comparison, our classification of the protozoa will continue to improve.

Trypanosoma brucei complex

Trypanosomes are flagellates within the class Kinetoplastida. Included within the *Trypanosoma brucei* complex are *T. brucei brucei*, the causative agent of nagana in animals, as well as *T. b. gambiense* and *T. b. rhodesiense,* which cause human trypanosomiasis, also called African sleeping sickness.

Distribution and prevalence Distribution is patchy in sub-Saharan Africa. *T. b. gambiense* is found in West and Central Africa. *T. b. rhodesiense* and *T. b. brucei* are mostly found in east and southern Africa (Figure 1). The at-risk human population is estimated at over 60 million. Fewer than 10,000 new human cases per year have been noted recently, but accurate numbers are difficult to obtain because of the large number of unreported cases.

Hosts and transmission *T. b. brucei* is mainly a parasite of native ruminants, many of which serve as asymptomatic carriers. If infected, introduced livestock such as cattle and horses are at high risk for symptomatic disease. Humans are not infected (Figure 2). Both *T. b. gambiense* and *T. b. rhodesiense* cause African trypanosomiasis in humans. Native African wildlife serves as reservoirs for *T. b. rhodesiense*, whereas *T. b. gambiense* appears to be primarily a human parasite. All members of the *T. brucei* complex rely on tsetse flies (genus *Glossina*) for vector transmission. The infective stage for mammals, the metacyclic trypomastigote, is introduced into the blood of the host as the fly feeds. *T. b. brucei* and *T. b. rhodesiense* are transmitted by *G. morsitans*, *G. pallidipes*, and *G. swynnertoni*, whereas *T. b. gambiense* is transmitted by *G. palpalis* and *G. tachinoides*.

Pathology Trypanosomes multiply in the blood causing a febrile illness. Other symptoms include headache, weakness, and cramping. In animals with nagana, paralysis and emaciation may ensue within a few days. Although many animals die within two or three weeks, the disease course depends on the susceptibility of the host species. Infection with *T. b. rhodesiense* in humans usually causes acute disease, with swollen lymph nodes and rapid weight loss. A chronic condition with neurological involvement, causing lethargy and coma, is associated with *T. b. gambiense* infection. Symptoms occur in cyclic waves, moderated by host immune response. The vast majority of African trypanosomiasis in humans is caused by *T. b. gambiense*.

Diagnosis Infection is traditionally diagnosed by observation of parasites in blood or cerebrospinal fluid (Figure 3). Various serological tests detect antibodies against *T. b. gambiense*. PCR can be used to amplify trypanosome DNA, permitting species or strain differentiation using certain genetic markers.

Treatment Pentamidine and suramin are effective prior to neurological involvement. Pentamidine is preferred for *T. b. gambiense* infections because it is less toxic, but it is ineffective against *T. b. rhodesiense*. These drugs do not enter the central nervous system. Thus, melarsoprol is usually used in late-stage *T. b. gambiense* infections with neurological involvement, although toxicity of this arsenic-based drug is high. Newer drugs used in *T. b. gambiense* infections include nifurtimox and eflornithine.

Control Control is largely based on vector eradication using insecticides, traps, or clearing vegetation. Some success has been achieved in breeding resistant livestock such as the ndama breed of cattle.

Did you know? The word "nagana" comes from the Zulu word for "depressed."

References

Barnett MP (2006) The rise and fall of sleeping sickness. *Lancet* **367**:1377–1378.

Simarro PP, Diarra A, Ruiz Postigo et al (2011). The human African trypanosomiasis control and surveillance programme of the World Health Organization 2000–2009: the way forward. *PLoS Negl Trop Dis* **5(2)**:e1007 (doi:10:1371/journal.pntd.0001007).

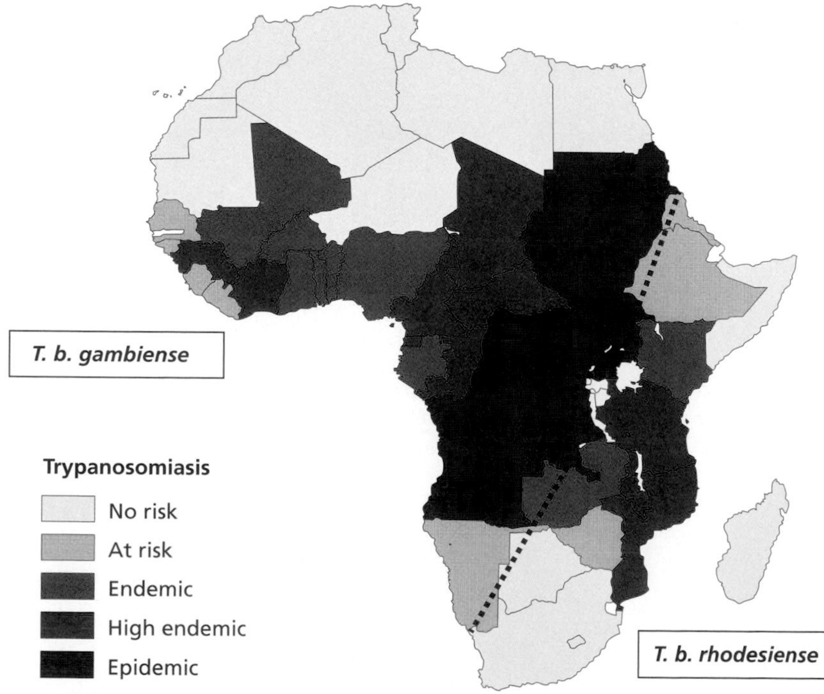

Figure 1 The distribution of *Trypanosoma* species in Africa. The distribution of *T. b. gambiense* and *T. b. rhodesiense* (to the left and right, respectively, of the dotted line) largely reflects the distribution of the *Glossina* species used as vectors. In general, those vector species harboring *T. b. gambiense* tend to be found in more dense vegetation. *Glossina* species used by *T. b. rhodesiense* are more commonly associated with open savannah. The distribution of *T. b. brucei* is similar to that of *T. b. rhodesiense* because the two species use the same vector species. (From WHO report on global surveillance of epidemic-prone infections diseases- African trypanosomiasis. With permission from WHO.)

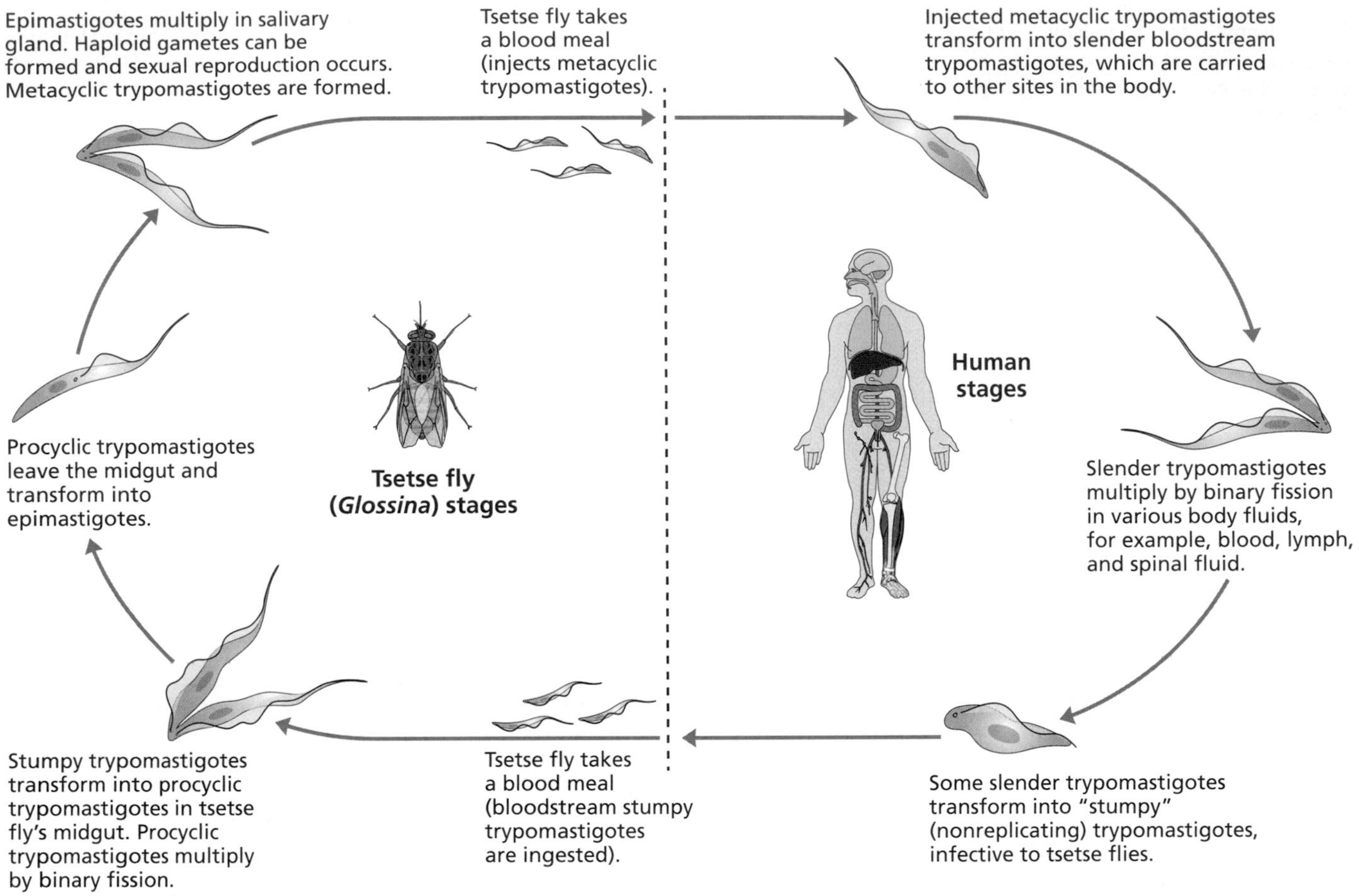

Epimastigotes multiply in salivary gland. Haploid gametes can be formed and sexual reproduction occurs. Metacyclic trypomastigotes are formed.

Tsetse fly takes a blood meal (injects metacyclic trypomastigotes).

Injected metacyclic trypomastigotes transform into slender bloodstream trypomastigotes, which are carried to other sites in the body.

Procyclic trypomastigotes leave the midgut and transform into epimastigotes.

Tsetse fly (*Glossina*) stages

Human stages

Slender trypomastigotes multiply by binary fission in various body fluids, for example, blood, lymph, and spinal fluid.

Stumpy trypomastigotes transform into procyclic trypomastigotes in tsetse fly's midgut. Procyclic trypomastigotes multiply by binary fission.

Tsetse fly takes a blood meal (bloodstream stumpy trypomastigotes are ingested).

Some slender trypomastigotes transform into "stumpy" (nonreplicating) trypomastigotes, infective to tsetse flies.

Figure 2 The life cycle of *T. brucei gambiense*. (Courtesy of CDC.)

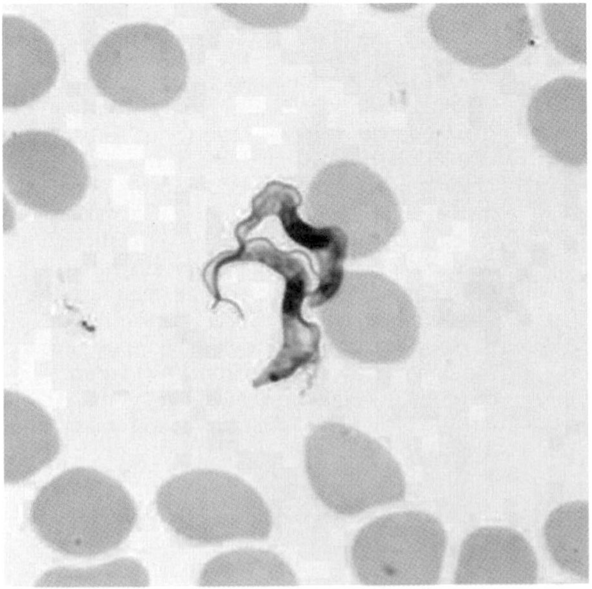

Figure 3 *Trypanosoma brucei rhodesiense* trypomastigotes in a blood smear of an infected human. Other members of the *T. brucei* species complex have an identical morphology. Typically only trypomastigotes are observed in the blood of infected individuals. (Courtesy of CDC/ Blaine Mathison.)

Trypanosoma evansi

Morphologically indistinguishable from *Trypanosoma brucei*, *Trypanosoma evansi* causes disease (called "surra" in Asia and "murrina" in South America) in many wild and domestic animals, including horses, camels, elephants, and dogs.

Distribution and prevalence *T. evansi* is widely distributed across northern Africa, central, south and Southeast Asia, and South America (Figure 1). Prevalence varies widely over this distribution.

Hosts and transmission *T. evansi* is thought to be originally a parasite of camels. It also is thought to have been introduced into the Americas by the Spanish in the sixteenth century in infected horses. Transmission is primarily mechanical via horse flies (genus *Tabanus*) (Figure 2). There is no development of the parasite while on the mouthparts of its arthropod vector.

Pathology Disease is most serious in dogs, horses, and elephants, but less serious in camels, where it generally causes a chronic condition. Cattle may remain asymptomatic. Symptoms of infection, such as severe anemia and weight loss, are similar to those for *T. brucei brucei*.

Diagnosis Infection can be demonstrated by the presence of trypanosomes in the blood (Figure 3). PCR and agglutination tests are also available.

Treatment Treatment is similar to that of nagana, relying on drugs such as isometamidium chloride, homidium bromide, and diminazene aceturate. In animals such as dogs and horses, untreated surra is almost 100% fatal.

Control Control principally focuses on management to reduce or eliminate the mechanical vectors. Throughout most of its distribution, little or no money is allocated to control efforts.

Did you know? In South America, vampire bats often serve as vectors for *T. evansi*.

Related species

T. congolense: This is the most common trypanosome of large mammals in South Africa. Its life cycle and pathogenesis are the same as *T. brucei*.

T. vivax: Similar to *T. brucei*, it is also found in South America, where transmission is mechanical via tabanid flies.

T. equiperdum: Transmitted sexually, it causes venereal disease in equines known as dourine.

References

Brun R, Hecker H & Lun ZR (1998) *Trypanosoma evansi* and *T. equiperdum*: distribution, biology, treatment and phylogenetic relationship (a review). *Vet Parasitol* **79**:95–107.

Desquesnes M, Dargantes A, Lai D et al (2013) *Trypanosoma evansi* and surra: a review and perspectives on transmission, epidemiology and control, impact and zoonotic aspects. *Biomed Research Int* Article ID 321237. http://dx.doi.org/10.1155/2013/321237.

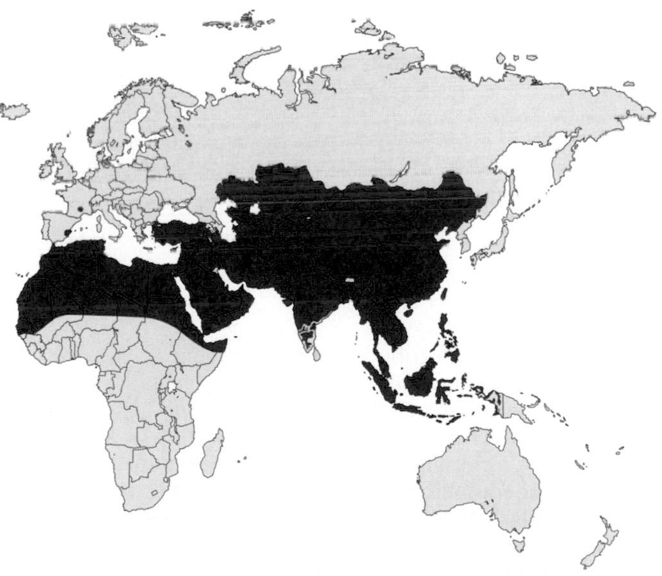

Figure 1 The distribution of *Trypanosoma evansi* in Africa and Asia. This parasite was also introduced into the Americas by the Spanish, in infected horses. (From Desquesnes M, Holzmuller P, Lai DH et al [2013] *Biomed Res Int* Article ID 194176.)

Figure 2 A horse fly. These biting flies, members of the family Tabanidae, serve as mechanical vectors of *Trypanosoma evansi*. Parasites adhere to the fly's mouthparts as it feeds. No parasite development occurs in association with its mechanical vector. (Courtesy of Dennis Ray, CC BY-SA 2.5.)

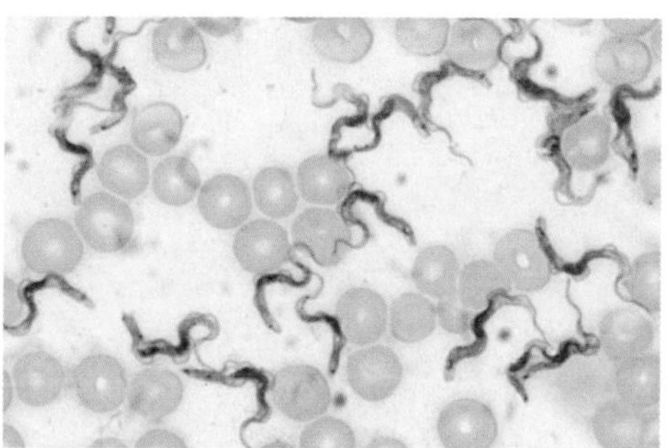

Figure 3 *Trypanosoma evansi* trypomastigotes. This blood smear was taken from a horse suffering from surra. The pale, round cells are erythrocytes. (Courtesy of Alan Walker, CC BY-SA 3.0.)

Trypanosoma cruzi

Trypanosoma cruzi is the causative agent of new world trypanosomiasis, which is also known as Chagas' disease.

Distribution and prevalence Between 7 and 8 million people in Mexico, Central America, and South America are believed to be infected, and many of those are asymptomatic (Figure 1). A number of infected mammals, as well as occasional human cases, have been observed in the southwestern and southeastern United States where 11 different species of kissing bugs occur.

Hosts and transmission Many domestic and wild mammals can become infected and serve as reservoirs (Figure 2). Those animals living in close association with humans, such as dogs and cats, are important sources of infection for humans. The opossum is a frequent wild reservoir. *T. cruzi* is transmitted by kissing bugs in the family Reduviidae (subfamily Triatominae), principally members of the genus *Triatoma* in the southern part of *T. cruzi* distribution and the genus *Rhodnius* in northern South America.

Pathology An acute phase occurs during the first weeks of infection. Mild symptoms include fever, body aches, and intestinal disorders. The most recognized indication is Romana's sign (Figure 3). The infection then enters a chronic phase, which is asymptomatic in between 60%–80% of infected individuals. Others develop life-threatening disorders, including cardiomyopathy. About two-thirds of those with chronic symptoms also experience a dilation of the digestive tract (megacolon and megaesophagus). Digestive problems can lead to malnutrition. Occasionally patients experience nervous system disorders. Most fatalities result from heart damage.

Diagnosis *T. cruzi* can be detected by microscopic examination of blood smears. Diagnosis is also possible with PCR. Various immunoassays distinguish among *T. cruzi* strains.

Treatment Chemotherapy, using benznidazole or nifurtimox, is most effective during the acute phase, during which parasites are eliminated in about 60%–85% of treated adults. Treatment becomes more problematic during the chronic phase, especially in adults, although benznidazole has been shown to slow the rate at which heart disease progresses.

Control Because the vectors are found in poor quality housing, control efforts focus on the elimination of thatched roof houses or other housing material in which reduviid bugs thrive. In endemic areas, blood is generally screened to prevent transfer during blood transfusion.

Did you know? Diagnosis is still occasionally made by xenodiagnosis, in which uninfected reduviid bugs are allowed to feed on a patient. The bugs are then examined for the presence of parasites in their guts.

References

Rassi A, Rassi A & Marin-Neto JA (2010). Chagas disease. *Lancet* **375(9723)**:1388–1402.

Kierzenbaum P (2005) Where do we stand on the autoimmunity hypothesis of Chagas disease? *Parasitol Today* **1**:4–6.

Figure 1 Distribution of *Trypanosoma cruzi* in the Americas , emphasizing areas where human infections are commonly found. The distribution largely reflects the distribution of the reduviid vectors. In the southern part of its distribution, the principal vectors are members of the genus *Triatoma*. In northern South America and in Central America, members of the genus *Rhodnius* are the most important *T. cruzi* vectors. Kissing bugs also commonly occur in the southern tier of United States and are infected with *T. cruzi* , but autochthonous human cases are rarely reported there.

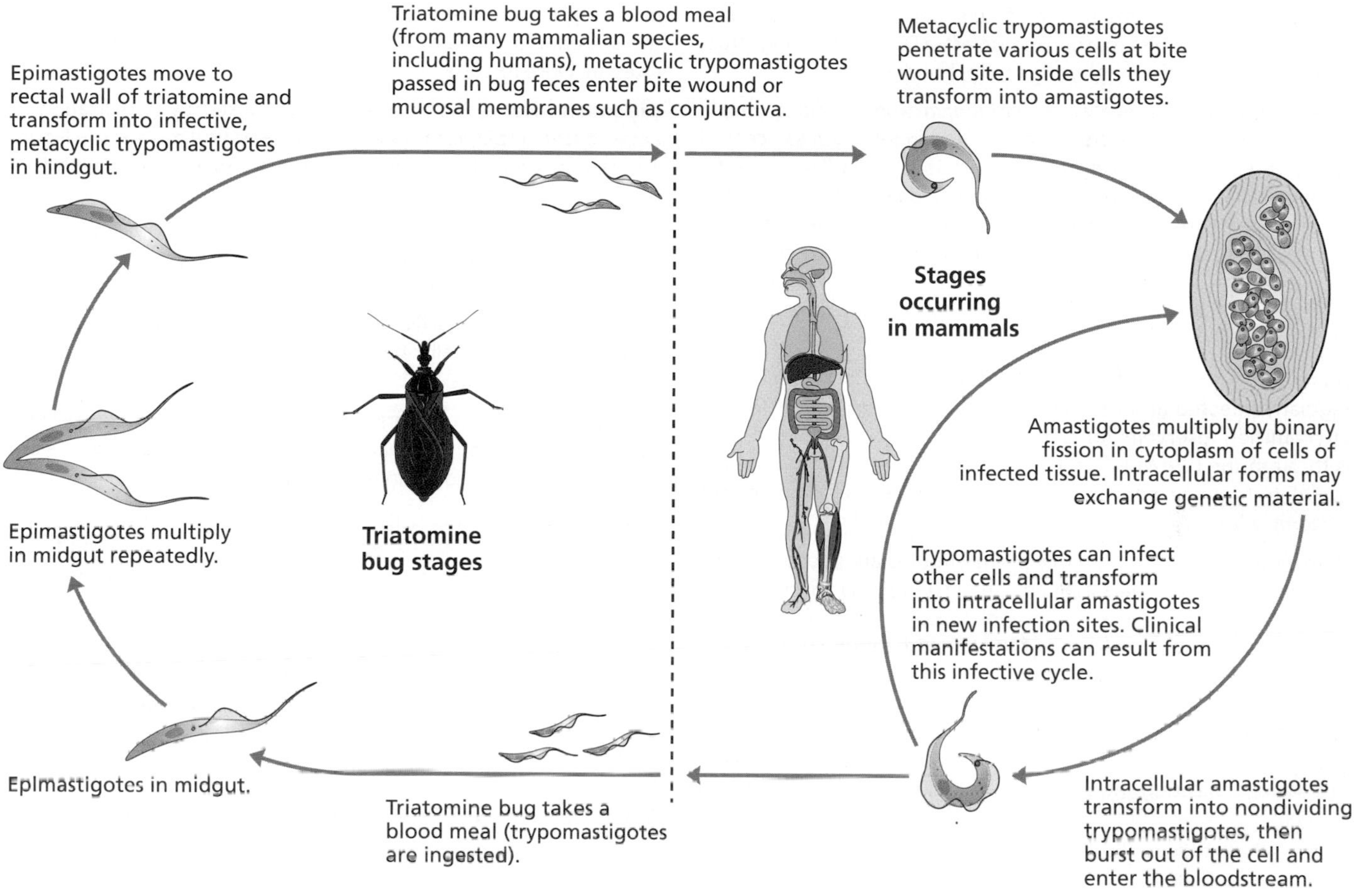

Epimastigotes move to rectal wall of triatomine and transform into infective, metacyclic trypomastigotes in hindgut.

Triatomine bug takes a blood meal (from many mammalian species, including humans), metacyclic trypomastigotes passed in bug feces enter bite wound or mucosal membranes such as conjunctiva.

Metacyclic trypomastigotes penetrate various cells at bite wound site. Inside cells they transform into amastigotes.

Stages occurring in mammals

Amastigotes multiply by binary fission in cytoplasm of cells of infected tissue. Intracellular forms may exchange genetic material.

Trypomastigotes can infect other cells and transform into intracellular amastigotes in new infection sites. Clinical manifestations can result from this infective cycle.

Intracellular amastigotes transform into nondividing trypomastigotes, then burst out of the cell and enter the bloodstream.

Triatomine bug takes a blood meal (trypomastigotes are ingested).

Epimastigotes in midgut.

Epimastigotes multiply in midgut repeatedly.

Triatomine bug stages

Figure 2 The life cycle of *Trypanosoma cruzi*. Mammals, Including humans, serve as the definitive host, while reduviid insects act as biological vectors and as intermediate hosts. (Courtesy of CDC.)

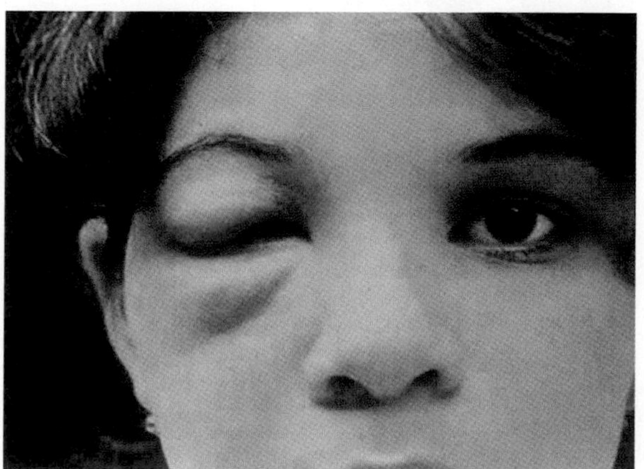

Figure 3 Romana's sign. The swollen eyelid known as Romana's sign is indicative of infection with *Trypanosoma cruzi*. The swelling is in response to feeding by the reduviid bug vector, which frequently feeds on the eyelids. Following feeding, the insect defecates, releasing infective metacyclic trypomastigotes in its feces. Irritation from the bite causes the person to scratch the eyelid, inoculating the parasite into the wound made by the feeding insect, or into the conjunctiva of the eye. Inflammation at the site causes the characteristic sign. (Courtesy of WHO/TDR.)

Genus *Leishmania*

Like trypanosomes, members of this genus are in the class Kinetoplastida, named for the kinetoplast, a DNA-containing structure found within the single, large mitochondrion.

Distribution and prevalence *Leishmania tropica* and *L. major*, causing cutaneous leishmaniasis, are found across central Africa, the Mideast, South America and South Asia (Figure 1). An estimated 0.7–1.3 million people are infected annually. *L. mexicana* causes cutaneous leishmaniasis in Central America and Mexico. Visceral leishmaniasis, caused by *L. donovani*, has an estimated incidence of 0.2–0.4 million cases annually. Most cases are in South Asia and East Africa. Overall, about 30 species of *Leishmania* are known.

Hosts and transmission Humans, dogs, and various rodents are regularly infected and can serve as reservoirs of infection. Gerbils are common reservoirs for *L. donovani*. Transmission to mammals is through the bite of infected sand flies (Figure 2) in the subfamily Phlebotominae: genus *Phlebotomus* (Africa and Asia) and genus *Lutzomyia* (Americas).

Pathology Infective metacyclic promastigotes are injected as the vector feeds. These parasites are phagocytized by macrophages and transform into amastigotes, which multiply in infected cells. The tissue affected is primarily the skin and submucosa in cutaneous leishmaniasis, which results in characteristic skin ulcerations (Figure 3). Such lesions may heal after several months, although secondary infections are common. Symptoms of visceral leishmaniasis (kala-azar) can be life threatening. Severe disease occurs as replicating intracellular amastigotes kill large numbers of phagocytes, causing the spleen and bone marrow to undergo increased phagocyte production. Erythrocyte production declines in consequence, resulting in severe anemia and weight loss, and the spleen and liver become grossly enlarged. Untreated fatality rates are close to 100%.

Diagnosis Visceral leishmaniasis is diagnosed by visualization of amastigotes in splenic aspirates or by serological means (see Figure 10.11). Cutaneous leishmaniasis requires a biopsy often with PCR from the ulcer.

Treatment Cutaneous leishmaniasis may not require treatment, as it often heals on its own, leaving only unpigmented scars. Visceral leishmaniasis has generally been treated with injected antimony-based drugs. Recent use of liposomal amphotericin B or miltefosine, taken orally, has shown efficacy in the treatment of visceral leishmaniasis.

Control Based on an integrated approach, combining early diagnosis and treatment, residual insecticide spraying to control sand flies, and the control of rodent reservoirs.

Did you know? People in the Mideast historically inoculated themselves with material from active cutaneous lesions onto normally unexposed body parts such as the buttocks. If infected later in life, immunity developed to the first inoculation often prevented new lesions from forming on exposed regions such as the face.

Related species

L. braziliensis: This species of *Leishmania* causes a mucocutaneous form of the disease in South America.

Reference

Santos DO, Coutinho CER, Madeira MF et al (2008) *Leishmaniasis* treatment—a challenge that remains: a review. *Parasitol Res* **103**:1–10.

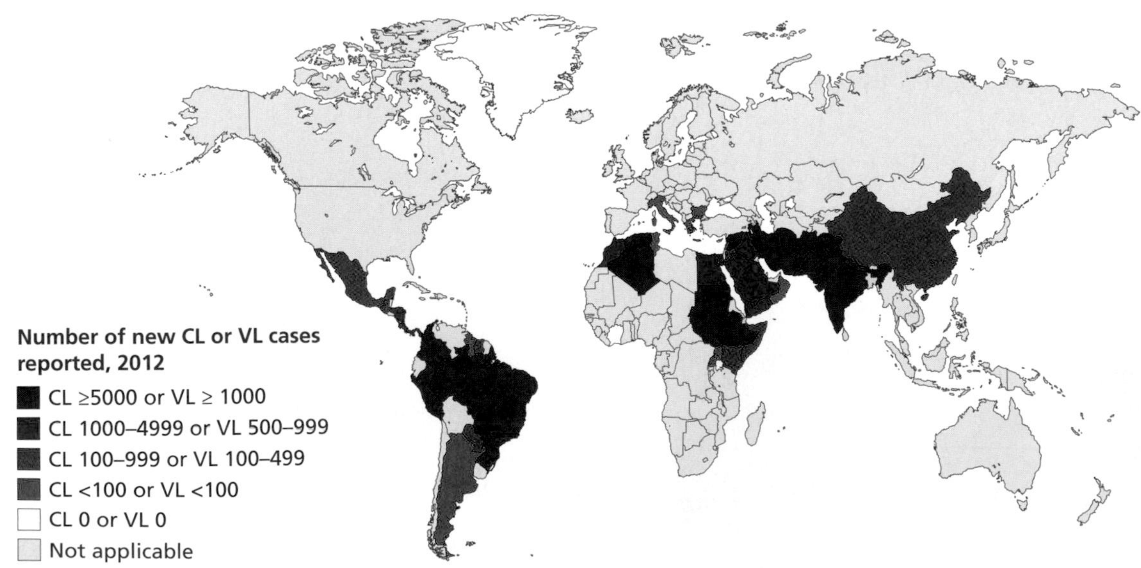

Figure 1 Distribution of cutaneous and visceral leishmaniasis. Leishmaniasis is endemic in 88 countries across four continents. The number of reported cases for both cutaneous and visceral leishmaniasis in 2012 is shown. (Adapted from WHO Leshmaniasis control programme. Annual country reports, 2012. With permission from WHO.)

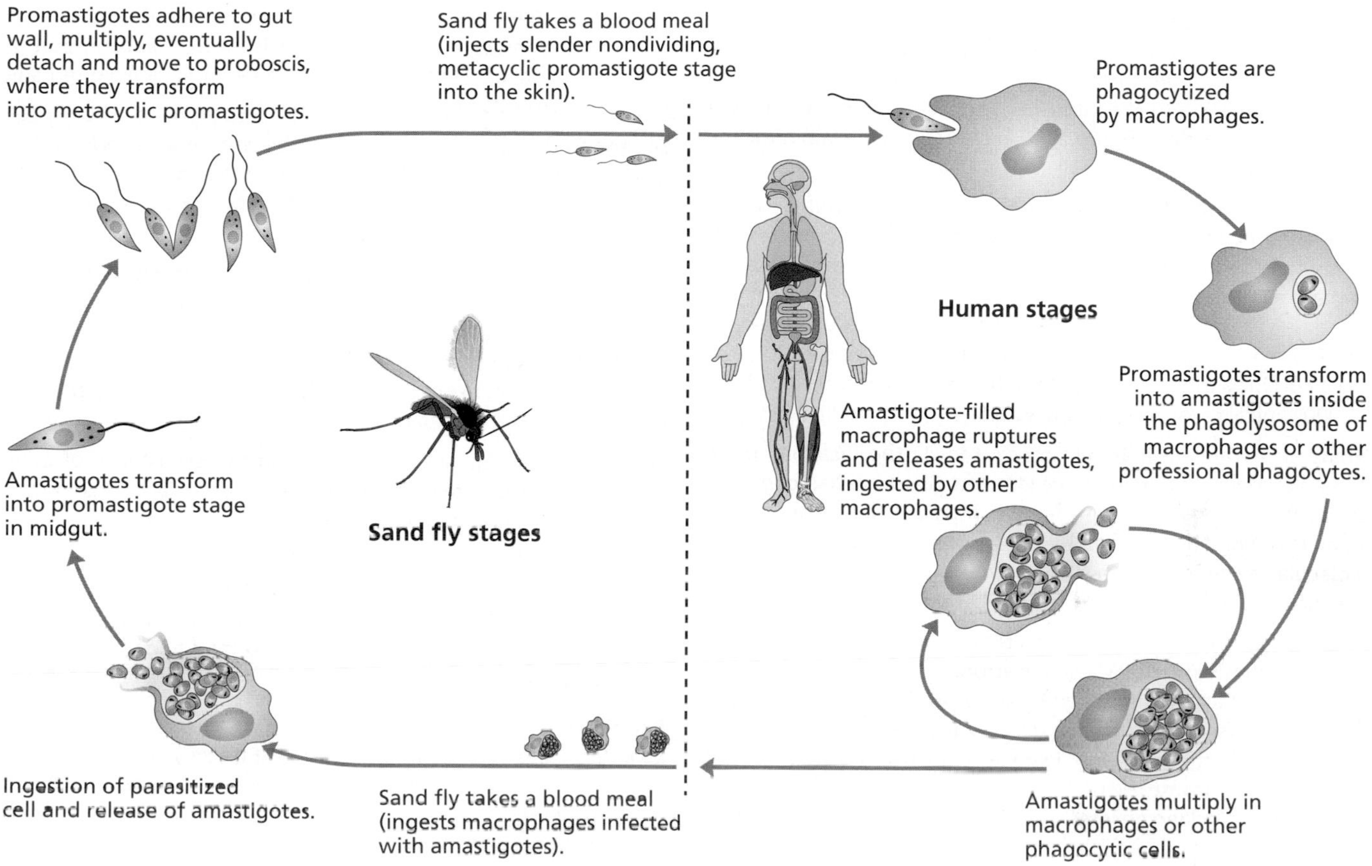

Promastigotes adhere to gut wall, multiply, eventually detach and move to proboscis, where they transform into metacyclic promastigotes.

Sand fly takes a blood meal (injects slender nondividing, metacyclic promastigote stage into the skin).

Promastigotes are phagocytized by macrophages.

Human stages

Promastigotes transform into amastigotes inside the phagolysosome of macrophages or other professional phagocytes.

Amastigote-filled macrophage ruptures and releases amastigotes, ingested by other macrophages.

Amastigotes transform into promastigote stage in midgut.

Sand fly stages

Ingestion of parasitized cell and release of amastigotes.

Sand fly takes a blood meal (ingests macrophages infected with amastigotes).

Amastigotes multiply in macrophages or other phagocytic cells.

Figure 2 The life cycle of *Leishmania*. (Courtesy of CDC.)

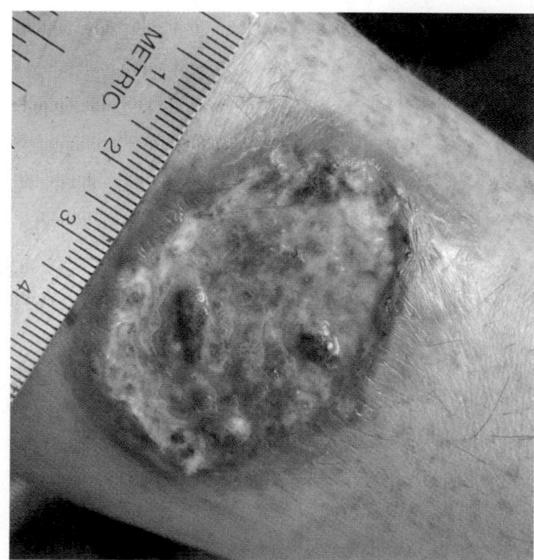

Figure 3 Cutaneous leishmaniasis. A typical lesion on the arm of an infected individual. Cutaneous leishmaniasis is the most common manifestation of leishmaniasis, often resolving on its own. In certain cases, however, cutaneous leishmaniasis can progress to a more diffuse cutaneous form or to visceral or mucocutaneous forms of the disease.

Giardia lamblia

Giardia lamblia (also known as *G. duodenalis* or *G. intestinalis*) belongs to the Excavata lineage, within the phylum Metamonada. It inhabits and reproduces in the upper portion of the small intestine, adhering to the intestinal epithelium with a prominent ventral adhesive disc (Figure 1). *G. lamblia* is responsible for the disease giardiasis.

Distribution and prevalence Although it is distributed worldwide, *G. lamblia* Is generally more common in developing countries where sanitation is poor. In developed countries, children in daycare centers, residents of nursing homes, and hikers are at elevated risk. Prevalence is hard to specify but about 2% of adults and 6%–8% of children in developing countries are positive. It is the most common intestinal parasite of humans in the US, and about 20,000 clinical cases per year are noted.

Hosts and transmission Transmission is via the fecal-oral route (Figure 2). Many mammals can be infected. Giardiasis has long been considered to be a zoonotic infection, but such transmission from animals to humans has recently been called into question. Recent molecular evidence suggests that some strains infecting humans may be largely distinct from those infecting other mammals.

Pathology Adherence of trophozoites to the intestinal epithelium via their adhesive disc may result in atrophy of the villi and reduced intestinal surface area for nutrient absorption. Infection may also cause a hypersecretion of chloride ions and water, which, in combination with malabsorption, leads to diarrhea. The trophozoites themselves aggravate the problem by covering portions of the epithelium and blocking absorption. Infection is often asymptomatic. Clinical giardiasis ranges from acute, often self-resolving diarrhea to chronic symptoms that also include nutritional disorders and weight loss. Besides diarrhea, other common symptoms include abdominal cramps, nausea, greasy stools (from malabsorption of fats), and anorexia. Infected, asymptomatic individuals often still pass infective cysts in their feces (Figure 3).

Diagnosis Detection of antigens on the surface of *Giardia* organisms from stool specimens is the most sensitive method of diagnosis. Examination of fecal samples on multiple days may be necessary to detect trophozoites or cysts.

Treatment Metronidazole is an effective treatment, but it cannot be used by pregnant women because it can increase the risk of birth defects. Tinidazole is equally effective and has fewer side effects. Paromomycin, a broad-spectrum anti-protozoan and anthelmintic protein synthesis inhibitor is also effective, and it is recommended for pregnant woman.

Control Its widespread distribution and the persistence of cysts in the environment make control difficult, especially in countries with inadequate sanitation. Most efforts focus on improved health education, sanitation, personal hygiene, avoidance, and boiling or filtering of drinking water potentially contaminated with cysts.

Did you know? Because giardiasis is sometimes confused with celiac disease, a duodenal biopsy is performed to obtain a differential diagnosis.

References

Monis PT, Caccio SM & Thompson RC et al (2009) Variation in *Giardia*: towards a taxonomic revision of the genus. *Trends Parasitol* **25**:93–100.

Buret AG (2008) Pathophysiology of enteric infections with *Giardia duodenalis*. *Parasite* **15**:75–80.

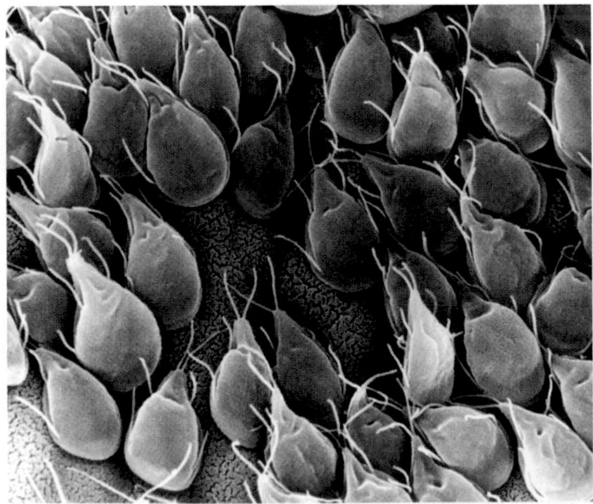

Figure 1 *Giardia lamblia* trophozoites. A dense mass of *G. lamblia* trophozoites adhering to the epithelium in the duodenum of their host. (Courtesy of CDC.)

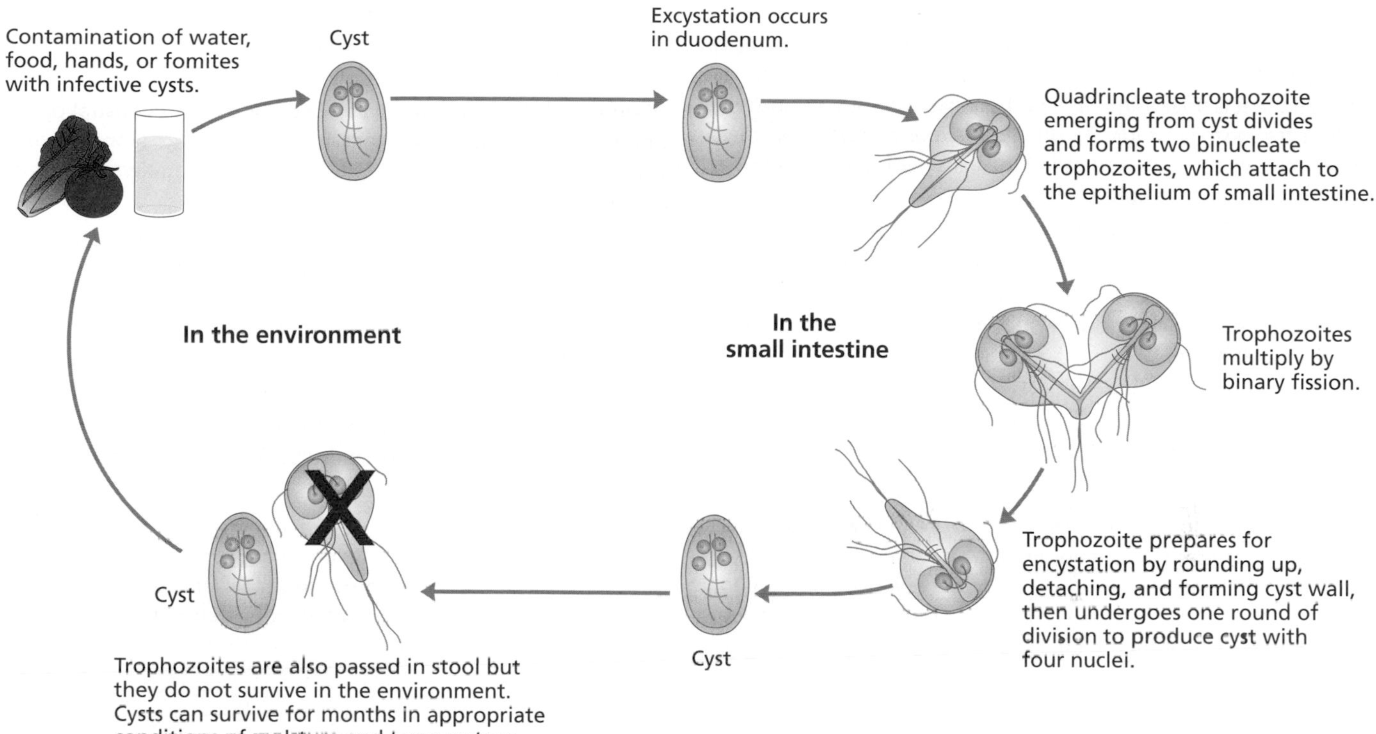

Figure 2 The life cycle of *Giardia lamblia*. Transmission is via the fecal–oral route. The highly resistant cyst is the infective stage, although pathology is generally associated with the feeding and growing trophozoites. Reproduction within the host is via binary fission. Ironically, asymptomatic carriers are more likely than patients showing symptoms to be shedding cysts in their feces and are therefore more likely to serve as a source of infection to other, uninfected, individuals. Those suffering from acute giardiasis are more likely to be passing noninfective trophozoites in their feces. Physiological cues in the host intestine determine whether a trophozoite continues to divide into additional trophozoites or undergoes cyst formation. (Courtesy of CDC.)

Figure 3 A *Giardia* cyst. This highly resistant infective stage can persist in the environment for months. (Courtesy of Joel Mills, CC BY-SA 3.0)

Trichomonas vaginalis

Trichomonas vaginalis is the causative agent of trichomoniasis. It is in the class Parabasalia, within the Excavata lineage. It is one of two members of the *Trichomonas* genus that infects humans and the only one that is unambiguously pathogenic. The life cycle includes only a trophozoite stage (Figure 1). Asexual reproduction is via binary fission. A central rod called the axostyle is thought to function in attachment to epithelial cells.

Distribution and prevalence *T. vaginalis* is distributed throughout the world and is the most common nonviral sexually transmitted pathogen in the world. It is estimated that there are 167 million new cases each year worldwide and up to 7.4 million cases in the United States alone.

Hosts and transmission Trichomoniasis is primarily an infection of the urogenital tract. Transmission is via sexual or genital contact. Humans are the only reservoir for *T. vaginalis*.

Pathology Most pathology is believed to result from irritation of the epithelium by the parasite. Secreted parasite proteases may also contribute to symptomology. Women are more likely to exhibit symptoms than men, and the severity of symptoms appears to reflect the number of parasites present. The most common symptom in women is inflammation of the vagina (vaginitis), which may be accompanied by a discharge and a burning or itching sensation. Most men are asymptomatic, but some experience urethral discharge or painful urination (dysuria).

Diagnosis Infections are diagnosed by the demonstration of trophozoites in vaginal, urethral, or prostatic secretions (Figures 2 and 3). Because such direct visualization is not always reliable, *in vitro* culture is often used to confirm diagnosis.

Treatment Trichomoniasis is usually highly treatable as long as sexual partners are treated simultaneously to prevent re-infection. Metronidazole and tinidazole are effective and are usually well tolerated. Fewer than 5% of all cases are resistant to some degree.

Control Control measures used for other sexually transmitted diseases, such as the limitation of the number of sexual partners and the use of condoms, are appropriate for *T. vaginalis* as well. Health education can alter behavior and therefore reduce the likelihood of infection.

Did you know? Although not life threatening itself, infection with *T. vaginalis* is associated with other conditions, such as an increased risk of cervical or prostate cancer and of HIV transmission. Infected women who are pregnant often deliver premature or low-birth-weight babies.

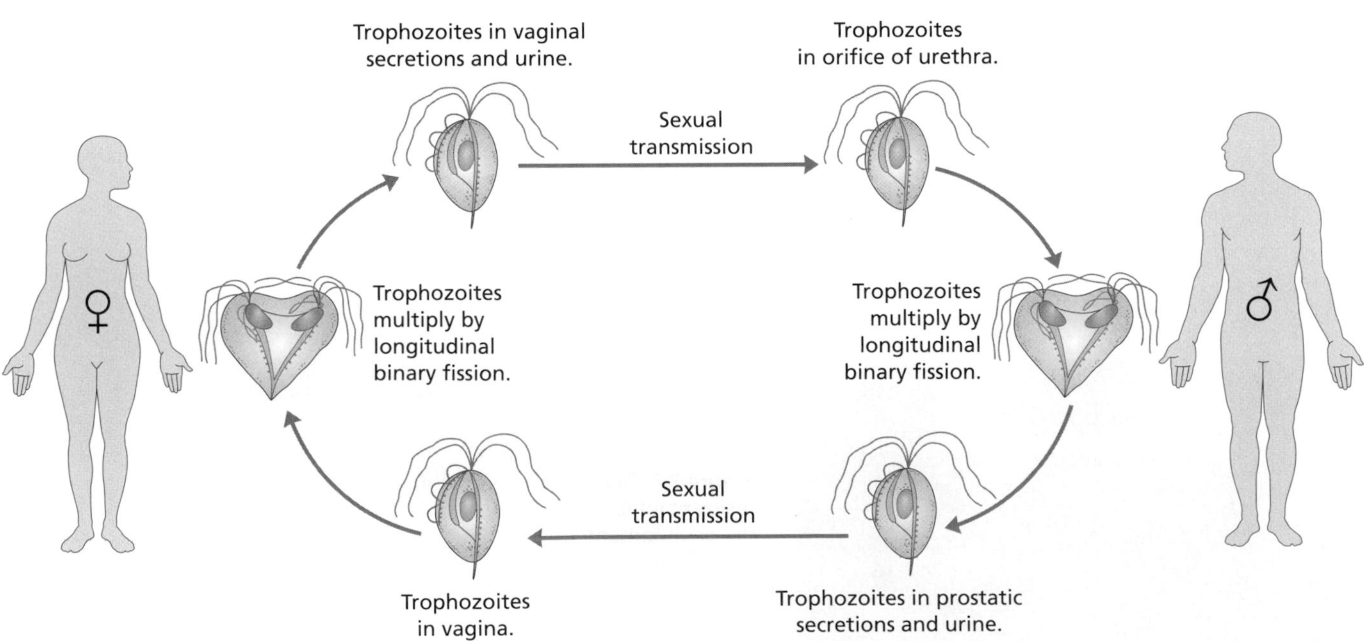

Figure 1 The life cycle of *T. vaginalis*. Unlike other protozoa, such as *Giardia*, in which infective stages are exposed to the external environment, *T. vaginalis* is transmitted sexually and has no cyst stage. (Courtesy of CDC.)

Related species

T. tenax: This species is a commensal of the human mouth, and is especially common in those with poor oral hygiene.

Pentatrichomonas hominis: *P. hominis* is a nonpathogenic species that inhabits the human colon.

Tritrichomonas foetus: *T. foetus* is a serious, sexually transmitted pathogen of cows and other bovines that can result in spontaneous abortion.

Histomonas meleagridis: Infects many birds, including a wide range of poultry. It inhabits the cecum where it can cause extensive tissue necrosis. Transmitted by the nematode *Heterakis gallinarum*.

References

Johnston VJ & Mabey DC (2008) Global epidemiology and control of *Trichomonas vaginalis*. *Curr Opin Infect Dis* **21**:56–64.

Hoots BE, Peterman TA, Torrone EA et al (2013) A trich-y question: should *Trichomonas vaginalis* be reportable? *Sexually Transmitted Diseases* **40**: 113–116.

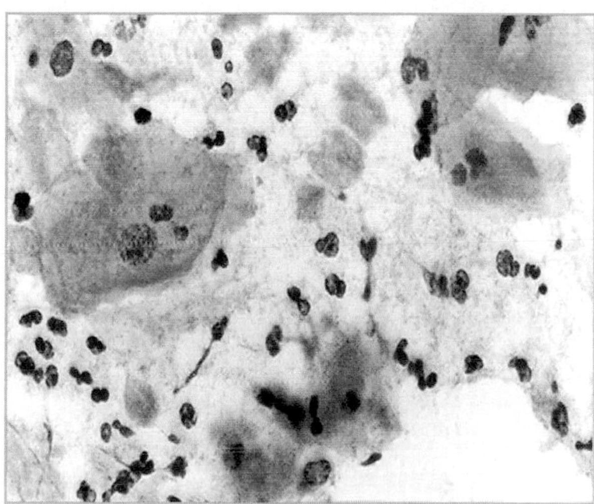

Figure 2 Diagnosis of *Trichomonas vaginalis* infection. A PAP smear showing infection with *T. vaginalis* trophozoites (dark purple). (Courtesy of Alex Brollo, CC BY-SA 3.0.)

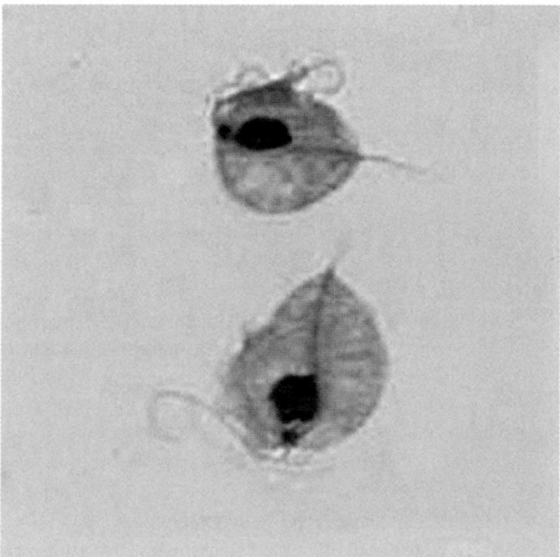

Figure 3 *Trichomonas vaginalis* trophozoites. The axostyle, which is composed of microtubules, is seen as the slender, dark line running down the cell's midsection. It is believed to function in attachment to the host epithelium. Note the trophozoites also have flagella and an undulating membrane for locomotion. (Courtesy of CDC.)

Entamoeba histolytica

Entamoeba histolytica is the agent of amebic dysentery. The organisms are anaerobic and lack Golgi bodies. They posess mitosomes, which are probably highly reduced mitochondria.

Distribution and prevalence *E. histolytica* is distributed worldwide, with most cases occurring in tropical and subtropical developing countries where good sanitation is often lacking. The estimated prevalence is 50 million worldwide. It was once believed that prevalence was as high as 0.5 billion, prior to the recognition that most of these infections were caused by the morphologically identical but nonpathogenic *E. dispar*.

Hosts and transmission Humans and other primates serve as hosts. Transmission is via the fecal–oral route (Figure 1), explaining the association of amebic dysentery with poor sanitation.

Pathology Infected individuals often remain asymptomatic or display mild symptoms. In approximately 10% of patients, amebas enter the intestinal epithelium resulting in an invasive infection (Figure 2). Damage to host cells by reproducing trophozoites (Figure 3) causes tissue necrosis and bloody stools. Ulceration may progress until amebas penetrate the serous membrane and muscle, resulting in a perforated bowel and peritonitis. Secondary bacterial infections are common. Entry into the circulatory system results in dissemination to other organs, such as the liver, causing additional foci of tissue necrosis. The reasons for initial invasion of the intestinal mucosa remain unclear but are thought to involve both parasite and host factors (see Chapter 5, pages 165–166 for a discussion).

Diagnosis Illness is generally diagnosed by the demonstration of amebas in fecal samples. *E. histolytica* is easily confused with nonpathogenic *E. dispar*, although PCR and monoclonal antibodies can be used to distinguish between these two species.

Treatment Several effective drugs are available to treat infections. Metronidazole is usually the drug of first choice because of its relatively low toxicity and effectiveness against both intestinal and invasive trophozoites.

Control Infections are controlled principally through improved sanitation. The treatment of asymptomatic infected individuals acting as reservoirs can reduce parasite transmission.

Did you know? In HIV+ patients also infected with *E. histolytica*, amebas sometimes consume virally infected cells. The virus remains viable within the ameba, but there is no evidence that amebas can serve as a mode of transmission for HIV.

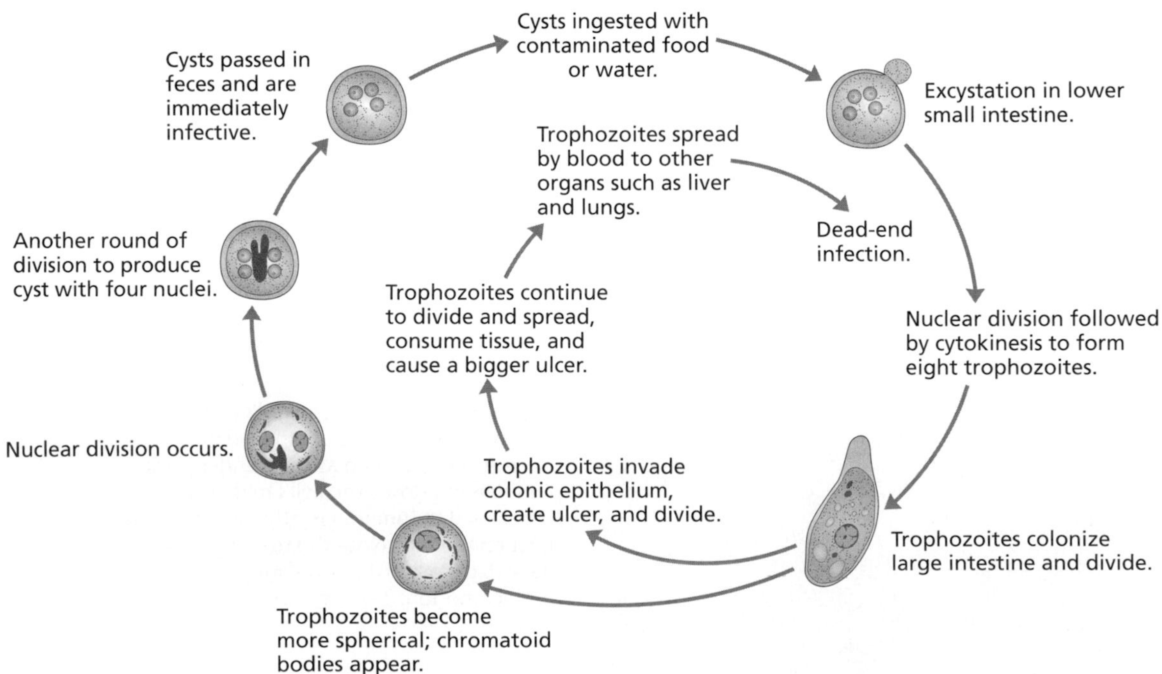

Figure 1 The life cycle of *Entamoeba histolytica*. Asymptomatic infected individuals shed primarily infective cysts in their feces, whereas symptomatic patients are more likely to shed non-infective trophozoites. Thus asymptomatic individuals acting as reservoirs play a disproportionate role in the epidemiology of amebiasis. (Courtesy of CDC.)

Related species

Entamoeba coli: This species is a nonpathogenic commensal, similar in appearance to *E. histolytica*.

E. gingivalis: This species, which lacks the cyst stage, is a commensal ameba of the mouth.

E. polecki: E. polecki generally infects pigs and monkeys and rarely causes human infections.

Endolimax nana: This species, which feeds on bacteria, is a human commensal of the colon.

Iodamoeba büetschlii: Although mainly a commensal of pigs, *I. buetschlii* may also infect humans.

References

Brown M, Reed S, Levy JA et al (1991). Detection of HIV-1 in *Entamoeba histolytica* without evidence of transmission to human cells. *AIDS* **5(1)**:93–96.

Petri WA Jr & Hague R (2014) *Entamoeba* species, including amebic colitis and liver abscess. In Mandell, Douglas, and Bennett's Principles and Practice of Infectious Diseases (Bennett JE, Dolin R, Mandell GL, eds.). 8th ed. Elsevier Saunders.

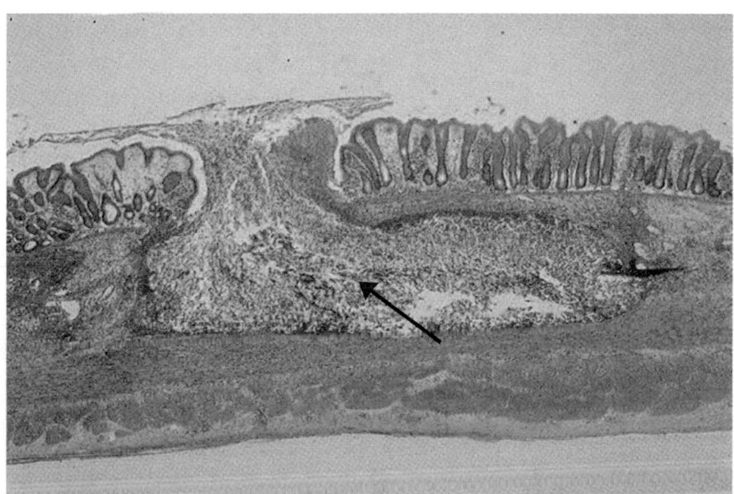

Figure 2 Pathology caused by *Entamoeba histolytica*. As amebas become invasive, they feed upon and cause tissue necrosis in the intestinal epithelium. A flask-shaped lesion of necrotic tissue beneath the epithelium is typical. The lumenal surface of the gut is at the top of the figure, the outer surface of the gut at the bottom. The arrow indicates the center of the lesion. (Courtesy of CDC/Mae Melvin.)

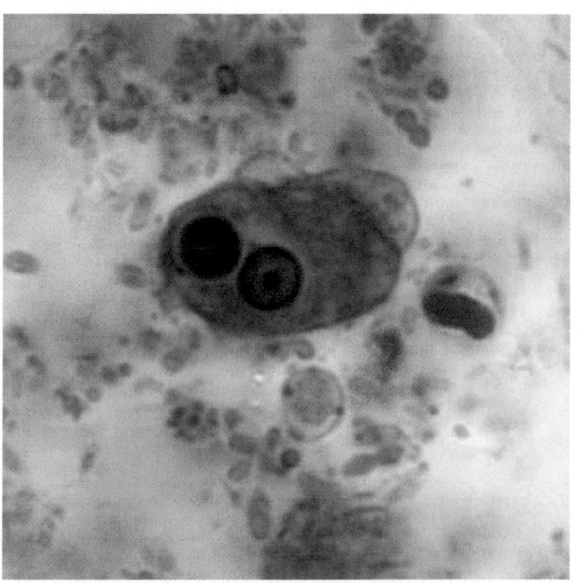

Figure 3 An *Entamoeba histolytica* trophozoite. Cysts are typically 10–20 microns across. Trophozoites usually are 20–30 μm across. The darkly stained inclusion is an ingested erythrocyte. The nucleus shows the typical thin layer of peripheral chromatin and a centrally placed endosome. Note the blunt pseudopod in the upper right of the organism. (Courtesy of CDC.)

Genus *Cryptosporidium*

The Apicomplexan genus *Cryptosporidium* consists of intracellular parasites of the intestinal epithelia that are found in many vertebrates. The taxonomy of the genus is unresolved, and the most widespread species, *C. parvum*, is probably a cluster of closely related species which, although morphologically identical, tend to infect different hosts. Members of the genus cause the diarrheal disease cryptosporidiosis.

Distribution and prevalence *Cryptosporidium* species are found worldwide, although they are more commonly observed in developing countries because of increased opportunities for transmission. There are an estimated 750,000 cases of cryptosporidiosis each year in the United States.

Hosts and transmission Different *Cryptosporidium* species infect a wide range of vertebrates. In humans, over 90% of infections are caused by *C. parvum*, which also infects cattle, and *C. hominis*, which is typically restricted to humans. The life cycle is direct. Transmission occurs via the fecal-oral route (Figure 1).

Pathology Mild to severe watery diarrhea is the most common symptom. Symptoms may last from several days up to four weeks. Many cases resolve on their own, but relapses are common. In immunocompromised patients, symptoms are far more severe, long-lasting, and potentially life threatening. Before the advent of modern chemotherapy for HIV+ patients, *Cryptosporidium* species caused frequent opportunistic infections and were a common cause of death. The underlying reasons for the diarrhea are unclear, but most likely it arises from inflammation and disruption of salt–water balance of the intestinal epithelium, as the parasite replicates in epithelial cells (Figure 2).

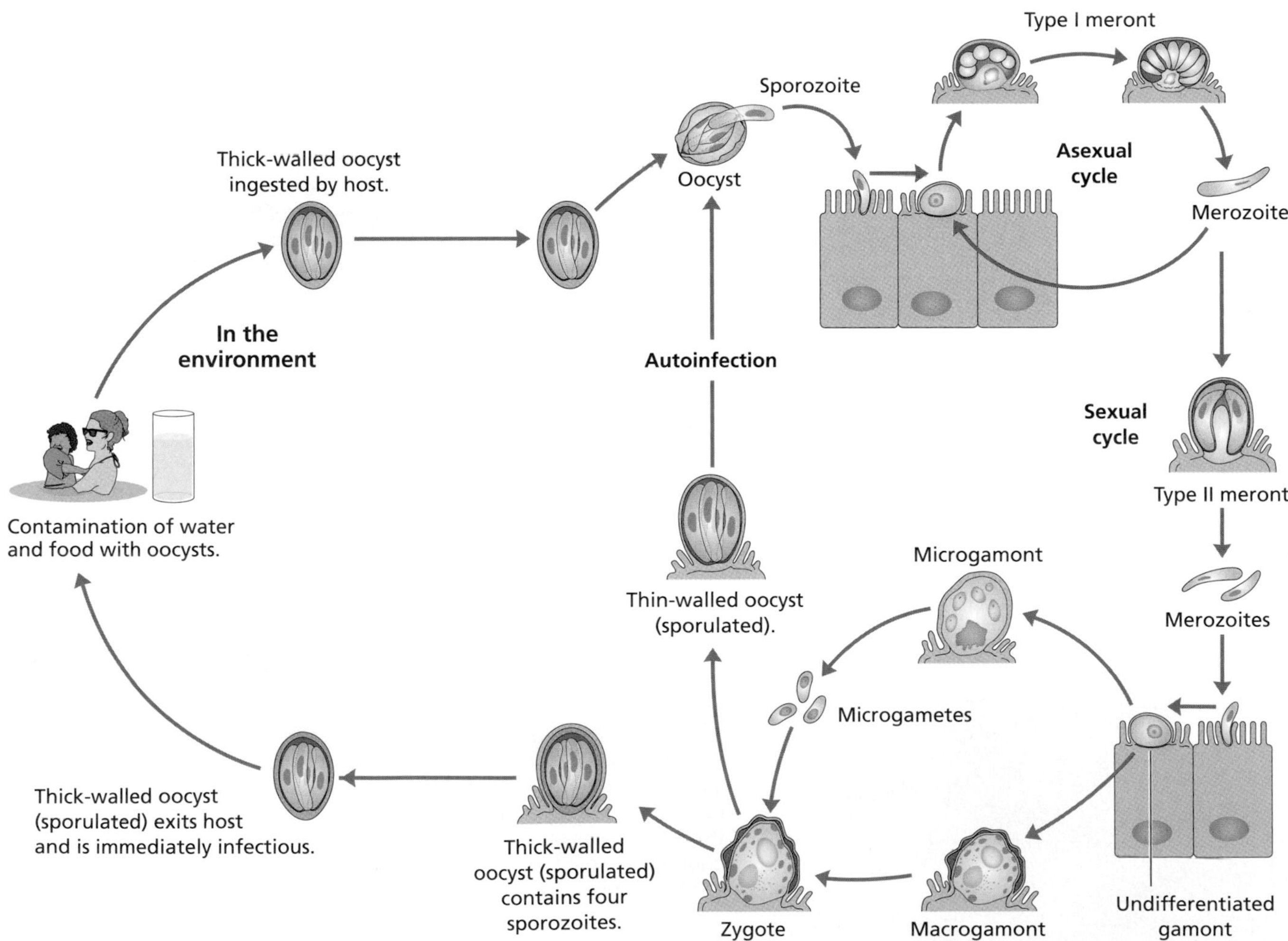

Figure 1 The life cycle of *Cryptosporidium*. Sporulated oocysts are shed in the feces. Transmission is via the fecal–oral route. Following ingestion, excystation occurs. Released sporozoites invade the gastrointestinal epithelial cells, where asexual reproduction (schizogony, also called merogony) occurs. This is followed by sexual reproduction (gametogony), producing microgamonts (male) and macrogamonts (female). Following fertilization, oocysts are produced. Oocysts may be thick-walled, which are commonly passed in the feces, or thin-walled, which are mainly involved in autoinfection. Since oocysts are sporulated, they are immediately infective for the next host. Millions of oocysts may be released in a single bowel movement. (Courtesy of CDC.)

Diagnosis Cryptosporidiosis is generally diagnosed based on oocyst detection in acid-fast stained stool samples (Figure 3). Examination of samples collected over several days may be required.

Treatment There is no uniformly effective anti-*Cryptosporidium* drug. Most treatment consists of amelioration of symptoms. Nitazoxanide may have some clinical value and can be used to treat immunocompetent patients. Its effectiveness for immunosuppressed patients is unclear.

Control Control is similar to that for other diseases transmitted via the fecal–oral route. Improved sanitation, protection of the water supply, and proper water treatment all reduce the likelihood of infection. Oocysts are resistant to normal chlorination levels in drinking water and swimming pools.

Did you know? In 1994 approximately 400,000 people in Milwaukee and surrounding areas developed symptomatic cryptosporidiosis. Water from Lake Michigan was the likely source of infective oocysts. Molecular analysis showed that the involved species was *C. hominis*, suggesting that a breakdown in sewage treatment was the immediate cause of the epidemic.

Related organisms

Eimeria tenella: This is the most important eimerian, which parasitizes chickens and causes the serious condition cecal coccidiosis. There are thousands of other species within this genus.

Cyclospora cayetanensis: This organism was first recognized as a human parasite in the 1990s. It causes diarrhea when infection occurs, via the consumption of contaminated produce.

References

Abrahamsen MS, Templeton TJ, Enomoto S et al (2004). Complete genome sequence of the apicomplexan, *Cryptosporidium parvum*. *Science* **304(5669)**:441–445.

Davies AP & Chalmers RM (2009) Cryptosporidiosis. *BMJ* **339**:963–967.

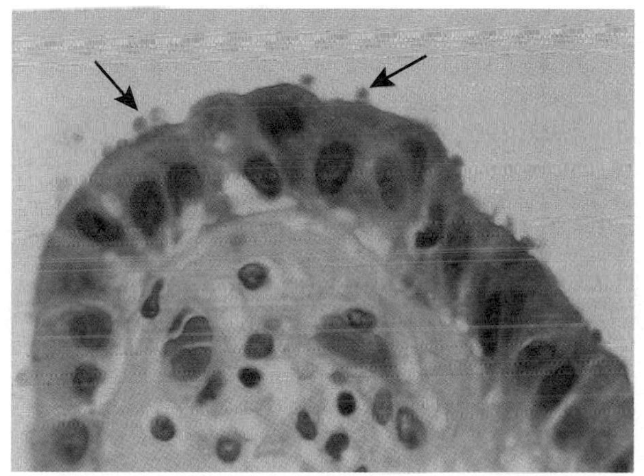

Figure 2 *Cryptosporidium* merozoites. Two merozoites, seen as tiny red spots on the surface of intestinal cells, are indicated by the arrows. The parasites attach to the surface of the epithelial membrane and subsequently become intracellular as the microvilli fuse around them. Merogony takes place immediately beneath the plasma membrane. Mature merozoites are then released into the gut lumen, where they may infect new cells. (Courtesy of HAD Lindquist, US EPA.)

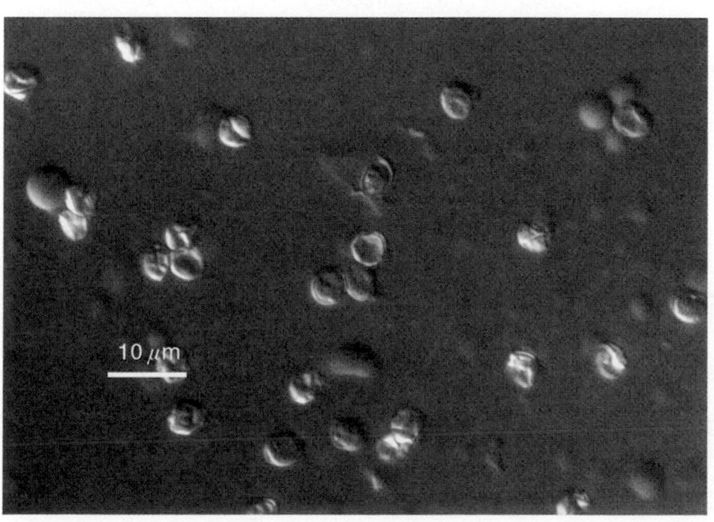

Figure 3 *Cryptosporidium parvum* oocysts recovered from mouse feces. The tiny oocysts are the sphere-shaped objects. Each contains four banana-shaped sporozoites, all of which cannot be seen in one plane of focus. (Courtesy of CDC.)

Toxoplasma gondii

Toxoplasma gondii is an apicomplexan parasite, within the class Conoidasida and the subclass Coccidia. It infects many mammalian species, including humans, as well as birds. It is the causative agent of toxoplasmosis.

Distribution and prevalence Although it tends to be more common in the tropics, *T. gondii* has a worldwide distribution, with the exception of very dry or cold climates. Prevalence in humans is high but varies with geographic location. In the United States, prevalence is estimated to be between 15% to 40%. In some parts of Latin America and Europe prevalence approaches 80%.

Hosts and transmission Toxoplasmosis is transmitted trophically from intermediate hosts (small mammals and birds) to the definitive host (domestic cats and other felids). Intermediate hosts become infected by consuming infective sporulated oocysts that are shed in the cat's feces. Large animals such as humans may consume oocysts and are usually dead-end hosts, unless they happen to be consumed by large felid species. Humans and other mammals can also become infected by consuming raw or undercooked meat, which harbors the cyst stages (containing bradyzoites) in other intermediate hosts. The complex life cycle consists of intestinal, tIssue, and external phases (Figure 1). Human fetuses can become infected via vertical transmission.

Pathology Symptoms of acute infection are generally lacking or mild in immunocompetent hosts. In immunocompromised

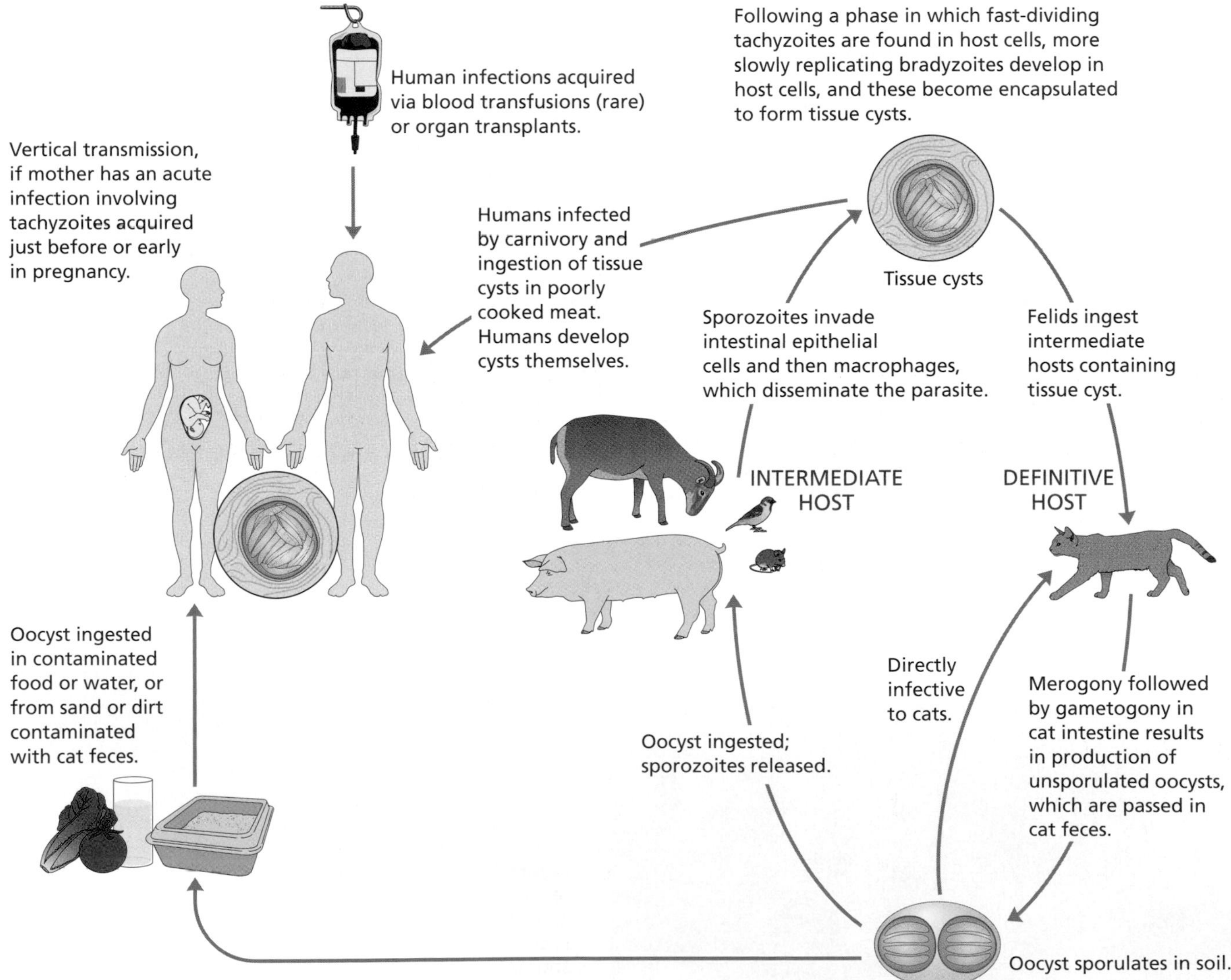

Figure 1 The life cycle of *Toxoplasma gondii*. Unsporulated oocysts are shed in an infected cat's feces. Sporulation takes from one to several days. Intermediate hosts become infected after consuming sporulated oocysts. Ingested oocysts release sporozoites, which invade epithelial cells and reproduce asexually. These rapidly reproducing cells are called tachyzoites.Tachyzoites localize in neural and muscle tissue and develop into tissue cysts containing bradyzoites. Cats become infected after consuming intermediate hosts harboring tissue cysts. Humans can become infected through the consumption of undercooked meat containing tissue cysts, or by inadvertently consuming sporulated oocysts in the environment. Transmission can also occur via blood transfusion, or vertically, from mother to fetus. (Courtesy of CDC.)

individuals, pathology can be severe, resulting in necrotic areas in the brain (Figure 2) as tissue cysts in the brain rupture, releasing bradyzoites that develop into tachyzoites and invade neighboring cells (Figure 3). Symptoms can include seizures, speech and coordination disorders, and blurred vision. Because the parasite can be transmitted vertically, it poses a special risk to the fetus if the pregnant woman is infected for the first time during pregnancy. Birth defects attributed to congenital toxoplasmosis include mental retardation, visual and auditory defects, and microcephaly. If a woman has become infected prior to pregnancy, there is little risk to the fetus because immunity developed during the initial infection will block vertical transmission during subsequent infections.

Diagnosis Toxoplasmosis is diagnosed using serological tests to detect anti-*Toxoplasma* antibodies.

Treatment Toxoplasmosis is usually untreated in immunocompetent individuals. Pregnant women with a primary infection can be treated with spiramycin to prevent vertical transmission. Immediate treatment is crucial in the immunocompromised. Several drugs, including pyrimethamine, clindamycin, and trimethoprim combined with a sulfa drug are effective, although relapses are common. Prophylaxis for immunocompromised patients is indicated until the immunosupression resolves.

Control Control is focused on avoidance of sources of infection, especially cat feces and undercooked meat. Such avoidance is especially necessary in previously uninfected pregnant women.

Did you know? Since sporulation of oocysts in cat feces requires a minimum of 24 hours to occur, daily litter box cleaning greatly reduces the risk of infection for cat owners.

Related organisms

Sarcocystis: *Sarcosystis* species have a similar life cycle to *T. gondii*. Humans are rarely infected, but they can serve as either the definitive or intermediate host, depending on the source of infection.

Neospora caninum causes neurological disease in dogs, which serve as definitive hosts. This organism is a major cause of spontaneous abortion in cattle, which may become infected if they consume sporulated oocysts that are passed in an infected dog's feces.

References

Boothroyd JC (2009). *Toxoplasma gondii*: 25 years and 25 major advances for the field. *Int J Parasitol* **39(8)**:935–946.

Hill DE et al (2005) Biology and epidemiology of *Toxoplasma gondii* in man and animals. *Anim Health Res Rev* **6**:41–61.

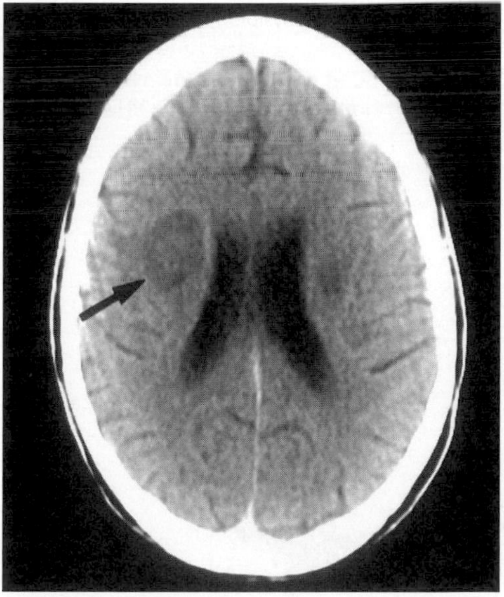

Figure 2 Toxoplasmosis. An MRI in which lesions in the brain caused by a reactivation of a latent *Toxoplasma gondii* infection are visible. The patient was HIV+ and therefore immunocompromised. (Image reprinted with permission from Medscape Reference.)

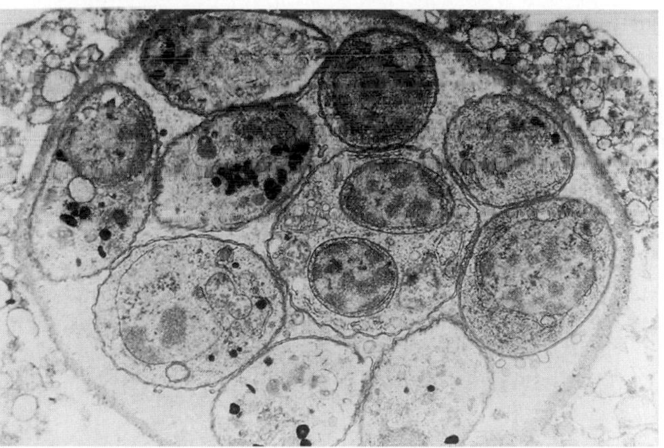

Figure 3 A *Toxoplasma gondii* tissue cyst delimited by the thick border seen in the upper right of the photograph. Ten bradyzoites are visible developing within the cyst. Typical cyst size is from 5 to 50 μm. They are most commonly seen in the brain, as well as skeletal and cardiac muscle. (Courtesy of CDC.)

Plasmodium falciparum

Plasmodium falciparum is the most virulent of the *Plasmodium* species infecting humans and the one associated with malignant tertian malaria.

Distribution and prevalence *P. falciparum* has an overlapping distribution with *P. vivax* across large parts of the tropics and subtropics (Figure 1). It is most prevalent in sub-Saharan Africa, where it is the most common causative agent of malaria. More than 75% of malaria cases are caused by this species in most sub-Saharan countries. It is estimated that there are over 100 million cases of falciparum malaria each year, 98% of which are in Africa. Most cases (as high as 85%) are in young children under five years of age.

Hosts and transmission Humans become infected when bitten by an infected *Anopheles* mosquito (Figure 2). In Africa, *A. gambiae* is the principal vector. Unlike *P. vivax*, *P. falciparum* is unable to form hypnozoites in the liver. Also, parasite-encoded proteins, found on the surface of infected erythrocytes, bind glycoproteins on the endothelium of capillaries. This process, called **sequestration**, results in very few circulating infected erythrocytes in which parasites are undergoing merogony. Gametocytes (Figure 3) do not code for these proteins. Consequently, gametocyte-infected erythrocytes return to circulation, where they may be ingested by a feeding mosquito.

Pathology Common symptoms include fever, chills and headache. Like other *Plasmodium* species, much of the pathology results from the release of inflammatory cellular debris and parasite waste products during lysis of infected erythrocytes. Because invasive stages use numerous receptors to attach and enter erythrocytes, they infect more cells and cause a much greater parasitemia than *P. vivax*, which uses only the Duffy antigen to bind erythrocytes. The sequestration of infected erythrocytes prevents infected cells from reaching the spleen, where they might be removed. Furthermore, sequestration can lead to blockage of circulation to the brain, resulting in hypoxia and the death of brain tissue known as **cerebral malaria**. This serious complication is exacerbated by inflammation initiated by the recognition of parasite molecules by TLRs on local antigen presenting cells, as described in Chapter 5 (page 177). Production of darkened urine (blackwater fever) is a hallmark of falciparum malaria.

Diagnosis Demonstration of parasites in fresh blood smears is the most common diagnostic tool. Diagnosis may also be made based on clinical symptoms and patient history. DNA probes specific for *P. falciparum* are available, as is a dipstick method for detecting parasite antigen.

Treatment Artemisinin-based combination therapies (ACTs) are recommended for uncomplicated falciparum malaria. The choice of ACT in an area will be based on the level of resistance to the constituents in the combination. In severe cases, intravenous or intramuscular artesunate is recommended. Quinine can be used as an alternative.

Control Control relies on the use of residual insecticide spraying of homes, the use of insecticide impregnated bed nets, and the reduction of mosquito breeding sites. Drugs, used either prophylactically or for treatment, also reduce transmission to vectors.

Did you know? Food shortages caused by epidemic falciparum malaria in agricultural areas around Rome in the first century BC contributed to population decline and the fall of the Roman Empire.

References

Clark IA & Cowden WB (2003) The pathophysiology of falciparum malaria. *Pharmacol Ther* **99**:221–260.

Perkins DJ, Were T, Davenport GC et al (2011). Severe malarial anemia: Innate immunity and pathogenesis. *Int J Biol Sci* **7(9)**:1427–1442.

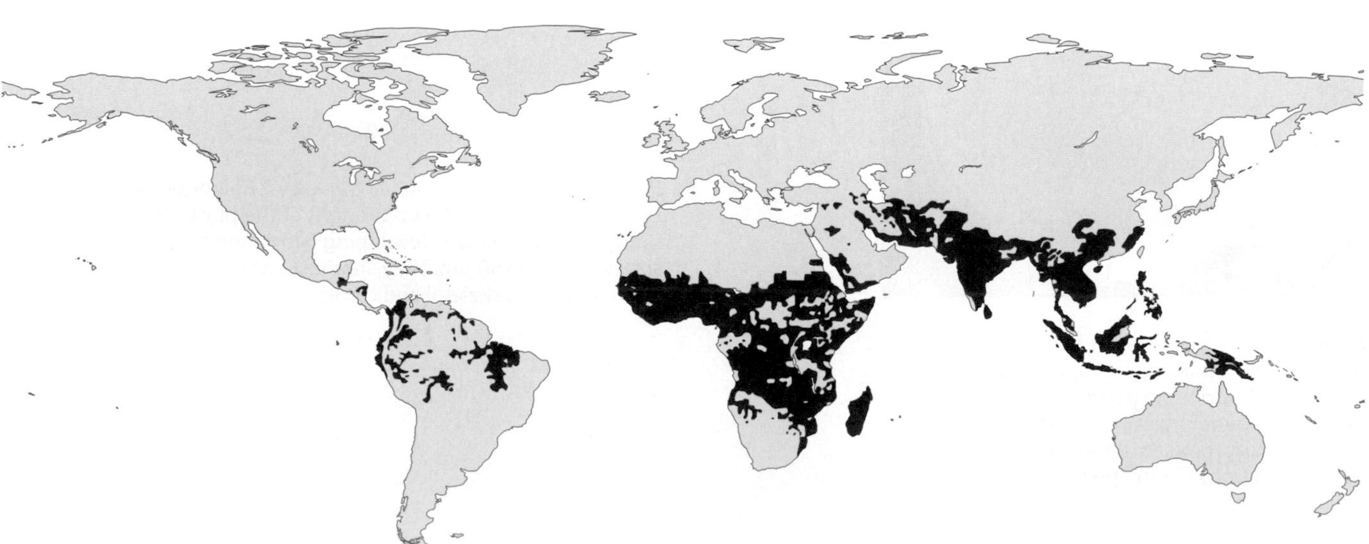

Figure 1 The distribution of malaria caused by *Plasmodium falciparum*. Most cases of falciparum malaria (up to 98% of all cases) occur in sub-Saharan Africa. (From Guerra CA, Snow RW & Hay SI [2005] *Trends Parasitol* 22:353–358. With permission from Springer Science.)

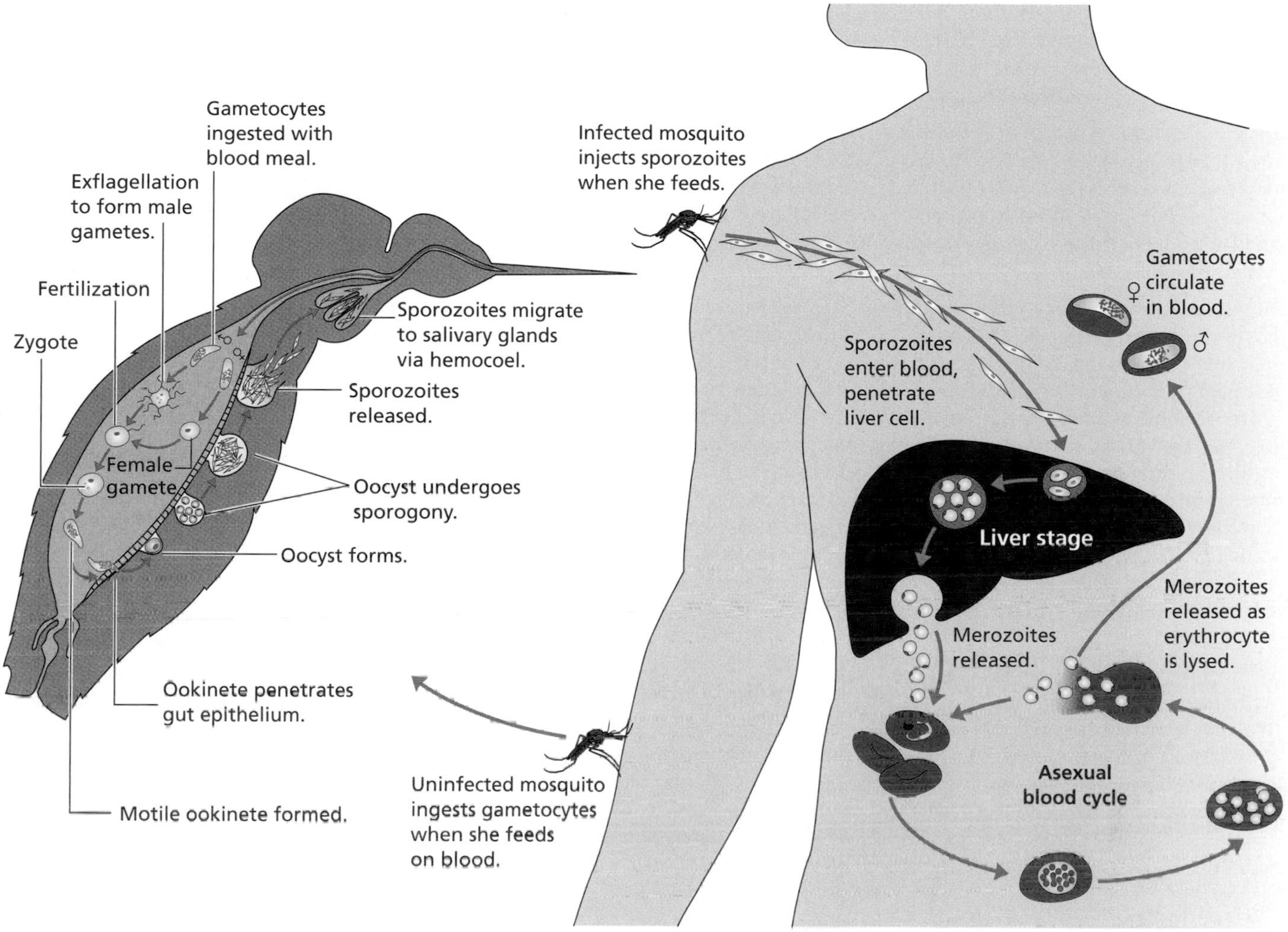

Figure 2 The life cycle of *Plasmodium falciparum*. Like other *Plasmodium* species, *P. falciparum* has a complex life cycle in which developmental changes and reproduction occur in both the human and mosquito hosts. Because union of gametocytes and therefore sexual reproduction occurs in the mosquito, the mosquito is considered to be the definitive host. Humans, as the host in which asexual reproduction occurs, serve as intermediate hosts.(Adapted from Su X, Hayton K & Wellems TE [2007] *Nature Reviews Genetics* 8:497–506. With permission from Macmillan Publishers Ltd.)

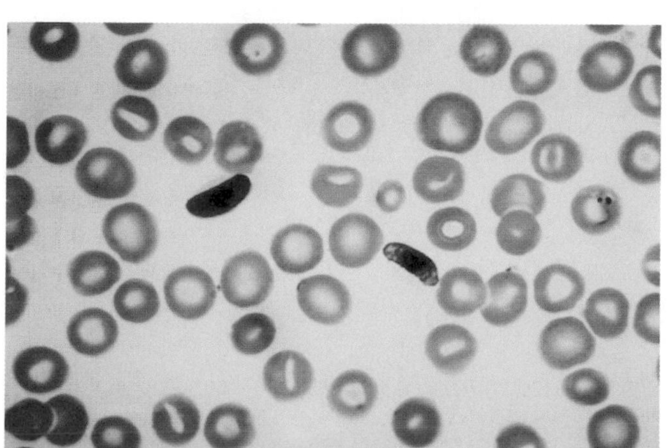

Figure 3 Gametocytes in a blood smear from an individual infected with *P. falciparum*. The gametocyte stage (the two banana-shaped darkly stained cells in the photo), unlike the merozoite stage, does not encode proteins that bind to glycoproteins on capillary epithelia. Consequently, gametocytes are not sequestered, but remain circulating in the blood, where a feeding mosquito may ingest them. (Courtesy of CDC/Mae Melvin.)

Plasmodium vivax

Plasmodium vivax is one of the four principal species of *Plasmodium* that infects humans, and it is the most common cause of recurring malaria. Although less virulent than *P. falciparum*, it is increasingly recognized as causing severe and potentially fatal disease.

Distribution and prevalence *P. vivax* is the most widely distributed human *Plasmodium* species and the most common species observed in temperate regions (Figure 1). It is the major cause of malaria outside of Africa, accounting for 65% of cases in Asia and Latin America. There are an estimated 70–80 million cases of *P. vivax* malaria annually.

Hosts and transmission Infective sporozoites reach human hosts through the bite of an infected female *Anopheles* mosquito. Sporozoites invade hepatocytes, where they either enter a dormant hypnozoite stage or undergo merogony (Figure 2). The ability to form hypnozoites results in relapses that can occur in infected individuals months or years post-infection. Merozoites invade erythrocytes, where they undergo additional rounds of merogony or until they undergo gametogony. Microgametocytes and macrogametocytes are ingested by a mosquito, where fertilization occurs. The zygote develops into a motile ookinete that invades the gut wall and undergoes sporogony. The developing sporozoites penetrate into the hemocoel and migrate to the salivary glands. See *P. falciparium*, Figure 2 for an illustration of the generalized *Plasmodium* life cycle.

Pathology Pathology occurs in tandem with the rupture of infected erythrocytes and the release of merozoites, parasite products such as hemozoin, and cellular debris. Proinflammatory cytokines such as TNFα are important in the onset of symptoms. Anemia caused by the destruction of erythrocytes is common and can be severe and life-threatening, especially in infants. Merogony, resulting in lysis of infected erythrocytes and the subsequent fever paroxysm, occurs in cycles every other day—hence the name tertian malaria. Fever episodes (Figure 3) are preceded by severe headache and chills, followed by profuse sweating.

Diagnosis Infections can be diagnosed using antibody strip fast test or visualized in blood smears of blood stages.

Treatment Chloroquine remains a common treatment, but resistance to this drug is high in some areas, especially in Asia. In such areas, artesunate is the drug of choice. Artemisinin-based combination therapy is also used. A high proportion (over a third) of patients suffer relapses unless primaquine is used to eradicate hypnozoites in the liver. Prophylaxis is recommended for visitors to areas where malaria is endemic.

Control Control efforts focus on residual insecticide spraying of dwellings, use of insecticide-treated bednets, as well as the use of antimalarial drugs to reduce transmission.

Did you know? Over 50% of West Africans are resistant to *P. vivax* infection because they lack the necessary Duffy blood group antigens on the surface of erythrocytes, needed by merozoites to invade these cells.

References

Greenwood BM, Fidock DA, Kyle DE et al (2008) Malaria: progress, perils, and prospects for eradication. *J Clin Invest* **118**:1266–1276.

Gething PW, Elyazar IR, Moyes CL et al (2012) A long neglected world malaria map. *Plasmodium vivax* endemicity in 2010. *PLOS Negl T rop Dis* **6**(9) e1814.doi:10.1371/journal.pntd.0001814.

Douglas NM, Anstey NM, Buffet PA (2012) The anemia of *Plasmodium vivax* malaria. *Malaria Journal* **135**: www.malariajournal.com/content/11/1/135.

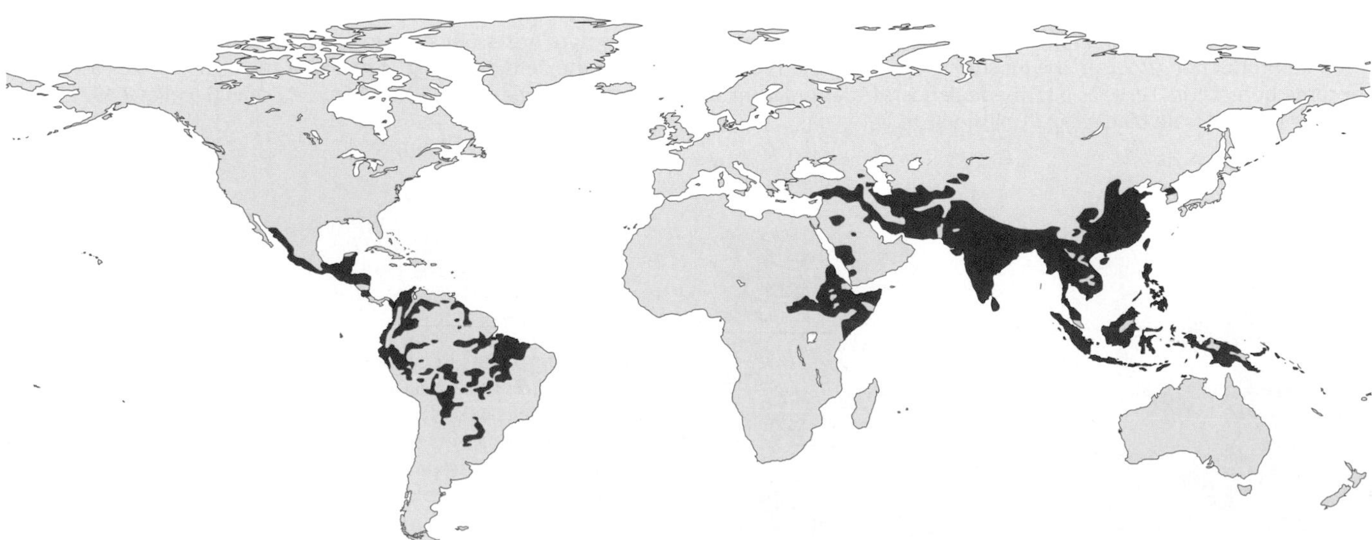

Figure 1 The distribution of *Plasmodium vivax*. (From Guerra CA, Snow RW & Hay SI [2005] *Trends Parasitol* 22:353–358. With permission from Springer Science.)

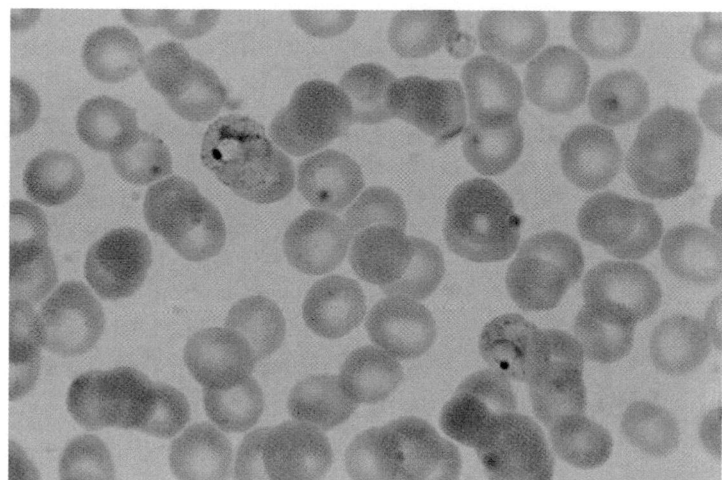

Figure 2 *Plasmodium vivax* ring stage. Two such stages are seen in this blood smear. Following erythrocyte invasion, merozoites develop into young trophozoites, called the ring form. As the trophozoite ingests host cytoplasm, it forms a large food vacuole, giving it its ring-like form. The nucleus is seen at the edge of the ring. The trophozoite will next develop into a schizont, during which time the parasite will undergo asexual reproduction (merogony). The newly formed merozoites are released upon host cell rupture. (Courtesy of CDC.)

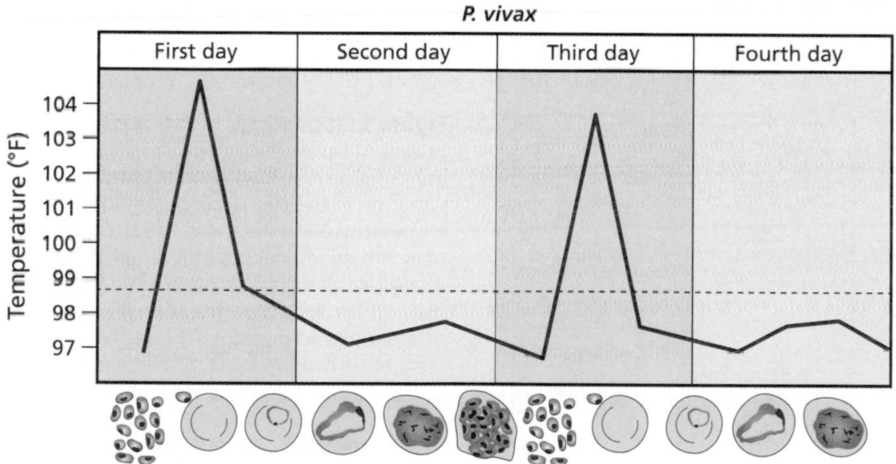

Figure 3 Fever periodicity in malaria caused by *Plasmodium vivax*. The onset of fever during an infection with *P. vivax* corresponds to the release of merozoites from red blood cells, which occurs every 48 hours. Such malaria is called tertian malaria, because, following the ancient Roman tradition of designating the day of an event as "day 1", the first fever bout occurs on day 1, the second on day 3, etc. *P. falciparum* merozoites are also released approximately every 48 hours, although the release is more drawn out, and the patient does not experience substantial relief between bouts of fever. Consequently, falciparium malaria is also referred to as tertian, but, because it is more severe than vivax malaria, it is sometimes called malignant tertian malaria. In contrast, infection with *P. vivax* is called benign tertian malaria. The major difference in the life cycle of these two parasites is the ability of *P. vivax* to form quiescent hypnozoites in the liver, which may begin to replicate months or years after the initial infection. For this reason, *P. vivax* can cause relapses, whereas *P. falciparum* cannot. For an overview of the *Plasmodium* life cycle, see the Rogues' Gallery entry for *P. falciparum*. (From Wiser MF [2011] Protozoa and Human Disease. Garland Science.)

Plasmodium malariae and *Plasmodium ovale*

Plasmodium malariae and *P. ovale* are two additional *Plasmodium* species responsible for malaria in humans.

Distribution and prevalence *P. malariae* is found in sub-Saharan Africa, South and Southeast Asia, and South America. In endemic regions, prevalence is from under 4% to over 20%. *P. ovale* is found primarily in West Africa and in parts of East Africa and Southeast Asia. In West Africa, prevalence may be 10% or more. Elsewhere, it is generally under 5% In endemic areas.

Hosts and transmission Like other human *Plasmodium* species, both *P. malariae* and *P. ovale* are transmitted by certain species of *Anopheles* mosquitoes. Humans are the only vertebrate host for *P. ovale*. *P. malariae* can also infect certain primates, including chimpanzees. See *P. falciparum*, Figure 2 for the generalized *Plasmodium* life cycle.

Pathology Infections caused by *P. malariae* and *P. ovale* are much less severe than for *P. vivax* and especially *P. falciparum*. Tertian malaria caused by *P. ovale*, like that caused by *P. vivax*, results in fever paroxysms every 48 hours, corresponding to the rupture of infected erythrocytes and the release of merozoites. Furthermore, sporozoites infecting liver cells do not necessarily undergo schizogony but may form hypnozoites, which may result in a relapse of disease up to 20 months after the initial infection. The slower growing *P. malariae* causes disease with paroxysms every 72 hours, known as quartan malaria. Although it does not enter a latent, hypnozoite stage, *P. malariae* can persist at low, subclinical levels for many years in infected individuals, who serve as reservoirs. It can later reach clinical significance again, a phenomenon known as recrudescence.

Diagnosis Different human *Plasmodium* species can be distinguished based on their different morphology during erythrocytic stages and in the manner in which they modify host erythrocyte morphology (Figure 1). Other immunodiagnostic and PCR-based tests are also available.

Treatment Chloroquine remains the standard treatment for both *P. ovale* and *P. malariae* infections. In the case of *P. ovale*, primaquine is added to the treatment regimen as it eliminates liver stages, including hypnozoites.

Control As with all malaria, control centers on reducing the number of vectors through use of residual insecticides in dwellings and removal of breeding sites, along with reducing human–mosquito contact through the use of insecticide impregnated bed nets.

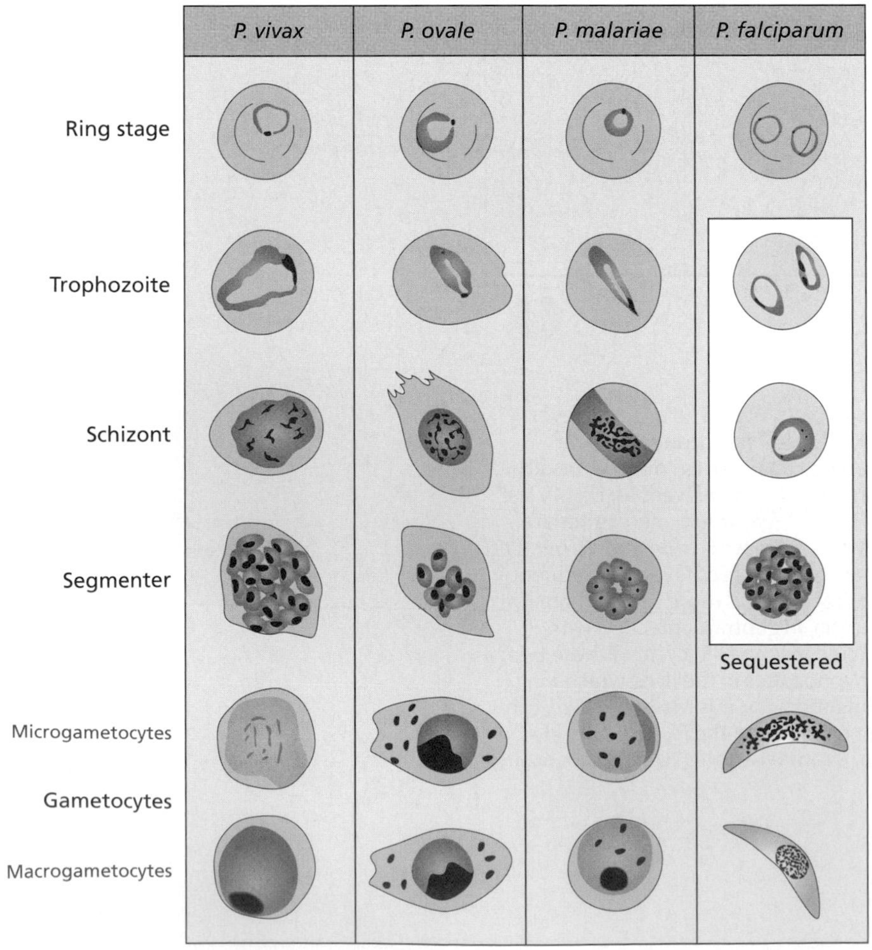

Figure 1 Blood stages of *Plasmodium*. Stages of the four principal species infecting humans are shown. Often variation in the morphology of different stages is used to diagnose the species of *Plasmodium* in a patient. Ring forms of all species are similar and difficult to distinguish. Trophozoites of *P. vivax* are often ameboid in shape. Those of *P. ovale* tend to be more compact, whereas those of *P. malariae* are very compact. Asexual forms of *P. falciparum* are not common in the blood so are delimited here in a box labeled "Sequestered". The number of merozoites produced per schizont varies between species. Typical numbers are 8–12 for *P. malariae*, 6–18 for *P. ovale*, 16–36 for *P. falciparum*, and 14–24 for *P. vivax*. Gametocytes for *P. falciparum* are crescent-shaped. In other species, they are round or oval. Gametocytes can be distinguished from trophozoites by their large size (nearly filling the erythrocyte) and a single nucleus. Mature microgametocytes tend to stain lighter than macrogametocytes and have a more diffuse nucleus. (From Wiser MF [2011] Protozoa and Human Disease. Garland Science.)

Did you know? The terms tertian and quartan to describe paroxysms every 48 and 72 hours, respectively, comes from the Roman custom of referring to the day on which an event occurs as day one. Thus, 48 hours later is day three and 72 hours later is day four.

Related organisms

Plasmodium knowlesi: P. knowlesi has traditionally been considered to be a cause of malaria in nonhuman primates. It is increasingly recognized as a cause of human malaria in Southeast Asia (Figure 2).

Haemoproteus columbae: This organism is a common parasite of pigeons that is vectored by biting flies (Hippoboscidae). There is limited or no pathology in otherwise healthy birds.

Leucocytozoon simondi: This organism parasitizes ducks and geese. It is vectored by black flies (family Simuliidae) and is highly pathogenic, especially in young birds. Severe anemia is a major symptom.

References

Franken G, Muller-Stover I, Holtfielter MC (2012) Why do *Plasmodium malariae* infections sometimes occur in spite of previous antimalarial medication? *Parasitol Res* **111**: 943-946.

Roucher C, Rogier C, Sokhna C et al (2014) A 20-year longitudinal study of *Plasmodium ovale* and *Plasmodium malariae* prevalence and morbidity in a West African population. *PlosONE* **9(2)**: e87169.doi:10.1371/journal.pone.008169.

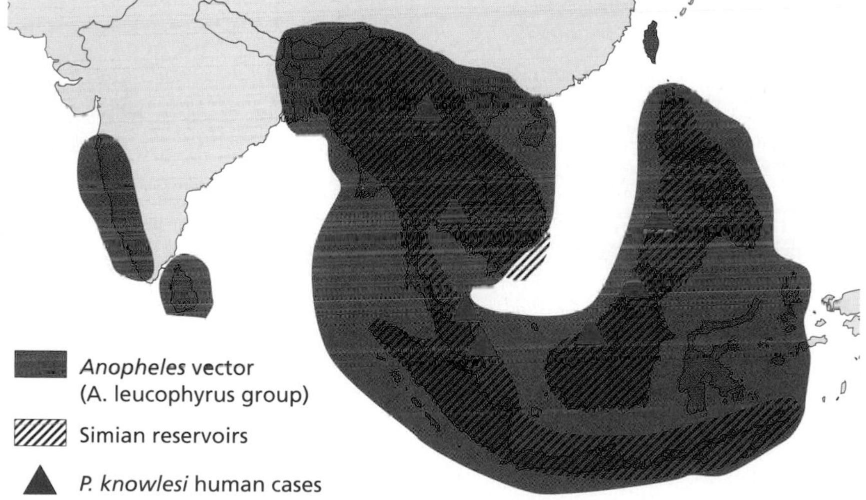

Figure 2 *Plasmodium knowlesi.* Long believed to be primarily a parasite of nonhuman primates, *P. knowlesi* is now recognized as an increasingly important cause of malaria in humans. The map highlights the distribution of competent vectors, important primate reservoirs, and places where human cases have occurred. Primate reservoirs include *Macaca fascicularis* (the long-tailed macaque) and *M. nemestrina* (the pig-tailed macaque). Deforestation, which drives macaques into closer contact with humans, may increase the likelihood of transmission to humans. (From Morbidity and Mortality Weekly Report [2009] 58:229–232. Courtesy of CDC.)

Babesia bigemina

Babesia bigemina is the apicomplexan blood parasite responsible for babesiosis or Texas red-water fever, a serious disease in cattle.

Distribution and prevalence Distributed wherever the ticks serving as vectors are found. These ticks, members of the genus *Rhipicephalus* (formerly *Boophilus*), inhabit the Americas, Africa, Asia, Australia, and southern Europe. Prevalence is generally low in developed countries and other places where effective tick control is practiced. In less developed areas, prevalence is variable and may be as high as 30%.

Hosts and transmission Many ruminants including deer, cattle, and water buffalo can serve as vertebrate hosts. Transmission occurs through the bite of infected ticks (genus *Rhipicephalus*, Figure 1). In North America, the most important vector species is *R. annulatus*.

Pathology Destruction of erythrocytes from intracellular asexual parasite reproduction causes severe anemia (Figure 2). Up to 75% of erythrocytes may be destroyed in fatal cases. Released hemoglobin and its waste products produce jaundice, and excess hemoglobin excreted in the urine give the urine a characteristic red color (hence the name red-water fever). Infected animals may remain weak and listless for several weeks, but once animals recover they are generally immune to subsequent disease. Disease is more severe in adults, with a mortality rate of approximately 50%–90% in untreated cases. Calves less than a year old rarely suffer serious disease.

Diagnosis Babesiosis may be suspected based on symptoms and the presence of *Rhipicephalus* ticks. Diagnosis is confirmed by observation of parasites in Giemsa-stained blood smears.

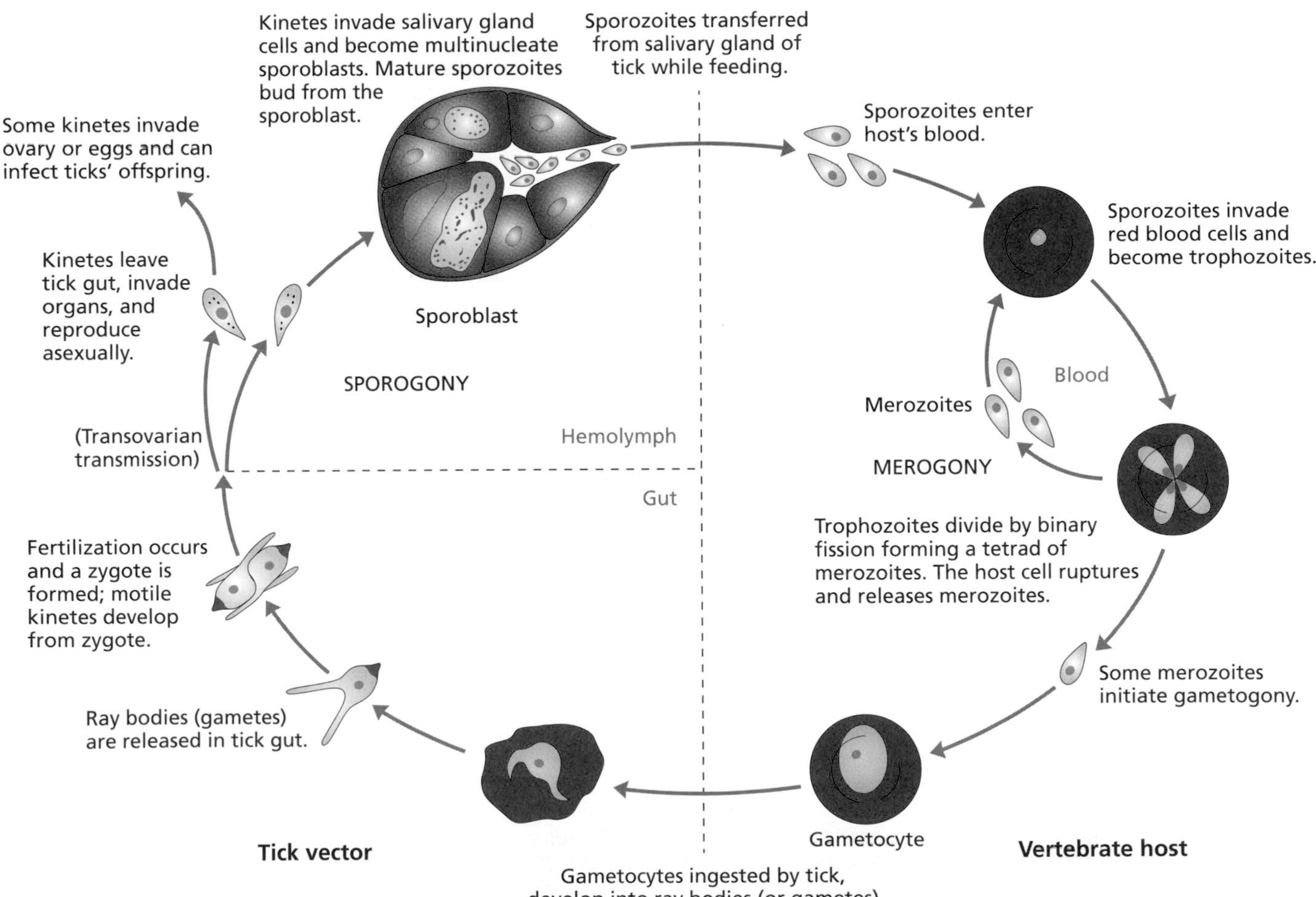

Figure 1 The life cycle of *Babesia bigemina*. Infective sporozoites are introduced into the mammalian host through the bite of an appropriate tick vector (genus *Rhipicephalus*). Unlike *Plasmodium*, there is no exoerythrocytic asexual reproduction. Sporozoites invade erythrocytes, where they develop into trophozoites. Binary fission produces merozoites, which are released upon erythrocyte lysis. Repeated rounds of erythrocyte invasion by merozoites continue, resulting in a rapid rise in parasitemia. Alternatively, some merozoites develop into gametocytes, which are ingested by a tick. Gametocytes develop into gametes called ray bodies in the tick's gut. The fusion of two ray bodies results in zygote formation, which then develops into a kinete. The kinete penetrates the gut wall to enter the hemolymph, from which it migrates to various organs, including the salivary glands. Here, the kinete forms a multinucleated structure called a sporoblast, from which sporozoites bud when the tick feeds. Transovarian transmission occurs when kinetes migrate to the ovaries and eggs. (From Wiser MF [2011] Protozoa and Human Disease. Garland Science.)

Treatment Diminazene (Berenil), imidocarb, or trypan blue (so-called because it also kills trypanosomes) are effective for the treatment of babesiosis.

Control Regular dipping of cattle in solutions of pyrethroids or organophosphates that kills ticks, effectively controls or even eliminates the disease in cattle or other ruminant livestock. Such tick control was used to eradicate babesiosis in the United States and elsewhere.

Did you know? *Rhipicephalus annulatus* (Figure 3), the principal vector in the Americas, is a rare example of a one-host tick that serves as a vector. One-host ticks feed, mature, and mate on a single animal. However, *B. bigemina* can pass from infected female ticks to developing eggs in the ticks' ovaries. Thus **transovarian transmission**, in which newly hatched larval ticks are themselves infected, explains how an arthropod that feeds on only one host can serve as a vector.

Related organisms

Theileria microti: Transmitted by *Ixodes scapularis, T. microti* parasitizes rodents. Since 1969, it has been identified as a cause of theileriosis in humans, with hundreds of reported cases in the United States.

Theileria parva: T. parva is the causative agent of East Coast fever, an important tick-borne disease of cattle in parts of Eastern and Southern Africa. It has largely been eradicated from Southern Africa.

References

Hunfeld KP, Hildebrandt A & Gray JS (2008). Babesiosis: Recent insights into an ancient disease. *Int J Parasitol* **38(11)**:1219–1237.

Uilenberg G (2006) *Babesia*—a historical overview. *Vet Parasitol* **138**:3–10.

Vannier E & Krause PJ (2012) Human Babesiosis *NEJM* **366**: 2397–2407.

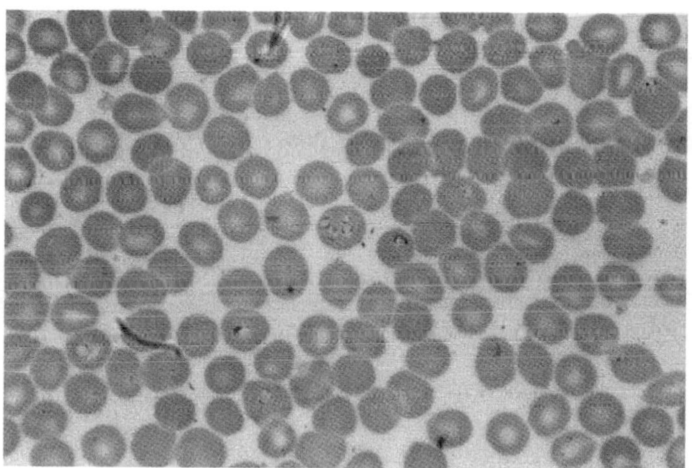

Figure 2 Blood smear from a cow infected with *Babesia bigemina*. After they have invaded red blood cells, trophozoites form merozoites (seen here in dark red) and undergo asexual reproduction within the host cells. The cell lysis caused by the intracellular parasite explains the anemia that is typical in heavily infected cattle. (Courtesy of CDC/Steven Glenn.)

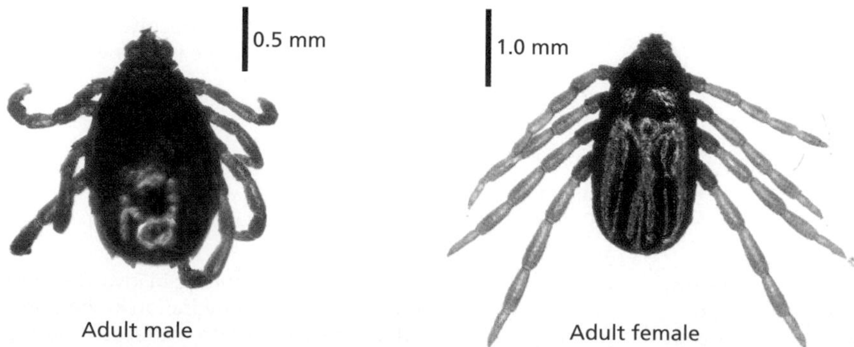

Figure 3 *Rhipicephalus annulatus*. Both an adult male and female of this one-host tick are shown. *R. annulatus* is the principal vector of *Babesia bigemina* in the Americas. (Courtesy of Christina Berry, University of Bristol.)

Balantidium coli

Balantidium coli is an alveolate eukaryote. It is the only known ciliate parasite in humans.

Distribution and prevalence *B. coli* is distributed worldwide but is more common in the tropics and in environments with poor sanitation. Prevalence is typically less than 1% of the population. Because of its high prevalence in pigs (20%–100%), prevalence is higher in those who live in close association with pigs. In Papua New Guinea, where pigs are very common, prevalence can be greater than 25% among pig farmers and those working in slaughterhouses.

Hosts and transmission Many mammals, including many primates and especially pigs, can serve as hosts for *B. coli*. Transmission is via the fecal–oral route (Figure 1), with the ingestion of the *B. coli* cyst (Figure 2).

Pathology Although pigs are generally asymptomatic, symptoms in humans can range from asymptomatic to severe. Many individuals experience mild symptoms, including diarrhea, nausea, and headache. These symptoms are believed to result from inflammation of the colonic mucosa. Chronic cases can persist for weeks, with diarrhea developing into dysentery. Occasionally trophozoites become invasive, causing ulceration of the colon, with symptoms similar to those seen in invasive amebiasis. Severe symptoms are most common in those with other underlying conditions such as immunodeficiency or malnutrition. Acute and severe infections can occasionally be life threatening.

Diagnosis Trophozoites of this large and ciliated protozoan (Figure 3) are identified in fecal examination and are unlikely to be confused with other intestinal protozoans.

Treatment Infections are typically treated with tetracycline. Results with metronidazole have been inconclusive, with efficacy in some but not all cases.

Control As with other parasites transmitted via the fecal–oral route, control relies on provisioning of safe food and drinking water and the prevention of contamination with fecal material. Those working with pigs should take special precautions.

Did you know? *B. coli*, like other ciliates, is unusual in that it has two nuclei. The large polyploid macronucleus is involved with metabolism and other general cell function. The macronucleus

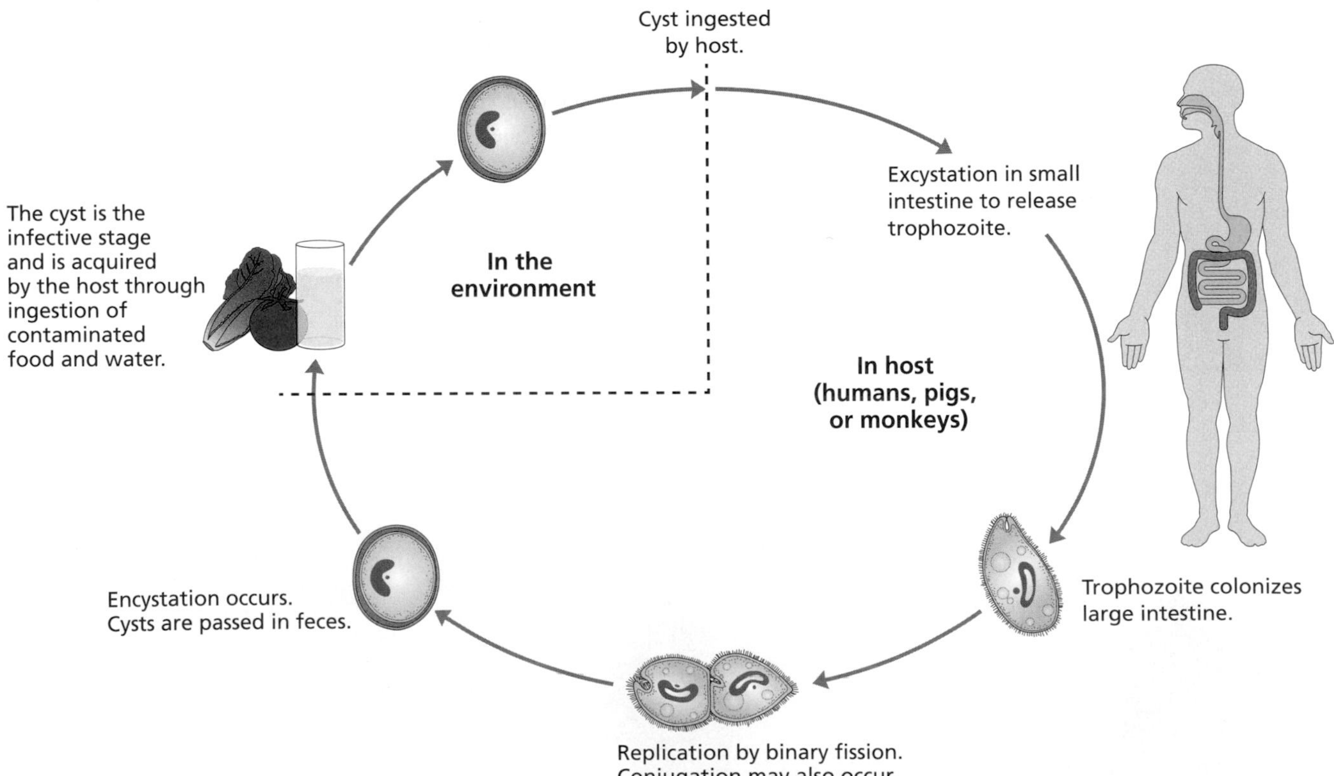

Figure 1 The life cycle of *Balantidium coli*. The infective stage, the cyst, is ingested, usually via the consumption of contaminated food or water. Excystation occurs in the small intestine, releasing the trophozoite. Trophozoites pass to the colon, where they feed on the intestinal mucosa and on bacteria. Some trophozoites may invade the epithelium of the colon resulting in invasive disease. Trophozoites respond to increasing dehydration in the colon by forming new cysts, which are passed with the feces. (Courtesy of CDC.)

is generated from the smaller diploid micronucleus by genome amplification. Many genes that do not function during the asexual portion of the life cycle are not present in the macronucleus. The micronucleus functions in sexual reproduction.

Another parasitic ciliate

Ichthyophthirius multifiliis: A common ectoparasite in freshwater fish, *I. multifiliis* causes the disease ich or white spot, often a serious problem for tropical fish enthusiasts.

References

Ramachandran A (2003) Introduction. The Parasite: *Balantidium coli*. The Disease: Balantidiasis. Stanford University. http://www.stanford.edu/group/parasites/ParaSites2003/Balantidium/Morphology.htm

Schister F L & Ramirez-Avila L (2008) Current world status of *Balantidium coli*. *Clin Microbiol Rev* **21(4)**:626–638.

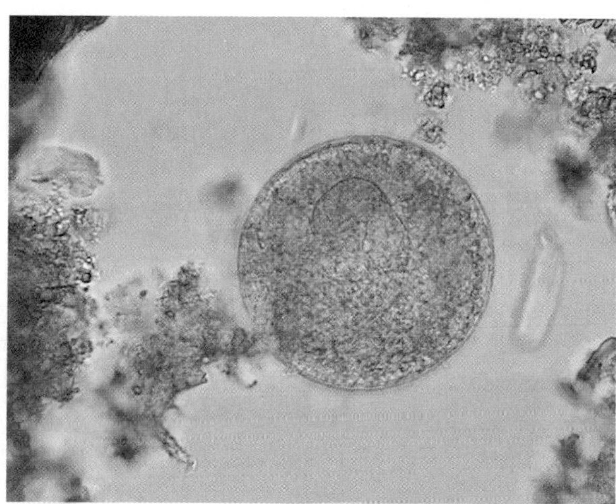

Figure 2 The cyst stage of *Balantidium coli*. Note its more rounded shape compared to that of the trophozoite in Figure 3. (Courtesy of CDC/LLLA Moore Jr.)

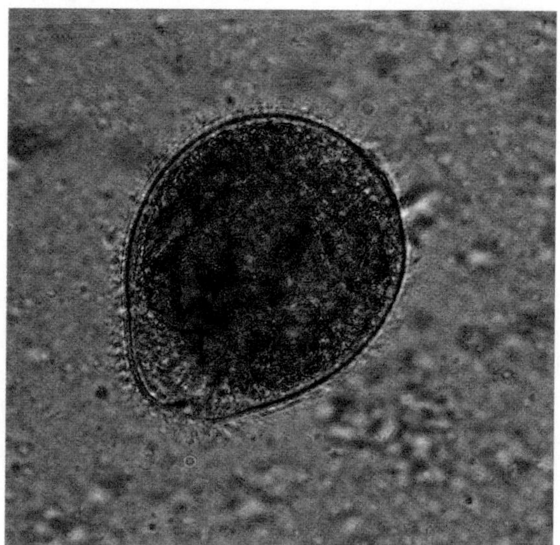

Figure 3 A *Balantidium coli* trophozoite. Note the cilia surrounding the cell. (Courtesy of Euthman, CC BY-SA 2.5.)

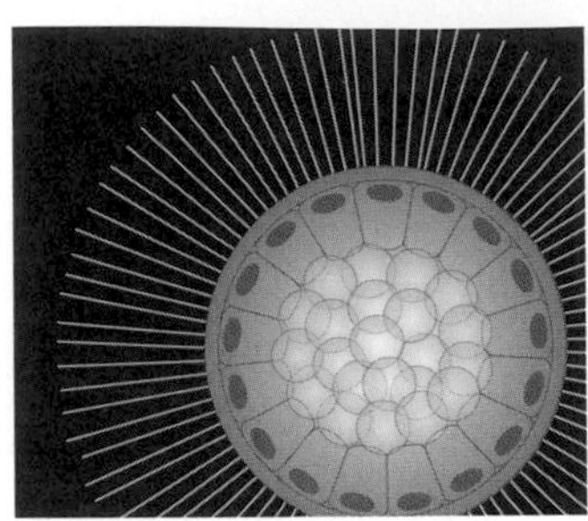

Phylum Platyhelminthes

Members of this large, lophotrochozoan phylum are commonly called flatworms. Familiar examples are planarians, flukes, and tapeworms. There are an estimated 55,000 species, with the majority being parasitic (~44,000 species). They exhibit bilateral symmetry and possess the three major germinal layers (ectoderm, mesoderm, and endoderm), but they lack a body cavity, or coelom, and are said to be **acoelomate**. They lack true segments. They exhibit cephalization (have an anterior end with a brain) and a ladderlike nervous system. Platyhelminths have no specialized circulatory or respiratory systems and rely on diffusion for gas exchange, hence their typical dorso–ventrally flattened appearance. Most have a mouth and gut, but they usually lack an anus. They have an osmoregulatory system consisting of flame cells and tubules that regulate water content and eliminate wastes. They are usually hermaphroditic.

Resolving relationships among platyhelminths is an ongoing endeavor. Today, the phylum Platyhelminthes is usually divided into the Catenulida and the Rhabditophora, the former consisting of a relatively small group of non-symbiotic species. Among the Rhabditophora are several families traditionally called turbellarians. These are mostly free-living predators but include a smattering of over 200 species belonging to 35 families that live in a permanent symbiotic association with other invertebrates, including several that are overtly parasitic.

The Rhabditophora also includes the Neodermata, a monophyletic group, and it is here that the major lineages of parasitic flatworms are found. Neodermatans have lost a ciliated epithelium in most life-cycle stages and instead are enveloped in their parasitic phases in a syncytial tegument called the **neodermis**. A remarkable feature of the neodermis is that the nuclei that govern its activities are sunken beneath the muscles that underlie the tegument. The neodermis is believed to be a key adaptation to parasitism in flatworms, both allowing for uptake of nutrients and for defense from host immune systems. The major neodermatan lineages are the Trematoda, discussed first below, followed by the Monogenea and Cestoda.

Class Trematoda The class Trematoda is divided into two groups: the Aspidogastrea and the Digenea. The Aspidogastrea, also called aspidobothrians, is a small group of 80 species that parasitize freshwater and marine molluscs and for which cartilaginous or bony fish and turtles serve as either facultative or obligatory hosts. Aspidogastrean adults are conspicuous for having a large ventral disc that is often divided into multiple compartments. None are of medical or commercial significance. Their biology holds important clues for the evolution of parasitism.

The Digenea are also known as digeneans, digenetic trematodes, or flukes. There are an estimated 18,000 species, making it one of the most diverse groups of metazoan endoparasites. A few species are essentially ectoparasites, living under the scales of fish. Digeneans have complex life cycles, almost always involving a mollusc (usually a snail) as an intermediate host, in which they undergo asexual reproduction, and a vertebrate as a definitive host, in which they undergo sexual reproduction. They thus exhibit an alternation of asexual and sexual generations, hence the designation as digenetic trematodes. Adults are usually hermaphroditic and typically have an oral and ventral sucker. Most species inhabit the intestine as adults but certain other species colonize just about any habitat within the vertebrate host. Many digenean species can cause serious veterinary or medical problems.

References

Hahn C, Fromm B & Bachmann L (2014) Comparative genomics of flatworms (Platyhelminthes) reveals shared genomic features of ecto- and endoparasitic Neodermata. *Genome Biol Evol* **6**:1105–1117.

Littlewood DTJ (2008) Platyhelminth systematics and the emergence of new characters. *Parasite* **15**:333–341.

Hymenolepis nana

Hymenolepsis nana, a member of the family Hymenolepididae, is also called *Vampirolepis nana*. It is known as the dwarf tapeworm and causes hymenolepiasis in humans and rodents.

Distribution and prevalence *H. nana* is cosmopolitan and is most common in temperate climates. It is probably the most common human tapeworm, with a prevalence of over 20% in some communities.

Hosts and transmission Figure 1 shows the life cycle, which is unusual because of the possibility of autoinfection.

Pathology This species can be problematic in children, causing enteritis, abdominal pain, diarrhea, anal pruritis, restlessness, headache, weakness, loss of appetite, nausea, and vomiting.

Diagnosis Diagnosis is by finding the characteristic eggs in the stool. Oval eggs (30–47 μm) have a thin outer shell and a thicker inner shell with polar thickenings that bear filaments.

Treatment Praziquantel, niclosamide, and nitazoxanide are effective in killing adult worms.

Control Treatment of infected people and others in the same environment, and prevention of the ingestion of eggs in food or water are effective, as are rodent and insect control.

Did you know? Because of autoinfection, a mouse infected with 10 cysticercoids may eventually contain more than 1500 adult tapeworms.

Other related species

H. diminuta: This species is larger than *H. nana* (the adult is up to 60 cm) and is primarily a parasite of rats, although it occasionally infects people. The scolex of this species is unarmed (lacks hooks), and autoinfection does not occur. The eggs lack polar filaments. Cysticercoids occur in stored grain beetles. Because of the ease of maintenance in the laboratory, this is one of the most intensively studied tapeworms.

Dipylidium caninum: A member of the family Dipylidiidae (Figure 2), *D. caninum* is the double-pored tapeworm, so named because it has two complete sets of reproductive organs per proglottid. It commonly infects dogs, cats, and other flea-infested animals. It occasionally infects people, usually children, if a flea from a pet containing a cysticercoid is accidentally ingested.

References

Chero JC, Saito M, Bustos JA et al (2007) *Hymenolepis* infection: symptoms and response to nitazoxanide in field conditions. *Trans R Soc Trop Med Hyg* **101**:203–205.

Soares Magalhães RJ, Fançony C, Gamboa D et al (2013) Extending helminth control beyond STH and schistosomiasis: The case of human hymenolepiasis. *PLoS Negl Trop Dis* **7(10)**: e2321. doi:10.1371/journal.pntd.0002321.

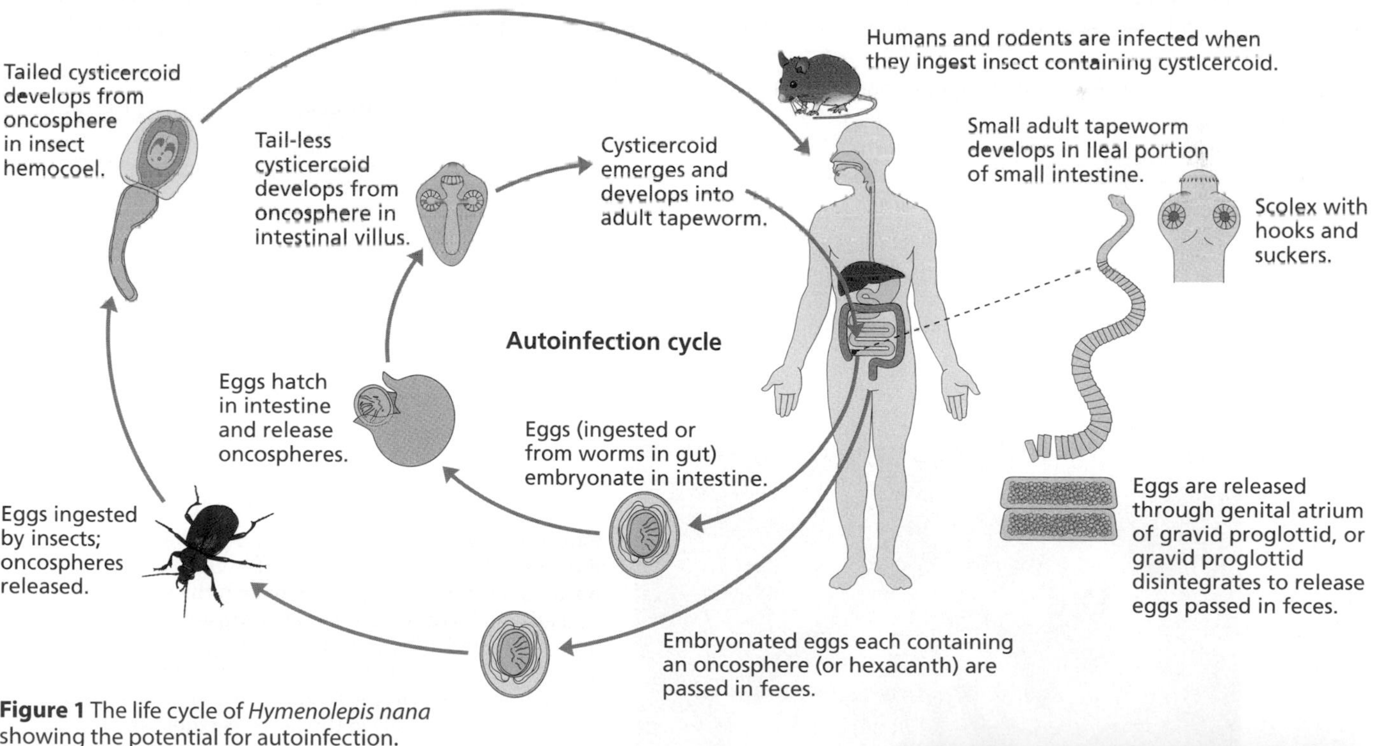

Figure 1 The life cycle of *Hymenolepis nana* showing the potential for autoinfection. (Courtesy of CDC.)

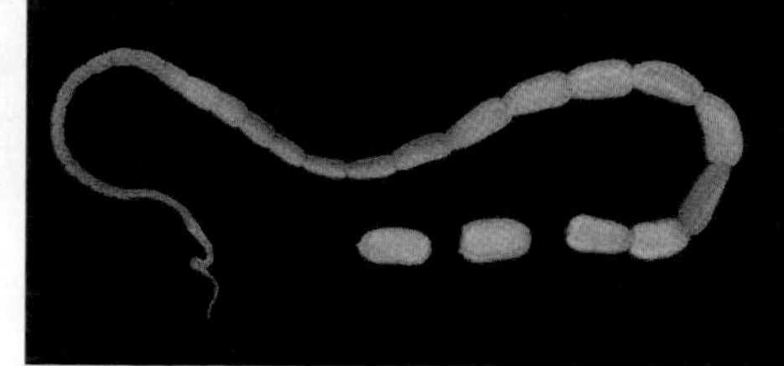

Figure 2 Adult of *Dipylidium caninum*, a common tapeworm of dogs and cats that also infects people. Note the almond-shaped proglottids. (Courtesy of CDC.)

Schistosoma mansoni

Schistosoma mansoni is a member of the digenean family Schistosomatidae that is characterized by having separate male and female adults (Figure 1). The adults live *in copula* in blood vessels, hence their name blood flukes. *S. mansoni* is one of the causative agents of human intestinal and hepatosplenic schistosomiasis. Schistosomiasis is also known as bilharzia or snail fever.

Distribution and prevalence *S. mansoni* is the most widely distributed of the human schistosomes, but most infections occur in sub-Saharan Africa. There are also large endemic areas in South America and in Southwest Asia. The estimated worldwide prevalence is about 83 million people.

Hosts and transmission The definitive hosts are primarily humans and occasionally other primates such as baboons; rodents are also infected. Characteristic lateral-spined eggs (Figure 2) are passed into water bodies and hatch to release miracidia. The miracidia infect intermediate hosts, which are particular species of freshwater pulmonate snails of the genus *Biomphalaria* (Planorbidae) (Figure 2). After a period of development and asexual reproduction in snails, many fork-tailed cercariae emerge, swim through the water, penetrate human skin, and initiate new infections.

Pathology Most pathology results from the many eggs that become trapped in the tissues. They Incite complex granuloma responses resulting in the eventual destruction of the eggs. The liver, spleen, and intestinal wall are all likely to be damaged by granulomas. The flow of blood in the liver can be impeded, resulting in hypertension of blood in the hepatic portal system. Enlargement of the spleen and liver (Figure 3) and formation of collateral veins are also consequences of infection. A distended abdomen with ascites, anemia, fatigue, and diarrhea with dysentery are all common symptoms. Children may exhibit impaired growth and cognitive development.

Diagnosis Schistosomiasis is diagnosed by finding the characteristic lateral-spined eggs in the feces or by detection of anti-schistosome antibodies or of circulating worm or egg antigens indicative of active infections.

Treatment The one widely available drug is the acylated quinolone-pyrazine, praziquantel, which results in a 70%–100% cessation of egg production after a single oral treatment. Adult worms are susceptible, whereas juvenile worms are not. Artemisinin is effective against immature worms.

Control Control relies primarily on treatment with praziquantel. In some locations, control of snails using chemical molluscicides or biological methods may be feasible. Improved sanitation, provision of safe water, and changing behavior to avoid excretion of eggs into snail habitats should be pursued and if so, can result in elimination, assuming reservoir hosts are not a factor in maintaining transmission. Continued progress in control is dependent on worms remaining fully susceptible to praziquantel. Development of safe alternatives is a high priority. Integration of schistosomiasis control with control of other helminths is an effective way forward.

Did you know? Adult *S. mansoni* can live for more than 30 years in some circumstances. The 363-mbp genome of *S. mansoni* has been sequenced and is estimated to contain about 12,000 known genes. Immunological studies suggest that humans slowly develop immunity to infection that is mediated by increased levels of IgE against adult worm antigens, eosinophilia, interleukin-5 (IL-5) and low levels of IgG4 antibodies. Development of a schistosome vaccine remains an important goal for long-term control prospects.

References

Berriman M, Hass BJ, LoVerde PT et al (2009) The genome of the blood fluke *Schistosoma mansoni*. *Nature* **460**:352–U65.

Colley DG, Bustinduy AL, Secor WE et al (2014) Human schistosomiasis. *Lancet* **383:**2253-2264.

Figure 1 Adults of *Schistosoma mansoni*, with the anterior end of the worms to the left and the posterior ends to the right. The male is thick-bodied, and has a slender female protruding from either end of its ventrally-placed groove-like gynecophoric canal. The dark, rust-colored material visible down the length of the worms is ingested blood, mostly in the female's digestive tract. (Photo courtesy of Melissa Sanchez and Martina Laidemitt.)

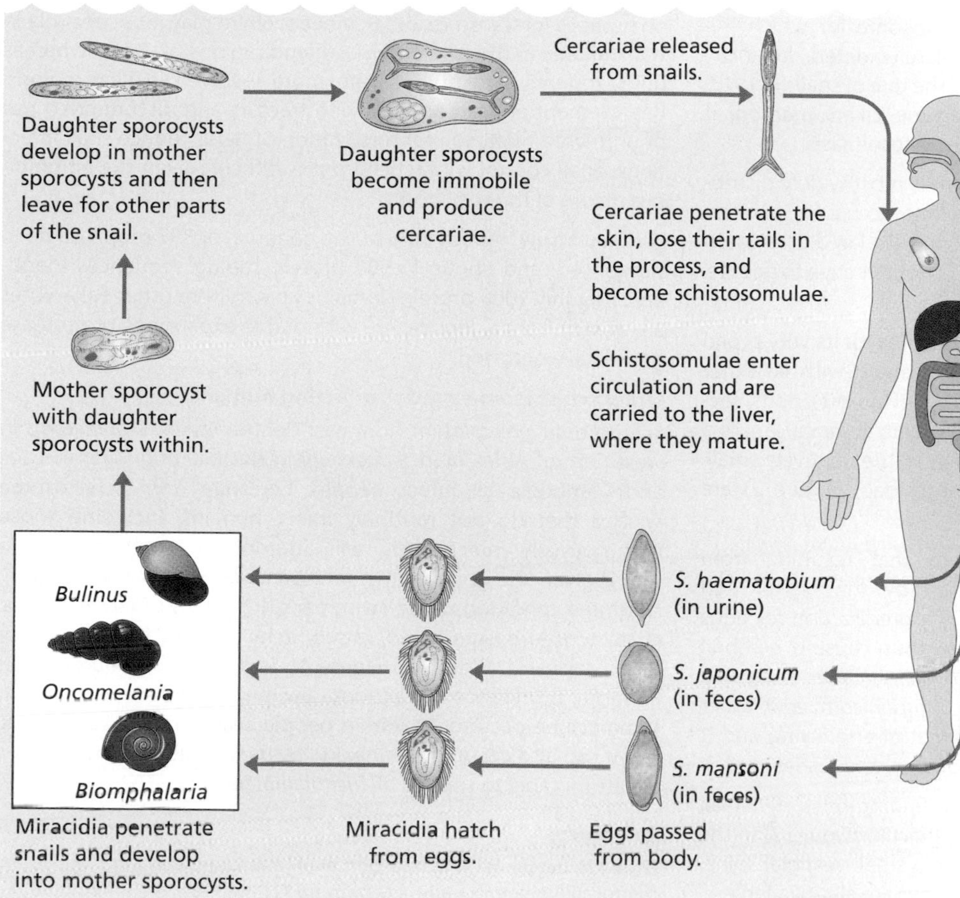

Daughter sporocysts develop in mother sporocysts and then leave for other parts of the snail.

Daughter sporocysts become immobile and produce cercariae.

Cercariae released from snails.

Cercariae penetrate the skin, lose their tails in the process, and become schistosomulae.

Schistosomulae enter circulation and are carried to the liver, where they mature.

Worms pair in liver and migrate: For *S. japonicum* and *S. mansoni*, worms pair in the liver and migrate to the mesenteric venules around the intestine where eggs are produced.

♀ ♂

The eggs then work their way through the intestinal wall and are passed in the feces. For *S. haematobium*, adults move to the veins surrounding the bladder and their eggs are passed in the urine.

Mother sporocyst with daughter sporocysts within.

Bulinus

Oncomelania

Biomphalaria

S. haematobium (in urine)

S. japonicum (in feces)

S. mansoni (in feces)

Miracidia penetrate snails and develop into mother sporocysts.

Miracidia hatch from eggs.

Eggs passed from body.

Figure 2 The life cycle of a schistosome. In the case of *Schistosoma mansoni*, adult worms are found in the mesenteric veins surrounding the intestine. They produce lateral-spined eggs that pass across the intestinal wall and into the feces. If the eggs are voided into freshwater, they hatch to release miracidia that infect freshwater snails (of the genus *Biomphalaria* in the case of *S. mansoni*). A miracidium transforms into a mother sporocyst that produces daughter sporocysts. The daughter sporocysts eventually migrate to the snail's digestive gland and ovotestis, where they produce cercariae. Eventually the fork-tailed cercariae leave the snail host, swim through the water, and penetrate human skin. The tail is lost and the remaining body (called a schistosomulum at this point) migrates from the skin to lungs and eventually to the liver, growing as it goes. Worms mature and pair in the liver and migrate to the mesenteric veins to produce eggs. (Courtesy of CDC.)

Figure 3 A child with a *Schistosoma mansoni* infection exhibits a swollen abdomen with ascites and an enlarged liver and spleen.

Schistosoma japonicum

Schistosoma japonicum was the first schistosome for which the involvement of snails in the life cycle was elucidated. Japanese investigators Miyairi and Suzuki described the role of snails in 1913. *S. japonicum* is called the oriental schistosome. Like *S. mansoni*, it also causes intestinal and hepatosplenic schistosomiasis.

Distribution and prevalence *S. japonicum* is most widely distributed in China and the Philippines, and it also occurs in Sulawesi. Historically it occurred in Japan, where it was declared eradicated in 1996. It occurs in Taiwan but, curiously, is not infective to people. About 2 million people are infected.

Hosts and transmission *S. japonicum* is unusual for its very broad host range in mammals, infecting humans along with 40 other species, commonly including water buffaloes (Figure 1), dogs, several species of rodents, and pigs. Infection with *S. japonicum* is a zoonosis. It occurs where susceptible snails of the relatively small, amphibious genus *Oncomelania* (Pomatiopsidae) (Figure 2) are available to support its transmission.

Pathology The pathology is similar to that resulting from *S. mansoni* infection but is exacerbated by the higher egg production rates of *S. japonicum*. There is a predilection for eggs of *S. japonicum* to appear in the brain, thus causing cerebral schistosomiasis. An acute form of schistosomiasis called Katayama fever is seen following infection with *S. japonicum*, and other schistosomes. It is associated with the onset of egg laying and is marked by fever, chills, nausea, abdominal pain, and rashes.

Diagnosis Infection can be diagnosed by finding relatively small (70–100 × 50–70 μm) ovoid eggs of *S. japonicum* (Figure 3) in the feces. The eggs either lack a spine or have a small vestigial spine. As noted for *S. mansoni*, immunological assays are also available.

Treatment Praziquantel is again the drug of choice (see *S. mansoni*).

Control Given its zoonotic nature, control of *S. japonicum* is rendered more difficult than for species more exclusively dependent on humans for transmission. Bovines seem to play more of a role in transmission in the lake and marshland regions of China, whereas dogs, rodents, and humans are more involved in hilly regions. Replacement of draft animals with tractors and discontinued use of untreated night soil for fertilization of fields reduce snail infections. Snail control is practiced but is difficult given the amphibious nature of the snail hosts.

Did you know? About 90% of the genome, or 397 mbp, has been sequenced, and about 13,500 protein-coding sequences identified. Roughly 1000 protein domains known from other eukaryotes are absent from *S. japonicum*, yet notable expansions of protease genes have occurred.

Other schistosome species infecting humans

Schistosoma intercalatum from east Central Africa, *S. guineensis* in west Central Africa, and *S. mekongi* in the Mekong River in Laos and Cambodia all infect people. Cercariae from schistosome species that do not routinely infect humans, including those that normally infect birds, will attempt to penetrate human skin if given the opportunity. In most cases they die in the skin, causing a condition called swimmer's itch, or cercarial dermatitis, characterized by reddened, raised, itchy papules that persist for a few days and disappear (Figure 4). Secondary exposures result in more pronounced reactions. Swimmer's itch is a frequent summertime occurrence where people swim in freshwater lakes, and it can also occur in marine and estuarine habitats. Many cases are attributable to species of *Trichobilharzia*.

References

Zhou XN, Bergquist R, Leonardo L et al (2010) *Schistosomiasis japonicum*: control and research needs. *Adv Parasitol* **72**:145–178.

Zhou Y et al (2009) The *Schistosoma japonicum* genome reveals features of host–parasite interplay. *Nature* **460**:345–U56.

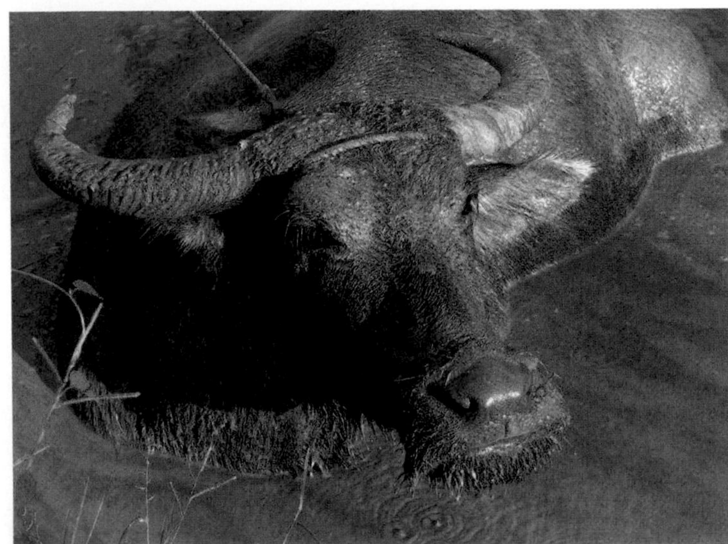

Figure 1 A water buffalo, one of many species able to serve as a definitive host for *Schistosoma japonicum*. (Courtesy of Cabajar, CC BY-SA 3.0)

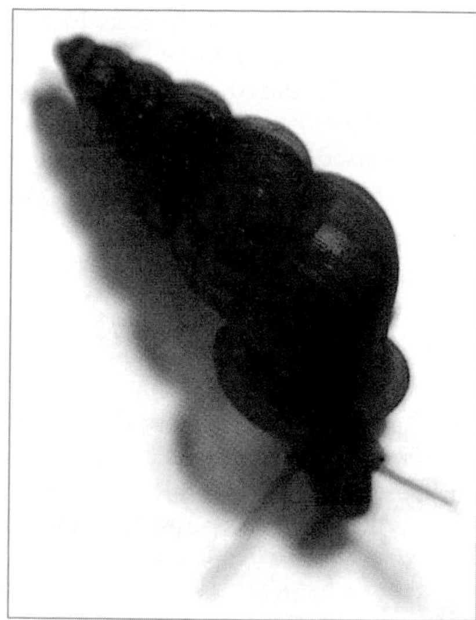

Figure 2 *Oncomelania hupensis*, the prominent snail intermediate hosts for *Schistosoma japonicum* in China. Unlike *Biomphalaria* or *Bulinus*, which are aquatic snails, *Oncomelania* is amphibious, often living on muddy surfaces. Oncomelania snails are only a few millimeters in length. (From Lewis FA, Liang Y, Raghavan N & Knight M [2008] *PLoS Negl Trop Dis* 2(7):e267.)

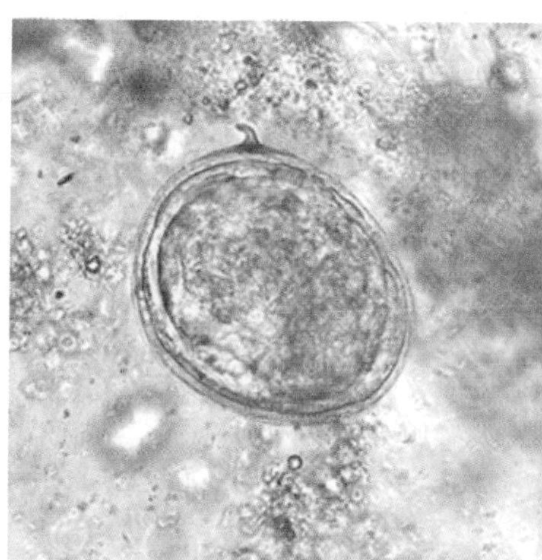

Figure 3 The relatively small eggs (70–100 × 50–70 μm) of *Schistosoma japonicum*, in this example having a vestigial spine, are passed in the feces of the definitive host. (Courtesy of CDC.)

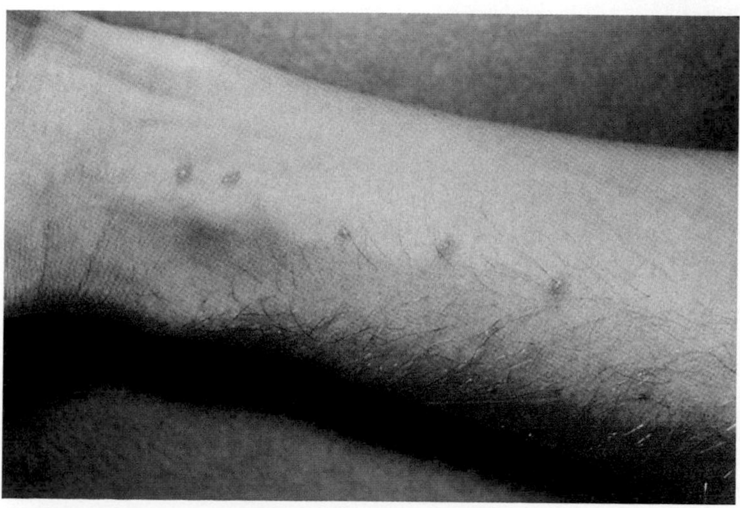

Figure 4 Arm of an individual suffering from swimmer's itch, or cercarial dermatitis. Each papule represents the site of entry of a cercaria of an avian schistosome. (Courtesy of CDC.)

Schistosoma haematobium

Schistosoma haematobium, which was first seen by Theodor Bilharz while performing autopsies on Egyptian peasants in 1851, is the cause of urinary schistosomiasis. Adult worms are found in the veins surrounding the urinary bladder.

Distribution and prevalence This helminth is mostly found in sub-Saharan Africa, with some foci in Southwest Asia, Yemen in particular. Historically, it is also known from Portugal and surprisingly, human cases have been acquired in Corsica in recent years. It is the most common of the schistosomes infecting humans, with an estimated 114 million people infected.

Hosts and transmission *S. haematobium* is the schistosome species most specific to humans. Other primates are occasionally infected. Intermediate hosts are certain species of freshwater pulmonate snails of the genus *Bulinus* (Planorbidae) (Figure 1). Reservoir hosts are not believed to play any significant role in transmission.

Pathology As with *S. mansoni*, the pathology caused by *S. haematobium* results from eggs trapped in the tissues, in this case primarily in the bladder wall. Infections are often accompanied by hematuria (blood in the urine) (Figure 2). Long-term infection can incite squamous cell carcinoma in the bladder, a rare example of cancer induced by a eukaryotic infectious agent. Damage to the ureters can obstruct the flow of urine and cause hydronephrosis (dilation of the pelvis and calyces of the kidneys). Pyelonephritis, or bacterial infection in the kidney, is another frequent consequence of infection. Damage to the glomeruli of the kidneys resulting from deposition of immune complexes can also occur. Infection can result in lesions in the lower female genital tract, likely facilitating the spread of sexually transmitted diseases, including HIV.

Diagnosis The presence of terminal-spined eggs (Figure 3) in the urine, often accompanied by hematuria, is the hallmark of *S. haematobium* infection. Use of dipsticks to detect hematuria provide a good way to quickly identify individuals likely to be infected. Immunologically based detection of antibodies to soluble worm or egg antigens or detection of circulating schistosome antigens can also be used.

Treatment Praziquantel is effective in treating urinary schistosomiasis: adult *S. haematobium* worms are more susceptible to the drug than other species and have shown no signs of development of resistance.

Control Control of *S. haematobium* is somewhat easier than for other schistosomes because transmission foci can often be identified by the presence of children with hematuria. Population-based treatment with praziquantel forms the cornerstone of present-day control programs. Efforts to educate local children not to urinate in the water are considered an important adjunct to control. Focal application of molluscicides to eliminate pockets of snails from irrigation systems have been effective in curtailing transmission in highly focal transmission sites as in Morocco. Improved sanitation and provision of safe water are useful long-term goals. Reservoir hosts are not a factor in maintaining transmission.

Did you know? Egyptian mummies have been found to contain the calcified eggs of *S. haematobium*. The time required from infection to first passage of eggs is greater for this species than for other human schistosomes: 56 days and longer.

References

Dent A & King CH (2007) *Schistosoma haematobium*: the urinary parasite. *Infect Med* **24**:489–496.

Young ND Jex AR, Li B, et al (2012) Whole-genome sequence of *Schistosoma haematobium*. *Nat Genet* **44**:221–225.

Figure 1 *Bulinus globosus*, an important snail intermediate host for *Schistosoma haematobium* in sub-Saharan Africa. Snail length 1cm. (Courtesy of Martina Laidemitt.)

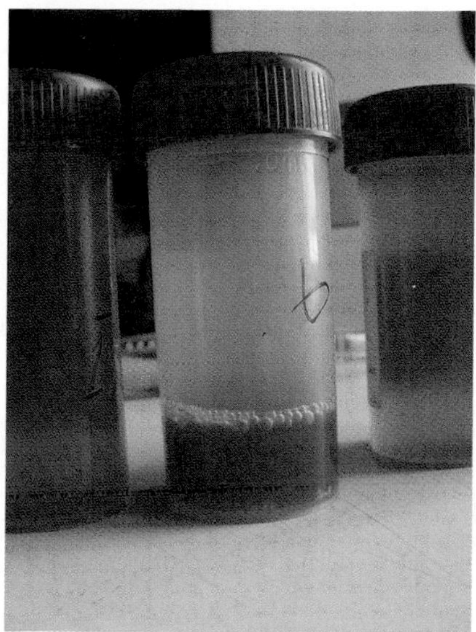

Figure 2 A recently collected urine sample (center) from a child with hematuria.

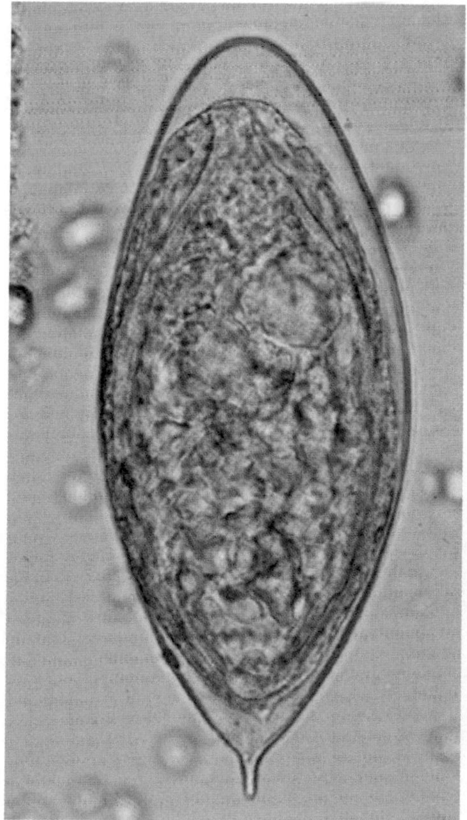

Figure 3 The terminally spined egg (112–170 × 40–70 μm) that is passed in the urine. Note the well-formed miracidium within the egg. (Courtesy of CDC.)

Fasciola hepatica and *Fasciola gigantica*

Large flukes of the family Fasciolidae live in the liver of sheep and cows, but they also increasingly inhabit humans (Figure 1). They cause fascioliasis. In the 1880s, Thomas and Leuckart described the life cycle of *Fasciola hepatica*, the first digenean to have its life cycle revealed.

Distribution and prevalence Fascioliasis has the widest distribution of any disease transmitted by snails. *F. hepatica* is present in Europe, Oceania, and the Americas, whereas in Africa and Asia both species are present. Human fascioliasis occurs in the Andes, Cuba, Egypt, Europe, the Caspian area, and Vietnam. Prevalence in people is between 2.4 to 17 million cases. Spread has been favored by adaptation to new lymnaeid snails hosts and the movement of livestock.

Hosts and transmission Definitive hosts include sheep, cattle, rabbits, camels, marsupials, and other herbivorous mammals, as well as humans. Adults in the gall bladder and bile ducts pass eggs in the feces, which embryonate in freshwater (Figure 2). Miracidia hatch from eggs and penetrate freshwater snails (Lymnaeidae). *Lymnaea truncatula* and *Lymnaea bulimoides* are hosts for *F. hepatica*, and *Radix natalensis* and *R. auricularia* are hosts for *F. gigantica*. Following a sporocyst generation, mother and daughter rediae are produced, and eventually cercariae are released and encyst on aquatic vegetation or at the water's surface. Metacercariae are ingested; excysted juveniles penetrate the gut wall and then the liver capsule to reach the bile ducts.

Pathology The extent of pathology depends on the worm burden. Juveniles do little damage until they penetrate the liver capsule and burrow through the parenchyma. Anemia can result. Worms in the bile ducts cause inflammation and stimulate fibrosis of duct walls, which can interfere with or obstruct the flow of bile, leading to impaired liver function and jaundice. The gall bladder may be damaged and its wall eroded.

Diagnosis Often adult worms are found upon slaughter of infected animals. Unembryonated eggs are passed in the feces. ELISA tests can detect antibodies to fasciolids. Stool antigen tests have been developed. Sequence-based methods may be required to distinguish the two species.

Treatment Triclabendazole is the drug of choice; praziquantel is ineffective.

Control Infection can be controlled by preventing ingestion of raw aquatic plants coming from localities that contain susceptible snails and that are contaminated with feces from infected definitive hosts. Drinking of water containing floating metacercariae should be prevented. Treatment to kill adult worms is also desirable.

Did you know? Hybrids between *F. hepatica* and *F. gigantica* occur.

Related species

Fasciolopsis buski: F. buski is the largest fluke (2–10 cm) to routinely infect humans, inhabiting the small intestine, causing fasciolopsiasis (Figure 3). *F. buski* also infects pigs. About 10 million human infections occur in tropical Asian countries. The life cycle is like that of *F. hepatica* and involves planorbid snails such as *Hippeutis*. Infections are acquired by eating raw plants containing the encysted metacercariae. Worms may cause diarrhea, severe abdominal pain, eosinophilia, and low vitamin B12 levels, and they may produce toxic and allergenic metabolites.

Fascioloides magna: *F. magna*, the giant liver fluke or deer fluke, occurs in North America and Europe. Adults are 4–10 cm long. In cervids, pairs of worms are often recovered from liver cysts with eggs passing from cysts to bile and then to the feces. Infections in domestic animals typically do not yield eggs. The life cycle is similar to that of *F. hepatica* and also employs lymnaeid snails hosts.

Reference

Mas-Coma S, Bargues MD & Valero MA (2005) Fascioliasis and other plant-borne trematode zoonoses. *Int J Parasitol* **35**:1255–1278.

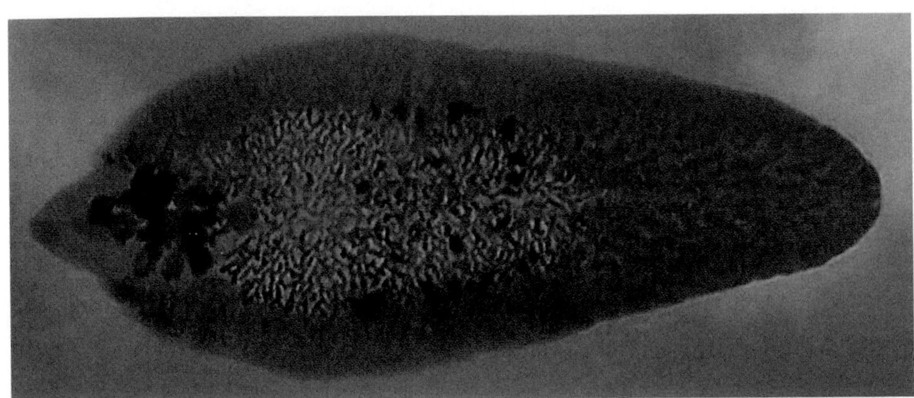

Figure 1 Adult of *Fasciola hepatica*. Note the pronounced cephalic cone and shoulders on the adult worm.

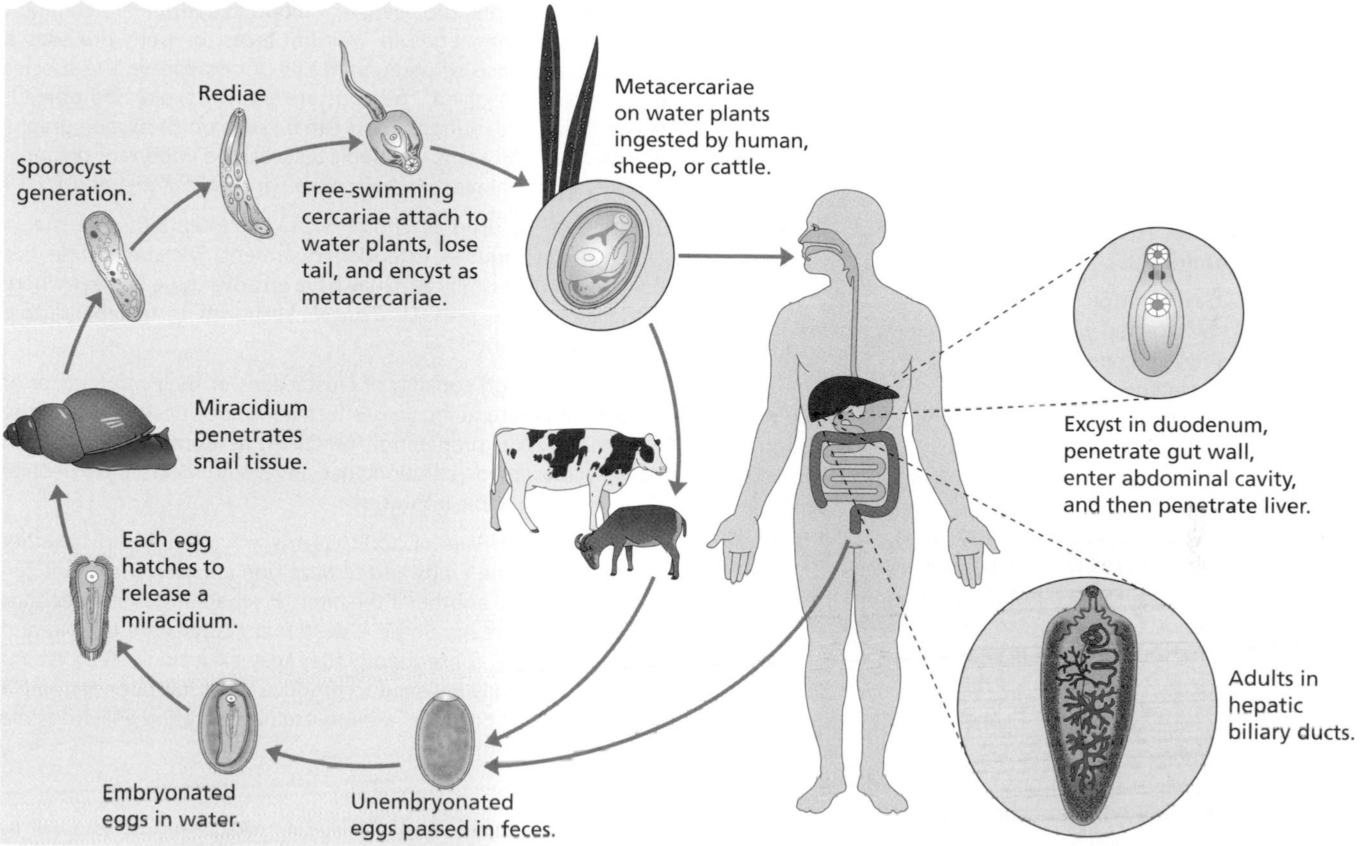

Figure 2 The life cycle of *Fasciola hepatica* and *F. gigantica*. (Courtesy of CDC.)

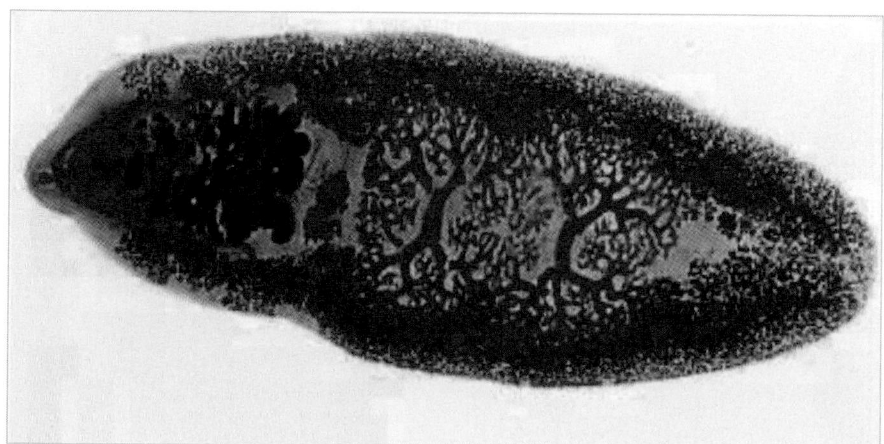

Figure 3 Adult of *Fasciolopsis buski*. This species lacks the pronounced cephalic cone and shoulders. Note the highly branched testes in the central third of the worm. Unlike *F. hepatica*, this worm lives in the small intestine.

Paragonimus westermani

Paragonimus westermani is the most important of many species of lung-inhabiting flukes (Figure 1) of the genus *Paragonimus* (family Troglotrematidae). It causes paragonimiasis. This is the only group of human helminths with adults that reside in the lung.

Distribution and prevalence *P. westermani* is the most widely distributed member of the genus, ranging from Japan throughout Southeast Asia to India. Other species of *Paragonimus* occur in Asia, Africa, and North, Central, and South America. About 22 million people are infected.

Hosts and transmission Crab-eating mammals, including humans, become infected with these spinose worms, which have the shape and size of coffee beans and live in cysts in the lungs (Figure 2). Operculated eggs (80 μm × 120 μm) are either expectorated or passed in the feces, embryonate in freshwater, and hatch to infect snails (*Semisulcospira* and others). Sporocyst and rediae stages occur, and short-tailed cercariae are produced that eventually penetrate and encyst in freshwater crabs (Figure 3) or crayfish as second intermediate hosts. Definitive hosts become infected when raw, pickled, or poorly cooked crabs are ingested. Worms fail to develop to maturity in wild boars, which serve as paratenic hosts. Both humans and tigers can become infected by eating uncooked flesh from these paratenic hosts.

Pathology Early-stage paragonimiasis is marked by abdominal pain, fever, diarrhea, eosinophilia, chest pain, and fatigue. Established pulmonary paragonimiasis causes cough and recurrent blood in sputum (hemoptysis), rust-colored sputum (from the presence of eggs), weakness, chronic bronchitis and wheezing cough, and dyspnea (shortness of breath). **Ectopic** paragonimiasis (manifestations resulting from the presence of parasite in a location other than the lungs) usually affects the brain. *Paragonimus* usually causes limited morbidity and rarely death.

Diagnosis Cough and presence of blood in sputum are common symptoms. Finding eggs in sputum, feces, or both provides a definitive diagnosis of paragonimiasis but may leave the species responsible unresolved. The eggs are brown, ovoid, and operculate, with opercular ridges. Cysts can be seen on chest radiographs or with CT or MRI scans. Serologic tests can be important diagnostic tools. Paragonimiasis is often misdiagnosed as tuberculosis.

Treatment The drug of choice is praziquantel. Bithionol is also effective but requires extended treatment. Triclabendazole has been used successfully and may have an advantage in areas where fascioliasis is also present. (*Fasciola* responds to triclabendazole but not praziquantel.)

Control Thorough cooking of crustaceans or their products or of meat from paratenic hosts is effective, as is education to change customs in food preparation practices, as exemplified in South Korea. Widespread pollution that adversely affects intermediate hosts may play a role in control.

Did you know? Pairs of adult worms are often found together in cysts within the lungs and fertilize one another. One adult can more readily find another if the interval separating their infections is less than 12 weeks. Some *P. westermani* adults are triploid and they self-fertilize, consequently they may exist singly in a cyst. An adult in the lung is believed to produce what has been termed "a zone of immune privilege" around itself by secreting enzymes able to digest host antibodies.

Reference

Procop GW (2009) North American paragonimiasis (caused by *Paragonimus kellicotti*) in the context of gobal paragonimiasis. *Clin Microbiol Rev* **22**:415–446.

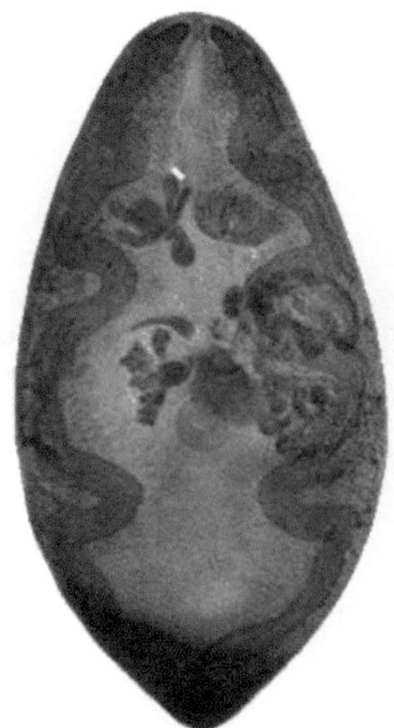

Figure 1 Adult of *Paragonimus westermani*. These worms are about the same size and shape as a coffee bean and live in cysts within the lungs. (Courtesy of CDC.)

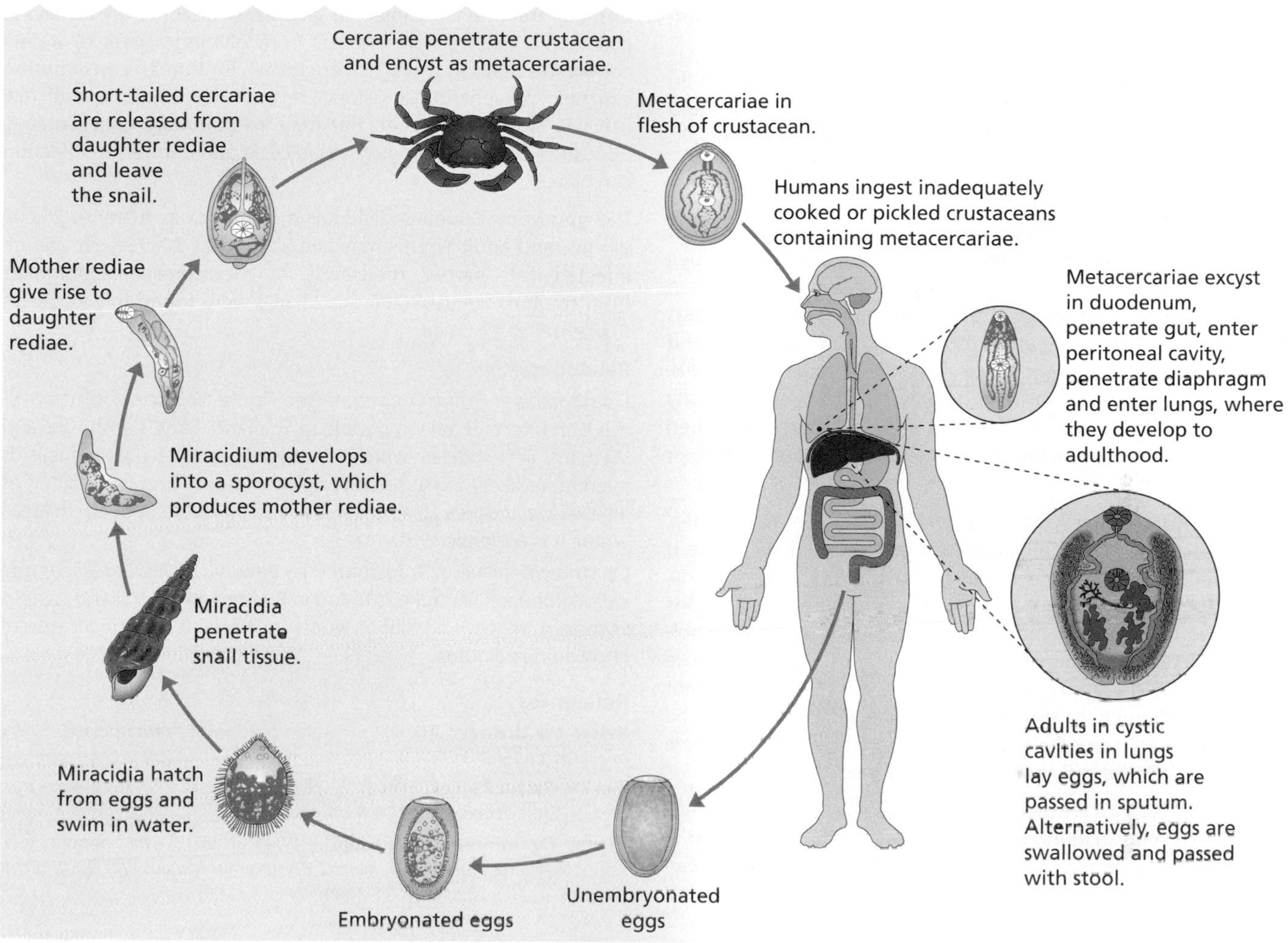

Cercariae penetrate crustacean and encyst as metacercariae.

Metacercariae in flesh of crustacean.

Humans ingest inadequately cooked or pickled crustaceans containing metacercariae.

Metacercariae excyst in duodenum, penetrate gut, enter peritoneal cavity, penetrate diaphragm and enter lungs, where they develop to adulthood.

Adults in cystic cavities in lungs lay eggs, which are passed in sputum. Alternatively, eggs are swallowed and passed with stool.

Unembryonated eggs

Embryonated eggs

Miracidia hatch from eggs and swim in water.

Miracidia penetrate snail tissue.

Miracidium develops into a sporocyst, which produces mother rediae.

Mother rediae give rise to daughter rediae.

Short-tailed cercariae are released from daughter rediae and leave the snail.

Figure 2 The life cycle of *Paragonimus westermani*. (Courtesy of CDC.)

Figure 3 Freshwater crabs serve as second intermediate hosts for *Paragonimus westermani*, harboring the metacercaria stage in the life cycle. People become infected by eating raw or poorly cooked crabs containg metacercariae. (Courtesy of CDC.)

Clonorchis sinensis

The Chinese or oriental liver fluke (Figure 1), a member of the family Opisthorchiidae, causes clonorchiasis.

Distribution and prevalence *Clonorchis sinensis* is found in regions of Asia, including China, Korea, Taiwan, and Vietnam, and formerly in Japan. About 35 million people are infected globally, with about 15 million of these cases occurring in China.

Hosts and transmission Fish-eating mammals including dogs, cats, and humans (and likely other species) serve as definitive host (Figure 2), with adult worms living in the bile ducts or gall bladder, or rarely the pancreatic duct. Eggs pass in the feces and are eaten by freshwater snails of the genus *Parafossarulus* and many others (including *Bithynia* and *Semisulcospira*) in which sporocysts and rediae develop. Tailed, swimming cercariae emerge from snails, encyst as metacercariae in many freshwater fish species (mostly Cyprinidae) or freshwater shrimp, and then are transmitted when the fish or shrimp are eaten raw, undercooked, dried, salted, or pickled.

Pathology Light infections (<100 worms) may cause diarrhea, eosinophilia, and jaundice, which may occur if the flow of bile is obstructed. In medium infections (<1000 worms), fever, anorexia, and abdominal distention may occur, and in heavy infections (>1000 worms), portal hypertension, abdominal pain, and cholangitis (bacterial infection of bile duct) may occur. Adults cause hyperplasia of the epithelium of the bile ducts, and malignant cholangiocarcinoma (cancer of the bile duct), which is also caused by *C. sinensis* (see also *Opisthorchis viverrini* below). This cancer is difficult to diagnose early and when found is usually fatal.

Diagnosis Infections are diagnosed through the detection of small characteristic eggs (Figure 3) in feces of people who have lived in endemic areas or eaten raw fish from such areas. Ultrasound can detect fibrosis in bile ducts and computer tomography can reveal occluded bile ducts.

Treatment Clonorchiasis is most commonly treated with praziquantel; cures are also achieved using albendazole.

Control Treatment of infected people is effective in achieving control. Preventing human feces from gaining access to waters containing snails in which fish are grown for human consumption can help prevent transmission. Preventing consumption of raw fish will stop transmission but may be culturally unacceptable. Education of children about the life cycle and dangers of infection can help.

Did you know? Some individuals may harbor as many as 25,000 worms, and adult worms may live as long as 20–25 years. Some infected fish harbor thousands of metacercariae. Mummies from the Han dynasty (206 BC–23 AD) were found infected with *C. sinensis*.

Related species

Opisthorchis viverrini: O. viverrini (see Figure 1) causes opisthorchiasis and infects 9 million people in Thailand, Laos, Cambodia, and Vietnam. This species, which is transmitted by *Bithynia* snails, is even more likely to cause cholangiocarcinoma and is classified as a type I carcinogen. This cancer is a leading cause of death in areas where it is endemic.

Opisthorchis felineus: Transmitted by *Bithynia* snails, *O. felineus*, the cat or Siberian liver fluke, is found in Russia and central and eastern Europe. It infects 1.5 million people in Russia. It may be a cause of cholangiocarcinoma.

References

Keiser J & Utzinger J (2005) Emerging foodborne trematodiasis. *Emerg Infect Dis* **11**:1507–1514.

Lun ZR, Gasser RB, Lai DH et al (2005) Clonorchiasis: a key foodborne zoonosis in China. *Lancet Infect Dis* **5**:31–41.

Petney TN, Andrews RH, Saijuntha W et al (2013) The zoonotic, fish-borne liver flukes *Clonorchis sinensis*, *Opisthorchis felineus* and *Opisthorchis viverrini*. *Int J Parasitol* **43**:1031–1046.

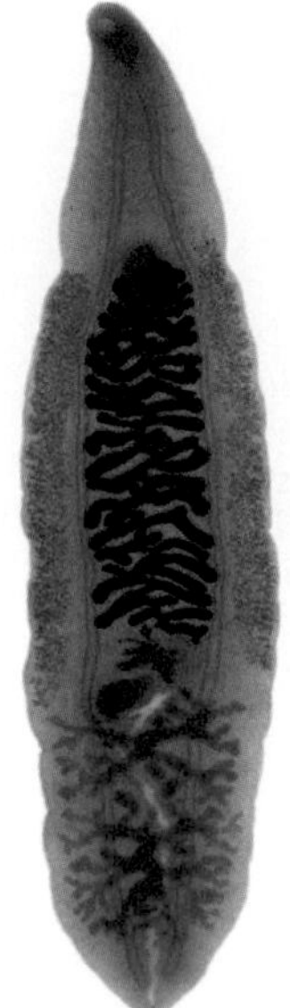

Figure 1 Adults of *Clonorchis sinensis* (left) and *Opisthorchis viverrini* (right). Note that the adults can be easily differentiated on the basis of the shape of the tandemly arranged testes located in the posterior part of the worm; the testes are highly branched for *C. sinensis* and condensed and lobed for *O. viverrini*. (From Sripa B, Kaewkes S, Sithithaworn P et al [2007] *PLoS Medicine* 4(7): e201.)

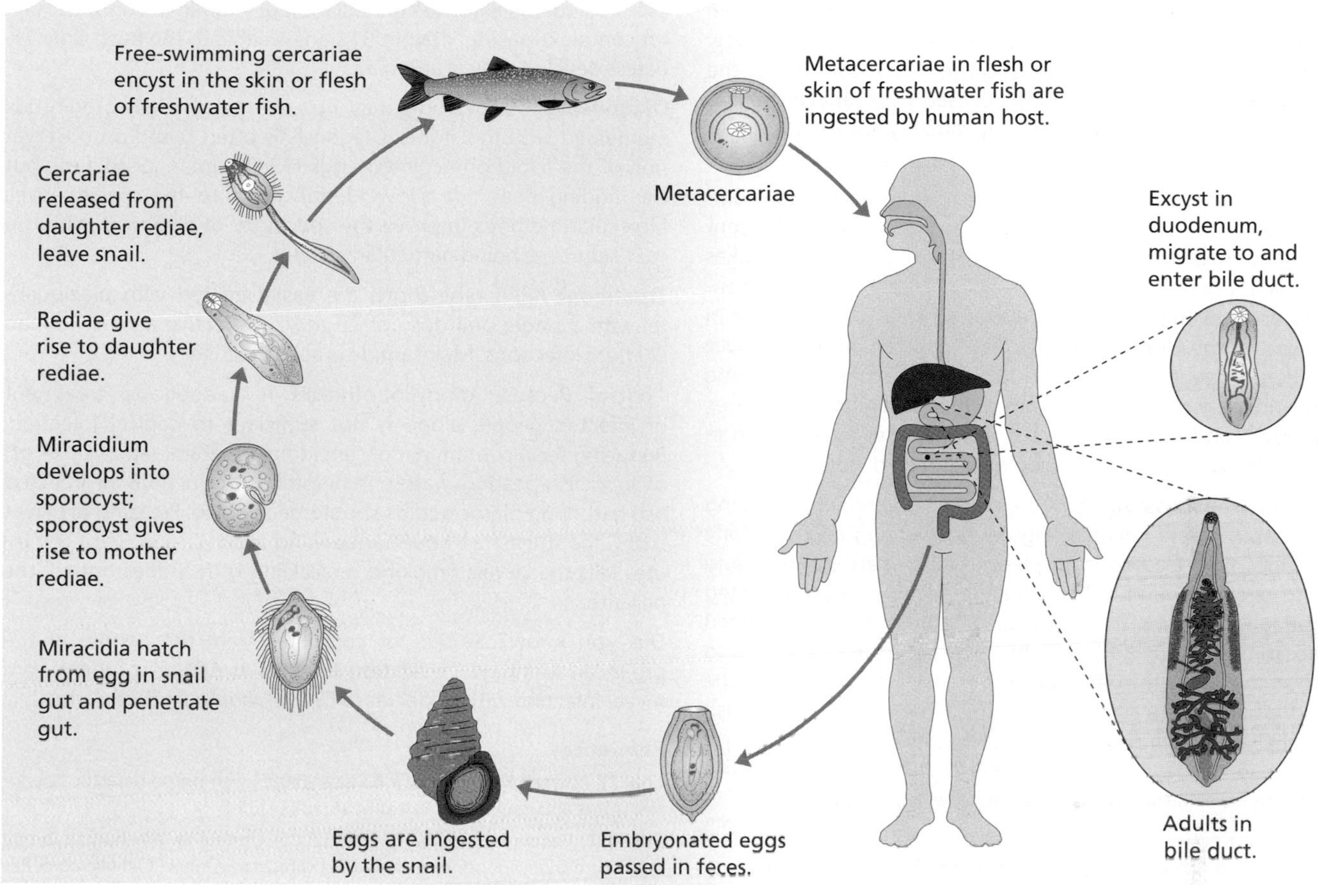

Free-swimming cercariae encyst in the skin or flesh of freshwater fish.

Metacercariae in flesh or skin of freshwater fish are ingested by human host.

Metacercariae

Excyst in duodenum, migrate to and enter bile duct.

Cercariae released from daughter rediae, leave snail.

Rediae give rise to daughter rediae.

Miracidium develops into sporocyst; sporocyst gives rise to mother rediae.

Miracidia hatch from egg in snail gut and penetrate gut.

Eggs are ingested by the snail.

Embryonated eggs passed in feces.

Adults in bile duct.

Figure 2 The life cycle of *Clonorchis sinensis*. (Courtesy of CDC.)

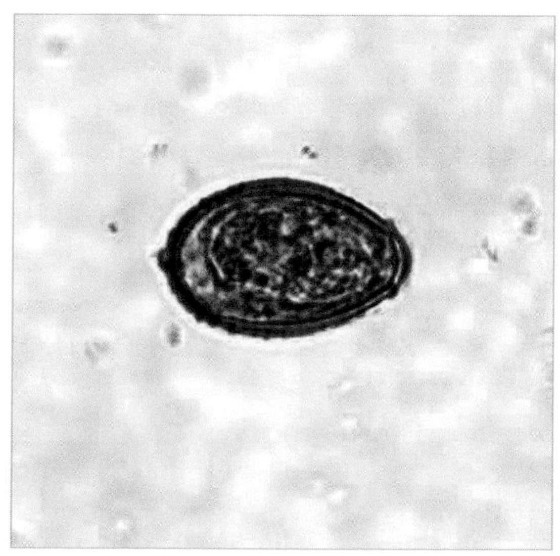

Figure 3 The relatively small (27–35 μm × 12–20 μm) operculated egg of *Clonorchis sinensis*. (Courtesy of CDC.)

Diphyllobothrium latum

A eucestode of the family Diphyllobothriidae, *Diphyllobothrium latum* (Figures 1 and 2) along with related species collectively known as broad or fish tapeworms, cause diphyllobothriasis, the most important fish-borne zoonosis caused by a tapeworm. It is the largest parasite of humans, usually ranging from 2–15 m, but can be up to 25 m in length.

Distribution and prevalence This zoonosis occurs most commonly in countries where raw or marinated fish are regularly consumed. It is present in North America (especially the Great Lakes region and Alaska) and Europe (especially Finland), and cases continue to occur even in the most developed areas. It is common in Russia east of the Urals and is regularly reported in Japan, South Korea, Peru, Argentina, and Chile. More work is needed to more precisely identify the species involved in each region. Freshwater, anadromous, and to a lesser extent marine fishes are involved in transmission. Up to 20 million people are infected worldwide.

Hosts and transmission Many fish-eating mammals, including people, can become infected (Figure 3) by eating raw or poorly cooked fish. Eggs passed in the feces embryonate in water and hatch to release swimming coracidia larvae, which are ingested by copepods in which they develop as procercoids. An infected crustacean is eaten by a fish in which the procercoid develops to a plerocercoid, usually in the body cavity. A large fish may eat the smaller fish and serve as a paratenic host. The most common fish hosts for *D. latum* are perch, pike, burbot, and walleye. Infections mature in 15–45 days in the definitive host, and an adult worm may produce a million eggs per day and live for decades.

Pathology Infections are usually asymptomatic, but abdominal pain, fatigue, diarrhea, itching, headaches, depression, or anxiety may occur. Prolonged infection may cause megaloblastic anemia as a result of the parasite's dissociation of vitamin B12-intrinsic factor complex, making vitamin B12 unavailable to the host. Only 2% or less develop clinical anemia.

Diagnosis The tapeworm may first be noted when proglottids are passed with the stools. Diagnosis is often based on observation of the ovoid operculated eggs (35–80 μm × 25–65 μm), but this finding does not allow identification to the species level. Molecular methods improve the specificity of diagnosis, with the *cox1* sequence being particularly useful.

Treatment Adult tapeworms are easily treated with praziquantel with a single oral dose of 25 mg/kg effective against human *D. latum* infections. Niclosamide is also effective.

Control Because diphyllobothriasis is a zoonosis, treatment of infected people alone is not sufficient to control infection. Reducing fecal contamination (including contamination by dogs) of food fish-bearing waters is helpful. Introductions of infected fish harboring plerocercoids should be avoided. Freezing of fish at −18°C for at least 24 hours or cooking at 55°C for at least 5 minutes kills the larvae. Smoking or pickling of fish does not kill the parasite.

Did you know? Studies of coprolites from the people of the pre-Incan Chiribaya civilization (600–1476 AD) in southern Peru reveal infection with a species of *Diphyllobothrium* tapeworm.

References

Chai JY, Murrell KD, Lymbery AJ et al (2005) Fish-borne parasitic zoonoses: status and issues. *Int J Parasitol* **35**:1233–1254.

Scholz T, Garcia H, Kuchta R et al (2009) Update of the human broad tapeworm *Diphyllobothrium*, including clinical relevance. *Clin Microbiol Rev* **22**:146–160.

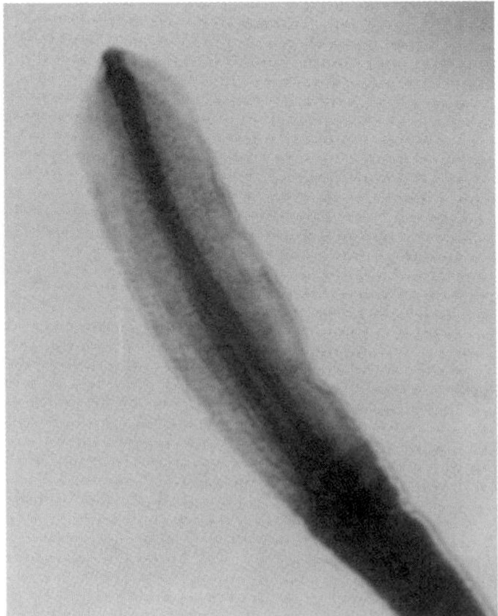

Figure 1 Scolex of *Diphyllobothrium latum*. The scolex of this pseudophyllidean tapeworm does not possess hooks or suckers, but rather two slit-like bothria, or grooves that run along either side of the scolex. These are used to adhere to the host's intestinal epithelium.

Figure 2 Proglottids of *Diphyllobothrium latum*, showing the centrally placed and convoluted dark rosette-shaped uterus full of eggs found in each proglottid. (Courtesy of CDC/Mae Melvin.)

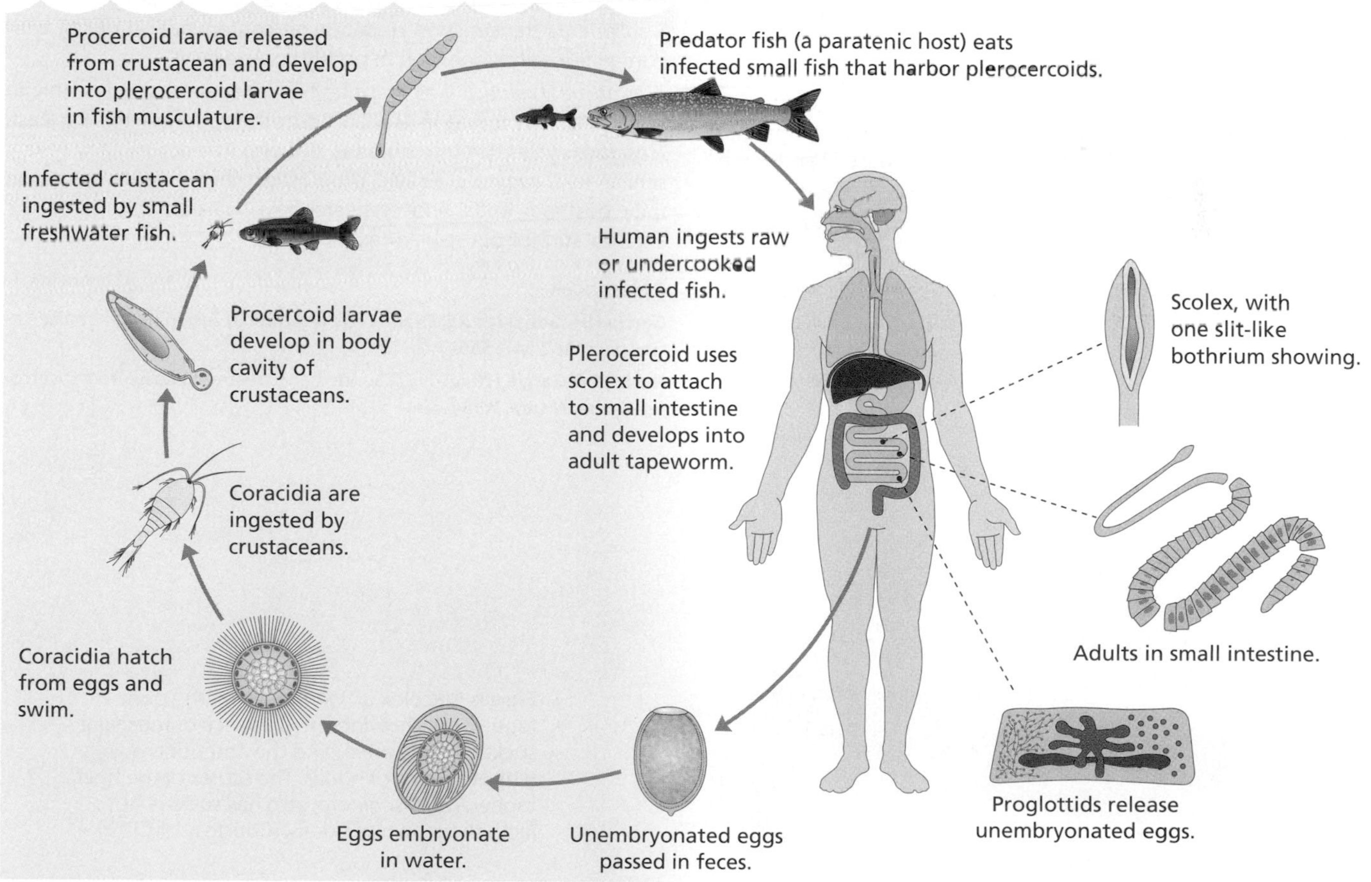

Procercoid larvae released from crustacean and develop into plerocercoid larvae in fish musculature.

Predator fish (a paratenic host) eats infected small fish that harbor plerocercoids.

Infected crustacean ingested by small freshwater fish.

Human ingests raw or undercooked infected fish.

Procercoid larvae develop in body cavity of crustaceans.

Plerocercoid uses scolex to attach to small intestine and develops into adult tapeworm.

Scolex, with one slit-like bothrium showing.

Coracidia are ingested by crustaceans.

Adults in small intestine.

Coracidia hatch from eggs and swim.

Proglottids release unembryonated eggs.

Eggs embryonate in water.

Unembryonated eggs passed in feces.

Figure 3 The life cycle of *Diphyllobothrium latum*. (Courtesy of CDC.)

Taenia solium

A member of the family Taeneidae, adults (Figure 1) live in the human small intestine causing taeniasis. Infection with the cysticercus larval stage causes cysticercosis.

Distribution and prevalence *Taenia solium* has a cosmopolitan distribution, but it is more prevalent in areas with poor sanitation and where pigs are free ranging and have access to consumption of human feces. It is common in Latin America, Eastern Europe, sub-Saharan Africa, Asia, and Oceania. About 400,000 people have neurocysticercosis in Latin America alone, with 1.7–3 million cases of epilepsy occurring worldwide per year. It is responsible for 50,000 deaths annually.

Hosts and transmission Adult tapeworms are found only in humans. Eggs or proglottids are passed in the feces (Figure 2) and are consumed by pigs, the only intermediate hosts of significance. The eggs hatch in the pig small intestine, burrow across the gut wall, gain access to the bloodstream, and are distributed to the tissues, where each larva grows into a bladder worm called a **cysticercus**. When raw or poorly cooked infected pork is ingested, the cysticerci will develop into adult tapeworms in the small intestine. Fecal contamination by an infected person (for example, due to lack of handwashing following a restroom visit) may result in eggs being ingested by nearby people, resulting in cysticerci in human tissues.

Pathology People with adult tapeworms suffer little damage (diarrhea, abdominal pain, or nausea) from the adult tapeworms but are at high risk for cysticercosis. Lodging of cysticerci in neural tissue (neurocysticercosis) is especially grave (Figure 3) and can be lethal. It is the world's greatest cause of acquired epilepsy; epileptic seizures occur in 50%–80% of patients with parenchymal brain cysts. Some cysts can grow unusually large.

Diagnosis Taeniasis is diagnosed by finding of typical taeneid eggs in the feces of people with a history of eating raw pork. Gravid proglottids can facilitate identification as *T. solium*. Detection of antigens in the stool by ELISA, and CT and especially MRI scans are effective for diagnosis of cysticercosis.

Treatment Niclosamide and praziquantel both kill adult worms. Niclosamide is preferred because it does not affect asymptomatic brain cysts and thereby provoke neurological symptoms. Steroids relieve symptoms of neurocysticercosis, and drug treatment (albendazole or praziquantel) may be called for. Antiepileptic drugs perform well in treating seizures.

Control Treatment of individuals with adult worms limits contamination of the environment and prevents cysticercosis both of the infected person and others living nearby. Practicing good personal hygiene, preventing access of pigs to human feces, and eating well-cooked pork are all important.

Did you know? Each gravid proglottid of an adult worm contains between 30,000 and 50,000 eggs. In one study, 37% of the people surveyed in Peru showed signs of having been infected at some time.

Related species

T. saginata: Adults of *T. saginata*, the beef tapeworm, also occur exclusively in humans, but cysticerci are found in the musculature of cattle. The adult worm is larger (up to 12 m) than for *T. solium*, and the scolex has suckers but lacks the double crown of hooks found on the *T. solium* scolex. The beef tapeworm does not cause cysticercosis in people, and so is considerably less dangerous than *T. solium*. Its transmission depends on contamination of soil with human feces and ingestion of poorly cooked beef.

T. asiatica: T. asiatica was identified as a separate species only in the 1980s. It is found in Taiwan and other parts of the Far East. *T. asiatica* cycles through humans and pigs. Its morphology is very similar to *T. saginata*. Gravid proglottids have protuberances on their posterior ends, and cysticerci have wartlike bumps on the external surface of the bladder.

References

Garcia HH, Gonzalez AE, Evans CAW et al (2003) *Taenia solium* cysticercosis. *Lancet* **362**:547–556.

Nash TE & Garcia HH (2011) Diagnosis and treatment of neurocysticercosis. *Nat Rev Neurol.* **7**:584–594.

Figure 1 Scolex of *Taenia solium*, the pork tapeworm, showing the presence of muscular suckers (there are 4) and the anterior crown with two rows of hooks. The scolex of the beef tapeworm, *T. saginata*, also has suckers but lacks the crown of hooks. (Courtesy of CDC.)

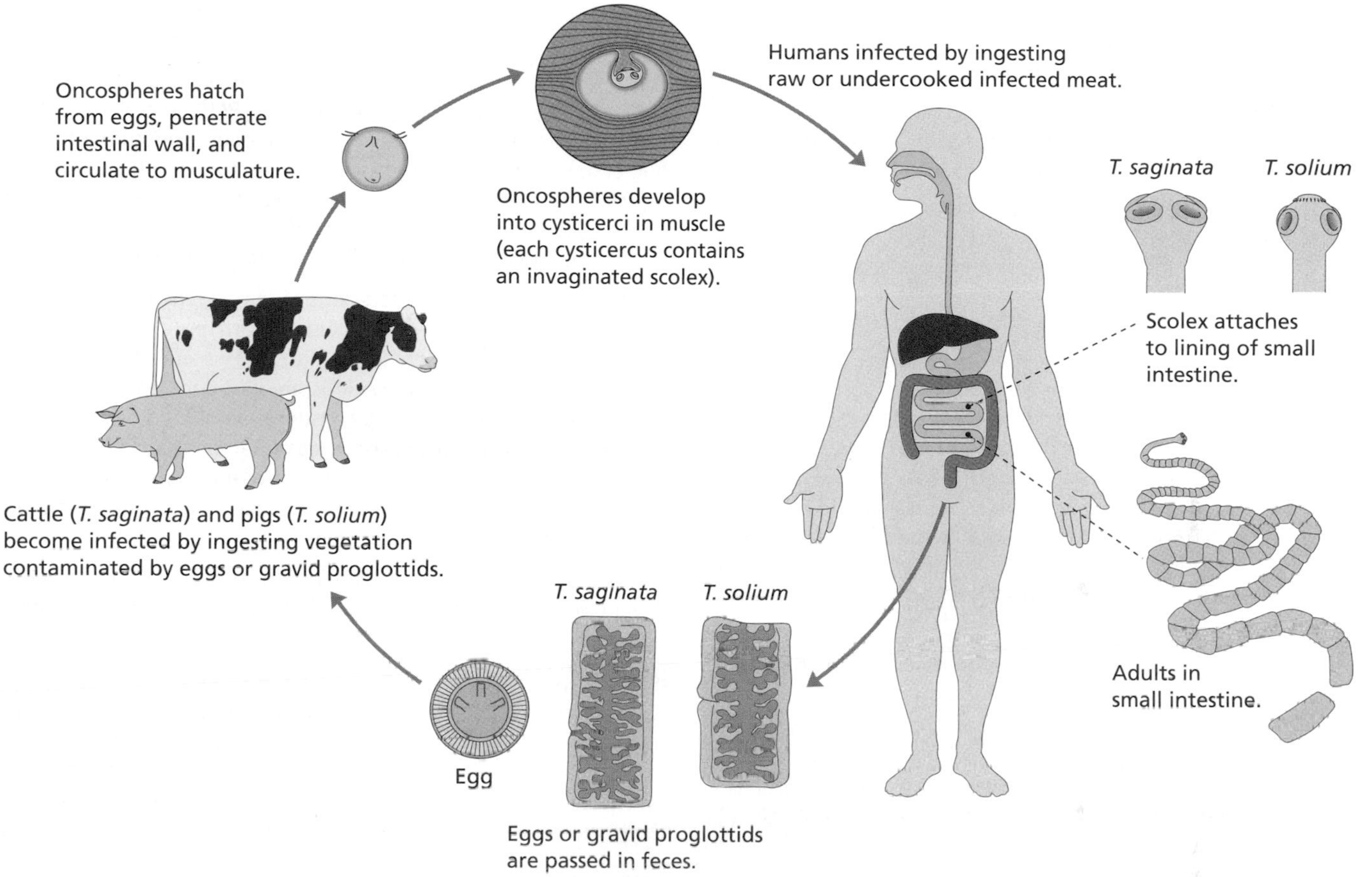

Oncospheres hatch from eggs, penetrate intestinal wall, and circulate to musculature.

Oncospheres develop into cysticerci in muscle (each cysticercus contains an invaginated scolex).

Humans infected by ingesting raw or undercooked infected meat.

T. saginata

T. solium

Scolex attaches to lining of small intestine.

Adults in small intestine.

Cattle (*T. saginata*) and pigs (*T. solium*) become infected by ingesting vegetation contaminated by eggs or gravid proglottids.

T. saginata

T. solium

Egg

Eggs or gravid proglottids are passed in feces.

Figure 2 The life cycle of *Taenia solium* and *T. saginata*. For *T. solium*, the cysticerci are found in swine, which serve as typical intermediate hosts. Humans may become infected with cysticerci, leading to cysticercosis. Adults of *T. solium* produce gravid proglottids with 7–13 uterine branches per side. For *T. saginata*, the intermediate hosts are cattle, and cysticercosis does not occur in people. Gravid proglottid have 15–20 uterine branches per side. Both species produce eggs with thick walls that contain numerous pores and cannot readily be distinguished from one other. (Courtesy of CDC.)

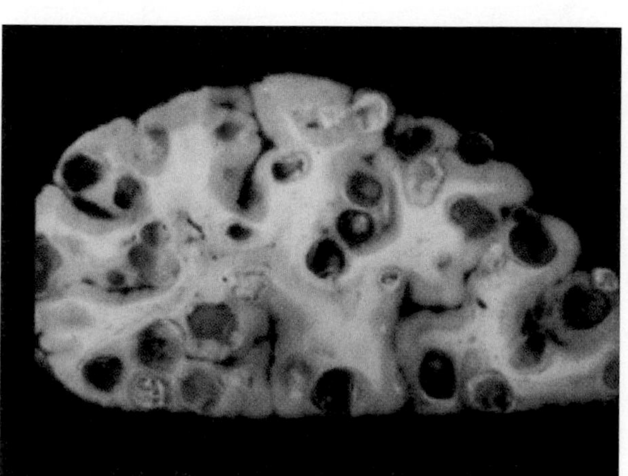

Figure 3 A case of neurocysticercosis showing numerous cysticerci of *Taenia solium* in the brain of an infected patient. It is likely such a patient also harbored an adult worm. (Courtesy of NIH.)

Echinococcus granulosus

Canids serve as definitive hosts for *Echinococcus granulosus*. The larval stage is problematic as it develops into an asexually proliferating unilocular hydatid cyst and causes echinococcosis, or hydatid disease.

Distribution and prevalence Distribution is cosmopolitan, and *E. granulosus* is found on all continents and in at least 100 countries.

Hosts and transmission The tiny (2–7 mm) adult tapeworms inhabit the small intestine of dogs and other canids, where they produce taeneid-type eggs that are passed in the feces (Figures 1 and 2). Eggs are ingested by grazing animals (sheep, goats, cattle, pigs, horses, and many others, and sometimes people), which serve as intermediate hosts. Following hatching of the egg in the gut, the oncosphere crosses the gut wall, enters the blood, and is carried to the liver, lungs, or other organs, where it will develop into a fluid-filled hydatid cyst, 1–20 cm in diameter. The hydatid cyst contains a protective wall, and inside it is a germinal layer that can produce thousands of protoscolices, each of which can develop into an adult tapeworm should a canid ingest the cyst. There are at least nine strains of *E. granulosus*, with some cycling through domestic dogs and ungulates, whereas others parasitize wild carnivores and ungulates.

Pathology Adult worms cause little or no harm to their canid hosts. The hydatid cysts occupy various organs, usually liver (Figure 3) and lungs but sometimes the bones or the brain of the intermediate host. They can grow to large size and cause morbidity or mortality. Hydatid cysts can cause abdominal pain, hepatomegaly, ascites, coughing, lung abcesses, and neurological symptoms, depending on their location. Cysts can rupture, leaking fluid that causes anaphylaxis.

Diagnosis Eggs can be found in canid fecal samples but cannot be differentiated from in the eggs of other taeneid species. ELISA based on antigens in feces can be used to detect infections in dogs. Diagnosis of echinococcosis in people relies on imaging techniques such as ultrasonography, but also computed tomography and X-rays. Serum antibodies can also be detected.

Treatment Infected dogs are treated with praziquantel. In people, cysts may be removed by surgery. The PAIR procedure is often implemented and consists of percutaneous puncture of cysts guided by ultrasonography, aspiration of most cyst fluid, replacement of cyst fluid by injection of ethanol to kill the germinal layer of the cyst, and then reaspiration of the ethanol. Cysts may be inactivated by chemotherapy with albendazole or mebendazole.

Control Treatment of infected dogs to remove the source of eggs is critical. Avoidance of close contact with potentially infected dogs, prevention of access of dogs to raw viscera of slaughtered animals, and control of stray dogs are all important. Practice of good basic sanitation to avoid consumption of food or water contaminated with canid feces and eggs of *E. granulosus* is important. Slaughterhouse inspections revealing hydatid cysts are indicative of a problem. Vaccination with recombinant oncosphere antigen EG95 can protect sheep.

Did you know? One of the *E. granulosus* strains is endemic on Isle Royale in Lake Superior, where it cycles between the indigenous moose and wolf populations.

Related Species

E. multilocularis: This Holarctic species infects foxes as definitive hosts and rodents (especially voles) as intermediate hosts. It produces a multilocular or alveolar hydatid cyst. The cysts can be very invasive and tumorlike in growth. If humans become infected with the cysts, it is one of the most lethal of human helminth infections.

Reference

Eckert J & Deplazes P (2004) Biological, epidemiological, and clinical aspects of echinococcosis, a zoonosis of increasing concern. *Clin Microbiol Rev* **17**:107–135.

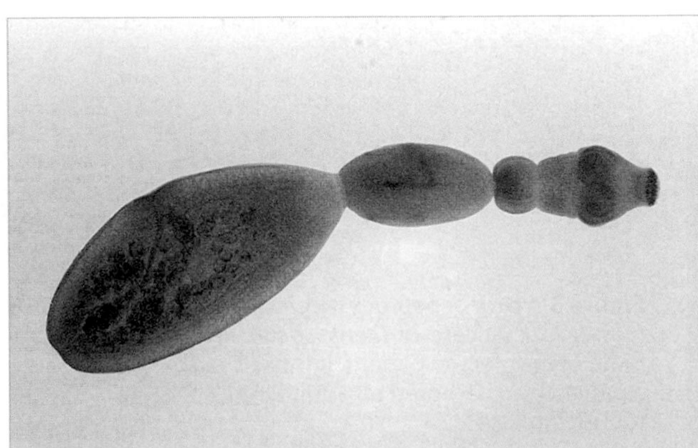

Figure 1 Adult of *Echinococcus granulosus* showing small size (2–7 mm long) with body composed of three segments (immature, mature, and gravid proglottids). (Courtesy of SJ Upton, Kansas State University.)

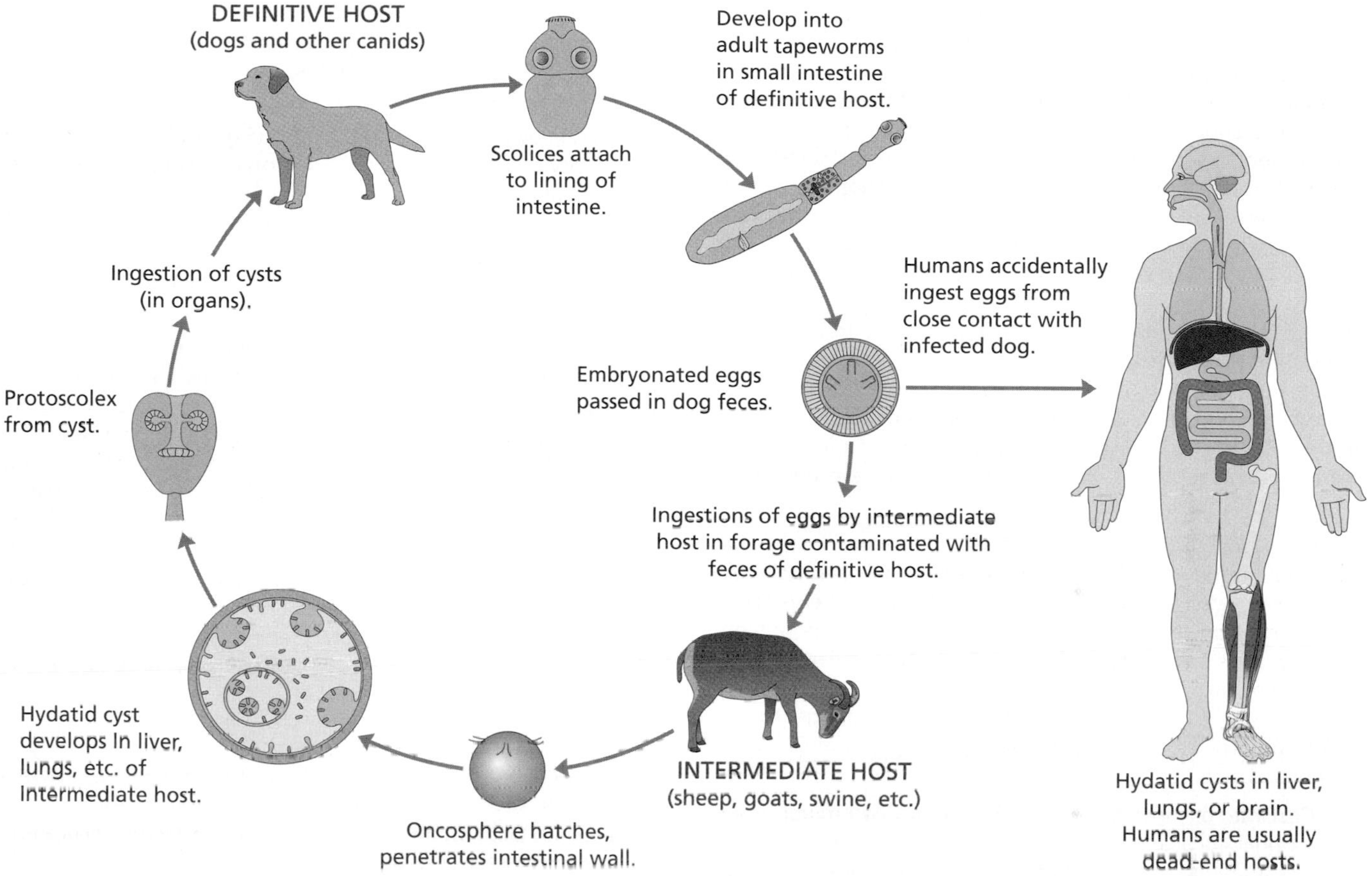

DEFINITIVE HOST
(dogs and other canids)

Scolices attach
to lining of
intestine.

Develop into
adult tapeworms
in small intestine
of definitive host.

Humans accidentally
ingest eggs from
close contact with
infected dog.

Embryonated eggs
passed in dog feces.

Ingestion of cysts
(in organs).

Protoscolex
from cyst.

Ingestions of eggs by intermediate
host in forage contaminated with
feces of definitive host.

Hydatid cyst
develops in liver,
lungs, etc. of
intermediate host.

INTERMEDIATE HOST
(sheep, goats, swine, etc.)

Oncosphere hatches,
penetrates intestinal wall.

Hydatid cysts in liver,
lungs, or brain.
Humans are usually
dead-end hosts.

Figure 2 The life cycle of *Echinococcus granulosus*. Note the involvement of canids as definitive hosts and grazing animals such as sheep as intermediate hosts. Humans are dead-end hosts for the parasite as their remains are unlikely to be consumed by canids, at least in the present world. (Courtesy of CDC.)

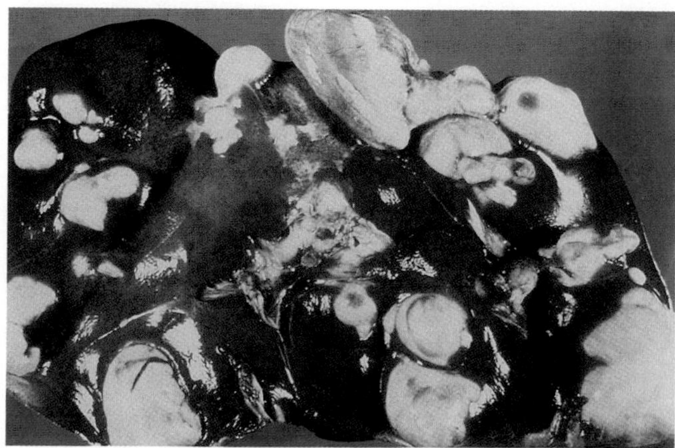

Figure 3 A sheep liver containing many hydatid cysts of *Echinococcus granulosus*. (Courtesy of CDC.)

Class Monogenea and Class Cestoda

Monogeneans are usually ectoparasitic on the gills or external surfaces of cold-blooded vertebrates, especially fish. There are also several species that are endoparasitic, colonizing locations such as the oral cavity, urinary bladder, or the orbit of the eye. The adults have conspicuous anterior and posterior holdfast structures that enable them to secure their position onto their hosts' surfaces. The posterior attachment structure, called an **opisthaptor**, is prominent, diverse in form among species and armed with suckers, various kinds of hooks, or both (Figure 1). Depending on the species, the adults feed either on the mucus of the host's skin or on blood. They have a direct life cycle with the hermaphroditic adult producing eggs in most species. The egg hatches to release a ciliated, short-lived **oncomiracidium** that swims to a new host and attaches and grows directly into an adult worm. They are usually host specific and often site specific as well. Some species are viviparous and can produce sequential generations of offspring without releasing them, and up to 20 generations have been found nested within one individual.

Monogeneans colonize freshwater, estuarine, and marine habitats and are found on bony fish, sturgeons and paddlefish, elasmobranchs, ratfish, amphibians, reptiles, squid, and crustaceans. There is even one species inhabiting the eye of the hippopotamus. Monogeneans can be life threatening to fish, particularly in hatcheries or when fish are farmed under high density. None are known to infect people. It is debated as to whether the monogeneans are a monophyletic group or **paraphyletic** (including some but not all the descendents of a common ancestor).

The Cestoda, or tapeworms, are divided into two minor groups, the Amphilinidea and Gyrocotylidea, and a major lineage, the Eucestoda. All are endoparasitic, with complex life cycles usually involving two hosts: usually an arthropod in which larval development occurs and a vertebrate definitive host in which the hermaphroditic adult worm is found. Cestodes lack a mouth and gut in all life-cycle stages. Tapeworms have a syncytial tegument featuring **microtriches**, which are tubelike microvilli that have an electron dense cap. The increased surface area of the tapeworm tegument facilitates uptake of nutrients.

The eight known species of amphilinids (Figure 2) are found in the body cavity of fishes and turtles. Gyrocotylids (Figure 3) are found in the spiral valves of ratfish. There are 10 species known. Adults of both groups are **monozoic**, meaning their bodies have one set of reproductive organs. Both amphilinideans and gyrocotylideans produce a ciliated larval stage called a lycophore that has 10 hooks at its posterior end.

Far more prominent are the Eucestoda, which are found as adults in all classes of jawed vertebrates. Eucestodes have a scolex, or anterior holdfast structure, containing grooves, hooks, or suckers that serve to anchor the adult worm to the wall of the host's small intestine. They also usually have **polyzoic** bodies that feature a serial repetition of gonads, usually each set of gonads separated from another in a segment. The segments most distal to the scolex are the oldest. They can produce prodigious numbers of eggs, each containing a six-hooked larva referred to as a **hexacanth**. Following development of a cestode larva in an intermediate host, the definitive host ingests the intermediate host and the larva will then grow into an adult tapeworm. Over 50 species of tapeworms have been reported from humans.

References

Bakke TA, Cable J, Harris PD et al (2007) The biology of gyrodactylid monogeneans: the Russian doll killers. *Adv Parasitol* **64**:161–376.

Buchman K & Lindenstrom T (2002) Interactions between monogenean parasites and their fish hosts. *Int J Parasitol* **32**:309–319.

Tsai IJ, Zarowiecky M, Holroyd N et al (2013) The genomes of four tapeworm species reveal adaptations to parasitism. *Nature* **496**:57–63.

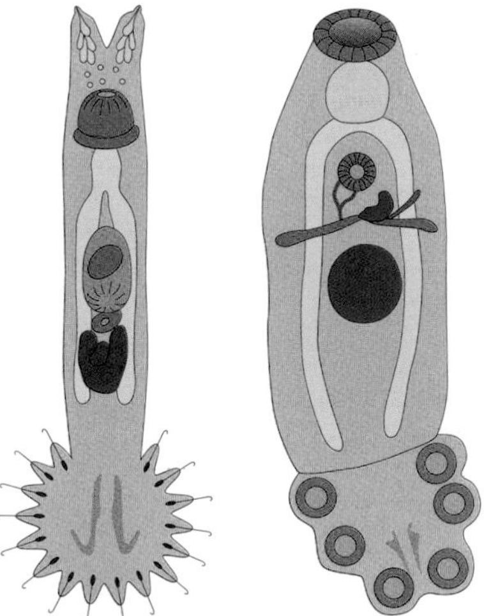

Figure 1 Two different kinds of monogeneans, one with an opisthaptor with several peripheral hooks, and one with with six prominent suckers.

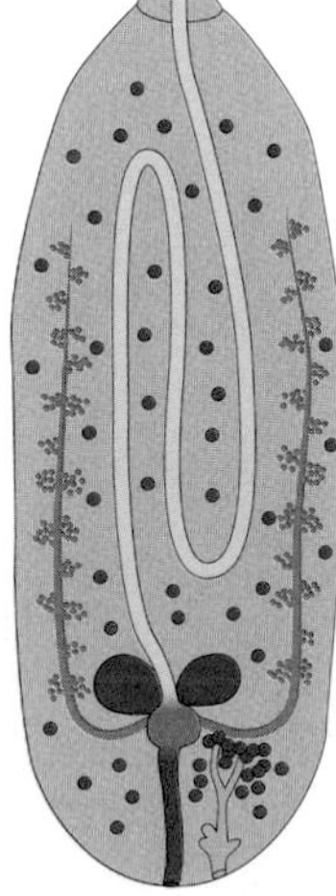

Figure 2 Adult amphilinid showing typical N-shaped uterus. The lycophore either penetrates or is eaten by a crustacean intermediate host in which it sheds its hooks and awaits ingestion by the definitive host.

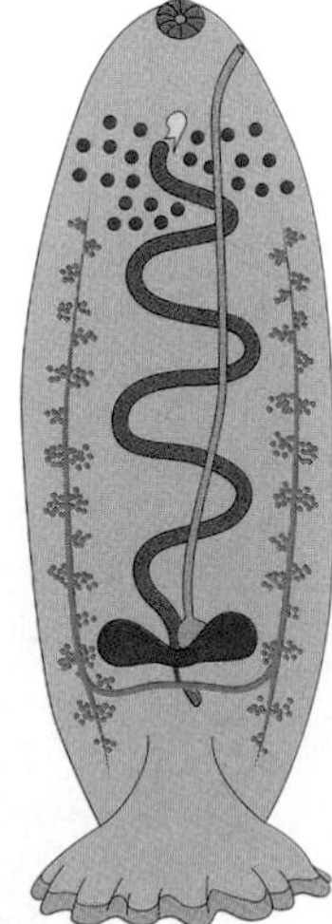

Figure 3 Adult *Gyrocotyle* showing characteristic rosette organ at posterior end. It is not clear if there is an intermediate host or if the cycle is direct.

(Figures 1, 2 and 3 adapted from Heinz Mehlhorn (eds) (2008) In Encyclopedia of Parasitology, 2nd ed. (Monogenea). With permission from Springer Science + Business Media.)

Phylum Nematoda

The phylum Nematoda is part of the Ecdysozoa or molting clade of protostome animals. There may be as many as a million species of nematodes (commonly known as roundworms) when they are all tallied, only about 23,000 species of which have been thus far described. Many species of nematodes are free-living and are common inhabitants of soil or rotting vegetation, or they are found in the sediments of marine and freshwater habitats. Some nematodes are parasites of plants and cause significant agricultural problems and may even serve as vectors of other plant diseases. Many species are parasites of invertebrates and can be exploited as biological control agents for insect pests. In addition, many are parasites of vertebrates. Some nematodes such as the filarial worms have life-cycle stages found in both an invertebrate (such as a mosquito) and a vertebrate host.

Nematodes have a complete gut with a mouth, pharynx, intestine, and anus. They lack circular muscles and so are incapable of changing their body diameter as does an earthworm. The lack of circular muscles gives nematodes a characteristic lashing appearance as they move. Nematodes have a fluid-filled body cavity that is incompletely bounded by tissue of mesoderm origin and is considered to be a **pseudocoelom**. Nematodes are usually dioecious, with the female being larger than the male, and the male having a characteristic spicule for copulatory purposes. Most nematodes are a few millimeters or less in length, but as a group they are surprisingly diverse morphologically. Some species such as the Guinea worm *Dracunculus medinensis* can grow to a meter in length and *Placentonema gigantissimum*—a nematode parasitizing the placenta of whales—is a whopper in the nematode world, reaching 8.5 m in length yet with a diameter of only 0.3 mm.

One of the most characteristic nematode features is their stereotypical developmental pattern. An L1 larva or juvenile is produced within an egg. The L1 will hatch from the egg and will proceed through a series of stages (L2, L3, L4, and adult), each separated by a period of growth followed by a molt. During molting, the old cuticle is replaced from underneath by a new one produced by the hypodermis. For some species, under unfavorable conditions, during the second molt, development is diverted to form a non-feeding **dauer-stage** larva that can eventually be reactivated to resume the life cycle. Nematodes have developed an astounding number of variations on this basic theme, as the examples presented in the following pages attest.

Key to increasing our overall understanding of nematodes and of animals in general has been the ascendance of *Caenorhabditis elegans* as a model organism. This humble nematode living on rotting fruit was nonetheless the first multicellular organism to have its genome completely sequenced. It has a conveniently transparent body and fast development time. Many *C. elegans* mutants have been identified, and genetic screens devised that have greatly facilitated functional studies of animal organ systems.

Phylogenetic studies indicate three major clades within the nematodes: the Chromadoria, the Enoplia, and the Dorylaimia. The Chromadoria make up a huge group of nematodes that includes the Rhabditina (containing hookworms, *C. elegans*, and invertebrate parasites); the Tylenchina (most notable for inclusion of *Meloidogyne* parasites of plants); and the Spirurina, where many familiar human parasites fall (*Ascaris*, the filarial worms, the pinworms, and *Dracunculus* among them). The Enoplia include mostly free-living nematodes, but the plant parasitic genus *Xiphinema* also is found here. Finally, the Dorylaimia include *Trichuris* and *Trichinella* and some additional parasites of vertebrates, invertebrates, and plants. Nematode parasitism of animals has arisen at least five or six separate times and of plants at least three times.

References

Blaxter M (2011) Nematodes: the worm and its relatives. *PLoS Biology* **9(4):** e1001050. doi:10.1371/journal.pbio.1001050

Lee D L (2002) The biology of nematodes. Taylor and Francis.

Trichuris trichiura

The whipworm (Figure 1), which is 30–50 mm in length, lives in the large intestine (especially caecum) of humans. It causes trichuriasis. The worm is named for the whiplike appearance of the adults.

Distribution and prevalence Distribution is cosmopolitan in the tropics, with 600–800 million people infected in Asia, Africa, and South America. Eggs become infective within two to three weeks after deposition on soil, and their survival is favored by shady, moist, warm conditions. *Trichuris trichiura* is grouped as a soil-transmitted helminth along with *Ascaris lumbricoides* and the hookworms.

Hosts and transmission Humans become infected by ingestion of soil, food, or water contaminated with feces containing embryonated eggs (Figure 2). The eggs hatch in the small intestine. Worms undergo four molts and complete their development in the mucosa of the large intestine and cecum, and produce eggs about three months after ingestion of the eggs. The thin anterior portion of the worm threads through the mucosa and the thicker posterior end hangs into the intestinal lumen. Females produce 3000–20,000 eggs per day. Each egg has a characteristic plug at either end. Adults live 1.5–2 years and possibly up to 5 years. Dogs harbor a distinctive species of whipworms, *T. vulpis*, and cats harbor two different whipworm species, *T. campanula* or *T. serrata*.

Pathology Most cases have relatively few worms and are asymptomatic. People with heavy infections, usually children from 5–15 years of age, may experience pain upon defecation and pass bloody or watery stools. In some cases, chronic dysentery and rectal prolapse occurs. Clubbing of the fingers may occur. Heavy infections in children are of concern because children suffer iron deficiency anemia potentially accompanied by impaired growth and cognitive development and poor school attendance.

Diagnosis Infection is diagnosed by finding the thick-shelled eggs (Figure 3) with a characteristic polar plug at either end. A method to concentrate the eggs may be required.

Treatment Mebendazole is more effective than albendazole, and multiple doses of either are probably required. Efficacy may be improved by using either in combination with ivermectin. Anemic children may require iron supplementation.

Control As for contorl of other soil-transmitted helminths, treatment to reduce adult worm populations and thereby minimize egg-production is important. Improved sanitation to prevent contamination of soil with eggs, preventing use of night soil as a fertilizer, and health education emphasizing the importance of personal hygiene and hand washing are also important. Vegetables and fruit should be washed, peeled, or cooked before consumption.

Did you know? The iceman Ötzi, a 5300 year old mummy found in the Alps near the border between Austria and Italy, was infected with *T. trichiura*. This parasite is frequently recovered from ancient latrines. Reasoning that helminths provoke anti-inflammatory responses, a *Trichuris* species from pigs, *T. suis*, is being used in a therapeutic context to treat people with Crohn's disease, ulcerative colitis, and other inflammatory disorders.

References

Bethony J, Booker S, Albonico M et al (2006) Soil-transmitted helminth infections: ascariasis, trichuriasis, and hookworm. *Lancet* **367**:1521–1532.

Knopp S, Mohammed KA, Speich B et al (2010) Albendazole and medendazole administered alone or in combination with ivermectin against *Trichuris trichiura*: a randomized controlled trial. *J Clin Dis* **51**:1420–1428.

Figure 1 Adults of *Trichuris trichiura*. Note the thick posterior and long slender anterior end. (Courtesy of James Liberatos, Louisiana Tech University.)

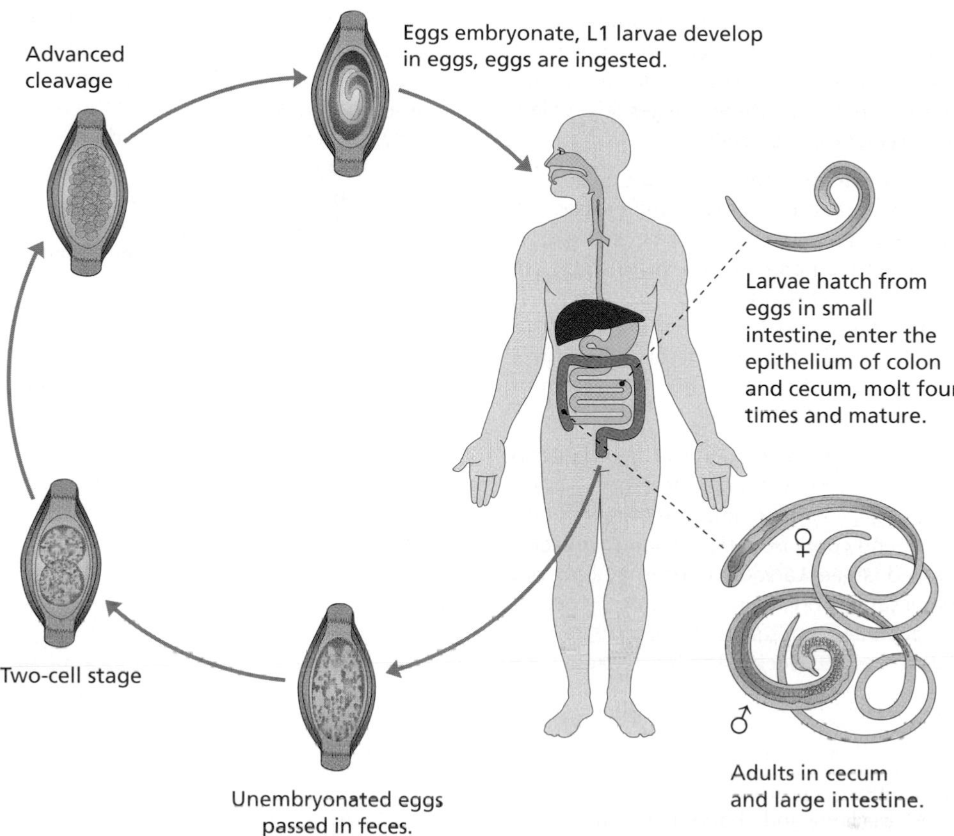

Figure 2 The direct, fecal–oral life cycle of *Trichuris trichiura*. (Courtesy of CDC/BG Partin.)

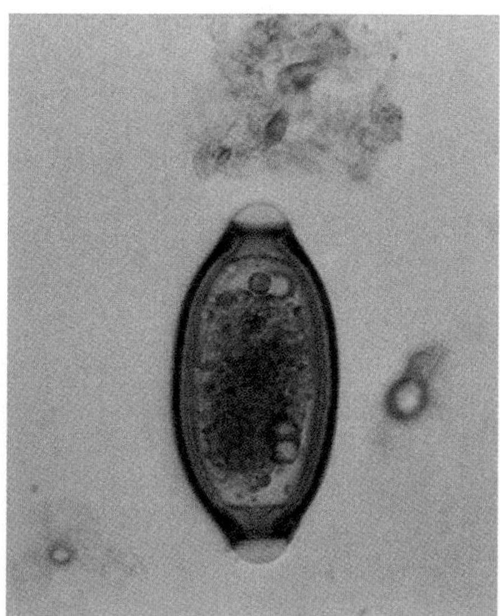

Figure 3 The egg of *Trichuris trichiura* with the characteristic plug at either end, measuring 50–54 μm × 22–23 μm. (Courtesy of CDC.)

Trichinella spiralis

This worm, known informally as the trichina worm (Figure 1), is responsible for causing trichinosis, or trichinellosis. It has the smallest adults (1.4–4.0 mm) of any nematode infecting humans. The worms live in the small intestine, where females release larvae that eventually colonize skeletal muscle cells.

Distribution and prevalence Trichinellosis occurs in about 55 countries where collectively about 10,000 human cases occur annually, with a fatality rate of 0.2%. Although once common in North America and Western Europe, *Trichinella spiralis* is now more common in countries of Central and Eastern Europe. Representatives of the other nine species and three as yet unnamed lineages of *Trichinella* are present on all continents (except Antarctica) where, except for *T. spiralis*, they are mostly parasites of wild animals.

Hosts and transmission *T. spiralis* infects domestic pigs and rats. A sylvatic cycle also occurs involving feral pigs and other carnivores. Transmission to people (Figure 2) is by ingestion of raw or undercooked meat (most frequently pork from free-ranging or backyard pigs) containing encysted larvae. Larvae enter the mucosa of the small intestine, molt four times, and begin to mate in just two days. By five to seven days postinfection, the female begins to release larvae (about 1500 larvae over 16 weeks). The larvae cross the gut wall, enter the circulation, and penetrate and encyst in skeletal muscle cells, where they can survive 40 years. The same individual first serves as definitive and then as intermediate host.

Pathology Pathology is directly related to how many infective larvae have been ingested. Diarrhea and abdominal pain may occur when adults derived from ingested larvae produce their own larvae that then cross the gut wall. As larvae begin to penetrate muscle cells (Figure 3), headache, fever, intense muscle pain (myalgia), especially of the eye muscles, facial and eyelid edema, and inflamed blood vessels are prominent. Although the larvae do not reside in cardiac muscle, myocarditis is noted in 5%–20% of cases. Eosinophilia and the presence of muscle enzymes in the blood are noted. In chronic infections, numbness and impaired muscle strength occur.

Diagnosis Ideally, trichinellosis is diagnosed early, before the female worms have a chance to produce larvae. Unfortunately, there are no unambiguous early signs of infection, so diagnosis typically occurs after invasion of the muscles. Muscle pain, fever, edema, eosinophilia, and elevated blood muscle enzyme levels are consistent with trichinellosis. Larvae may be seen in muscle biopsy, and epidemiological investigation may reveal a likely source of infection. Common source epidemics, as when a cluster of cases results from ingestion of infected meat at a social gathering, are not uncommon.

Treatment Treatment with albendazole or mebendazole as soon as possible is desirable to kill adults and larvae and to limit colonization of the muscles. Glucocorticosteroids to reduce swelling and preparations to compensate for electrolyte imbalances are also important.

Control Education regarding the risk of eating raw, poorly cooked, dried, or smoked meat, either from domestic animals (free-ranging pigs, horses) or wild carnivores can prevent infection. Farming practices that deny pigs access to raw meat of any kind will keep pigs uninfected. Cooking of meat to a core temperature of 71°C for at least one minute and solid freezing (−15°C for at least three weeks) will kill *T. spiralis* larvae. Larvae of other species of *Trichinella* can resist freezing.

Did you know? Once a *T. spiralis* larva infects a skeletal muscle cell, the myofibrils break down, the larva is encapsulated, and a capillary network surrounds the muscle cell, which at this point is called a nurse cell. *Trichinella* are among the largest of all intracellular parasites.

References

Gottstein B, Pozio E, Noeckler K et al (2009) Epidemiology, diagnosis, treatment, and control of trichinellosis. *Clin Microbiol Rev* **22**:127–145.

Pozio E (2014) Searching for *Trichinella*: not all pigs are created equal. *Trends Parasitol.* **30**: 3–11.

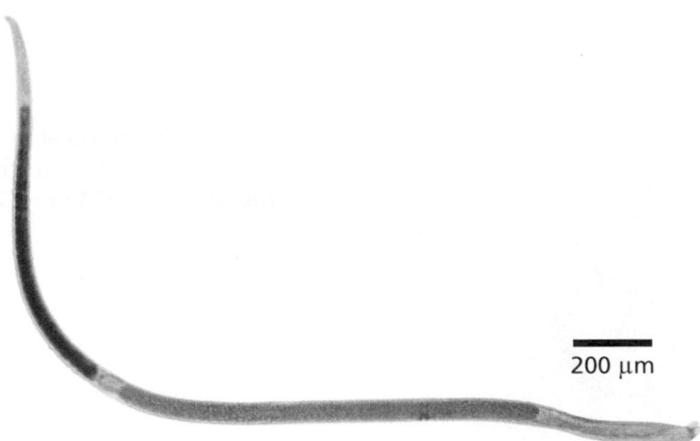

Figure 1 Adult male of *Trichinella spiralis*. (Courtesy of John Sullivan, University of San Francisco

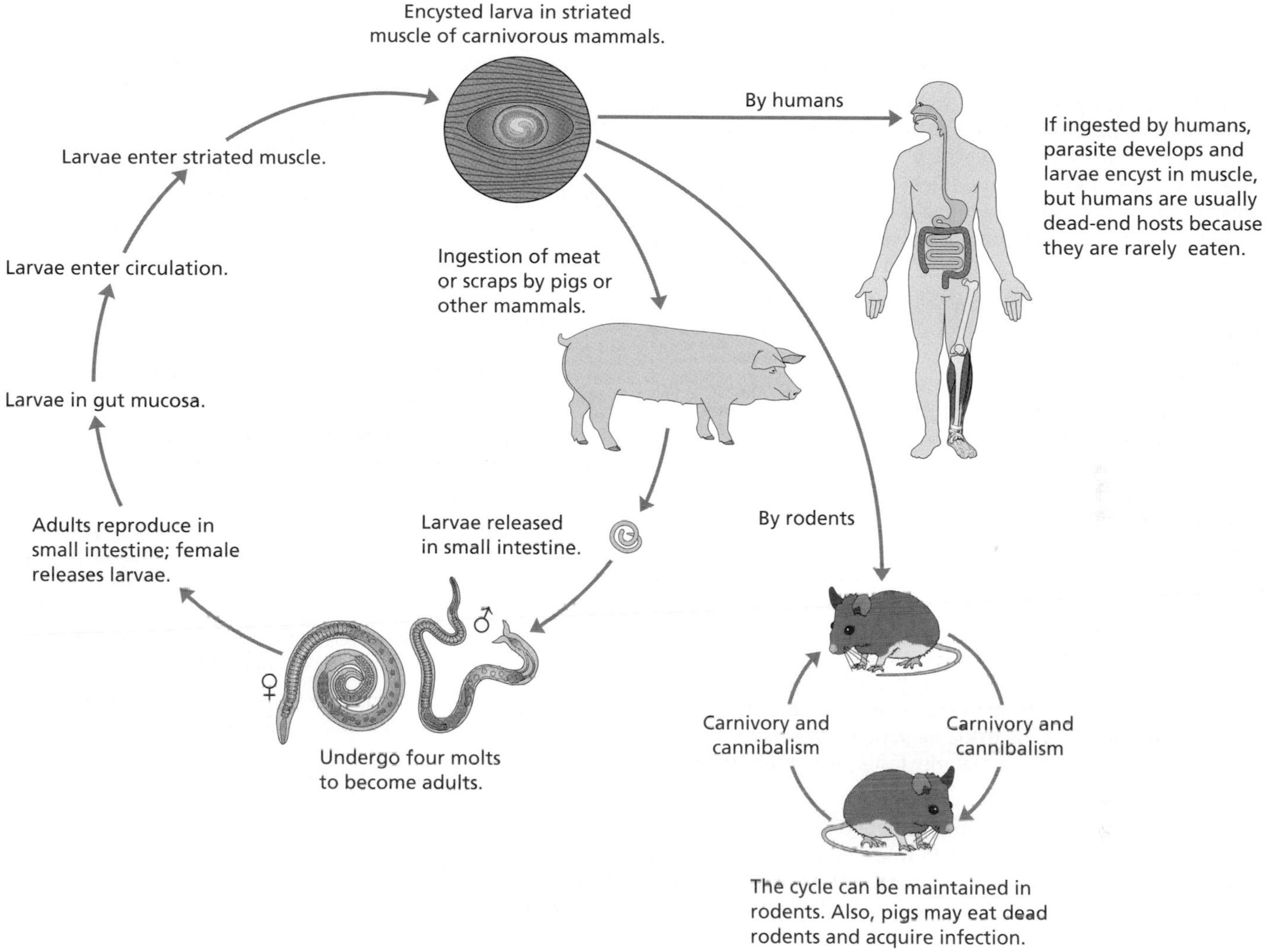

Encysted larva in striated muscle of carnivorous mammals.

By humans

If ingested by humans, parasite develops and larvae encyst in muscle, but humans are usually dead-end hosts because they are rarely eaten.

Larvae enter striated muscle.

Larvae enter circulation.

Ingestion of meat or scraps by pigs or other mammals.

Larvae in gut mucosa.

Adults reproduce in small intestine; female releases larvae.

Larvae released in small intestine.

By rodents

♂

♀

Undergo four molts to become adults.

Carnivory and cannibalism

Carnivory and cannibalism

The cycle can be maintained in rodents. Also, pigs may eat dead rodents and acquire infection.

Figure 2 The domestic life cycle of *Trichinella spiralis*. Carnivorism plays a prominent role, and a single individual first serves as definitive and then as intermediate host. Pigs can acquire infection by eating infected rats, though rats probably do not serve as reservoirs of infection for pigs. Traditionally, humans become infected by eating improperly cooked pork. (Courtesy of CDC.)

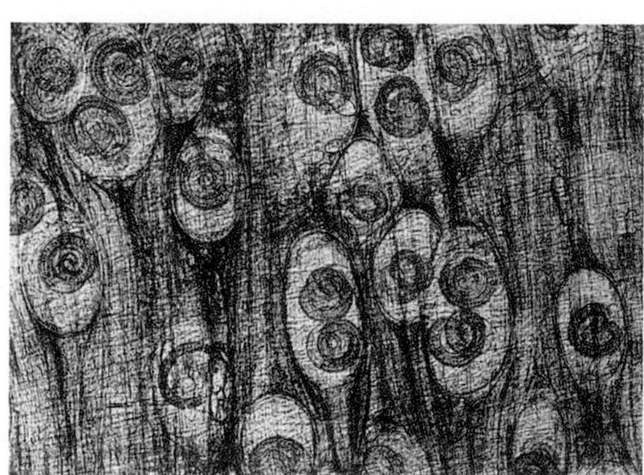

Figure 3 A heavy infection showing many encysted larvae of *Trichinella spiralis* in skeletal muscle. (Courtesy of CDC.)

Ascaris lumbricoides

Ascaris lumbricoides, the large human intestinal roundworm (Figure 1), resides in the small intestine and is the causative agent of ascariasis. Males are 15–31 cm and females 20–49 cm long.

Distribution and prevalence *A. lumbricoides* is one of the most common of all parasites, infecting 0.8–1.2 billion people, or about one sixth of humanity. It is prevalent in tropical and subtropical locations with poor sanitation and hygiene, including sub-Saharan Africa, the Americas, China, and East Asia. It is grouped with *Trichuris trichiura* and the hookworms as a soil-transmitted helminth.

Hosts and transmission Humans, especially children, are the usual host for *A. lumbricoides*. Pigs can rarely be infected, just as humans are occasionally infected with the closely related roundworm of pigs, *A. suum*. The two species are close relatives but gene flow is limited between them. Infection is acquired by ingestion of soil, food, or water contaminated with feces containing *Ascaris* eggs (Figure 2). An L3 larva covered by the L2 cuticle emerges from the egg in the small intestine, penetrates the gut, and moves to the liver where the L2 cuticle is lost. The L3 moves to the lungs, is coughed up and swallowed and molts to L4 and then adulthood in the small intestine. Adult worms begin egg production at nine to 11 weeks after egg ingestion and live for about a year; a female can produce 200,000 eggs per day. The exceptional life span of the egg (up to 15 years) contributes greatly to this species' success. Why do *Ascaris* worms bother to leave the gut to undertake their pulmonary migration, only to return to where they started in the gut? The answer is not clear, but the tissue migration may simply allow the worms to grow and achieve maturity faster than if they remained in the intestine.

Pathology Most cases are asymptomatic, and the likelihood of suffering adverse effects increases with worm burden. Children 5–15 years of age usually have the most intense infections. Lung inflammation, difficulty in breathing, eosinophilia, and fever may occur as the larval worms undergo their migration through the lungs. Adult worms can cause abdominal discomfort, distension, nausea, and diarrhea. They also can cause intestinal obstructions, especially in young children, and gut perforation can occur. Adult worms can obstruct the bile duct and can migrate from the small intestine into the nasopharynx or to the anus. Chronic infections are associated with appetite loss, impaired weight gain and growth, and cognitive impairment, though the latter effect may be confounded by poverty, lack of psychological stimulation, and general poor health. The larvae of a related species, *Toxocara canis* of dogs and cats, can cause visceral larval migrans in people following accidental ingestion of eggs.

Diagnosis Infection is diagnosed by finding the characteristic eggs with bumpy shells (Figure 3) passed in the feces. A method to concentrate the eggs may be required. Elongate, unfertilized eggs may also be passed. Passed adults may also provide a dramatic indication of infection.

Treatment Both albendazole and mebendazole are effective treatments and cause few side effects. These drugs bind to β-tubulin and inhibit microtubule polymerization.

Control Treatment of infected people helps reduce egg-producing adult worm populations. Provision of basic sanitation prevents contamination of soil with eggs, and use of untreated human wastes (night soil) as fertilizer should be discouraged. Good personal hygiene, hand washing, and the washing and cooking of raw vegetables and fruit should be emphasized.

Did you know? Ascaris eggs are legendary for their resistance to harsh chemicals: they can embryonate in 2% formalin and 50% hydrochloric acid. *Ascaris* infections tend to be overdispersed in people, meaning that some individuals are "wormy" and bear heavier infections than most. These same individuals tend to reacquire heavy infections following treatment.

References

Crompton DWT (2001) *Ascaris* and ascariasis. *Adv Parasitol* **48**:285–375.

Dold C & Holland CV (2011) *Ascaris* and ascariasis. *Microbes Infect* **13**:632–637.

Figure 1 A number of adult *Ascaris lumbricoides* worms passed by an African child following treatment. (Courtesy of CDC/Henry Bishop.)

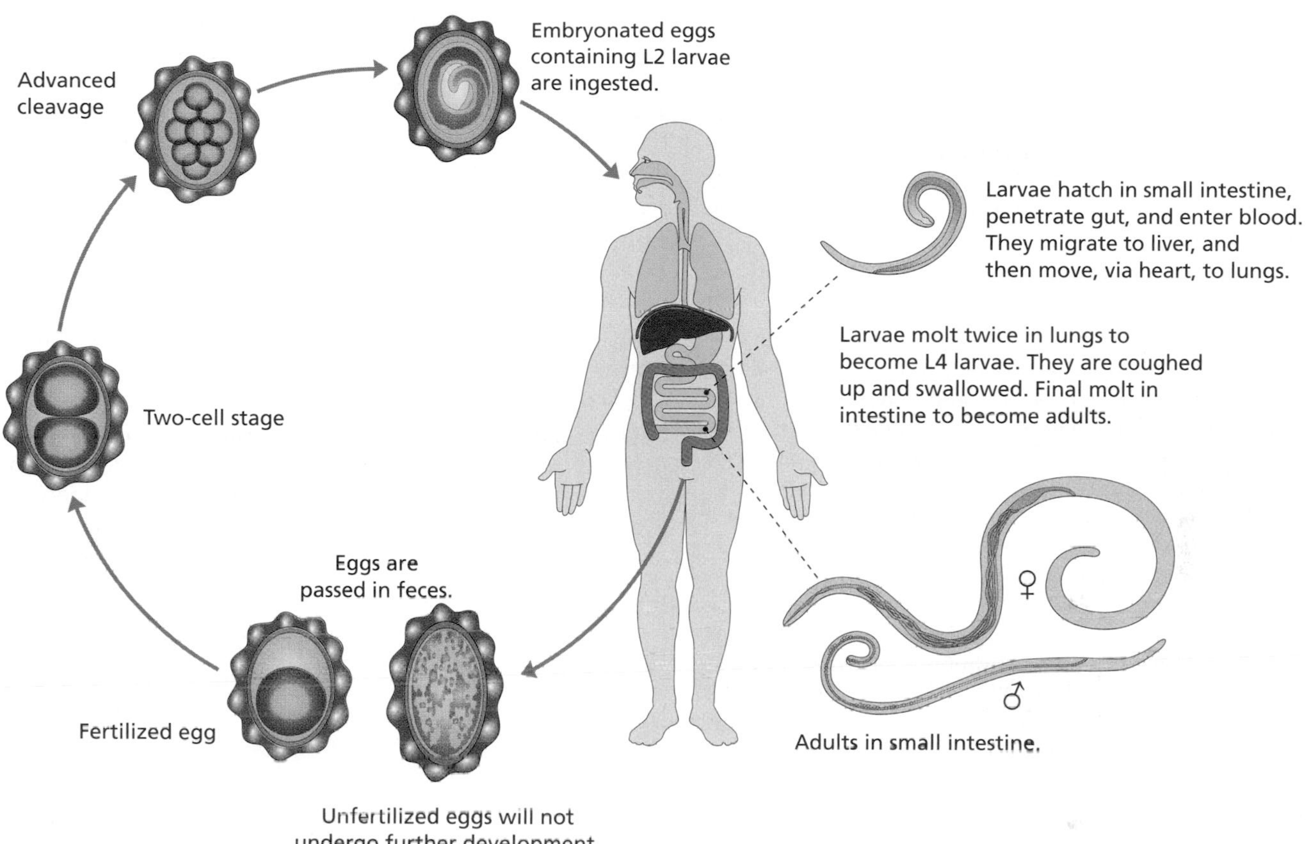

Figure 2 The life cycle of *Ascaris lumbricoides*. Note that the worms, once hatched from the egg, penetrate the small intestine. Larvae are coughed up and swallowed, and return to the small intestine, where they develop to maturity. (Courtesy of CDC.)

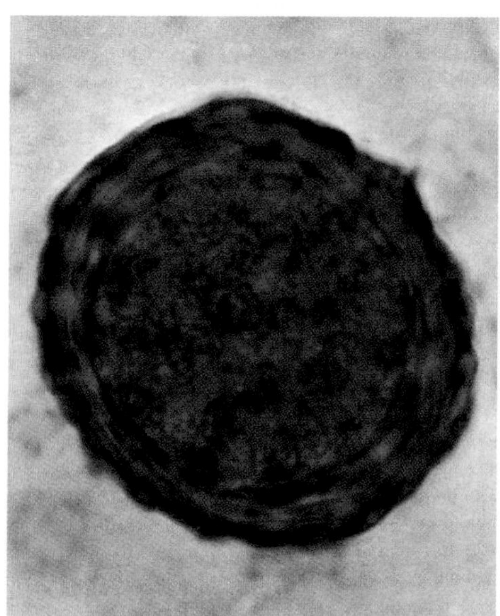

Figure 3 The egg of *Ascaris lumbricoides*, measuring 45–75 μm × 35–50 μm. (Courtesy of Graham Colm, CC BY-SA 3.0.)

Necator americanus and *Ancylostoma duodenale*

The hookworms (Figure 1), adults of which are 5-11 mm and 8-13 mm in length, respectively, live in the upper small intestine of humans.

Distribution and prevalence These soil-transmitted helminths infect 576–740 million people. Hookworms thrive in warm, moist locations with well-drained soils that support egg hatching and survival of their free-living and infective larval stages. *Necator americanus* is common in Latin America, sub-Saharan Africa, and Asia, whereas *Ancylostoma duodenale* is usually found in the Middle East, India, North Africa, and in Europe, and is better adapted to cooler conditions.

Hosts and transmission Larvae develop on soil contaminated with feces and benefit from shade and moisture to survive (Figure 2). Humans become infected when the infective L3 filariform larvae penetrate skin, or in the case of *A. duodenale*, may also be acquired through the mouth. Larvae pass via the vasculature to the lungs, are coughed up and swallowed, and molt twice to become adults. It takes from five to nine weeks for eggs to be produced. Adults of *N. americanus* produce 5000–10,000 eggs per day whereas those of *A. duodenale* produce 10,000–30,000 per day. The adults can live five to seven years. Larvae of *A. duodenale* may remain dormant in the tissues, reactivating if their host becomes pregnant; they may then pass to the mammary glands and be transmitted by lactation. *A. braziliense* infects cats, but its larvae may penetrate human skin and cause cutaneous larval migrans. Dogs are infected with *A. caninum*, a species that can infect people.

Pathology Children and women with heavy infections are especially prone to develop iron-deficiency anemia and protein deficiency as a result of the blood-feeding habits of adult hookworms. Blood loss in milliliters per worm per day is 0.03 for *N. americanus* and 0.15–0.23 for *A. duodenale*. The intensity of infection may increase with age, possibly related to fecal contamination of work environments. Facial edema, eosinophilia, and pica accompany iron-deficiency anemia, and emaciation, ascites, and cardiac failure can occur in extreme cases. Children with chronic infection suffer impaired physical and cognitive development and frequent school absenteeism. Infected mothers have premature deliveries and babies with low birth weight.

Diagnosis Thin-shelled oval eggs (Figure 3) are found on stool examination. Persistent eosinophilia and evident anemia may also aid diagnosis.

Treatment Efficacy of a single dose treatment is 72% for albendazole and 15% for mebendazole. Treatment of infected, pregnant women (but not in the first trimester) is strongly advocated. Iron supplementation is also warranted in cases of anemia.

Control Control is as for other soil-transmitted helminths: anthelmintic treatment to prevent morbidity and reduce contamination, sanitation, and health education. Wearing shoes and otherwise avoiding contact of soil contaminated with infective larvae, as might occur on the damp floor of an outhouse, is also important.

Did you know? Hookworm infection was present in inhabitants of South America at least 7000 years ago. It is generally considered these tropical parasites could not have survived the cold temperatures of Beringia. So how did they get to the American tropics? Possibly the parasites were transmitted by people who followed coastal migration routes or who arrived in South America by a trans-Pacific route.

References

Araújo A, Reinhard KJ, Ferreira LF et al (2008) Parasites as probes for prehistoric human migrations? *Trends Parasitol* **24**:112–115.

Bethony J, Brooker S, Albonico M et al (2006) Soil-transmitted helminth infections: ascariasis, trichuriasis, and hookworm. *Lancet* **367**:1521-1532.

A

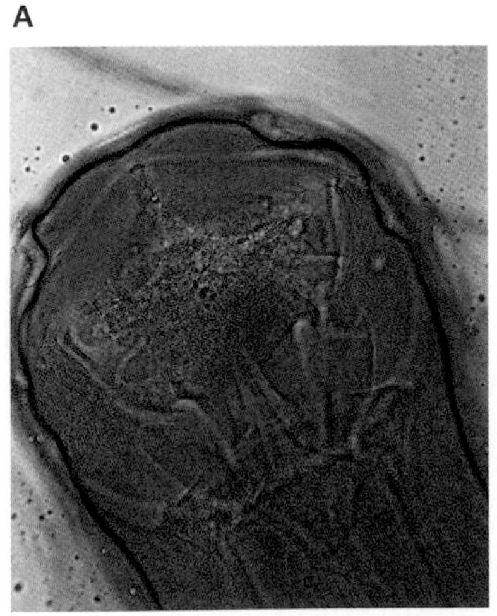

B

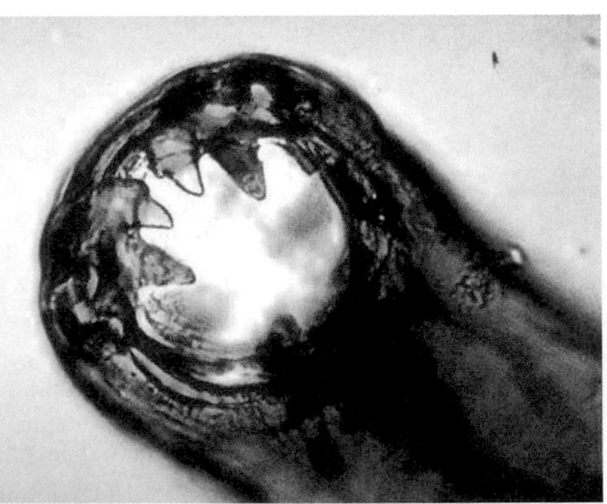

Figure 1 The buccal capsule of *Necator americanus* (left) showing cutting plates (light colored lines), and of *Ancylostoma duodenale* (right) with two pairs of conspicuous teeth. (A, courtesy of CDC.)

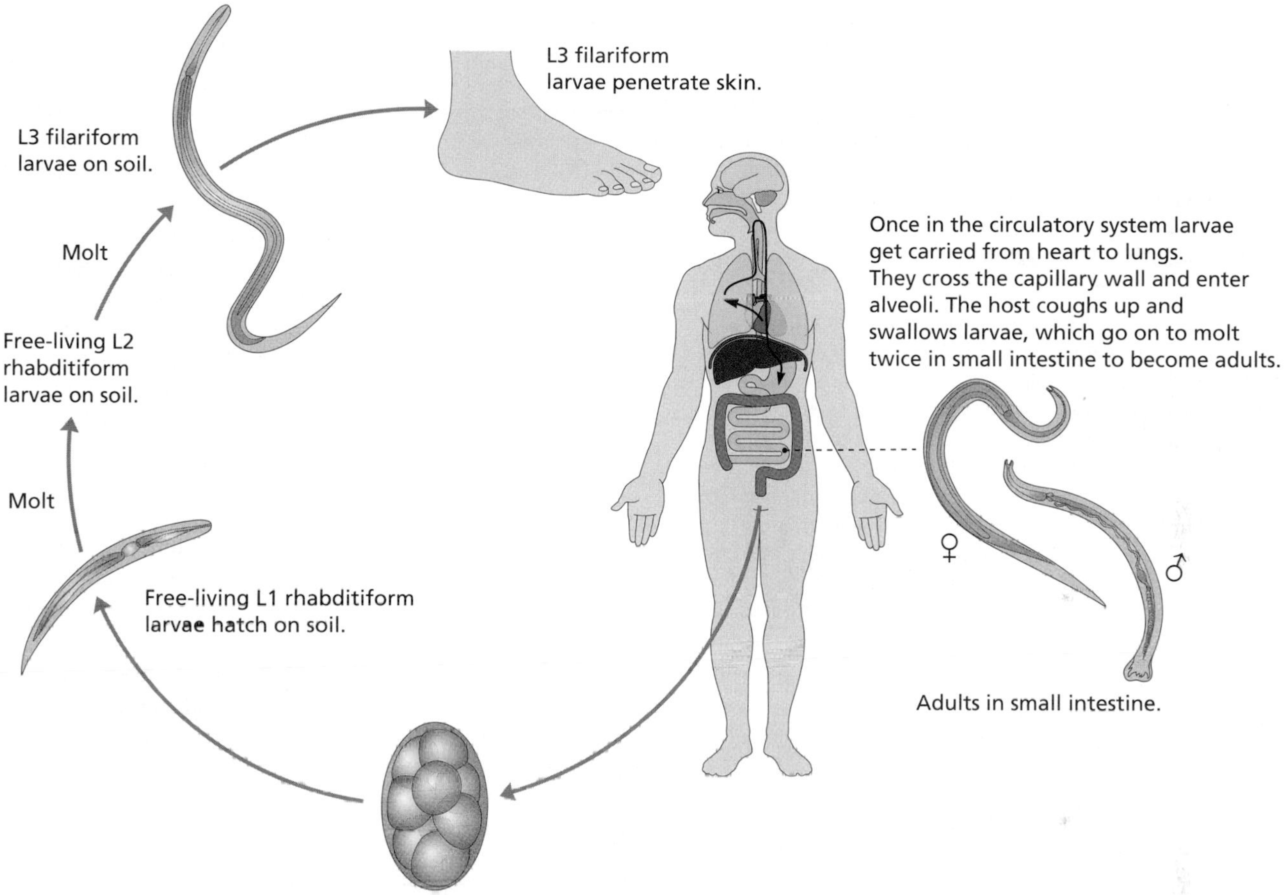

L3 filariform larvae penetrate skin.

L3 filariform larvae on soil.

Molt

Free-living L2 rhabditiform larvae on soil.

Molt

Once in the circulatory system larvae get carried from heart to lungs. They cross the capillary wall and enter alveoli. The host coughs up and swallows larvae, which go on to molt twice in small intestine to become adults.

♀

♂

Adults in small intestine.

Free-living L1 rhabditiform larvae hatch on soil.

Eggs in feces.

Figure 2 The life cycle of hookworms. Note its involvement of rhabditiform and filariform larvae living on the soil, with the latter eventually penetrating human skin. (Courtesy of CDC.)

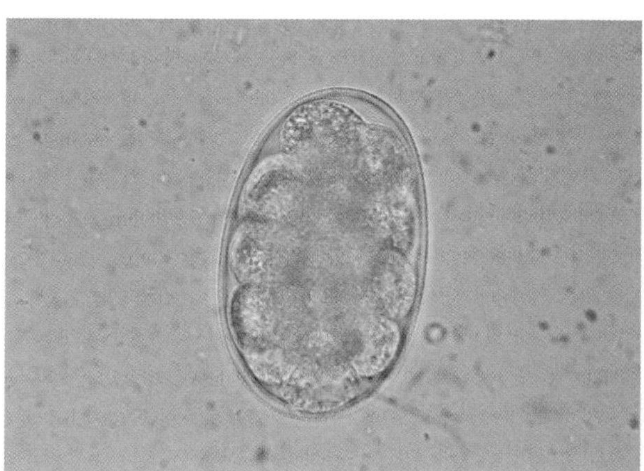

Figure 3 Hookworm egg with characteristic thin shell, measuring 65–75 μm × 36–40 μm. The eggs of *Necator americanus* are slightly larger than those of *Ancylostoma duodenale*, but they are hard to distinguish from one another. (Courtesy of Joel Mills, CC BY-SA 3.0.)

Wuchereria bancrofti

Wuchereria bancrofti, along with *Brugia malayi* and *B. timori*, causes lymphatic filariasis. The adults (females 80–100 mm and males 40 mm long; Figure 1) reside in the lymphatic vessels. Filarial worms are arthropod-transmitted worms of the superfamily Filarioidea.

Distribution and prevalence Lymphatic filariasis occurs in 73 countries, where it infects over 120 million people, with about 40 million disfigured and with over 1.4 billion people at risk of acquiring infection. *W. bancrofti* is responsible for 90% of the cases of lymphatic filariasis, with *B. malayi* more important in Southeast Asia and *B. timori* in southeastern Indonesia.

Hosts and transmission Humans become infected (Figure 2) when bitten by infected *Aedes*, *Anopheles*, or *Culex* mosquitoes. There are no important reservoir hosts. L3 larvae are transferred from the mosquito's proboscis to the skin, where they enter the wound. Over about one year, the larvae migrate to the afferent lymphatic system and undergo two molts to become adults. The adults live for five to 10 years. Females produce thousands of sheathed microfilariae (mff) per day. The sheath is a retained egg membrane. The mff exhibit nocturnal periodicity. When ingested by mosquitoes, the mff unsheathe, penetrate the gut, enter the thoracic muscles, and undergo two molts, and by 10–12 days the L3 larvae move to the proboscis.

Pathology Most infections are asymptomatic, but about a third of infected people exhibit clinical signs including fever, swelling of lymph nodes, lymphedema (obstruction by adult worms of lymph flow followed by swelling), and hydrocele (accumulation of fluid in the scrotum). Chyluria (lymph in the urine) may occur. Inflammatory responses to *Wolbachia* symbionts found in adults may contribute to pathology. Impaired lymphatic function makes it harder to fight skin and lymphatic infections. The most well-known manifestation is elephantiasis (Figure 3), namely the thickening and hardening of skin and underlying tissue, especially in the legs and feet. Permanent disability can result and disfigured people may be shunned. Tropical pulmonary eosinophilia syndrome marked by cough, wheezing, high levels of IgE, and eosinophilia may result.

Diagnosis Blood smears taken at night will reveal presence of sheathed microfilariae, but sensitivity is low. A more sensitive filariasis card test detecting worm antigens is now the diagnostic method of choice, but it is costly. Ultrasonography can detect adult worms.

Treatment Treatment is complicated and must be exercised with care. A multiday course of DEC (diethylcarbamazine), which kills both adults and microfilariae, in combination with albendazole is recommended. Contraindications exist for use of DEC in areas where onchocerciasis occurs and for both DEC and ivermectin where loaiasis occurs. For mass drug administration programs, usually single annual dose combinations of ivermectin and albendazole or DEC and albendazole are used. Hygienic skin care is useful in treating patients with elephantiasis.

Control Many mosquito bites over several months are required to become infected. Reducing mosquito exposures helps to minimize risk. A global effort to eliminate lymphatic filariasis is well underway, featuring annual mass drug administration in entire communities.

Did you know? The mff of *W. bancrofti* circulate in the peripheral blood only at night, mostly between 10pm and 2am, a time when their mosquito hosts are likely to bite. In areas where transmission is by day-biting mosquitoes, the mff are found in the peripheral blood during the day.

References

Knopp S, Steinmann P, Hatz C et al (2012) Nematode infections: filariases. *Infect Dis Clin North Am* **26**:359–381.

Rebollo M & Bockarie MJ (2013) Toward the elimination of lymphatic filariasis by 2020: treatment update and impact assessment for the endgame. *Expert Rev Anti Infect Ther* **11**:723–731.

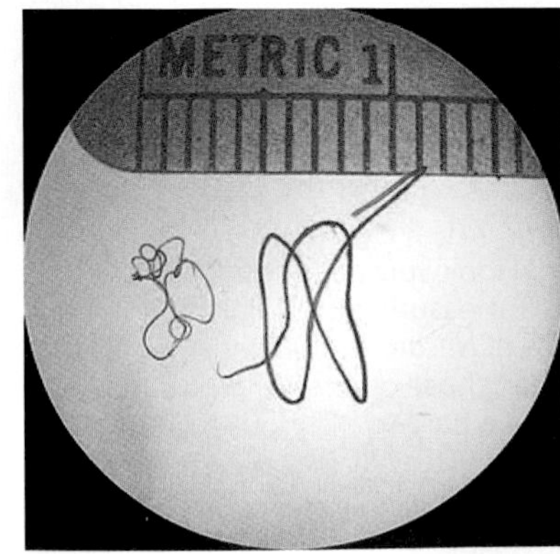

Figure 1 Adults of *Wuchereria bancrofti*, with the female (right) larger than the male. Scale bar in millimeters. (Courtesy of CDC.)

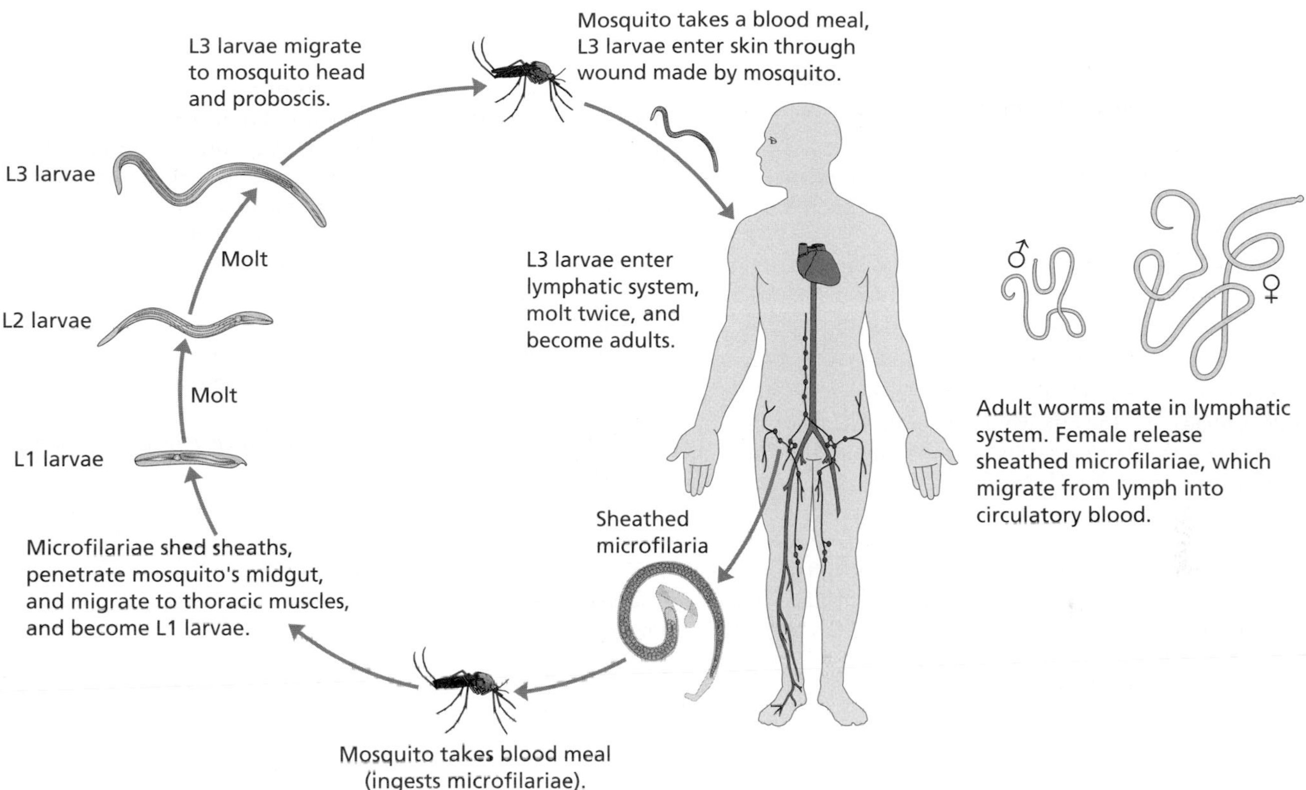

Figure 2 The life cycle of *Wuchereria bancrofti*. Note the involvement of a mosquito vector, which must bite twice, once to acquire the infection and a second time to transmit it to a new host. (Courtesy of CDC.)

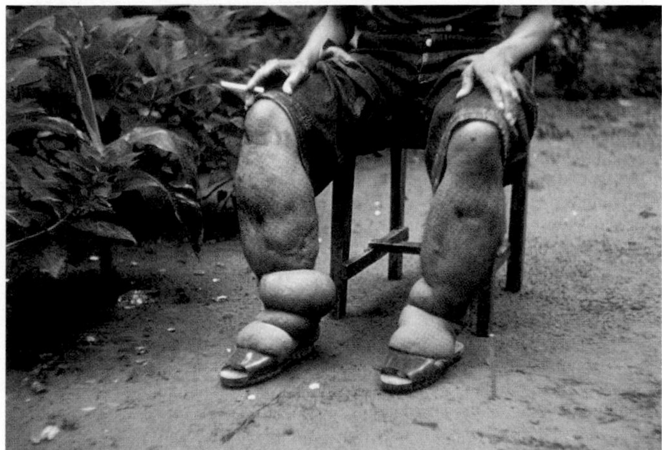

Figure 3 Elephantiasis caused by *Wuchereria bancrofti*. (Courtesy of CDC.)

Onchocerca volvulus

Onchocerca volvulus is the filarial worm that causes onchocerciasis, or river blindness, in people. The large females (350–700 mm long) and males (200–500 mm long) live subcutaneously. Onchocerciasis is second only to trachoma as an infectious disease responsible for blindness.

Distribution and prevalence Onchocerciasis is endemic in 27 sub-Saharan African countries, in Yemen, and in six Latin American countries. About 37 million people are infected and over 100 million people are at risk, with about 99% of the cases occurring in Africa. People in West Africa living close to rivers and streams that support populations of black fly vectors have been most at risk.

Hosts and transmission Humans are infected (Figure 1) when an infected black fly (*Simulium*) bites and transfers infective L3 larvae to the skin. Larvae enter the bite wound, migrate to subcutaneous tissue, molt twice to become adults, and then mate. They take 6 to 12 months to mature. Adult females live for 9 to 14 years and can produce 1500 unsheathed microfilariae (mff) per day. The mff migrate to the dermis and some enter the eyes. They can survive up to two years in the human body, with up to 2000 mff found per milligram of skin. They are ingested by feeding black flies, penetrate the gut, move to the thoracic muscles, and undergo two molts to become infective L3 larvae, which then move to the proboscis for transmission about two weeks after ingestion.

Pathology Adults are found in subcutaneous nodules on the head, arms, shoulders, hips, or chest, but otherwise they cause few problems. The mff they produce are responsible for most of the problems associated with onchocerciasis, which usually become evident in older people. Mff die in the skin, causing intense itching and inflammation, and in chronic infections this results in depigmentation, hyperpigmentation, elephantiasis, and loss of skin elasticity. A condition called hanging groin is one manifestation of damaged skin. In chronic infections, death of mff in the eye with the release of *Wolbachia* symbionts leads to inflammation and can result in corneal opacity and damage to the retina and optic nerve (Figures 2 and 3). The intense itching and blindness associated with onchocerciasis have strong negative effects on socioeconomic development.

Diagnosis Skin snips may reveal the unsheathed mff, but the sensitivity is low. A skin patch test is available that involves placing DEC (diethylcarbamazine) cream on the skin. DEC kills mff in the underlying skin predictably causing a rash to occur 24 hours following application of the cream.

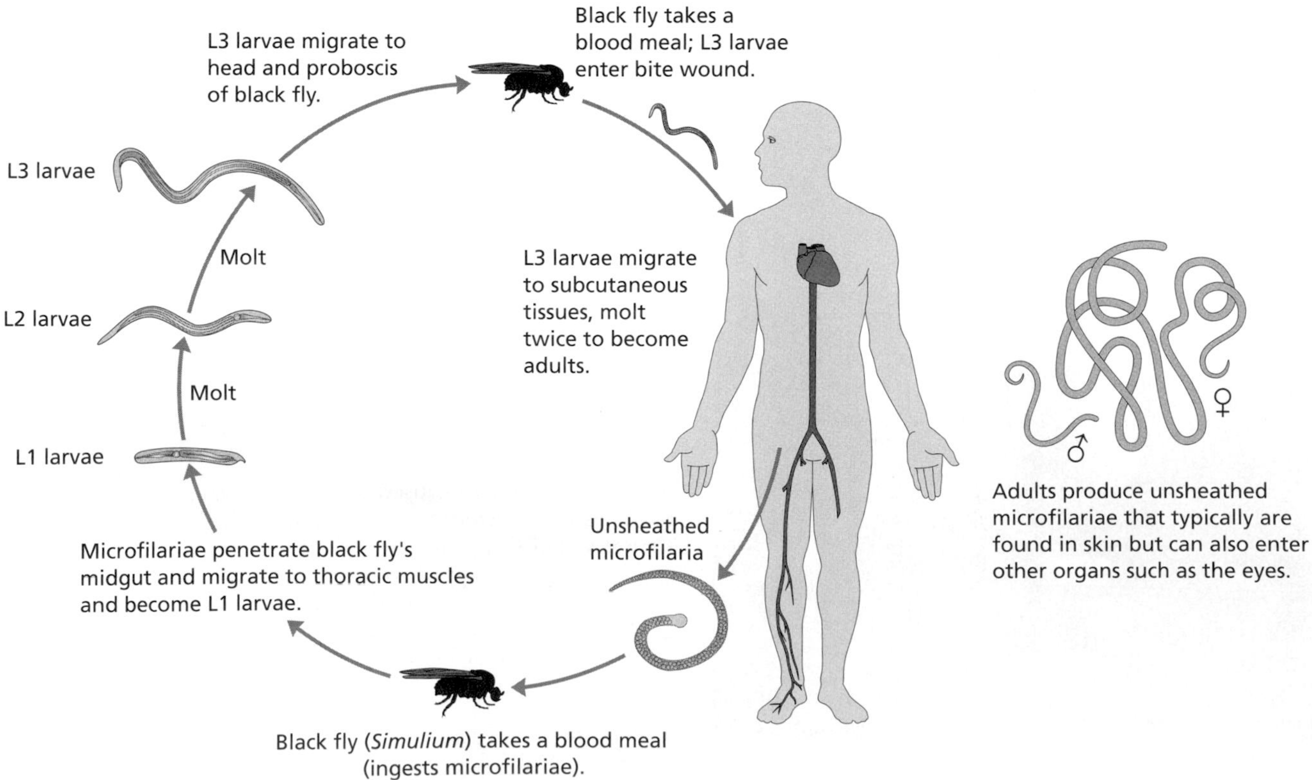

Figure 1 The life cycle of *Onchocerca volvulus*. Female black flies (*Simulium*) serve as vectors. (Courtesy of CDC.)

Treatment Treatment with ivermectin (Mectizan, donated by Merck) administered twice annually kills mff. Although the death of mff may cause itching and a transient fever, treatment removes them as a concern for causing progressive ocular damage. Ivermectin also temporarily halts production of mff, thereby affecting transmission. Ivermectin is given to entire communities at risk to infection. DEC is not used because it causes death of mff that can lead to strong inflammatory responses in the eyes. Prolonged doxycycline courses will kill *Wolbachia* and cause sterilization of females.

Control Prevention of bites by black flies, which mostly bite by day, is important in prevention. Control programs for onchocerciasis have been under way in Africa since 1974; these programs first targeted black flies for control and later have emphasized ivermectin treatment of people.

Did you know? A peculiar form of epilepsy called nodding disease affecting children has been associated with *O. volvulus* infection. Underlying reasons for this connection are not known.

References

Crump A, Morel CM, Omura S et al (2012) The onchocerciasis chronicle: from the beginning to the end? *Trends Parasitol* **28**:280-288.

Knopp S, Steinmann P, Hatz C et al (2012) Nematode infections: filariases. *Infect Dis Clin North Am* **26**:359–381.

Figure 2 Note the corneal opacity resulting from chronic *Onchocerca volvulus* infection.

Figure 3 Historically, in villages in West Africa located near streams supporting *Simulium* populations, many people blinded by *Onchocerca volvulus* would be led from place to place by younger people yet to be affected. (Courtesy of WHO.)

Dracunculus medinensis

The Guinea worm (Figure 1), sometimes called the fiery serpent, is responsible for causing dracunculiasis in humans. The female worm ranges from 600–1000 mm in length whereas the male is much smaller, 15–40 mm long.

Distribution and prevalence In 2013, there were only 148 reported cases of dracunculiasis, with 131 of these coming from South Sudan and the remainder from Ethiopia, Mali, and Chad. In 1986, *Dracunculus medinensis* was endemic in 20 countries in Africa and Asia and infected 3.5 million people.

Hosts and transmission Humans become infected (Figure 2) by drinking water containing copepods (*Cyclops*) with the infective L3 larvae of the parasite. The larvae are released upon digestion of the copepod in the gut, and L3 larvae penetrate the wall of the small intestine and enter subcutaneous tissues where they molt twice to become adults. Mating occurs 60–90 days after infection. The fertilized females continue to migrate subcutaneously to an extremity, usually the foot or lower leg. There, about one year post-infection, they provoke the formation of a blister, which breaks and leaves an exit hole for the female. When the affected area is immersed in water, a portion of the female's uterus is extruded through the wound and L1 larvae are released into the water. The L1 larvae, which are produced in the hundreds of thousands, are infective for about three days in the water. They are ingested by copepods, penetrate into the hemocoel, and undergo two molts to become L3 larvae. Transmission is often seasonal.

Pathology The characteristic blister is a response to larvae emerging from the female. The blister is accompanied by a burning sensation, which is often the first visible sign of infection. Once the blister has broken, the underlying wound is susceptible to infection and can become inflamed and painful. Sepsis or tetanus can also result. The wound may interfere with the ability to work or walk. The wound may interfere with the infected person's livelihood, especially if infections occur during an important planting or harvesting season. Infected children are likely to miss school. A person may become infected year after year.

Diagnosis Diagnosis usually occurs with the dramatic appearance of a blister and the emergence of portions of the female worm.

Treatment The time-honored means of slowly winding the extruded female guinea worm onto a stick is still practiced. Use of ice can accelerate removal and relieve the burning sensation. Keeping the wound clean and treatment with antibiotic ointments are helpful.

Control Control measures have been very effective and eradication within a decade seems possible. The measures include filtration of drinking waters to remove infected copepods, provision of clean borehole water, treatment of water in transmission foci with Temephos to kill copepods, education to inform people how infection can be prevented, prevention of female worms from expelling progeny into water bodies, monitoring and case control efforts, and use of media and local celebrities to help promote control efforts. Control has been facilitated by the lack of reservoir hosts. The work of the Carter Center to spearhead the Global Guinea Worm Eradication Program with involvement by the WHO, CDC, UNICEF, and others has been exemplary.

Did you know? Dracunculiasis is predicted to be the first disease caused by a eukaryotic parasite to be eradicated, and the first disease to be eradicated without the use of a vaccine.

References

Hopkins DR (2013) Disease eradication. *N Engl J Med* **368**:54–63.

Hopkins DR, Ruiz-Tiben E, Weiss A et al (2013) Dracunculiasis eradication: and now, South Sudan. *Am J Trop Med Hyg* **89**:5–10.

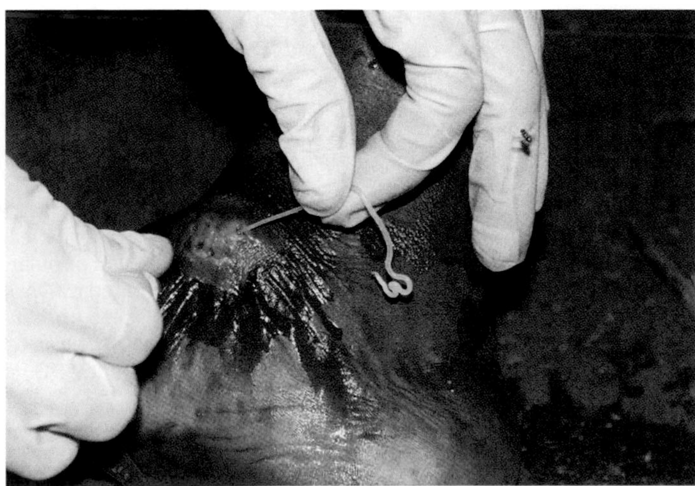

Figure 1 A female *Dracunculus medinensis* protruding through a hole in the ankle of this infected person. (Courtesy of CDC.)

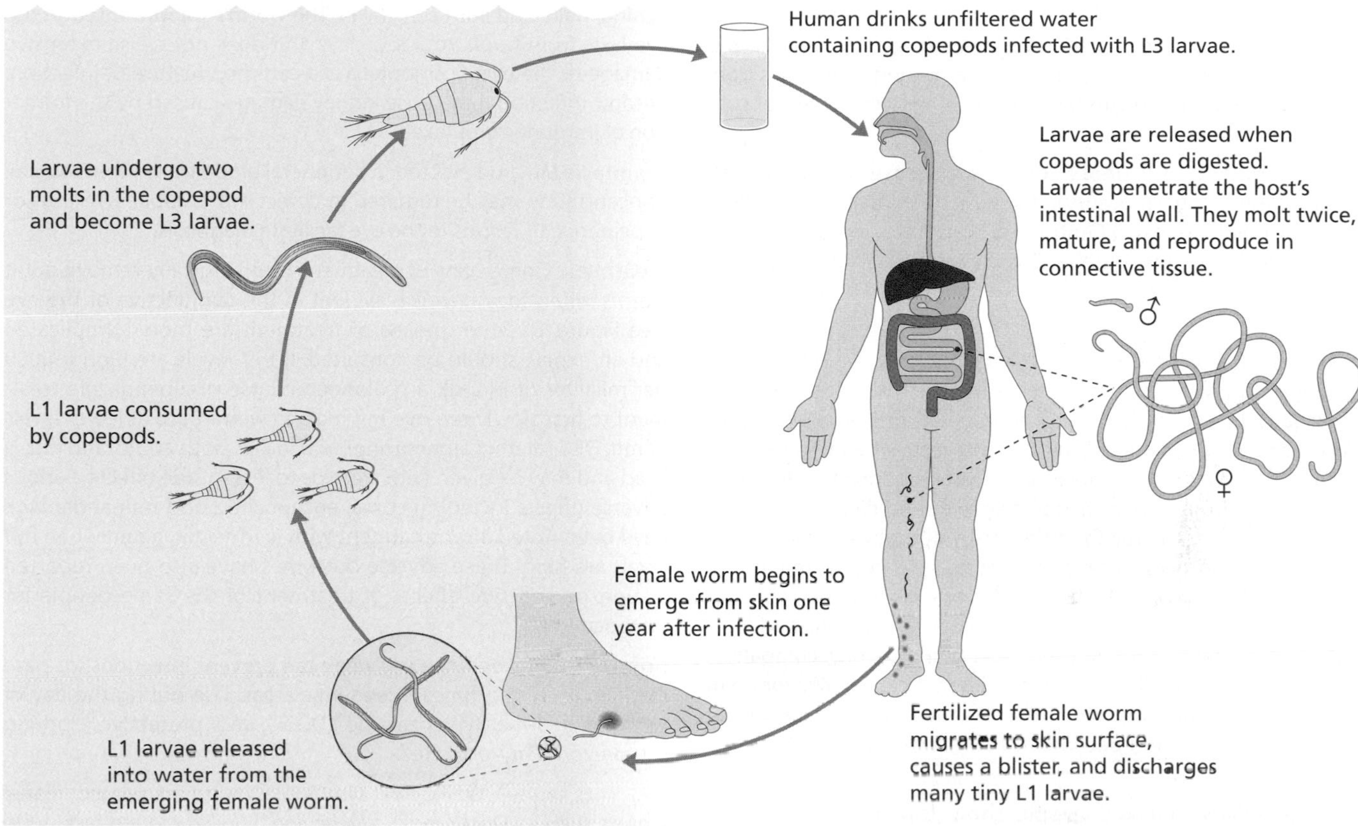

Figure 2 The life cycle of *Dracunculus medinensis*. Copepods (*Cyclops*) play an important role as intermediate hosts. (Courtesy of CDC.)

Figure 3 Filtering of drinking water to remove copepods is a simple yet effective way to prevent Guinea worm infection. (© Carter Center/L Gubb.)

Loa loa

Loa loa, a filarial worm called the African eye worm (Figure 1), is responsible for causing loiasis in the rainforests of Central and West Africa. Females range from 20–70 mm long, and males from 20–35 mm long. The worms live and move between layers of subcutaneous tissues.

Distribution and prevalence Loiasis is a disease with a relatively circumscribed range; it occurs in the equatorial rainforests (including rubber plantations) of Central and West Africa, extending as far east as Uganda, as far south as Zambia, and as far north as 10°N. It is endemic in about 11 countries and afflicts an estimated 3–13 million people.

Hosts and transmission Humans are the only definitive hosts, although closely related worms occur in other primates. Infections are acquired through the bite of an infected mango fly (*Chrysops silacea* and *C. dimidiata* are prominent vectors) as people work during the day in open areas in or around forests (Figure 2). Infections are common in the rainy season, and flies are attracted to wood smoke. L3 larvae from the fly's mouthparts enter the bite wound, migrate to subcutaneous tissues, molt twice, and after a minimum of five months become adults. Adults can live up to 17 years. Females produce thousands of sheathed microfilariae (mff) per day. Mff circulate in the peripheral blood by day, compatible with the day-biting habits of mango flies. Mff can survive for up to one year in the body. Ingested mff cross the fly's gut, enter the fat bodies, molt twice, and become infective L3s, which move to the proboscis of the fly.

Pathology Most infected people show few if any symptoms. Localized and transient subcutaneous itchy, swellings called Calabar swellings appear, which correspond to the presence in the underlying subcutaneous tissues of migrating adults releasing mff. The swellings are usually found around the wrists or ankles and can make movement painful. Adults sometimes pass into the conjunctiva and cornea of the eyes, hence the common name eyeworm (see Figure 1). Presence of the adult worms in the eye causes itching, pain, and light sensitivity. The worm's sojourn through the eye lasts from hours to a few days and does not cause extensive damage to the eye. Eosinophilia is a common feature of infection. Chronic infection may cause kidney damage caused by the formation of immune complexes.

Diagnosis Mff are detected in peripheral blood from 10am to 2pm. Concentration may be required to detect mff. Calabar swellings or appearance of worms in the eye facilitate diagnosis.

Treatment One means of treatment is to surgically remove adult worms when they become evident in the conjunctiva of the eye (see Figure 1). Other means of treatment are more complicated and an expert should be consulted. If mff levels are high (>8000 per milliliter of blood), a prolonged course of albendazole treatment to first slowly remove mff may be warranted. At lower levels of mff, DEC (diethycarbamazine), which kills both adults and mff, is used and may be given with steroids to lessen side effects. Serious adverse effects including coma, encephalitis, and retinal damage have been noted after treatment with ivermectin, again when mff levels are high. These adverse outcomes have also been reported as unexpected side effects of treatment of the same people for onchocerciasis.

Control DEC taken once per week can prevent infections for people living for long times in endemic areas. Use during the day of *N,N*-diethyl-3-methylbenzamide (DEET) and protective clothing can prevent *Chrysops* bites.

Did you know? Wolbachia symbionts found in related filarial worms such as *Onchocerca*, *Wuchereria*, and *Brugia* are lacking in *Loa loa*.

References

Holmes D (2013) *Loa loa*: neglected neurology and nematodes. *Lancet Neurol* **12**:631–632.

Padgett JJ & Jacobsen KH (2008) Loiasis: African eye worm. *Trans R Soc Trop Hyg* **102**:983–989.

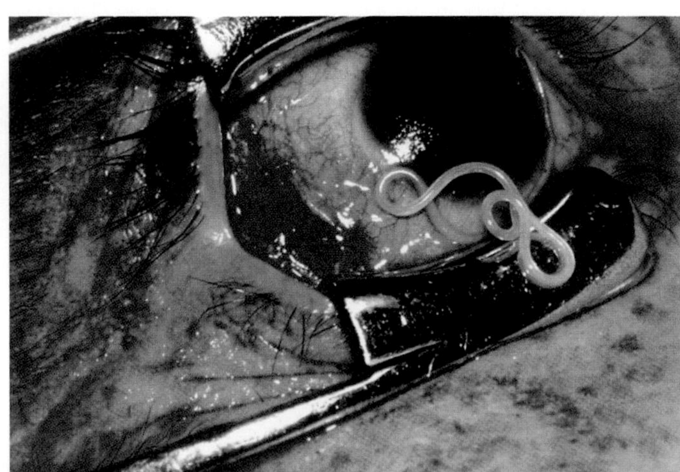

Figure 1 Removal of an adult *Loa loa* from the conjunctiva of the eye.

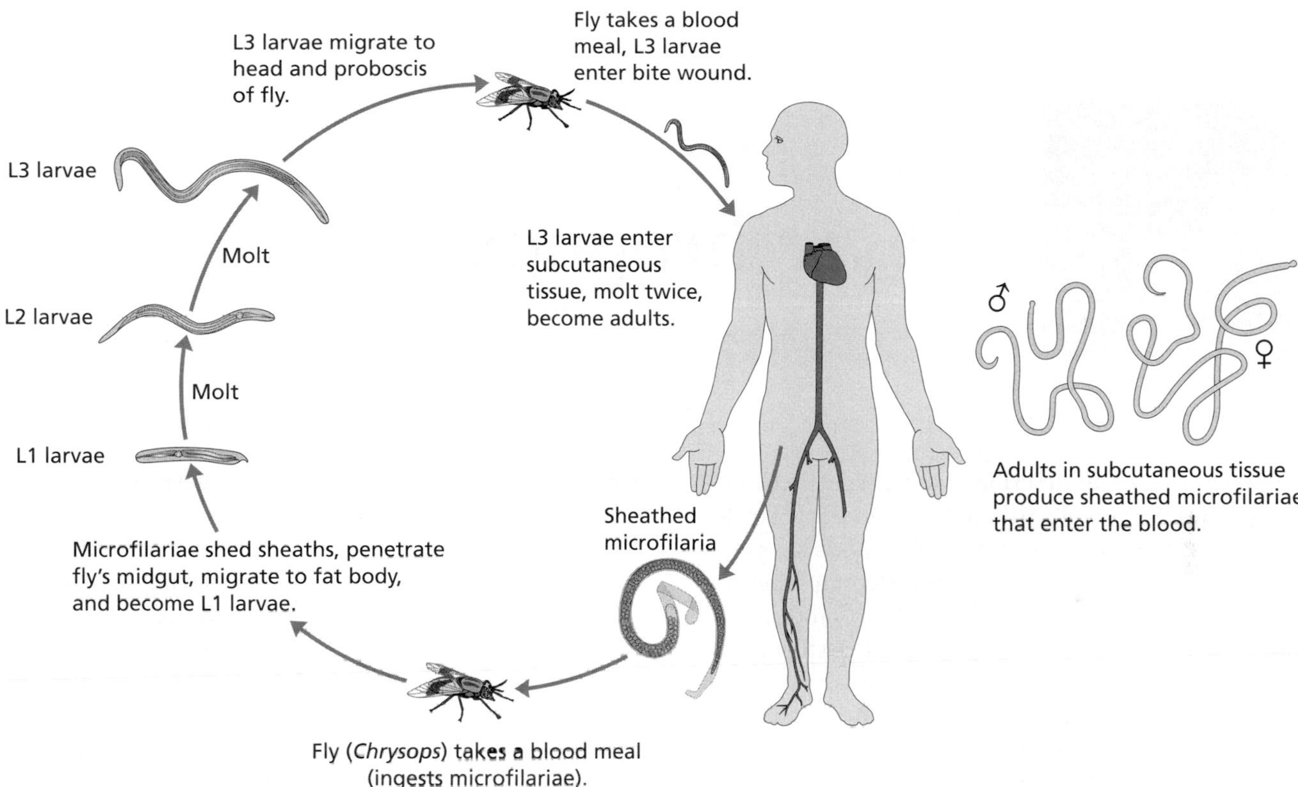

Figure 2 The life cycle of *Loa loa*, showing the involvement of the biting mango fly, *Chrysops* sp. as a vector. (Courtesy of CDC.)

The Arthropods

Approximately 80% of all described animal species belong to the phylum Arthropoda. Although considerable strides have been made, the taxonomy of the arthropods has yet to be fully resolved. Among protostome invertebrates they are placed in the superphylum Ecdysozoa: those protostomes that molt as a normal part of their development. Arthropods are divided according to some taxonomic schemes into six classes, three of which, Arachnida (spiders, scorpions, pseudoscorpions, mites, and ticks), Insecta (the insects), and Crustacea (lobsters, shrimp, crabs, barnacles, copepods, and others) predominate, making up over 95% of all described arthropod species. Many recent phylogenetic analyses consider the insects and the crustaceans to be sister taxa within the subphylum Pancrustacea. Other pancrustacean arthropods include the noninsect hexapods (the Collembola and the Protrura) and the Branchiopoda. According to many of these same analyses, the arachnids are placed in the subphylum Chelicerata, along with the Merostomata (the horseshoe crabs), class Pycnogonida (the sea spiders), and the Myriapoda (millipedes and centipedes).

Arthropods display a **metameric** body plan: a segmented body in which structures arising from the ectoderm and mesoderm are repeated in each segment. Unlike the annelids, which are also metameric and once believed to be closely related to the arthropods, most modern arthropods have greatly reduced the number of segments by fusing segments into fewer, more specialized segments. Such specialization of segment groups is called **tagmatization** and each specialized group of fused segments is called a **tagma**. The crustaceans and insects have bodies composed of three tagmata: head, thorax, and abdomen. In the arachnids, the head and thorax are fused to form the prosoma. Arthropods are also typified by having an exoskeleton, composed largely of the polysaccharide chitin, and jointed appendages. Their internal organs are found in a cavity called a **hemocoel**, which also serves as a cavity for blood circulation. Arthropods have a pair of ventral nerve cords, with paired ganglia in each segment. The arthropod brain consists of fused ganglia around the esophagus in the head.

Although many arthropods are free-living, there are numerous parasitic species. Many of the typical arthropod features described above have been secondarily reduced or lost in these organisms as adaptations to parasitism. For example, the pentastomids have lost all appendages, and their only chitinous structures are those used to adhere to their preferred site within the host—generally the respiratory system of a vertebrate. Likewise, the loss of wings in fleas is thought to be an adaptation to parasitism, allowing them greater freedom of movement on their hosts' bodies.

Arthropods act as both endo- and ectoparasites in their own right and can frequently impose significant morbidity on their hosts. They also often serve as both definitive and intermediate hosts for other parasites, including protozoa, flatworms, nematodes, and on occasion even other arthropods. Most important vectors of viral, prokaryotic, and eukaryotic parasites are arthropods, making them a major focus of those interested in human and animal parasitology.

Sand flies (subfamily Phlebotominae)

These dipterans (the true flies) comprise the subfamily Phlebotominae, which along with subfamily Psychodinae make up the family Psychodidae. Certain species of sand flies transmit *Leishmania* in both the Old and New World and *Bartonella bacilliformis*, the bacterial agent of Carrion's disease, in South America. Others transmit viruses such as the sand fly fever virus in the Old World. There are about 900 species of sand flies.

Distribution and prevalence Sand flies (Figure 1) have a widespread distribution and are typically found between approximately latitudes 50°N and 40°S, although they are absent from certain oceanic islands. Because they require high humidity and organic material for reproduction and larval development, they are typically found in caves, rock crevices, animal burrows, or underneath logs and other vegetation. In the Old World, members of the genus *Phlebotomus* are responsible for *Leishmania* transmission. In the Americas, members of the genus *Lutzomyia* transmit both leishmaniasis and Carrion's disease.

Hosts and transmission Only female phlebotomines take blood meals, which are required for egg production (Figure 2). Blood feeding is by **telmophagy** (feeding from blood that has pooled in small wounds made by the fly). One blood meal can support the production of about 100 eggs, which are laid singly in humid organic matter. Males feed on nectar. Adult females feed mainly at dawn, at dusk, or at night. Host specificity varies according to species. Although some species tend to be host specific on various mammals, birds, or reptiles, others are generalists. Species serving as *Leishmania* vectors become infected when they ingest *Leishmania* amastigotes. After migrating through the fly's body, reproducing asexually, and undergoing sequenced developmental changes (see *Leishmania* life cycle), promastigotes arrive in the esophagus and pharynx, which they tend to clog. As feeding sand flies pump esophageal contents repeatedly, attempting to remove the obstruction, promastigotes are inoculated into the next host. Transmission can also occur if the host crushes the sand fly against a mucus membrane or a break in the skin.

Pathology See genus *Leishmania* for a description of pathology in leishmaniasis. Sand fly bites cause little pain, but they develop into small, itchy, inflamed bumps that may last several days.

Treatment See genus *Leishmania* for treatment of leishmaniasis. Bites can be treated with a variety of topical anti-inflammatories.

Control Control of sand flies is practiced in areas endemic for *Leishmania*. Methods include use of pyrethroid insecticides for residual spraying, insecticide impregnated bed nets, space spraying, and use of repellents on skin or fabrics for personal protection. Because dogs are an important reservoir host for *Leishmania*, spraying in animal shelters and the use of repellent-impregnated dog collars is also useful. As breeding sites for sand flies are so difficult to find and access, control measures against larval sand flies are often not feasible, although laboratory studies suggest the utility of several biological and chemical agents for larval control.

Did you know? Phlebotomine sand flies have been used forensically. Ingested blood can be recovered from the flies and DNA analysis can be used to identify the person from whom the blood was taken. Bite marks can be used to place a person in an area where the flies were collected.

References

Alexander B & Maroli M (2003) Control of phlebotomine sand flies. *Med Vet Entomol* **17**:1–18.

Ready PD (2013). Biology of phlebotomine sand flies as vectors of disease agents. *Annu Rev Entomol* **58**: 227–250.

Figure 1 An adult sand fly. A female *Phlebotomus papatasi* taking a blood meal. Note the elevated position of the wings while at rest, characteristic of sand flies. (Courtesy of CDC/ Frank Collins.)

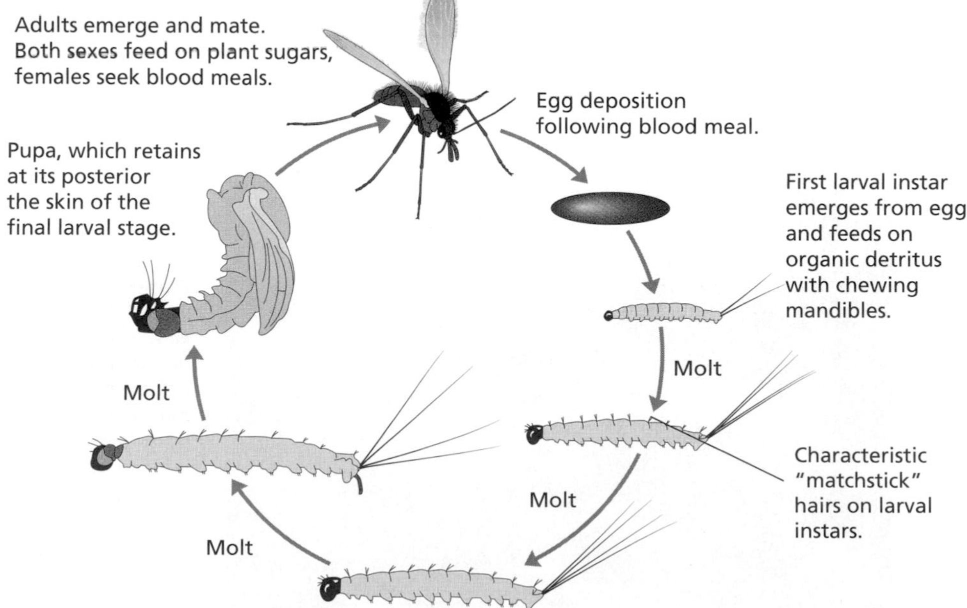

Figure 2 The life cycle of phlebotomines. Following a blood meal, females lay up to 100 eggs in humid, organic material in caves, rock crevices, animal burrows and other dark places. There are four instar stages, followed by the formation of a pupa, from which the adult emerges. Larvae do not develop in water bodies, posing difficulties for control efforts. (Courtesy of CDC.)

Lice (*Pediculus humanus, Pthirus pubis*)

The most important species of sucking louse (order Anoplura) occurring on humans exists in two distinct subspecies: *Pediculus humanus humanus*, the body louse, and *P. humanus capitis*, the head louse (Figure 1). Both are blood feeding, as are all anoplurans. Body lice live on host's clothing, except while feeding. Eggs (nits) are cemented to clothing. Head lice generally remain in the host's hair, cementing their nits to hair (Figure 2).

Distribution and prevalence Distribution is worldwide. Infestations with *P. humanus capitis* are far more common. With the exception of follicle mites, head lice are recognized as the most common human ectoparasite. In the United States, 6–12 million cases (mostly children) are treated annually. Other countries experience similar high infestation rates. All socioeconomic groups are affected, and prevalence is not closely related to levels of hygiene. In contrast, the prevalence of body lice is closely tied to levels of cleanliness and hygiene. Infestations are rare in the developed world, except among the homeless. The likelihood of infestations is greatest during wartime or when other events preclude personal cleanliness and clothes washing. It is most common in cold regions requiring heavy clothing where there is also significant poverty.

Hosts and transmission Both subspecies are exclusively ectoparasitic on humans, and both are transmitted via close physical contact, or shared clothing or hair brushes. Body lice transmit epidemic or louse-borne typhus, trench fever, and louse-borne relapsing fever, caused by *Rickettsia prowazeki*, *Bartonella quintana*, and *Borrelia recurrentis*, respectively. Head lice, previously thought not to vector diseases, are now known to carry *B. quintana*.

Pathology Unless they transmit an infectious disease, a louse infestation (pediculosis) is not life threatening. Feeding lice cause very itchy red papules, which often lead to dermatitis. Chronic infestation leads to thickened, darkened skin, a condition often called vagabond's disease (Figure 3).

Diagnosis Diagnosis is generally by observation of either adult lice or nits.

Treatment A variety of commercial chemical treatments are available to treat lice infestations. Such treatments often have no effect on eggs. Heated air has been used to desiccate head lice and their eggs.

Control Good personal hygiene and regular clothes washing are effective in the control and prevention of body louse infestation. For head lice, keeping hair tied up may prevent infestation. Lice combing may be repeated for an entire family once a week while children are school age.

Did you know? Molecular evidence suggests that body lice and head lice diverged around 100,000 years ago, allowing an approximation of when humans began to wear clothing.

Other related species

Pthirus pubis: This species known as the crab louse infests pubic hair in humans. Crab lice are not prominently involved as disease vectors.

References

Kittler R, Kayser M & Stoneking M (2003) Molecular evolution of *Pediculus humanus* and the origin of clothing. *Curr Biol* **13**:1414–1417.

Burgess IF (2004) Human lice and their control. *Annu Rev Entomol* **49**:457–481.

Ascunce MS, Toups MA, Kassu G et al (2013) Nuclear genetic diversity in human lice (*Pediculus humanus*) reveals continental differences and high inbreeding among worldwide populations. *PLoSONE* **8**(2): e57619. doi:10.1371/journal.pone.0057619.

Figure 1 *Pediculus humanus capitis*. An adult head louse adheres to a human hair using its powerful clawed appendages. The morphology of *P. humanus humanus* and *P. humanus capitis* is very similar. Head lice are generally slightly smaller (about 1.25 mm for males and about 1.9 mm for females) compared to body lice (about 2.5 mm and 3.0 mm for males and females, respectively). (Courtesy of Gilles San Martin, CC BY-SA 2.0.)

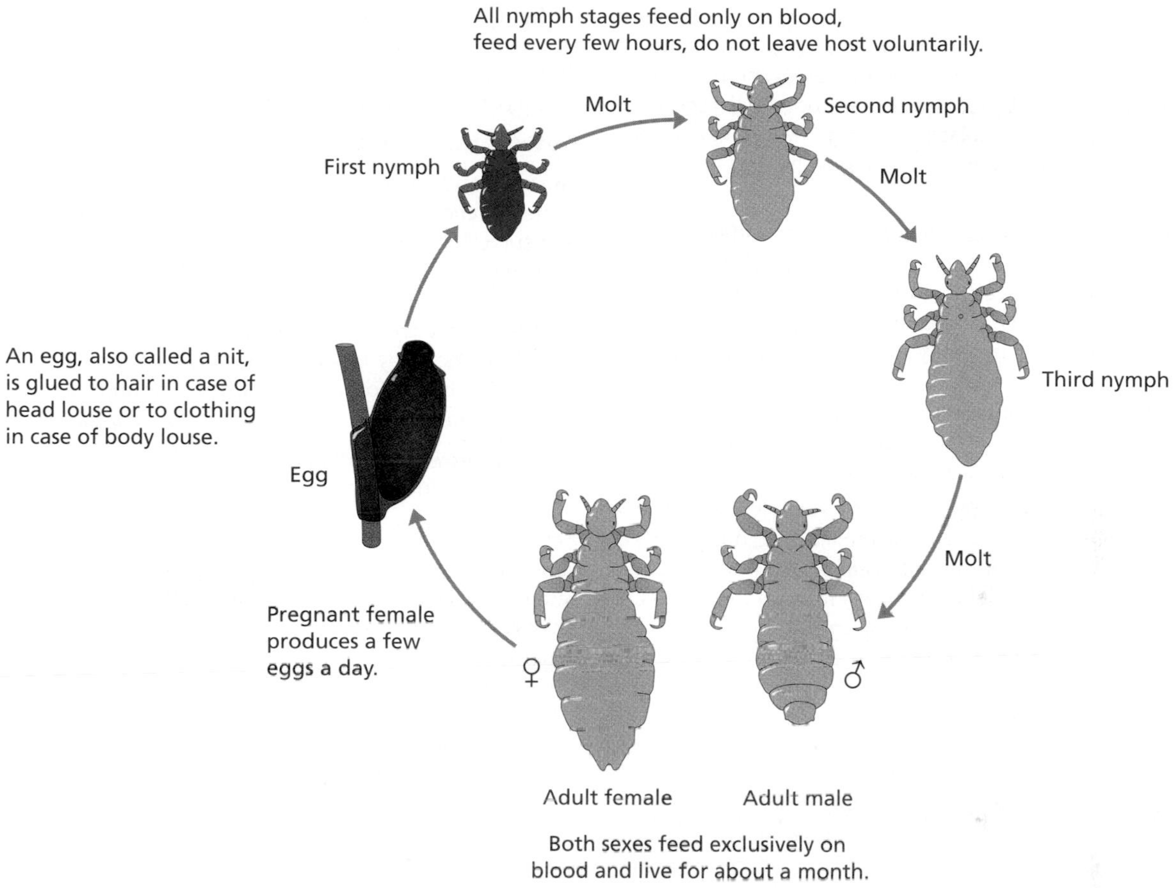

All nymph stages feed only on blood,
feed every few hours, do not leave host voluntarily.

Molt

Second nymph

First nymph

Molt

An egg, also called a nit,
is glued to hair in case of
head louse or to clothing
in case of body louse.

Third nymph

Egg

Molt

Pregnant female
produces a few
eggs a day.

♀

♂

Adult female

Adult male

Both sexes feed exclusively on
blood and live for about a month.

Figure 2 The life cycle of *Pediculus humanus*. Eggs of body lice and head lice are attached to clothing or hair fibers, respectively, and hatch in about a week. First-stage nymphs pass through two subsequent nymph stages and reach adulthood in about nine days, although cooler temperatures may lengthen the time necessary to complete the life cycle. (Courtesy of CDC.)

Figure 3 Pediculosis caused by an infestation of body lice. Note the large number of adult lice adhering to the pulled-down sock and the dark, thickened skin (known as vagabond's disease), caused by chronic louse feeding.

Kissing bugs (subfamily Triatominae)

The subfamily Triatominae consists of blood-sucking insects found within the family Reduviidae and the order Hemiptera (Figure 1). These insects are commonly called conenose or kissing bugs. Most or all of the approximately 155 species are potential vectors of *Trypanosoma cruzi*, the kinetoplastid causing Chagas' disease, but only those species that are adapted to living with humans are considered important vectors.

Distribution and prevalence Triatomines are widespread in South and Central America and much of North America (Figure 2). Various species tend to be found in different sites, such as vegetation, animal burrows, rock piles or human dwellings. Important Chagas' disease vectors are typically those that infest human habitations. The primary vectors are *Triatoma infestans* and *Rhodnius prolixus* in southern and northern South America, respectively, and *T. dimidiata* in Central America. *Panstrongylus megistus*, *T. sordida*, and *T. brasiliensis* are important vectors in Brazil.

Hosts and transmission Triatomines will generally feed on any vertebrate. Those species living in close proximity to humans feed readily on humans and other domestic animals such as dogs and cats. Those living in other habitats typically feed on those wildlife species that are available. Many mammalian species can serve as reservoir hosts for *T. cruzi*. Domestic dogs and cats are probably most important to human health. Transmission of infective *T. cruzi* metacyclic trypomastigotes to the mammalian host occurs when parasites are passed in the insect's feces following feeding. The mammal may then scratch or rub the bite site, inoculating the parasite into the wound or mucous membrane. Reservoir hosts can also become infected by eating infected kissing bugs.

Pathology Triatomines tend to bite humans in the soft tissues around the lips and eyes (hence the name kissing bugs). Generally these bites are painless. If parasites are introduced, an acute inflammatory reaction occurs at the site of infection (see *T. cruzi*).

Treatment See *T. cruzi*.

Control Residual insecticides are effective for kissing bug control as are insecticide-treated bednets. Control efforts include community surveillance followed by focal insecticide application. Paint containing insecticides can be effective, especially in house interiors (Figure 3). Triatomine populations in human dwellings can be dramatically reduced by eliminating insect hiding places such as wall cracks, or by replacement of thatched roofing with metal roofing. Elimination of wood piles, garbage, and excess vegetation from the areas surrounding housing is also helpful.

Did you know? In Brazil, kissing bugs are called "barbeiros", the Portuguese word for barbers.

References

Bern C, Kjos S, Yabsley MJ et al (2011) *Trypanosoma cruzi* and Chagas' disease in the United States. *Clin Microb Rev* **24(4)**: 655-681.

Abad-Franch F, Ferraz G, Campos C et al (2010) Modeling disease vector occurrence when detection is imperfect: Infestation of Amazonian palm trees by Triatomine bugs at three spatial scales. *PLoS Negl Trop Dis* **4(3):** e620. doi:10.1371/journal.pntd.0000620.

Zeledón R & Rabinovich M (1981) Chagas' disease: an ecological appraisal with special emphasis on its insect vectors. *Annu Rev Entomol* **26**:101–133.

Figure 1 The life cycle stages of *Rhodnius prolixus*. Like other triatomines, *R. prolixus* and other kissing bugs undergo incomplete metamorphosis. A wingless, first instar nymph hatches from an egg, passing through second through fifth instars, before molting into an adult possessing two pairs of functional wings. All nymph stages and adults of both sexes take blood, and all stages of both sexes can become infected with *T. cruzi*. (Courtesy of Thierry Heger, CC BY-SA 3.0.)

Figure 2 Distribution of important triatomine vectors in South America. The two most important vectors of Chagas' disease are *Triatoma infestans*, commonly found in southern South America, and *Rhodnius prolixus*, concentrated in the northern part of the continent.

Figure 3 Housing conducive to triatomine infestation. Features such as a thatched roof, dirt floors, and cracked, adobe walls provide ideal habitat for important human vectors of Chagas' disease. (Courtesy of WHO.)

Black flies (genus *Simulium*)

There are about 1800 species in the dipteran family Simuliidae (the black flies), many of them in the genus *Simulium*. Members of this genus transmit *Onchocerca volvulus*, the nematode responsible for onchocerciasis (river blindness), and the apicomplexan *Leucocytozoon*, a malaria-like parasite of birds.

Distribution Members of the genus are found worldwide. Black flies are found mainly in close proximity to fast-flowing waterways, where eggs are laid and where larvae develop (Figures 1, 2, and 3). The most important vector species in Africa are *S. damnosum* and *S. neavei*. In the Americas, *S. callidum*, *S. exiguum*, *S. metallicum*, and *S. ochraceum* are major vectors.

Hosts and transmission Only female black flies (Figure 4) take blood meals and are involved in parasite transmission. They tend to be host specific in terms of where they take their blood meals. The species involved in *O. volvulus* transmission readily feed on humans. Female flies are **telmophages**. Unlike some other blood-feeding arthropods called **solenophages**, they do not introduce their mouthparts directly into a blood vessel. Rather, they cut through skin and blood vessels causing blood to pool on the surface, which they then ingest. *Onchocerca* vectors become infected when they ingest microfilariae in this pooled blood. Within the black fly, microfilariae eventually develop into infective L3 larvae, which migrate to the fly's mouthparts, and are able to infect a new human host when the fly takes its next blood meal.

Pathology See *O. volvulus* for a description of the pathology observed in onchocerciasis. Black flies can be vicious biters and can cause significant problems for livestock, poultry, and wildlife. Repeated biting by swarms of black flies can cause anemia and even death. Bites usually elicit a local inflammatory response causing red, itchy lesions. A large number of bites can cause a systemic response, probably from extensive histamine release. In humans, this reaction is called black fly fever. Headaches, fever, nausea, general dermatitis, and asthma are typical symptoms. Especially severe reactions can be fatal.

Treatment A variety of anti-inflammatory topical medications are available to relieve the itching and inflammation associated with insect bites. In severe cases of black fly fever, supportive care is required.

Control Black flies may be controlled through the use of larvicide spraying of fast-flowing rivers. Individuals can protect themselves from black flies with repellants such as DEET (*N,N*-diethyl-3-methylbenzamide).

Did you know? In 1923, *Simulium colombaschense* killed 16,000 horses, mules, and cattle in Europe. From 1944 until 1948, *S. venustum* killed more than a thousand cattle annually in western Canada.

References

Adler PH & Crosskey RW (2009) World black flies (Diptera: Simuliidae): a comprehensive revision of the taxonomic and geographical inventory. Natural History Museum, London.

Crump A, Morel CM, Omura S et al (2012) The onchocerciasis chronicle: from the beginning to the end. *Trends Parasitol* **28**: 280–288.

Lakwo TL, Garms R, Rubaale T et al (2013) The disappearance of onchocerciasis from the Itwara focus, western Uganda, after elimination of the vector *Simulium neavei* and 19 years of annual ivermectin treatments. *Acta Tropica* **126**: 218–221.

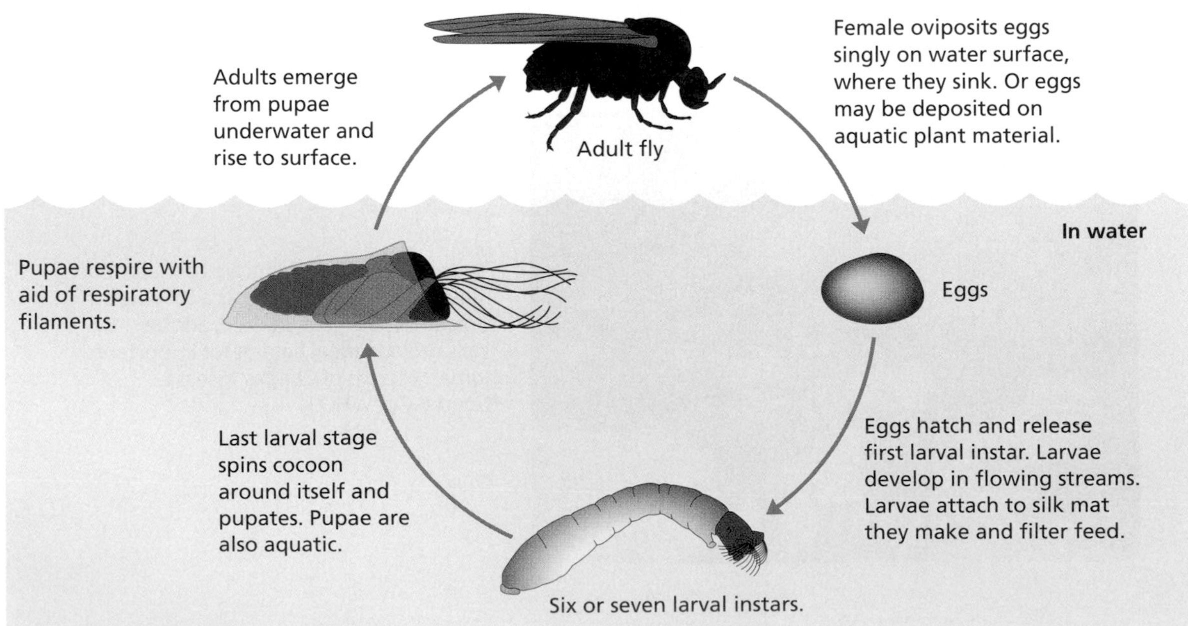

Figure 1 The life cycle of *Simulium*. Eggs are deposited on vegetation in or near fast-flowing streams or rivers. There are typically six or seven larval instars. The final stage larva spins a cocoon around itself in which pupation occurs. Adult flies emerge from the aquatic pupa and float to the surface. The life cycle can take from just over two weeks to six weeks, depending on temperature and food availability. (Courtesy of CDC.)

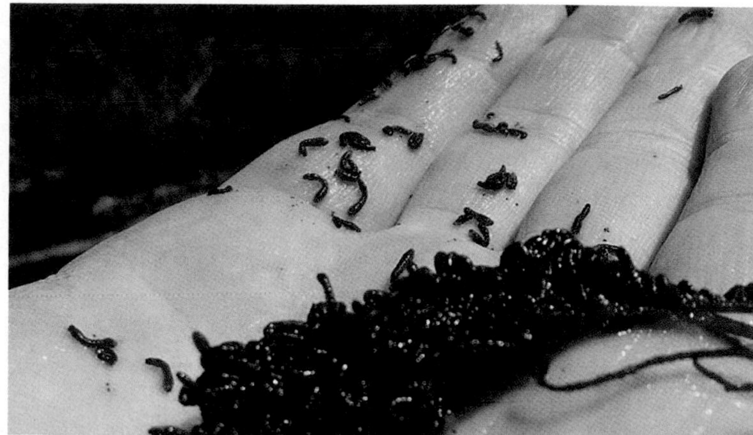

Figure 2 Black fly larvae. *Simulium* larvae collected on only a few blades of vegetation in a fast-flowing North American stream. Some rivers can produce millions of flies per kilometer per day. (Courtesy of Clemson University.)

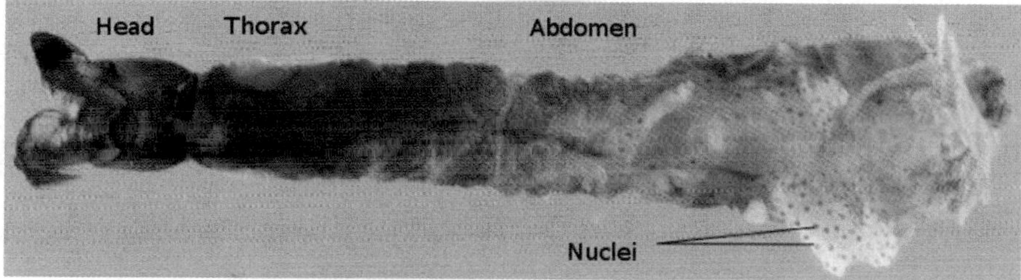

Figure 3 Larval *Simulium*. The nuclei at the posterior end contain giant polytene chromosomes. Following hatching, the larva uses modified salivary glands to spin a silken mat, which remains attached to an underwater object. With its head then floating downstream, it utilizes rake-like projections around the mouth to filter feed. (Courtesy of Clemson University.)

Figure 4 A female black fly. Note the small area of pooled blood from which the fly is feeding. (Courtesy of F Christian Thompson, USDA.)

Mosquitoes (family Culicidae, genera *Culex*, *Anopheles*, and *Aedes*)

Mosquitoes comprise the family Culicidae, within the insect order Diptera (the true flies). Mosquitoes are the most important vectors of parasitic diseases. Various mosquito species take blood meals from a wide variety of vertebrate hosts, including mammals, birds, reptiles, amphibians, and even fish. Although some species are host specific, others have broad host ranges, feeding for instance, on both mammals and birds. There are approximately 3500 described species of mosquitoes that make up two principal subfamilies within family Culicidae. The Culicinae consists of many genera, including two of the most important medically: *Culex* and *Aedes*. The Anophelinae consists of three genera, the most medically important of which by far is *Anopheles*.

Distribution and prevalence Mosquitoes are found in all parts of the world with the exception of Antarctica. In warmer areas of the world, especially in humid tropical areas, mosquitoes may be active throughout the year. In temperate areas, inseminated females hibernate through the winter months in appropriate microhabitats. Eggs from temperate species are also capable of sustained survival at low temperatures. Although details vary across species, all mosquito life cycles possess the same life-cycle stages (Figure 1). Mosquito populations in a particular area can be enormous, and although not all mosquitoes feed on blood, they are the most common blood-feeding arthropods.

Hosts and transmission Mosquito host preference has a genetic basis but commonly shows a high level of plasticity, influenced by the abundance of potential host species. Those species characterized by relatively high host specificity, including the most medically important species in the genera *Anopheles*, *Aedes*, and *Culex* (Figure 2), are often the most important vectors. For example, *Anopheles gambiae*, an important malaria vector, shows a strong preference for humans. Likewise, *Aedes aegypti*, an important vector of dengue and yellow fever viruses, as well as other pathogens, is also anthropophilic. *Culex quinquefasciatus*, alternatively, a known vector of West Nile virus, may feed primarily on mammals, primarily on birds, or widely on both birds and mammals, depending on local conditions. Some of the other medically important parasites transmitted by members of these three genera are: *Culex:* various encephalitis viruses, including western equine encephalitis and St. Louis encephalitis viruses, avian malaria, and various filarial worms, including *Wucherieria bancrofti*. *Aedes:* dengue, yellow fever, numerous other viral pathogens, and several nematode parasites, including the canine heartworm *Dirofilaria immitis*. *Anopheles:* various viruses and filarial worms, *Plasmodium* species in mammals.

Pathology Pathology associated with eukaryotic parasites transmitted by mosquitoes is described elsewhere in the Rogue's Gallery and in the text. Reaction to mosquito bites varies across individuals, with most experiencing only mild, itchy, and inflamed swelling at the bite site. This inflammation is initiated by compounds found in mosquito saliva. Scratched or traumatized bites may be vulnerable to secondary bacterial infection. In places where mosquito numbers are exceedingly high, such as in the high Arctic during the brief summer mosquito season, animals may suffer extensive blood loss from swarms of feeding mosquitoes. A caribou, for instance, may lose up to 300 ml of blood daily during this time.

Treatment Mosquito bites are often untreated and resolve on their own within a few days. Washing with soap and warm water may lessen the likelihood of a secondary bacterial infection. Mosquito bites may itch intensely, but scratching these bites

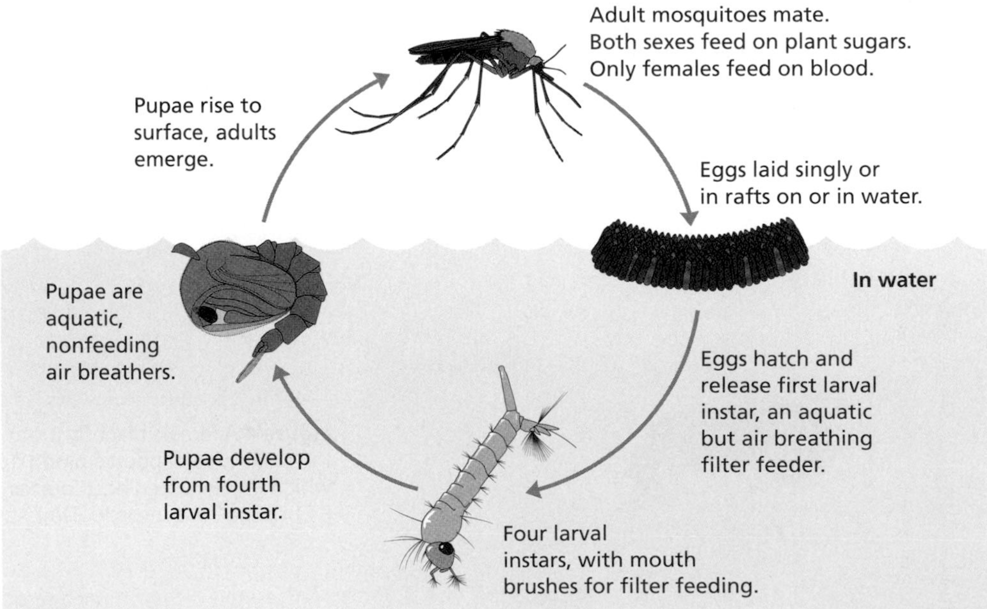

Figure 1 The life cycle of the mosquito. All mosquitoes display complete metamorphosis with egg, larval, pupal, and adult stages. Eggs may be laid in rafts (genus *Culex*), singly (*Anopheles*) on water, or on soil near the water's edge to await flooding before hatching (*Aedes*). Most larvae are found suspended at the water surface by a breathing siphon. Larvae pass through four instars prior to pupation. In most species, the pupa also remains at the water surface, breathing through a pair of respiratory trumpets. Adult females may live from about two weeks to several months. In blood-feeding species, females take a blood meal prior to egg deposition. Males typically live only about a week and feed on nectar. (Courtesy of CDC.)

often exacerbates the itching. The scratching itself can result in increased histamine release, which causes increased inflammation and consequently, increased itching. Scratching also increases the likelihood of breaks in the skin and subsequent bacterial infection. The use of an over-the-counter antihistamine compound such as diphenhydramine can reduce itching. Alternatives exist for those allergic to diphenhydramine. The use of an anti-itching topical cream is advisable. The application of an ice pack or a cold washcloth may also relieve inflammation and intense itching.

Control Many methods of control have been used to reduce mosquito numbers. Larvicides are frequently applied directly to mosquito larval habitats to kill aquatic larvae. Larvae have also been controlled biologically with *Gambusia*, but this technique still generates controversy (see Chapter 9). The removal of standing water in objects such as old tires or buckets is valuable for the control of certain mosquito species, especially those in the genus *Aedes*, which are adapted to lay their eggs in small amounts of standing water (Figure 3). During outbreaks of mosquito-vectored diseases, aerial spraying of pesticides such as malathion can be used to reduce numbers of adult mosquitoes. See Chapter 9 for a discussion of more recent experimental strategies for mosquito control such as the use of transgenic mosquitoes or the introduction of bacterial symbionts that interfere with mosquito reproduction. Simple measures such as the use of bed nets, window screening, appropriate clothing, spraying of houses with residual insecticides, and insect repellents are all effective.

Did you know? Female mosquitoes in search of a blood meal are attracted to potential hosts by compounds in animal breath and sweat such as carbon dioxide and lactate. They also sense body heat. Females can detect such cues from up to 75 feet away. The mosquito follows concentration gradients, flying back and forth through the chemical plume in the direction of increased concentration until the host is located.

Because it is such an excellent malaria vector, and because it prefers to feed on humans and breed in habitats formed by human activity, *Anopheles gambiae* is arguably the world's most dangerous animal.

References

Becker N, Petric D, Zgomba M et al (2010) Mosquitoes and Their Control, 2nd ed. Springer.

Dickinson K & Paskewitz S (2012) Willingness to pay for mosquito control: How important is West Nile Virus risk compared to the nuisance of mosquitoes? *Vector-Borne and Zoonotic Diseases* **12**: 886–892.

Spielman A & D'Antonio M (2001) Mosquito: A Natural History of Our Most Persistent and Deadly Foe. Hyperion.

Takken W & Verhulst NO (2013) Host preferences of blood-feeding mosquitoes. *Ann Rev Entomol* **58**:433–453.

Arensburger P, Megy K, Waterhouse RM et al (2010) Sequencing of *Culex quinquefasciatus* genome establishes a platform for mosquito comparative genetics. *Science* **330**: 86–88.

A

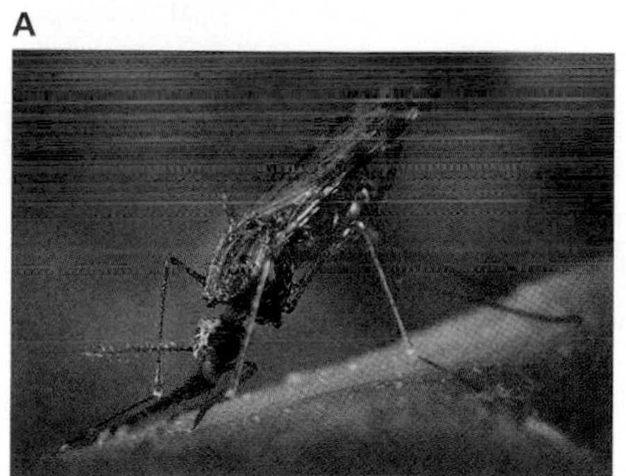

B

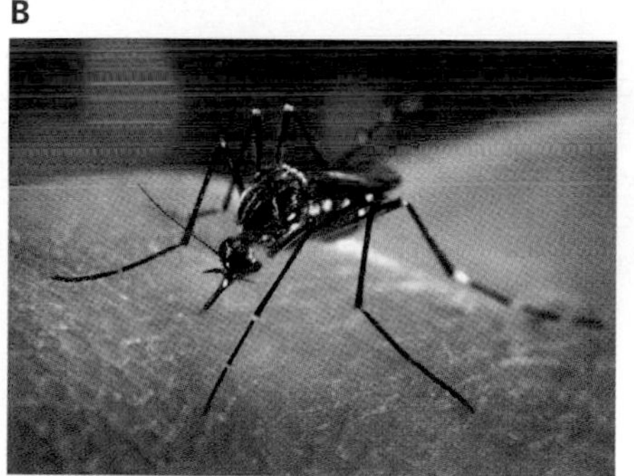

C

Figure 2 Blood feeding mosquitoes. Female mosquitoes in the process of taking a blood meal. (A) *Anopheles albimanus*, (B) *Aedes aegypti*, (C) *Culex quinquefasciatus*. Members of these three genera are the most important in terms of parasite transmission. Different adult mosquito species are identifiable by their size, patterns of coloration, antennae, and other anatomical features. Furthermore, mosquitoes may be identified by certain characteristics such as the typical resting posture of *Anopheles* mosquitoes, in which the abdomen is elevated. (Courtesy of CDC/James Gathany.).

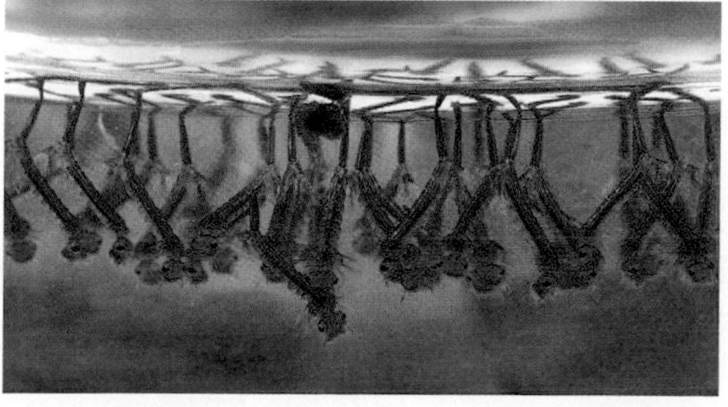

Figure 3 *Culex* larvae. A single pupa is also seen in the center of the photo. *Culex* larvae typically hang suspended in the water, using their abdominal breathing siphon to attach to the surface and for respiration. Larvae of other genera may adopt different postures, allowing their identification. *Anopheles* larvae, for example, generally lie parallel to the surface rather than with head down, as seen here. Pupae are highly active and will quickly swim to the bottom if disturbed at the surface. (Courtesy of CDC/James Gathany.).

Tsetse flies (genus *Glossina*)

Large blood-feeding flies commonly known as tsetse flies are in the family Glossinidae (order Diptera), which contains one genus, *Glossina*. The genus includes 23 species, 20 of which are able to transmit trypanosomes to mammals.

Distribution and prevalence *Glossina* are restricted to Africa (Figure 1), except for a few sites on the Arabian Peninsula. Mainly confined to the tropics, tsetse flies inhabit much of the continent between the Sahara Desert to the north and the Kalahari Desert to the south. Of the most important species in terms of human and animal health, *G. palpalis*, *G. tachinoides*, and *G. fuscipes* are found in tropical portions of West and Central Africa, mainly along rivers in areas of heavy vegetation. *G. morsitans*, *G. pallidipes*, and *G. swynnertoni* inhabit open woodlands and savanna in East Africa.

Hosts and transmission Both males and females take blood meals (Figure 2) on a wide range of hosts, including wildlife, domestic animals, and humans. Each species has characteristic feeding patterns; some species are likely to feed on any available mammalian host. Others feed preferentially on specific hosts, such as cattle, native African ruminants, pigs, or humans. *G. palpalis*, *G. tachinoides*, and *G. fuscipes* are the principal vectors for *Trypanosoma brucei gambiense*, the parasite responsible for chronic African trypanosomiasis. Acute trypanosomiasis, caused by *T. b. rhodesiense* is primarily vectored by *G. morsitans*, *G. pallidipes*, and *G. swynnertoni*. It is likely that all of these *Glossina* species, and other species of tsetse flies as well, are competent vectors for *T. b. brucei*, the causative agent of nagana. Because *T. b. brucei* uses African ruminants as reservoirs and because such animals are most common in East Africa, it is likely that the most important vectors of nagana are those that also transmit *T. b. rhodesiense*.

Pathology See the entry on the *Trypanosoma brucei* complex for the pathology associated with African trypanosomiasis. The tsetse fly (Figure 3) can have a very painful bite, as the stout labium has tooth-like structures that penetrate the skin.

Treatment The tsetse bite itself will resolve on its own, but immediate treatment should be sought if symptoms of trypanosomiasis occur. These symptoms include a red, inflamed chancre at the site of the bite, generally appearing one to two weeks after the infection.

Control As tsetse flies are attracted to colors such as navy blue, turquoise, or red, the risk of bites can be reduced by wearing neutral colored clothing. Repellents containing *N,N*-Diethyl-meta-toluamide (DEET) are generally ineffective. Numbers of tsetse flies can be effectively reduced by clearing vegetation where adult flies rest. Because flies deposit larvae on soil under brush, such measures also reduce fly reproduction. Spraying with insecticides is effective, as are traps designed to attract tsetse flies. The release of irradiated, sterile males has been effective in certain areas, such as Zanzibar.

Did you know? A biblical passage (Isaiah 7:18) is thought to refer to tsetse flies: "And it will come about in that day that the Lord will whistle for the fly that is in the remotest part of the rivers of Egypt."

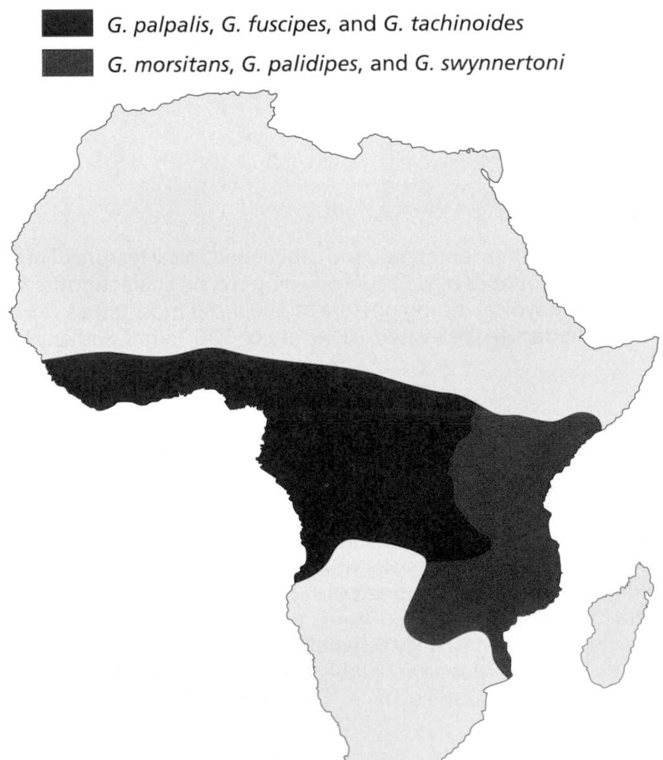

Figure 1 Distribution of *Glossina* species in Africa. The fly belt in Africa lies roughly between 15°N and 25°S. The vectors of chronic trypanosomiasis, caused by *Trypanosoma brucei gambiense*, are often called the *G. palpalis* group; these vectors are typically found along heavily wooded waterways in Central and West Africa. The vectors of acute trypanosomiasis caused by *T. b. rhodesiense*, are often called the *G. morsitans* group; these vectors are found in open savannah regions of East Africa. (Courtesy of University of California, Los Angeles.)

References

Goodling RH & Krafsur ES (2005) Tsetse genetics: contributions to biology systematics and control of tsetse flies. *Ann Rev Entomol* **50**:101–123.

Hunt RC (2004) Microbiology and Immunology On-line, trypanosomiasis page. University of South Carolina. http://pathmicro.med.sc.edu/lecture/trypanosomiasis.htm.

Weitz B (1963) The feeding habits of *Glossina*. *Bull World Health Organ* **28**:711–729.

International Glossina Genome Initiative. (2014). Genome sequence of the tsetse fly (*Glossina morsitans*): vector of African trypanosomiasis. *Science* **344**: 380–386.

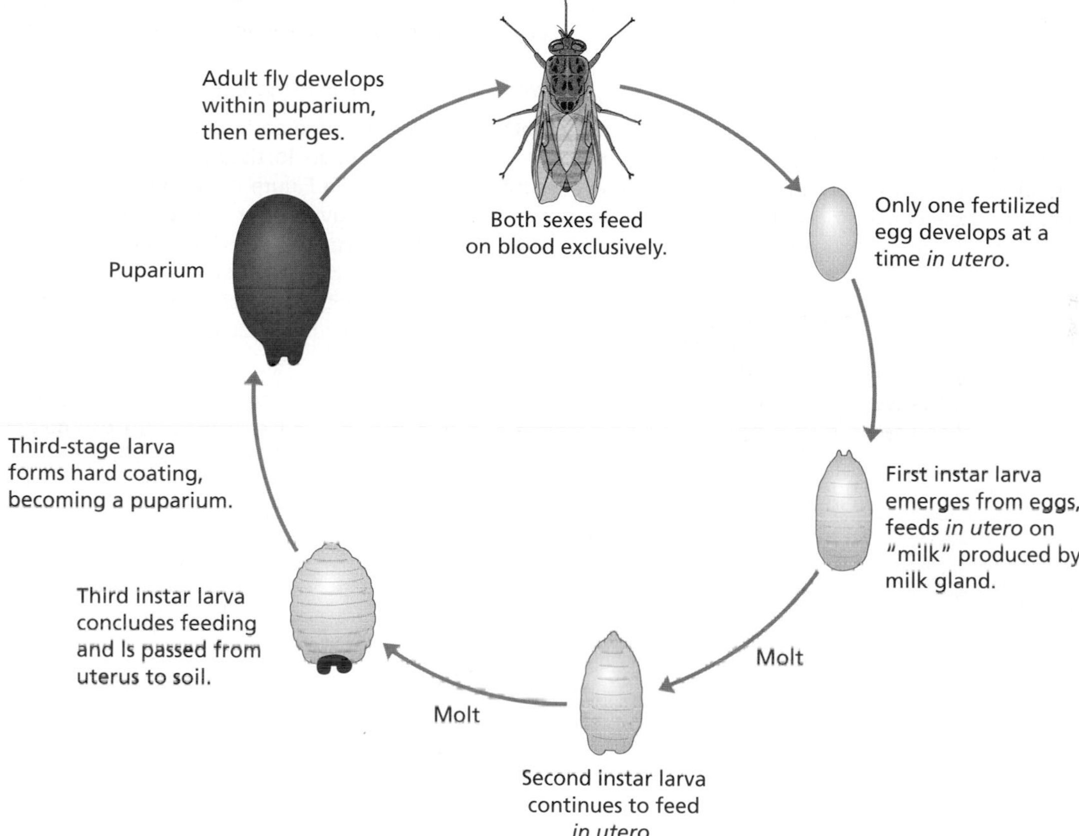

Figure 2 The life cycle of the tsetse fly. Tsetse flies have an unusual life cycle in which only a single egg is fertilized at a time. The egg hatches within the uterus of the female into a first instar larva. Larvae feed on secretions from specialized glands. All development through the third instar occurs in the uterus. Third instar larvae, which are nearly as large as adult females, are deposited on loose soil, generally under vegetation. Within an hour, the integument hardens to form the puparium. Tsetse flies are totally dependent on blood for their nutrition and harbor specialized symbionts to provide needed vitamins lacking in blood.

Figure 3 An adult tsetse fly. This species, *Glossina morsitans*, is a primary vector for chronic trypanosomiasis in West and Central Africa. Tsetse flies are recognizable by their wings, which fold on top of each other when the fly is at rest, and by the proboscis, which protrudes straight out in a resting fly. (Courtesy of Alan R Walker, CC BY-SA 3.0.)

Bot flies (family Oestridae)

The family Oestridae consists of dipterans that are commonly called bot flies or warble flies. Adults are free-living and their larvae parasitize mammals. Their life cycles vary according to species. The larvae of skin bots develop subepidermially in the host and maintain a hole in host's skin through which they respire. Stomach bots parasitize the digestive tract. Important genera include *Cuterebra*, skin bot flies of small mammals; *Dermatobia*, the human skin bot fly; *Hypoderma*, skin bot flies that primarily infect cattle; and *Gasterophilus*, stomach bot flies of large mammals.

Distribution and prevalence *Cuterebra* spp. are found in temperate and tropical regions of the Americas. *Dermatobia hominis* (Figure 1A) is a New World species that is common in forested areas across most of Central and South America (Figure 1B). *Hypoderma* spp. are widely found in the Northern Hemisphere, where they can be locally common. *Gasterophilus intestinalis*, the horse bot fly, originally an Old World species, has been introduced into the Americas, where it can be common.

Hosts and transmission *Cuterebra* females lay eggs near the mouth, nose, eye, or anus of rodents or lagomorphs. After hatching, larvae enter the body and burrow under the skin, where they feed. *D. hominis* uses **phoresy** to reach its (often human) host. Adult females catch other insects (for example, mosquitoes) and glue eggs to them. Eggs reach the mammalian host when the carrier insect lands on the host. Eggs hatch on the skin, and larvae penetrate the skin, developing in the subepidermis. Both *Hypoderma* and *Gasterophilus* lay their eggs directly on their hosts. *H. bovis* eggs are laid primarily on the hind legs. When larvae emerge, they penetrate the skin, migrating ultimately to the animal's back, where they develop. Pupating larvae emerge from the skin and then fall to the soil, where they remain as pupae until the adult emerges. Female *G. intestinalis* attach their eggs to the legs of horses (Figure 2A). When the horse licks its hair, the eggs are stimulated to hatch. Larvae (Figure 2B) penetrate the tongue and burrow to the stomach. Following development, they pass out with feces to pupate in the soil.

Pathology Skin bots are an important cause of **myiasis** (wound formation and damage to tissue caused by fly larvae). Lesions, such as that shown in Figure 3, can be painful and may interfere with normal host activities. For example, *Hypoderma* infection can result in weight loss and reduced milk production in cattle. Horses with light stomach bot infections may be asymptomatic. Heavy infections can damage the gastrointestinal epithelium, causing significant damage and even death. Gastrointestinal obstruction is also a possibility.

Diagnosis Stomach bot fly eggs are visible on the hair of domestic animals, primarily on the legs. Larvae can also be detected in the feces of infected animals. *D. hominis* signals its presence in humans by the formation of a cutaneous nodule containing a single larva.

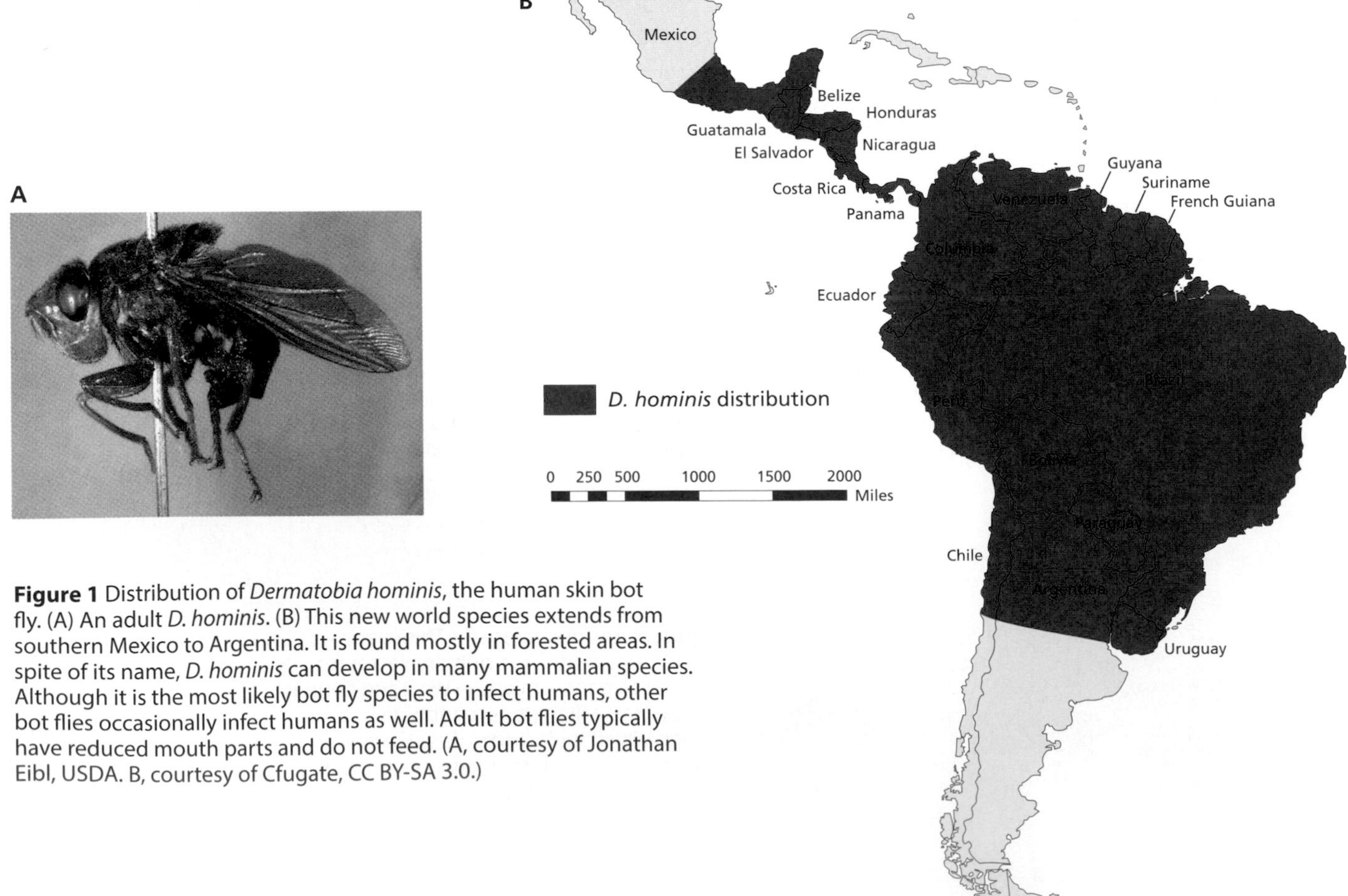

Figure 1 Distribution of *Dermatobia hominis*, the human skin bot fly. (A) An adult *D. hominis*. (B) This new world species extends from southern Mexico to Argentina. It is found mostly in forested areas. In spite of its name, *D. hominis* can develop in many mammalian species. Although it is the most likely bot fly species to infect humans, other bot flies occasionally infect humans as well. Adult bot flies typically have reduced mouth parts and do not feed. (A, courtesy of Jonathan Eibl, USDA. B, courtesy of Cfugate, CC BY-SA 3.0.)

Unlike a mosquito bite, the nodule discharges blood or serum. There is often itching and occasional shooting pain. As the larva grows, movement of the nodule can be observed.

Treatment Skin bots are often treated by sealing the larval breathing tube with nail polish or Vaseline. The larvae are removed by making a small incision near the larval breathing hole and expressing the larva. Bot fly infection in horses is treated with an anthelmintic such as ivermectin, which is also effective against many other arthropods.

Control Fly sprays can be used to reduce the likelihood of infection. In horses, eggs can be removed from the hair on the legs with a razor blade. Horses and other domestic livestock can also be treated prophylactically with ivermectin.

Did you know? Cattle can become highly agitated by ovipositing *Hypoderma* females. They may run wildly to avoid the flies, an activity called gadding. This behavior has given rise to the term "gadfly".

Other related organism

Oestrus ovis: O. ovis parasitizes sheep and goats. Larvae are deposited in the nostrils, and development occurs in the sinuses.

References

Mullen G & Durden L (eds) (2009) Medical and Veterinary Entomology. Academic Press.

Pape T (2001) Phylogeny of Oestridae (Insecta: Diptera). *Syst Entomol* **26**:133–171.

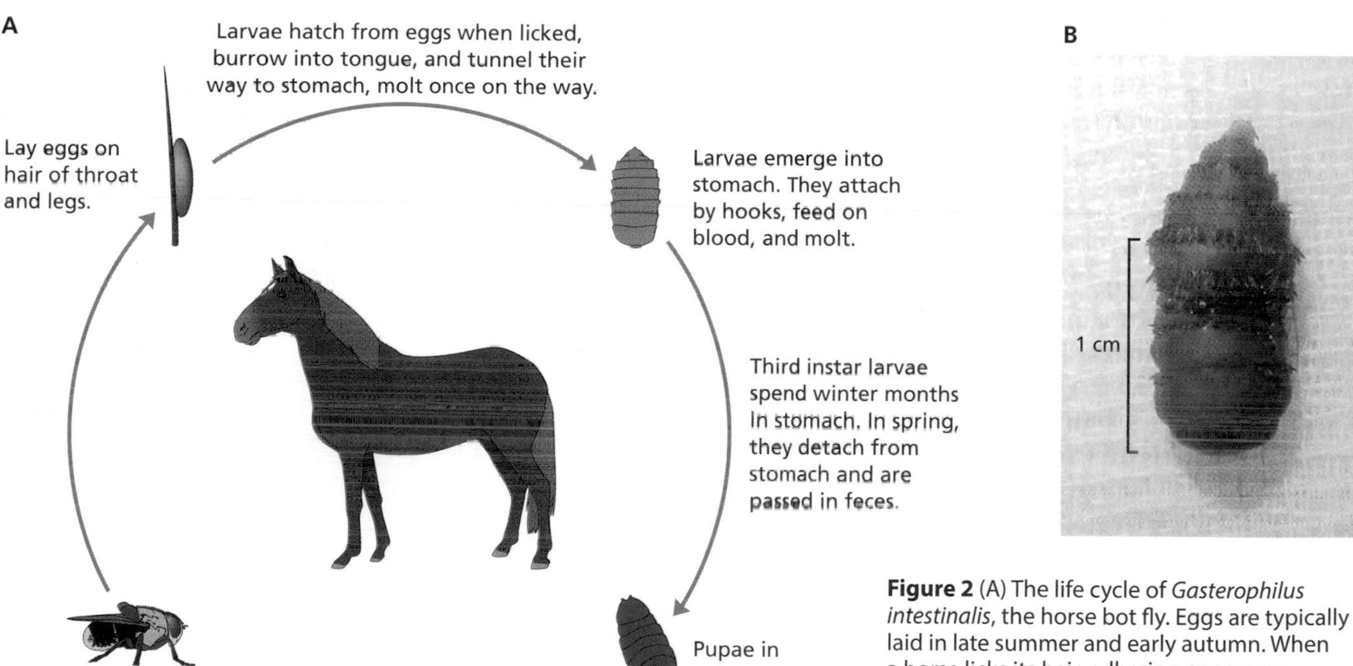

Figure 2 (A) The life cycle of *Gasterophilus intestinalis*, the horse bot fly. Eggs are typically laid in late summer and early autumn. When a horse licks its hair, adhering eggs may be picked up on the tongue. The tongue's warmth and moisture induce hatching, and the emerging first-stage larva penetrates the tongue. It then burrows to the stomach, molting along the way, where it attaches to the stomach lining with anterior hooks and feeds on blood. An additional molt follows. The following spring or summer the third-stage larva detaches and passes out with the feces, pupating in the soil. The adult emerges roughly one month later. (B) A *G. intestinalis* larva. (B, Courtesy of Kalumet, CC BY-SA 3.0.)

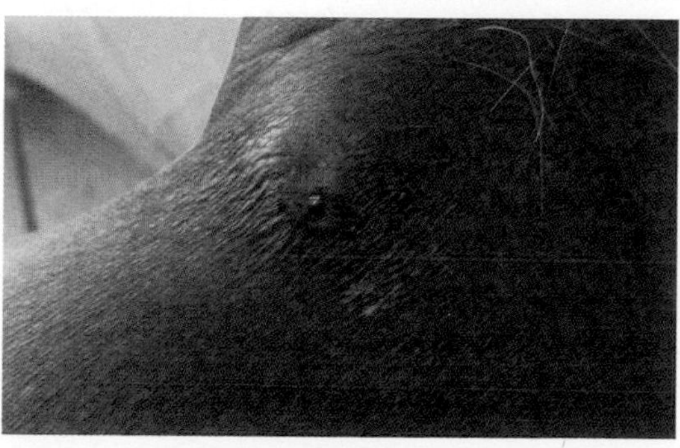

Figure 3 *Dermatobia hominis* infection. Myiasis caused by *D. hominis*. The hole through which the larva respires is visible in the center of the lesion. Such lesions can be quite painful. After deposition by a phoretic insect, the egg hatches in response to warm skin temperature. The emerging larva penetrates either unbroken skin or via the hole made by its phoretic mosquito host, and develops. Following development lasting about six weeks, the larva emerges, falls to the ground, and pupates.

Fleas (order Siphonaptera)

The fleas, small insects in the order Siphonaptera, are ectoparasitic on mammals and birds. There are approximately 2500 species. Their bodies are bilaterally flattened, and they have lost wings as an evolutionary adaptation to parasitism (Figure 1). Adult males and females have mouthparts adapted for piercing skin and sucking blood. Larvae are free-living (Figure 2).

Distribution and prevalence Various species of fleas are found worldwide. Individual species may have a more restricted distribution. For example, the cat flea *Ctenocephalides felis* is common across most of North America, but it does not occur in parts of the Rocky Mountain region. Part of the reason many flea species are so widespread is that, although they may be susceptible to environmental extremes such as temperature and low humidity, fleas often remain in microhabitats such as animal burrows, where conditions are compatible with survival.

Hosts and transmission Although they often have preferred hosts, most fleas are not highly host specific, readily transferring between hosts of different species. For instance, the cat flea *C. felis* readily uses both cats and dogs, as well as humans and other mammals as hosts. Various flea species transmit a variety of viral, bacterial, and eukaryotic parasites. The most important flea-borne disease is bubonic plague, caused by the bacterium *Yersinia pestis*. The most important vector species is probably the oriental rat flea *Xenopsylla cheopis*, but other species, including *Pulex irritans*, the human flea, are capable of transmitting plague. *Rickettsia typhi*, the bacterium responsible for murine typhus, is transmitted by several flea species. Myxoma virus, a pathogen in rabbits, is transmitted by several arthropods including fleas. *C. felis*, *C. canis*, and *P. irritans* all serve as intermediate hosts of the tapeworm *Dipylidium caninum* (see page 459). Trypanosomes in the subgenus *Herpetosoma*, and tapeworms in the genus *Hymenolepis* may also be vectored by fleas.

Pathology Flea bites usually result in **pulicosis**, slightly raised, swollen, and itchy lesions with puncture points in the center (Figure 3). The bites often are in clusters and can remain bothersome for up to several weeks. Hair loss may occur as a result of frequent scratching by the host. In extreme cases, anemia is possible. Fleas and their feces are allergenic for sensitive individuals. The ability of house dust to cause allergies may be in part attributable to flea allergens.

Treatment Various medications including anti-itch cream containing antihistamines or hydrocortisone are effective in treating flea bites.

Control Various insecticides such as diflubenzuron (inhibits egg development) and methoprene (results in infertile eggs) are useful indoors. Flea collars or oral medications may be used to inhibit flea activity on pets. Keeping areas in the home and outside, where livestock are kept, free from debris is beneficial. Regular vacuuming, for instance, can eliminate most adult fleas in the home. Baking soda can be used to dehydrate and kill fleas.

Did you know? The oriental rat flea *Xenopsylla cheopis* can jump more than 100 times its body length.

References

Dobler G & Pfeffer M (2011) Fleas as parasites of the family Canidae. *Parasit Vectors* **4**:139 (doi:10.1186/1756-3305-4-139).

Order Siphonaptera—Fleas—BugGuide.Net. http://edis.ifas.ufl.edu/IG087.

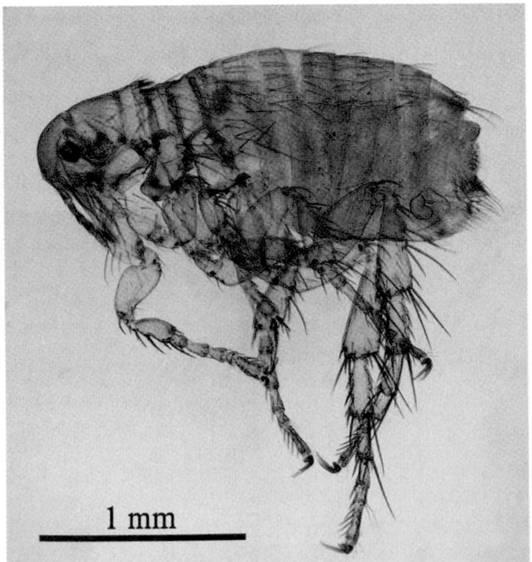

Figure 1 An adult dog flea, *Ctenocephalides canis*. The dog flea uses both dogs and cats as hosts, as well as other mammals. This species serves as an intermediate host for the tapeworm *Dipylidium caninum*. The mammal serving as definitive host becomes infected when it inadvertently consumes an infected adult flea during grooming. (Courtesy of Kalumet, CC BY-SA 3.0.)

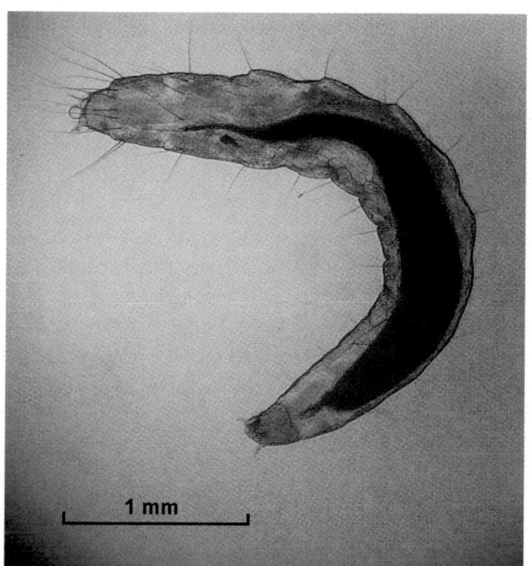

Figure 2 A flea larva. As holometabolous insects, fleas undergo complete metamorphosis. Typically there are three larval instars, followed by pupation at about 12 days. Adults may remain quiescent within the pupal cocoon for months until a host appears. Larvae are free-living, feeding on any available organic material , including fecal material from adult fleas. (Courtesy of Luis Fernandez-Garcia, CC BY-SA 3.0.)

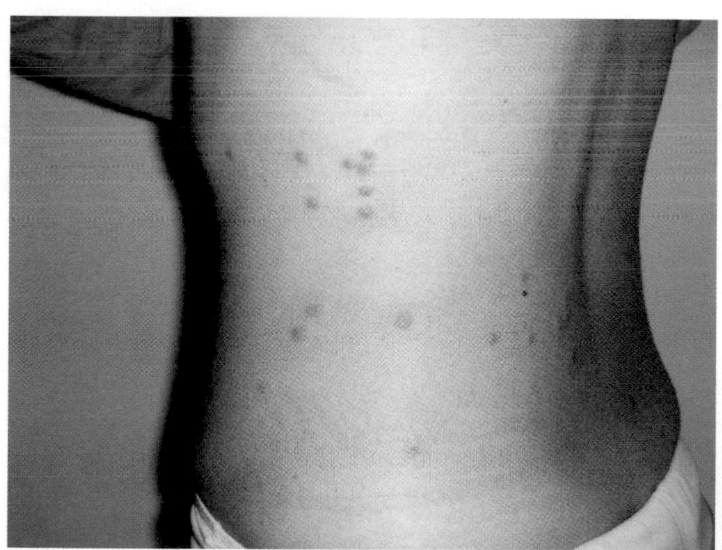

Figure 3 Flea bites. Although a flea bite is painless and may go unnoticed, skin irritation begins minutes later. After two or three days, a more severe rash may form. The skin condition that results is known as pulicosis.

Ticks (order Ixodida)

The Ixodida (ticks), along with mites, comprise the subclass Acari within the class Arachnida. Ticks are blood-feeding ectoparasites in the order Ixodida. There are three families of ticks: the Ixodidae or hard ticks (over 700 species), characterized by the presence of a hard dorsal shield (Figure 1), the Argasidae, or soft ticks (approximately 200 species), in which, unlike the ixodids, the capitulum bearing the mouthparts is concealed beneath the body (Figure 2), and the Nuttalliellidae, consisting of a single species found in southern Africa.

Distribution and prevalence Ticks, in general, are found widely across the world, although individual species may have discrete distributions. Ixodid ticks are most common where conditions are warm and humid. Low temperatures inhibit egg hatching, while some humidity is necessary for development. When necessary environmental conditions are satisfied and where host species are abundant, tick numbers can be extremely high. Argasid ticks are more common in areas of low humidity.

Hosts and transmission Although they may have preferred hosts, most ixodid ticks feed opportunistically on a wide range of hosts (Figure 3). Because they are generalist feeders and because they are competent vectors for many pathogens, they are important in the transmission of many viral, bacterial, and protozoan pathogens to humans and domestic animals. A few of the diseases transmitted by ixodid ticks as well as the principal genera of vectors include: Lyme disease, *Theileria microti* and ehrlichiosis (transmitted by members of genus *Ixodes*); tularemia and Rocky Mountain spotted fever (genus *Dermacentor* and genus *Amblyomma*); East Coast fever (*Theileria parva*) in cattle (genus *Rhipicephalus*); Crimean-Congo hemorrhagic fever (genus *Hyalomma*); and *Babesia bigemina*, the cattle disease red water fever (genus *Rhipicephalus (Boophilus)*). Argasid ticks in the genus *Ornithodoros* may transmit relapsing fever (*Borrelia hermsi*).

Pathology In addition to their role as vectors, ticks can cause anemia from heavy blood loss in severe infestations. Tick bites can additionally cause inflammation and ulceration, exacerbated by components of the tick's saliva. Dermatitis in the ears (**otoacariasis**) is problematic for many animals. Toxic components in tick saliva can result in **tick paralysis**, when the bite is near the base of the skull. The paralysis is reversed when the tick is removed.

Treatment Tick bites are usually treated by first physically removing the tick and then disinfecting the bite. Topical anti-inflammatory creams can be used to reduce swelling and itching. For removal, a tick should be grasped as close to the skin as possible with forceps, and then steadily pulled out, to avoid leaving the mouthparts behind. Such mouthparts contribute to the dermatitis often associated with tick bites.

Control Efforts to eradicate ticks have been generally unsuccessful. Ticks can be controlled locally in livestock by dipping, in which the animals are forced to walk through a liquid mixture of insecticides and acaricides to protect against ectoparasites, including ticks. Topical compounds are available for domestic pets, although some of these compounds may cause adverse reactions.

Did you know? Guinea fowl are effective tick control agents. Just two guinea fowl can eliminate ticks from two acres in one year. People previously bitten by ticks may develop allergies to red meat.

References

Magnarelli LA (2009) Global importance of ticks and associated infectious disease agents. *Clin Microbiol News* **3**:33–37.

Nicholson WL, Sonenshine DE, Lane RS et al (2009) Ticks (Ixodida). In Medical and Veterinary Entomology (Mullen G & Durden L eds), pp. 483–532. Academic Press.

Figure 1 A female adult *Ixodes scapularis*. Also known as the black-legged tick or the deer tick, this species is the primary vector in the eastern United States for *Borrelia burgdorferi*, the bacterial agent responsible for Lyme disease. Like all hard ticks, the anterior body region, known as the capitulum, bears the mouth parts, which are used to pierce the skin of hosts and ingest blood. A sclerotized plate called the scutum covers most of the dorsal surface of male hard ticks, but only the anterior part of the dorsum of a female hard tick, as seen in this photograph. (Courtesy of Scott Bauer, USDA.)

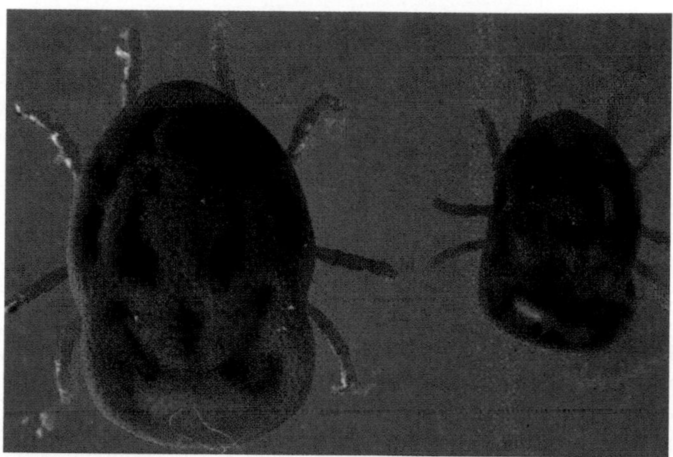

Figure 2 *Otobius megnini*, the spinose ear tick. Note the lack of the scloerotized dorsal scutum in this soft tick. Note also that, unlike hard ticks, the capitulum is not visible from a dorsal view. The larvae and nymphs of this species commonly infest the ears of many mammals, including dogs, cattle, and sheep. Many other soft ticks are primarily ectoparasitic on birds. (Courtesy of Matt Pound, CC BY-SA 3.0.)

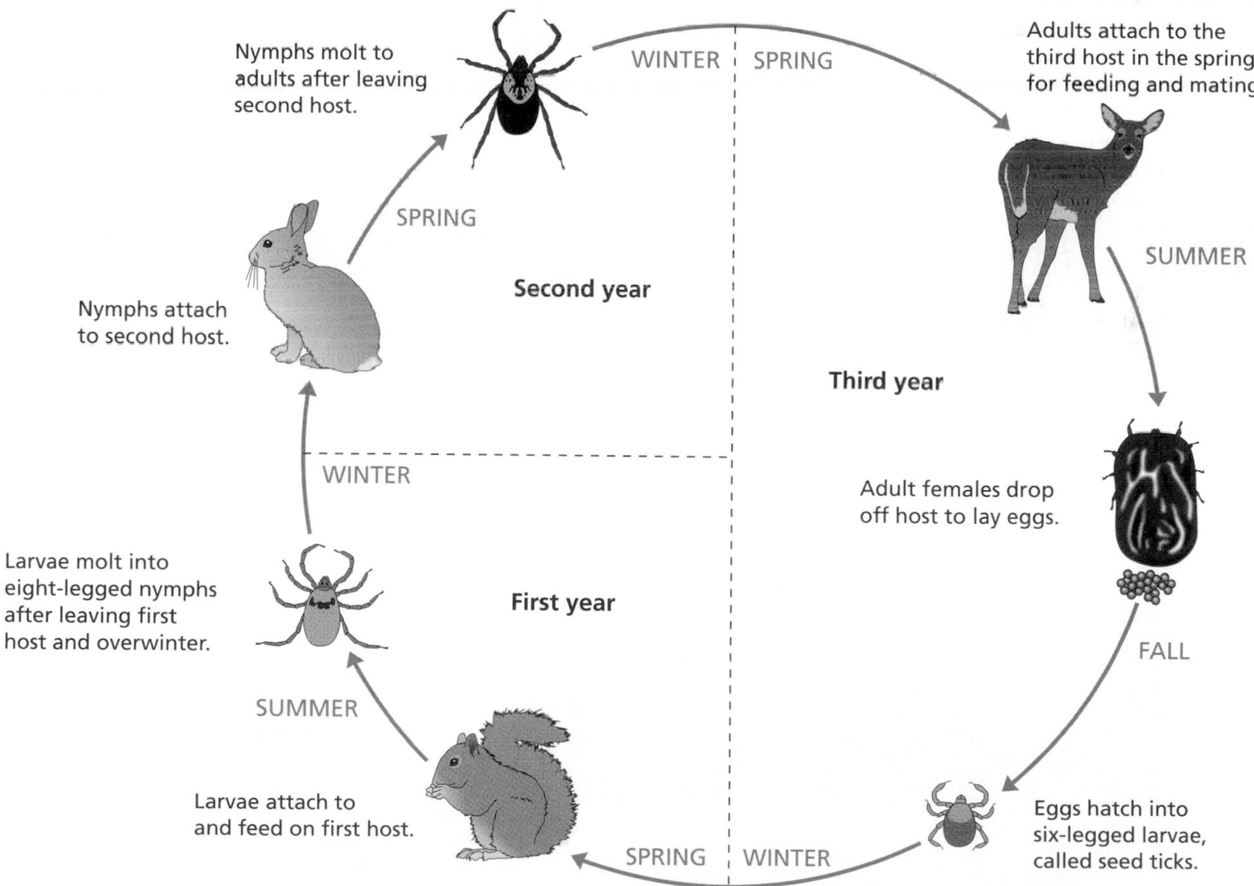

Figure 3 The life cycle of *Ixodes scapularus*. All ticks have four basic life-cycle stages: the egg, larva, nymph, and adult. A blood meal by the female following mating is required for egg production. An engorged female drops off the host and deposits eggs in the soil or in organic material. Larvae have six legs as opposed to nymphs and adults, which have eight legs. *I. scapularus*, like most hard ticks, is an example of a three-host tick, in that the larva, nymph, and adult all feed on different hosts. Other ticks may use only a single host or two hosts and are referred to as one-host or two-host ticks (see Figure 3.13). Argasid ticks have a multihost life cycle. The larval stage, each of from 2 to 8 nymphal instars, and adults of both sexes of soft ticks require blood meals. Unlike hard ticks, which stay attached to their hosts for up to several days, soft ticks feed rapidly for about an hour and subsequently leave the host. (Courtesy of CDC.)

Mites (subclass Acari)

Like ticks, mites are arthropods in the class Arachnida and the subclass Acari. There is no universal taxonomic consensus below the subclass level, but there are probably several orders of mites, based traditionally on morphological features, especially of the respiratory system. Mites may be free-living or parasitic on vertebrates, invertebrates, or plants.

Distribution and prevalence Over 48,000 species have been described; many more await discovery and naming. Because they are so diverse, mites in general have a worldwide distribution. Free-living species are found in water or soil, where each species has a characteristic distribution based on abiotic factors. The distribution of parasitic species is often dictated by the distribution of required hosts. The numbers of mites found in a preferred habitat can be high. House dust mite mites such as *Dermatophagoides pteronyssinus* can reach numbers of between 100 and 500 per gram of dust. Most if not all of people over 18 years of age may harbor one or two species of *Demodex* follicle mites (Figure 1).

Hosts and transmission Parasitic mites use a very wide range of hosts, depending on species. Some transmit pathogens of medical, veterinary, or agricultural importance. For example, members of the family Halarachnidae are parasitic in mammals and are usually found in the respiratory system. Members of the family Dermanyssidae parasitize various birds and mammals. Members of the family Trombiculidae may transmit *Rickettsia tsutsugamushi*, the causative agent of scrub typhus, to humans (Figure 2). Varroa mites (two distinct species) parasitize honey bees, negatively affecting the bees' ability to pollinate crops (see Box 1.1).

Pathology Mite infestations can cause pathology in various ways. Parasitic mites can be a source of annoyance and can cause dermatitis in hosts, which in some cases can be severe. The chicken mite (*Dermanyssus gallinae*), for example, can reach numbers high enough to kill birds, or it may cause hens to abandon their nests, resulting in the death of chicks. Various proteins (a cysteine protease released in mite feces or structural proteins in the exoskeleton) are powerful allergens, causing hay fever, eczema, or asthma. Sarcoptic mange or scabies is caused by infestation with *Sarcoptes scabiei* (Figure 3). Following mating, female mites burrow through the skin causing severe itching, rash, and crusting in severe cases.

Treatment Medications such as antihistamines can alleviate allergies caused by exposure to mite antigens. Medications with anti-arthropod activity are used topically to treat dermatitis or scabies caused by mites. Topical treatments may be combined with injected ivermectin to treat sarcoptic mange in animals (Figure 3).

Control Mites are difficult to control. Methods vary according to the type of mite under consideration. House dust mites can be controlled through regular washing of mattresses and blankets with hot water followed by drying at high temperatures. Varroa mites can be controlled with products such as thymol oil, the vapor of which is allowed to build up in beehives killing mites. Animals with mange should be isolated from other animals to avoid transmission.

Did you know? About 1200 species of spider mites (family Tetranychidae) live on plant leaves. They damage the plant, including many crop species, by puncturing the plant cells to feed. They are so named because many species spin protective silk webs.

References

Brooks GD & Bush RK (2009) Allergens and other factors important in atopic disease. In Patterson's Allergic Diseases, 7th ed (Grammer LC & Greenberger PA eds), pp. 73-103. Lippincott.

Halliday RB, O'Connor BM & Baker AS (2000) Global diversity of mites. In Nature and Human Society: The Quest for a Sustainable World: Proceedings 1997 Forum on Biodiversity (Raven PH & Williams T, eds), pp. 192–212. National Academies Press

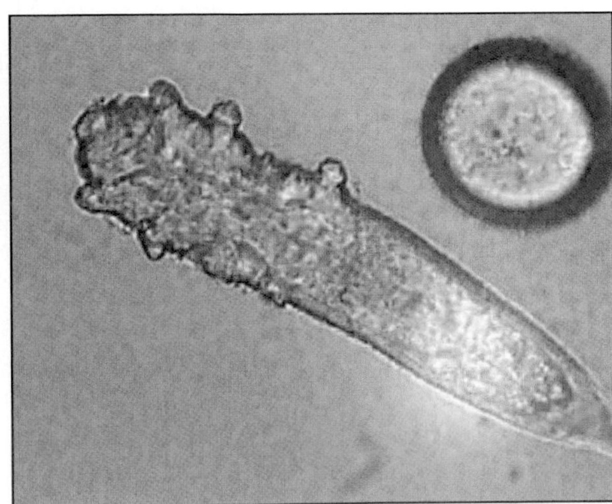

Figure 1 Humans commonly harbor two species of follicle mites, *Demodex brevis* which inhabits sebaceous glands associated with hairs, and *D. folliculorum* which inhabits hair follicles above the level of the sebaceous glands. Given the number of hair follicles found on the human body (>5 million), and the number of humans alive today (>7 billion), the number of habitats potentially exploitable by these mites is immense.

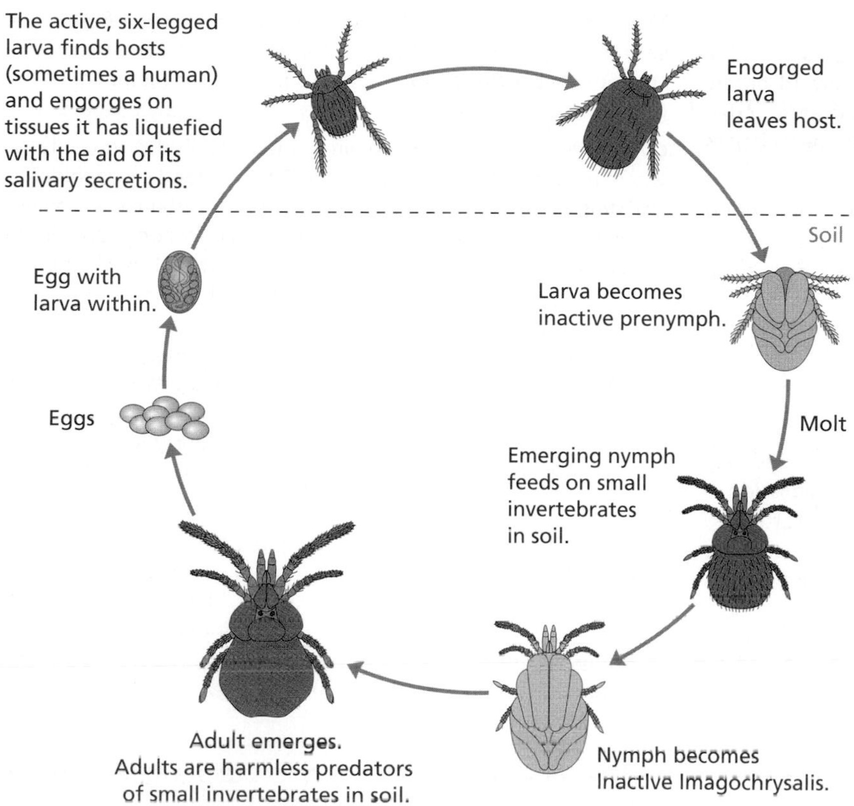

Figure 2 The life cycle of a chigger mite. These mites, members of the family Trombiculidae, are common in many tropical and temperate regions. Among parasitic mites, chiggers are unusual as only the larva is parasitic. Following hatching, the larvae seek out a host on which to feed. The mouthparts of the larva are used to penetrate the skin, after which the larva injects saliva containing proteolytic enzymes, which digest host cells. The parasite then ingests the cellular material, along with interstitial fluid. The engorged larva then leaves the host and becomes an inactive prenymph. Emerging nymphs feed on small soft-bodied invertebrates or insect eggs. There is a subsequent inactive stage from which the adult emerges. A number of species parasitize humans and although some humans experience few or no symptoms, others have a severe, itchy reaction to feeding larvae. Other species affect a variety of livestock and poultry. (Courtesy of CDC.)

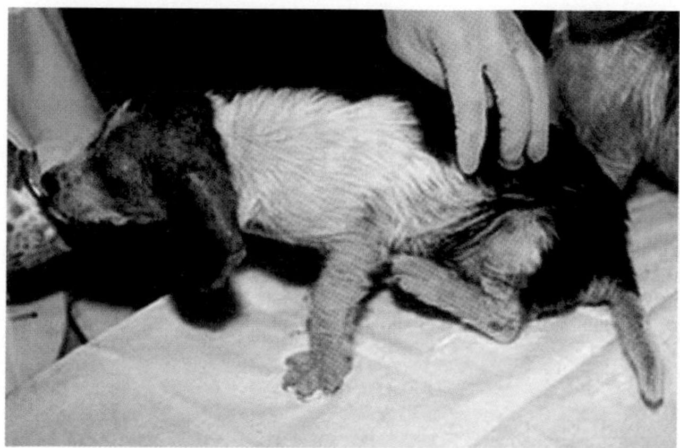

Figure 3 A puppy suffering from sarcoptic mange. A highly contagious condition caused by *Sarcoptes scabiei canis*, mange can also affect cats, horses, and other animals. The condition in humans, commonly called scabies, is usually caused by a closely related subspecies. The observed pathology is in response to the burrowing activity of female mites. Several other kinds of mites can also cause mange in animals.

Parasitic Crustaceans

The Crustacea (considered a subphylum within the phylum Pancrustacea or a class within phylum Arthropoda) is a highly diverse taxon. Many, but not all, of the approximately 14 orders contain parasitic members. Prominent among parasitic crustaceans are those in the subclass Copepoda and the subclass Cirripedia.

Distribution and Prevalence Parasitic crustaceans are found mainly but not exclusively in aquatic habitats. Individual species may have discrete distributions, determined by both host availability and abiotic factors.

Hosts and Transmission Some species use invertebrates while others use vertebrates as hosts (Figure 1). Amphipods in the family Hyperiidae, for example, parasitize jellyfish. Pentastomes (tongue worms) parasitize the respiratory system of vertebrates. Parasitic copepods may use aquatic invertebrates or vertebrates (Figure 2), depending on species. In most cases transmission is achieved when various life-cycle stages actively seek and attach to an appropriate host. Parasites in the subclass Cirripedia parasitize marine invertebrates, commonly other crustaceans (Figure 3).

Pathology Light infections often cause little pathology, but heavy infections can have serious consequences for the host. The copepod *Lernaea cyprinacea* (the anchor worm), for example, parasitizes freshwater fish. The adult embeds in the fish's flesh, held in place by a large anterior process. The host suffers damage to its scales, skin, and underlying muscle tissue. Inflammation, ulceration, and secondary infections may follow. Members of the family Pennellidae are common parasites of marine fish and mammals. Several species invade the circulatory system, most commonly a large vessel, which can result in blockage. Among the Cirripedia, the most important order is the Rhizocephala. After invading their crustacean host, these parasites develop a network of rootlike processes that extend throughout the host's tissues. These processes are used to absorb nutrients and often cause castration. See Chapter 7 for a description of the bizarre behavioral changes that occur in crabs serving as hosts for members of the genus *Sacculina*.

Control Preventative measures are sometimes used in aquaculture, where parasites may pose a problem. The copepod *Lepeophtheirus salmonis*, (commonly called a sea louse), for instance, is a major problem in commercial salmon operations. Infection is prevented by removal of dead and sick fish, and separating fish of different ages. Parasite numbers can also be reduced through the introduction of cleaner fish, which feed on parasites attached to the body of commercially important species. Various anti-parasitic drugs, including organophosphates and pyrethroids, are available for use in aquaculture operations. For control of the copepod intermediate hosts of *Dracunculus medinensis*, see the entry for this nematode parasite. Prevention and control of other parasitic crustaceans using aquatic invertebrates or wild fish is rarely if ever practiced.

Did you know? The pentastomid *Linguatula arctica* uses reindeer as its definitive host, in which it parasitizes the nasal sinuses. Damage to the nasal epithelium causes blood to pool in the sinuses. It has been suggested by some that pooled blood, which may impart a reddish color to the nasal area, is the basis of the Christmas story, *Rudolf the Red-Nosed Reindeer*.

References

Heckmann R (2003) Other ectoparasites infesting fish: copepods, branchiurans, isopods, mites, and bivalves. *Aquaculture* **22**:44–56.

Kabata Z (1980) Crustaceae as enemies of fish. In Diseases of Fish (Sniezko SF & Axelrod HR eds) TFH Publications.

Glover KA, Stolen AB, Messmer A et al (2011) Population genetic structure of the copepod *Lepeophtheirus salmonis* throughout the Atlantic. *Mar Ecol Prog Ser* **427**: 161–172.

Figure 1 A female *Cyamus boopis*, a species of whale louse specific to humpback whales, *Megaptera novaeangliae*. Despite the name, whale lice are not lice but amphipod crustaceans. They are external parasites, found in skin lesions, genital folds, nostrils, and eyes of marine mammals of the order Cetacea (whales and dolphins). Unlike free-living amphipods, which have distinct lateral flattening, *C. boopis* is dorsoventrally flattened as an adaptation to its ectoparasitic life style. The highly developed appendages are adapted for grasping to the host. (Courtesy of Hans Hillewaert, CC BY-SA 3.0.)

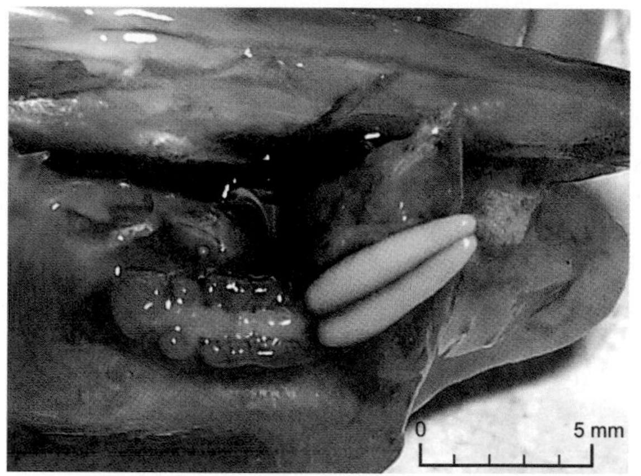

Figure 2 Female *Acanthochondria limandae*. Like many parasitic copepods, this species, an ectoparasite of bottom-dwelling marine flatfish, displays a morphology highly adapted to its parasitic lifestyle and looks quite unlike free-living copepod species. Typically, parasitic copepods have complex adaptations for adhesion to the host and a reduction in locomotor appendages. The reproductive structures are often greatly increased in size (note the two elongated white egg sacs), and there is often a loss of external segmentation. There is typically a reduction in sense organs. The parasite here is attached to the base of the gill arches. (Courtesy of Hans Hillewaert, CC BY-SA 3.0)

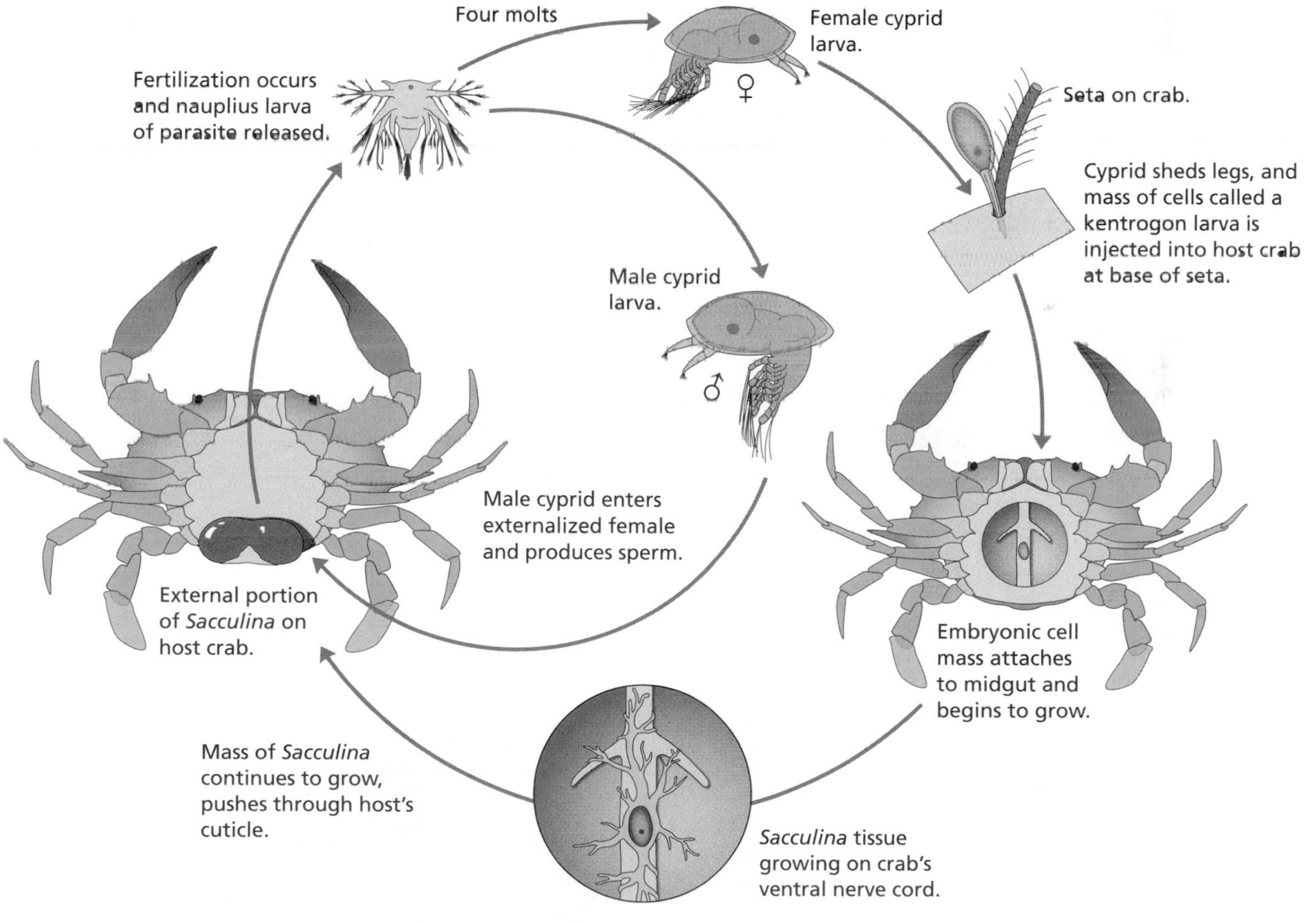

Four molts
Female cyprid larva.
♀
Seta on crab.
Fertilization occurs and nauplius larva of parasite released.
Cyprid sheds legs, and mass of cells called a kentrogon larva is injected into host crab at base of seta.
Male cyprid larva.
♂
Male cyprid enters externalized female and produces sperm.
External portion of *Sacculina* on host crab.
Embryonic cell mass attaches to midgut and begins to grow.
Mass of *Sacculina* continues to grow, pushes through host's cuticle.
Sacculina tissue growing on crab's ventral nerve cord.

Figure 3 The life cycle of *Sacculina*. These parasites are highly modified barnacles (subclass Cirripedia) that parasitize marine crabs. As adults they display a highly modified morphology that lacks a digestive system and appendages. Like many crustaceans, their initial larval stage is the nauplius. The nauplius stage undergoes four molts to develop into a cyprid. The female cyprid seeks out an appropriate crab, whereupon it uses its antennules to attach to the crab. Swimming legs and associated musculature are then shed; the remaining larva is called a kentrogon. The kentrogon consists only of undifferentiated cells. This cell mass enters the crab's hemocoel at a site where the crab's exoskeleton is thin, such as the base of a seta. The cell mass migrates to a site below the ventral nerve cord and begins to grow. As the parasite continues to grow, it presses against the ventral exoskeleton, which eventually breaks open, exposing the female *Sacculina*'s gonads. Male cyprids are attracted to the now exposed gonads. The male enters receptacles in the female and produces sperm within these receptacles. Fertilization results in a new generation of nauplius larvae.

Microsporidians (*Encephalitozoon cuniculi*)

The Microsporidia are a phylum of intracellular, fungal parasites. There are approximately 1500 known species of these unicellular fungi, all of which parasitize animals. *Encephalitozoon cuniculi*, common in many mammals, provides a representative example of veterinary importance. It is especially common in rabbits (Figure 1).

Distribution and prevalence *E. cuniculi* is distributed worldwide. Its natural prevalence is difficult to determine because many cases are subclinical. Prevalence of up to 42% has been reported for humans with a history of tropical diseases or a stay in tropical countries. Some surveys in European rabbits estimate prevalence as high as 23% of healthy rabbits and 85% of rabbits with neurological symptoms.

Hosts and transmission As in other microsporidians, the infective stage is the unicellular spore. Following consumption, osmotic pressure rises within the spore, resulting in the release of the polar filament, which penetrates nearby cells in the gut. The ameba-like sporoplasm then passes into the cell of the host. The sporoplasm undergoes extensive multiplication and ultimately gives rise to new spores, which are released in the urine.

Pathology In some cases, most commonly in rabbits, *E. cuniculi* can undergo replication whereupon it may reach the central nervous system and other organs. In the brain (Figure 2), it induces granuloma formation, which results in granulomatic encephalitis. A common symptom in rabbits is torticollis (head tilt) (Figure 3). Other symptoms include partial or complete paralysis, loss of coordination, and seizures. Rare human fatalities have occurred in individuals with advanced HIV infection.

Diagnosis Diagnosis is usually conducted serologically, as infected animals produce high levels of anti-*E. cuniculi* antibodies. Enzyme-linked immunosorbent assay (ELISA) using spore antigens is commonly used.

Treatment Various benzimidazole-type drugs are suggested as treatments, including the anthelmintics albendazole, oxibendazole, and fenbendazole. Some studies suggest that there is enhanced efficacy when one of these drugs is administered in combination with enrofloxacin (Baytril®), a bacteriocidal drug commonly used in veterinary medicine.

Control Because the infective stage, the spore, is common in the environment and because these spores remain infective for extended periods of time, there is little or no attempt at control in wild rabbits. In rabbit colonies, animals displaying symptoms should be quickly isolated and treated. Newborn rabbits to infected mothers should also be treated.

Did you know? Historically, *E. cuniculi* was considered as a possible cause of rabies and polio.

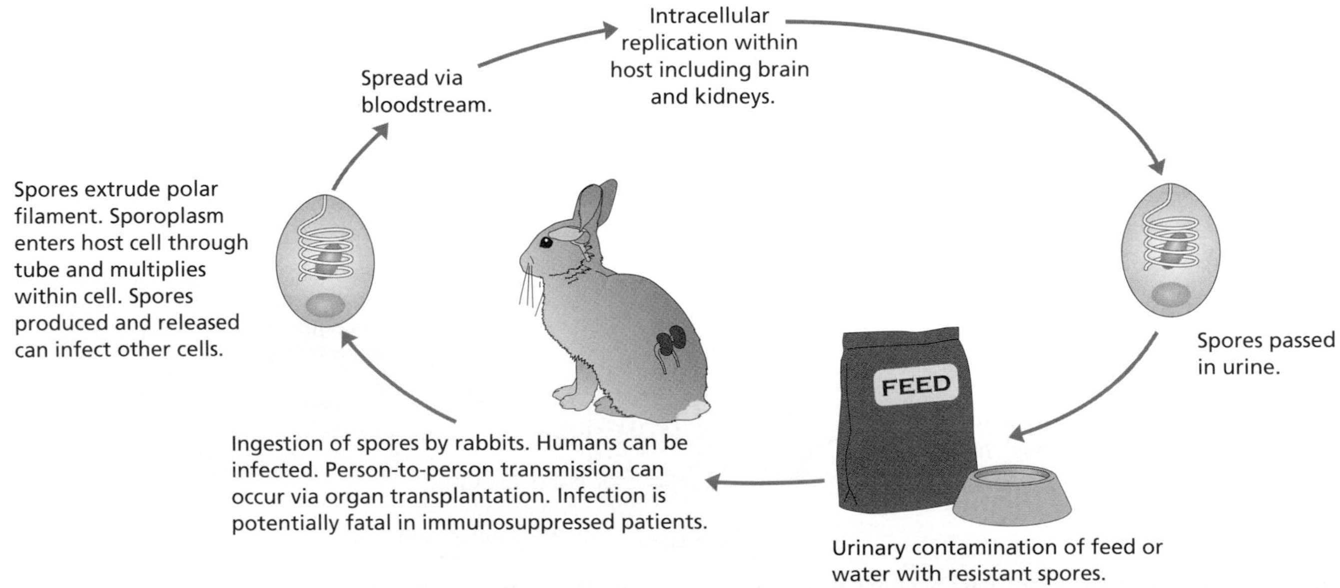

Figure 1 The life cycle of *Encephalitozoon cuniculi*. Transmission occurs when spores passed in the urine are consumed inadvertently by other animals. Following ingestion, the spore extrudes its polar filament, allowing invasion of cells of the gut wall by the sporoplasm. Spores are produced that can spread to other parts of the body, including the brain and kidneys, where granulomas form in response to parasite reproduction. Vertical transmission from a mother rabbit to her offspring is also common (not shown).

Related organisms

Nosema apis: This organism is a common parasite of honeybees.

N. bombycis: A parasite of silkworm larvae, *N. bombycis* was perhaps the first microscopic infectious organism identified. It was discovered by Louis Pasteur in 1870.

References

Mathis A, Weber R & Deplazes P (2005) Zoonotic potential of the Microsporidia. *Clin Microbiol Rev* **18**:423–445.

Suter C, Muller-Doblies UU, Hatt J-M & Deplazes P (2001) Prevention and treatment of *Encephalitozoon cuniculi* in rabbits with fenbendazole. *Vet Rec* **148**:478–480.

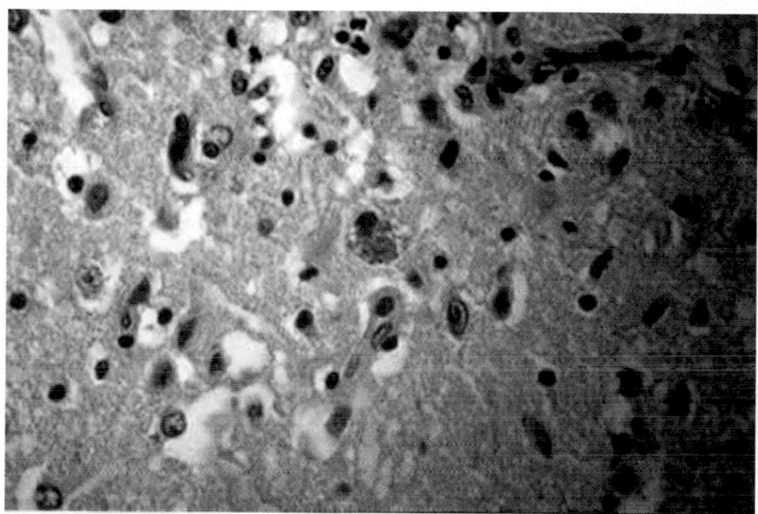

Figure 2 Neuropathology caused by *Encephalitozoon cuniculi*. A cross section of brain tissue from a rabbit that died of *E. cuniculi*. (Courtesy of Tracy Bartick.)

Figure 3 Head tilt. Rabbit with torticollis (head tilt) from cerebral infection with *Encephalitozoon cuniculi*. (Courtesy of Buckeye House Rabbit Society)

Myxozoans (*Myxobolus cerebralis*)

Myxobolus cerebralis is an animal parasite of salmonid fishes in the phylum Myxozoa. It is the causative agent of whirling disease.

Distribution and prevalence *M. cerebralis* was originally a parasite of brown trout (*Salmo trutta*) native to central Europe. As brown trout were introduced elsewhere, *M. cerebralis* was likewise introduced into other parts of Europe and North and South America (Figure 1). Typically, infected *S. trutta* are asymptomatic, but even this species can experience severe symptoms if heavily infected. Other species of salmonids, especially rainbow trout (*Oncorhynchus mykiss*), are particularly vulnerable, and in some places steep declines or even complete elimination of entire populations has occurred.

Hosts and Transmission The life cycle of *M. cerebralis* requires two hosts, a salmonid fish and an aquatic annelid worm (*Tubifex tubifex*). As shown in Figure 2, fish are infected by contact with a stage called the triactinomyxon (TAM) which facilitates penetration and colonization via the fish's skin. Alternatively, a fish may eat an infected *T. tubifex*, whereupon penetration is across the gut epithelium. Spores, released into the water when an infected fish dies, are consumed by the annelid. In the annelid, TAMs develop within cysts and are ultimately released with the worm's feces.

Pathology Following penetration of the fish's epithelium, the sporoplasm is released from the TAM. The sporoplasm divides into individual cells, which digest cartilage as they spread through the host. The skeletal deformities and erratic swimming (whirling) in which the fish chases its tail are caused by damage to cartilage as well as the spinal cord. The fish's age and the number of infecting TAMs affect the severity of the infection. Skeletal deformities are most pronounced in young fish, as their skeletons are not completely ossified, and there is correspondingly more cartilage upon which the parasite can feed.

Diagnosis Infection with *M. cerebralis* is often presumed based on appearance (Figure 3) and behavioral abnormalities. Definitive diagnosis usually relies on finding spores upon histological examination of cartilage. Infection can also be confirmed serologically or by amplifying a portion of the *M. cerebralis* 18S rRNA gene.

Treatment There is no effective treatment for *M. cerebralis* infections. Impoundments in which infected fish have been found can be decontaminated by drainage, followed by treatment with calcium cyanamide or quicklime to destroy spores.

Control Because some species or even strains within species seem to have increased resistance, using more resistant fish in aquaculture can reduce incidence. Likewise, the elimination of *Tubifex* worms, required in the *M. cerebralis* life cycle as definitive hosts, can reduce infection rate in fish. Rearing young fish in facilities that are known to be parasite-free can help reduce severity of disease and mortality rate. Certain chemical compounds can be used to interfere with spore development, reducing subsequent infection. Fishermen can help prevent spread by carefully cleaning fishing equipment between fishing trips.

Did you know? *Ceratomyxa shasta* is another pathogenic myxozoan parasite of salmonids. It is also an example of a parasite that

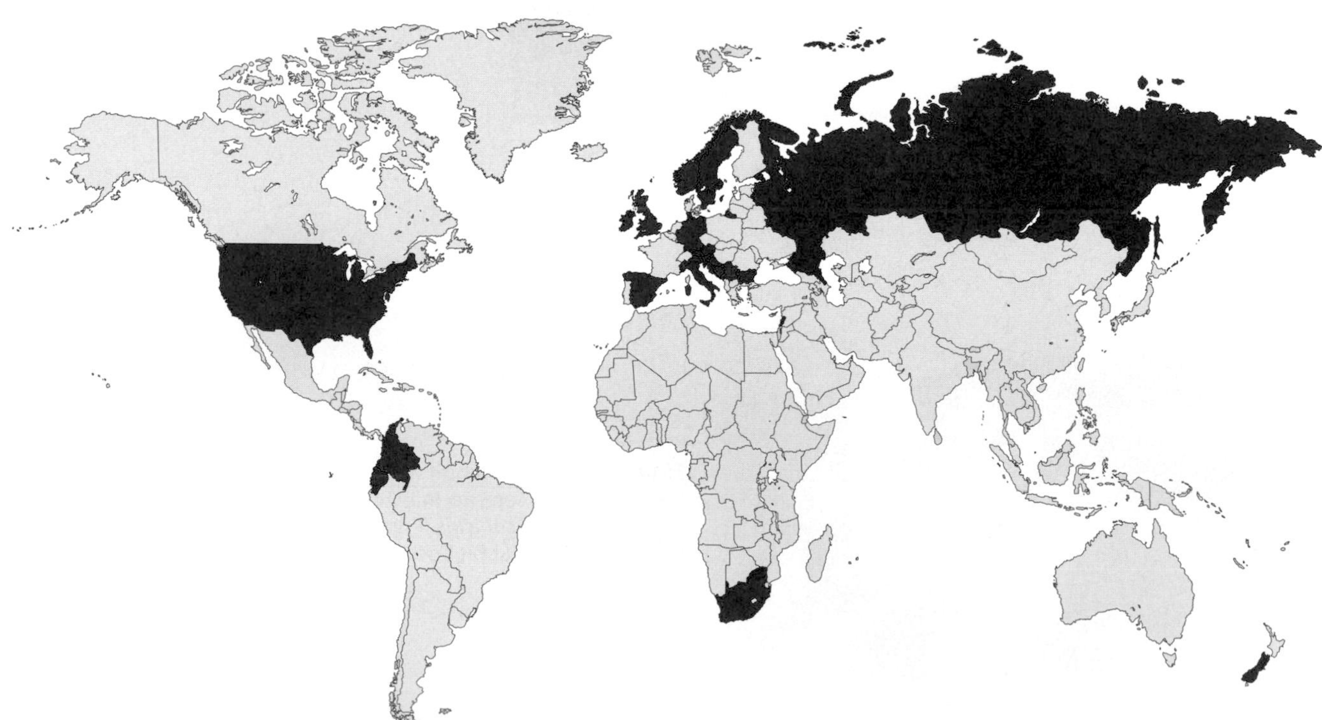

Figure 1 The worldwide distribution of *Myxobolus cerebralis*. Originally found in central Europe, the parasite has spread mainly through the introduction of European brown trout to other locations. (Adapted from Hoffman GL [1990] *J Aquat Anim Health* 2:30–37.)

continues to develop after the death of its host. Once fully developed, spores are released from the body of the dead fish, which then infect freshwater polychaete worms.

References

Gilbert MA & Granath WO Jr (2003) Whirling disease and salmonid fish: life cycle, biology, and disease. *J Parasitol* **89**(4):658–667.

Granath WO (2014) Effects of habitat alteration on the epizootiology of *Myxobolus cerebralis,* the causative agent of salmonid whirling disease. *J. Parasitology* **100**(2):157–165.

http://en.wikipedia.org/wiki/JSTOR.

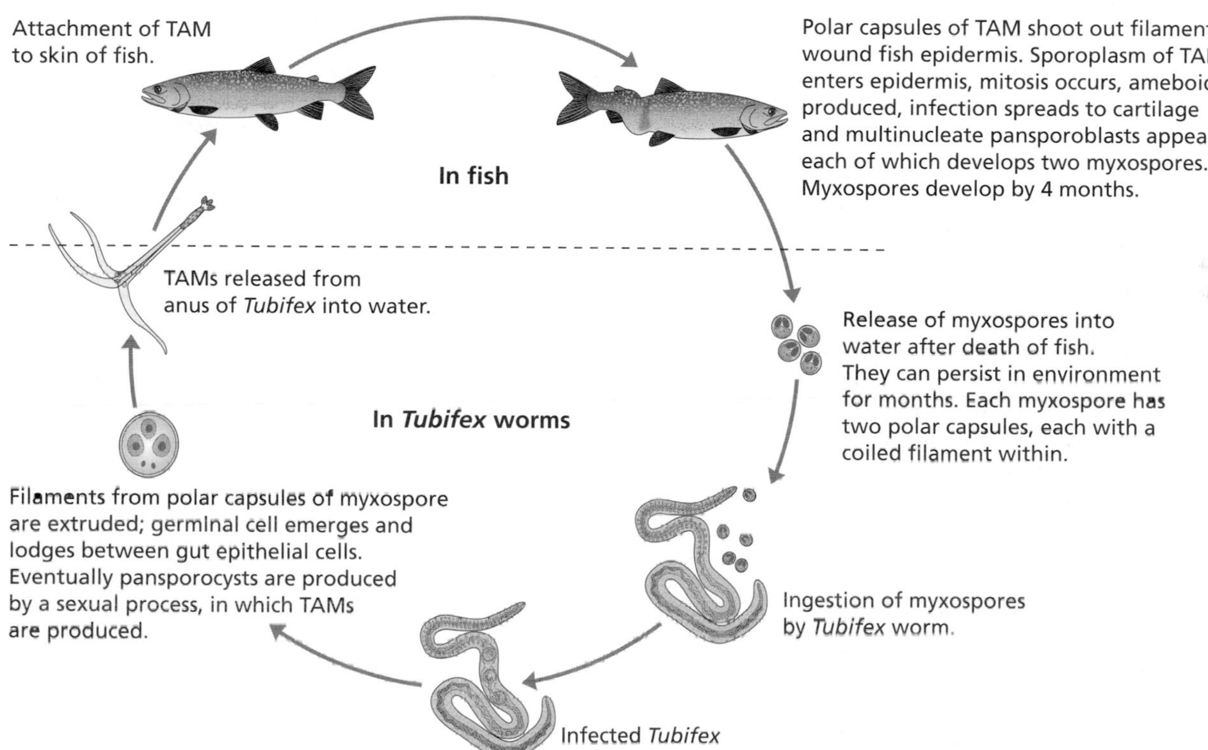

Figure 2 The life cycle of *Myxobolus cerebralis*. Two hosts, a salmonid fish and an aquatic annelid (*Tubifex tubifex*), are required for the life cycle. The infective stage for fish, called a triactinomyxon (TAM), is released with the feces from the worm. Alternatively, a fish can become infected by eating an infected *T. tubifex*. TAMs passively contact a fish and, upon the rapid release of filaments from each of their three polar capsules, release a mass of germ cells called a sporoplasm, which enter the fish through the wound caused by the filaments. The sporoplasm then divides into individual cells that spread through the fish's body. Asexual reproduction ultimately gives rise to myxospores, which are released when the fish dies and decomposes or when it is eaten. Myxospores are consumed by *T. tubifex*. The myxospore also employs polar capsules and filaments to create a wound in the gut epithelium, allowing germ cells to penetrate. In the intestinal epithelium, a new generation of TAMs is formed, which may then be released to complete the life cycle.

Figure 3 A deformed brook trout infected with *Myxobolus cerebralis*. As a rule, brook trout (*Salvelinus fontinalis)* and rainbow trout (*Oncorhynchus mykiss*) tend to suffer heavier infections and more severe pathology than brown trout (*Salmo trutta)*, which may have been the original host for *M. cerebralis*. The disease has the most severe effects on young fish less than five months old, as their skeletons are not yet ossified, making them more susceptible to deformities.

Horsehair worms (phylum Nematomorpha)

Approximately 350 species of nematomorphs, also called horsehair or Gordian worms, have been described. Phylogenetically, nematomorphs appear to be close relatives of the nematodes, and they are commonly confused with mermithid nematodes. Although they are of no medical or veterinary importance, they are presented here because of their inherent interest and because, owing to the conspicuous nature of the adults, they are frequently observed (see also Box 2.3).

Distribution and prevalence Nematomorphs (Figure 1) have a worldwide distribution, but they are generally restricted to areas near water. Five described species are marine. The remaining species are found in freshwater habitats. Approximately a dozen species occur in the United States. The most commonly encountered and perhaps most closely studied species is *Gordius robustus*. Prevalence of this species in its insect hosts is probably more common than usually expected. For example, this species can infect large proportions of hosts such as the Mormon cricket (*Anabrus simplex*), where prevalence over 90% in some habitats has been reported.

Hosts and transmission Adults are free-living in aquatic environments, whereas larvae are parasitic in arthropods (Figure 2). Common hosts include crickets, cockroaches, beetles, mantids, grasshoppers, and occasionally spiders. Marine species utilize crustaceans as hosts. Following mating, females release gelatinous strings containing prodigious numbers of eggs. The eggs develop into characteristic larvae in water. The larva has a proboscis with three rows of hooks. The larvae are ingested by a variety of invertebrate paratenic hosts in which they encyst. It is believed these paratenic hosts are eventually consumed, perhaps after their death, by a host like a cricket, beetle, or grasshopper. The emerging larva penetrates the gut wall and enters the body cavity. Here the larva develops into a large adult, with a length many times exceeding that of the host. By the time the worm is mature, it fills most of the host's body cavity, except for the head and appendages. After about two or three months, the now adult worms exit the host through an opening that forms near the anus (Figure 3). It is believed that behavioral changes in the host induced by the parasite cause the host to seek out water, releasing the adult into an appropriate aquatic habitat.

Pathology Although infected insects harbor remarkably large nematomorphs relative to their own body size, and they are typically castrated while infected, some insect hosts do survive following exit of the nematomorph, and in some cases can even commence reproduction.

Treatment and control As these parasites do not affect humans or animals of veterinary importance, treatment or control measures are not used.

Did you know? Although they pose no threats to humans, people are often alarmed when they discover the long (usually 30–40 cm) adults in swimming pools or toilet bowls. Infected insects are drawn to such sites, apparently by the water-seeking behavior induced by the parasite.

References

Hanelt B, Thomas F & Schmidt-Rhaesa A (2005) Biology of the phylum Nematomorpha. *Adv Parasitol* **59**:244–305.

Sato T, Watanabe K, Kanaiwa M et al (2011) Nematomorph parasites drive energy flow through a riparian ecosystem. *Ecology* **92**: 201–207.

Thorne G (1940) The hairworm, *Gordius robustus* Leidy, as a parasite of the Mormon cricket, *Anabrus simplex* Haldeman. *J Wash Acad Sci* **30**:219–231.

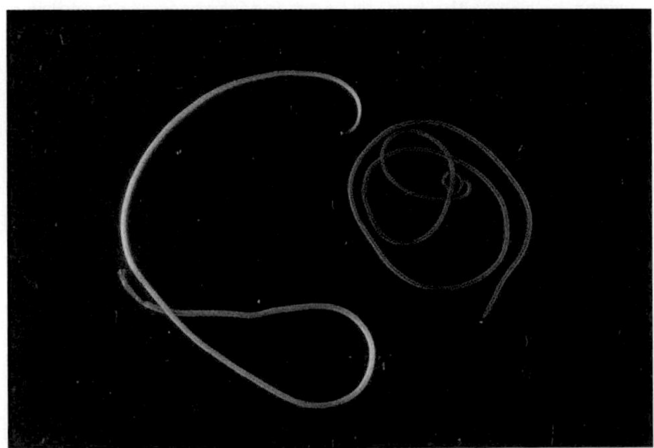

Figure 1 Adult *Paragordius tricuspidatus*. These large worms can reach lengths of over 15 cm. Adults are often observed in dense, tangled clusters, giving rise to the common name Gordian worms, after the famous Gordian knot of Greek mythology. Like nematodes, nematomorphs are dioecious (except for a single described monoecious species) and rely on internal fertilization. Also like nematodes, they are ecdysozoans because they molt their exoskeleton as they grow. They have nonfunctional digestive tracts and lack specialized respiratory, circulatory, and excretory systems. (Courtesy of Bildspende von D. Andreas Schmidt-Rhaesa, CC BY-SA 3.0.)

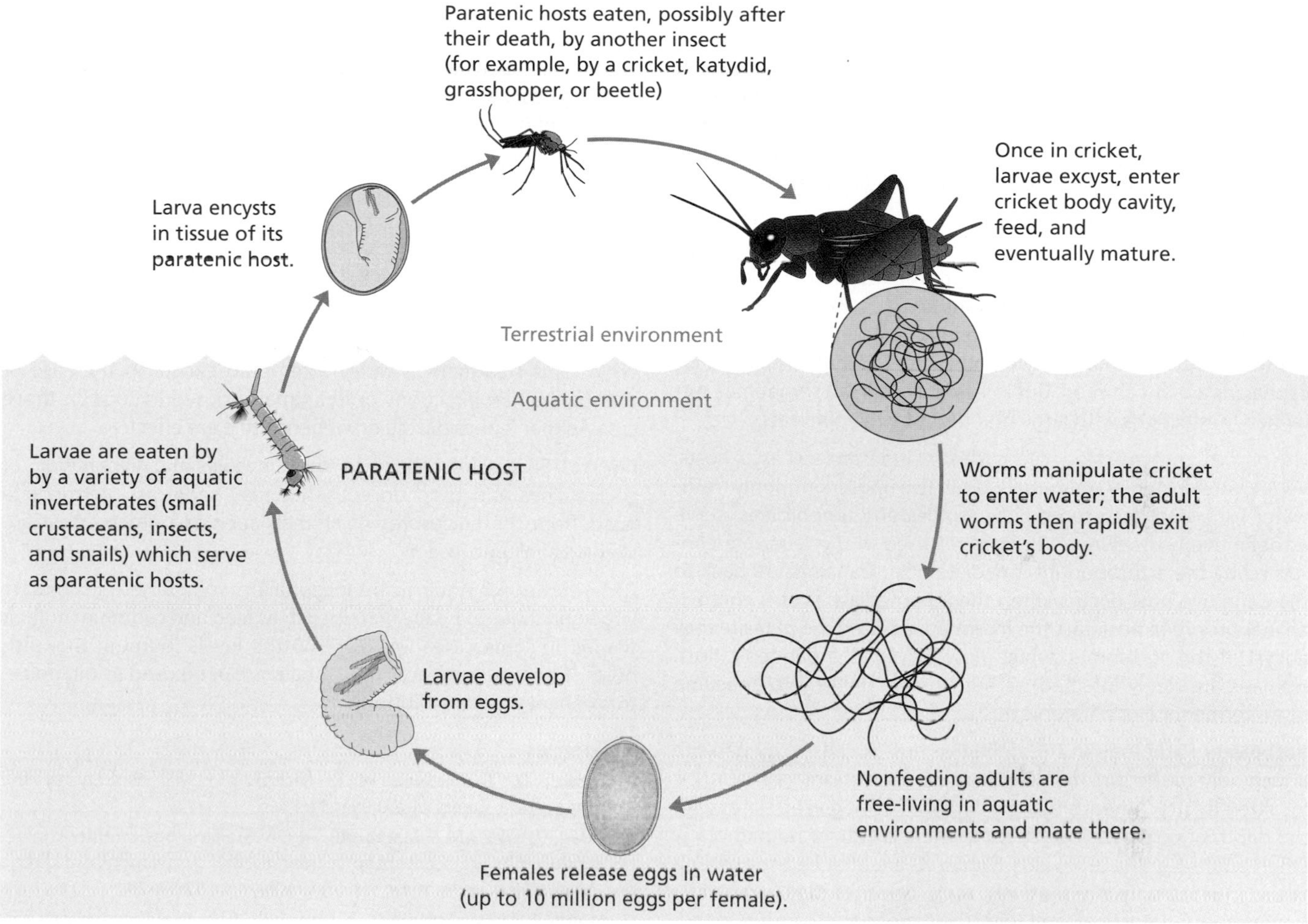

Paratenic hosts eaten, possibly after their death, by another insect (for example, by a cricket, katydid, grasshopper, or beetle)

Once in cricket, larvae excyst, enter cricket body cavity, feed, and eventually mature.

Larva encysts in tissue of its paratenic host.

Terrestrial environment

Aquatic environment

PARATENIC HOST

Larvae are eaten by by a variety of aquatic invertebrates (small crustaceans, insects, and snails) which serve as paratenic hosts.

Worms manipulate cricket to enter water; the adult worms then rapidly exit cricket's body.

Larvae develop from eggs.

Nonfeeding adults are free-living in aquatic environments and mate there.

Females release eggs in water (up to 10 million eggs per female).

Figure 2 A typical life cycle of a nematomorph. The requirement for a paratenic host for all nematomorph species requires further study. (Courtesy of Ben Hanelt, University of New Mexico.)

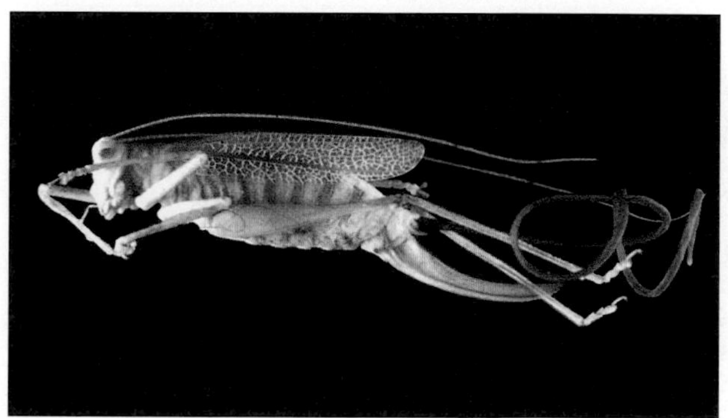

Figure 3 *Spinochordodes tellinii.* The adult worm is seen emerging from its katydid host. Whether the insect host (the katydid in this example) is a true definitive host is a matter for debate. Although the worms develop to near maturity in the katydid, they actually mate and expel progeny only in the external environment. (Courtesy of Bildspende von D. Andreas Schmidt-Rhaesa, CC BY-SA 3.0.)

Phylum Acanthocephala

There are approximately 1200 species in the phylum Acanthocephala, which are also known as spiny-headed worms because of their characteristic spine-studded proboscis that is used to penetrate into the intestinal lining of their definitive host (Figure 1). Molecular analysis suggests that they are closely related to rotifers (Figure 2.19). Further study may show that they actually are rotifers, highly modified for their parasitic life style.

Distribution and prevalence Distribution is worldwide, although individual species may have discrete geographic ranges that are dictated primarily by host distribution. Acanthocephalans are most common in freshwater environments, as freshwater fish are the most commonly used adult hosts. Compared with other helminth parasites, such as platyhelminths and nematodes, acanthocephalans are uncommon, but individual hosts can sometimes suffer heavy infections with large numbers of adult parasites.

Hosts and transmission Life cycles include at least two hosts. Vertebrates, including mammals, birds, fish (most commonly freshwater fish), and, less commonly, reptiles and amphibians, serve as definitive hosts (Figure 2). Eggs released in the feces are consumed by the arthropod intermediate host. Transmission back to the definitive host occurs when the intermediate host is eaten. If an inappropriate host eats the intermediate host, the parasite may encyst in the vertebrate, which now serves as a paratenic host. Humans are rarely infected, as humans are unlikely to consume raw intermediate or paratenic hosts.

Pathology Pathology in the definitive host is caused by parasite attachment in the gut. In low or moderate infections, pathology is localized to attachment sites. The extent of damage correlates with the depth of proboscis penetration. There is little or no pathology when adult acanthocephalans (Figure 3) are attached to the epithelial mucosa only. Alternatively, some parasites may completely penetrate the intestinal wall with their proboscis, causing extensive granuloma formation and fibrosis. In many cases, the depth of penetration depends on the species of definitive host. In birds and other definitive hosts, pathology is generally most severe in young animals. Mortality is rare unless intensity is heavy. Some species of acanthocephalans are known to alter the behavior of their intermediate hosts to facilitate transmission (see Chapter 6 for ecological examples involving acanthocephalans).

Diagnosis Diagnoses are not often attempted except in zoos, in aquaculture, or for research purposes. In such cases, diagnosis is made by the detection of eggs in fecal samples or visual detection of adult worms in the intestine of the definitive host. The structure of the proboscis, as well as the number, shape, and arrangement of hooks on the proboscis, may be used to aid in species identification.

Treatment Treatment is rarely attempted except in the case of unusual human infections or in animals housed in zoos. In those cases, either fenbendazole or mebendazole are effective.

Control Successful control in zoos involves the elimination of cockroaches and other insects that may serve as intermediate hosts from the enclosures of animals such as primates that are likely to consume insects.

Did you know? Adult acanthocephalans sometimes accumulate large quantities of heavy metals such as lead and cadmium in their tissues, in some cases up to X2700 the levels found in their fish hosts. Consequently, acanthocephalans can be used as bioindicators of heavy metal pollution.

References

Crompton D, Thomasson W & Nickol B (1985) Biology of the Acanthocephala. Cambridge University Press.

La Sala LF, Perez AM & Martorelli SR (2013) Pathology of enteric infections induced by the acanthocephalan *Profilicollis chasmagnathi* in Olrog's gull, *Larus atlanticus*, from Argentina. *J. Helminthol* **87**:17–23.

Gismondi E, Cossu-Leguille C & Beisel J (2012) Acanthocephalan parasites: help or burden in gammarid amphipods exposed to cadmium? *Exotoxocol* **21**:1188–1193.

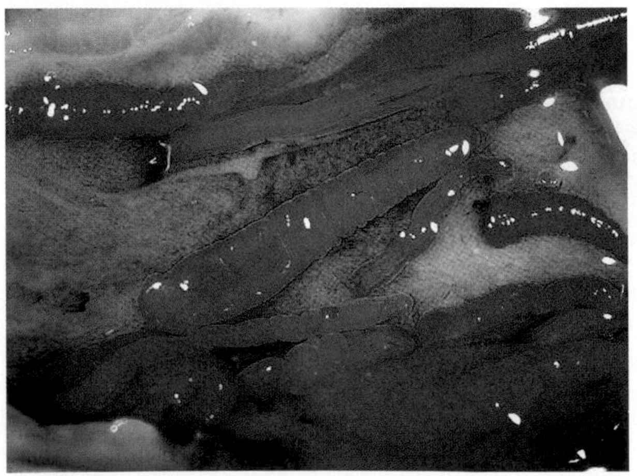

Figure 1 Adult acanthocephalans in the definitive host. These adults, members of the genus *Pomphorhynchus*, have their proboscises imbedded in the intestinal lining of their definitive host, a bluefish (*Pomatomus saltatrix*). Most of the pathology due to acanthocephalans is the result of inflammation in response to damage caused by the proboscis, which is used to attach to the gut wall. With no dedicated digestive system, acanthocephalans are absorptive feeders, similar to tapeworms.

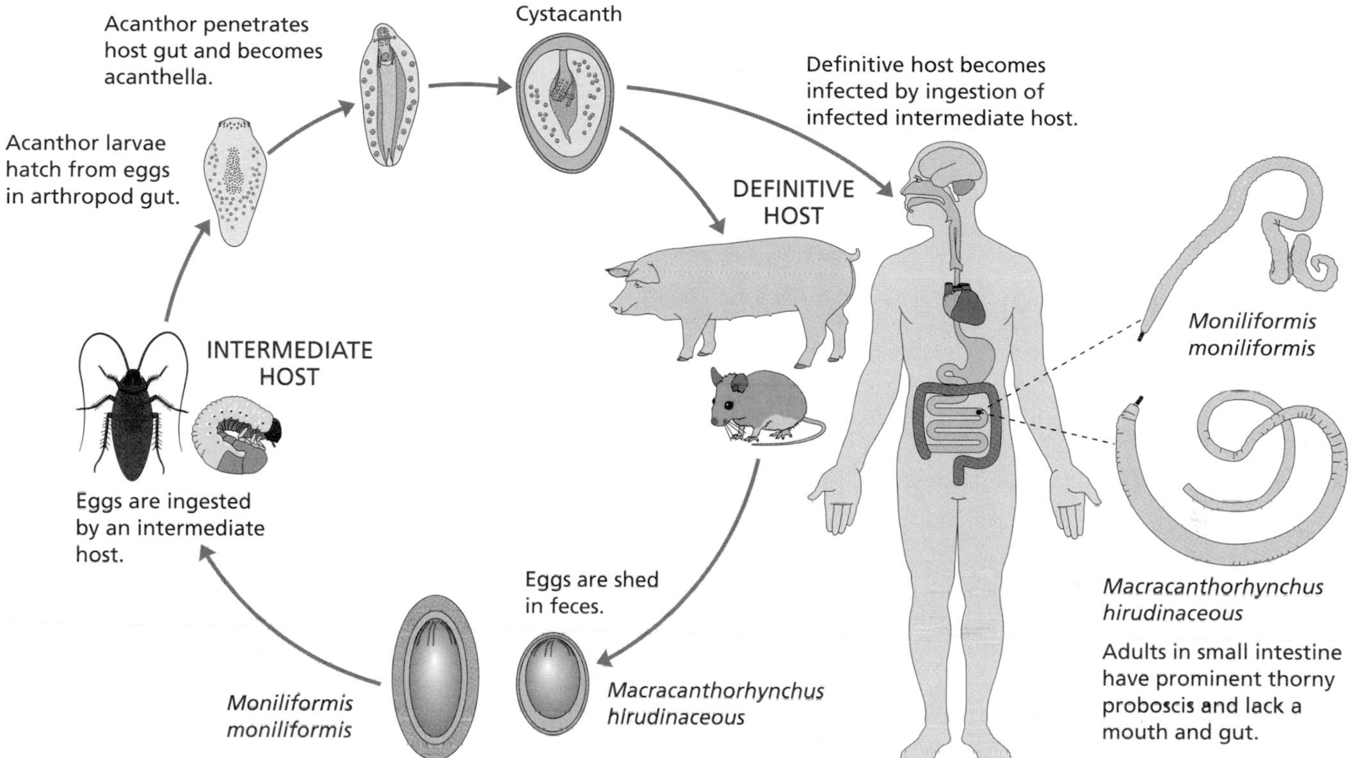

Figure 2 The life cycle of acanthocephalans. The life cycle for the two species most likely to cause occasional infections in humans, *Macracanthorhynchus hirudinaceous* and *Moniliformis moniliformis*, are shown. Eggs are shed in the feces of definitive hosts, most commonly rats for *M. moniliformis*, and pigs for *M. hirudinaceous*. Humans and other mammals may serve as accidental hosts. Eggs contain a larval stage known as the acanthor. The eggs are consumed by an insect, which serves as intermediate host. Within the hemocoel of the insect, the acanthor develops into the second larval stage called an acanthella, which ultimately develops into the infective stage called a cystacanth. Following consumption by the definitive host, the cystacanth attaches to the lining of the small intestine, where it matures into an adult. (Courtesy of CDC.)

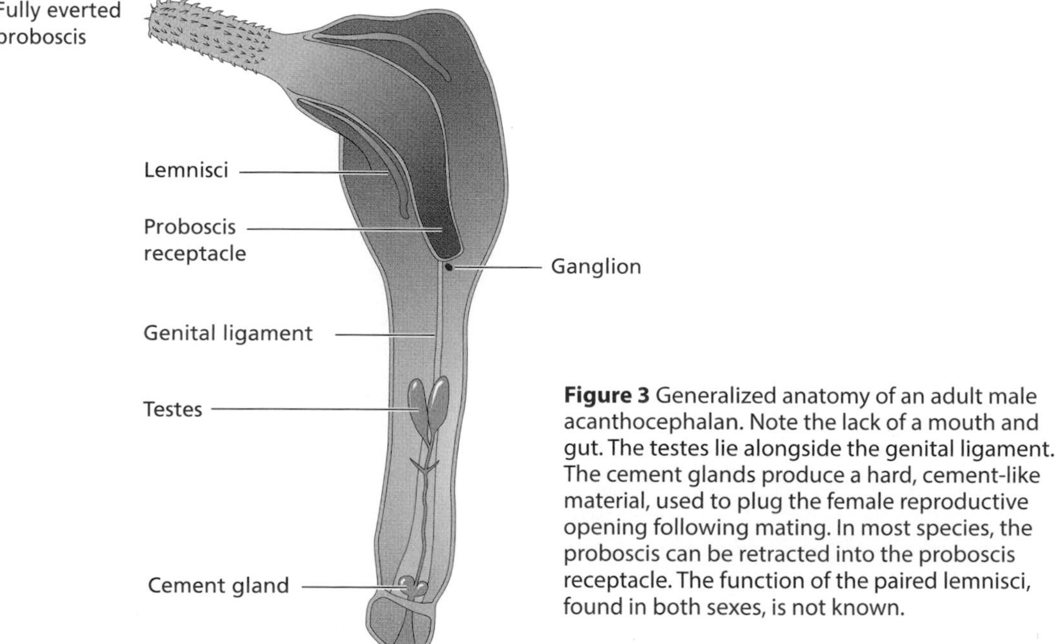

Figure 3 Generalized anatomy of an adult male acanthocephalan. Note the lack of a mouth and gut. The testes lie alongside the genital ligament. The cement glands produce a hard, cement-like material, used to plug the female reproductive opening following mating. In most species, the proboscis can be retracted into the proboscis receptacle. The function of the paired lemnisci, found in both sexes, is not known.

GLOSSARY

acetabula
The adhesive sucker found on the ventral surface of trematodes or on the scolex of cestodes.

acoelomate
Those organisms such as platyhelminths that lack a body cavity known as a coelom, which arises in many animals during embryological development.

adaptive immune system
Components of the immune system, mediated by lymphocytes, in which these cells respond to specific antigen in a manner that becomes stronger with repeated exposure to the same antigen.

adaptive immunity
Immunity to infection conferred by an adaptive immune response.

adelphoparasites
Parasites that use hosts to which they are closely related, often in the same family or genus. Also called agastoparasites.

adjuvants
Compounds that increase the immune response to the antigens with which they are mixed.

agastoparasites
see adelphoparasites.

alarm cytokines (or alarmins)
A group of cytokines released by intestinal epithelial cells in response to infection by intestinal nematodes; the cytokines help skew the immune response toward a Th-2 response.

allee effect
In small populations, individual fitness increases as population density increases.

allelic exclusion
Individuals with two or more alleles for a particular gene express only one of those alleles at a time.

allochronic speciation
A type of sympatric speciation in parasites in which populations remain separate because they have different periods of transmission.

allogrooming
The act of grooming an individual other than oneself.

alloparasites
Parasites that use taxonomically unrelated organisms as hosts.

allopatric speciation (or vicariant speciation)
Speciation that occurs when biological populations of the same species become vicariant, or isolated from each other to an extent that prevents or interferes with genetic interchange.

allopolyploidy
A phenomenon in which chromosomes are contributed by different parental species to form a new species.

alternatively activated macrophages
A type of macrophage activated by Th-2 cells that promotes smooth muscle contraction, tissue repair, and other responses to infection with intestinal helminth parasites. Also called M2 macrophages.

amebapores
Proteins produced by amebas that are used to kill bacteria within food vacuoles. They may play a role in the pathogenicity of *Entamoeba histolytica.*

amebic dysentery
A condition caused by *Entamoeba histolytica* that occurs when amebas penetrate and feed on the intestinal epithelium of the host and that is characterized by frequent bloody bowel movements.

amoebozoa
A proposed subdomain, within the domain Eukarya, which includes lobose amebas and slime molds.

Andrews binding energy
An estimated binding energy that a molecule would establish with a hypothetical molecular partner if all functional groups contributed to binding.

anergy
A state in which B and T lymphocytes become unresponsive to their specific antigen.

angiosperms
Flowering plants. Seed plants that produce flowers as part of their reproductive cycle.

annotation
A process in genome analysis in which the location and function of specific genes within the genome is assessed.

antibody capping
An immune evasion tactic used by *Entamoeba histolytica* in which antibodies bound to surface antigens are first aggregated and then sloughed off the surface of the parasite.

antibody-dependent cell-mediated cytotoxicity
The killing of antibody-coated target cells by cells with Fc receptors for the bound antibody. Used by natural killer cells, as well as by eosinophils and mast cells.

antigenic variation
An immune evasion strategy used by various parasites. Surface antigens are altered from one generation to the next, protecting the new generation from antibodies against the previous generation.

antigens
The portion of a molecule recognized by either antibody or T-cell receptor that stimulates an adaptive immune response.

apical complex
A structure typical of apicomplexans. The apical complex, which gives apicomplexans their name, is used to recognize, attach to, and penetrate host cells.

apicoplast
A structure found in most apicomplexans that contains its own circular, plastid genome. It is believed to be the remnant of an algal chloroplast that was acquired by ancestoral apicomplexans by secondary endosymbiosis.

apparent competition
A type of intraspecific competition in which the negative impact on the competing organisms is an indirect effect, mediated by a third party.

Archaeplastida
A proposed subdomain within the domain Eukarya, which includes red algae, green algae, and plants.

arms race
In parasitology, it is the coevolution between host and parasite in which an evolutionary adaption in one results in selection pressure and subsequent adaptation in the other.

artemisinin-based combination therapy
A standard antimalarial drug regimen in which artemisinin is taken in combination with one or more other antimalarial drugs.

ascites
A buildup of tissue fluid in the mesenteries and abdominal cavity.

autogrooming
The act of grooming oneself.

autoinfection
Reinfection by a larval-stage parasite of the same host in which it was produced, which occurs without leaving the host.

B1 B cell
A type of atypical B cell with a significantly less diverse antigen–receptor repertoire compared to conventional B cells, which seems to respond to polysaccharide antigens without the assistance of $CD4^+$ T cells.

basic reproductive rate
R_0. The average number of new infections generated by a single infected host when all members of the host population are considered to be susceptible to infection.

bioinformatics
The application of computer science and information technology to biology, typically to aid in analysis of large biological data sets, such as all the genes in a particular organism.

biological indicators
Species whose presence, abundance, and biological roles can be used to reveal and monitor the extent of ecosystem integrity.

biological species concept
The view that a biological species is a group of individuals with similar properties that are able to interbreed with one another and produce fertile offspring and that do not regularly interbreed with other species.

biological vector
An invertebrate that transports a parasite to the next vertebrate host in its life cycle, in which the parasite multiplies, undergoes developmental changes, or both.

bradyzoites
The small, slowly reproducing stage in various coccidian life cycles.

broadcast reproduction
A reproductive strategy in which reproductive propagules are released into the environment to be dispersed by water or wind.

brood parasites
Birds that surreptitiously deposit their eggs into a nest of the same or a different species of bird, so that the foster parents will rear the progeny to fledgling.

cardiomyopathy
A disease state characterized by the deterioration of the heart's musculature.

castrator
Parasites that consistently eliminate the reproductive activities of their hosts, diverting these host resources instead to their own development and reproduction.

CD1
Proteins similar to MHC 1. Unlike conventional MHC 1 proteins, which present peptides to $CD4^+$ T cells, CD1 is used to present glycolipids to $CD4^+$ T cells.

cercariae (singular: cercarium)
A larval life-cycle stage typical of digenetic trematodes, produced by asexual reproduction within either rediae or sporocysts, depending on species.

cerebral malaria
A complication that may result from infection with *Plasmodium falciparum*, characterized by elevated temperature, severe headache, and coma, and that may cause death.

chitin
A polysaccharide commonly found in the cell wall of fungi and in the exoskeleton and other hard parts of many invertebrate animals, composed of polymers of *N*-acetyl glucosamine.

cholangiocarcinoma
A cancer of the bile duct, which can be the result of a chronic infection with the digenetic trematode *Opisthorchis viverrini*, the Southeast Asian liver fluke.

classical biological control
A type of biological control in which an effective biological control agent drives the pest to a low, stable equilibrium population.

cleaning symbiosis
A mutualism between two organisms in which one organism removes and consumes ectoparasites from the body of the other organism.

cleptoparasite
Organism that regularly steals food or other resources obtained by another organism.

clonal deletion
A process that occurs during lymphocyte development in which immature lymphocytes that recognize self-antigen are removed, allowing for self-tolerance.

coevolution
Reciprocal evolution between two species in which each species imposes selection on the other.

combination therapy
The use of more than one drug simultaneously to prevent the development of resistance by the disease-causing organism.

commensal
A participant in a symbiotic relationship that benefits from the symbiosis, while the other participating organism is neither benefited nor harmed.

compatibility filter
The principle that a parasite may successfully infect many hosts, but it can only thrive and complete its life cycle in hosts that are compatible in terms of the host environment and nutrients that they provide the parasite.

competition-dependent regulation
A type of density-dependent population regulation in which intraspecific competition increases as population increases, ultimately causing a decrease in population.

competitive exclusion
A possible outcome of interspecific competition in which one of the two competing species is driven to extinction as the result of the competition.

component population
In parasitology, all the parasites of a given species in a particular ecosystem that are found within all the hosts of a particular species.

concomitant immunity
A type of immunity that is dependent on the presence of a particular parasite to stimulate immunity against additional challenges by infective stages of this same parasite species. Also called premunition.

concomitant predation
Predation that occurs when a predator consumes prey and also consumes the parasites living in prey. In such cases the predator may derive some nutrition from the parasites.

congeners
Members of the same genus.

conjugate vaccines
An acellular type of vaccine consisting of specific antigens only rather than whole organisms. Antigens are selected by their ability to stimulate both B-cell and T-cell immunity.

connectance
The ratio of the number of observed links to the number of potential links in a food web.

conoid
A set of microtubules arranged in a spiral configuration that forms part of the apical complex found in apicomplexans.

conservation biology
The study of the Earth's biodiversity, with the goal to preserve as many species as possible within as natural a setting as possible.

constitutive defenses
Defense mechanisms that are always present rather than induced by exposure to a parasite. Examples include barriers to the entry of parasites such as the skin or stomach acid in animals.

contig
A set of overlapping cDNA segments that provide longer stretches of sequence and thus help to identify particular genes.

conventional macrophages
A class of phagocytic cells important in both the innate immune response and in antigen presentation to $CD4^+$ T cells. Following their activation, conventional macrophages release various inflammatory cytokines.

cospeciation
A phenomenon in parasitology in which parasites respond to speciation in their host by forming new species themselves.

CRISPR
A mechanism found in many prokaryotes that is used to defend against invasion by viruses or foreign plasmids. An acronym for "clustered regularly interspaced short palindromic repeats."

cryptic species
Closely related organisms that are so similar morphologically that they have frequently been classified as a single species. Genetic analysis reveals differences among members of this group such that they merit the status of separate species.

cutaneous larval migrans
A condition that occurs when larval hookworms penetrate an inappropriate host. Rather than reach the circulatory system, the larvae wander through the skin leaving a trail of inflammation.

cyclodevelopmental transmission
Transmission using a biological vector in which the parasite undergoes developmental changes but does not multiply.

cyclopropagative transmission
Transmission using a biological vector in which the parasite both undergoes developmental changes and multiplies.

cyst
A general term for a resistant, long-lived, and relatively metabolically inactive stage in the life cycle of many parasites. The cyst is typically the infective stage for the next host in the parasite life cycle.

cystatins
Cysteine protease inhibitors produced by various helminths that interfere with host MHC 2 presentation, thus decreasing the likelihood of parasite destruction by the host immune system.

cysticercoid
The life-cycle stage that develops in the onchospere in the intermediate host in many cestodes life cycles.

cysticercus
A larval stage typical of tapeworms in the family Taeniidae, found in the tissues of the intermediate host.

danger hypothesis
The idea that certain molecules associated with pathogens or released by injured cells act as danger signals, evoking a response from certain cells of the immune system.

dauer-stage
A nematode larval stage in which development is arrested when conditions are unsuitable, resuming when conditions improve.

daughter sporocysts
The asexually produced generations, derived from the mother sporocyst, found in the life cycle of some digenetic trematodes. They are typically formed within the body of the first intermediate host.

decision-dependent regulation
A type of density-dependent population regulation in which parasites avoid hosts that are already infected.

definitive host
A host in which the parasite develops into a sexually mature adult. In most cases, sexual reproduction also takes place within this host.

delayed-type hypersensitivity response (DTH)
A type of cell-mediated immunity initiated by antigen. It is called "delayed" because the reaction occurs hours or days after exposure to the antigen.

deterministic models
A type of mathematical model designed to predict the outcome of a

biological phenomenon, which will always produce the same result if provided with the same starting parameters.

diapedisis
The process by which phagocytic cells cross an epithelial border by squeezing through spaces between epithelial cells.

dilution effect
A reduction in the transmission of a parasite as a result of the complexity of the environment in which the parasite is found. Such complexity may include the presence of potential predators that consume the parasite.

dioecious (or gonochoristic)
An organism in which adults are divided into separate male and female individuals.

diplostomulum metacercaria
A life-cycle stage in certain digenetic trematodes. It develops from the mesocercaria in the newly infected definitive host. Following migration within the host, it develops into the adult trematode.

direct life cycle (or monoxenous life cycle)
A parasitic life cycle in which only a single host is used.

disease
A pathological condition with symptoms that set it apart from a normal body state.

disseminated cutaneous leishmaniasis
A form of cutaneous leishmaniasis characterized by disfiguring lesions, in which the parasite is spread over a wide area by infected macrophages.

DNA vaccines
Vaccines consisting of only pathogen DNA. Such DNA may be taken up by host cells and subsequently expressed. An immune response against these expressed foreign antigens may be protective against a later challenge by the pathogen.

Dollo's law
The concept that parasites evolve from free-living forms and not vice versa because, as they evolve, parasites lose genes and structures needed to return to a free-living state and they gain specialized adaptations for parasitism.

drowning on arrival
A situation following a cospeciation event in which the parasite becomes extinct in one of the new host lineages.

duplication
The formation of two closely related parasite species as a result of sympatric species that inhabit the same host. Also called synxenic speciation.

ecosystem services
The ability of natural environments to provide food and clean water, to regulate climate and infectious diseases, to support nutrient cycles, and to supply other tangible benefits.

ectoparasites
Parasites that live on the external surface of their host.

ectopic
In parasitology, refers to an infection in which the parasite is found in a location other than the normal location.

effector-triggered immunity (ETI)
A defensive mechanism in plants in which parasite-encoded "effector proteins" are detected by plants and used to activate signal transduction pathways that lead to downstream immune responses.

emerging infectious diseases
An infectious disease with elevated incidence in recent years. Such diseases may be new to science or known diseases that have recently increased in incidence or spread to new geographic areas.

empirical methodology
A strategy of drug development in which many compounds are tested against a particular infectious agent. Those compounds that display efficacy against the agent are researched further.

encounter filter
The idea that parasites may never use otherwise biologically compatible hosts if factors such as behavior prevent parasite and host from ever coming in contact with each other.

endemic
A disease that maintains a steady presence in a particular host population without the need for input of infected individuals from other populations.

endoparasites
Parasites that live within the body of their host.

enemy release hypothesis
The idea that if an invading host species has managed to leave its parasites behind in its original habitat, then it may have greater success in its new environment.

epidemic
Any unusual and unexpected outbreak of disease above that expected based on recent experience.

epidemiology
The study of the behavior of disease in human populations.

epigenetic
Heritable changes in gene expression or phenotype caused by mechanisms other than a change in the DNA's nucleotide sequence.

epizootiology
The study of the distribution and determinants of disease in animals.

eradication
In the context of disease control, a permanent global reduction to zero of the incidence of new infections with no danger of reintroduction.

esophageal sarcomas
Neoplasms initiated by cysts containing adult worms, found in the esophagus of dogs infected with *Spirocerca lupi*.

eukaryotes
Organisms with cells that have a defined nucleus and membrane-bound organelles. Eukaryotes include unicellular protozoans, fungi, plants, and animals.

evolutionary species concept
The view that a biological species is a group of organisms composed of a single lineage with a shared evolutionary trajectory.

exaptation
A process in which a trait evolves for one purpose but later is used for another.

excavata
A proposed subdomain within the domain Eukarya, which includes the diplomonads, the kinetoplasts, and the parabasalids.

facultative parasites
Organisms that are fully able to live a free-living existence but that may become parasitic if the opportunity or necessity to do so arises.

Fahrenholz's rule
A parasite and host phylogenies will be congruent as a consequence of repeated cospeciation events.

failure to speciate
The speciation of a particular host that is not followed by a speciation of its parasites.

Fecal–oral
A type of parasite transmission in which propagules are released in the feces, where they may contaminate food or water, to be consumed by the next host in the parasite's life cycle.

filarial worms
A group of nematode parasites, all of which dwell in the tissues of the definitive host and all of which rely on arthropod as intermediate hosts.

fitness
A measure of the success of an individual in passing on its genes to future generations. Fitness is influenced by the individual's ability to survive and to reproduce.

food web
A graphic representation of the trophic interactions that occur among the members of a biological community.

functional response
A result of biological competition, in which the population of one or both of the competing populations alters its niche as a consequence of the competition.

gametogony
The formation of gametes in the apicomplexa.

geographical information systems (GIS)
Systems designed to capture, maintain, analyze, and display various kinds of information, such as rainfall, satellite images of vegetation, or parasite prevalence records in a spatially explicit way.

geohelminth
Nematodes that are primarily transmitted via soil contaminated with infectious eggs or larval stages.

glochidium
A larval stage typical of many fresh water bivalves that attaches to the gill filaments of fish, where it lives as an ectoparasite. Following metamorphosis, glochidia drop off and settle to the bottom of the water body.

glycosomes
A membrane-bound organelle found in the cells of kinetoplastids in which the enzymes involved in glycolysis are sequestered.

guard concept
A defensive mechanism of plants in which R proteins recognize parasite-damaged host molecules as "altered self," triggering a protective response.

hard polytomy
A feature of some phylogenetic trees in which a cluster of closely related species diverged from each other relatively recently, making it difficult to resolve the exact phylogenetic relationships of these species.

haustoria
The invasive roots of parasitic angiosperms that invade the roots or shoots of other land plants.

haustorium
The tip of the long, slender process produced by some parasitic plants that penetrates the host plant and gives the parasite access to the host's interior.

heirloom parasite
A parasite that was originally acquired by an ancestral host organism and that is now found in the descendents of that ancestor.

helminth
A common nontaxonomic term used to describe a parasitic worm. Helminths include trematodes, tapeworms, nematodes, acanthocephalans, and nematomorphs.

helminth therapy
The deliberate infection of patients with specific intestinal nematodes to alleviate symptoms of certain clinical conditions such as chronic inflammatory bowel disease.

heme detoxification protein (HDP)
A protein produced by *Plasmodium* that is thought to catalyze the dimerization of heme into hemozoin.

hemiparasites
Parasitic plants that retain the ability to perform photosynthesis.

hemocoel
A body cavity typical of arthopods. Serves as a cavity for blood circulation.

hexacanth
The larval stage emerging from the egg of many cestodes. Also called an oncosphere.

historical biogeography
The study of the manner in which historical aspects of geology, ecology, or climate affect the present or past distribution of species.

holoparasites
Parasitic plants that are no longer capable of their own photosynthesis and that obtain all their nutrients from the host plant through their haustoria.

homoplasy
Phenotypic similarities between two species resulting from convergent evolution rather than common ancestry.

horizontal gene transfer
The transfer of genetic information between organisms in a manner other than conventional modes of reproduction.

host switching
A phenomenon in which one or a few parasites colonize a new host species, essentially forming a peripheral population.

host-death dependent regulation
A type of density-dependent parasite population regulation in which the most heavily infected hosts succumb to their infections and die.

hydrogenosome
An organelle found in some protozoans that generates ATP from pyruvate while giving off hydrogen (H_2) as a by-product.

hygiene hypothesis
The idea that early exposure to microorganisms is necessary for proper development of the immune system and that, without such exposure, immune dysfunction may lead to allergies or autoimmunity later in life.

hyperparasite
An organism that is a parasite of another parasite.

hyperparasitism
A phenomenon in which a parasite uses a different parasite as a host.

hyperplasia
An increase in cell number. Other than their accelerated proliferation, hyperplastic cells appear normal.

hypersensitive response (HR)
A phenomenon in which plant cells at the site of parasite infection are induced to undergo programmed cell death.

hypertrophy
Accelerated cell division in response to an exogenous stimulus.

hypertrophy
An increase in the size of an organ or tissue resulting from an enlargement of its component cells.

hyphae
Branching filaments of filamentous fungi that absorb nutrients from the surrounding environment. Collectively the hyphae make up the densely branched fungal body or mycelium.

hypobiosis
The cessation of development, during which the organism remains able to reinitiate development in response to specific environmental stimuli.

immune system
The molecules, cells, and tissues involved in innate and adaptive immunity.

immunocompetence handicap hypothesis
The idea that sexual selection in males to develop a new adaptation that makes them more successful in securing mates might impose a cost on immunocompetence, rendering males carrying the adaptation more vulnerable to infection with parasites.

immunological memory
The capacity of the vertebrate immune system to respond faster and more effectively to second and subsequent exposure to a specific antigen.

immunoparasitology
The immune interactions between parasites and their hosts.

immunopathology
The damage to the host that occurs as a result of an inappropriate or excessive immune response, rather than from any direct effect of a parasite.

inbreeding
Reproduction by mates who are closely related, which leads to homozygosity, loss of genetic diversity, and increased chances that offspring can be affected by deleterious traits.

inbreeding depression
The loss of fitness in a population resulting from the breeding of related individuals.

incidence
The number of new cases of a disease acquired in some unit of time, usually expressed as a proportion or rate.

indirect life cycle (or heteroxenous life cycle)
A type of parasitic life cycle that requires more than a single host.

inducible defense
Innate immune processes that are stimulated to occur following infection, as opposed to constitutive defensives such as anatomical barriers, which are always present.

infection
The entry, development, and/or multiplication of an infectious agent in the body.

infectious agent
An organism or suborganismal entity such as a virus that is capable of producing an infection or infectious disease.

infectious disease
The clinical manifestations resulting from an infection.

infestations
The development or multiplication of an ectoparasite on the surface of its host. In contrast to infections, which typically refer to endoparasites within the host's body.

infrapopulation
The number of parasites of a particular species harbored by a specific individual host at a given point in time.

innate immune memory
A phenomenon observed in some invertebrates in which a heightened immune response occurs upon second exposure to the same pathogen.

innate immune system
The molecules, cells, and tissues involved in those defensive mechanisms considered to be inborn or innate.

innate immunity
The various inborn resistance mechanisms that are encountered by a pathogen, before adaptive immunity is induced.

integrated drug discovery
Partnerships in which academic institutions, industry, and the public health community work together to facilitate the discovery and development of new drugs against parasites.

interactive site segregation
A possible outcome of biological competition in which one or both of the competing species will react to the presence of the other by occupying a different segment of its habitat and so diminish contact with the other species.

interference competition
A type of intraspecific competition in which active aggression between interacting individuals is observed.

intermediate host
A host in which a parasite undergoes a required developmental step and possibly asexual reproduction but does not reproduce sexually.

interspecific competition
Biological competition between two different species.

intraguild predation
A type of predation in which one member of a guild (a group of species that uses the same resources) attacks and consumes another member of the guild.

intraspecific competition
Biological competition between members of the same species.

intraspecific variation
The genetic diversity found within a particular species.

invariant natural killer T cells (iNKT cells)
A type of lymphocyte bearing T-cell receptors of limited diversity that recognizes glycolipid antigens presented by CD1 molecules.

invasional meltdown
As the number of introductions of exotic species increases, the ease with which they and other exotics become established also increases.

isolate
A sample of a parasite species derived from a particular host at a particular time.

keystone species
A species found in a biological community that is interconnected with many other species in that community and, if removed, causes significant changes in the structure of the community.

killed vaccines
Vaccines consisting of inactivated virus or dead organisms. Also called inactivated vaccines.

kinetoplast
A structure found within the single large mitochondrion of excavates such as *Trypanosoma* and *Leishmania*. The kinetoplast contains its own DNA, called kDNA.

knock-on effects
The idea that introductions of parasites, of their hosts, or even of nonhost species that affect indigenous transmission cycles may have further, more cryptic, downstream effects that could affect entire communities of organisms.

Lamarckian inheritance
A trait acquired during an organism's lifetime that can subsequently be passed on to offspring.

lead molecule
An initial molecule found to have potential as an antiparasite drug. This molecule must undergo substantial modification and optimization before its use can be considered.

lectins
Carbohydrate-binding proteins.

live, attenuated vaccines
Vaccines that contain a weakened strain of a particular parasite, which although able to infect and replicate in the host, is unable to cause disease.

local adaptation
A situation in which parasites are more successful in hosts from the same geographic area than in hosts from a different geographic area.

local maladaptation
The opposite of local adaptation, when parasites are more successful in hosts with allopatric distributions than they are with sympatrically distributed hosts.

localized cutaneous leishmaniasis
A form of cutaneous leishmaniasis in which infected macrophages do not move far beyond the initial site of inoculation and in which lesions are generally self-healing.

M2 macrophages
see alternatively activated macrophages.

macroevolution
Large-scale evolutionary patterns and processes at or above the species level that occur on an evolutionary time scale, typically measured over thousands of generations and millions of years.

macrogamete
The female gamete in the apicomplexan life cycle.

magic traits
Genetically determined traits that are under divergent selection pressures, one of which is imposed by parasites and one of which affects mate choice, and consequently can increase the likelihood of host reproductive isolation and speciation.

mass drug administration
The administration of drugs to whole populations in high-transmission areas, whether or not individuals were infected at the time.

mechanical vector
A vector that picks up a pathogen from one host and transfers it to another, with the pathogen undergoing no development or multiplication within or on the vector.

merogony
A type of asexual reproduction observed in apicomplexans in which multiple fission produces merozoites.

merozoites
An invasive stage in the life cycle of most apicomplexans, produced as a result of merogony.

mesocercaria
A larval stage in the life cycle of some digenetic trematodes, arising from the cercaria and giving rise to the metacercaria.

metagenomics
The characterization of genetic material recovered directly from a particular environment without the need to culture the organism present.

metameric
A body plan consisting of repeated segments, in which structures arising from the mesoderm and ectoderm are repeated in each segment.

metaplasia
A generally reversible condition in which differentiated cells of one type are replaced by cells of a different type, often as a result of chronic irritation by a foreign substance.

microbiome
The collection of all microorganisms that typically live on or in the body of a host organism.

microevolution
The evolutionary process as it occurs below the species level, particularly at the level of populations within a species, with an implied ecological time scale.

microfilariae
The first-stage larvae of ovoviviparous filarial nematodes.

microgamete
The male gamete in the apicomplexan life cycle.

micronemes
Slender, convoluted secretory structures that are connected to the rhoptries within the apicoplast of apicomplexans.

microparasites
Microscopic parasites (viruses, bacteria, or protozoans) that establish infections in hosts for which it is difficult or impossible to quantify the actual numbers of individual infectious entities present.

micropredator
Small invertebrates, such as mosquitoes, ticks, or leeches, that take meals from a given host but then leave that host once the meal is complete. As with predators, the source of each meal may be a different individual.

microsatellite markers
Short DNA nucleotide repeat sequences (usually 2–6 nucleotides long) that undergo rapid change in the number of times they are duplicated and that provide a good way to discriminate among individuals of the same species.

microtriches
Small microvilli arising from the tegument of tapeworms, which increase the surface area of the tegument.

miracidium (plural: miracidia)
The first stage larva of digenetic trematodes.

missing the boat
A situation in which a particular species of host diverges into two new species, and a parasite of the original host species fails to colonize one of the newly diverged host species.

mitosomes
A reduced version of the mitochondrion found in *Giardia*. Although they retain the mitochondrial double membrane, they lack mitochondrial DNA. They are incapable of aerobic respiration, although they are involved in ATP production.

mode of transmission
The manner in which a parasite moves from one host to another.

molecular mimicry
A similarity between certain parasite and host antigens, such that an immune response against the parasite results in a reaction against host tissues bearing the similar antigen. May be involved in some autoimmune responses.

monoecious
Hermaphroditic. An organism that contains both male and female reproductive systems.

monophyletic
A phylogenetically related group of organisms, all derived from the most recent hypothetical common ancestor of that group.

monoxenous
Parasites requiring only a single host for the completion of their life cycles.

monozoic
Describes organisms bearing one set of reproductive organs.

mother sporocyst
A life-cycle stage in many digenetic trematodes. It arises from the miracidium, following entry of the miracidium into the first intermediate host.

mutualism
Symbiosis between two organisms in which the fitness of both organisms is enhanced as a result of the symbiosis.

mycelium
The network of hyphae in a filamentous fungus.

mycology
The study of fungi.

mycorrhizal epiparasitism
A phenomenon in which mycorrhizal fungal mycelia become conduits that funnel nutrients away from host plants to other plants that are unambiguously parasitic.

myiasis
The infestation of tissues of a live host by larvae of flies that then feed on the living tissue, causing the damage to the host.

nagana
A disease of African ungulates and livestock, resulting from infection with *Trypanosoma brucei brucei*.

natural helper cells (or nuocytes)
A type of lymphocyte lacking T-cell receptors. In response to chemical mediators released by epithelial cells, natural helper cells release various cytokines that help skew an immune response toward a Th-2 response.

natural killer cells (or NK cells)
Lymphocytes lacking the specificity of T or B cells that are important in the early response to pathogens, before the activation of adaptive immunity. An important early source of gamma interferon.

NB-LRRs (nucleotide-binding leucine-rich repeat proteins)
Proteins produced in response to plant parasite effector proteins. Once produced they are involved in the activation of signal transduction pathways with downstream immune responses, collectively called effector-triggered immunity.

neglected tropical diseases
Any of various parasitic diseases typically found in tropical areas that historically have received relatively low priority in the public health community.

neodermis
The syncytial tegument, characteristic of the parasitic stages in the life cycle of many flat worms.

neoplasia
An abnormal proliferation of cells in response to an external stimulus and that continues even after cessation of the stimulus.

numerical response
A manifestation of biological competition in which the population of one or both of the competing populations is negatively affected as a consequence of the competition.

nurse cell
A metaplasia resulting from infection with *Trichinella*. *Trichinella* modifies skeletal muscle cells to become nurse cells, which secrete a protective collagen capsule around both the cell and the parasite.

obligate epiparasitism
A phenomenon in which one parasitic plant forms a required haustorial attachment to another parasitic plant.

obligatory parasite
A parasite that is unable to complete its life cycle in the absence of required hosts.

old friends hypothesis
A refinement of the hygiene hypothesis, proposing that animal hosts rely on their normal microbial flora to encode certain gene products necessary for proper immune function.

oncomiracidium
The larval stage released from the egg, typical of the monogenean life cycle.

oncosphere
The larval cestode that emerges from the egg.

oocyst
The stage in the apicomplexan life cycle following union of the macro and microgametocyte, in which sporozoites are produced.

opisthaptor
The posterial attachment structure, typical of adult monogeneans.

opisthokonta
A proposed subdomain with the domain Eukarya, consisting of animals, fungi, and choanoflagellates.

opportunistic parasite
A parasite that takes advantage of particular circumstances to

initiate an infection in a host that it normally does not infect or in which it does not normally cause disease.

orthologous genes
Genes in different species that originated by vertical descent from a single gene of the last common ancestor.

osmotrophic
A heterotrophic mode of nutrient acquisition in which nutrients are absorbed from the surrounding environment rather than consumed. Typical of fungi.

otoacariasis
Dermatitis of the ears, caused by an infestation of ticks or mites.

pandemic
An epidemic of an infectious disease that has spread across large regions such as continents or even across the entire world.

paraphyletic
A taxonomic group of organisms, including the most recent common ancestor of those organisms, except for a small number of monophyletic groups descended from the same common ancestor.

parasite
An organism that infects or infests a host, where it does some level of damage, while the parasite itself benefits in some way from its association with the host.

parasite-mediated sexual selection hypothesis
The view that parasites indirectly influence a male's ornamentation and consequently help determine female mate choice.

parasitemia
A quantitative measure of parasites in the blood. May also refer to the parasite load in the host and provide an indication of the degree of an active parasitic infection.

parasitism
The mode of existence used by parasites.

parasitoid
An organism that spends a significant amount of its life attached to or within a single host, often sterilizing or killing it and sometimes fully consuming it in the process.

parasitophorous vacuole
The membrane-bound vacuole in which an intracellular parasite may be found.

paratenic hosts (or transport hosts)
A host in which the parasite does not undergo further or necessary development but that bridges a trophic gap in a life cycle, thus making transmission more probable.

pathogen
An infectious agent capable of causing a state of disease in a host.

pathogen pollution
The introduction of pathogenic parasites into a new or naive host species or population.

pathogen-associated molecular patterns (PAMPs)
Common and conversed molecules associated with particular groups of pathogens. Recognized by pattern recognition receptors to help initiate an induced response to infection.

pattern recognition receptors (PRR)
Receptor molecules capable of binding specific pathogen-associated molecular patterns (PAMPs). Once such receptors are bound, induced immune responses are initiated.

periodicity
The emergence of certain parasitic larvae from a particular host at specific times, coordinated to coincide with the diurnal activity pattern of the next host, to better insure contact with that host.

peripatric speciation (or peripheral isolates speciation)
A type of speciation that occurs when a small population at the periphery of a large geographic range becomes isolated sufficiently, such that genetic drift or inbreeding causes it to diverge from the other individuals in the broader range.

peritonitis
Inflammation of the peritoneal cavity. It can be the result of invasive intestinal parasites, such as *Entamoeba histolytica*.

philopatry
A tendency on the part of individuals of a particular population to associate only with other individuals from the same population in preference to individuals from other populations. It limits dispersal of the philopatric individuals.

phoresy
The association of one organism with another for the purpose of transport from one place to another with no implied physiological dependence.

phylogenetics
The use of evolutionary trees to make and evaluate hypotheses about historical patterns of descent.

pleiotropy
A phenomenon in which a single gene has multiple phenotypic effects.

plerocercoid
A larval stage characteristic of certain tapeworms with aquatic larval stages. It is often found encysted in the skeletal muscles of a second intermediate host and serves as the infective stage for the definitive host.

plerocercoid growth factor (PGF)
A molecule produced by the plerocercoid stage of the tapeworm *Spirometra mansonoides* that mimics the effects of mammalian growth factor.

polar rings
Electron dense structures of unknown function that form part of the apical complex, found in apicomplexans.

pollutogens
Infectious agents that originate outside a particular ecosystem and are able to develop within a host found in that ecosystem, yet they do not require that host for reproduction.

poly-IgA receptor
A molecule involved in the translocation of IgA antibody from the basal to the apical side of surface epithelium. Once transport of IgA across the epithelium is complete it is released extracellularly.

polyclonal activation
The activation of lymphocytes regardless of antigen specificity. This phenomenon leads to the activation of many lymphocyte clones with different antigen specificities.

polytomies
A term used in phylogenetics to describe nodes in the phylogenetic tree that are not completely resolved to dichotomies.

polyzoic
The body plan of organisms bearing serial repetition of gonads.

portal of exit
The host anatomical structure through which parasite propagules leave the host.

premunition
A phenomenon in which regular and repeated exposure to a particular pathogen is required to maintain some level of immunological memory to that pathogen. Also called concomitant immunity.

prevalence
The percentage of host individuals infected with a particular parasite species within a defined geographic area.

primary endosymbiosis
The engulfment of a bacterium by a primitive eukaryotic cell, which eventually, over evolutionary time, gives rise to a mutualistic interaction in which both cells lose the capacity for independent living.

primary vectors
Biological vectors with sufficient vector capacity to maintain the life cycle of a specific parasite, in the absence of other vector species.

proglottid
A set of reproductive organs in a cestode, generally corresponding to a single segment in a chain of segments called the strobila.

propagative transmission
A type of vector transmission in which the parasite replicates within the vector but does not undergo any developmental progression.

prophylaxis
The use of drugs in otherwise healthy individuals to prevent infection in the first place.

protelean parasite
An organism that is parasitic when immature but free-living as an adult.

protozoans or protists
Phylogenetically non-valid terms used commonly to describe eukaryotic, unicellular organisms that are microscopic. Many are motile.

pseudocoelom
A body cavity that forms during embryological development, which is not completely surrounded by tissue derived from the mesoderm.

PTI (pattern-triggered immunity)
The series of induced immunological changes that occur in response to the binding of a pattern recognition receptor to its PAMP. Also called PAMP-triggered immunity.

pulicosis
The skin condition that may result due to flea bites.

radiation
A relatively sudden expansion in the number of species, typically resulting from elaboration of an innovation. Among parasites, examples of such innovations may be a switch to new host lineage or new form of immune evasion.

rational drug design
The process of developing new medications based on a prior knowledge of the biological target of the proposed drug.

Red Queen hypothesis
The idea that biotic interactions are a fundamental driver of evolution and that for interacting species, particularly for antagonistic interactions such as a parasite and its host, a change in one is likely to select for a change in another.

reduction strategies
Techniques to control vectors of infectious disease that are based on the introduction of vectors with reduced fitness. The overall goal is to reduce vector populations and thereby reduce disease transmission.

replacement strategies
Vector control based on the introduction of genetically altered vectors that are resistant to infection with a specific parasite. If such vectors are able to replace wild type vectors in the field, there will be an overall decline in parasite transmission.

reservoir host
The host in which a parasite normally dwells and where it can be reliably maintained, even when not being actively transmitted, and from which it can then colonize other hosts.

resistan-like molecule β (RELM β)
A molecule produced by intestinal epithelial cells in response to IL-4. It is involved in various antihelminth effector responses, which may include interference with parasite feeding and sensory reception.

resource competition (or exploitative competition)
Density-dependent biological competition for resources such as food or space.

respiratory burst
Metabolic changes that occur in neutrophils and macrophages following phagocytosis, in which toxic metabolites that kill the ingested parasite are produced.

rhoptries
Elongated bodies with secretory function extending within the polar rings of the apical complex, found in apicomplexans.

sacculinization
The evolution of simplified bodies in many parasites as compared to their free-living relatives. Often these parasites do not appear to be related to their free-living relatives.

SAR (systemic acquired resistance)
Immunological changes that occur in plants in response to effector triggered immunity. SAR can provide long-term protection to not only the parasite that initiated the response, but to others as well.

SAR (Stramenopila-Alveolata-Rhizaria)
A proposed subdomain with the domain Eukarya, which includes the apicomplexans, the ciliates, and the giant kelps.

secondary endosymbiosis
The type of symbiosis that occurs when the product of primary endosymbiosis is itself engulfed and retained by another free-living eukaryote. Obligate secondary endosymbionts become dependent on their symbiont and are unable to survive in their absence.

secondary vectors
Biological vectors, which, although they can transmit a specific parasite, do not have sufficient vector capacity to maintain transmission in the absence of the primary vector.

selective site segregation
A situation in which two potential competitors do not overlap but occupy different parts of their habitat.

selective sweep
A situation in which a particular phenotypic and genetically determined trait is replaced by new trait that increases to fixation.

selective toxicity
A concept in drug development in which the targeted parasite is adversely affected by a drug, while the host remains unaffected.

Selective toxicity is achieved by attacking parasite targets that are lacking or significantly different in the host.

selfing
Self-fertilization that occurs when a hermaphroditic individual produces sperm that fertilize eggs that it also has produced.

sensitive
In the context of disease diagnosis, refers to the ability of a diagnostic test to accurately detect the patients who are truly positive. A highly sensitive test thus does not result in false negative results.

sequestration
The adherence of erythrocytes infected with *Plasmodium falciparum* to the endothelium of capillaries, which prevents their transport to and subsequent elimination in the spleen.

sexual selection
Natural selection imposed by either competition for mates between males or when one sex, usually the female, exercises choice in mate selection.

short-sighted hypothesis
Competition between parasites infecting the same host in which the more rapidly reproducing parasites come to predominate. As a consequence of this accelerated reproduction, more damage is done to the host, which may die as a result.

sink
An organism that becomes infected with a particular parasite but does not allow the parasite to complete its development. Such an organism, by acting as a sponge for infective parasite stages, lowers transmission of the parasite to its competent host.

site specificity
The requirement of many parasites to arrive at a specific anatomical location within a particular host in order to grow, reproduce, or otherwise complete its life cycle.

social immunization
A process in certain ants in which ants infected with a fungal pathogen are sought out by uninfected ants, which are subsequently exposed to low levels of infection. This activates immunity to the fungus, preventing a more serious infection.

social parasites
Organisms that invade or lay their eggs in the nest of a host organism, allowing their young or immature stages to develop on food provided in that nest.

soft polytomy
In the analysis of phylogenetic trees, an unresolved node (polytomy) resulting from an inadequate amount of sequence data to enable full resolution of the node. Particularly likely to occur at deeper (older) nodes in the tree.

solenophages
Blood-feeding arthropods that introduce their mouthparts directly into a blood vessel.

souvenir parasite
A parasite found on a particular host organism that was acquired when the parasite underwent a host shift from its original host to this more recently acquired host.

species
One of the basic units of biological classification and a taxonomic rank. According to the biological species concept, species is often defined as the largest group of organisms capable of interbreeding and producing fertile offspring.

specific
In the context of disease diagnosis, a diagnostic test that does not give positive results if the patient is not infected or is infected with some other species of parasite.

spondylosis
Degenerative osteoarthritis of the joints between the spinal vertebrae. A symptom of infection with the nematode *Spirocerca lupi* in dogs.

sporogony
The asexual reproduction that occurs in the life cycle of apicomplexans, following sexual reproduction, which gives rise to sporozoites.

sporozoite
An invasive stage in the life cycle of apicomplexans, produced by sporogony.

sterile-male technique
A type of vector control in which large numbers of laboratory-reared sterile males are released into foci of disease transmission. These males mate with wild females, which then produce unfertile eggs. The wild target vector population consequently declines.

stochastic models
A mathematical model designed to predict the outcome of a biological phenomenon, into which an element of chance has been added such that the same outcome will not necessarily occur, even if provided with the same starting parameter values.

straggling
A situation in which a particular parasite manages to colonize a new host but the association is accidental and temporary.

strain
An intraspecific group of parasites that differs from other such groups in one or more traits, including traits that might be relevant to control or treatment.

subclinical infection
An infection with a parasite that has no accompanying signs or symptoms of disease.

subspecies
A distinctive group of organisms within a species that may occupy a particular region and that can interbreed with other subspecies but typically do not because of their isolation or some other reason.

subunit vaccine
A type of acellular vaccine consisting only of particular immunostimulatory antigens.

superspreaders
Infected members of a population that transmit a parasite to other uninfected members at a rate significantly higher than R_0 for the parasite.

suprapopulation
All the parasites of a given species, in all stages of development, both free-living or in all supportive host species, that are found in a given ecosystem.

symbiosis
An intimate association between organisms of two different species. Types of symbiosis include parasitism, mutualism, and commensalism.

sympatric speciation
A type of speciation that occurs without any physical or

geographical barriers being present to isolate the populations undergoing speciation.

synapomorphies
Phenotypic characteristics that are shared by members of a monophyletic group as well as by the group's immediate common ancestor.

synxenic speciation
In parasitology, it takes place when sympatric speciation occurs within a particular host species, resulting in two closely related parasite species occupying the same host species.

systematics
The study of the diversification of life on Earth, including the relationships among organisms over time.

systems biology
A field of biology that studies complex biological interactions by taking a holistic rather than a reductionist approach.

tachyzoites
A life-cycle stage in *Toxoplasma gondii*, characterized by rapid proliferation.

tagma
In segmented animals, a specialized group of segments that are fused together.

tagmatization
The process by which body segments in segmented animals are fused together into specialized units for a particular purpose.

taxonomy
The science of identification, description, and naming of organisms.

telmophages
Blood-feeding arthropods that drink blood that has pooled on the surface of the host due to damage caused by the arthropod.

telmophagy
Feeding from blood that has pooled in small wounds made by a blood-feeding arthropod.

therapeutic index
The amount of a drug that is toxic to host as compared to the amount that will kill the parasite. The higher the therapeutic index, the greater the ability of the drug to destroy the parasite without harming the host.

threshold of disease
The minimum number of parasites in a host required to cause detectable signs and symptoms in an infected individual. When parasite numbers are below this threshold value, the infection is subclinical.

tick paralysis
A paralytic condition in the host, caused by toxic compounds found in the saliva of certain ticks.

tolerance
The idea that in some host-parasite interactions it is advantageous to the host to devise strategies for minimizing the harmful consequences of infection without limiting the presence of the parasite population.

tomites
The trophozoite stage in *Ichthyophthirus multifiliis*.

trade-off hypothesis
In reference to the evolution of virulence in parasites, the idea that an increase in virulence may result in decreased transmission, possibly from more rapid mortality of the host.

transcriptome
The set of all RNA molecules from a cell, a set of cells, or an organism.

transgenesis
The deliberate introduction of exogenous genetic material into a living organism, altering the phenotype of that organism in some desired manner.

transmission
The passage of a parasite to a new host.

transovarian transmission
The passage of a parasite to progeny of an infected female, from the transmission of the parasite to eggs.

tree of life
A commonly used metaphor in which a tree illustrates the common origin of all life, followed by diversification of lineages of life as represented by twigs on this growing tree.

trophically transmitted
A type of parasite transmission in which transmission is achieved when an infected host is eaten by the next host in the parasite's life cycle.

trophozoite
The feeding and replicating stage in the life cycle of many protozoa.

unilocular hydatid cyst
Developmental stage of *Echinococcus granulosus* in the intermediate host. The germinal layer within the cyst gives rise to protoscoleces, the infective stage for the definitive host.

variant surface glycoprotein (VSG)
An antigenic coat protein found on the surface of certain African trypanosomes. Different VSGs are expressed by different generations of trypanosomes, explaining the antigenic variation observed in these parasites.

vector capacity
A measure of the relative ability of a competent biological vector to transmit a specific parasite.

vector competence
The ability of a particular vector to transmit a specific parasite. Any vector able to transmit the parasite is said to be competent. Opposed to vector capacity, which measures the relative ability of competent vectors to transmit the parasite.

vertical transmission
Transmission of a parasite from one generation to the next through the egg, the developing fetus, or larvae within the female's body.

zoonosis
An animal disease that can be transmitted to and cause disease in humans.

INDEX

Page references with the suffixes B, F and T indicate that the topic is covered in a Box, Figure or Table only on that page. Relevant non-text material on pages where a text treatment is already suitably indexed has not been distinguished in this way.

F

J

K

L

M

T

U

V